Studies in Surface Science and Catalysis 85

ADVANCED ZEOLITE SCIENCE AND APPLICATIONS

Studies in Surface Science and Catalysis

Advisory Editors: B. Delmon and J.T. Yates

Vol. 85

ADVANCED ZEOLITE SCIENCE AND APPLICATIONS

Editors

J.C. Jansen
Laboratory of Organic Chemistry and Catalysis, Delft University of Technology, Delft, The Netherlands

M. Stöcker
SINTEF SI, Department of Hydrocarbon Process Chemistry, P.O. Box 124, Blindern, Oslo, Norway

H. G. Karge
Fritz Haber Institute of the Max Planck Society, Faradayweg 4–6, D-14195 Berlin, Germany

J. Weitkamp
Institute of Chemical Technology I, University of Stuttgart, D-70550 Stuttgart, Germany

ELSEVIER   Amsterdam — London — New York — Tokyo   1994

ELSEVIER SCIENCE B.V.
Sara Burgerhartstraat 25
P.O. Box 211, 1000 AE Amsterdam, The Netherlands

ISBN: 0-444-82001-9

Transferred to digital printing 2005
Printed and bound by Antony Rowe Ltd, Eastbourne

Preface

Development in zeolite science is shown by the yearly increase in the number of publications and scientists involved. Frequently held meetings in different forms are therefore imperative.

In order to improve the communication and understanding between newcomers and experienced scientists in the field, the concept of a three-day school preceding the International Zeolite Conference was adopted. The school on "Introduction to Zeolite Science and Practice" was given before the start of the 8th and 9th IZC.

The organizers of the 10th IZC came to the conclusion that for this Summer School lectures at an advanced level rather than introductory courses are preferred.

The main outlines of the programme are the following:
(i)   details on important issues of zeolite science and technology
(ii)   subjects which are underexposed at the IZC
(iii) new and future applications of molecular sieves.
A broad variety of subjects has been chosen to lecture on. In some cases, a deliberate choice has been made to present a subject from different points of view.

Altogether, the Summer School is intended as a forum for open, detailed and firm discussions on zeolite science and applications.

The contents of this book entitled" Advanced Zeolite Science and Applications" reflect in an extended way the lectures given at the 10th IZC Summer School. The editors are convinced that the text will be useful to experienced workers from academic institutions and industry. In particular the combination of science and applications offers useful information for almost all readers interested in molecular sieves.

We would like to express our gratitude to the authors who, despite their heavy workloads, have been willing and enthusiastic to co-operate. In particular, Mrs. Mieke van der Kooij-van Leeuwen is acknowledged for her help in processing the manuscript of this book in such a short time.

Koos Jansen, Michael Stöcker, Hellmut G. Karge and Jens Weitkamp

April 1994

List of contributors

C. Baerlocher
Crystallography, ETH-Zentrum, CH-8092 ZüRICH, Switzerland

H. van Bekkum
Department of Organic Chemistry and Catalysis, Delft University of Technology, Julianalaan 136, 2628 BL DELFT, The Netherlands

G. Bellussi
Eniricerche S.p.A., via F. Maritano 26, 20097 San Donato, MILANO - Italy

N.P. Blake
Department of Chemistry, University of California at Santa Barbara, SANTA BARBARA, California 93106, U.S.A.

J.L. Casci
ICI Katalco RT&E Department, P.O. Box 1, Billingham, Cleveland, TS23 12B, United Kingdom

E.G. Derouane
Facultés Universitaires N.D. de la Paix, Laboratoire de Catalyse, 61, Rue de Bruxelles, B-5000 NAMUR, Belgium

A. Erdem-Senatalar
Department of Chemical Engineering, Istanbul Technical University, 80626 Maslak-ISTANBUL, Turkey

F. Fajula
CNRS - URA 418 - Laboratoire de Chimie Organique, Physique et Cinétique Appliquées, School of Chemistry, 8, Rue de l'Ecole Normale, 34053 MONTPELLIER Cedex, France

E.R. Geus
Department of Organic Chemistry and Catalysis, Delft University of Technology, Julianalaan 136, 2628 BL DELFT, The Netherlands

H. Gies
Institut für Mineralogie, Ruhr-Universität Bochum, Germany

S. Gonthier
Department of Chemical Engineering, Worcester Polytechnic Institute, 100 Institute Road, WORCESTER, Massachusetts 01609, U.S.A.

I.I. Ivanova
Moscow State University, MOSCOW, Russia

P.A. Jacobs
Centrum voor Oppervlaktechemie en Katalyse, KU Leuven, Kardinaal Mercierlaan 92, B-3001 HEVERLEE, Belgium

J.C. Jansen
Department of Organic Chemistry and Catalysis, Delft University of Technology, Julianalaan 136, 2628 BL DELFT, The Netherlands

D. Kashchiev
Institute of Physical Chemistry, Bulgarian Academy of Sciences, ul. Akad. G. Bonchev 11, SOFIA 1113, Bulgaria

V.B. Kazansky
Zelinsky Institute of Organic Chemistry, MOSCOW B-339, Russia

H. Kessler
Laboratoire de Matériaux Minéraux, URA CNRS 428, Ecole Nationale Supérieure de Chimie de Mulhouse, Université de Haute-Alsace, 3 rue Alfred Werner, 68093 MUL-HOUSE Cedex, France

H.W. Kouwenhoven
Laboratorium für Technische Chemie, Eidgenössische Technische Hochschule Zürich, Technisch-Chemische Laboratorium, Universitätsstraße 6, CH-8092 ZüRICH, Switzerland

J. Livage
Chimie de la Matière Condensée, Université P. et M. Curie, 4 Place Jussieu, 75252 PARIS, France

D. Markgraber
Department of Chemistry, University of California at Santa Barbara, SANTA BARBARA, California 93106, U.S.A.

J.A. Martens
Centrum voor Oppervlaktechemie en Katalyse, KU Leuven, Kardinaal Mercierlaan 92, B-3001 HEVERLEE, Belgium

L.B. McCusker
Crystallography, ETH-Zentrum, CH-8092 ZüRICH, Switzerland

H. Metiu
Department of Chemistry, University of California, SANTA BARBARA, California 93106, U.S.A.

J. Patarin
Laboratoire de Matériaux Minéraux, URA CNRS 428, Ecole Nationale Supérieure de Chimie de Mulhouse, Université de Haute-Alsace, 3 rue Alfred Werner, 68093 MUL-HOUSE Cedex, France

D. Plee
CECA Adsorption, Groupement de Recherches de Lacq, B.P. 34, 64170 LACQ,
France

M.S. Rigutto
Department of Organic Chemistry and Catalysis, Delft University of Technology,
Julianalaan 136, 2628 BL DELFT, The Netherlands

D.R. Rolison
Naval Research Laboratory, Surface Chemistry Branch, Code 6170, WASHINGTON,
DC 20375-5000, U.S.A.

R.A. van Santen
Eindhoven University of Technology, Inorganic Chemistry and Catalysis, P.O. Box
513, 5600 MB EINDHOVEN, The Netherlands

C. Schott-Darie
Laboratoire de Matériaux Minéraux, URA CNRS 428, Ecole Nationale Supérieure de
Chimie de Mulhouse, Université de Haute-Alsace, 3 rue Alfred Werner, 68093 MUL-
HOUSE Cedex, France

G. Schulz-Ekloff
Institut für Angewandte und Physikalische Chemie, Universität Bremen, D-28334
BREMEN, Germany

S.T. Sie
Delft University of Technology, Chemical Process Technology, Julianalaan 136,
2628 BL DELFT, The Netherlands

V.I. Srdanov
Department of Chemistry, University of California, SANTA BARBARA, California
93106, U.S.A.

M. Stöcker
SINTEF S1 - Forskningsvelen 1, P.O. Box 124, Blindern, N-0314 OSLO, Norway

G.D. Stucky
Department of Chemistry, University of California, SANTA BARBARA, California
93106, U.S.A.

R.W. Thompson
Department of Chemical Engineering, Worcester Polytechnic Institute, 100 Institute
Road, WORCESTER, MA 01609, U.S.A.

Contents

Chapter 9. Theory of Brønsted acidity in zeolites
R.A. van Santen

Chapter 10. Analysis of the guest-molecule host-framework interaction in zeolites with NMR-spectroscopy and X-ray diffraction
H. Gies

Chapter 11. The preparation and potential applications of ultra-large pore molecular sieves: A review
J.L. Casci

Chapter 12. Advances in the in situ ^{13}C MAS NMR characterization of zeolite catalyzed hydrocarbon reactions
I.I. Ivanova and E.G. Derouane

Chapter 13. Practical aspects of powder diffraction data analysis
C. Baerlocher and L.B. McCusker

Chapter 14. Review on recent NMR results
M. Stöcker

Chapter 15. Supported zeolite systems and applications
H. van Bekkum, E.R. Geus and H.W. Kouwenhoven

J.C. Jansen, M. Stöcker, H.G. Karge and J. Weitkamp (Eds.)
Advanced Zeolite Science and Applications
Studies in Surface Science and Catalysis, Vol. 85
© 1994 Elsevier Science B.V. All rights reserved.

Sol-Gel Chemistry and Molecular Sieve Synthesis

J. Livage

Chimie de la Matière Condensée, Université Pierre et Marie Curie
4 place Jussieu - 75252 Paris - France

Sol-gel chemistry provides a new approach to the preparation of oxide materials [1]. Starting from a solution, a solid network is progressively formed via inorganic polymerization reactions. The term sol-gel chemistry could actually be used in a broader sense to describe the synthesis of inorganic oxides by wet chemistry methods such as precipitation, coprecipitation or hydrothermal synthesis. Two routes are currently used depending on the nature of the molecular precursor. The inorganic route with metal salts in aqueous solutions and the metal-organic route with metal alkoxides in organic solvents. In both cases the reaction is initiated via hydrolysis in order to get reactive M-OH groups. This reaction can be simply performed by adding water to an alkoxide or by changing the pH of an aqueous solution [2]. Condensation then occurs leading to the formation of metal-oxygen-metal bonds. Condensed species are progressively formed from the solution leading to oligomers, oxopolymers, colloids, gels or precipitates. Oxopolymers and colloidal particles give rise to sols which can be shaped, gelled, dried and densified in order to get powders, films, fibres or monolithic glasses [3].

This paper describes the basic chemical reactions involved in sol-gel syntheses from both inorganic and metal-organic precursors.

1. METAL CATIONS IN AQUEOUS SOLUTIONS

Molecular sieves are usually prepared via hydrothermal methods from aqueous solutions [4]. This route has already been used for a long time in industry for the synthesis of catalysts and ceramic powders. Literature provides many data describing the hydrolysis of metal cations in dilute solutions [5] but very little is known about the formation of polynuclear species at concentrations greater than about 1 mol.l^{-1}, although these conditions are generally relevant to the synthesis of solid phases. One of the main problem arises from the fact that water behaves both as a ligand and a solvent. A large number of oligomeric species are formed simultaneously. They are in rapid exchange equilibria and it is not obvious to predict which one would nucleate the solid phase. The key parameter is usually the pH of the aqueous solutions but anions, cations or templates are often added in order to obtain the desired product [4][6].

The whole synthesis occurs in an aqueous medium in which water exhibits very specific properties, both as a molecule and as a solvent.
. The water molecule has a high dipolar moment, $\mu=1.84$ Debye, and liquid water exhibits a high dielectric constant ($\epsilon=80$). Water is therefore a good solvent for most ionic compounds. It breaks polar bonds (ionic dissociation) and dipolar water molecules solvate both cations and anions.
. The water molecule is a Lewis base via its $3a_1$ molecular orbital. It reacts with metal cations M^{z+} (Lewis acids) via acid-base reactions giving OH_2, OH^- or O^{2-} ligands. These hydrolysis equilibria are responsible for the behavior of metal cations toward condensation. It is therefore very important to know the chemical nature of aqueous species as a function of parameters such as pH, temperature or concentration.

The Partial Charge Model

The Partial Charge Model (PCM) will be used as a guide to describe the aqueous chemistry of metal cations [7]. It is based on the electronegativity equalization principle stated by R.T. Sanderson as follows: "when two or more atoms initially different in electronegativity combine, they adjust to the same intermediate electronegativity in the compound" [8]. The main consequence is that both the electronegativity χ_x of a given atom X and its partial charge δ_x vary when the atom is chemically combined. These two parameters must be related. A linear relationship is usually assumed as follows :

$$\chi_x = \chi_x^0 + \eta_x \delta_x \tag{1}$$

where η_x is the hardness of atom X as introduced by Pearson. Hardness is related to the softness $\sigma_x=1/\eta_x$ which provides a measure of the polarisability of the electronic cloud around X. Softness increases with the size of the electronic cloud, i.e. with the radius r of X. Therefore hardness varies as $1/r$. According to the Allred-Rochow scale, electronegativity is proportional to Z_{eff}/r^2 [9]. Hardness may then be approximated as :

$$\eta = k\sqrt{\chi^0} \tag{2}$$

where k is a constant that depends on the electronegativity scale, k=1.36 when Pauling electronegativities are expressed in the frame of Allred-Rochow's model (Table 1).

The total charge "z" of a given chemical species is equal to the sum of the partial charges of all individual atoms $z = \Sigma \delta_i$. This together with equations (1) and (2) leads to the following expressions for :

the mean electronegativity

$$\chi = \frac{\Sigma_i \sqrt{\chi_i^0} + 1.36 z}{\Sigma_i 1/\sqrt{\chi_i^0}} \qquad (3)$$

and the partial charge

$$\delta_i = (\chi - \chi_i^0)/1.36\sqrt{\chi_i^0} \qquad (4)$$

Eq. (4) can also be writen as $\quad \delta_i = \sigma_i(\chi - \chi_i^0) \qquad (5)$

where

$$\sigma_i = (1.36\sqrt{\chi_i^0})^{-1} \qquad (6)$$

H 2,10 ;507																	He 3,20 ;411
Li 0,97 ;747	Be 1,57 ;587											B 2,02 ;517	C 2,50 ;465	N 3,07 ;420	O 3,50 ;393	F 4,10 ;363	Ne 5,10 ;326
Na 1,01 ;732	Mg 1,29 ;647											Al 1,47 ;606	Si 1,74 ;557	P 2,11 ;506	S 2,48 ;467	Cl 2,83 ;437	Ar 3,50 ;393
K 0,91 ;771	Ca 1,04 ;721	Sc 1,23 ;663	Ti 1,32 ;640	V 1,56 ;589	Cr 1,59 ;583	Mn 1,63 ;576	Fe 1,72 ;561	Co 1,75 ;556	Ni 1,80 ;548	Cu 1,75 ;556	Zn 1,66 ;571	Ga 1,82 ;545	Ge 2,00 ;520	As 2,20 ;496	Se 2,50 ;465	Br 2,69 ;448	Kr 3,10 ;418
Rb 0,89 ;779	Sr 0,99 ;739	Y 1,19 ;674	Zr 1,29 ;647	Nb 1,45 ;611	Mo 1,56 ;589	Tc 1,67 ;569	Ru 1,78 ;551	Rh 1,84 ;542	Pd 1,85 ;541	Ag 1,68 ;567	Cd 1,60 ;581	In 1,49 ;602	Sn 1,89 ;535	Sb 1,98 ;523	Te 2,15 ;501	I 2,33 ;482	Xe 2,60 ;456
Cs 0,87 ;788	Ba 0,97 ;747		Hf 1,36 ;631	Ta 1,50 ;600	W 1,59 ;583	Re 1,88 ;536	Os 1,99 ;521	Ir 2,05 ;514	Pt 2,00 ;520	Au 2,02 ;517	Hg 1,80 ;548	Tl 1,60 ;581	Pb 1,92 ;531	Bi 2,03 ;516	Po 2,12 ;505	At 2,28 ;487	Rn 2,30 ;485
Fr 0,86 ;793	Ra 0,95 ;754																

Table 1. Electronegativities χ_i and softness σ_i of atoms X_i according to [7]

The Partial Charge Model provides an easy way to work out the mean electronegativity of chemical species and the charge distribution on each atom. In the case of the water molecule for instance equations (3) and (5) lead to $\chi(H_2O)=2.49$, $\delta_H \approx +0.2$ and $\delta_O \approx -0.4$.

Hydrolysis of metal cations

The word "hydrolysis" is used here to describe those reactions of metal cations with water that liberate protons and produce hydroxy or oxy species [5]. In aqueous solutions this reaction results from the solvation of positively charged cations by dipolar water molecules. It leads to the formation of $[M(OH_2)_N]^{z+}$ species. Water is a Lewis base and the formation of a $M{\Leftarrow}OH_2$ bond with the metal cation (Lewis acid) draws electrons away from the bonding σ molecular orbital of the water molecule. This electron transfer weakens the O-H bond and coordinated water molecules behave as stronger acids than the solvent water molecules. Spontaneous deprotonation then takes place as follows :

$$[M(OH_2)_N]^{z+} + h\,H_2O \Rightarrow [M(OH)_h(OH_2)_{N-h}]^{(z-h)+} + h\,H_3O^+ \qquad (7)$$

Where "h" is called the hydrolysis ratio. It indicates how many protons have been removed from the solvation sphere of the metal cation. The acidity of coordinated water molecules increases as the electron transfer within the M-O bond increases. In dilute solutions this leads to a whole set of more or less deprotonated species ranging from aquocations $[M(OH_2)_N]^{z+}$ (h=0) to neutral hydroxydes $[M(OH)_z]^0$ (h=z) or even oxyanions $[MO_{N'}]^{(2N'-z)-}$ corresponding to the case when all protons have been removed from the coordination sphere of the metal. The electron transfer within the $M{\Leftarrow}OH_2$ bond increases with the oxidation state of the metal cation M^{z+} and coordinated water molecules become more acidic as z increases (Table 2).

Table 2. Partial charge distribution in $[M(OH_2)_6]^{z+}$ species

M^{z+}	χ	δ_M	δ_O	δ_H
Mg^{2+}	2.625	+0.86	-0.34	+0.27
Al^{3+}	2.754	+0.78	-0.29	+0.33
Ti^{4+}	2.875	+0.76	-0.25	+0.39
V^{5+}	3.023	+0.51	-0.19	+0.47
W^{6+}	3.283	+0.31	-0.09	+0.60

A convenient rule-of-thumb shows that the hydrolysis ratio h of a given precursor $[M(OH)_h(OH_2)_{N-h}]^{(z-h)+}$ mainly depends on two parameters, the pH of the solution and the oxidation state of the metal cation M^{z+}. A charge-pH diagram can then be drawn in order to show which aqueous species predominate (Fig.1). Two lines corresponding to h=1 and h=2N-1 respectively separate three domains in which H_2O, OH or O^{2-} ligands are observed [10].

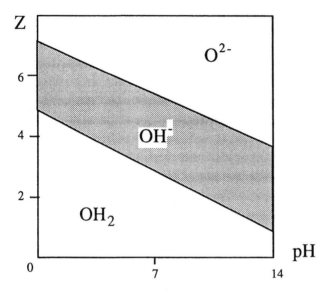

Figure 1. Charge-pH diagram showing the evolution of hydrolyzed species [10]

Determination of the hydrolysis ratio.

In very dilute aqueous solutions, metal cations exhibit several hydrolyzed monomeric species in the pH range 0-14. The problem is then to know whether it is possible to predict the chemical nature of these aqueous species at a given pH.

Following the electronegativity equalization principle it can be stated that deprotonation (eq.7) goes on until the electronegativity χ_h of hydrolyzed species $[M(OH)_h(OH_2)_{N-h}]^{(z-h)+}$ becomes equal to the mean electronegativity χ_{aq} of the aqueous solution. According to thermodynamics, the chemical potential of protons varies linearly with pH :

$$\mu_{H+} = \mu^\circ_{H+} - \beta pH \qquad \text{where } \beta = 2.3RT \qquad (8)$$

Proton exchange reactions between H_3O^+ and H_2O species are very rapid in aqueous solutions so that it can be assumed that $\chi_{H+}=\chi_{aq}$. As a consequence a linear relationship can also be established between the mean electronegativity χ_{aq} and the pH of an aqueous solution.

$$\chi_{aq} = \chi_{aq}° - \lambda pH \tag{9}$$

$\chi_{aq}°$ and λ depend on the reference state for proton and the electronegativity scale. Choosing $[H_5O_2]^+$ as a reference for proton at pH=0 and 2.49 as the mean electronegativity of water at pH=7 leads to [7]:

$$\chi_{aq} = 2.732 - 0.035pH \tag{10}$$

The mean electronegativity χ_h of hydrolyzed precursors can then be calculated as a function of pH and the hydrolysis ratio h is deduced from the basic equations (3) and (5) of the Partial Charge Model.

The charge of the hydrolyzed species $[M(OH)_h(OH_2)_{N-h}]^{(z-h)+}$ is :

$$z-h = \delta_M + N\delta_O + (2N-h)\delta_H \tag{11}$$

$$h = [z-\delta(M(OH_2)_N)]/[1-\delta_H] \tag{12}$$

The hydrolysis ratio h can be expressed as a function of the partial charges $\delta_i = \sigma_i(\chi_{aq} - \chi_i)$ which depend on the mean electronegativity χ_{aq} of the aqueous solution. According to (10), this means that h can be expressed as a function of pH leading to the following expressions :

at pH = 0 $\qquad$ $h = 1.47z - 0.50N - 1.08(2.732 - \chi_M)/\sqrt{\chi_M}$ $\qquad$ (13)

at pH = 14 $\qquad$ $h = 1.08z + 0.37N' - 0.79(2.242 - \chi_M)/\sqrt{\chi_M}$ $\qquad$ (14)

This shows that at a given pH, h mainly depends on the oxidation state "z" and the coordination number "N" of the cation M^{z+}.

Hydrolysis of Si^{IV} and Al^{III}

The Partial Charge Model, applied to Si^{IV} ($\chi_{Si}=1.74$, N=4) leads to:

$$h = (2.088 + 0.217pH)/(0.679 + 0.018pH) \tag{15}$$

Four different hydrolyzed precursors $[H_nSiO_4]^{(4-n)-}$ can be found in aqueous solutions, ranging from $[Si(OH)_3(OH_2)]^+$ (h=3) at pH=0 to $[SiO_2(OH)_2]^{2-}$ (h=6) at pH=14 (Fig.2).

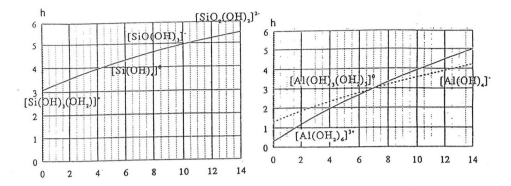

Figure 2. Hydrolysis of Si^{IV} (N=4) and Al^{III} (N=6, N'=4) as a function of pH

Aluminium provides a more interesting example for which the coordination of the metal cation decreases from octahedral to tetrahedral as pH increases. This coordination change occurs around pH≈6.

Let us consider the M-OH$_2$ bond. The negatively charged oxygen atom gives electrons to both the metal M and the hydrogen : $M^{\delta+} \Leftarrow O^{\delta-} \Rightarrow H^{\delta+}$. As deprotonation goes on, electrons are more attracted by the metal. The partial charge δ_M becomes less positive and the M-O bond becomes less polar (Table 3), leading to covalent anions such as $[AlO_4]^{5-}$. Octahedral aluminium $[Al(OH_2)_6]^{3+}$ species are observed at low pH whereas tetrahedral aluminates $[Al(OH)_4]^-$ are formed at high pH (Fig.2).

Table 3. Partial charge in hydrolyzed Al^{III} species

precursor	χ	δ_{Al}	$\delta(H_2O)$
$[Al(OH_2)_6]^{3+}$	2.75	+0.78	+0.37
$[Al(OH)(OH_2)_5]^{2+}$	2.67	+0.73	+0.26
$[Al(OH)_2(OH_2)_4]^+$	2.59	+0.68	+0.14
$[Al(OH)_3(OH_2)_3]^0$	2.49	+ 0.62	0

Stability of M-OH bonds

The order of preference for ligands is usually $H_2O < OH^- < O^{2-}$ as the (charge/radius) ratio of the cation increases (Fig.1). However M-OH bonds are not always stable in aqueous solutions. The high dielectric constant of water can lead to the dissociation of polar bonds. The stability of $M^{\delta+}\text{-}O^{\delta-}\text{-}H^{\delta+}$ bonds toward ionic dissociation then depends on the electronegativity of the metal.

. Basic dissociation is observed with low-valent metal cations such as Na^+ or Ba^{2+}. Their electronegativity is so small ($\chi_M < 1$) that electrons are strongly attracted toward the OH group leading to the dissociation of the polar $M^{\delta+}\text{-}OH^{\delta-}$ bond : $M\text{-}OH_{aq} \Rightarrow M^+_{aq} + OH^-_{aq}$.
NaOH is a strong base and $[Na(OH_2)_N]^+$ cannot be deprotonated under usual conditions.

. Acid dissociation is observed with electronegative elements at the upper right of the periodic table (P^V, S^{VI}, Cl^{VII}) or with highly charged "d^0" metal cations such as Mn^{VII} that give covalent oxyanions. These cations are more electronegative than hydrogen ($\chi_H = 2.1$). Electrons are attracted toward the metal, increasing the positive charge $H^{\delta+}$ and leading to the dissociation of the highly polar $MO^{\delta-}\text{-}H^{\delta+}$ bond : $MO\text{-}H_{aq} \Rightarrow MO^-_{aq} + H^+_{aq}$.
$HMnO_4$ is a very strong acid and $[MnO_4]^-$ species cannot be protonated even at low pH.

. M-OH bonds are not stable either when transition metal ions have a high oxidation state and empty "d" orbitals. Strong $d\pi\text{-}p\pi$ transfers favor the formation of M=O double bonds which decrease the positive charge of the metal ion. Spontaneous internal proton transfer is then observed between two OH groups bonded to the same metal atom. High-valent small cations (V^{IV}, V^V, Mo^{VI}, W^{VI}) give rise to oxo-aqueous species such as $[VO(OH_2)_5]^{2+}$ or $[VO_2(OH_2)_5]^+$ at low pH rather than $[V(OH)_2(OH_2)_4]^{2+}$ or $[V(OH)_4(OH_2)_2]^+$ [11].

2. CONDENSATION OF HYDROLYZED PRECURSORS

The "charge-pH" diagram is a very useful guide for sol-gel chemistry. Condensation becomes possible when at least one stable M-OH bond is formed i.e. in the intermediate domain in figure 2.
. Condensation is usually initiated via pH modification, by adding a base to low-valent aquo-cations or an acid to high-valent oxy-anions. Large condensed species leading to precipitates or gels are obtained around the Point of Zero Charge (PZC) while less condensed solute species known as polyanions or polycations can be formed above or below the PZC.
. Condensation could also be initiated via redox reactions. MnO_2 gels for instance cannot be formed from Mn^{IV} precursors which are usually not soluble in aqueous solutions. They have been prepared via the reduction of permanganate salts $[MnO_4]^-$ by fumaric acid.

The enthalpy change for the first hydrolysis reaction is positive and often close to the enthalpy of dissociation for water (13.3 kcal/mole). The tendency of metal cations to hydrolyze therefore increases with temperature, a phenomenon which is widely used for the hydrothermal synthesis of molecular sieves or monodispersed colloids [4][12].

Olation and oxolation.

The two main mechanisms for condensation are called olation and oxolation. In both cases, polynuclear species are formed via the elimination of water molecules from precursors containing at least one M-OH group.

Olation corresponds to the nucleophilic addition of a negatively charged OH group onto a positively charged hydrated metal cation. As aquo-cations usually exhibit their maximum coordination number, the formation of an "ol" bridge requires the departure of one molecule of water as follows :

$$M\text{-}OH^{\delta-} + M^{\delta+}\text{-}OH_2 \Rightarrow M\text{-}OH\text{-}M^{\delta+}\text{-}OH_2^{\delta+} \Rightarrow M\text{-}OH\text{-}M + H_2O \tag{16}$$

This reaction should involve the formation of bridging $H_3O_2^-$ ligands. Such species could play a fundamental role in the structure of these primary hydrolysis products of metal ions. Their existence was first demonstrated by single crystal X-ray diffraction in the solid state and then confirmed in concentrated solutions by differential anomalous X-ray scattering. An OH ligand coordinated to a metal atom may form a short and symmetrical hydrogen bond to an H_2O ligand of another metal thereby forming a hydrogen-oxide bridging ligand ($H_3O_2^-$) with a characteristic distance of about 5Å between metal atoms [13].

$$>\!M\text{-}\underset{H}{\overset{H}{O}} + HO\text{-}M< \Rightarrow >\!M\text{-}\underset{H}{O}\ldots H\text{-}O\text{-}M< \Rightarrow >\!M\text{-}\overset{H}{O}\text{-}M< + H_2O \tag{17}$$

Oxolation involves the condensation of two OH groups to form one water molecule which is then removed giving rise to an "oxo" bridge as follows :

$$>\!M\text{-}OH + HO\text{-}M< \Rightarrow >\!M\text{-}O\text{-}M< + H_2O \tag{18}$$

This reaction could be described as the dehydration of olated species. It requires the formation of one water molecule between two hydrogen bonded OH groups. Oxolation is usually slower than olation. However it is not completely clear at the present time why some compounds form oxo bridges rather than double hydroxo bridges.

Olation and oxolation require at least one negatively charged nucleophilic OH group in the coordination sphere of the metal cation. However a survey of experimental data suggests that this is not enough and condensation does not seem to occur at room temperature when the positive charge of the metal cation is too small ($\delta_M \leq +0.3$).

Two mononuclear protonated precursors have to be taken into account for Cr^{VI} in aqueous solutions, namely $[CrO_2(OH)_2]^0$ and $[CrO_3(OH)]^-$. They correspond to h=6 and h=7 respectively. In both cases, OH groups have a negative partial charge so that oxolation is possible as follows :

$$. h = 7 \qquad 2[CrO_3(OH)]^- \Rightarrow [Cr_2O_7]^{2-} + H_2O \qquad (19)$$

Condensation stops as hydroxy groups are no more available in the dimer.

$$. h = 6 \qquad 2[CrO_2(OH)_2]^0 \Rightarrow [(HO)O_2Cr\text{-}O\text{-}CrO_2(OH)]^0 + H_2O \qquad (20)$$

OH groups are still available in the dimer but they are positively charged ($\delta_{OH}=+0.04$) and condensation stops at this stage. This means that hydrogen atoms have a strong positive partial charge so that the dimer exhibits acidic properties.

$$[Cr_2O_5(OH)_2] \Rightarrow [Cr_2O_7]^{2-} + 2\,H^+ \qquad (21)$$

Owing to the small size of Cr^{VI} coordination expansion is not possible. Only dimeric bichromate species are therefore observed in aqueous solutions despite the negative partial charge of oxo groups.

Pure orthophosphoric acid $PO(OH)_3$ is a colorless crystalline solid. In aqueous solutions, hydrolysis gives orthophosphoric species $[H_x(PO_4)]^{(3-x)-}$. The phosphate anion $(PO_4)^{3-}$ is a strong base ($pK_a=12.4$) which could only be observed at very high pH. Under usual conditions, only three protonated species (x=1, 2, 3) have to be taken into account as potential precursors for condensation. The condensation of orthophosphoric species would lead to pyrophosphates $(P_2O_7)^{4-}$ as follows :

$$2\,PO(OH)_3 \Rightarrow (OH)_2\text{-}OP\text{-}O\text{-}PO(OH)_2 + H_2O \qquad (22)$$

This oxolation reaction can be described as a nucleophilic substitution. It is mainly governed by the negative charge of the OH group and the positive charge of the metal atom. However the charge conditions are not sastisfied ($\delta_{OH}<0$ and $\delta_P>0.3$) and condensation does not occur. The corresponding values (Table 4) show that in acid solutions, P-OH groups are not nucleophilic enough (H_3PO_4 is a strong acid) whereas at higher pH the positive charge on phosphorus is too small. As a consequence, polyphosphates cannot be formed in aqueous solutions at room temperature via the condensation of monomeric precursors. They are usually prepared by thermal dehydration. Polyphosphates are not stable toward hydrolysis in aqueous solutions. Hydrolysis rates increase as the number of P atoms increases and pH decreases leading to the formation of H_3PO_4.

Table 4. Partial charge on phosphate precursors $[H_x(PO_4)]^{(3-x)-}$

precursor	χ	δ_P	δ_{OH}
$[H_3PO_4]^0$	2.71	+0.35	0.0
$[H_2PO_4]^-$	2.49	+0.19	-0.2
$[HPO_4]^{2-}$	2.18	+0,03	-0.48

Condensation of a trivalent cation, Al^{III}

According to the charge-pH diagram, Al^{3+} gives aquo-cations $[Al(OH_2)_6]^{3+}$ at low pH (pH<3). Condensation can be performed by increasing the pH. It follows an olation mechanism via the nucleophilic addition of negatively charged OH groups onto positively charged metal cations. This leads to the departure of coordinated water molecules and the formation of "ol" bridges. Olated polycations are formed during the first steps of condensation [14]. Hydroxides or oxy-hydroxides are precipitated around the point of zero charge [15].

The trimer $[Al_3(OH)_4(OH_2)_{10}]^{5+}$ was characterized by potentiometric titration but NMR measurements (^{1}H, ^{17}O, ^{27}Al) have recently shown that the di-μ_2-hydroxo cation $[(H_2O)_4Al(OH)_2Al(OH_2)_4]^{4+}$ is not formed in detectable amounts upon neutralization. A sulfate dimer can be crystallized from sulfuric acid solutions [16] but, when such crystals are dissolved in water, ^{27}Al NMR shows that the dimer disproportionates giving $[Al(OH_2)_6]^{3+}$ monomers and a highly hydrolyzed (h≈2.5) trimeric polycation of unknown structure [17]. As shown by dilution experiments such a polycation transforms directly and rapidly into the $[Al_{13}O_4(OH)_{24}(OH_2)_{12}]^{7+}$. This well known tridecamer "Al$_{13}$" exhibits a Keggin-like structure and corresponds to an hydrolysis ratio h ≈ 3 [18]. When hydrolyzed under acidic conditions Al^{3+} solutions slowly convert to boehmite (γ-AlOOH).

12

The first condensation steps of trivalent aquo-cations $[M(OH_2)_6]^{3+}$, were clearly evidenced with Cr^{3+} which gives inert complexes because of crystal field stabilization effects arising from its electronic d^3 configuration [19]. The first step begins with the h=1 $[Al(OH)(OH_2)_5]^{2+}$ precursor (Fig.3). It gives rise to edge sharing dimers as follows :

$$2 \, [Al(OH)(OH_2)_5]^{2+} \Leftrightarrow [(H_2O)_4\text{-}Al \overset{OH}{\underset{OH}{<\quad>}} Al\text{-}(OH_2)_4]^{4+} + 2H_2O \qquad (23)$$

Such edge sharing dimers have been evidenced by X-ray diffraction in the basic sulfate $[Al_2(OH)_2(OH_2)_8][SO_4]_2,2H_2O$ [14].

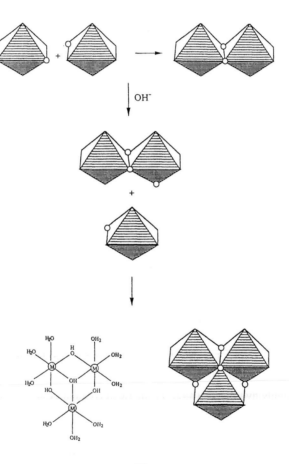

Figure 3. Early stages of the hydrolysis of M^{III} cations (M=Cr, Al, Fe). o = OH group

Further hydrolysis leads to trimeric species formed by adding one hydrolyzed monomer (h=2) to the previous dimer. It could be described as follows:

$$[Al_2(OH)_2(OH_2)_8]^{4+} + [Al(OH)_2(OH_2)_4]^+ \Rightarrow [Al_3(OH)_4(OH_2)_9]^{5+} + H_2O \qquad (24)$$

Such trimeric species are often observed in aqueous solutions of polyanions or polycations [20]. They exhibit a compact cyclic structure in order to minimize electrostatic repulsions between cations in edge-sharing adjacent $[MO_6]$ octahedra. $[Al_3(OH)_4(OH_2)_9]^{5+}$ trimers cannot be isolated in aqueous solutions, but their citric derivative has been precipitated as single crystals [15]. Al^{3+} ions are bonded to a central μ_3-OH group with an sp^3 hybridized oxygen atom. Three of these sp^3 orbitals are involved in M-O bonds while the fourth one gives the O-H bond. The acidity of the μ_3-OH hydrogen atom therefore depends on electron transfers within the three Al-O bonds.

Electron transfer toward the small and highly polarizing Al^{3+} ion with empty d (and even p) orbitals is rather strong, making the μ_3-OH group highly acidic and therefore easily deprotonated.

$$[Al_3(OH)_4(OH_2)_9]^{5+} + H_2O \Rightarrow [Al_3O(OH)_3(OH_2)_9]^{4+} + H_3O^+ \qquad (25)$$

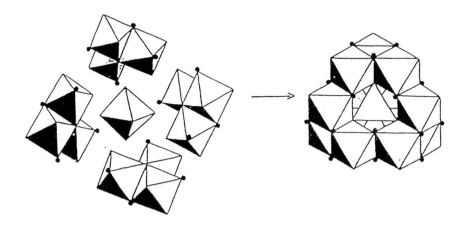

Figure 4. Suggested mechanism for the formation of $[Al_{13}O_4(OH)_{24}(OH_2)_{12}]^{7+}$

A nucleophilic oxygen atom is formed which can react with positive $[Al(OH_2)_6]^{3+}$ species. Such a reaction is limited only by difffusion rates and should be rather fast. The nucleophilic addition of four deprotonated trimers onto a single aquo-ion leads to the formation of the well known Keggin-like $[Al_{13}O_4(OH)_{24}(OH_2)_{12}]^{7+}$ polycation in which one tetrahedral $[AlO_4]$ is surrounded by twelve $[AlO_6]$ octahedra (Fig.4) [21]. Such a reaction path is in agreement with NMR data showing that the structure of this polycation remains the same in the aqueous solution as in crystalline sulfates or selenates. Moreover, they show that small oligomeric species (which may be trimers) directly condense upon hydrolysis into "Al_{13}" with no detectable intermediates.

The neutralization of an aqueous solution of an Al^{3+} salt (h>2.6) mainly leads to neutral solute precursors $[Al(OH)_3(OH_2)_3]^0$ from which $Al(OH)_3$ solid phases are formed via rapid olation reactions :

$$[Al(OH)_3(OH_2)_3]^0 \quad \Rightarrow \quad Al(OH)_3 \; + \; 3\,H_2O \tag{26}$$

Oxolation then leads to oxo-hydroxydes which upon heating transform into alumina.

$$2\,Al(OH)_3 \; \Rightarrow \; 2\,AlO(OH) + 2\,H_2O \; \Rightarrow \; Al_2O_3 + H_2O \tag{27}$$

When heated (T>80°C) aluminium hydroxide $Al(OH)_3$ readily transforms into oxy-hydroxides $AlOOH$, mainly boehmite, γ-$AlOOH$. Diaspore, α-$AlOOH$, is obtained via hydrothermal synthesis. The preferential formation of boehmite rather than diaspore may be due to the acidity of Al_3OH bridges. Increasing the temperature probably dissociates hexameric $[Al_6(H_3O_2)_{18}]^0$ oligomers leading to more compact $[Al_3(OH)_9(OH_2)_4]^0$ trimers. Moreover, temperature also enhances the intrinsic acidity of Al_3OH bridges.

Condensation can proceed one step further via the addition of a neutral monomer $[Al(OH)_3(OH_2)_3]^0$ onto the compact trimer via two corner-sharing $[AlO_6]$ octahedra. A third condensation reaction could occur either with the μ_3-OH group at the center of the trimer or with the μ_2-OH groups of the adjacent $[AlO_6]$ sites. The μ_3-OH bridge is more acidic so that condensation occurs preferentially with this deprotonated μ_3-O$^-$ via the elimination of one water molecule. This gives rise to a skewed tetramer $[Al_4O(OH)_{10}(OH_2)_5]^0$ with one μ_4-O and four μ_2-OH bridges. These tetrameric species could be considered as Secondary Building Units (SBU) for the formation of boehmite γ-$AlOOH$. Further condensation via olation and oxolation then leads to the corrugated sheets typical of the boehmite γ-$AlOOH$ phase. Hydrogen bonds between these sheets finally lead to a three dimensional rigid structure (Fig.5).

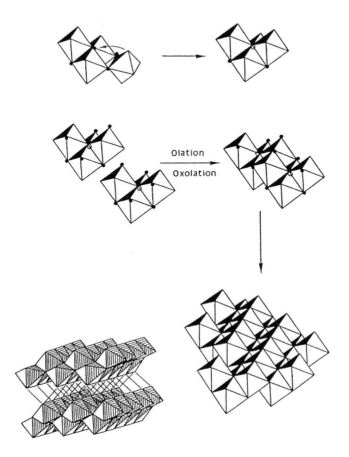

Figure 5. Suggested mechanism for the formation of boehmite γ-AlOOH from tetrameric building blocks.

Around room temperature (T<80°C) crystalline hydroxides $Al(OH)_3$ are precipitated from aqueous solutions, either by increasing the pH of a nitrate solution or upon acidification of a sodium aluminate solution. These hydroxides exhibit lamellar structures based on hexagonal rings linked through μ_2-OH bridges (Fig.6). Depending on the packing of these $Al(OH)_3$ sheets, gibbsite (cubic packing) or bayerite (hexagonal packing) structures are formed. The formation of these planar sheets from aqueous precursors could be described following two different pathways (Fig.6) :

16

i) The first one occurs when a base is added to an Al^{3+} aqueous solution. It involves the octahedral $[Al(OH)_3(OH_2)_3]^0$ neutral precursor. This precursor can also be written as $[Al(H_3O_2)_3]^0$ where $H_3O_2^-$ groups behave as bidentate chelating ligands. Changing the coordination mode from chelating to bridging would first lead to cyclic oligomers. Because of the steric hindrance of $H_3O_2^-$ the smallest one would correspond to hexameric species with twelve $H_3O_2^-$ bridges. Tetramers would be too small for eight $H_3O_2^-$ bridges. Once formed, this embryo could easily collapse into a critical nucleus. Water elimination via olation leads to stronger μ_2-OH bridges. Condensation of $[Al(H_3O_2)_3]^0$ monomers or aggregation of hexameric units then leads to solid particles.

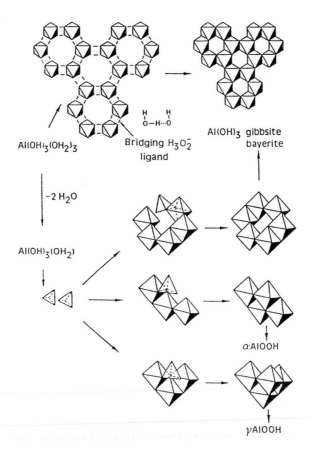

Figure 6 : Suggested mechanism leading to aluminum hydroxides $Al(OH)_3$ or oxy-hydroxides AlOOH from either acidic solutions (octahedral $Al(OH_2)_6^{3+}$) or basic solutions (tetrahedral $Al(OH)_4^-$).

ii) The second pathway occurs upon acidification of an aluminate solution. Tetrahedral $[Al(OH)_3(OH_2)]^0$ neutral precursors are formed according to :

$$[Al(OH)_4]^- + H_2O \Rightarrow [Al(OH)_3(OH_2)]^0 + OH^- \tag{28}$$

Tetrahedral monomeric aluminate ions have been clearly evidenced by ^{27}Al NMR. After proton transfer from a water molecule, a $[Al(OH)_3(OH_2)]^0$ neutral precursor is formed. Owing to its low coordination number, very fast addition reactions can lead to octahedral oligomers such as trimers, tetramers or hexamers. The hexameric oligomer is obviously a critical nucleus for bayerite or gibbsite. Smaller oligomers could act as critical nuclei for oxy-hydroxides phases.

Condensation of a tetravalent cation, Si^{IV}.

The aqueous chemistry of Si^{IV} is dominated by the fact that it remains tetrahedrally coordinated over the whole range of pH. Condensation occurs upon acidification of an aqueous solution of silicates. Condensed phases are formed of corner sharing $[SiO_4]^{4-}$ tetrahedra in order to minimize electrostatic repulsions between cations [22].

As shown in the first part, only four protonated species $[H_nSiO_4]^{(4-n)-}$ have to be taken into account as precursors for condensation (Table 5).

Table 5. Partial charge calculation on $[H_nSiO_4]^{(4-n)-}$ precursors

precursor	hydrolysis ratio	pK	χ	δ_{Si}	δ_{OH}
$[H_5SiO_4]^+$	h = 3	< 0	2.74	+0.56	0
$[H_4SiO_4]^0$	h = 4	9.9	2.58	+0.47	-0.12
$[H_3SiO_4]^-$	h = 5	13	2.3	+0.35	-0.30
$[H_2SiO_4]^{2-}$	h = 6		2.10	+0.20	- 0.55

. The $[H_2SiO_4]^{2-}$ precursor (h=6) is a strong base while $[H_5SiO_4]^+$ (h=3) is a strong acid. They can only be observed in very basic or acid aqueous solutions respectively .

. The h=5 precursor $[H_3SiO_4]^-$ leads to the formation of a large number of oligomeric species around pH≈12. About 20 different species ranging from monomers to decamers have been evidenced by ^{29}Si NMR in aqueous solutions of potassium silicate. A survey of these data shows that above pH=9, Q^4 species are never observed and that most polysilicates are highly branched or cyclic. Linear oligomers are no more observed beyond four Si atoms (Fig.7).

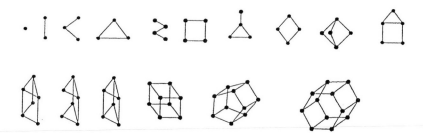

Figure 7. Some oligomeric silicate species characterized by ^{29}Si NMR

This can be qualitatively understood if we consider a linear trimer $[Si_3O_{10}]^{8-}$ (Fig.8). Bridging oxygen atoms which give electrons to two Si atoms are less negatively charged ($\delta_{O1}=-0.74$) than terminal oxygen atoms ($\delta_{O3}=-1.0$). Q^2 Si atoms are more positively charged ($\delta_{Si1}=0.30$) than Q^1 Si atoms ($\delta_{Si2}=+0.17$). Therefore condensation occurs via the nucleophilic addition of a terminal Si-O$^{\delta-}$ atom of one trimer onto the middle Si$^{\delta+}$ atom of another one, leading to branched oligomers. Cyclisation usually occurs, so that only rather small oligomers are formed in alkaline medium.

The chemical nature of counter cations in the solution appears to have a strong effect on the formation of these oligomeric species. Cage polysilicates for instance are formed in the presence of alkylammonium ions such as $[N(CH_3)_4]^+$ whereas cycles are observed with Na^+. This is a key point for the synthesis of zeolites in the presence of templates [4].

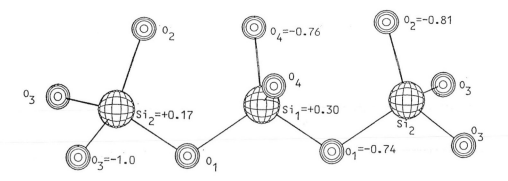

Figure 8. Partial charge distribution in a $[Si_3O_{10}]^{8-}$ trimer.

In the pH range 3≤pH≤9, neutral species $Si(OH)_4$ are predominant. They lead, via oxolation reactions, to the formation of amorphous hydrated silica : $Si(OH)_4 \Rightarrow SiO_2 + 2H_2O$. Around the Point of Zero Charge (pH≈3), gelation is very slow but reaction rates increase significantly by changing the pH.

. Base catalysis occurs at higher pH. It is governed by the nucleophilic addition of negatively charged Si-O onto positively charged Si atoms. This leads to branched species which give rise to dense colloidal silica particles (Fig.9).

. Acid catalysis (pH<2) leads to the protonation of the leaving silanol group. The more negative OH groups are then involved, i.e. terminal Si-OH rather than bridging Si-OH-Si. This leads to the formation of chain polymers which give rise to polymeric gels. (Fig.9)

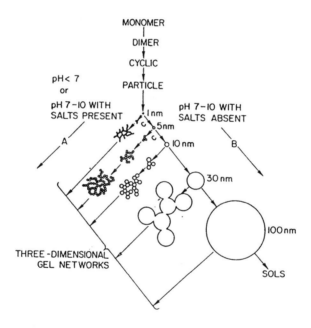

Figure 9. Polymerization behavior of silica in basic (B) or acid (A) solutions (ref.22).

Condensation of a pentavalent cation, V^V

A large variety of V^V species can be found in aqueous solutions. At room temperature they mainly depend on vanadium concentration and pH. V^V is a highly charged cation so that oxo-anions $[VO_4]^{3-}$ in which vanadium is surrounded by four equivalent oxygen atoms are formed in highly alkaline aqueous solutions (pH≈14). Protonation occurs as pH decreases giving rise to hydrolyzed monomeric species $[VO_{4-n}(OH)_n]^{(3-n)-}$ (0≤n≤3) in very diluted solutions ($c<10^{-4}$M). There are no water molecules in the coordination sphere of vanadium so that olation is not possible. Condensation occurs via oxolation. The mean negative charge of these anionic species decreases as n increases (Table 5), down to the Point of Zero Charge (pH≈2). Below that pH value, monomeric cationic species $[VO_2]^+$ are observed in which V^V is six-fold coordinated. Polyanionic species such as pyrovanadate $[V_2O_7]^{4-}$, metavanadate $[V_4O_{12}]^{4-}$ or decavanadate $[V_{10}O_{28}]^{6-}$ are formed in more concentrated solutions (Fig.10) [20].

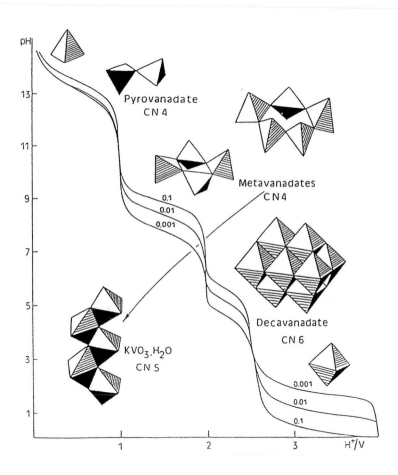

Figure 10. V^V species formed in aqueous solutions [20]

Vanadium pentoxide gels V_2O_5,nH_2O are obtained around the point of zero charge (pH≈2). They exhibit a fibrous structure made of vanadium pentoxide double chains in which $[VO_6]$ octahedra share edges and corners. These gels can be conveniently synthesized via the acidification of metavanadate aqueous solutions by proton exchange resins [23]. Their formation from the neutral (h=5) precursor $[VO(OH)_3]^0$ can be described as follows (Fig.11):

Table 6. Partial charge distribution in $[VO_{4-n}(OH)_n]^{(3-n)-}$ precursors

precursor	hydrolysis ratio	pK	χ	δ_V	δ_{OH}
$[VO_4]^{3-}$	h = 8	13.5	1.58	+0.01	
$HVO_4]^{2-}$	h = 7	7.5	2.05	+0.29	-0.71
$[H_2VO_4]^-$	h = 6	3.8	2.38	+0.48	-0.30
$[H_3VO_4]^0$	h = 5	3.2	2.61	+0.62	-0.09

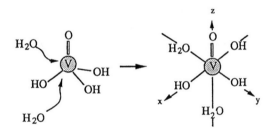

$$[VO(OH)_3]^° + 2H_2O \Rightarrow [VO(OH)_3(OH_2)_2]^°$$

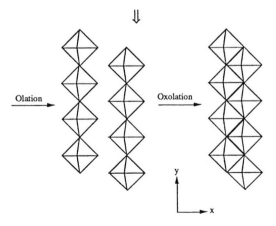

Figure 11. Suggested mechanism for the formation of V_2O_5 gels from $[VO(OH)_3]^0$ precursors.

Coordination expansion of the neutral precursor from 4 to 6 occurs via the nucleophilic addition of two water molecules. This gives hexacoordinated species $[VO(OH)_3(OH_2)_2]^0$ in which one water molecule lies along the z axis opposite to the short V=O double bond while the second one is in the equatorial plane opposite to an OH group. One $V-OH_2$ and three V-OH bonds are formed in this plane so that x and y directions are not equivalent. Fast olation reactions along the $HO-V-OH_2$ direction give rise to olated corner-sharing chain polymers. Slower oxolation reactions then lead to the formation of edge-sharing double chains. This results in a layered structure made of water molecules intercalated between V_2O_5 ribbons [23].

3. COMPLEXATION OF CATIONIC PRECURSORS

The formation of condensed phases from aqueous solutions is mainly governed by the pH of the solution. However most aqueous systems also contain counterions introduced during the dissolution of inorganic salts in water. Foreign anions (or cations) are also often added in the solution in order to obtain the desired products. Such ions, even when added in small amounts, can change the morphology, the structure and even the chemical composition of the resulting solid phases [12]. Spherical hematite particles ($\alpha-Fe_2O_3$) are obtained when aqueous solutions of $FeCl_3$ are heated under reflux while acicular particles are formed with $Fe(NO_3)_3$ in the presence of phosphate ions. Goethite ($\alpha-FeOOH$) is precipitated from $Fe(NO_3)_3$ while Akaganeite ($\beta-FeOOH$) is obtained with $FeCl_3$. The thermo-hydrolysis of $Ti(SO_4)_2$ aqueous solutions leads to anatase TiO_2 while the rutile phase is formed in the presence of chloride or nitrate .

These examples show that anions (or cations) can play an important role in the formation of condensed phases from aqueous solutions. They have not yet been taken into account in our description of hydrolysis and condensation reactions. However, negatively charged anions X^{x-} can react with positively charged aquo-cations (and vice versa) to give complexes as follows :

$$[M(OH_2)_N]^{z+} + aX^{x-} \Leftrightarrow [M(X)_a(OH_2)_{N-\alpha a}]^{(z-ax)+} + \alpha aH_2O \tag{29}$$

α corresponds to the number of water molecules which are replaced by one anionic ligand X. $\alpha=1$ for monodentate ligands such as Cl^- and $\alpha=2$ for bidentate ligands such as $(SO_4)^{2-}$. The stability of such complexes is given by the complexation constant:

$$k = [MX_a^{(z-ax)+}]/[M^{z+}][X^{x-}]^a \tag{30}$$

The complexation of metal cations by anionic species leads to the formation of a new precursor and the whole hydrolysis-condensation process is modified. Basic salts in which both X^- anions and OH groups are present can even be obtained when the complexing anion is not removed during the nucleation and growth of solid phases. Many data can be found in the literature about the formation of monomeric complexes but very little is known about the complexation of oligomeric species. It would therefore be very useful to be able to describe complexation within the frame of the Partial Charge Model.

Complexation and electronegativity.

Let us consider the complexation of aquo-cations by a monodentate and monovalent ($\alpha=1$, $x=1$) anion X^-. Complexation occurs in an aqueous medium where water can play a double role:
. Water is a solvent with a high dielectric constant ($\varepsilon \approx 80$). It leads to the dissociation of ionic species:

$$[M(X)(OH_2)_{N-1}]^{(z-1)+} \Rightarrow [M(OH_2)_N]^{z+} + X^-_{aq} \qquad (31)$$

. Water is a σ donor molecule so that anions also have to compete with aquo ligands. Complexation can be described as the nucleophilic substitution of water molecules by anions in the coordination sphere of metal cations. However, as complexation proceeds in an aqueous medium in the presence of a large excess of water molecules, the reverse reaction could also occur. Water is a Lewis base which can substitute less nucleophilic ligands leading to the hydrolytic dissociation of metal complexes as follows :

$$[M(X)(OH_2)_{N-1}]^{(z-1)+} + H_2O \Rightarrow [M(X)(OH_2)_N]^{(z-1)+}$$
$$\Rightarrow [M(HX)(OH)(OH_2)_{N-1}]^{(z-1)+} \Rightarrow [M(OH)(OH_2)_{N-1}]^{(z-1)+} + HX \qquad (32)$$

For complexes to remain stable in aqueous solutions, M-X bonds have to avoid both ionic dissociation by the aqueous solvent and hydrolytic dissociation via the nucleophilic substitution of HX by water molecules. Both reactions depend on the extent of electron transfer between metal and anion in the M-X bond. The complexing ability of a given anion X^- toward a metal cation M^{z+} then depends on charge transfers within the M-X bond, *i.e.* on the electronegativities of the complexed precursor (χ_p), the anion X^- (χ_x) and its protonated form HX (χ_{HX}). Two cases can be found:
i) The anion is more electronegative than the precursor ($\chi_x > \chi_p$). Electron density is withdrawn from the metal toward the anion (M$\Rightarrow$X). The negative charge of the anion increases

giving a highly polar M-X bond and ionic dissociation could occur. For a monovalent anion, the limiting partial charge would correspond to $\delta_x = -1$ in the $[M(X)(OH_2)_{N-1}]^{(z-1)+}$ precursor. ii) The anion is less electronegative than the precursor ($\chi_x < \chi_p$). The $M \Leftarrow X$ electron transfer gives a more covalent bond which remains stable toward ionic dissociation. However, as the electronegativity of X decreases, substitution by a water molecule could occur if HX becomes a positively charged leaving group (eq.32). The limiting partial charge would then be $\delta_{HX} = 0$ in the $[M(XH)(OH)(OH_2)_{N-1}]^{(z-1)+}$ precursor.

Two critical electronegativities, χ_d for ionic dissociation and χ_h for hydrolytic dissociation can be deduced from these two partial charge conditions ($\delta_x = -1$, $\delta_{HX} = 0$). Equalization of the mean electronegativity χ_p of the metal precursor with the electronegativities χ_x and χ_{HX} of the basic and acid forms of the anion in an aqueous solution leads to an electronegativity range $\chi_h \leq \chi_x \leq \chi_d$ where complexation is expected to occur. For a given cationic precursor $[M(OH)_h(OH_2)_{N-h}]^{(z-h)+}$ an electronegativity range can be defined in which anions are able to form complexes. Outside this range ionic ($\chi_x \geq \chi_d$) or hydrolytic ($\chi_x \leq \chi_h$) dissociation prevails in the aqueous medium and complexation does not occur.

Complexation and pH.

The above conditions for complexation ($\chi_h \leq \chi_x \leq \chi_d$) can only be applied to a given precursor at a given pH. Protonation or deprotonation reactions take place when changing the pH. The following side reactions have then to be also taken into account :

. Cationic precursors are deprotonated and their positive charge decreases when pH increases. Their mean electronegativity decreases and ionic dissociation would occur for less electronegative anions.

$$[M(OH)_h(OH_2)_{N-h}]^{(z-h)+} + q\ OH^- \Leftrightarrow [M(OH)_{h+q}(OH_2)_{N-h-q}]^{(z-h-q)+} + q\ H_2O \qquad (33)$$

. Anions are protonated when pH decreases giving rise to their acid forms.

$$X^{x-} + q\ H_3O^+ \Leftrightarrow [H_qX]^{(x-q)-} + q\ H_2O \qquad (34)$$

In both cases the observed complexation constant decreases and "M-X" complexes remain stable over a limited range of pH only. As a general rule, the pH at which complexes can be formed shifts toward higher values when the charge of the cation M^{z+} decreases, when the mean electronegativity of the anion decreases or when α increases. Some anions can then be complexing at low pH, at the beginning of the condensation process and become non complexing during the nucleation and growth of solid particles.

Complexation of Zr^{IV} precursors.

When dissolved in water, $ZrCl_4$ is readily hydrolyzed giving rise to h=2 species $[Zr(OH)_2(OH_2)_6]^{2+}$. OH groups are negatively charged ($\delta_{OH}=-0.06$) so that olation leads to the well known cyclic tetramer, $[Zr_4(OH)_8(OH_2)_{16}]^{8+}$ in which Zr atoms are eightfold coordinated by four bridging OH groups and four terminal water molecules [24].
The mean electronegativity of this tetramer is $\chi=2.7$ and partial charge calculations lead to the following critical electronegativity values for complexation (Fig.12):
. $\chi_h=2.49$ and $\chi_d=2.69$ for monodentate anions ($\alpha=1$)
. $\chi_h=2.50$ and $\chi_d=2.80$ for bidentate anions ($\alpha=2$)

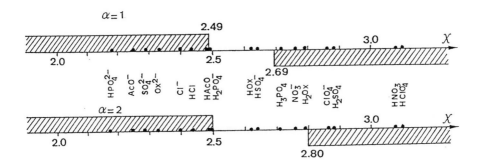

Figure 12. Electronegativity range for complexation of precursors by monodentate ($\alpha=1$) or bidentate ($\alpha=2$) anions.

Anions such as ClO_4^- and Cl^- are not able to give complexes with this h=2 precursor.
. ClO_4^- is a highly electronegative anions ($\chi=2.86>\chi_d$). It would give very polar ClO_4-Zr bonds in which the partial charge of the perchlorate anion would become more negative than -1. Ionic dissociation would occur in aqueous solutions and complexes cannot be formed.
. Cl_{aq}^- ($\chi=2.3<\chi_h$) is not electronegative enough to give monodentate complexes with the Zr^{IV} precursors below pH=5. Only outer sphere complexes can be formed at lower pH, in agreement with X-ray diffraction studies showing that $ZrOCl_2,nH_2O$ crystals are made of tetrameric cationic species $[Zr_4(OH)_8(OH_2)_{16}]^{8+}$ and Cl^- anions [25]. There is no Zr-Cl bond (Fig.13). Monoclinic zirconia can be obtained at low pH when an aqueous solution of $ZrOCl_2$ is heated under reflux. Cl^- anions are released in the solution and are not involved in the

26

polymerization process described by A. Clearfield [26]. Adding a base leads to the precipitation of amorphous hydrous zirconia ZrO_2,nH_2O. However, Cl^- anions become complexing during the process, between pH=5 and pH=9. They are released at higher pH but a significant amount of anions remain strongly adsorbed in the amorphous oxide. They can be removed by careful washing. The amount of Cl^- retained by the precipitate depends on the pH. It decreases as the final pH of precipitation increases.

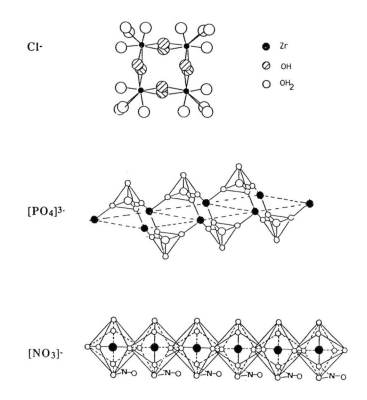

Figure 13. Structure of zirconium salts formed in the presence of anions.

Phosphates exhibit several protonated forms $[H_nPO_4]^{(3-n)-}$ which are strongly complexing over a large range of pH ($\chi(H_2PO_4)^- = 2.5$). They remain bonded to Zr^{IV} during the formation of the solid phase giving rise to the precipitation of phosphates at low pH. Phosphoric acid reacts with aqueous solutions of zirconium salts giving an amorphous gelatinous precipitate which crystallizes into α-$Zr(HPO_4)_2,H_2O$ when the solution is heated under reflux [27]. This phosphate is built up of negatively charged $[Zr(PO_4)_2]_n^{2n-}$ layers in which phosphate groups are bonded to three $[ZrO_6]$ octahedra (Fig.13).

Weakly complexing anions such as nitrates ($\chi=2.76$) can form bidentate complexes ($\alpha=2$) with tetrameric zirconium precursors below pH≈2, but not monodentate complexes ($\alpha=1$). A zirconium hydroxide nitrate $Zr(OH)_2(NO_3)_2,4.7H_2O$ has been precipitated from concentrated zirconium nitrate solutions at low pH. Its crystal structure was recently determined from X-ray powder diffraction data [28]. This basic salt is made up of $[Zr(OH)_2(OH_2)_2(NO_3)]^+$ chains in which zirconium atoms are eightfold coordinated by four OH groups, two water molecules and one chelating nitrate (Fig.13). These chains are held together by hydrogen bonds through additional water molecules and non complexing nitrate groups located between the chains. Nitrates can obviously be involved during the early stages of the hydrolysis-condensation process and basic salts are precipitated at low pH. However these anions are then released at higher pH so that amorphous hydrous zirconia is obtained at high pH when a base is added.

Sulphates $(H_nSO_4)^{(2-n)-}$ form complexes with Zr^{IV} precursors up to pH≈11. They can displace OH groups from hydrolyzed Zr^{IV} species and form sulphate bridges between zirconium atoms. Zirconium sulphate, $Zr(SO_4)_2,4H_2O$, is readily crystallized from sulphuric acid solutions. Many basic zirconium sulphates have been described in which $(SO_4)^{2-}$ groups behave as bridging ligands between zirconium atoms [30][31]. Moreover sulphate anions can be used to control the precipitation of zirconia. Polydispersed ZrO_2 particles are obtained when an acidic aqueous solution of $ZrOCl_2,8H_2O$ is refluxed whereas monodispersed spherical ZrO_2 particles are formed in the presence of K_2SO_4. Complexation with $(SO_4)^{2-}$ anions leads to the in-situ generation of sulphato-complexes which behave as precursors for nucleation. Protons are formed during the process and non complexing $(HSO_4)^-$ anions are released at a further stage leading to the precipitation of zirconia free of sulphate.

This example shows that multivalent anions, which exhibit several protonated forms, can be conveniently used to control the formation of solid phases. This is not possible with monovalent anions which lead to the nucleation and growth of basic salts as soon as they have some complexing behavior. Polyvalent anions are very important in the synthesis of solids from aqueous solutions. They prevent, or slow down, condensation reactions and can be used to tailor the size, morphology and structure of solid particles. Their precise role is not easy to describe and would require an accurate characterization of all the intermediate species formed during the nucleation and growth of the solid phase.

Formation of alumino-silicates

Like phosphates, silicate species $[H_nSiO_4]^{(4-n)-}$ are strongly complexing toward most cations. In the presence of high-valent transition metal cations such as W^{VI} or Mo^{VI}, they lead to the formation of the so-called heteropolyoxometalates [20]. These condensed species are usually made of trimeric $[MO_6]_3$ octahedral building blocks (M=W, Mo, ...) surrounding a central $[SiO_4]$ tetrahedron as in the well known Keggin ion $[SiW_{12}O_{40}]^{4-}$.

Silicate anions react readily with Al^{III} precursors to give alumino-silicate compounds. These reactions are currently used for the synthesis of zeolites. In most of the synthesis of zeolite molecular sieves, the first event that occurs upon mixing the silica and alumina sources is the precipitation of a gel.

. In alkaline solutions aluminium gives negatively charged tetrahedral $[Al(OH)_4]^-$ species, a structure consistent with its incorporation into a zeolite framework. Because of the lower charge of Al^{III} compared to Si^{IV}, counter cations such as Na^+ or $[N(CH_3)_4]^+$ have to be added for charge compensation. They are known to play a major role in the formation of the zeolite structure.

. At lower pH, when F^- is used as a mineralyzer instead of OH^- for instance, aluminium precursors are positively charged octahedral species. The formation of alumino-silicates can then be described as the complexation of cationic Al species by silicate anions.

4. ALKOXIDE PRECURSORS

The sol-gel synthesis of glasses and ceramics is mainly based on the hydrolysis and condensation of alkoxide precursors. These alkoxides are not soluble in water and a common solvent such as the parent alcohol has to be used. Water is typically diluted in an alcohol and added slowly to the alkoxide solution in order to prevent precipitation. The metal-organic route is much more versatile than the inorganic one. Many chemical parameters, other than pH can be used to control the reactions (solvent, hydrolysis ratio, alkoxy group, acid or base catalysis). However alkoxides are rather expensive and highly reactive toward moisture. Therefore they are not currently used in industry and most research is performed in universities. The sol-gel chemistry of alkoxides is somewhat different from the chemistry of aqueous solutions. However, as soon as an excess of water is added, most organic groups are removed and chemical reactions become similar to those observed in aqueous solutions.

Hydrolysis and condensation of metal alkoxides

The so-called sol-gel process is based on the hydrolysis and condensation of molecular precursors such as metal alkoxides $M(OR)_z$ where R is typically an alkyl group (R= CH_3, C_2H_5, ...).

$$\text{Hydrolysis :} \quad >M\text{-}OR + H_2O \Rightarrow >M\text{-}OH + ROH \tag{35}$$
$$\text{Condensation :} \quad >M\text{-}OH + RO\text{-}M< \Rightarrow >M\text{-}O\text{-}M< + ROH \tag{36}$$

These alkoxides have been known for a long time and have been extensively described in a book published almost twenty years ago [32]. They can be synthesized directly via the reduction of alcohols by strongly electropositive metals or via substitution reactions of metal salts such as chlorides with alcohols or alkaline alkoxides.

Alkoxy groups are rather hard π-donor ligands. They stabilize the highest oxidation number of the metal. Therefore alkoxides of main group elements and d^0 transition metals are well known while those corresponding to soft d^n late transition metals have been much less studied. The number and stability of metal alkoxides decrease from left to right across the periodic table [33].

For coordinatively saturated metals the hydrolysis and condensation of metal alkoxides $M(OR)_z$ corresponds to the nucleophilic substitution of alkoxy ligands by hydroxylated species XOH as follows:
$$M(OR)_z + x\ XOH \longrightarrow [M(OR)_{z-x}(OX)_x] + x\ ROH \tag{37}$$

where X stands for Hydrogen (hydrolysis), a Metal atom (condensation) or even an organic or inorganic Ligand (complexation).

These reactions can be described by a SN_2 mechanism :

$$
\begin{array}{c}
H \\
\backslash \\
O^{\delta-} + M^{\delta+}\text{-}O^{\delta-}\text{-}R \\
\diagup \\
X
\end{array}
\Rightarrow
\begin{array}{c}
H^{\delta+} \\
\backslash \\
O\text{-}M\text{-}O^{\delta-}\text{-}R \\
\diagup \\
X
\end{array}
\Rightarrow XO\text{-}M\text{-}
\begin{array}{c}
H^{\delta+} \\
\diagup \\
O \\
\backslash \\
R
\end{array}
\Rightarrow XO\text{-}M + ROH
\qquad (38)
$$

The reaction starts with the nucleophilic addition of negatively charged $HO^{\delta-}$ groups onto the positively charged metal $M^{\delta+}$, leading to an increase of the coordination number of the metal atom in the transition state. The positively charged proton is then transferred toward an alkoxy group and the positively charged protonated ROH ligand is finally removed.

The chemical reactivity of metal alkoxides toward hydrolysis and condensation mainly depends on the positive charge of the metal atom δ_M and its ability to increase its coordination number "N". As a general rule, the electronegativity of metal atoms decreases and their size increases when going toward the bottom left of the periodic table (Table 7). The corresponding alkoxides become progressively more reactive toward hydrolysis and condensation. Silicon alkoxides are rather stable while cerium alkoxides are very sensitive to moisture. Alkoxides of electropositive metals must be handled with care under a dry atmosphere otherwise precipitation occurs as soon as water is present. Alkoxides of highly electronegative elements such as $PO(OEt)_3$ cannot be hydrolyzed under ambient conditions, whereas the corresponding vanadium derivatives $VO(OEt)_3$ are readily hydrolyzed into vanadium pentoxide gels.

Table 7. Electronegativity "χ", partial charge "δ_M", ionic radius "r" and maximum coordination number "N" of some metal alkoxides

alkoxide	χ	δ_M	r(Å)	N
$Si(OPr^i)_4$	1.74	+0.32	0.40	4
$Ti(OPr^i)_4$	1.32	+0.60	0.64	6
$Zr(OPr^i)_4$	1.29	+0.64	0.87	7
$Ce(OPr^i)_4$	1.17	+0.75	1.02	8
$PO(OEt)_3$	2.11	+0.13	0.34	4
$VO(OEt)_3$	1.56	+0.46	0.59	6

Sol-gel chemistry of silicon alkoxides.

The sol-gel chemistry of silicon alkoxides has been extensively described [1]. It is not very different from the chemistry of silicates in aqueous solutions. The reaction of silicon tetrachloride with water or alcohol for instance could be written as follows:

$$SiCl_4 + 2H_2O \Rightarrow Si(OH)_4 + 4HCl \tag{39}$$
$$SiCl_4 + 4ROH \Rightarrow Si(OR)_4 + 4HCl \tag{40}$$

The hydrolysis of $Si(OR)_4$ then leads to $Si(OH)_4$ and the formation of SiO_2 [34].

Si^{IV} is fourfold coordinated ($N=z=4$) in the precursor as well as in the oxide so that coordination expansion does not occur and silicon alkoxides $Si(OR)_4$ are always monomeric. The electronegativity of Si is rather high ($\chi=1.74$) and its positive charge ($\delta\approx+0.3$) quite small. Silicon alkoxides are therefore not very sensitive toward hydrolysis. Gelation can take several days. Their reactivity decreases when the size of alkoxy groups increases. This is mainly due to steric hindrance which prevents the formation of hypervalent silicon intermediates.

As for aqueous solutions, hydrolysis and condensation rates of silicon alkoxides can be enhanced by acid or base catalysis. Inorganic acids reversibly protonate negatively charged alkoxide ligands and increase the reaction kinetics by producing better leaving groups. Basic catalysis provides better nucleophilic OH^- groups for hydrolysis whereas deprotonated silanol groups $Si-O^-$ enhance condensation rates. Acid catalysis mainly increases hydrolysis rates whereas basic catalysis enhances condensation [1].

However catalysis does not only increase reaction rates. As in aqueous solutions, it leads to polymeric species of different shapes. The negative charge of OR groups increases as the electron-providing power of oxo and alkoxo ligands increases. The ease of protonation of OR groups therefore decreases as the connectivity of the adjacent Si atom increases. Acid-catalyzed condensation is directed preferentially toward the ends of oligomeric species resulting in chain polymers (Fig.14a). The positive partial charge δ_{Si} increases with its connectivity so that nucleophilic addition of $Si-O^-$ is directed preferentially toward the middles of oligomers leading to more compact, highly branched species (Fig.14b).

The chemical reactivity of silicon alkoxides can also be increased by nucleophilic activation in the presence of chemical species such as DMAP (dimethylaminopyridine), NaF or n-Bu$_4$NF which behave as Lewis bases. A pentavalent intermediate is reversibly formed with F^- that stretches and weakens the surrounding Si-OR bonds. The positive charge δ_{Si} increases rendering the silicon atom more prone to nucleophilic attack. This nucleophilic activation is efficient for both hydrolysis and condensation reactions [35].

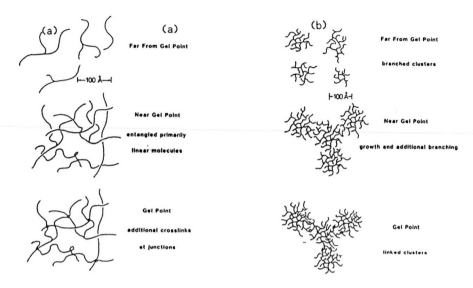

Figure 14. Formation of oligomeric silica species via acid (a) or base (b) hydrolysis [1]

Complexation of metal alkoxides and chemically controlled condensation

Most metal alkoxides, other than silicon, are highly reactive toward hydrolysis and condensation. Precipitation occurs as soon as water is added. Therefore their chemical reactivity has to be tailored in order to avoid uncontrolled precipitation. This is now currently performed via the chemical modification (or complexation) of metal alkoxides [2]. The strong electronegativity of oxygen (χ_O=3.5) makes the M-OR bond strongly polar and the metal atom highly prone to nucleophilic reagents [33]. Metal alkoxides react with hydroxylated ligands XOH such as carboxylic acids or β-diketones giving $[M(OR)_{z-n}(OX)_n]$ species. Such reactions lead to important modifications of the molecular structure of the alkoxide. As for aqueous solutions, a new precursor is formed which exhibits different chemical reactivity and functionality. Complexed alkoxides are usually much less sensitive to hydrolysis and nucleophilic chemical additives are currently employed in order to stabilize highly reactive metal alkoxides and control condensation reactions [36].

It has been observed for instance that adding acetic acid to titanium alkoxides prevents the precipitation of TiO_2 and increases the gelation time [37]. Acetate groups (OAc = CH_3COO) behave as bidentate bridging ligands. The coordination number of titanium increases from four to six and oligomeric species $[Ti(OPr^i)_3(OAc)]_n$ (n=2 or 3) are formed for a molar ratio

AcOH/Ti=1 (Fig.15a). Acetate ligands are less easily hydrolyzed than alkoxy groups therefore slowing down condensation reactions.

Similar reactions have been observed with acetyl acetone (acacH=CH_3-CO-CH_2-CO-CH_3) which reacts with metal alkoxides as a chelating ligand [38]. Oligomers are not readily formed and for a stoichiometric acac/Ti=1 ratio, nucleophilic substitution leads to [Ti(OPr^i)$_3$(acac)] monomers in which Ti is only fivefold coordinated (Fig.15b). This strongly chelating ligand cannot be removed upon hydrolysis (unless pH<2) even in the presence of a large excess of water. Condensation is then prevented and only small oligomers are formed.

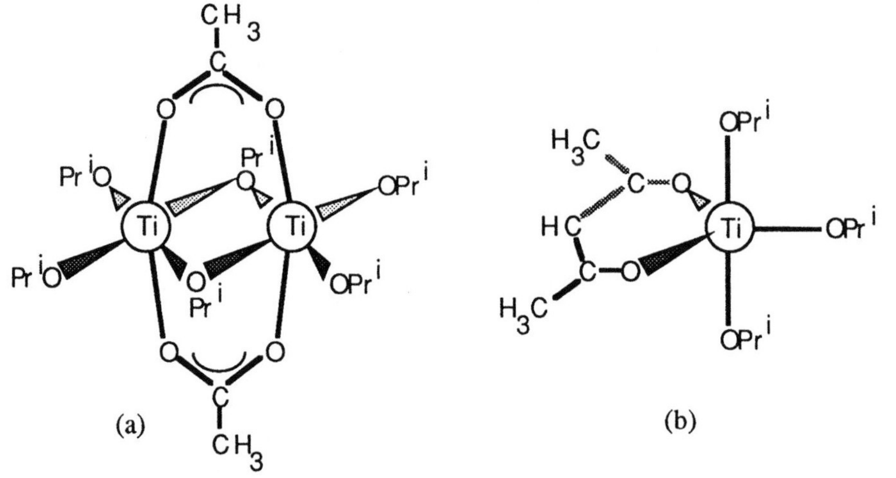

Figure 15. Chemical modification of Ti(OPr^i)$_4$ with :
a) acetic acid [Ti(OPr^i)$_3$(OAc)]$_2$ and b) acetylacetone [Ti(OPr^i)$_3$(acac)]

The hydrolysis of metal alkoxides gives reactive M-OH bonds which lead to condensation and favor the formation of larger species. Complexation leads to M-OX bonds which act as polymerization lockers and favor the formation of smaller species. A large variety of oligomeric species can then be obtained upon hydrolysis and condensation. Molecular clusters or colloidal particles can be synthesized depending on the relative amount of hydrolysis (h=H_2O/M) and complexation (x=X/M).

Titanium ethoxide Ti(OEt)$_4$ is highly reactive toward hydrolysis and leads to the uncontrolled precipitation of TiO_2 when water is added. However, for a very low hydrolysis ratio (h<1), condensation is mainly governed by the formation of μ-oxo and alkoxo bridges.

Oligomeric oxo-alkoxides such as $Ti_7O_4(OEt)_{20}$ (h=0.6) [39], $Ti_{10}O_8(OEt)_{24}$ (h=0.8) [40] or $Ti_{16}O_{16}(OEt)_{32}$ (h=1) [41] are obtained (Fig.16). They are structurally related to the well-known polyoxoanions in aqueous solutions [20].

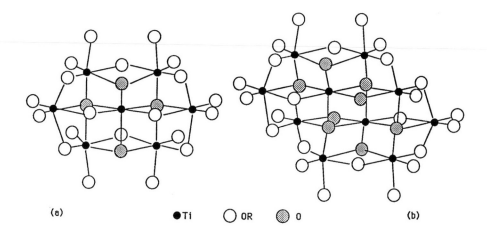

Figure 16. Molecular structure of titanium oxo-alkoxides
a) $Ti_7O_4(OEt)_{20}$ b) $Ti_{10}O_8(OEt)_{24}$

The hydrolysis of titanium alkoxides complexed by acetylacetone leads to different oligomeric species depending on the respective values of x and h. [TiO(acac)]$_2$ is formed upon hydrolysis of Ti(acac)$_2$(OR)$_2$. It is made of dimers with sixfold coordinated Ti atoms linked through oxygen bridges. Strongly complexing acac ligands cannot be hydrolyzed easily and condensation does not go any further. Larger molecular species such as $Ti_{18}O_{22}(OBu)_{26}(acac)_2$ are obtained for smaller amounts of acetylacetone (x=0.1, h=1.2) [42]. They are built of eighteen [TiO$_6$] octahedra sharing edges or corners whereas complexing acac ligands and non hydrolyzed OR groups remain outside of the [$Ti_{18}O_{22}$] core of the molecule (Fig.17a).

Similar experiments have been reported with zirconium alkoxides [43]. In the presence of a small amount of water (h<1), only few alkoxy groups are removed leading to molecular oligomeric species. Hydrolyzed alkoxy groups give rise to μ-oxo or μ-OH bridges, while complexing ligands remain bonded to the metal atom. They are located outside the oxo rich core of the molecular species (i.e. at the surface) and act as termination ligands preventing further condensation (Fig.17b, c).

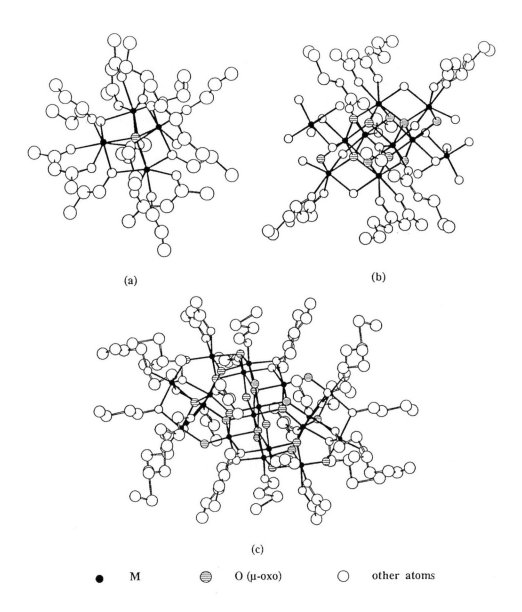

(a)

(b)

(c)

● M   ⊜ O (μ-oxo)   ○ other atoms

Figure 17. Molecular structure of oxo-alkoxides

a) $Zr_4O(OPr^n)_{10}(acac)_4$, b) $Zr_{10}O_6(OH)_4(OPr^n)_{18}L_6$, c) $Ti_{18}O_{22}(OBu)_{26}(acac)_2$

Polymeric sols are obtained when more water is added to zirconium alkoxide solutions. They are made of amorphous zirconium oxo-species in which most acac ligands remain bonded to zirconium preventing condensation and the formation of an oxide network. Crystalline zirconia particles are never obtained when hydrolysis is performed at room temperature. For crystallization to occur, zirconium bonded complexing ligands have to be removed in order to allow the formation of Zr-O-Zr bonds. A crystalline oxide network is formed upon heating as for the thermo-hydrolysis of aqueous solutions of zirconium oxychloride. However in order to avoid aggregation reflux has to be performed in a highly acid medium, in the presence of paratoluene sulfonic acid (PTSA). The reversible dissociation of Zr-acac bonds then occurs followed by preferential recoordination at the surface of the growing cluster. These surface acetylacetonato ligands, in equilibrium with free acac in the solution, slow down the growth process of the colloidal particles and prevent their aggregation via steric hindrance effects. Kinetically stable non-aggregated crystalline zirconia nano-particles some nanometers in diameter are then obtained [44].

Hetero-alkoxides and molecular building blocks.

In covalent alkoxides $M(OR)_z$ metal atoms usually exhibit a lower coordination than in the oxide $MO_{z/2}$. Coordination expansion is therefore a general tendency of the sol-gel chemistry of metal alkoxides, leading to the formation of oligomeric alkoxides or oxoalkoxides [32].

Oligomerization occurs via the nucleophilic addition of two alkoxide molecules and the formation of alkoxy bridges. This reaction is similar to the olation reaction described previously but, as coordination expansion is possible, there is no need for a leaving group. As for OH, OR groups can be bonded to two (μ_2-OR) or three (μ_3-OR) metal atoms. Most metal alkoxides exhibit oligomeric molecular structures $[M(OR)_z]_n$. Their molecular complexity increases with the size of the metal atom, its electropositive character and the "N-z" difference. It therefore increases when going down the periodic table (Table 8).

Table 8. Molecular complexity "n" of some tetravalent metal alkoxides [32]

M	Si (OEt)$_4$	Ti (OEt)$_4$	Zr(OEt)$_4$	Hf(OEt)$_4$	Th(OEt)$_4$
n	1	2.9	3.6	3.6	6

The molecular complexity of metal alkoxides also depends on the steric hindrance of alkoxy groups. Bulky secondary or tertiary alkoxy groups tend to prevent oligomerization. Trimeric species $[Ti(OEt)_4]_3$ have been evidenced in pure liquid titanium ethoxide (Fig.18b) whereas titanium iso-propoxide $Ti(OPr^i)_4$ remains monomeric (Fig.18a). This is no more the case for zirconium iso-propoxide which is dimeric because of the larger size of Zr^{IV}. Moreover, solvent molecules can also be used for coordination expansion leading to solvated dimers $[Zr(OPr^i)_4(Pr^iOH)]_2$ when the alkoxide is dissolved in its parent alcohol (Fig.18c).

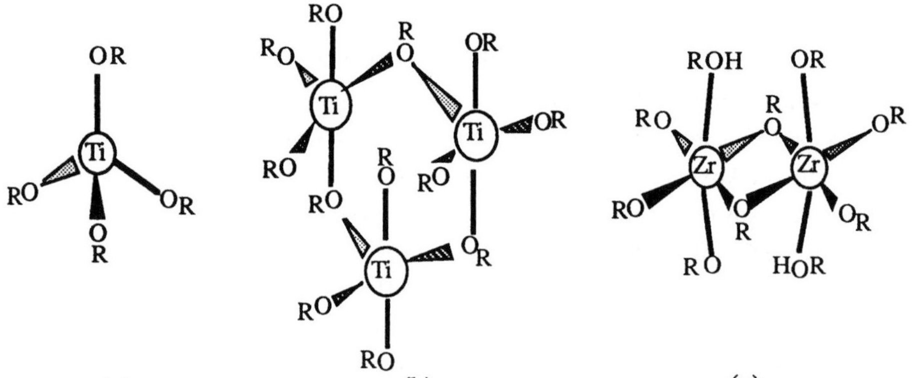

(a)   (b)   (c)

Figure 18. Molecular structure of transition metal alkoxides.
a : $Ti(OPr^i)_4$, b : $[Ti(OEt)_4]_3$, c : $[Zr(OPr^i)_4(Pr^iOH)]_2$

Most advanced ceramics or molecular sieves are multicomponent materials having two or more types of cations in the lattice. Since alkoxide precursors are mixed at the molecular level in the solution, a high degree of homogeneity can be expected. However a major problem in forming homogeneous multicomponent gels is the unequal hydrolysis and condensation rates of the metal alkoxides. This may result in phase separation during hydrolysis or thermal treatment, leading to higher crystallization temperatures or even undesired phases. It is therefore necessary to prepare gels of high homogeneity in which cations of various kinds are uniformely distributed at an atomic scale through M-O-M' bridges. Several approaches have been attempted to overcome this problem, including the partial prehydrolysis of the less reactive alkoxide or matching of hydrolysis rates by the complexation of the most reactive alkoxide.

In order to prepare crystalline materials at low temperatures it would be necessary to design metal-organic precursors such that the metal ions are dispersed at the molecular level and the ligands undergo facile elimination during the transformation from molecular to bulk material. The development of precursor solutions which consist of polynuclear complexes with the metals in the stoichiometry of the desired oxide product will have a beneficial effect by lowering processing temperatures and times. Heterometallic alkoxides, which contain two or more different metal atoms linked by μ-OR bridges are claimed to provide best precursors for the sol-gel synthesis of multicomponent materials [45].

The formation of heteroalkoxides is governed by coordination expansion and acid-base properties. The nucleophilic addition of alkoxide groups between two different alkoxides can be described as a Lewis acid-base reaction. It is favored by a large difference in the electronegativity of metal atoms. Heteroalkoxides are therefore easily formed by simple mixing of two alkoxides of low and high electronegativity. Heteroalkoxides incorporating an alkali metal such as $LiNb(OEt)_6$ represent so far the most important class of heteroalkoxides. They are often used as "building blocks" for the sol-gel synthesis of multicomponent ceramics. The $LiNbO_3$ perovskite or the spinel $MgAl_2O_4$ for instance have been obtained via the hydrolysis of the bimetallic alkoxides $[LiNb(OEt)_6]$ and $[Mg(Al(OR)_4)_2]$ respectively [46][47]. Heterometallic alkoxides are not limited to two metals. A wide range of polymetallic alkoxides containing up to five different metals in the same molecular species have been reported. This explains the emphasis on heterometallic alkoxides. However polar M-OR-M' bonds can be broken upon hydrolysis and the question arises as to how the sol-gel synthesis really leads to the formation of an homogeneous oxopolymer or oxide network rather than in nanophase materials.

Oxo-alkoxides would therefore be better precursors than heterometallic alkoxides. Condensation, via ether elimination, leads to the formation of oxo-bridges [45]. This normally occurs upon heating and μ-oxo complexes are often formed during the purification of metal alkoxides via distillation. Condensation via ether elimination is favored by the smaller size of oxo ligands and their ability to exhibit higher coordination numbers, up to six, favoring coordination expansion of the metal atom. Oxo-alkoxides are more stable than the corresponding alkoxides and less reactive toward hydrolysis and condensation. The tendency to form oxo bridges increases with the size and charge of metal ions. Large electropositive metals are known to give oxoalkoxides such as $Pb_4O(OEt)_6$, $Pb_6O_4(OEt)_4$, $Bi_4O_2(OEt)_8$, $Y_5O(OPr^i)_{13}$ or $Nb_8O_{10}(OEt)_{20}$ [48]. They are usually built of edge sharing $[MO_6]$ octahedra and their molecular structure is close to that of the corresponding polyoxoanions formed in aqueous solutions [20].

As for homometallic alkoxides, oxo-bridges can also be formed when two different metal alkoxide solutions are refluxed leading to heterometallic oxo-alkoxides. $Pb_4O(OEt)_6$ for instance undergoes complete dissolution in ethanol when $Nb(OEt)_5$ is added giving rise to $[Pb_6O_4(OEt)_4][Nb(OEt)_5]_4$. Hetero-alkoxides can also provide molecular precursors with the correct M'/M stoichiometry in which some M-O-M' bonds are already formed. Whereas alkoxy bridges are usually hydrolyzed during the sol-gel synthesis, oxo bridges are strong enough not to be broken. The crystallization of the perovskite $BaTiO_3$ phase synthesized via the sol-gel route occurs around 600°C when $Ti(OPr^i)_4$ and $Ba(OPr^i)_2$ are used as precursors. This crystallization temperature decreases to 60°C when the mixture of alkoxides is refluxed prior to hydrolysis [49]. This should be due to the in-situ formation of the bimetallic oxo-alkoxide $[BaTiO(OPr^i)_4(Pr^iOH)]_4$. Such a compound has been isolated as a single crystal and characterized by X-ray diffraction [50]. This bimetallic precursor exhibits the right Ba/Ti stoichiometry and Ba-O-Ti bonds are already formed in the solution.

Synthesis of alumino-silicates from alkoxides

Molecular sieves are often formed of an alumino-silicate network. The problem is then to make Si-O-Al bonds in the molecular precursor and to keep them during the whole hydrolysis and condensation process. However because of the large difference in reactivity of aluminium and silicon alkoxides towards hydrolysis and condensation, it appears to be rather difficult to prepare homogeneous SiO_2-Al_2O_3 gels.

By far the most used synthetic approach is to partially pre-hydrolyze silicon tetraethoxide (TEOS) with a small amount of acidic water in order to form $Si(OEt)_{4-x}(OH)_x$ species $(x\neq0)$ which could then react with $Al(OR)_3$ to form Si-O-Al bonds. Commercially available double alkoxides such as $(Bu^sO)_2$-Al-O-Si$(OEt)_3$ have also been used. Si-O-Al bonds are already formed in this molecular precursor and it could be expected that they would remain in the gel. However, ^{29}Si and ^{27}Al NMR experiments show that Al-OBu^s groups are hydrolyzed much faster than Si-OEt. groups. Condensation then occurs with Al-OBu^s rather than with Si-OEt leading to the formation of Al-O-Al bonds instead of Al-O-Si bonds [51].

Moreover Al^{III} exhibits several coordination numbers, from four to six, so that coordination expansion tends to occur as soon as water is added for hydrolysis. The tetrahedral coordination of aluminium can be preserved either via complexation or at high pH.

Cordierite $Mg_2Al_4Si_5O_{18}$ is a typical example of the sol-gel synthesis of a silica based multicomponent compound in which four-fold coordinated Al^{III} behaves as a network former. When aqueous solutions of aluminum nitrate $Al(NO_3)_3$ are used as precursors together with TEOS and magnesium nitrate, six-fold coordinated Al^{III} species are formed in the solution. This coordination state remains in the gel that can be described as a silica gel in which Al^{III} and Mg^{II} ions are dispersed. No Si-O-Al bonds are formed at this stage. During a thermal treatment, Al^{III} reacts with Mg^{II} via solid-state reactions leading to the spinel phase $MgAl_2O_4$. The μ-cordierite phase is formed at higher temperature, and still remains the main phase at 1000°C [52].

The formation of the spinel phase can be prevented when aluminium alkoxides are used as precursors [53]. The μ-cordierite phase is then obtained directly around 900°C and completely transforms into α-cordierite at 1000°C. However $Al(OBu^s)_3$ is very viscous and highly reactive toward moisture. Better results are obtained with aluminium alkoxides modified with acetylacetone or ethylacetoacetate (etac). The hydrolysis of $[Al(OBu^S)_2(etac)]_n$ leads to the formation of Al-OH groups, but their condensation with $Al-OBu^S$ groups is partly prevented by the etac ligand. Hydroxylated Al-OH groups can then react with $Si(OR)_4$ to give Al-O-Si bonds [54]. $[Al(OBu^S)_2(etac)]_n$ has been co-hydrolyzed with TEOS using an aqueous solution of magnesium acetate. ^{27}Al NMR spectra show the formation of a transient species characterized by a sharp peak at $\delta \approx 51ppm$ due to tetracoordinated Al atoms in a highly symmetric environment which could be assigned to $Al\{O[Si(OR)_3]\}_4$ species. ^{29}Si NMR also shows a peak due to $Si(OAl)(OR)_3$. Such species could behave as nuclei for the growth of a silico-aluminate network [53]. Si-O-Al bonds are not broken when the gel is heated. The spinel phase $MgAl_2O_4$ is not formed and the pure α-cordierite phase is obtained at 1000°C.

These sol-gel preparations of cordierite clearly demonstrate the importance of the formation of heterometallic Si-O-Al bonds in the solution. Aluminum alkoxides provide reactive species containing Al-OH groups which can easily react with Si-OEt groups to built an alumino-silicate network. An interesting point is the fact that Al remains four-fold coordinated. This seems to be due to the complexation of Al^{III} by silicate species. All SiO_4 tetrahedra seem to be bonded to one or even two or three, AlO_4 tetrahedra.

5. CONCLUSION

All chemical mechanisms reported in this paper could be applied to the synthesis of molecular sieves from aqueous solutions or alkoxide precursors. They have been described in the frame of the partial charge model. The chemical reactivity of molecular precursors was described as a function of parameters such as electronegativity, pH, hydrolysis ratio, complexation, steric hindrance. However many other parameters could also be taken into account. May be the two most important ones would be temperature and templates.

. Hydrolysis can be described as a Lewis acid-base reaction :

$$>M-OH_2 + H_2O \Leftrightarrow >M-OH + H_3O^+ \tag{41}$$

The enthalpy change for this reaction ΔH is positive and often close to the enthalpy of dissociation for water (13.3 kcal/mole). Hence the tendency of cations to hydrolyze increases with temperature. This has been widely used to perform the homogeneous hydrolysis of metal cations via thermohydrolysis in order to obtain monodispersed particles [6].
. Heating leads to a decrease of the viscosity of water allowing the faster diffusion of solute species. The dielectric constant of water increases with pressure but decreases drastically when

the temperature increases. This point is very important, water looses its properties as a good solvent toward ionic species. Ionic dissociation becomes less important so that ion-pairs and neutral complexes are more often observed.

. Templates are also often used in order to obtain the desired phase. Their role has not been described in this paper. It could be related to the complexation of molecular precursors but weak interactions (van de Waals, hydrogen bond, hydrophilic-hydrophobic interactions, solvation...) should be involved rather than true chemical bonds as in the formation of metal complexes.

REFERENCES

1. C.J. Brinker, G.W. Scherer, Sol-Gel Science, Academic Press, New York (1989)
2. J. Livage, M. Henry, C. Sanchez, Progress in Solid State Chem., 18 (1988) 259
3. L.C. Klein, Sol-Gel technology, Noyes Pub., Park Ridge (1988)
4. R.M. Barrer, Hydrothermal Chemistry of Zeolites, Academic press, London (1982)
5. C.F. Baes, R.E. Mesmer, Hydrolysis of Cations, J. Wiley, New York (1976)
6. E. Matijevic, Ann. Rev. Mater. Sci., 15 (1985) 483
7. M. Henry, J.P. Jolivet, J. Livage, Structure and Bonding, 77 (1992) 153
8. R.T. Sanderson, Science, 114 (1985) 670
9. A.L. Allred, E. Rochow, J. Inorg. Nucl. Chem., 5 (1958) 264
10. C.K. Jorgensen, Inorganic Complexes, Academic Press, London (1963)
11. D.L. Kepert, The Early Transition Metals, Academic Press, London (1972)
12. E. Matijevic, Acc. Chem. Res., 14 (1981) 22
13. M. Ardon, A. Bino, Structure and Bonding, 65 (1987) 1
14. G. Fu, L.F. Nazar, Chem. Mater., 3 (1991) 602
15. G. Sposito, The Environmental Chemistry of Aluminium, CRC Press, Boca Raton (1989)
16. G. Johansson, Acta Chem. Scand., 16 (1962) 403
17. J.W. Akitt, B. Milic, J. Chem. Soc. Dalton Trans., (1984) 981
18. G. Johansson, Arkiv Kemi, 20 (1962) 321
19. H. Stünzi, L. Spiccia, F.P. Rotzinger, W. Marty, Inorg. Chem., 28 (1989) 66
20. M.T. Pope, Heteropoly and Isopolyoxometallates, Springer Verlag, Berlin (1983)
21. J.W. Akitt, J.M. Elders, J. Chem. Soc. Dalton Trans., (1988) 1347
22. R.K. Iler, The Chemistry of Silica, John Wiley & Sons, New York (1979)
23. J. Livage, Chem. Mater., 3 (1991) 578.
24. A. Clearfield, P.A. Vaughan, Acta Cryst., 9 (1956) 555
25. D.B. McWhan, G. Lundgren, Acta Cryst., 16 (1963) A36

26. A. Clearfield, Rev. Pure and Appl. Chem., 14 (1964) 91

27. A. Clearfield, Inorganic Exchange Materials, CRC Press, Boca Raton (1981)

28. P. Bénard, M. Louer, D. Louer, J. Solid State Chem., 94 (1991) 27

29. D.B. McWhan, G. Lundgren, Inorg. Chem., 5 (1966) 284

30. P.J. Squattrito, P.R. Rudolf, A. Clearfield, Inorg. Chem., 26 (1987) 4240

31. M. El Brahami, J. Durand, L. Cot, Eur. J. Solid State Inorg. Chem., 25 (1988) 185

32. D.C. Bradley, R.C. Mehrotra, D.P. Gaur, Metal Alkoxides, Academic Press, London (1978)

33. L.G. Hubert-Pfalzgraf, New J. Chem., 11 (1987) 663.

34. L.L. Hench, J.K. West, Chem. Rev. 90 (1990) 33

35. V. Belot, R. Corriu, C. Guérin, B. Henner, D. Leclercq, H. Mutin, A. Vioux, Q. Wang, Mater. Res. Soc. Symp. Proc., 180 (1990) 3

36. C. Sanchez, J. Livage, New J. Chem., 14 (1990) 513

37. S. Barboux-Doeuff, C. Sanchez, Mat. Res. Bull., 29 (1994) 1

38. A. Leaustic, F. Babonneau, J. Livage, Chem. Mater., 1 (1989) 240

39. K. Watenpaugh, C.N. Caughlan, Chem. Comm., 2 (1967) 76

40. V.W. Day, T.A. Ebersacher, W.G. Klemperer, C.W. Park, F.S. Rosenberg, J. Am. Chem. Soc., 113 (191) 8190

41. A. Mosset, J. Galy, C.R. Acad. Sci. Fr., 307 (1991) 8190

42. P. Toledano, M. In, C. Sanchez, C.R. Acad. Sci. Fr., 313 (1991) 1247

43. P. Toledano, M. In, C. Sanchez, C.R. Acad. Sci. Fr., 311 (1990) 1161

44. M. Chatry, M. Henry, C. Sanchez, J. Livage, Mater. Res. Bull. (in the press)

45. K.G. Caulton, L. Hubert-Pfalzgraf, Chem. Rev., 90 (1990) 969.

46. D.J. Eichorst, D.A. Payne, S.R. Wilson, K.E. Howard, Inorg. Chem., 29 (1990) 1458

47. K. Jones, T.J. Davies, H.G. Emblem, P. Parkes, Mater. Res. Soc. Symp. Proc., 73 (1986) 111

48. L. Hubert-Pfalzgraf, R. Papiernik, M.C. Massiani, B. Septe, Better Ceramics through Chemistry IV, Mater. Res. Soc. Symp. Proc., 180 (1990) 393

49. K.S. Mazdiyasni, R.T. Dolloff, J.S. Smith, J. Am. Ceram. Soc., 52 (1969) 523

50. A.I. Yanovsky, M.I. Yanoskaya, V.K. Limar, V.G. Kessler, N.Y. Turova, Y.T. Struchkov, J. Chem. Soc. Chem. Comm., (1986) 1605

51. J.C. Pouxviel, J.P. Boilot, A. Dauger, L. Hubert, Mater. Res. Soc. Symp. Proc., 73 (1986) 269.

52. U. Selvaraj, S. Komarneni, R. Roy, J. Am. Ceram. Soc., 73 (1990) 3663

53. L. Bonhomme-Coury, F. Babonneau, J. Livage, Chem. Mater., 5 (1993) 323

54. L. Bonhomme-Coury, F. Babonneau, J. Livage, J. Sol-Gel Sci. Techn. (in press)

J.C. Jansen, M. Stöcker, H.G. Karge and J. Weitkamp (Eds.)
Advanced Zeolite Science and Applications
Studies in Surface Science and Catalysis, Vol. 85
1994 Elsevier Science B.V.

Effects of Seeding on Zeolite Crystallisation, and the Growth Behavior of Seeds

Sylvie Gonthier and Robert W. Thompson

Department of Chemical Engineering, Worcester Polytechnic Institute
100 Institute Road, Worcester, Massachusetts 01609, U.S.A.

It is common to add seed crystals of desired molecular sieve zeolites to synthesis batches to promote crystallisation of certain zeolite phases, and to increase the rate of crystallisation. However, the effect of seed crystal addition on the crystallisation mechanism(s) has not been thoroughly assessed or understood.

The fundamental influence of the addition of zeolite seed crystals to a molecular sieve zeolite crystallisation batch is evaluated. The crystallisation rate enhancement frequently observed as a result of seed crystal addition to batch synthesis mixtures is reviewed and interpreted. The similarities between adding "macroscopic" sized seed crystals, adding "directing agents," and aging a synthesis solution are pointed out.

Lastly, by evaluating seeded synthesis experiments, it is shown that a great deal can be learned regarding apparent intergrowths of crystals frequently observed in zeolite syntheses. Numerous rather random aggregations of crystals can be understood as **un-ordered overgrowths** which, at best, are very loosely connected ensembles of single crystals which came together during their growth stages.

1. INTRODUCTION

The crystallisation of molecular sieve zeolites generally involves mixing an aluminate source with a silicate source in a basic medium to form an amorphous aluminosilicate gel, followed by hydrothermal treatment at some predetermined elevated temperature. Syntheses are usually done in batch systems, and, therefore, the amount of crystalline zeolite increases with time from near zero initially to essentially 100% at later times, while the amorphous gel dissolves simultaneously to replenish the reagents in solution as they are consumed by crystallisation. Successful syntheses produce essentially pure crystalline zeolite. The per cent zeolite plotted against time typically gives a sigmoid shaped curve, commonly referred to as the "crystallisation curve" for that synthesis. Details of synthesis procedures and applications of molecular sieve zeolites can be found in two classic treatises on the subject [1,2].

Mechanisms of crystallisation have been investigated for decades, but most crystallisation processes occurring in solutions have long been agreed to involve a nucleation step followed by crystal growth. Nucleation, that is the creation of the smallest entity which could be recognized as having the crystalline atomic structure, may occur by one or more of several mechanisms, which are well summarized in most up-to-date texts on crystallisation [e.g., 3]. Crystal growth in solution, the continued evolution of the nuclei to larger, macroscopic sizes, is generally agreed to involve the assimilation of dissolved solute into the ordered crystalline atomic form at the crystal surface, and may involve a diffuse boundary near the surface in which the ordering process takes place.

The issue of most interest in the debate over the mechanisms of molecular sieve zeolite crystallisation in hydrothermal systems is the nucleation mechanism. Of the myriad alternatives which have been proposed (homogeneous, heterogeneous, autocatalytic, solid-solid transformation,...), there has been little agreement on the governing nucleation mechanism(s), even in unseeded systems. The prevailing nucleation mechanism(s) in zeolite crystallisation systems will have to be understood before meaningful progress can be made in predicting or controlling the outcome of such crystallisation processes. In addition, it is known that several quite different mechanisms of secondary nucleation have been observed in solution crystallisations: microattrition, fluid shear, needle breeding, and initial breeding. The characteristic feature of these nucleation mechanisms is that the nucleation rate is catalysed by the presence of macroscopic parent crystals, including seed crystals. The standard text on this subject is sufficiently enlightening that further discussion of the details of these mechanisms is not warranted [3].

It has been known since at least the late 1960's that adding seed crystals of the desired zeolite phase to a synthesis batch increases the "rate of crystallisation," defined loosely as the slope of the crystallisation curve previously described, and shortens the time required for the crystallisation to be completed [4,5]. Numerous other studies have reported on crystallisation rate enhancement for other zeolite synthesis systems [6-11]. The mechanism of this rate enhancement has not been clearly delineated, but two explanations have been offered: i) the added surface area of the seed crystals results in the more rapid consumption of reagents, reducing the supersaturation, even to the extent that nucleation of new crystals is prohibited, or ii) seeds promote nucleation by some secondary nucleation mechanism [3], as yet not clearly specified. The additional area resulting from this enhanced nucleation, then, results in the faster consumption of reagents.

Several other examples of this behaviour when seeds were added have been reported. Han, et al [12] reported that adding seeds of ZSM-5 to a ZSM-5 synthesis batch enabled them to separate the nucleation and growth stages of the process. Ueda and Koizumi [13] added analcime seeds to an analcime batch synthesis solution, and showed that a new population of crystals was nucleated, even though the addition of seeds might have relieved the supersaturation by growth on the seed crystal surfaces and completely suppressed nucleation of a new population. Chi and Sand [14] reported on the seeded and unseeded synthesis of clinoptilolite, and concluded that systems seeded with clinoptilolite crystals proceeded more quickly than unseeded systems, and produced purer products.

It also has been found that, on occasions, adding seed crystals of a particular zeolite phase forces the solution to produce that particular phase [5-7,15], rather than the phase the solution would otherwise form. For example, Mirskii and Pirozhkov [6] added zeolite NaA to a hydroxysodalite synthesis solution, producing zeolite NaA. They added hydroxysodalite to a zeolite NaA synthesis solution, producing hydroxysodalite. Lastly, they added mordenite crystals to a zeolite NaX synthesis solution, producing mordenite. Warzywoda and Thompson [15] reported that adding zeolite NaA crystals in sufficient quantity to a zeolite G synthesis solution resulted in the rapid formation of zeolite NaA, in remarkably shorter time than the zeolite G otherwise would have formed. The particle size distribution resulting from this study suggested that the seeds promoted nucleation of a new zeolite NaA population.

Curiously, however, there are several examples of zeolite crystallisations which can only be accomplished in the presence of added seed crystals [16]. In addition, under the conditions of the study, Culfaz and Sand [7] noted that adding zeolite X seeds to their zeolite X synthesis batches did not change the induction times or the crystallisation rates. These two examples might be viewed as the extreme limits of seeded syntheses in which on the one hand seeds are absolutely required, while on the other hand the presence of seeds makes absolutely no difference in the system.

In a series of studies on the effects of adding Silicalite seeds to an Al-free NH_3-ZSM-5 system [17-19], it has been shown rather convincingly that nucleation enhancement in this system is by an initial breeding mechanism [3]. That is, nuclei, or nuclei breeding entities, exist on the seed crystal surfaces and become dislodged from these crystal surfaces when added to the synthesis batch to promote crystallisation; this mechanism is so dominant that the population of new crystals is significantly smaller than those in the absence of added seed crystals [18]. It also was shown that these entities could be separated from the seed crystal surfaces by slurrying the seed crystals in either water or mild NH_4OH solution followed by filtration [19]. Maped. et al [20], showed that seeds accelerate the crystallisation of high silica zeolites, but that in order to do so they must have the stabilizing organic cation within the crystal framework. This result is at least qualitatively confirmed by Hou's results [17].

In addition to understanding the role of seed crystals in batch syntheses, it is essential to understand the behaviour of the seed crystals themselves in the growth environment of the solution. From the series of studies just noted [17-19] the seed crystals generally appeared to be covered with a population of new crystals growing from the surface in random orientations, giving what might be called a "porcupine" morphology, as illustrated in the photographs of the seed crystals shown in Figure 1. It appeared, therefore, that some of the aforementioned initial bred nuclei dislodged from the surface to generate a new population of crystals, while some of these entities remained on the seed crystal surface and grew from there. The cavities on the seed crystal surface in Figure 1b appear to have been left by crystallites which were previously on the surface, but which popped off; the cavities indicate that the seed crystal was growing around them. New results on the surface-bound crystallites will be presented later in this text.

Numerous studies have been conducted in which the mixed amorphous gel solution was aged at some lower temperature prior to reacting at elevated temperatures [21-31]. The role of aging some amorphous gel solutions appears to provide a time during which the solution can form nuclei, though even at dramatically reduced rates, which can then become activated at elevated temperatures. These steps are sometimes necessary for successful crystallisation, as in the case of zeolite Y [22-24], can sometimes result in essentially complete crystallisation at room temperature [25], can at times result in a bimodal distribution of crystal sizes [29,30], or can at other times produce no effect at all [31]. The magnitude of the temperature dependent parameters will determine the outcome of the aging process. However, when nuclei are formed during room temperature aging, the solution begins its high temperature period of the process as though one had added seeds. Thus, one notes the strong similarity between physically adding seed crystals, aging solutions which are known to form nuclei at room temperature, and adding nuclei-sized entities created elsewhere (sometimes called "directing agents") to a fresh gel solution which were the result of a brief hydrothermal treatment sufficient only to create nuclei-sized entities [32].

Figure 1a. Silicalite crystallites growing on Silicalite seed crystal surfaces [42].

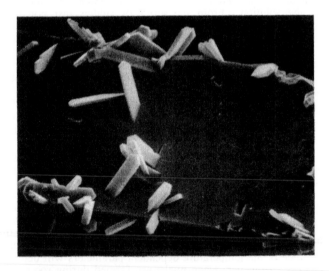

Figure 1b. Close-up of seed crystal surface, showing cavities where crystallites likely became dislodged [42].

The final aspect which will be considered here is the understanding of the apparent twinning or intergrowth behaviour of some zeolite crystals, which has been gleaned from observing the growth behaviour of the Silicalite seed crystals in batch synthesis solutions. At the current time, the appearance of these assemblages seems to be a random occurence, although it also is possible that our understanding of the mechanism of their formation is incomplete. Examples of this behaviour are illustrated in Figure 2, showing several different zeolite phases which seem to be crystal twins or intergrowths. It is noteworthy that these apparent intergrowths appear to be randomly oriented with respect to their crystal axes, and that the centers of the adjoining crystals do not appear to overlap, suggesting that they probably were nucleated in different locations within the synthesis solution. Similar observations have been made in the literature recently for other zeolite phases [33-36], even one sample which was synthesized in the microgravity environment of space [35]. Two examples have been reported recently in which ordered overgrowths have been described [37,38]. These arrangements are somewhat different than the **randomly** oriented configurations shown here, and noted elsewhere [33-36], and result from crystallographic orientations. Observing the growth behaviour of seed crystals will be shown to provide some insight to this phenomenon.

2. THEORETICAL FOUNDATIONS

In order to understand the anticipated effects of adding seeds to a zeolite synthesis system, recent modelling results will be utilized [39]. The basis of that model was the presumption that discrete entities exist in the synthesis gel as formed which are responsible for zeolite crystal nucleation, a presumption based on another recent study reported in the literature [31]. Further, these entities were assumed to be located in the outer regions of the gel phase, and were presumed to be completely activated by the time 10-15% of the gel had been converted, consistent with observations [40].

The mathematical model used previously, and extended here, was based on the population balance methods developed by Randolph & Larson [3] and studied extensively over the last 30 years. The partial differential equation describing the batch crystallizer is easily converted to a system of ordinary differential equations (known as the moment equations), as shown previously [39], resulting in the following:

$$\frac{dm_0}{dt} = B(t) \qquad [1]$$

$$\frac{dm_1}{dt} = Q\, m_0 \qquad [2]$$

$$\frac{dm_2}{dt} = 2Q\, m_1 \qquad [3]$$

$$\frac{dm_3}{dt} = 3Q\, m_2 \qquad [4]$$

where the moments $(m_0 \ldots m_3)$ correspond to the cumulative number of crystals, the cumulative length of all the particles, the cumulative crystal surface area, and the

48

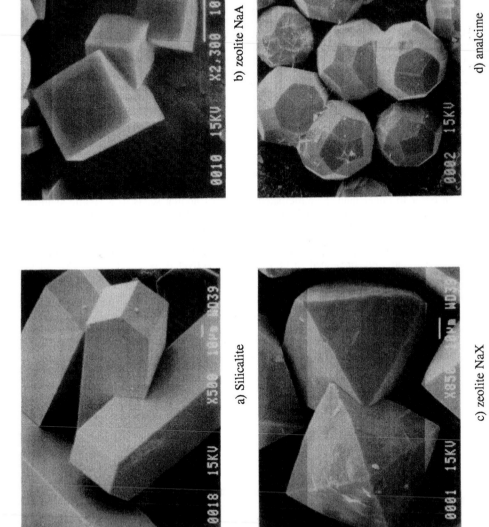

a) Silicalite

b) zeolite NaA

c) zeolite NaX

d) analcime

Figure 2. Examples of zeolite crystals which appear to be "intergrown."

cumulative crystal volume (or mass), per unit volume of the batch system, respectively. The solution of these four differential equations would describe the time evolution of these four properties of the crystal size distribution in the batch crystallizer. It is seen, for example, that the total number of particles increases as a result of the nucleation event, B(t), while the other three moments change in proportion to the crystal growth rate, Q, times some other factors.

The birth rate is described as a function of the rate of gel consumption, since most of the nucleating entities are assumed to be in the gel phase, multiplied by a weighting factor which serves to locate the nucleating entities near the outer regions of the gel particles:

$$B(t) = - N_A \frac{dG}{dt} \exp\{-k*(1-\frac{G}{G_0})\} \qquad [5]$$

where G_0 and N_A are the initial gel concentration and nuclei density, respectively, and G represents the gel concentration at some later time, t. The exponential term is an empirical function which serves the purpose of the nuclei becoming activated at early conversion levels, that is, the quantity $(1-G/G_0)$ is equal to the conversion level of the amorphous gel.

The functions describing the gel and solution phase material balances are given by the following:

$$\frac{dG}{dt} = -k_1 G + k_2 G G* \qquad [6]$$

$$\frac{dG*}{dt} = -\frac{dG}{dt} - \frac{\sigma\phi Q}{2} m_2 \qquad [7]$$

where k_1 and k_2 are rate constants for the forward and reverse gel dissolution steps, σ is the crystal density, ϕ is a crystal area shape factor, and $G*$ is the solution phase concentration. The crystal growth rate, Q, is independent of crystal size, as suggested by previous reports [2,21,25,26], and is given by a linear driving force as:

$$Q = k_4(G* - G*_{eq}) \qquad [8]$$

where the subscript "eq" designates the equilibrium value of the solution concentration at which the driving force for crystal growth would be zero, that is where the supersaturation drops to zero.

In the previous work [39], the initial values of the moments reflected only sufficient particles of "zero size" to initiate the process, since the majority of the nuclei resided in the amorphous gel phase. The initial values of the moments, m_1-m_3, in that study were "near zero," since the nuclei were presumed to be angstroms in dimension. The number of nuclei initially in the liquid phase, m_0, represented about 27% of the total. In the present study, we wish to examine the effects of adding seeds to the synthesis batch, which would be simulated by using non-zero initial values. That is, seeds of certain sizes (micrometers in dimension, for example) might be added in certain amounts (mass %, for example), and these additions would be included by specifying that the initial values of the moments are positive, non-zero values; e.g., $m_3(0) = 2.0E10$. The mass per cent of seeds added in the simulations is reported as:

$$\% \text{ seeds} = \frac{(\text{mass of seeds added})}{(\text{mass of seeds} + \text{gel})} \times 100\% \qquad [9]$$

Aging amorphous aluminosilicate gels at room temperature are thought to promote nucleation at a relatively slow rate, followed by the normal nucleation step once at elevated temperature [21,25,26,28,30]. In later simulations, the effects of aging the amorphous gel prior to reacting at elevated temperature were simulated by doubling the values of the nuclei initially in the solution, N_0, and the gel-bound nuclei, N_A. Other degrees of aging could be investigated by choosing per cent increases other than 100% (doubling), but this is left for another time.

The simulations reported below were carried out on the DECsystem network at WPI using the DSS2 differential equation solver software package. No particular stiffness was noted in solving these equations simultaneously, as long as extremely large or small numbers were scaled.

2.1. Simulation Results

Figures 3-6 show the results of simulations with no seeds added compared to the predicted results when 0.1% and 1.0% seeds, having characteristic dimension of 1.0 micrometer, were added to the solution. A rather complete picture can be obtained by examining the crystallisation curves (related to the third moment), the birth rate histories (related to the zeroth moment), and the evolution of the crystal sizes (computed by integrating the growth rate expression, $dL/dt = Q(t)$).

Figure 3, for example, shows the predicted crystallisation curves for these simulations. The unseeded system, curve (a), shows that this typical synthesis proceeded to completion in about 13-15 hours, following an "induction period" of about 4 hours. Adding 1.0 micrometer seeds in the amount of 0.1% caused the process to proceed more quickly, finishing in about 11 hours, for example. The process was completed in about 7 hours when using 1.0% seeds of 1.0 micrometer dimension. The reason for the enhancement is that the added seeds provide more cumulative surface area for the assimilation of reagents from solution, and, as noted in equation [4], this feature causes the third moment (proportional to the degree of crystallinity) to increase more rapidly.

Figure 4 shows the birth rate histories for these three simulations, based on the rate at which dormant nuclei become activated by their release from the outer regions of the amorphous gel phase. The unseeded system, curve (a), proceeds to a maximum at about 4 hours, and is complete in about 8 hours. As seeds are added in increasing amounts, curves (b) and (c), the nuclei, being the same in number, become activated more quickly, again due to the increased surface area contributed by the added seeds. The nuclei become active sooner in the process since the seeds consume solute more quickly, and the gel dissolves more quickly.

Figures 5 and 6 show the evolution of the largest crystals in the new crystal population and the seed crystal population, respectively. Curve (a) indicates that the crystals proceed at a constant rate to a limit of about 6.7 micrometers in the unseeded

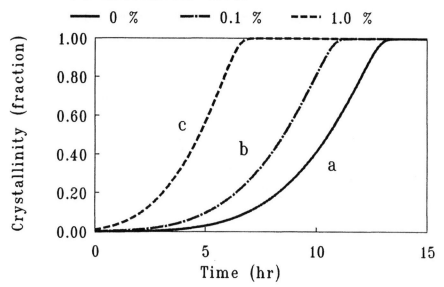

Figure 3. % Crystallinity vs. Time
1.0 micrometer seeds

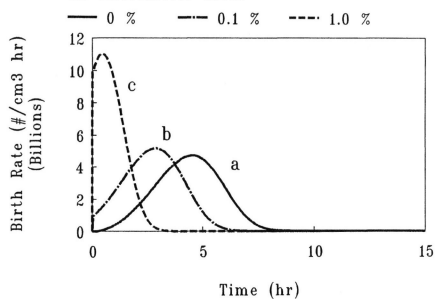

Figure 4. Birth Rate vs. Time
1.0 micrometer seeds

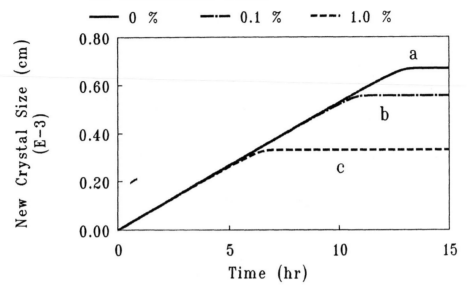

Figure 5. New Crystal Size vs. Time
1.0 micrometer seeds

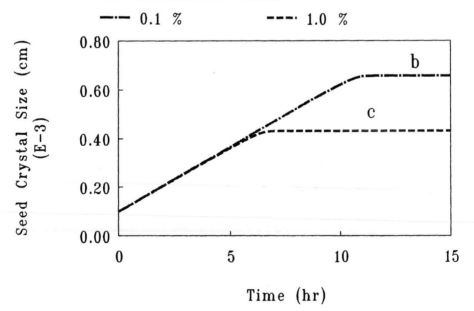

Figure 6. Seed Crystal Size vs. Time
1.0 micrometer seeds

system, while the new crystals stop growing at about 5.6 and 3.5 micrometers, respectively, in curves (b) and (c), that is when 0.1% and 1.0% seeds are used. These results can be understood to stem from the competition for reagents between new crystals (being the same in number in all three cases) and the added seed crystals. The seed crystals grow from their starting size of 1.0 micrometer to about 6.5 micrometers when 0.1% were added, but only to about 4.5 micrometers when 1.0% were added. Again, the smaller increase in size results from the competition between larger numbers of crystals when 1.0% were added compared to addition of 0.1% seeds.

Figure 7 shows the predicted crystallisation curves when 1.0% seeds were added, but with seed crystals having characteristic dimensions of 1.0, 4.0, and 7.0 micrometers. One can see that the increase in the "crystallisation rate" is more pronounced as the seed crystal size is reduced. This is to be expected, because the same mass of smaller crystals has more cumulative surface area than large crystals, and equation [4] indicates that the cumulative surface area, that is the second moment, m_2, is one of the key factors in enhancing the "crystallisation rate."

Figure 8 shows the birth rate histories arising from the 1.0% seed addition, using the different sized seeds. As expected, the nucleation proceeds more quickly when using smaller seed crystals in the same amount, again due to the higher cumulative surface area per gram, and proceeding along the same arguments as before related to gel consumption rates.

Aging amorphous gel mixtures has been suggested to promote nuclei formation either in the solution phase or the gel phase during the aging step [21,25,26,28,30]. This step of the zeolite crystallisation was simulated by assuming that a certain aging step

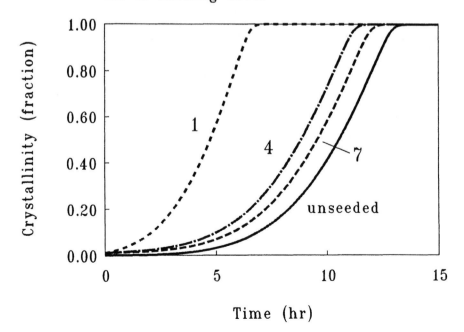

Figure 7. % Crystallinity vs. Time 1.0 % seeding level

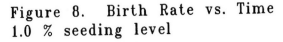

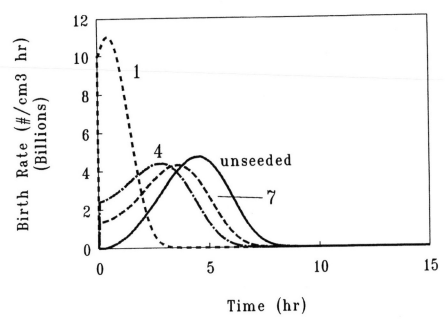

Figure 8. Birth Rate vs. Time
1.0 % seeding level

would double the number of seed nuclei in the solution phase and double the dormant nuclei in the gel phase, although other enhancement factors could easily be simulated. Figure 9 shows the predicted crystallisation curves resulting from these assumptions. As expected, the process occurs at a faster rate and is completed sooner. This enhancement also results from the increased cumulative surface area due to the larger number of crystals. The rate enhancement is observed somewhat later in the process, that is only after the crystals grow to macroscopic size and the differences in surface area are significant.

The predicted birth rate histories are shown in Figure 10, and it is observed that nucleation in the aged solution proceeds (and finishes) more quickly, because more crystals were created in the system (given by the area under the curves). The final crystal sizes resulting from the two simulations were 6.7 micrometers and 5.3 micrometers in the unaged and aged solutions, respectively, as would be expected from systems in which more nuclei were created.

It should be mentioned here that creating nuclei in a synthesis batch through an aging step can be viewed as quite similar to adding "nuclei" in the form of "directing agents" [32]. In cases in which nuclei-sized entities are formed by a separate hydrothermal step, separate from the main synthesis step, and added to a synthesis batch, the consequence is the same as creating nuclei via a low temperature aging process. As such, the results shown in Figures 9 and 10 can be viewed as qualitatively simulating the behaviour of adding "directing agents," as observed [32]. Introducing "initial-bred nuclei" from seed crystal surfaces [17-19] may be viewed in the same manner.

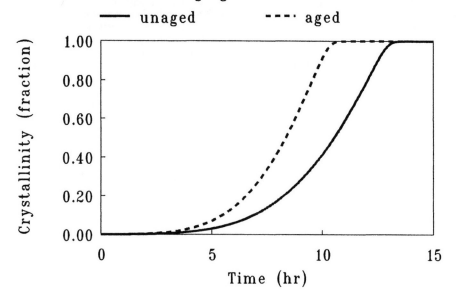

Figure 9. % Crystallinity vs. Time
effect of aging

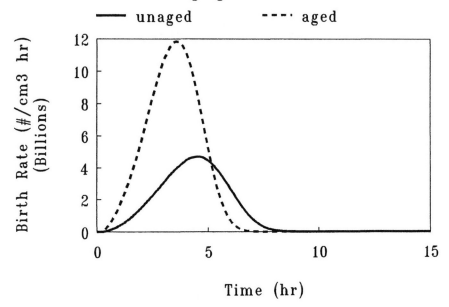

Figure 10. Birth Rate vs. Time
effect of aging

2.2. Comments on the Number of Nuclei Formed

It has been assumed in these simulations that a finite number of discrete entities exist in the unseeded, unaged synthesis batch, which ultimately will give rise to the formation of nuclei. In a recent study of zeolite NaX synthesis it was concluded that these entities entered the system in the silica source used, and could be correlated with the levels of several impurities in the silica sources [31]. It has been assumed here that adding seeds to the solution to create additional nuclei did not prohibit the activation of any of these preformed entities. That is, they all were activated ultimately. In a recent study in which Silicalite seeds were added to a fresh Silicalite synthesis batch [41] it was not demonstrated that activation of the nuclei otherwise formed in the solution was suppressed by the addition of the seeds. On the other hand, the same study reported that the absolute number of nuclei formed at higher temperatures was less than at lower temperatures, suggesting that there is a mechanism whereby the presence of the growing Silicalite crystals suppressed further nucleation. Thus, whether the addition of seeds will suppress subsequent nucleation, as would be expected if the nucleation process were solely supersaturation dependent, is currently unresolved. Models of the sort described here could include such effects, but this matter deserves further investigation.

3. SEEDED SYNTHESIS RESULTS

Several series of experiments were conducted to illucidate further the role of seed crystals in molecular sieve zeolite syntheses, but especially to determine their effects on zeolite crystal nucleation. The ammonium-Silicalite system was used for these studies. Seeds, when added, were expressed as a percentage of the amorphous silica mass used in the synthesis batch.

The synthesis system to be described here in which the effects of adding seed crystals were investigated has been described previously [17-19], and involved the following "standard" batch composition:

$$60 \ (NH_4)_2O - 90 \ SiO_2 - 6 \ (TPA)_2O - 994.4 \ H_2O$$

where TPA represents tetrapropylammonium ions. The ingredients used were ammonium hydroxide (30% in water, Baker), Ludox AS-40 (Du Pont), tetrapropylammonium bromide (Fluka), and triple exchanged and deionised water (Barnstead Nanopure II). All syntheses were carried in sealed Teflon-lined Morey type autoclaves without agitation, placed in a laboratory convection oven at 180°C, and at autogenous pressure. Autoclaves were removed and quenched in cold water at the appropriate synthesis time. The unseeded, unaged system produced coffin-shaped "Al-free" ZSM-5 (Silicalite) crystals in seven days at temperature, having a narrow size distribution, with crystals having a characteristic length of approximately 300 micrometers, as shown in Figure 11. It is observed that there is a relatively small degree of twinning or intergrowth of particles in this sample, but that the sample is predominantly single crystals, characteristic of this system.

This system was noted to be quite fluid (water-like) at room temperature, but formed a very viscous, almost paste-like, mass after about 30 minutes at reaction temperature [42]. This observation resulted in several interesting advantages in this study, since, after this gelling time, particles had very little mobility in this system. This effect

Figure 11. Coffin-shaped Silicalite crystals from "standard" batch composition.

also offered the advantage that seeds could be isolated from the gelatinous mass (e.g., seeds on the bottom) after selected reaction times.

3.1. Results from Early Times with Seeds

The standard batch composition was seeded with 280 micrometer Silicalite seed crystals at levels of 3%, 7.6%, 9.5%, and 16.1%. Samples were removed from the convection oven after two hours of reaction to note the degree of free crystal formation and crystallite formation on the seed crystal surfaces.

Figure 12 shows illustrative samples of the seed crystals, collected from the bottoms of the autoclaves, from the preparations having the four different seeding levels. It is not clear that at this relatively short reaction time there were new crystallites formed on the seed crystal surfaces, until 9.5% seeds was used in the preparation. On the other hand, it is obvious (except in the 3% seeding case) that new crystals had formed in the gel phase around the seed crystals, as observed by the 5-7 micrometer crystals shown in the optical photograph of the gel mass from the 7.6% seed level experiment in Figure 13.

The "porcupine" morphology, noted previously, was not observed in these results with seeding levels of 3 % and 7.6 %, and only marginally after two hours with 9.5 % and 16.1 % seeding levels. Therefore, this experiment was the first indication that it is extremely unlikely that any nuclei became activated and grew directly from the seed crystal surfaces, but that some of them adhered to the surfaces at some later time. That is, the extra nuclei which formed when seeds were added either were not initially bound to the seed crystal surfaces, or essentially all became dislodged from the seed crystal surfaces prior to their subsequent incorporation there.

58

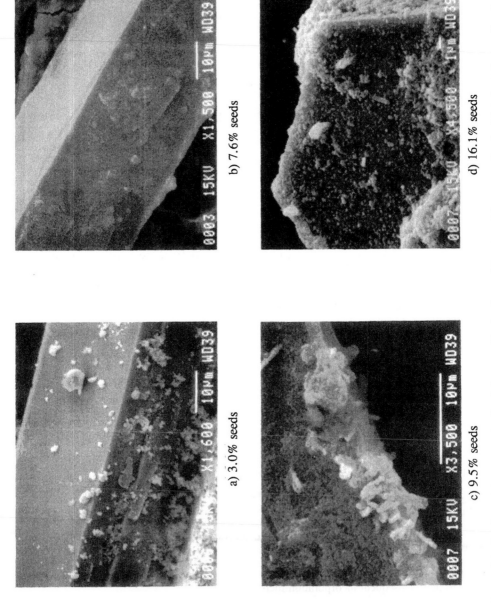

a) 3.0% seeds

b) 7.6% seeds

c) 9.5% seeds

d) 16.1% seeds

Figure 12. Seed crystals removed after 2 hours of hydrothermal treatment in a new standard batch at the seeding level specified.

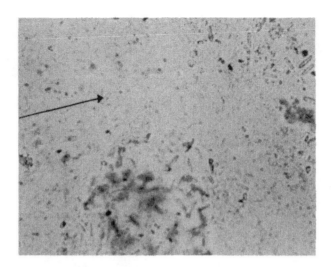

Figure 13. Optical photograph of gel and crystals formed after 2 hours of hydrothermal treatment of a standard batch at 7.6% seeding level.

3.2. Small Seed Crystal Additions

The previous experiment, intended to observe the early stages of new crystal formation on seed crystal surfaces, was followed by a similar experiment in which crystals having 5-7 micrometer characteristic length were used. Reactions were carried out for 1.5, 3, and 48 hours at temperature, using a 4.3% seeding level.

Figure 14 shows SEM photomicrographs representative of the samples at 1.5, 3, and 48 hours of reaction time, showing that there were no noticeable examples of crystallites emerging from the seed crystal surfaces, even though new Silicalite crystals were detected in the bulk after two hours of reaction. There are a few noticeable cavities in the seed crystal surfaces at 48 hours, indicating that there may have been crystallites there for a brief time, but their abundance was quite small.

This was the second indication that the new population of nuclei was activated in the solution, or in the amorphous gel phase, but not on the seed crystal surface directly. It appeared from these two experiments that the new population of crystals, known to exist at early times in the bulk, adhered to the seed crystal surfaces given sufficient seed crystal surface area or reaction time. The issues of how the crystallites came into contact with the seed crystal surfaces, and how they became attached there were yet unresolved.

3.3. Dried Gel Experiments

Several experiments were conducted in which a partially reacted "dried gel" was used as the silica source. The dried gel was formed by reacting an unseeded standard batch for a few hours at temperature, such that the conversion was less than 10%, and Silicalite crystals in the mixture were on the order of 10-15 micrometers. This partially

a) after 1.5 hours

b) after 3.0 hours

c) after 48 hours

Figure 14. Small seed crystals
treated hydrothermally in a standard
batch at 180°C for the times designated.

reacted mass was dried at 70°C to be used in subsequent seeded syntheses. The dried gel converted to Silicalite crystals when reacted alone at 180°C in an NH_4OH medium, as shown in Figure 15.

In one such experiment with seeds and dried gel, for example, 0.0245 grams of 165 micrometer seeds and 0.0213 grams of dried gel were added to 6.84 grams of 30% NH_4OH in water solution and reacted at 180°C for one week. This dried gel originally contained Silicalite crystals of about 19 micrometers characteristic length. The estimated growth in the dried gel-bound crystals was about:

$$L_{max,final} - L_{max,initial} = 71.6 - 19.3$$
$$= 52.3 \text{ micrometers}$$

where the L's represent the maximum crystal sizes initially and at the end of the experiment. It was presumed that the Silicalite crystals distributed throughout the system grew at a rate independent of particle size, an observation reported recently for Silicalite [43]. The growth of the seed crystals themselves during this time, estimated from SEM observations, was about 42 micrometers, which is noted to be less than the dried gel-bound crystals, perhaps due to diffusional transport limitations in the fluid phase. In spite of this difference, however, any crystals observed in the final sample smaller than 52 micrometers should have been from nuclei formed during the synthesis, while crystals larger than 52 micrometers must have been those which already existed in the dried gel particles.

Figure 16 shows two examples, of many observed, in which crystals, ca. 66-68 micrometers in length, are seen to be encapsulated by the much larger seed crystals.

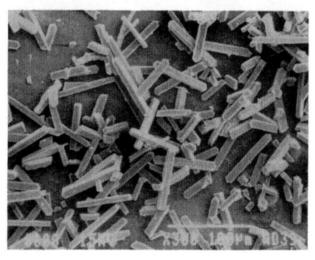

Figure 15. Silicalite crystals produced from "dried gel" treated hydrothermally at 180°C in NH_4OH solution for three days.

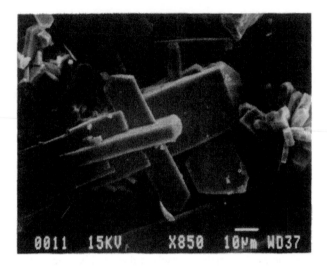

a) entrapped crystal approximately 65.9 micrometers.

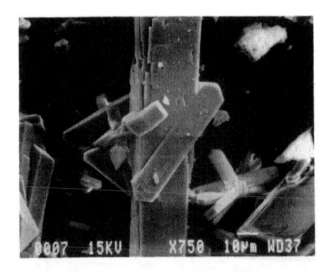

b) entrapped crystal approximately 68.3 micrometers.

Figure 16. Silicalite crystals originally formed in the "dried gel" trapped in larger growing seed crystals after hydrothermal treatment at 180°C for 7 days.

These results provide convincing evidence that crystals formed within the dried gel, and clearly quite separate from the seed crystals, became attached and encapsulated by the seed crystals. Their migration was probably due to gravitational settling in this case, since the 30% NH_4OH solution remained quite fluid. From these observations, it appeared likely that crystals which came into contact, or close proximity, were joined as shown in the SEM photographs included here.

3.4. Fluid Solution Syntheses

It was noted that the standard NH_4+ batch composition became very viscous after about 30 minutes at temperature. To examine the effect of this viscosity increase, a batch composition, based on $Na+$, which remained fluid over the entire synthesis was used to further demonstrate the results shown in the previous experiment. The batch composition used for this experiment was given by:

$$2.2 \, Na_2O \quad Al_2O_3 \quad 66 \, (TPA)_2O \quad 20 \, SiO_2 \quad 6000 \, H_2O$$

a mole oxide composition which produces 100 micrometer long coffin-shaped Silicalite crystals after 12 days at 180°C. These 100 micrometer crystals were used as seeds in the subsequent experiment, using the same mole oxide composition as from which the seeds were created.

Figure 17 shows examples typical of these experiments in the more dilute, less viscous fluid synthesis systems. It is observed that the seed crystals have grown to about 165-170 micrometers, and that there are noticeably fewer small particles adhering to the seed crystal surfaces compared to those from the more viscous solutions. In addition, more of the crystals on the seed crystal surfaces appear to be laying flat on the surfaces, which would be expected if the new particles had arrived on the seed crystal surfaces by gravitational settling in a low viscosity fluid. That is, many of the crystals entrapped by the seed crystals do not seem to have the same random orientations in three dimensions as in the previous experiments using the standard batch composition, producing the "porcupine" morphology, but seem to have faces oriented along the seed crystal faces, suggesting that they arrived from the solution by settling rather than nucleating and growing from the seed crystal surfaces. The "porcupine" morphology, then, can be suggested to result from the manner in which crystals come into contact. The new crystals, of size 20-25 micrometers, are significantly smaller than those produced without seeds, again suggesting that nucleation was promoted by the addition of the seed crystals.

3.5. Different Sized Seed Crystals

Several experiments were conducted with two different sized seed crystal populations in the same gel mixture. These experiments were intended to illucidate definitively that the crystals could become "overgrown" after their formation, and how crystals become attached to one another in such syntheses. The sizes of the seed crystal populations were chosen so that they could easily be distinguished at the end of the experiments.

Figure 18 shows typical particle agglomerations from an experiment in which the seeds originally having linear dimensions of 60 and 250 micrometers were added to the bottom of an autoclave, which was then filled with a standard batch composition, and reacted at 180°C for 15 hours. Any crystals nucleated during that time were well smaller than 60 micrometers, avoiding any confusion in discerning their origins. It is

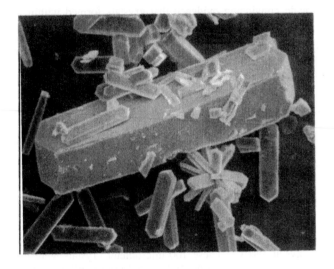

a) seed crystal has grown to approximately 165 micrometers.

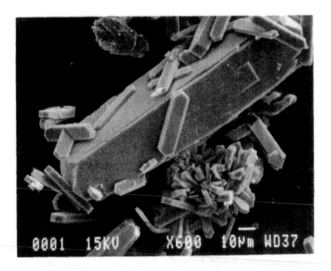

b) seed crystal has grown to approximately 170 micrometers.

Figure 17. Silicalite product from Na+ based ZSM-5 batch composition, which remains fluid over the entire hydrothermal synthesis time; 12 days at 180°C.

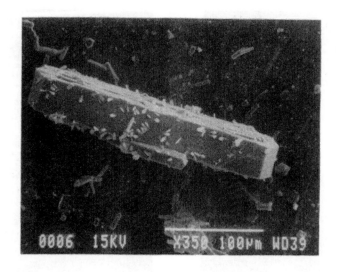

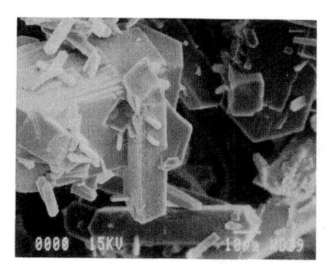

Figure 18. Two examples of 60 micrometer seed crystals trapped in 250 micrometer seed crystals after hydrothermal treatment for 15 hours at 180°C.

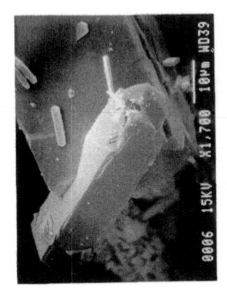

b) after 7 hours

c) after 8 hours

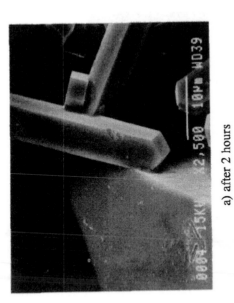

a) after 2 hours

Figure 19. Progressive entrapment
of 100 and 200 micrometer seed
crystals after different periods
of hydrothermal treatment at 180°C.

unmistakeable that several 60 micrometer seed crystals were trapped in larger, 250 micrometer seed crystals. Thus, it is demonstrated that entrapped crystals, as well as the crystallites growing from the seed crystal surfaces, result from physical contact, followed by crystals growing around one another. The reason why this effect appeared to be more significant in the NH_4^+-Silicalite system than the Na^+-Silicalite system was probably the paste-like medium which restricted particle mobility compared to the low viscosity medium, in which even mild thermal gradients could keep particles moving on the micrometer scale. In the ammonium system, the smaller crystallites were probably overgrown by the seed crystals, because they were not free to move away from the advancing seed crystal face.

A similar experiment was conducted with seeds having linear dimensions of 50 and 200 micrometers. Equal amounts of these seed populations were added to a standard batch to make up 2.4% seeding, and the mixture reacted for 2, 3, 7, and 8 hours. Observations at different times were intended to show increased degrees of penetration of the two seed crystal sizes, recognizing the extent of variability even within a single sample.

Figure 19 shows SEM photographs of typical intergrowth behaviour from three such samples, and provides evidence that the extent of entrapment appeared to increase as time went on. The photograph at 2 hours, for example shows two crystals which appear to have just come into contact, with perhaps minor levels of bonding occurring between them. At 7 and 8 hours, there is no doubt that the smaller crystals have been grown over by the larger crystals. It appears from these observations that both particles continued to expand by growth along their normal crystallographic axes, as noted previously by Hay, et al. [37], forming randomly configured overgrowth patterns determined by their orientation upon contact.

Figure 20 shows an example of a larger seed crystal growing over, and around, a smaller crystal, taken from one of the many experiments performed during this study [42].

Figure 20. Example of a seed crystal growing over and around a smaller crystal, beginning to completely engulf the smaller crystal.

The larger crystal has grown well over one micrometer beyond the entrapped crystal, creating a slit in itself afterwards.

3.6. Zeolite NaA and NaX Seeds

Several experiments were conducted using Silicalite seeds and seed crystals of another zeolite phase to ascertain the effect of extraneous solids in these systems, and to demonstrate the generality of the un-ordered overgrowth mechanism. Thus, for example, Silicalite and zeolite NaA crystals were placed on the bottom of an autoclave, gently mixed, and then the standard Silicalite batch solution was added to the autoclave. The autoclave was withdrawn from the oven after 21 hours at 180°C and quenched.

Figure 21 shows the samples thus obtained, from which several observations can be made. First, the zeolite NaA cubes are noted to be pocked, an indication that they were dissolving in this solution. Second, in spite of the cubes dissolving, it is noted that the Silicalite seed crystals were growing around the cubes. That is, the two crystal types had come into contact, and the Silicalite crystal was growing around the NaA crystal, to entrap it, whether the NaA cube was bonded, adsorbed, or just in physical contact with the Silicalite crystal. Thirdly, it is noted that a new population of highly twinned ZSM-5 crystals had formed in the bulk, undoubtedly due to the Aluminium which was released into the solution by the dissolving zeolite NaA crystals. Note, too, that some of these new ZSM-5 crystals also were entrapped by the growing Silicalite.

Similar experiments were conducted with Silicalite and zeolite NaX seed crystals, but the NaX crystals completely dissolved within the time frame of the experiment, so no further information was learned from that work.

4. CONCLUSIONS

It is well known that most hydrothermal zeolite syntheses will proceed in the absence of seed crystals, thus demonstrating that most amorphous aluminosilicate solutions are capable of creating nuclei and sustaining crystal growth, at least given an appropriate batch composition. Nucleation and crystal growth in unseeded zeolite synthesis systems, and their mathematical analysis, have been discussed previously [44]. As such, they were not the focus of this study, except to build on recent prior works [31,39,41], which showed that "nuclei" could be discrete entities in solution and near the edges of amorphous gel particles, and that they were activated at some temperature-dependent rate.

Seed crystals are not an essential element of most batch compositions, but are known to catalyse the zeolite crystallisation rate. This study has focused on the role of added seed crystals in molecular sieve zeolite syntheses. In particular, the results from studies using seed crystals, and specifically how the addition of seed crystals enhances zeolite nucleation and growth, has been examined.

A population balance model was developed and solved numerically to generate predictions of the behaviour of seed crystals added to a hydrothermal zeolite synthesis. Theoretical results were consistent with expectations, that is, added seed crystals provided more cumulative surface area for the assimilation of nutrient material from solution. As such, crystallisations were predicted to proceed faster with increasing amounts of seed crystal surface area added, either in the form of more seeds or the same amounts of smaller seed crystals. The size evolution of seed crystals and newly formed crystals was shown to proceed along similar trajectories with time, and the final average crystal sizes

Figure 21. Three examples of the "product" of hydrothermal treatment of a standard batch composition for 21 hours at 180°C with both Silicalite and zeolite NaA seeds present in the batch.

were shown to be reduced with increased amounts of seed crystal surface area. While not simulated specifically, it should be obvious that the simulations of aging or directing agent addition, consistent with the results noted above, could equally well represent the addition of initial bred nuclei dislodged from larger seed crystal surfaces; the effects predicted can be expected to be the same.

Examination of scanning electron photographs of Silicalite crystals from several experiments revealed that new crystallites, previously believed to initiate their growth from the seed crystal surfaces, were not observed unequivocally to be on large seed crystal surfaces at early times when new crystallites were observed to have already formed in the bulk around the seed crystals. Thus, it is unlikely that the new generation of nuclei, activated upon seed crystal addition, was due to crystallites which were adhering to the seed crystal surfaces after being placed in solution. The same observations were made when small seed crystals were used instead of the much larger ones just mentioned.

Experiments with seed crystals of different, but well defined, sizes demonstrated that single crystals can overgrow one another in random orientations, giving the appearance of true intergrowths. In the case of the dried gel and the fluid solution experiments, the low viscosity medium probably permitted the different sized crystals to come together by gravitational settling. In the fluid solution experiments more crystals appeared to come together in face-to-face contact, while there were somewhat more random orientations noted in the dried gel experiments, due to the crystals' random orientations in the dried gel particles. In the cases of the high viscosity medium systems, the mobility of the crystals was undoubtedly restricted after about 30 minutes at temperature, and the overgrowth morphology formation was likely due to the crystals' lack of mobility; they simply grew into one another. The ability of crystals to grow around one another was also demonstrated by using zeolite NaA seeds in a Silicalite batch. These results would lend credence to the mechanism, proposed by Sano, et al [45], by which growing zeolite crystals could envelope small active catalytic particles to form a molecular sieve barrier around them.

Thus, the formation of the "porcupine" morphology was demonstrated to be the result of randomly oriented crystal/crystal overgrowths, after many "initial bred nuclei" had dislodged from the seed crystal surfaces, if they were ever actually on the seed surfaces. Any such "nuclei" remaining on the seed crystal surfaces appeared to be inactive. Furthermore, the "porcupine" morphology was shown to have formed from the random orientation which the seed crystals and the new crystals had relative to one another as their growing faces advanced toward each other in the viscous medium which restricted their respective mobilities. The same mechanism of randomly oriented crystal/crystal overgrowths can be suggested to explain similar behaviour observed elsewhere in the zeolite literature, noted earlier in this text.

There are several unanswered questions surrounding this subject which should be mentioned before closing. The formation of the "initial-bred nuclei" is not clearly understood. At this point, it is not known if these entities are originally associated with seed crystal surfaces, which would mean that most zeolite crystals would bear them, if there are simply microcrystals contained in zeolite crystal products, or if these entities are formed by microattrition in the dry state from the "rough" manner in which crystals are commonly handled (as in filtering, drying, scraping, etc.). At this time it has not been demonstrated that secondary nucleation by initial-breeding occurs in other zeolite syntheses, but it is not unreasonable to suggest that mechanisms should be quite similar among many hydrothermal zeolite systems. Presuming that "initial-bred nuclei" truly exist on seed crystal surfaces in the dry state, it is not known why they would appear to leave the surface when in a synthesis solution, or remain inactive (as indicated by the results of this study), although this would certainly suggest a surface charge effect. Lastly, it has not yet been demonstrated whether adding seeds to a synthesis batch supresses the nucleation event which occurs in the absence of seeds, or whether the uncatalysed nucleation event occurs in spite of the presence of seeds, and the new crystals simply do not grow to a size at which they can be distinguished as a separate population.

5. ACKNOWLEDGEMENTS

The authors gratefully acknowledge the support of the National Science Foundation through grant CTS-9103357. The authors also gratefully acknowledge the assistance of Mr. George E. Schmidt, Jr., who reproduced the photographs in a rather short time.

6. REFERENCES

1. Breck, D. W., *Zeolite Molecular Sieves,* 1974, John Wiley & Sons, New York.

2. Barrer, R. M., *Hydrothermal Chemistry of Zeolites,* 1982, Academic Press, London.

3. Randolph, A. D. and Larson, M. A., *Theory of Particulate Processes*, second edition, 1988, Academic Press, London.

4. Kerr, G. T., **J. Phys. Chem.**, 1966, **70**, 1047-1050.

5. Kerr, G. T., **J. Phys. Chem.**, 1968, **72**, 1385-1386.

6. Mirskii, Ya. V. and Pirozhkov, V. V., **Russ. J. Phys. Chem.**, 1970, **44**, 1508-1509.

7. Culfaz, A. and Sand, L. B., ACS No. 121 (eds. Meier, W. M. and Utterhoeven, J.B.), 1973, 140-151.

8. Kacirek, H. and Lechert, H., **J. Phys. Chem.**, 1975, **79**, 1589-1593.

9. Lechert, H., Stud. Surf. Sci. & Catal., no. 18 (eds. Jacobs, P. A., Jaeger, N. I., Jiru, P., Kazansky, V. B., and Schuilz-Ekloff, G.), 1984, 107-123.

10. Narita, E., **Ind. Eng. Chem. Prod. Res. Dev.**, 1985, **24**, 507-512.

11. Narita, E., **J. Crstl. Gr.**, 1986, **78**, 1-8.

12. Han, S., Shi, S., Liang, Q., and Xu, R., **Gaed. Xue. Hua. Xue.**, 1983, **4**, 540-544.

13. Ueda, S. and Koizumi, M., **Amer. Miner.**, 1979, **64**, 172-179.

14. Chi, C.-H. and Sand, L. B., **Nature**, 1983, **304**, 255-257.

15. Warzywoda, J. and Thompson, R. W., **Zeolites**, 1991, **11**, 577-582.

16. Zhdanov, S. P., Adv. Chem. Ser. 101 (eds. Flanigen, E. M. and Sand, L. B.), 1971, 20-43.

17. Hou, L.-Y. and Thompson, R. W., **Zeolites**, 1989, **9**, 526-530.

18. Warzywoda, J., Edelman, R. D., and Thompson, R. W., **Zeolites**, 1991, **11**, 318-324.

19. Tsokanis, E. A. and Thompson, R. W., **Zeolites**, 1992, **12**, 369-373.

20. Maped, N. F., Kubasov, A. A., Limova, T. V., and Burenhova, L. N., **Russ. J. Phys. Chem.**, 1985, **59**, 1582.

21. Zhdanov, S. P. and Samulevich, N. N., Proc. 5th Int. Conf. on Zeol. (ed. Rees. L. V. C.) 1980, 75-84.

22. Dewaele, N., Bodart, P., Gabelica, Z. and B.Nagy, J., **Acta. Chim. Hung.**, 1985, **119**, 233-244.

23. Dutta, P. K., Shieh, D. C. and Puri, M., **J. Phys. Chem.**, 1987, **91**, 2332-2336.

24. Fahlke, B., Starke, P., Seefeld, V., Wieker, W. and Wendlandt, K.P., **Zeolites**, 1987, **7**, 209-213.

25. Sand, L. B., Sacco, A., Thompson, R. W. and Dixon, A. G., **Zeolites**, 1987, **7**, 387-392.

26. Bronic, J., Subotic, B., Smit, I. and Despotovic, Lj. A., Stud. Surf. Sci. & Catal., no. 37 (eds. Grobet, P. J., Mortier, W. J., Vansant, E. F., and Shulz-Ekloff, G.) 1988, 107-114.

27. Ginter, D. M., Went, G. T., Bell, A. T. and Radke, C. J., **Zeolites**, 1992, **12**, 733-741.

28. Ginter, D. M., Bell, A. T. and Radke, C. J., **Zeolites**, 1992, **12**, 742-749.

29. Di Renzo, F., Dutarte, R., Espiau, P., Fajula, F., and Nicolle, M.-A., "Crystallisation of Zeolites in the Microgravity Environment of Space," 8th Euro. Sym. on Mater. & Fl. Sci. in Space, Euro. Space Ag. (sp. publ., ESA SP-333) 691-695.

30. Brock, A. A., Link, G. N., Poitras, P. S. and Thompson, R. W., **J. Mater. Chem.**, 1993, **3**, 907-908.

31. Hamilton, K. E., Coker, E. N., Sacco, A., Dixon, A. G. and Thompson R. W., **Zeolites**, 1993, **13**, 645-653.

32. Xu, R., Zhang, J., Pang, W., and Li, S., Report No. 421, Jilin University, P. R. C., 1983.

33. Edelman, R. D., Kudalkar, D. V., Ong, T., Warzywoda, J. and Thompson, R. W., **Zeolites**, 1989, **9**, 496-502.

34. Kornatowski, J., Kanz-Reuschel, B., Finger, G., Baur, W., Bulow, M., and Unger, K. K., **Collect. Czech. Chem. Commun.**, 1992, **57**, 756-766.

35. Sano, T., Mizukami, F., Kawamura, M., Takaya, H., Mouri, T., Inaoka, W., Toida, Y., Watanabe, M., and Toyoda, K., **Zeolites**, 1992, **12**, 801-805.

36. Hufton, J. R. and Danner, R. P., **AIChE J.**, 1993, **39**, 962-974.

37. Hay, D. G., Jaeger, H. and Wilshier, K. G., **Zeolites**, 1990, 10, 571-576.

38. de Vos Burchart, E., Jansen, J. C. and van Bekkum, H., **Zeolites**, 1989, **9**, 432-435.

39. Gonthier, S., Gora, L., Guray, I. and Thompson, R. W., **Zeolites**, 1993, **13**, 414-418.

40. Golemme, G., Nastro, A., B.Nagy, J., Subotic, B., Crea, F. and Aiello, R., **Zeolites**, 1991, **11**, 776-783.

41. Twomey, T. A. M., Mackay, M., Kuipers, H. P. C. E. and Thompson, R. W., **Zeolites**, 1994, in press.

42. Gonthier, S., M. S. thesis, 1993, Chemical Engineering, WPI.

43. Cundy, Lowe, Sinclair, **Fara. Disc.**, in press.

44. Thompson, R. W., "Population Balance Analysis of Zeolite Crystallization," in *Modelling of Structure and Reactivity in Zeolites,* ed. C. R. A. Catlow, Academic Press, 1992, 231-255.

45. Sano, T., Okabe, K., Kohtoku, Y., Shimazaki, Y., Saito, K., Takaya, H., and Bando, K., **Zeolites**, 1985, **5**, 194-196.

J.C. Jansen, M. Stöcker, H.G. Karge and J. Weitkamp (Eds.)
Advanced Zeolite Science and Applications
Studies in Surface Science and Catalysis, Vol. 85
© 1994 Elsevier Science B.V. All rights reserved.

The opportunities of the fluoride route in the synthesis of microporous materials

H. Kessler, J. Patarin and C. Schott-Darie

Laboratoire de Matériaux Minéraux, URA CNRS 428, Ecole Nationale Supérieure de Chimie de Mulhouse, Université de Haute-Alsace, 3 rue Alfred Werner, 68093 Mulhouse cedex, France

An overview is given on the synthesis of crystalline microporous materials by using fluoride-containing media. For silica-based materials the replacement of OH- by F- as a mineralizer makes it possible to obtain zeolites at pH values lower than 10-11. The synthesis of high-silica zeolites and clathrasils is described with emphasis on the effect of the presence of fluoride on the crystal size, the substitution of silicon by trivalent or tetravalent elements and the organic species leading to a given material. In a second part is addressed the preparation of aluminophosphate-based solids and gallophosphates. It is shown that besides already known phases, numerous novel ones could be obtained in the presence of F- anions. Fluorine is generally part of the framework as terminal or bridging species, or occluded in double-four-rings as in the LTA-type $AlPO_4$ and $GaPO_4$, and in the large-pore gallophosphate cloverite with 20-membered ring openings.

1. INTRODUCTION

The use of fluoride as a flux component for the crystal growth from a melt is well known. On the other hand the mineralizing role of fluoride in hydrothermal synthesis was known by the ancient mineralogists and the chemists [1], but until the end of the seventies it was little studied in the synthesis of microporous materials. The first clear example of the use of fluoride was for the crystallization of silicalite-1 in slightly alkaline media by Flanigen and Patton [2]. The fluoride route was then extensively investigated by our group for silica-based zeolites, alumino- and gallophosphates. Several others groups have reported on it too.

For the synthesis of silica-based zeolites which are normally prepared in alkaline medium, the replacement of the hydroxide anions by fluoride anions as mineralizers makes it possible to obtain zeolites even in slightly acidic media (pH ~5). At such pH values the solubility of silica for example increases significantly in the presence of fluoride because of the formation of hexafluorosilicate species [3]. Such species were observed in particular in the mother liquor of fluoride silicalite-1 by ^{19}F and ^{29}Si NMR [3,4]. No other silicon-containing species could be observed. It can be assumed that the hydrolysis of fluorosilicate anions yields polycondensable hydroxylated species whose condensation leads to the crystalline material. In the synthesis of borosilicalite-1, the presence of the hydroxyfluoroborate anions BF_3OH^- and $BF_2(OH)_2^-$ beside BF_4^- and SiF_6^{2-} was indeed evidenced by ^{19}F NMR [3].

Most of the syntheses employing the fluoride route have been carried out in aqueous medium. However, a non-aqueous fluoride medium including ethyleneglycol has been used for the preparation of microporous gallophosphates for example [5,6]. In fact, in most cases small amounts of water were present. The non-aqueous fluoride route was recently developed for the synthesis of very large crystals in the mm range. Thus, Kuperman *et al.* [7] reported on the synthesis of well defined single crystals of all-silica materials, aluminosilicates and aluminophosphates 0.4-5.0 mm in size. The systems included pyridine, HF, reagent quantities of water and optionally an organic template. The authors [7] conclude that HF-pyridine and HF-alkylamines are novel mineralizers that make it possible to control the amount of water present in the reaction medium.

More recently a third route by heating a dry powder of silica and NH_4F thus producing NH_3 and water vapor was found [8]. Apparently, the water vapor enables a local gas phase transport of silica which recrystallizes into silicalite-1 in the presence of tetrapropylammonium bromide.

The aim of this chapter is to give an overview on the synthesis of microporous materials in the presence of fluoride. The synthesis of silica-based materials and alumino- and gallophosphates will be addressed. Emphasis will be placed on the role of fluoride which besides its mineralizing-complexing role shows a structure directing and templating effect. The last two, which were first observed for octadecasil [9], are particularly apparent in the synthesis of microporous aluminophosphates and gallophosphates.

2. CRYSTALLIZATION OF SILICA-BASED ZEOLITES

2.1. Usual synthesis conditions

Typically the synthesis mixture is prepared by adding to a silica source (fumed, colloidal, precipitated silica,...) a source of element T (T=Al,Fe,Ga,B...) if necessary, an organic species and a fluoride source. Among the various fluoride sources used, HF and NH_4F are preferred because the calcination of the as-synthesized material leads directly to the H form of the zeolite. The temperatures of crystallization are similar to those of the syntheses without F⁻ but the crystallization time is usually higher.

2.2. Materials obtained in the presence of F⁻ anions

The various materials obtained in fluoride media are reported in Table 1. With the exception of the AST, CJS-1, CJS-2 and CJS-3 materials, all the solids have been previously obtained in the absence of fluoride anions.

Table 1
Silica-based materials (structure code) obtained in the presence of F⁻ anions

Acidic or near neutral fluoride media (pH = 5-11)
MFI, TON, MTT, FER GIS, SOD, BEA, MTW LEV, MEL, ZSM-48* (MTN, Nu-1*, AST, CJS-2*, CJS-3*)** CJS-1*

* No structure code   ** Clathrasil

2.3. MFI - type zeolites

MFI- type zeolites with a large framework composition [10-16] could be obtained in fluoride medium with $Pr_xNH_{4-x}^+$ templates (Pr = n-propyl, x = 1 to 4) according to the procedure developed in our laboratory [17]. More recently, other organic species such as ethylenediamine, diethylamine, choline, piperazine and 1,4-diazabicyclo (2,2,2) octane, were found to lead to MFI-type zeolites too in the presence of fluoride anions [18]. Nevertheless tetrapropylammonium seems to be the best template since, as shown in Table 2, in some cases, other structure types are obtained .

Table 2

Templating efficiency of $Pr_xNH_{4-x}^+$ in the absence and in the presence of Al substituting for Si

Template	Framework composition	
	T = Si	T = Si, Al
Pr_4N^+	MFI	MFI
Pr_3NH^+	MFI	MFI
$Pr_2NH_2^+$	MTT (MFI)	MFI (MTT)
$PrNH_3^+$	MTN (FER) (MFI)	FER (MFI)

( ) traces

The crystals are always of good quality and the size exceeds generally the values observed in alkaline-type synthesis.

The morphology is the same as for crystals obtained by the usual alkaline route.The length/width ratios decreases with x in the $Pr_xNH_{4-x}^+$ template and the substitution degree of Si. The substitution by trivalent elements leads generally to less flat faces which indicates that crystallization occurred in a more supersaturated medium.

2.3.1. Purely siliceous MFI-type zeolite

Typical synthesis examples leading to siliceous MFI-type samples using $Pr_xNH_{4-x}^+$ (x = 2 to 4) templates are reported in Table 3.

Among the templating cations which were used, Pr_4N^+ (x = 4) proved to be the most efficient for easy and fast crystallization without seeds.

The as-synthesized samples show different crystal morphologies (Figure 1). As mentioned above, the length/width ratio of the crystal decreases with x. The Pr_4N^+-containing crystals are rather elongated (80 x 40 x 20 μm^3), whereas the crystals obtained with Pr_3N (x = 3) and Pr_2NH (x = 2) show an isometric facies (30 x 25 x 20 μm^3 and 6 x 5 x 4 μm^3 respectively). It should be noted that under specific conditions (stirring, seeding) small cystals (1-5 μm) could be obtained at a lower temperature with Pr_4N^+ species (e.g., 80°C).

Table 3
Typical synthesis mixtures leading to siliceous MFI - type zeolites in the presence of $Pr_xNH_{4-x}^+$ templates.

Sample	x	Starting molar composition	T (°C)	Crystallization time (days)
A	4	$1 SiO_2$: $0.125 Pr_4NBr$: $0.5 NH_4F$: $30 H_2O$	170	15
B	3	$1 SiO_2$: $1 Pr_3N$: $2HF$: $30H_2O$: seeds *	170	15
C	2	$1 SiO_2$: $1 Pr_2NH$: $2HF$: $30H_2O$: seeds *	170	50

* 2 wt % per SiO_2

According to the powder X-ray patterns, all the as-synthesized samples are orthorhombic (space group Pnma [19]). The crystallographic characteristics [20] of the obtained siliceous MFI-type zeolites are given in Table 4.

Table 4
Unit cell formula and crystal parameters of the siliceous MFI-type samples [20]

Sample	Unit cell formula	a(Å)	b(Å)	c(Å)
A	$(Si_{96}O_{192})4Pr_4NF$	20.039(3)	19.928(3)	13.382(3)
B	$(Si_{96}O_{192})4Pr_3NHF \cdot 8H_2O$	20.048(2)	19.889(2)	13.383(3)
C	$(Si_{96}O_{192})4.8Pr_2NH_2F \cdot 6.5H_2O$	20.045(4)	19.886(3)	13.379(5)

Water molecules occupy the volume of the channels which is left empty by the template carrying less than four propyl groups. The cationic nature of the template was established by ^{13}C NMR spectroscopy.Whereas the Pr_4N^+ species occupy the straight and zigzag channels [21], the Pr_3NH^+ and $Pr_2NH_2^+$ cations are located preferentially in the zigzag channels.Such a conclusion could be drawn from the following main observations [20]:

(i) b (parallel to the straight channels) decreases, a increases and c stays constant on going from Pr_4N^+ to $Pr_2NH_2^+$ (see Table 4).

(ii) According to the differential thermal analysis under argon, the Pr_3NH^+ and $Pr_2NH_2^+$ species decompose at about the same temperature (535 and 515°C).Such a temperature is much higher than the decomposition temperature of $Pr_2NH_2^+$ occluded in the MTT structure (<400°C) in which this template is located in straight channels analogous to those of the MFI structure.

(iii) The ^{13}C chemical shift of the carbon β of $Pr_2NH_2^+$ in the MFI-type structure (δ = 18.2 ppm/TMS) is different from the corresponding value for $Pr_2NH_2^+$ in the MTT-type structure (δ = 20.2 ppm/TMS).

Figure 1. Scanning electron micrographs of siliceous MFI-type zeolites prepared with $Pr_xNH_{4-x}^+$ species as templates . Top : x = 4, sample A; middle : x = 3, sample B and bottom : x =2, sample C.

For the three as-synthesized samples, the molar ratio F$^-$/organic species is close to one. Price *et al.*[21] found from single-crystal X-ray diffraction data that the fluoride anions were located close to the Pr$_4$N$^+$ species at the intersection of the 10-ring channels of the framework.Whereas, more recently [22] by Rietveld refinement, the F$^-$anions were located between two five membered rings near the intersection of the channels.

The efficiency of the Pr$_4$NF species compared to the Pr$_3$NHF and Pr$_2$NH$_2$F species was confirmed by the determination of the standard enthalpies of formation of the as-synthesized samples [23].The deduced $\Delta H°$ of stabilization are -596.8 ± 230.3 KJ.mol^{-1}, + 431.6 ± 231.9 KJ.mol^{-1} and +508.0 ± 213.9 KJ.mol^{-1} for samples A, B and C respectively. It appears that the Pr$_4$NF species only contributes significantly to the stabilization of the MFI structure. This result agrees with the observed specificity of the three organic species with respect to the MFI structure type.The tetrapropylammonium species lead to the MFI structure type alone, whereas the two other species lead also to the structure MEL [24] and MTT [20] respectively.

2.3.2. MFI-type zeolites with Si partly substitued by T^{III} elements (T=B, Al, Fe, Ga).

By using a similar starting composition, but in the presence of a trivalent element source (salt, hydroxide,...), the fluoride route can be used to prepare MFI- type zeolites in which silicon is partly substituted by a T element (T = B, Al, Fe, Ga) [12-14,16] The Si^{IV}/T^{III} ratio in the zeolite varies with the synthesis conditions and is always larger than about 10. Nevertheless for a ratio lower than 20, some extraframework T^{III} species are often present in the final material. For T-poor samples (Si/T>100) the template cations (Pr$_4$N$^+$) occluded in the channels of the structure are essentially neutralizing the fluoride anions like in the siliceous MFI-type material. In T-richer samples, the organic cation compensates partially or completely the negative charge of the framework.

Large twinned crystals are obtained contrary to those produced by the usual alkaline route. A study of the distribution of the elements within the crystals was performed by X-ray emission mapping.The most homogeneous intra- and intercrystalline compositions are those with a Si^{IV}/T^{III} ratio between 20 and 30. This corresponds to nearly 4 T^{III} per unit-cell ; *i.e.*, 4 negative charges balanced by the 4 alkylammonium cations. When Si^{IV}/T^{III} increases,homogeneity decreases, the core of the crystals being richer in T^{III} than the outer shell. The concentration gradient decreases from Ga^{III} to Fe^{III} and Al^{III}. When Ga and Al are introduced in the same mixture, it can be shown by X-ray emission mapping of the respective T^{III} atoms that the core of the crystal is saturated with Ga and that Al follows Ga towards the outside (Figure 2). Nevertheless, the degree of substitution and the importance of the segregation will depend on both T$_{Al}$^{III}/T$_{Ga}$^{III} and T$_{Si}$^{IV}/(T$_{Al}$^{III}+T$_{Ga}$^{III}) ratios in the reaction mixture. When these ratios are lower than about 1 and 30 respectively only Ga will be present in the crystals.

By using a combined source of Si^{IV} and T^{III} elements, the concentration gradient of T^{III} in the crystal can be strongly reduced. This is illustrated in the case of the crystallization of Fe^{III}-containing MFI-type zeolites.

The iron distribution in two zeolite samples prepared from a ferrisilicate xerogel (obtained by the hydrolysis of tetraethoxysilane in the presence of a ferric salt

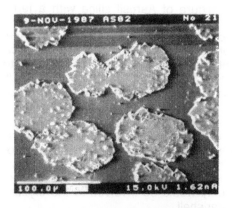

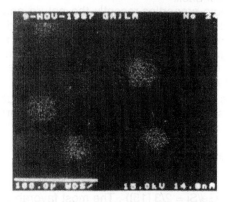

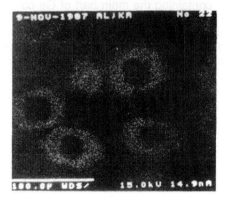

Figure 2. Section across (Ga, Al, Si) - MFI crystals: scanning electron micrograph (top) ; X-ray emission mapping of Ga (middle) and Al (bottom).

solution) and from a mixture of Aerosil silica with a ferric salt solution can be compared in Figure 3. A starting Si/Fe molar of 70 was used for both experiments and was found too by chemical analysis of the two crystalline samples. For the solid prepared by the first procedure, *i.e.*, homogeneous ferrisilicate xerogel, there is no concentration gradient in the crystal. The Si/Fe ratio determined by Si and Fe analysis performed on individual spots is similar to the Si/Fe ratio determined by chemical analysis. The contour of the FeKα X-ray emission map matches the contour of the crystals on the micrograph (Figure 3, left). For the solids prepared by the second procedure (Aerosil 130) the crystals show an iron-rich core with a Si/Fe molar ratio near 30 and an essentially siliceous rim (Figure 3, right). In this latter case, it can be considered that the silica particles are coated with ferric species that would dissolve preferentially. Iron-rich (Si/Fe = 30) nuclei would therefore be formed. During crystallization, the iron concentration decreases and in the last crystallization stage, silica species only would be involved in the crystal growth, leading to an essentially siliceous outer shell.

Concerning the acidic properties, recently the nature and strength of acid sites present in (Si, Al) MFI-type zeolites synthesized in OH⁻ and F⁻ medium have been compared [25]. Silica rich (Si/Al >25-30) H-MFI prepared in the presence of fluoride anions have for any given Si/Al ratio a number of acid sites similar to that of the corresponding H-MFI materials synthesized in alkaline medium. On the contrary, aluminium-rich H-MFI (Si/Al ≤20) samples exhibit very few acid sites compared to their OH counterparts. As mentioned before, such a result might be due to the presence in these samples of cationic extraframework aluminium species compensating part of the framework charges.

2.3.3. MFI-type zeolites with Si^{IV} substituted by Ge^{IV} or Ti
Germanium

By using the fluoride synthesis route, high substitution degrees could be reached with germanium (up to Ge/Si = 2/3 [15]) . The most favorable pH range lies between 9 and 11. Other oxides containing the main part of Ge co-precipitate with the MFI - type zeolites in neutral or acidic medium. In a very alkaline medium, the formation of soluble germanates restricts the incorporation of germanium in the zeolite. Moreover, if the pH is adjusted with alkali or ammonium hydroxide, the corresponding germanates admix with an almost purely siliceous MFI-type zeolite. These difficulties can be avoided if the pH is adjusted near to 10 with an amine such as CH_3NH_2. Under such conditions the presence of F⁻ allows crystallization to occur within 15 hours at 180°C with the following reaction mixture :

$$1\ GeCl_4 : 1 SiO_2 : 0.5\ Pr_4NBr : 1\ HF: 8\ CH_3NH_2 : 35\ H_2O.$$

After removal of the template by calcination at 550° C in air, the orthorhombic symmetry (as-synthesized sample) becomes monoclinic (calcined sample). The unit -cell parameters of the monoclinic materials are a function of the degree of substitution x (Ge per 96 tetrahedra). Besides, the monoclinic-orthorhombic transition temperature of the calcined samples increases linearly from ~80° C (x = 0) to ~250° C (x = 32.8). It should be noted that the reverse phenomenon is observed when trivalent elements (B, Al, Fe ...) are incorporated into MFI - type zeolites [26].

Two Ge-MFI samples (Si/Ge = 5 and 10) were characterized in their as -synthesized and calcined form by X-ray absorption spectroscopy [27]. In both cases, XANES

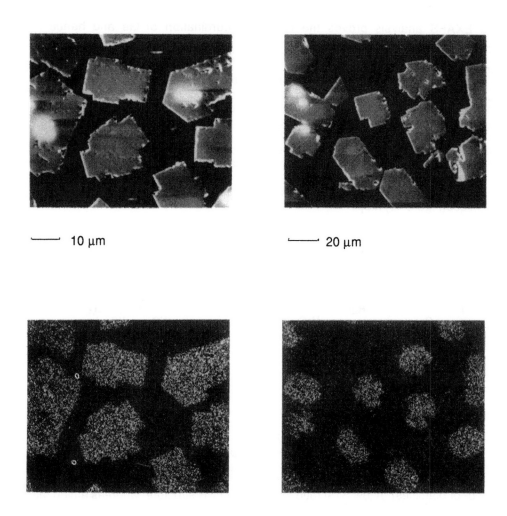

10 μm

20 μm

Figure 3. Section across (Si, Fe)-MFI crystals:scanning electron micrographs (top) and FeKα X-ray emission mapping (bottom). Left, homogeneous, and right, heterogeneous iron distribution.

and EXAFS analysis proves the four-fold coordination of Ge and hence its incorporation into the framework. The Ge-O bond distance is close to 1.74 and 1.72Å for the as-synthesized and calcined materials respectively.

Titanium

Titanium-containing MFI-type zeolite could be synthesized in the presence of fluoride anions too. Nevertheless the substitution degree is lower (Si/Ti ≥ 50).Titanium is incorporated into the framework, but a significant amount of Ti can also be present in the form of extracrystalline or intracrystalline inclusions of TiO_2 depending on the F/Ti ratio in the synthesis mixture. For the calcined samples, without TiO_2 impurities and in the absence of water, titanium was found in a four-fold coordination in the zeolite framework by X-ray absorption spectroscopy (Figure 4a). In the as-synthesized and calcined hydrated forms, strong interactions with water molecules result in a six-fold coordination for titanium,according to the same technique(Figure 4b) [28].

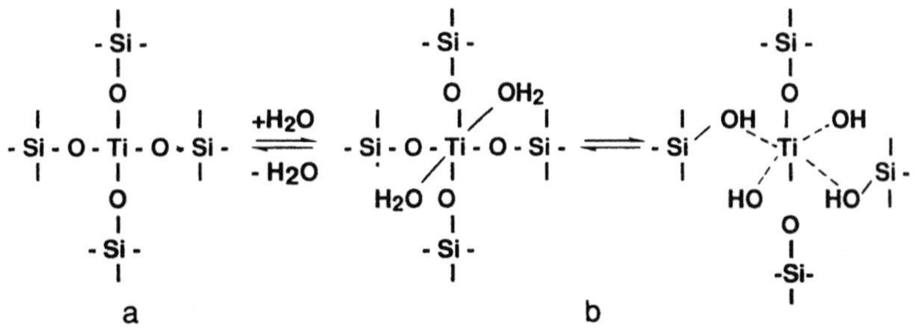

Figure 4. Proposed model for titanium coordination in MFI-type zeolite.
a : in the absence of water ; b : in the presence of water [28].

2.4. FER - ,TON - and MTT - type zeolites

Three pentasil zeolites with the FER, TON and MTT structure- types were obtained at low pH values (pH = 5 - 9) in the presence of fluoride anions and amines [29-31]. The first two are obtained with the same amine families (linear aliphatic mono- or diamines). A low Si/Al ratio leads to the FER structure, whereas a high ratio favors the TON structure. The influence of the carbon chain length x of the amine is shown in Figure 5.The composition of the reaction mixture was

$(1-y)SiO_2$: y Al_2O_3:$10H_2O$:$3[CH_3(CH_2)_{x-1}NH_2,HF]$ or $1[NH_2(CH_2)_xNH_2,2HF]$.

For the FER-type zeolite, y ~0.066 and the synthesis duration was 14 days at 170° C in a static mode. The TON-type zeolite was obtained for y ~0 and heating during 3 days at 170° C under agitation with 5wt % of seeds (with respect to silica).

By using monoamines, highly crystalline FER-type and TON-type materials were obtained for x equal to 3 or 4 and 4 or 5 respectively. With diamines, the best results were obtained with x equal to 4 or 5 for the FER-type and x equal to 5 or 6 for the TON-type materials. It should be noted that with diamines, MFI-type zeolite was obtained instead of FER-type zeolite for x = 6 and of TON-type zeolite for x = 7.

In addition to the above-mentioned templates two other amines, di-n pentylamine and 1-4 diamino n-pentane led also to the TON structure type in the presence of F⁻ anions.

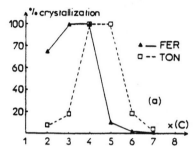 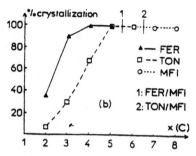

Figure 5. Crystallization (%) of FER- (▲) TON -(□) and MFI-(○) type materials as a function of the number x of carbons in the monoamines (left) and in the diamines (right) used as templates.

A MTT-type structure is obtained by using one of the three following amines, di-n-propylamine, isopropylamine or pyrrolidine under the same conditions as for the synthesis of TON-type zeolites.

The FER-type crystals are aggregates of platelets, the size of which can be larger than 200 μm (Figure 6, top). As mentioned in the introduction, it should be noted, that very large crystals (mm range) were obtained recently by Kuperman *et al.* in HF/pyridine media [7]. The TON and MTT structure-types show a fibrous aspect, the length being over 100 μm for a diameter below 1 μm (Figure 6, middle and bottom respectively).

From the chemical analysis and the ^{13}C NMR results, the template in these materials is always present in the cationic form (compensation of framework charges and / or F⁻ anions in the channels).

2.5. Silicalite- 2 and zeolite ZSM-48

Silicalite-2, the Al-free analog of ZSM-11 (MEL structure type) and ZSM-48 were synthesized in the presence of fluoride anions from systems containing either tetrabutylammonium(TBA),hexamethonium(HM) cations or 1,12-diaminododecane (DADD) or the combinaison of these organic species [32].

Crystallization occurred at 170°C during 20 days under static conditions from a gel (pH ~9 – 10) having the following molar composition :

$10NaF : (xTBABr : y HMBr_2 - z DADD) : 10 SiO_2 : 330 H_2O$, with x+y+z = 1.25.

At high TBA content in the system (x = 0.85-1.25), the product was amorphous or contained only small amounts of silicalite-2, pure silicalite-2 zeolite crystallizing in the

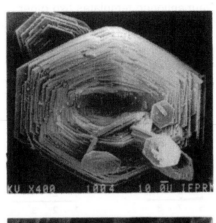

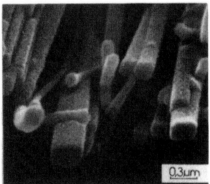

Figure 6. Scanning electron micrographs of zeolites FER-(top), TON-(middle) and MTT-(bottom) type zeolites prepared in fluoride medium.

region 0.35≤x≤0.85. With no hexamethonium ion in the reaction mixture, the crystallization time was extremely low. High concentrations of diamine or low concentrations of TBA (x = 0.0-0.32) favor the formation of zeolite ZSM-48. Without diamine or without fluoride anions no crystallization was observed. This result confirms the important role of F⁻ ions in the syntheses at low alkalinity. A maximum of 0.4 and 2 fluoride anions per unit-cell was found in the as-synthesized ZSM-48 and silicalite-2 samples respectively.

2.6. SOD-type zeolite

During the attempt to synthesize zeolite A (LTA) in the presence of fluoride anions, sodalite (SOD) was obtained as impurity and its amount increased when the fluoride content in the mixture increased. From the gel compositions reported in Table 5, aggregated small crystals (<1 µm) of fluorosodalite as main phase were obtained [33].

Table 5
Synthesis conditions leading to fluorosodalite [33]

Sample	Gel composition	T (°C)	Time (hrs)	Products
A	$1SiO_2 : 1 Al_2O_3 : 1.4 Na_2O : 5 NaF$	150	48	SOD + (CAN) *
B	$1SiO_2 : 1 Al_2O_3 : 1.4 Na_2O : 1 NaF$	200	48	SOD + (CAN)*

* (CAN) = traces of cancrinite.

In the two samples, the ratios Si/Al and Na/(Al+F) determined by chemical analysis are close to one. By analogy to hydroxysodalite and considering that the size of the fluoride anion (ionic radius=1.33 Å) is comparable to the size of the hydroxide anion (1.40 Å), the chemical formula of the fluorosodalite would be $Na_6 [Al_6 Si_6 O_{24}] 2 NaF \cdot x H_2O$. As determined from various samples, the number x of water molecules lies between 4 and 6.

The fluorosodalite samples were characterized in their as-synthesized, calcined, and calcined rehydrated form by ¹⁹F NMR spectroscopy [33].For the as-made fluorosodalite, the ¹⁹F NMR spectrum shows a sharp signal at -178.8 ppm (reference $CFCl_3$) with a small one at -174.3 ppm (Figure 7a). The main component was assigned to fluoride located in the sodalite cage. After calcination at 100°C without rehydration, two main peaks are observed at -179.0 and -184.7 ppm (Figure 7b). At higher temperatures (300° C and 600° C) no change in the powder XRD pattern was observed but only the last NMR signal (~–185 ppm) is present (Figure 7c). The dehydration process is reversible. After 24 hours of rehydration a similar spectrum to that of Figure 7b is obtained and after one month of rehydration, only one main component at -178.7 ppm is present on the NMR spectrum (Figure 7d). The main conclusions on this NMR study are the following :

- At 100° C, the fluoride anion can be found in two well defined states of hydration in the fluorosodalite,corresponding to the two main ¹⁹F NMR peaks at -179 and -184.7 ppm.

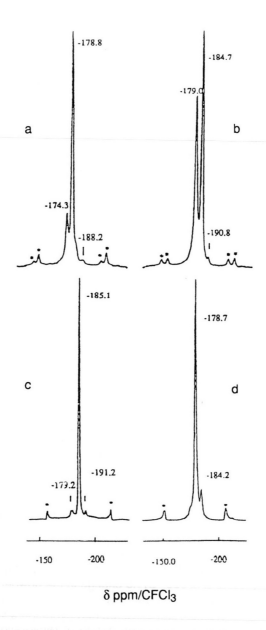

Figure 7. ^{19}F NMR spectra of fluorosodalite zeolite : a) as - synthesized ; b) after calcination at 100° C ; c) after calcination at 600° C ; d) after calcination at 600° C and complete rehydration * spinning side bands.

- At 600° C, the dehydration is complete and the peak at -185.1 ppm can be assigned to a fluoride ion in a sodalite cage containing no water molecule.
- In the as-synthesized sample the ^{19}F NMR signal at -178.8 ppm corresponds to a fluoride ion in a sodalite cage containing 2 to 3 water molecules.

2.7. BEA-, MTW- and LEV - type zeolites

2.7.1.BEA-type zeolite

Zeolite β is usually prepared from basic media (without F⁻ anions) in the presence of Na⁺ and TEA⁺ cations as the structuring agents. In fluoride medium, zeolite β crystallizes at 170° C under static conditions from a silica - alumina gel (pH ~7-9) containing 1,4-diazabicyclo(2,2,2)octane(DABCO) and methylamine as templates [34]. The molar ratios of the starting mixture are in the range 5 to 30 for Si/Al, 0.5 to 1 for F/Si as well as for templates/Si, and 10 to 20 for H_2O/Si. As shown in Figure 8, for high Si/Al molar ratios, or under stirring conditions, MTW-type zeolite or cristobalite are obtained. Complete crystallization of zeolite β occurs after 7 to 15 days of synthesis only in the presence of methylamine and by adding 2wt % of seeds (per SiO_2) in the reaction mixture. As a matter of fact, without seeding only partial conversion of the gel is observed. Such a result might be explained by the very large viscosity of the synthesis medium.

Crystals of zeolite β are larger than those prepared in a high alkaline medium (typically 5-10 μm). They are characterized by a truncated square bipyramidal morphology (Figure 9, top).The Si/Al molar ratios lie between 9 and 22 and the amount of organic species in the as- synthesized sample is close to 25 wt %.

As observed for MEL-type zeolite, the fluoride content is low (about 2 F⁻ per unit cell). In this case too, F⁻ plays a mineralizing role and might compensate partially the charge of the template molecules.

By ^{13}C NMR spectroscopy, two signals are observed at 46.1 and 54.7 ppm/TMS (Figure 10). The less intense signal corresponds to the dabconium cations (δ= 46.1 ppm), whereas the main peak was assigned to a cationic polymer of polyethylene piperazine (δ = 54.7 ppm) resulting from the polymerization of DABCO (Figure 11).

The unit-cell formula for the silica-richest and -poorest samples of zeolite Beta prepared in fluoride medium may be written as [34]

~ 2.4 [DABCO⁺] ~ [$C_{43}N_{12}N^+_{2.5}H_{88}$] , 2.1 F⁻, 61.2 SiO_2 , 2.8 AlO_2^- , ~8.5 H_2O
and
~ 2.6 [DABCO⁺] , ~ [$C_{47}N_{10.5}N^+_{5.5}H_{94}$] , 1.9 F⁻, 57.8 SiO_2 , 6.2 AlO_2^- , ~8.5 H_2O

In both cases, the DABCO molecule and the polymeric species are partially protonated.

2.7.2. MTW-type zeolite

As can be seen in Figure 8, MTW-type zeolite crystallizes from high-silica gels (Si/Al between 12 and 100) in the presence of DABCO as template with or without methylamine [35]. The molar composition of the gel (H_2O/Si, F/Si, DABCO/Si) is similar to that of zeolite β. Pure samples are obtained only under stirring conditions after 5 days at 170° C or 1 day at 200° C.

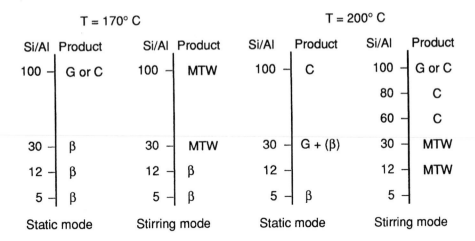

β = zeolite Beta, C = cristobalite, G = gel

Figure 8. Crystallization fields of BEA- and MTW-type zeolites in fluoride medium at two different temperatures with DABCO and methylamine.

MTW-type zeolite shows a fibrous aspect, the length of the fibers being over 100 μm for a diameter less than 1 μm (Figure 9, middle). The Si/Al molar ratio of the obtained products is in the range 30 to 90 and the amount of organic species is close to 10 wt %. The same polymeric organic species as found in zeolite β are present in the as-made materials (similar ^{13}C CP MAS NMR spectrum). The removal of the template by calcination is difficult. Indeed, a complete white sample, without no significant loss of crystallinity is obtained only after heating at 900° C during 2 hours.

The amount of F$^-$ anions in the as-made zeolite is low (0.1 to 0.6 wt %) and from ^{19}F NMR spectroscopy, the fluoride is found to be present mainly as SiF_6^{2-} species (sharp peak at -127 ppm / $CFCl_3$). The presence of such species could explain the low n-hexane adsorption capacity (2wt %) of the corresponding calcined samples compared to the values reported (4-10 wt %) for a classical MTW-type zeolite prepared in high alkaline medium. As a matter of fact, after calcination under air, the thermal decomposition of these SiF_6^{2-} species could lead to silicic residues, obstructing partially the channels of the structure.

2.7.3. LEV-type zeolite

In the presence of quinuclidine (Q) instead of DABCO the hydrothermal crystallization of silica-alumina gels with low Si/Al molar ratios (5 ≤ Si/Al ≤ 10) leads to LEV-type materials [36].When the Si/Al ratio increases to 10≤Si/Al≤100, a mixture of levyne and of the clathrasil octadecasil is obtained; pure octadecasil crystallizes from high silica gels (Si/Al>100) (see below).

Figure 9. Scanning electron micrographs : BEA - (top), MTW - (middle) and LEV-(bottom) type zeolites prepared in fluoride medium.

54.7

46.1

δ ppm/TMS

Figure 10. ^{13}C CP MAS NMR spectrum of as-synthesized zeolite β prepared in fluoride medium.

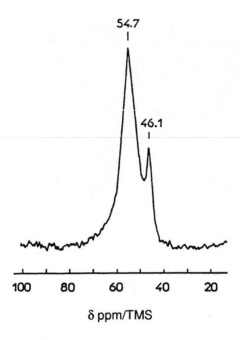

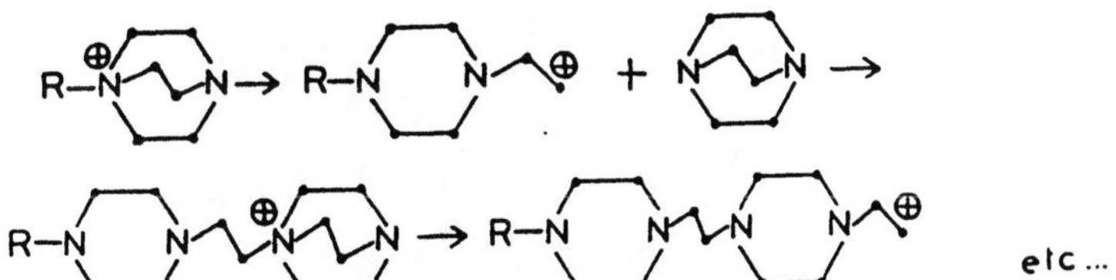

etc...

Figure 11. Proposed mechanism of cationic polymerization of DABCO : R being a proton or a methyl group resulting from the thermal decomposition of methylamine.

For instance, a pure LEV-type material is obtained after 4 days at 170 - 200° C under static conditions with the gel ratios F/Si and Q/Si equal to 0.5.

The samples are characterized by a Si/Al ratio close to10 and a low fluoride content (0.3 to 0.6, depending on the sample). In the absence of seeds, large spheric aggregates consisting of small platelets are obtained (Figure 9, bottom).A typical formula for the hexagonal unit-cell is

48.1 SiO_2 , 5.9 AlO_2^- , 6.9 Q^+ , 1.3 F^-, ~ 5.7 H_2O.

In this case, most of the fluoride anions are compensated by Q^+, the excess (0.3F) can be explained by the presence of SiF_6^{2-} impurities.They show up as a sharp signal at-128 ppm (reference $CFCl_3$) beside the line at -82 ppm assigned to the F^- anions neutralizing part of Q^+.

2.8. CJS-2, CJS-3 and octadecasil clathrasils

2.8.1. CJS-2 and CJS-3 clathrasils

Recently, two clathrasils named CJS-2 and CJS-3 were reported [18]. They were obtained by using the fluoride route at pH = 7 in the presence of pyrrolidine and N,N,N',N'-tetramethylethylenediamine respectively.The molar compositions of the starting mixtures were

0.5 R : 1.0SiO_2 : 0.5 HF : 40 H_2O (R = organic species).

After heating at 190°C for 23 (CJS-2) and 54 days (CJS-3) large crystals of these two materials were obtained. The topological structure of CJS-2 was found to be consistent with that of the previously published nonasil [37] but with a different space group. However, CJS-3 presents an original powder XRD pattern with a unit-cell parameter equal to 19.4795 (3) Å.

In both clathrasils, the organic species are protonated, they could not be completely removed even after calcination under O_2 at 900° C for 4 hours.

2.8.2. Octadecasil (AST-type structure)

As mentioned before (Section 2.7), octadecasil of the structure type AST [38] was synthesized in fluoride medium from high silica gels (Si/Al >100) in the presence of quinuclidine (Q) as template [9]. This phase was obtained too from a mixture containing tetramethylammonium cations (Me_4N^+) instead of quinuclidine. However, in that case other phases such as MTN and Nu-1 materials were often present in the final product.

Typically, with quinuclidine, the synthesis conditions were the following : 170° C ; 15 days ; F/Si and Q/Si molar ratios (always equal) between 0.5 and 2 ; H_2O/Si molar ratio between 16 and 20. The pH of the mixture was between 7 and 8. Whatever the template used (Me_4N^+ or Q), as shown in Figure 12 (left), the crystals have a octahedral shape with some truncatures when prepared with Me_4N^+ ions. Their size varies from about 20 to 100 μm (up to 200 μm in some cases). The silicon to aluminum molar ratio of the as-made samples is high (Si/Al >100). Actually, octadecasil was mainly obtained in its pure silica form.

 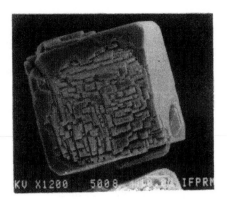

Figure 12. Scanning electron micrographs of octadecasil (left) and Nu-1 (right) prepared in fluoride medium.

The as-synthesized octadecasil crystallizes in the tetragonal symmetry (space group I4/m). The unit-cell formula and the crystallographic characteristics are given in Table 6.

Table 6
Unit-cell formula and crystal parameters of octadecasil samples [9]

Sample	Organic species used	Unit-cell formula	a (Å)	c (Å)
A	Me_4N^+	$[Si_{20}O_{40}]\ 2(Me_4NF)$	n.d.	n.d.
B	Q	$[Si_{20}O_{40}]\ 2(QF)$	9.194(2)	13.396(4)

n.d.: not determined

Quinuclidine was found protonated by [13] C NMR spectroscopy. In both samples, the fluoride anions are present as ion-pairs. As an example, the ^{19}F NMR spectrum of sample B is reported in Figure 13. An unusual very sharp line is observed at -38.22 ppm (reference $CFCl_3$). Such a chemical shift corresponds to a specific position of the F^- in the structure. As a matter of fact, from the single crystal structure analysis performed on sample B, the fluoride anions were located in the double-four membered rings (D4R units) of the structure ; 85 % of these units being occupied. It was the first case where F^- anions were found to play a structure orienting role beside their mineralizing role. Four highest maxima of the residual electron density map present inside the octadecahedral cages of the framework were assigned to the disordered protonated quinuclidine [9].

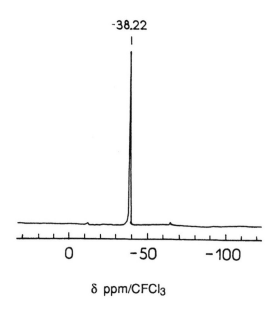

Figure 13. ^{19}F MAS NMR spectrum of as-made quinuclidinium octadecasil

After calcination at 500° C in air, the product is black and the structure becomes cubic. A complete white sample was obtained after calcination up to 980°C, but the structure was partly collapsed.

2.9. Zeolite P, ZSM-39 and Nu-1

Zeolite P (GIS-type structure) and ZSM-39 (MTN-type structure) were obtained in non alkaline medium with the same organic species (tetramethylammonium cations, Me$_4$N$^+$) [18], A low Si/Al ratio led to zeolite P, whereas a high Si/Al ratio favored ZSM-39 formation. In both cases, the crystals had a typical polyhedral shape and their size was close to 150 (zeolite P) and 400 µm (ZSM-39).

In Table 7 are reported the typical synthesis conditions leading to these two zeolites.

Table 7
Typical synthesis conditions leading to zeolite P and ZSM-39 [18]

Zeolite	Starting molar composition	T (°C)	Crystallization time (days)
P	1.0 Me$_4$N$^+$:1.0 SiO$_2$:0.1 Al$_2$O$_3$:1.0 NH$_4$F:42 H$_2$O	190	21
ZSM-39	1.1 Me$_4$N$^+$:1.0 SiO$_2$:1.8 NH$_4$F:40 H$_2$O	190	7

In more concentrated media, Nu-1 crystallizes [39] but in most cases with a MTN-type material.

After 5 days of crystallization at 170° C, a pure Nu-1 sample was produced from the following starting mixture (pH=9-10):

0.5 Me$_4$N$^+$: 1 SiO$_2$: 0.06 Al$_2$O$_3$: 1 NH$_4$F : 9 H$_2$O.

After a longer crystallization time (15-30 days), ZSM-39 was obtained as the main phase .

The Si/Al molar ratio of the Nu-1 sample was close to 15. The crystal size depends on the crystallization time and varies from 20 to 50 μm. In fact, as shown in Figure 12, each crystal consists of aggregates of small crystallites with a size close to 3 μm. As observed for MTN-type materials, the removal of the template by calcination is difficult. Indeed, after calcination in air at 950°C (weight loss ~ 9.3 %), the solid was black and no significant loss of crystallinity was observed. The structure of Nu-1 was recently solved from powder X-ray synchrotron data [40] and this material was ranged in the clathrasil family.

By ^{19}F NMR spectroscopy, a sharp signal was observed at -116.6 ppm (reference CFCl$_3$) besides a large one at -140.6 ppm [39]. The latter which was found too in the MTN-type sample corresponds to (NH$_4$)$_3$AlF$_6$ impurities always present in the final products. The first peak was assigned to a fluoride species which is typical of the Nu-1 material.

A ^{1}H liquid NMR study performed on the as-synthesized sample after dissolution in hydrofluoric acid solution, showed clearly that during the synthesis, part of the tetramethylammonium cations decomposed into trimethylamine [39]. Such a result could explain the difficulties to synthesize a pure material.

A thermal decomposition of the templates was also observed during the synthesis of a novel aluminosilicate zeolite named CJS-1 in non alkaline media [18].

CJS-1 was prepared by using three different heterocyclic compounds (piperazine, pyrrolidine and piperidine) as templates. The synthesis conditions were the following:

(0.5-4.0)organic species:1.0 SiO$_2$:(0.05-0.5)Al$_2$O$_3$:(0.5 -3.0)HF:(20-200)H$_2$O

T = 150-200 °C;crystallization time 6-10 weeks.

This new aluminosilicate shows an original XRD pattern and characteristic micropore adsorption properties.

3. SYNTHESIS OF ALPO$_4$-BASED MATERIALS

3.1. Usual synthesis conditions with fluoride

In contrast to the alkaline pH values used in the conventional synthesis of silica-based zeolites, the usual pH of the reaction mixture for the synthesis of AlPO$_4$-based materials is slightly acidic to slightly alkaline (typically, starting pH = 3-10) [41,42]. Therefore, the pH conditions for the synthesis of AlPO$_4$-based solids in the presence of fluoride are close to those that would be used in its absence. Thus, the mineralizing effect may not be so evident as in the case of silica-based zeolites. Nevertheless, various beneficial effects are observed in the presence of fluoride, the crystallization times are generally shorter and the crystals usually larger and well formed. Thus, in the synthesis of AFI-type materials, SAPO-34 and CoAPSO-34, Xu *et al.* [43] observed that when F⁻ is present, the induction time is smaller (divided by about 3) but the rate of crystal growth is slower than when it is absent.

They assumed that the presence of fluoride anions favors the fast production of fewer nuclei, after which crystal growth consumes preferentially but slowly the precursors. The stability of the fluorocomplexes must not be so high that further reaction involving them is inhibited.

Another beneficial effect of the presence of fluoride is the production of a number of phases which do not form in a fluoride-free medium, thus showing a structure directing role of it. For such phases, fluorine is generally part of the framework bonded to Al atoms as terminal or bridging species, or even trapped in double-four-ring units as it was observed in the clathrasil octadecasil [9].

3.2. Aluminophosphates and derived materials obtained as well in the presence as in the absence of fluoride

Up to now materials of the structure types AFI, AEL, CHA, FAU and SOD have been prepared in the presence of fluoride. In Table 8 are given the organic molecules used and the corresponding framework compositions.

Typically, $AlPO_4$-5 for example is obtained by heating a gel of molar composition

$$1Pr_3N : 1 \ Al_2O_3 : 1P_2O_5: 1 \ HF : 70H_2O$$

at 170 °C for 17 hrs with stirring (tumbling of the autoclave). Essentially the same conditions are used for the synthesis of SAPO-5, the starting SiO_2/Al_2O_3 ratio is typically in the range 0-0.6.

Fluorine is present in the as-synthesized materials (chemical analysis, ¹⁹F MAS NMR). In the $AlPO_4$-type solids with no framework charge, fluoride neutralizes the cationic organic species and the ratio F/organic species is generally equal to one, whereas in the substituted materials less fluorine is present because part of the organic cations compensate the negative framework charge. By calcination fluorine-containing species are removed together with the decomposition products of the organics. The resulting solid is essentially fluorine-free. In some cases the thermal stability of the material prepared in fluoride medium is increased with respect to that of the solid produced in its absence. This is presumably due to a higher crystallinity.

The location of the organic and F species has been determined by structure analysis [47] on a single crystal of $AlPO_4$-5 (80 μm x 500 μm) which was obtained by heating at 170 °C for 10 days a gel of the following molar composition :

$$2.2Pr_4NOH : 1Al_2O_3 : 1P_2O_5 : 1.7 \ NH_4F : 318H_2O.$$

The unit-cell formula is $(AlPO_4)_{12}Pr_4NF$ and the parameters a=13.740(5)Å, c=8.474(4)Å. The organic cations are located in the 12-ring channels and F⁻ in the 4-ring columns between two 4-rings (Figure 14). Thus, F⁻ is not located close to the tetrapropylammonium cation, contrary to OH⁻ which was found in the tripod-shaped Pr_4N^+ cation in the 12-ring channels of the AFI-type material prepared in the absence of fluoride and whose unit-cell formula is $(AlPO_4)_{12}Pr_4NOH$ [48].

Table 8
AlPO$_4$-type and derived materials obtained with or without fluoride

Structure type	Organic molecule	T elements	Ref.
AFI	Et$_3$N	Al,P;Al,P,Me(Me=Mg,Cr,Mn,Co,Ni)	44
		Al,P;Al,P,Me(Me=Mg,Co)	45
	Pr$_3$N	Al,P;Al,P,Si	46
		Al,P,Co	45
	Pr$_4$NOH	Al,P	46,47
AEL	n-Pr$_2$NH	Al,P;Al,P,Si	46
CHA	Et$_3$N	Al,P,Si;Al,P,Si,Co	43
	Et$_4$NOH	Al,P,Si;Al,P,Me(Me=Mg,Co)	46,45
	Morpholine	Al,P,Si	45
FAU	Me$_4$NOH + Pr$_4$NOH	Al,P,Si	45
SOD	Me$_4$NOH	Al,P,Me(Me=Co,Ni)	45

^{19}F MAS NMR confirmed the structure analysis, *i.e.*, one single line was observed (chemical shift = -119 ppm/CFCl$_3$). As the ^{19}F chemical shift in AlPO$_4$-11 (AEL) is very close to the value observed for the AFI-type phase, one may assume that the environment of fluorine is essentially the same in both structures, *i.e.*, that F is not close to the organic species (dipropylammonium cations) in this structure too.

3.3. Aluminophosphates obtained solely in the presence of fluoride

The aluminophosphates which were obtained in the presence of fluoride only are reported in Table 9.

AlPO$_4$-CJ2 is an ammonium hydroxyfluoroaluminophosphate with 6-membered ring channels containing NH$_4^+$ [50]. The crystal structure was redetermined recently by Férey *et al.* [52]. Fluorine is part of the framework as bridging and terminal species. The other materials of Table 9 will be discussed in more detail.

3.3.1. Tetravalent variant of AlPO$_4$-16

On attempting the synthesis of cloverite with the AlPO$_4$ composition and the usual template for cloverite, *i.e.*, quinuclidine [53], a tetragonal variant of AlPO$_4$-16 was obtained with a=9.3423(1)Å, c=13.4760(2)Å, space group I$\bar{4}$ [49], whereas the material prepared in the absence of HF is cubic with a=13.3832(6)Å, space group I4/m [54]. Thus, like in octadecasil [9] the presence of fluoride induces a tetragonal distortion of the structure. A Rietveld refinement showed that fluoride is located in the D4R's like in octadecasil.

On the removal of the organic species and fluorine the material transforms into the cubic form.

Table 9
Aluminophosphate obtained in the presence of fluoride only

Structure	Organic species (+HF)	Ref
Tetragonal variant of $AlPO_4$-16	Quinuclidine	49
Triclinic CHA-like	1-methylimidazole	6
	Morpholine	46
	Piperidine	6
	Pyridine	6
	N,N,N',N-tetramethylenediamine	45
$AlPO_4$-CJ2	Hexamethylenetetramine (decomposes)	50
LTA-type	Diethanolamine + tetramethylammonium	51

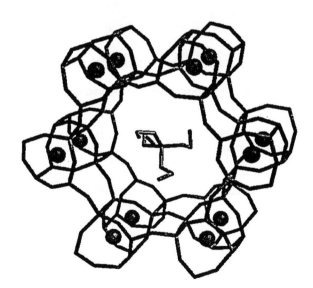

Figure 14. Plot of the 12-ring channel and the 4-ring columns with Pr_4N^+ and F location in $(AlPO_4)_{12}Pr_4NF$. One Pr_4N^+ position is shown . Balls : fluoride anions.

3.3.2. Triclinic CHA-like AlPO$_4$

A triclinic CHA-like AlPO$_4$ was produced in the presence of HF with morpholine as a template [46,55]. A mixture of molar composition :

1.5 Morpholine : 1Al$_2$O$_3$: 1P$_2$O$_5$: 1HF : 100 H$_2$O

was heated at 170 °C for 10 days. The obtained crystals are shown in Figure 15. By adding a silica source and keeping the Al/P ratio equal to one the same triclinic variant of SAPO-34 was obtained [46].

The crystal structure of the AlPO$_4$ form was refined by the Rietveld technique [56]. Two fluorine atoms were found to bridge two Al atoms of a 4-ring which connects two D6R's of the chabazite-like topology (Figure 16).

Figure 15. Scanning electron micrograph of the triclinic CHA-like AlPO$_4$[55].

The additional F atoms are neutralized by two protonated organic molecules located in each chabazite cage. Thus, the unit-cell formula is (AlPO$_4$)$_6$ (C$_4$H$_{10}$O)$_2$F$_2$. As reported in Table 9 this triclinic material was further obtained with 1-methylimidazole , piperidine, pyridine and N,N,N',N'-tetramethylenediamine. It is a precursor of AlPO$_4$-34 which is obtained after removal of the organic template and HF by calcination [6].

3.3.3. LTA-type AlPO$_4$ and derived materials

The silicoaluminophosphate materials ZK-21, ZK-22 and SAPO-42 of the LTA type have been reported [57] but phosphorus is only present in a small amount (typically 0.01$\leq$P/(P+Si+Al)$\leq$0.04).

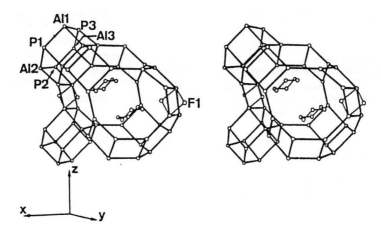

Figure 16. Stereoplot of the chabazite cage of the triclinic CHA-like AlPO$_4$ with the bridging fluorine atoms and the two occluded morpholinium cations. The three types of Al and P atoms are shown (From [56]).

LTA-type AlPO$_4$ was synthesized for the first time by using a fluoride medium containing tetramethylammonium cations and diethanolamine (DEA) as templates [51]. A starting mixture with the molar composition

0.078Me$_4$NCl : 0.93DEA : 1Al$_2$O$_3$: 1P$_2$O$_5$: 0.17HF : 40H$_2$O

was heated at 170 °C for 8.5 hrs. SAPO, CoAPO and MeAPSO (Me=Co,Zn) samples were also obtained with the same templates.

The crystal structure was determined on a CoAPSO single crystal [51]. The space group is Fm$\overline{3}$c, a=24.063(3)Å. There is Al, P alternation and some of them are substituted by Si or Co. The fluoride anions were found in the D4R's and the tetramethylammonium cations in the sodalite cages. The nitrogen atom of diethanolamine (essentially protonated) is located in the center of the 8-ring window with ethanol groups -CH$_2$-CH$_2$OH protruding in two adjacent α cages in a similar way as the propyl group of di-n-propylammonium in LTA-type GaPO$_4$ (see 4.2.2.).

4. SYNTHESIS OF GALLOPHOSPHATES

In 1985, a number of gallophosphates with a porous framework structure obtained by hydrothermal synthesis were reported by Parise [58]. This was followed by several studies with various organic species [59,60] and structural characterization on single crystals [61-66].

Compared to the aluminophosphates, the gallophosphates possess more complicated framework structures, which result from the variety in the coordination of the gallium atom.

The fluoride method was first used for the synthesis of gallophosphates in 1991. The results were exciting with the discovery of cloverite [53], which is the first molecular sieve with three-dimensional 20-membered ring channels, and the LTA-type gallophosphate [67]. Since then continuous interest has been shown for the Ga_2O_3-P_2O_5-HF-H_2O-R (R=organic species) system [68,69].

The new route using a non-aqueous medium [70-72] allowed to obtain in the Ga_2O_3-P_2O_5 system, with or without fluoride, new materials or already known phases with different templates [5-73].

4.1. Usual synthesis conditions for microporous gallophosphates in the presence of fluoride

4.1.1. In aqueous medium
A typical starting gel composition is the following :

$$zR : 1Ga_2O_3 : 1P_2O_5 : xHF : yH_2O$$

where   R is the organic species
$$0.2 \leq x \leq 2$$
$$40 \leq y \leq 300$$
$$1 \leq z \leq 6$$

Generally the amount of R is such that the starting pH value is in the range 3-7.

Various gallium sources have been used. The most employed have been the oxide hydroxide GaOOH, the oxide Ga_2O_3 or an amorphous gallium oxide produced by heating gallium nitrate at 250°C. Hydrated gallium sulfate $Ga_2(SO_4)_3 \cdot xH_2O$ was found to be the most reactive for the synthesis of cloverite.

The preferred and most used source of P_2O_5 is orthophosphoric acid H_3PO_4 as for the synthesis of aluminophosphates [42]. The source of fluoride was generally HF. It was found that when NH_4F was used as the fluoride source, the ammonium cation showed a strong templating effect and the ammonium gallophosphate $(NH_4)_{0.93}(H_3O)_{0.07}GaPO_4(OH)_{0.5}F_{0.5}$ analogous to $AlPO_4$-CJ2 [50] was obtained. Loiseau [74] suggested that in the presence of NH_4F the role of the amine is limited to a pH-controlling one.

The order of addition of the reactants is usually the following : H_3PO_4, water, gallium source, hydrofluoric acid and finally the organic species. The obtained gel is mixed at room temperature for about 2hrs and transferred into a PTFE-lined stainless steel autoclave. The crystallization temperature and heating time may vary respectively form 80°C to 220°C [75] and from a few hours to a few weeks.

Only a few studies on the framework substitution by Si or Al were performed. The synthesis procedure was usually the same as that used for the $AlPO_4$-derived materials [42].

4.1.2. In "non-aqueous" medium
The conditions of synthesis in non-aqueous medium are close to those used in aqueous medium, the difference is that alcohol substitutes for water, however the medium is not strickly non aqueous because the sources of phosphorus and fluorine were always introduced as H_3PO_4 85% and HF 40%. For the Ga_2O_3-P_2O_5 system only syntheses in ethyleneglycol were performed [5,6]. A typical gel composition is the following :

4R: $1Ga_2O_3$: $1.8P_2O_5$: $1.5HF$: $45EG$

which is sealed in a Teflon-lined stainless-steel autoclave and heated at 180°C for 3-10 days.

4.2. Gallophosphate materials obtained in the presence of fluoride

The gallophosphate type phases which were obtained with various organic species in the presence of HF are reported in Table 10.

Table 10
Gallophosphates obtained in aqueous medium in the presence of fluoride with various organic species

Organic species (+HF)	Phase obtained
Quinuclidine Methylquinuclidinium iodide 3-azabicyclo(3,2,2)nonane Piperidine Hexamethyleneimine	cloverite
Di-n-propylamine Pyridine ("non-aqueous" medium)	LTA
Dimethylamine Piperazine Pyrrolidine	$GaPO_4$-21
Pyridine 1-methylimidazole	tricl.$GaPO_4$ or one unknown phase
Diaminoethane Diaminopropane	ULM-4 ULM-3, ULM-4, ULM-6
Diaminobutane Diaminopentane	ULM-3
Diaminohexane (HMDA)	ULM-5 GaPO-HMDA-2 GaPO-HMDA-3
Diaminoheptane Diaminooctane	ULM-5
Diazabicyclo (2,2,2)octane	ULM-1, ULM-2 Other DABCO-containing gallophosphates

104

Except for GaPO$_4$-21 which was formed with dimethylamine, piperazine or pyrrolidine (Table 10) fluorine was found in the structure of all obtained materials. GaPO$_4$-21 has already been synthesized with isopropylamine [62], ethylenediamine [65] and ethanolamine [76] in the absence of fluoride.

4.2.1. -CLO-type GaPO$_4$

As mentioned above, a remarkable material obtained with the fluoride method in the Ga$_2$O$_3$-P$_2$O$_5$ system is cloverite. In Figure 17 is given a SEM picture of cloverite crystals. This exceptional structure with a three dimensional 20-membered ring channel system and large cages of 30 Å across [53] could not be obtained without fluoride. In the absence of fluoride, GaPO$_4$-a [60] was obtained with quinuclidine which was the first organic species leading to cloverite. The unusual shape of the window in the form of a four-leafed clover is due to the presence of terminal hydroxyl groups in the framework (Figure 18). Fluorine is occluded in the D4R's of the structure, a corresponding ^{19}F MAS NMR chemical shift at -68ppm was found. The D4R's are distorted in such a way that the F$^-$ ions are close to all four Ga atoms that become five coordinated. Recently, Barr et al. [77] showed the strong acidity resulting from the terminal hydroxyl groups in the structure. On the other hand, several studies on ^{31}P and [71]Ga NMR spectroscopy were undertaken [78-81].

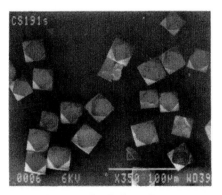

Figure 17. Scanning electron micrograph of cloverite.

The hydrothermal synthesis of cloverite is now possible with various templates, i.e., quinuclidine (large range of temperature, [75]), methylquinuclidinium iodide, 3-azabicyclo (3,2,2)nonane and piperidine [6]. It should be mentioned that Huo et al. [5] reported its synthesis in a non-aqueous medium with the last template, and more recently we obtained cloverite with hexamethyleneimine in aqueous medium [82]. These results are not so surprising considering the similarities between these molecules. In Figure 19 are shown the various molecules having been found to lead to cloverite.

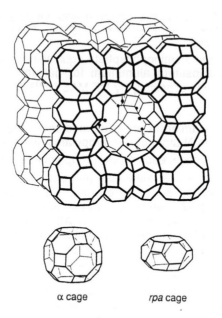

α cage   *rpa* cage

Figure 18. The framework topology of cloverite showing the cubic arrangement of α and rpa cages • Hydroxyl groups.

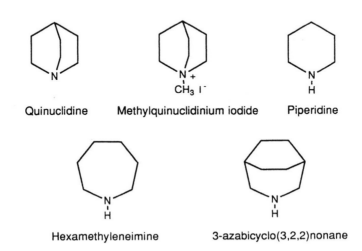

Quinuclidine   Methylquinuclidinium iodide   Piperidine

Hexamethyleneimine   3-azabicyclo(3,2,2)nonane

Figure 19. Organic molecules leading to cloverite.

106

The incorporation of silicon in the structure was not very sucessfull and only the presence of Si(4Si) islands resulting from a $2Si^{4+} = P^{5+}+Ga^{3+}$ substitution might possibly be considered [75].

In view of preparing the aluminophosphate equivalent of cloverite we studied the possibility to partially substitute aluminium for gallium in the structure. Partial substitution could be proven especially by ^{19}F MAS NMR [83].

4.2.2. LTA-type GaPO$_4$

The use of di-n-propylamine in aqueous medium in the presence of HF led to LTA-type GaPO$_4$ [67].The N-atom of di-n-propylammonium cation was found in the 8-membered ring of the α cage, each propyl group pointing towards the center of two adjacent α cages. The structure refinement [84] showed that the N-atom is shifted away from the center of the 8-membered ring. This results in a strong N-H---O hydrogen bond (d(N-O)=2.7Å) which was evidenced by ^{31}P MAS NMR. Ojo *et al.* reported a series of syntheses of AlPO$_4$ and AlPO$_4$-derived materials with di-n-propylamine and concluded to the preferential orientation towards exclusively channel-containing structures [85]. This shows the particularity of the Ga$_2$O$_3$-P$_2$O$_5$-HF-H$_2$O system which allowed to obtain a structure with cages. As mentioned above, fluorine is localized in D4R's as in cloverite.

The use of pyridine in non-aqueous medium containing ethylene glycol led to the LTA-type GaPO$_4$ too [6]. ^{19}F MAS NMR shows that fluorine is in the same environment as in the material obtained in aqueous medium. Seeing the differences between di-n-propylamine and pyridine, it will be of interest to determine the localization of the latter and also whether ethyleneglycol is occluded in the structure. The crystal morphology depends on the synthesis medium. In aqueous medium the crystals are cubic whereas in the non-aqueous one cuboctahedra are obtained (Figure 20).

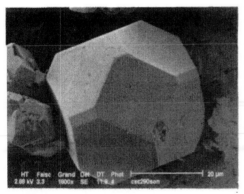

Figure 20. Scanning electron micrographs of GaPO$_4$-LTA obtained in ethyleneglycol (left) and in water with di-n-propylamine (right).

4.2.3. Triclinic CHA-like GaPO$_4$

The use of either pyridine or 1-methylimidazole as organic species in fluoride medium led to isotypes of a triclinic variant of chabazite which was already obtained in the AlPO$_4$ form [6]. The structure was determined on a single crystal of the phase containing 1-methylimidazole. The unit cell parameters are a=9.426(6)Å, b=9.168(6)Å, c=9.308(6)Å, α=90.38(5)°, β=103.75(5)°, γ=92.58(5)°, space group P$\overline{1}$.

As for the triclinic CHA-like AlPO$_4$, fluorine is part of the framework, the Ga-F distances are 1.959(5)Å and 1.991(6)Å [82].

Two gallium atoms per unit cell which belong to a four-ring are linked to four framework oxygens and to two symmetrically equivalent fluorine atoms, therefore two gallium atoms out of six are hexacoordinated. By ^{19}F MAS NMR, one single line is observed at a chemical shift of -98 ppm.

On removal of water, HF and the organic species by calcination, GaPO$_4$-34 which is isostructural to AlPO$_4$-34 was obtained. This form is stable up to at least 600°C under inert atmosphere, but as for many gallophosphates slow amorphization occurs in a moist atmosphere [82].

Both organic species (pyridine or 1-methylimidazole) led to another novel phase which is unidentified at this time [82]. The ^{19}F MAS NMR spectrum shows two lines at -111.1ppm and -131.5ppm which could correspond to fluorine in the structure. The line at -111.1ppm is close to that observed by Loiseau in GaPO$_4$-CJ2 and in ULM-4 for a terminal fluorine [74].

4.2.4. Gallophosphates ULM-3, ULM-4 and, ULM-6

By using diaminopropane, diaminobutane and diaminopentane Loiseau [74] obtained a new orthorhombic phase called ULM-3. The unit-cell parameters are a=10.154(1)Å, b=18.393(2)Å, c=15.773(2)Å, space group Pbca. The three-dimensional structure shows 10 membered ring channels along [100], they are interconnected with 8-membered ring channels along [101] and [10$\overline{1}$] [86]. Two types of F are bridging two gallium atoms of a GaO$_4$F$_2$ octahedron and a GaO$_4$F trigonal bipyramid. By ^{19}F MAS NMR, chemical shifts close to that observed for GaPO$_4$-tricl.-CHA were recorded at -93.1 ppm and -100.6 ppm. In the same system, with diaminopropane as organic species but for different pH values, adjusted by changing either the Ga$_2$O$_3$/P$_2$O$_5$ or the Ga$_2$O$_3$/R ratio, the same authors obtained two other new structures ULM-4 and 6. ULM-4 as well as ULM-6 shows a framework based on 8 MR channels and contains fluorine. The structure of ULM-4 is based on the same units arrangement as ULM-3, whereas that of ULM-6 consists of the connexion of PO$_4$ tetrahedra with GaO$_4$X (X=F,H$_2$O) trigonal bipyramids and GaO$_4$F$_2$ octahedra.

4.2.5. Gallophosphate ULM-5

The use of a linear diamine with a longer chain (H$_2$N(CH$_2$)$_n$NH$_2$, n=6-8) led to the first 16-membered ring material. Whereas we had obtained it as a powder only, by heating a gel of starting composition

2.8HMDA: 1Ga$_2$O$_3$: 1P$_2$O$_5$: 1HF: 80H$_2$O

at 180°C for 24hrs [6], Loiseau and Férey [87] prepared single crystals (n=6-8) and solved the structure by single crystal X-ray diffraction. In Figure 21 is given a SEM picture of our sample. The three-dimensional network is built up from three

Figure 21. Scanning electron micrograph of the gallophosphate ULM-5 obtained with hexamethylenediamine as the organic template.

types of basic building units: two hexameric Ga_3P_3 and one octameric Ga_4P_4, the latter as in cloverite. The framework delimits 16- and 6-membered ring channels along [100] and 8-membered ones along [010] (Figure 22).

The diprotonated amines are inserted in the 16-membered ring channels, whose free aperture is 12.20 x 8.34Å. For the first time, both bonding and encapsulated F atoms are present in the same structure. By ^{19}F MAS NMR we observed three chemical shifts at -67.8 ppm, -99.2ppm, -113 ppm [6]. The first one is close to that already observed for cloverite [88].

The study of the thermal stability of the material by high temperature X-ray diffraction (Guinier camera) showed a slight structure change at ~350 °C and a transformation into a mixture of quartz and cristobalite-type $GaPO_4$ at 650 °C [6]. The structure stable at 600 °C is retained on cooling to room temperature under a dry atmosphere. However, amorphization occurs when the calcined material is subjected to moisture.

The increase of the pH of the starting gel resulted in two different phases. At this time, these materials could only be obtained as a mixture with some ULM-5. The observation of low-angle powder diffraction lines ($2\theta < 6°$) might be indicative of a porous character [82].

4.2.6. DABCO-containing gallophosphates

When the bicyclic diamine DABCO was used as organic species in the Ga_2O_3-P_2O_5-HF-H_2O system, six different phases could be obtained [82]. The crystal structure of two of them has been previously solved by Loiseau et al.. The first one, which was named ULM-1, is based on 8-membered ring channels along [001], 6-

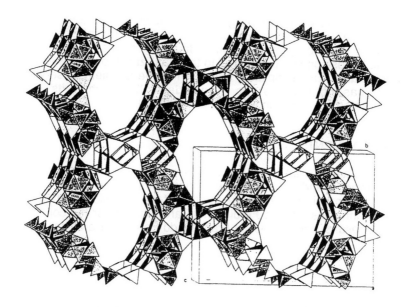

Figure 22. Perspective view of ULM-5 close to [100]. The amino groups have been omitted for clarity (G. Férey and T. Loiseau, private communication).

and 8-rings along [101] and [010] [68]. The cationic framework of ULM-2 is built up from the stacking of $[(12.3)^2.(12.3.8).(12.3.8)]^4$ gallium-phosphorus plane nets and phosphorus square plane nets along 4_1 helicoidal axes [69].

Modifications in heating time, temperature or water content led to four other phases which could only be obtained as a mixture [82]. Single crystals could be isolated from one of the mixtures and a structure analysis was performed. The structure, whose refinement is still in progress, shows Dabconium cations and discrete Ga-O-P D4R's hosting F⁻ anions. The question arises whether such a species is a precursor for a three-dimensional structure.

CONCLUSION

The use of fluoride media in the synthesis of microporous solids appears to offer many possibilities among which an increase of the number of possible phases. Besides known zeolites, some new silica-based materials such as CJS-1, CJS-2, CJS-3 and a AST-type phase were obtained. On the other hand, the crystal size of the known zeolites synthesized in the presence of fluoride is larger than that of the corresponding phases prepared in its absence. Moreover, the protonic form of the zeolite can be obtained directly by calcination of the as-synthesized solid. A more pronounced effect of the fluoride anions is observed in the systems Al_2O_3-P_2O_5 and Ga_2O_3-P_2O_5. Indeed, the fluoride route appears the most fruitful for these systems.

Besides its mineralizing-complexing role, F⁻ shows a structure directing and a templating effect. In most aluminophosphates and gallophosphates F is part of the framework or even occluded in double-four-ring cages as first observed for octadecasil. It thus contributes strongly to the stabilization of the structure as do the organic templates. This templating effect of fluoride is chiefly observed when the framework has no charges (as in octadecasil) or when its charges are autocompensated (as in alumino- or gallophosphates).

ACKNOWLEDGMENTS

The authors are grateful to Dr Philippe Caullet and Professor Jean-Louis Guth for fruitful discussions, and to Mrs Simone Einhorn, Mrs Sophie Garnier and Dr. Jean-Louis Paillaud for their help in the preparation of the manuscript. Professors Gérard Férey, Hermann Gies and Dr. Ferdi Schüth are thanked for the communication of unpublished results, and the Institut Français du Pétrole for a doctoral fellowship to Céline Schott-Darie.

REFERENCES

1. R.M. Barrer, Hydrothermal Chemistry of Zeolites, Academic Press, London, 1982.
2. E.M. Flanigen and R.L. Patton, US Patent No. 4073865 (1978).
3. F. Hoffner-Marcuccilli, Thesis, Mulhouse, 1992.
4. S.A. Axon and J. Klinowski, J. Chem. Soc., Faraday Trans., 89 (1993) 4245.
5. Q. Huo and R. Xu, J. Chem. Soc., Chem. Comm., (1992) 1391.
6. C. Schott-Darie, H. Kessler and E. Benazzi, in "Proceedings of the International Symposium on Zeolites and Microporous Crystals ' 93", Nagoya, 1993, in press.
7. A. Kuperman, S. Nadimi, S. Oliver, G.A. Ozin, J.M. Garcés and M.M. Olken, Nature, 365 (1993) 239.
8. F. Schüth, private communication.
9. P. Caullet, J.L. Guth, J. Hazm, J.M. Lamblin and H. Gies, Eur. J. Solid. State Inorg. Chem., 28 (1991) 345.
10. J.L. Guth, H. Kessler, M. Bourgogne, R. Wey and G. Szabo, Fr. Pat., Appl. No. 84/07773 (1984).
11. J.L. Guth, H. Kessler, M. Bourgogne, R. Wey and G. Szabo, Fr. Pat., Appl. No. 84/11521 (1984).
12. J.L. Guth, H. Kessler, R. Wey and A.C. Faust, Fr. Pat.,Appl. No. 85/07978 (1985).
13. J. Patarin, J.L. Guth, H. Kessler, G. Coudurier and F. Raatz, Fr. Pat.,Appl. No. 86/17711 (1986).
14. J.L. Guth, H. Kessler and J.M. Popa, Fr. Pat., Appl. No. 87/07187 (1987).
15. Z. Gabelica and J.L. Guth, Fr. Pat., Appl. No. 88/04367 (1988).
16. A. Seive, J.L. Guth, F. Raatz and L. Petit, Fr. Pat., Appl. No. 88/06509 (1988).
17. J.L. Guth, H. Kessler and R. Wey in "New Developments in Zeolite Science and Technology", Proceedings of the 7th International Zeolite Conference, Y. Murakami, A. Iijima and J.W. Ward, (eds.), Elsevier, Amsterdam, 1986, p. 121.

18. D. Zhao, S. Qiu and W. Pang in "Proceedings from the 9 th International Zeolite Conference", Part I, R. von Ballmoos, J.B. Higgins and M.M.J. Treacy (eds.), Butterworth - Heinemann, Stoneham 1993 p. 337.
19. D.H. Olson, G.T. Kokotailo, S.L. Lawton and W.M.J. Meier, J. Phys. Chem., 85 (1981) 2238.
20. J. Patarin, M. Soulard, H. Kessler, J.L. Guth and J. Baron, Zeolites, 9 (1989) 397.
21. G.D. Price, J.J. Pluth, J.V. Smith, J.M. Bennett and R.L. Patton, J. Am. Chem. Soc., 104 (1982) 5971.
22. B.F. Mentzen, M. Sacerdote - Peronnet, J.L. Guth and H. Kessler, C.R. Acad. Sci. Paris,313 (1991) 177.
23. J. Patarin, M. Soulard, H. Kessler J.L. Guth, and M. Diot, Thermochim. Acta, 146 (1989) 21.
24. Z. Gabelica, E.G. Derouane and N. Blom, Appl. Catal., 5 (1983) 109.
25. J.F. Joly, A. Auroux, J.C. Lavalley, A. Janin and J.L. Guth in "Proceedings from the 9th International Zeolite Conference", Part II , R. von Ballmoos, J.B. Higgins and M.M.J. Treacy (eds.), Butterworth - Heinemann, Stoneham, 1993, p.235.
26. D.G. Hay and H. Jaeger, J. Chem. Soc. Chem. Comm. (1984) 1433.
27. M.H. Tuilier, A. Lopez, J.L. Guth and H. Kessler, Zeolites, 11 (1991) 662.
28. A. Lopez, M.H. Tuilier, J.L. Guth, L. Delmotte and J.M. Popa, J. Solid State Chem. ,102 (1993) 480.
29. J.L. Guth, A.C. Faust, F. Raatz and J.M. Lamblin, Fr. Pat., Appl. No.86/16362 (1986) .
30. J. Patarin, J.M. Lamblin, A.C. Faust, J.L. Guth and F. Raatz, Fr. Pat., Appl. No.88/06841 (1988) .
31. J. Patarin, J.L. Lamblin, A.C. Faust, J.L. Guth, F. Raatz, Fr. Pat., Appl. No.88/08105 (1988) .
32. R. Mostowicz, A. Nastro, F. Crea and J.B. Nagy, Zeolites, 11 (1991) 732.
33. B. Feron, J. L. Guth and N. Mimouni-Erddalane, Zeolites, in press.
34. P. Caullet, J. Hazm, J.L. Guth, J.F. Joly, J. Lynch and F. Raatz, Zeolites, 12 (1992) 240.
35. J.F. Joly, P. Caullet, A.C. Faust, J. Baron and J.L. Guth, Fr. Pat., Appl. No. 90/16256 (1990).
36. P. Caullet, J.L. Guth, A.C. Faust, J.F. Joly and F. Raatz, Fr. Pat., Appl. No. 89/17163 (1989).
37. B. Marler, N. Dehnbostel, H. Ewert, H. Gies and F. Liebau, J. Inclusion Phenom., 4 (1986) 339.
38. J.M. Bennett and R.M. Kirchner, Zeolites, 11 (1991) 502.
39. J. Patarin, B. Marler, P. Caullet, A.C. Faust and J.L. Guth, to be published.
40. H. Gies, private communication.
41. S.T. Wilson, B.M. Lok, C.A. Messina and E.M. Flanigen, in "Proceedings of the 6th International Zeolite Conference", D. Olson and A. Bisio (eds.), Butterworth, Guildford, 1984, p.97.
42. S.T. Wilson, in "Introduction to Zeolite Science and Practice", Studies in Surface Science and Catalysis, Vol. 58, H. van Bekkum, E.M. Flanigen and J.C. Jansen (eds.), Elsevier, Amsterdam,, 1991, p. 137.
43. Y. Xu, P.J. Maddox and J.M. Couves, J. Chem. Soc., Faraday Trans., 86 (1990) 425.
44. Y. Xu, P.J. Maddox and J.M. Thomas, Polyhedron, 8 (1989) 819.

112

45. M. Goepper, Thesis, Mulhouse, 1990.
46. F.M. Guth, Thesis, Mulhouse, 1989.
47. S. Qiu, W. Pang, H. Kessler and J.L. Guth, Zeolites, 9 (1989) 440.
48. J.M. Bennett, J.P. Cohen, E.M. Flanigen, J.J. Pluth and J.V. Smith, in "Intrazeolite Chemistry", G.D. Stucky and F.G. Dwyer (eds.), ACS Symp. Ser., 218, American Chemical Society, Washington DC, 1983, p. 109.
49. C. Schott-Darie, J. Patarin, Y. Le Goff, H. Kessler and E. Benazzi, Microporous Materials, in press.
50. L. Yu, W. Pang and L. Li, J. Solid State Chem., 87 (1990) 241.
51. L. Sierra, C. Deroche, H. Gies and J.L. Guth, Microporous Materials, in press.
52. G. Férey, T. Loiseau, P. Lacorre and F. Taulelle, J. Solid. State Chem., 105 (1993) 179.
53. M. Estermann, L.B. McCusker, C. Baerlocher, A. Merrouche and H. Kessler, Nature, 352 (1991) 320.
54. J.M. Bennett and R.M. Kirchner, Zeolites, 11 (1991) 502.
55. H. Kessler, in "Synthesis, Characterization and Novel Applications of Molecular Sieve Materials", R.L. Bedard *et al.* (eds.), Materials Research Society, Pittsburgh, (1991), p.47.
56. A. Simmen, Thesis, Zurich, 1992.
57. G.D. Kühl and K.D. Schmitt, Zeolites, 10 (1990) 2.
58. J.B. Parise, J. Chem. Soc., Chem. Commun., (1985) 606.
59. R. Xu, J. Chen and S. Feng, in "Chemistry of Microporous Crystals", Studies in Surface Science and Catalysis. Vol. 60, I. Inui, S. Namba and T. Tatsumi (eds.) Elsevier, Amsterdam, 1991, p. 63.
60. S.T. Wilson, N.A. Woodward, E.M. Flanigen and H.G. Eggert, Eur. Patent Appl. 0226219 (1987).
61. J.B. Parise, Inorg. Chem., 24(1985) 4312.
62. J.B. Parise, Acta Crystallogr., C42 (1986) 144
63. J.B. Parise, Acta Crystallogr., C42 (1986) 670.
64 G. Yang, S.Feng and R. Xu, J. Chem. Soc., Chem. Commun., (1987) 1254.
65. S. Feng, R. Xu, G. Yang and H. Sun, Chem. J. Chinese Univ. (English edition), No.4 (1988)1.
66. T. Wang, G. Yang, S. Feng, C. Shang and R. Xu, J. Chem. Soc., Chem. Commun., (1989) 948.
67. A. Merrouche, J. Patarin, M. Soulard, H. Kessler and D. Anglerot, in "Zeolite and Pillared Clays Synthesis",Vol. 2, M. Occelli and H. Robson (eds.), Van Nostrand Reinhold, New York, 1992, p.384.
68. T. Loiseau and G. Férey, J. Chem. Soc., Chem. Commun., (1992) 1197.
69. T. Loiseau and G. Férey, Eur. J. Solid. State Inorg. Chem., 30 (1993) 369.
70. D.M. Bibby and M.P. Dale, Nature, 317 (1985) 157.
71. Q. Huo and R. Xu, J. Chem. Soc., Chem. Commun., (1988) 1486.
72. Q. Huo and R. Xu, J. Chem. Soc., Chem. Commun., (1990) 783.
73. Q.B. Kan, F.P. Glasser and R. Xu, Journal of Materials Chemistry, 3 (1993) 983.
74 T. Loiseau, Thesis, Le Mans (1994).
75. J. Patarin, C. Schott, A. Merrouche, H. Kessler, M. Soulard, L. Delmotte, J. L. Guth and J. F. Joly in "Proceedings from the 9th International Zeolite Conference", Vol. I, R. von Ballmoos, J. B. Higgins and M. M. J. Treacy (eds.), Butterworth-Heinemann, 1993, p.263.
76. G. Yang, S. Feng and R. Xu, Jiegou Huaxue, J. Struct. Chem., 7 (1988) 235.

77. T. L. Barr, J. Klinowski, H. Y. He, K. Alberti, G. Muller and J. A. Lercher, Nature, 365 (1993) 6445.
78 R. L. Bedard *et al.*, J. Am. Chem. Soc., 115 (1993) 2300.
79. H. M. Meyer zu Altenschildesche, H. J. Muhr and R. Nesper, Microporous Materials, 1 (1993) 257.
80. S. M. Bradley, R. F. Howe and J. V. Hanna, Solid State Nuclear Magnetic Resonance, 2 (1993) 37.
81. B. Zibrowius, M. W. Anderson, W. Schmidt, F. F. Schüth, A. E. Aliev and K. D. M. Harris, 13 (1993) 607.
82. C. Schott-Darie, H. Kessler, M. Soulard, V. Gramlich and E. Benazzi, 10th International Zeolite Conference, Garmisch-Partenkirchen, (1994), submitted.
83. C. Schott-Darie, L. Delmotte, H. Kessler and E. Benazzi, Solid State Nuclear Magnetic Resonance, in press.
84. A. Simmen, J. Patarin and C. Baerlocher in "Proceedings from the 9th International Zeolite Conference", Vol. I, R. von Ballmoos, J. B. Higgins and M. M. J. Treacy (eds.), Butterworth-Heinemann, (1993), p.433.
85. A. F. Ojo and L. B. McCusker, Zeolites, 11 (1991) 460.
86. T. Loiseau, R. Retoux, P. Lacorre and G. Férey, J. Solid State Chem., in press.
87. T. Loiseau and G. Férey, J. Solid State Chem., in press.
88. A. Merrouche, J. Patarin, H. Kessler, M. Soulard, L.Delmotte, J. L. Guth and J. F. Joly, Zeolites, 12 (1992) 226.

J.C. Jansen, M. Stöcker, H.G. Karge and J. Weitkamp (Eds.)
Advanced Zeolite Science and Applications
Studies in Surface Science and Catalysis, Vol. 85
© 1994 Elsevier Science B.V. All rights reserved.

Alkali Metal and Semiconductor Clusters in Zeolites

V. I. Srdanov, N. P. Blake, D. Markgraber, H. Metiu and G. D. Stucky

Department of Chemistry, University of California, Santa Barbara, California 93106, U.S.A.

1. INTRODUCTION

The properties of molecular clusters whose dimensions are sufficiently small that deviations from bulk behavior are observed is one of the most exciting areas of modern research . Numerous experiments on interaction-free gas phase molecular clusters continue to provide an important foundation for the theoretical interpretation of their properties. From this work it has become clear that structural, optical, electronic, and magnetic properties of matter consisting of nanocluster particles could be varied, at will, if cluster size and size distribution can be controlled. The recognition of the potential usefulness of such materials has influenced chemists and physicists to consider various approaches to large scale synthesis of molecular clusters.

Because of their uniform pore sizes and their ability to absorb molecular species, zeolites have, from very early on, been considered as promising hosts for synthesis of such clusters. There are two types of clusters that have received particular attention. The clusters formed from II-VI and III-V compound semiconductors have been intensively studied because of their potential for fabrication of new opto-electronic devices. In parallel, significant advances in both the synthesis and theoretical understanding of alkali metal clusters have been achieved. The ultimate simplicity of the electronic structure of alkali metal clusters serves as an excellent starting point for testing applicability of the existing and new theories into the size quantized regime.

Unlike interaction-free gas phase molecular clusters, the properties of small molecular clusters confined inside the zeolite pores strongly depend upon the nature of the cage/cluster interface and the electrostatic field within the cage itself. The nature of the cage electric field has been probed locally at selected sites, but has never been adequately defined for the cage structure as a whole. The best classical potential parameters currently used in modelling zeolite inclusion chemistry still have no universally accepted framework charge distribution.

Here we attempt to give a short background on size quantization phenomena in semiconductors followed by a critical evaluation of the current state of affairs on inclusion of semiconductors in zeolites. We accomplish this by focusing on available data on inclusion of II-VI compound semiconductors in zeolites where particular attention is given to the most commonly studied CdS/Zeolite heterostructure. The second part of this review is dedicated to synthesis of alkali clusters in zeolites with particular attention given to properties of solvated electrons in sodalites. The utmost simplicity of precisely defined sodalite cage structure allows for pioneering use of quantum-mechanical calculations in treating the interaction of the solvated electrons with the electrostatic potential of the sodalite framework. It is hoped that the outcome of these calculations would provide an important answer regarding the actual zeolite framework charge distribution and the cage potential surfaces.

2. SEMICONDUCTOR CLUSTERS IN ZEOLITES

2.1. Quantum size effects in semiconductors

Research on the inclusion of semiconductor particles in zeolite hosts has been carried out in numerous laboratories around the world. A partial list includes the groups of Bein[1] from Purdue University; Bogomolov[2] of the Ioffe Institute in St. Petersburg, Russia; Fox[3] from the University of Texas, Herron and Wang[4] at Dupont; Goto[5], Miyazaki[6] and Terasaki[7] of Japan, Ozin[8] at Toronto, Schulz-Ekloff[9] from the University of Bremen, Thomas[10] from Notre Dame, and our work[11].

The interest in small semiconductor clusters was triggered by the discovery of Ekimov[12] who noticed that absorption spectra of small CuCl particles significantly differ from the absorption spectra of CuCl bulk. This strange phenomenon was explained by Efros and Efros[13] to be the consequence of three-dimensional space quantization of the bound electron-hole pair. The electron-hole pair is formed when a photon with energy equal to or greater than the semiconductor energy gap ($h\nu > E_g$) excites an electron from the valence to the conduction band. The positive charge left behind is appropriately called a hole and, at low temperatures, it can be weakly bound to the electron over relatively large distances (~ 100 Å). This bound electron-hole pair is called an exciton. The average distance at which the electron-hole pair is still attracted by the Coulombic forces is called the exciton Bohr radius in close analogy to the hydrogen atom. Quantum confinement occurs when the individual clusters of the semiconductor are comparable or smaller than the exciton Bohr radius. The motion of the weakly bound electron-hole pair becomes quantized which gives rise to unusual optical and electronic properties.

The Quantum Size Effects (QSE) can occur in either one, two or all three dimensions. The corresponding structures are known as Quantum Wells (QW), Quantum Wires (QWR), and Quantum Dots (QD). The following most important QSE have been experimentally documented and theoretically explained.

a) Less delocalization of the electrons in the conduction band means less kinetic energy and an increase in the semiconductor band gap with decreasing particle dimensions. The absorption edge for direct transitions of nanosized semiconductor clusters is thus shifted to higher energies (blue shifted) in respect to the bulk parent material[12,13]. There is a corresponding increase of the exciton binding energy[14].

b) It has been shown that the refractive index of QS semiconductor clusters strongly depends on small changes in the incident light intensity. This nonlinear process manifests in an enhancement of the third-order optical susceptibility $\chi^{(3)}$ whose dependence on semiconductor particle size has clearly been demonstrated[15,16,17].

c) Upon saturation of the absorption transition, when the rate of filling is equal to the rate of depletion of the lowest energy band, any additional photons absorbed must have higher energy in order to promote electrons into the next available band. Photons with lower energy are then transmitted so that the nanosized semiconductor particle becomes transparent to light. This property can be used in optical switching[18], "hole burning"[19] and laser focusing[20].

d) There is an increase of the oscillator strength for exciton radiative recombination[21] resulting in a superirradiant material with fast optical responses which makes quantum confined materials particularly attractive for applications in opto-electronics.

These extraordinary physical properties were the driving force behind the recent research efforts focused on the development of a new generation of opto-electronic devices. The main attention of physicists and engineers has been given to direct gap III-V compound semiconductors (GaAs and others) because these can be grown by high performance Molecular Beam Epitaxy (MBE) and Metalo Organic Chemical Vapor Deposition (MOCVD) techniques. The II-VI compound semiconductors received considerable attention from chemists who invented several different methods of production of size quantized particles with reasonable narrow particle size distribution. Among them the most notable was the method of arrested precipitation[22,23] and directed molecular synthesis of capped clusters[24].

QSE in Group IV semiconductors including Si, the heart of current microelectronics technology, have received little attention. This is because silicon is an indirect gap semiconductor (which implies intrinsically small probability for radiative recombination of the electron-hole pair) leaving little hope for applications in opto-electronics. The recent discovery of strong light emission from porous silicon[25] however, resulted in considerable excitement in the semiconductor community leading to over 400 citations in the current contents data base. The origin of unexpected luminescence has been ascribed by various groups to surface related species such as siloxene[26], oxidized silicon[27], and hydride complexes[28]. Another opinion is that QSE in the sponge-like skeleton of the porous silicon may play a key role in its optical properties[26,29]. It has actually been proposed that the electronic structure of silicon bulk reverses in such a way that Si becomes a direct gap semiconductor as the size quantization regime is approached. The latest careful measurements of Brus and coworkers[30] however, indicates that the Si indirect gap is preserved in small isolated Si particles which adds new fuel to the already hot controversy about the true nature of porous silicon. At the very least, however, these results defined an opportunity to create quantum confinement in thin semiconducting walls of such porous materials as Si and certain metal sulfides.[31,32,33,34,35]

2.2. Relation between the cluster size and the energy gap

It should be re-emphasized that the theory of size quantization relies on the extension of the effective mass approximation, established for bulk semiconductors, to the size quantized regime. The validity of this assumption is constantly being challenged as documented by the story of "holy silicon". Electronic and transport properties of a bulk semiconductor directly depend on the dielectric constant of the material and the electron and hole effective masses. These widely differ in various semiconductors and are sensitive function of inter atomic distances and geometry. Two known crystalline forms of CdS, for example, differ markedly in electronic structure and transport properties.

Specificallly, the onset of QSE appears whenever size constrains become comparable to the effective exciton Bohr radius (a^*_B) of a semiconductor:

$$a^*_B = \hbar^2 \, \varepsilon / \mu_B{}^* e^2 \qquad\qquad (1)$$

Here $\mu^*_B = m^*_h \, m^*_e / (m^*_h + m^*_e)$, ε: dielectric constant, μ^*_B : effective reduced mass of the exciton, and m^*_h and m^*_e : effective masses of the hole and electron respectively. The effective mass of an electron in semiconductors is typically one or two orders of magnitude smaller than that of the free electron. This is the reason for very large exciton Bohr radius which in certain semiconductors such as InSb approaches microscopic dimensions; $a^*_B(InSb) =$ 830 Å. For semiconductors such as GaAs and CdS noticeable QSE occur for particles smaller than 100Å. The detection limit of X-ray diffraction and SEM techniques is often comparable to the dimension of size quantized particles which explains why spectroscopic data are frequently used as the only evidence for their existence.

Effective masses of the electrons and holes for common semiconductors are readily available in the literature. In order to quantitatively analyze QSE by using Eq. 2 derived by Brus[36], the following assumptions must be made: (i) negligible perturbation of the semiconductor electronic structure by the surrounding environment, (ii) preserved bulk crystalline lattice, (iii) narrow particle size distribution and (iv) spherical shape of the size quantized particles. If these are fulfilled, the onset of the "blue shifted" absorption, E (in eV), of the particles with radius R can be calculated and takes the form:-

$$E = \hbar(\pi/R)^2 - 1.786 \ e^2/\varepsilon R \qquad (2)$$

This equation predicts energy of the lowest bound exciton state which is only several meV smaller than the value of the semiconductor energy gap (Eg); the difference of the two determines the exciton binding energy. Thus Eq. (2) can be used, to a good approximation, to predict the onset of absorption of size quantized particles or to deduce the average particle size from spectroscopic measurements. One should be careful to ensure that no violation of the above assumptions is likely to occur in the system under the investigation. This can adversely effect the appearance of the absorption spectra leading to erroneous conclusions when the size of semiconductor particles is to be deduced.

2.3. Semiconductor clusters in zeolites: the dream and the reality

The best developed methods for creating QSE materials include MBE and MOCVD grown III-V semiconductor heterostructures. While the fabrication of QW structures has been nearly perfected, a number of problems remain to be solved in fabrication of QWRs and QDs. For example, use of electron beam lithography to shape QWRs or QDs starting from a QW heterostructure produces lateral resolution of only ~ 200 Å. The recent advance in the direct growth of QDs on strain superlattices[37] appears to overcome this problem, although, considerably broad particle size distribution and the lack of spatial ordering need further improvements.

The ultimate goal in fabrication of the QDs and QWRs is to create a three dimensional periodic array of nanoclusters with uniform dimensions in the range between 5 Å and 300 Å. The density of such arrays has to be sufficiently high to assure good performance in opto-electronic circuits. Bearing this in mind, it is quite difficult to escape from the intuitively beautiful picture of zeolites or mesoporous (e.g. MCM-41) lattices as hosts for such semiconductor clusters. If zeolite and molecular sieve cages of 6-30 Å diameter were successfully filled with semiconductor particles, a strong QSE should occur for most semiconductors. Zeolites are transparent in the visible and near UV region which allows unobstructed optical access to the QDs. Furthermore, a relatively large variety of zeolites with slightly different cage sizes permits fine tuning of the electronic properties of confined particles. Most importantly, the periodic crystalline lattice of the host provides an **ordered** three-dimensional array of semiconductor particles with **uniform** sizes. *A priori* it would appear that the major prerequisites for technological applications of quantum confined structures are met!

There are however, a number of factors which must be defined both experimentally and theoretically before 3-d porous surfaces are to be device useful. These include:

- size and topographical uniformity of nanoclusters
- 3-d homogeneity
- tunability with respect to atomic modification of
  - topography
  - cluster dimensions
  - surface states defined by the cluster/zeolite interface
  - inter-cluster coupling
- thermal and optical stability
- optical transparency
- processability into thin films or device structures

The first item above includes long range crystal order as well as order/disorder properties of the clusters relative to the cage or pore walls. In packaging clusters within a cage, it is clearly desirable if the cluster point group symmetry matches that of the cage. If the available cage size does not properly match the final cluster size, variations in cluster topography and size can result.

Two critical points involve surface states at the cluster/cage wall interface and the electronic coupling between clusters. Confined particles may behave as isolated QDs if the potential barrier separating them prevents significant overlap of the carrier wave functions. In an ideal case, when interaction of the electronic states of the cluster and the zeolitic host are small, a low temperature absorption spectrum would consist of a set of sharp lines. The onset of the absorption can be predicted by Eq. 2 *provided the perturbations caused by the zeolite matrix can be neglected* and the bulk crystalline structure is preserved.

At sufficiently high cluster concentrations, when there may be appreciable carrier wavefunction overlap between neighboring clusters, one is forced to consider the extended band structure of the semiconductor sub-lattice. Band structure calculations of this interesting hypothetical heterostructure may need to take into account periodic boundary conditions of two interleaved crystal lattices which possess two different periodicities. This situation is likely to occur in zeolite-based heterostructures because of the significant electron wave function overlap which is expected in the case of closely separated clusters[38]. To the best of our knowledge no band calculations of this kind have been attempted.

Such heterostructures can be synthesized by (a) direct synthesis, (b) ion exchange chemistry, (c) gas phase inclusion and subsequent nucleation within the host, (d) liquid phase inclusion of the guest, (e) gas or solution phase inclusion with one or more precursor guest reactants and topotactic synthesis of new guest species within the host, and (f) a combination of (b) and (e). The most common procedure for CdS inclusion starts with Cd++/2Na+ ion exchange followed by dehydration and exposure of Cd-Z (Cd exchanged zeolite) to the H_2S vapor to form CdS species. It is important to note that the ion exchange process can yield very different siting of cations depending on temperature, pH, solvent vs. melt ion inclusion, other extra-framework ions, calcination, and loading levels. This process must be systematically controlled along with the conditions for treatment with H_2S or H_2Se in order to obtain materials which can be consistently reproduced and which contain monosize clusters. Exposure to air and water vapor can result in acid decomposition of the zeolite framework and/or conversion of the CdS to Cd(O,OH) so that special care must be taken in order to obtain reproducible results.

Y. Wang and N. Herron[39] and subsequently Herron et al.[40] reported the synthesis of cubane like $(CdS)_4$ species located inside the sodalite cages of Z-Y. The change in the "blue shift" from 290 nm to 350 nm with the increase of CdS concentration was explained by the transition from an assemble of isolated $(CdS)_4$ clusters to completely filled "super cluster" structure also called a "super lattice"[1,3].

At a low loading level of CdS in zeolite Y, isolated Cd(S,O) molecular units are formed with an absorption peak around 290 nm and no emission even at 4.2 K. At higher loadings, X-ray single crystal and powder diffraction, EXAFS, and optical absorption data show that $[Cd(S,O)]_4$ clusters can be uniquely located within the sodalite cages. The discrete $[Cd(S,O)]_4$ cubes within the small sodalite units of the CdS zeolite Y structure begin to interconnect as the loading density within the zeolite rises. The Cd atoms point toward each other through the double 6-rings linking the sodalite units with a Cd-Cd distance of 6.2 Å. As this 3-dimensional interconnection proceeds, the corresponding changes in optical properties indicate a progression toward a *semiconductor supercluster*. If the clusters are loaded into the sodalite cages of zeolite A, clusters are oriented with their faces parallel across double 4 rings at a separation of ~7.3 Å[41]. The absorption edge is correspondingly blue shifted in respect to the previous structure. The transition from clusters to aggregates upon increasing CdS loading is not continuous but rather abrupt, as judged from optical absorption and emission spectra. This suggests that the formation of the superlattice from the individual clusters inside the zeolite may be a percolative process.

When fully loaded, the superlattice shows what is designated as an exciton shoulder near 340 nm whose radiative recombination was observed at low temperature. Neither annealing nor increasing CdS concentration further shifts the exciton shoulder to the red. In contrast to this, when CdS is deposited on the exterior surface of the zeolite crystallite a continuous red shift towards bulk band gap absorption is observed. A similar behavior was found in the size dependent absorption of CdS colloids.[42]

Of particular note here is a recent theoretical study of Catlow and co-workers[43] who used lattice simulation and quantum cluster calculations to address structural properties of CdO and CdS clusters in zeolite Y. The authors find that $(CdS)_4$ cubes are energetically favored over the CdS (CdO) diatomic species in zeolites in nice agreement with the early experiments of Herron et al.[40] They also find that the repulsive Coulomb interactions between the zeolite framework and the $[Cd(S,O)]_4$ clusters causes a reduction in the clusters size compared to "interaction free" clusters. Most importantly perhaps, the repulsive interaction between the $(CdS)_4$ clusters in the adjacent cages of ZY was calculated to be on the order of 1 eV, which is believed to be responsible for the experimental difficulties experienced in synthesis of a perfectly ordered semiconductor "superlattice".

Using a similar synthetic procedure Liu and Thomas[44] also reported formation of CdS clusters in Z-X and Z-A. The Na^+/Cd_2^+ exchange and the destiny of Cd^{2+} species in sodalite cages during the exposure to H_2S were carefully monitored by infrared spectroscopy. These results indicated formation of CdS particles inside the α-cages of Z-X and Z-A with no evidence for $(CdS)_4$ formation in sodalite cages. In fact, migration of Cd^{2+} from the sodalite to the α-cages of Z-A was observed whenever the Cd^{2+} deficiency in the α-cages occurred. The size of the CdS clusters is believed to be limited by the free radius of the α-cages and for Z-A was estimated to be on the order of $(CdS)_{4.2}$ in average. The onset of absorption at 380 nm in Z-A and 430 nm in Z-X is ascribed to the CdS clusters confined by the zeolite α-cages. It is important to note that an extra framework absorption band in the IR region in CdS/Z-A was correlated to the appearance of a weak and broad absorption in the UV-VIS spectrum with the onset at 856 nm. This was taken as an evidence for partial destruction of the Z-A framework which allows for formation of larger CdS clusters embedded in the zeolite matrix.

By using spectroscopic data from the CdS synthesis in Z-X, Schulz-Ekloff and coworkers[45] found that the calculated radius (Eq. 2) of the CdS particles exceeds the radius of the zeolite α–cages by at least factor of two. Since photocorrosion experiments indicated that CdS particles are located inside zeolite pores rather than at the zeolite surface, the authors conclude that "sulfide particles probably grow within the zeolite matrix by migration of smaller aggregates followed by coalescence . . . (and) the fragmentation of the zeolite framework."

In yet another study[46] on inclusion of series of semiconductors (CdS, PbS, CdSe and TiO2) inside different zeolite hosts, a similar conclusion was reached regarding the sizes of the semiconductor particles calculated by using Eq. 2. The largest discrepancy was found for semiconductors in Z-X where the size of the particles was estimated to be five times larger than the size of the α-cages. Results were explained by the tendency of semiconductor particles to aggregate and migrate outside the zeolite cages particularly when exposed to water vapor. It is noteworthy that the authors find a strong correlation between the calculated size of the particles and the zeolite host "entry aperture".

Another method used to control the individual cluster topology is through diffusion of organometallic precursors which are sterically restricted to large channels or cages, and then co-reacting these thermally [47] or photochemically [48] to accomplish cluster synthesis. For example, in zeolite Y the organometallic precursor molecules are too large to enter the small sodalite cages and cluster formation can selectively take place in the supercages. Either of these methods requires careful consideration of the role of the host cage or channel walls and their influence on the cluster structure. The importance of this was documented in studies[11,49] of adsorption isotherms of dimethyl zinc and dimethyl mercury conducted in zeolite Na-Y with Si/Al ratios of 3.2 and dealuminated Z-Y. The greater reactivity of dimethyl zinc with the molecular sieve framework is shown in the nearly complete irreversibility of its adsorption into Na-Y. Dimethylmercury in Na-Y shows a two step adsorption curve indicating that above about six molecules per supercage, which may correspond to a monolayer of surface coverage, adsorption becomes fully reversible. In dealuminated Y, which has essentially zero framework charge, both dimethyl zinc and dimethyl mercury adsorb reversibly. Opportunities clearly exist to use the chemical bonding properties of the molecular sieve internal surface to control both the self-assembly chemistry and the cluster geometry. This approach was first applied to the synthesis of GaP in Z-Y[50]

The synthetic route for inclusion of II-VI nanoclusters in the supercages of Z-Y involving MOCVD type precursors has recently been described in detail by Ozin and co-workers[51]. The reaction of the gas phase $Cd(CH_3)_2$ precursor with partially hydrogenated Z-Y ($H_xNa_{55-x}Y$) produces $Cd-CH_3$ species anchored at the Bronsted sites. Subsequent treatment with H_2S produces cubanelike $Cd_2(M_4S_4)^{4+}$ species in α-cages of Z-Y bounded by the ZO-M^{2+}-S bridging groups. All synthetic steps were carefully monitored by in situ mid-IR spectroscopy. Similarly to previous studies, a "blue shifted" absorption with the threshold at ~420 nm and the strong photoluminescence in the visible with $\lambda_{max} = 625$ nm were explained by "spatial and quantum confinement effects".

Photoluminescence (PL) spectra of semiconductor clusters in zeolites have been sought by several authors, but no clear sign of the sharp exciton emission at the edge of the absorption band has been found. Instead, the PL spectra are dominated by a broad emission in the visible region with λ_{max} significantly larger than the observed λ_{Eg}. This is typical for semiconductors containing high concentrations of defects such as ion and cation vacancies which serve as efficient traps for photo-generated carriers. The imperfections found in the Cd/Z-Y "superlattice"[40,41] are in agreement with the PL results.

This review is not meant to be comprehensive and has dealt primarily with II-VI cluster systems. The most recent studies on other semiconductors in zeolites such as PbS[52]; and SnO$_2$[53] show results similar to those described above for the CdS-Z systems. However, there have been several important breakthroughs in synthesis of other types of semiconductors in zeolites in which rather different synthetic approaches have been used. Attempts to use ion exchange as a route to the formation of III-V semiconductors in zeolite frameworks have failed. The reason is the loss of crystallinity of the material that occurs at very low pH values required to keep group III cations in solution as hydrated cations. Alternative methods of anhydrous nitrate and halide melts also failed to give the desired inclusion products, as did methylene chloride solutions of group III halides as precursors. Instead, it has been demonstrated that MOCVD technique can be successfully applied in synthesizing of GaP inside the pore structure of zeolite Y[47,54]. The course of the reaction of trimethyl gallium and phosphine was monitored by ^{31}P MASNMR and optical spectroscopy as a function of reaction temperature and loading for both NaY and the acid zeolite HY. In HY, trimethyl gallium molecules are first anchored at room temperature by the elimination of methane to give [(CH$_3$)$_2$Ga]-OZ, where OZ refers to the zeolite framework oxygen atoms.

GaP has both a direct and indirect band gap with Eg = 466 nm direct and 545 indirect. The absorption spectrum shows pronounced peaks at 350 nm or less which are blue shifted relative to bulk GaP. The supercage point group symmetry of the large cage of zeolite Y is 43m, the same as bulk GaP, which is expected to help in obtaining periodic and local ordering. Extended X-ray absorption fine structure (EXAFS) spectroscopy and synchrotron X-ray diffraction studies identified 26-28 atom (GaP)$_{13-14}$ clusters in the super cages which were ~11Å in diameter, corresponding to 3 coordination spheres around a central Ga or P atom in the bulk structure. However, a considerable disorder within the supercages as well as fluctuations in the cage to cage content have also been documented.

The possibility for synthesis of QWRs and QDs by using the gas phase inclusion of II-VII compound semiconductor PbI$_2$ inside the cages and channels of zeolites A, Y, L, and mordernite has also been demonstrated[55,56]. As in the CdS/Z-Y heterostructures, a marked changes in the absorption spectra with the increase of PbI$_2$ content have been found. Careful measurements revealed several distinct spectral features that could have been associated to the different contents of the Z-A α-cages varying between 1 and 5 PbI$_2$ molecular units. Again, the fully loaded PbI_2/Z-A heterostructure showed the absorption edge at 3.2 eV which is considerable blue shifted in respect to the bulk PbI_2 band gap absorption at 2.57 eV. If a given zeolite cage is exchanged with another cation, the band edge red shifts because more space is available to the PbI_2 clusters as evident from the absorption maxima of K (3.20 eV), Na (3.15 eV) and Mg (3.07 eV) exchanged Z-Y. The origin of a low energy absorption tail found in both PbI_2 loaded mordenite and zeolite L was tentatively associated to the wire-like PbI_2 structures that can be formed in the channels of these two zeolites.

It is essential that inclusion chemistry proceeds under thermodynamic equilibria. In less favorable cases the zeolite network can become clogged due to the diffusion limitations preventing homogeneous guest distribution to be attained. This is particularly true when dynamic MOCVD is used because gaseous byproducts often react with the cluster atoms. It is possible to overcome such problems by uniform loading of a gaseous precursor prior to photo-induced dissociation of the precursors in which case a uniform cluster growth can be achieved at relatively low temperature.

This approach has been nicely demonstrated in synthesis of WO inside the cages of Z-Y[57]. The volatile binary metal-carbonyls of the VIb group (Group 5, new notation) have small dimensions relative to the cage openings of the large pore zeolites. They can be easily diffused into the zeolite pores and photo-converted to the respective metal oxide with negligible carbon contamination. The intra-zeolite photo-oxidation chemistry of α-cage encapsulated hexacarbonyltungsten(0) in $Na_{56}Y$ and $H_{56}Y$, $n\{W(CO)_6\}$-$Na_{56}Y(H_{56}Y)$, with O_2 provides a synthetic pathway to a zeolite alpha-cage-located W(VI) oxide, $n(WO_3)$-$Na_{56}Y(H_{56}Y)$, (n = 0-32). This formulation represents the unit cell contents which has eight supercages so that, at full loading, there are four WO_3 units per supercage. At loadings below one WO_3/supercage the absorption edge was found at 3.5 eV. Further increase of the WO_3 concentration causes an abrupt shift of the absorption edge to a limiting value of 3.3 eV which remains constant to the highest loading composition of roughly four WO_3/supercage. FTIR, MASNMR, XPS, and EXAFS data suggest first the formation of $(WO_3)_2$ dimers, then $(WO_3)_4$ tetramers as the loading is increased.

To conclude, there have been a number of high quality experimental studies devoted to semiconductor clusters in zeolites. In most of the studies deviation from the optical properties of bulk semiconductors have been documented. There is, however, marked disagreement in the interpretation of the experimental results. One opinion is that the "blue shift" in absorption spectra is due to a new type of semiconductor-zeolite "superlattice" comprised of closely packed semiconductor clusters in zeolite cages. Favorable experimental evidence have been provided for the existence of CdS/Z-Y, PbS/Z-Y and CdO/Z-Y "superlattices" in this context. Exact calculations of the electronic band structure of such materials are lacking. Another opinion is that the absorption spectra of the II-VI cluster originate from semiconductor particles with an average radius of ~ 25 Å, which aggregate at the zeolite surface or are embedded within the damaged zeolite framework. All researchers agree that partial destruction of the zeolite framework during the synthesis of II-VI compound semiconductors can occur if appropriate care is not taken during the synthesis.

Serious shortcomings in the interpretation of these studies lies in the fact that little attention has been given to the requirements regarding the validity of the Eq. 2 when particle sizes were deduced from spectroscopic data. In particular, the possible influence of the zeolite matrix or the shape of the semiconductor particles on the absorption spectra has been neglected. Most importantly, the possibility of significant electron wavefunction overlap between the isolated clusters has been disregarded. As indicated earlier, this could be responsible for the observed 'anomalous' absorption spectra; which neither correspond to the bulk semiconductor band gap, nor to that expected for the isolated clusters.

Unfortunately, the inherent guest disorder in all of the above system leads to inhomogeneously broad spectral features in both absorption and luminescence. This, along with the requirements for chemical and thermal stability and easy processibility, appear to be limiting factors for future development of possible zeolite-based opto-electronic devices.

3. ELECTRON SOLVATION IN ZEOLITIC STRUCTURES: ALKALI METAL CLUSTERS IN ZEOLITES

3.1. Background

In the previous section we considered the issue of semiconductor clusters in pore zeolites. The zeolitic environment offers unique opportunities for the growth of small clusters within the cages of the lattice. In controlling the size of these clusters one can investigate phenomena of contemporary interest in physics and chemistry and especially quantum size effects. However, it must be emphasized that in effect one is using the interface chemistry of a three dimensional surface, as opposed to a two dimensional surface. The development of devices from two dimensional deposition processes has slowly evolved to its present, technologically useful, state through many years of extensive research and development. In addition to the new set of parameters which arise for 3-d surfaces in dealing with the differences in the substrate composition and structure of porous materials, one must also address the consequences of 3-d nano or meso pore confinement on diffusion, nucleation, and guest cluster growth. The problems associated with this are evident from the discussion in the previous section. In order to model and fundamentally understand nano- or meso-composite materials which contain confined charge carrier clusters, we have chosen to examine in more detail the relatively simple sodalite 60 atom cage structures in which the structural details of the host and guest can be well defined. In this context it is possible to quantitatively model fundamental properties such as cage electrostatic potential, interaction between the charge carriers, and guest and host structural changes with confinement.

Efforts to parameterize the potential parameters for inorganic matrices is considerably behind that for organic molecules so that *ab initio* quantum mechanical calculations of the interaction of an atom or molecule with a 3-d surface are in the initial stages of being tested against experiment[58,59,60,61,62,63,64,65,66,67]. For the specific case of zeolites, model calculations at the *ab initio* level have generally been restricted to simple silicate or aluminosilicate compounds. In all but a notable few studies, calculations have been restricted to systems of finite size, with terminating hydrogens to satisfy valence requirements of the 'dangling' O atoms. Some attempts have been made to use a periodic treatment for extended lattices, for example in the program CRYSTAL[68]; however, its application has been limited to simple zeolites with an all silica (SiO_2) framework.

The usefulness of the model compound calculations stems from the construction of zeolite force fields which can be used in classical molecular dynamics and Monte Carlo simulations to model lattice dynamics and sorbate-host interactions. In general this classical force field is constrained to take the form:

$$V_{total} = V_{bond\ stretch} + V_{bond\ angle\ bend} + V_{torsion} + V_{Lennard-Jones} + V_{electrostatic\ potential} \qquad (3)$$

The electrostatic interaction potential for the framework requires atomic charges as parameters, regardless of the form of potential that is used. Lattice dynamics will be dominated by the bond-interactions supplied by all but the last term in equation (1). For the so-called non-bonding interactions however, such as those responsible for sorption processes within the zeolite, it is the long-range electrostatic terms that are important. These are of primary importance in the cluster chemistry. In most present day descriptions the electrostatic potential force field is essentially a Born-Mayer field, which takes the following form for two charges i and j with an internuclear separation $\mathbf{r}$,

$$V_{electrostatic\ potential} = A\ e^{-r/\rho} + q_i q_j/r \qquad (4)$$

The exponential term is there to model the effects of Pauli exclusion at distances short enough for wavefunction overlap, and are generally only important when i and j are charges of opposite sign. Sometimes dispersion terms are also included, although these short range terms are relatively unimportant in comparison to the Coulombic and exponential repulsion terms. Successful description of sorption processes therefore requires realistic charges. However the charges, which are generally arrived at from Mulliken population analyses from *Ab Initio* calculations, are very sensitive to the level of calculation performed. Electronegativity equalization, CNDO and INDO semiempirical and SCF/Mulliken have given atomic charges for the silicon atom which range from 0.4 to 1.91[69,70,71,72]. For example, the variation in SCF /Mulliken charges for silicon in a given silicate is 0.69 to 1.1 with a corresponding variation in the potential field continuum. This large variation in charge reflects the unreliability of low-level CNDO and MNDO calculations for such systems. Basis set superposition error for the low level alumino-silicate calculations can be quite high and consequently the calculations in general over-estimate charge separation. Higher level STO-3G calculations are known to give much more reliable estimates for charge populations[73].

Another indirect way in which one may extract electrostatic charges is by the modelling of experiments which probe the electric field in the cavities of the zeolite. Experimentally, it is well established that the electrostatic field within zeolite or molecular sieve cavities can be large and used in dramatic ways to modify the chemistry[74,75,76,77,78]. Some of the earliest estimates of these fields at cation sites in zeolites were made by Dempsey [79]. The magnitude of these fields was used to explain the reason why (a) NO disproportionates to NO^+ , NO_2 and N_2O, even though the net free energy for this reaction under standard state conditions is positive; and (b) the ease with which NaCl can be introduced into the zeolite cage 300°C below its melting point[80]

A more recent example which demonstrates the importance of cage potential on inclusion chemistry is shown by the ability to form cation radicals of the polyenes within Na-ZSM-5 by Ramamurthy and co-workers[81]. They report that arenes may be photo-ionized to produce the arene cation and a trapped electron at a p-type defect. The most probable source of this effective potential is cationic Lewis defect sites formed by dehydroxylation of the zeolite framework. An important point here is that local site potential not only influences self organization and assembly within a zeolite cage, but also may severely perturb the electronic structure of small molecules.

Perhaps the most striking example of how the electric field of the zeolite can modify both physically and chemically the sorbate is provided by the alkali metal clusters, and it is to this topic that we dedicate the remainder of this chapter. Uptake of the alkali atoms by the zeolite, leads to the introduction of color into an otherwise colorless zeolitic host. To adequately describe these optical properties it is not possible to resort to a picture of an isolated alkali atom perturbed by the electric field of the zeolite, since the electric field autoionises the valence electron. This electron is solvated through the presence of other alkali ions, denoted M, and the physical and chemical properties of the M-zeolite system take on the character of cluster of type $M_n^{(n-1)+}$. As expected, the optical properties of the valence electrons are very sensitive to the electric field supplied by the zeolite framework. The size of these clusters and their spatial distribution is defined by the geometry and size of the channels and cavities of the zeolite. Variation of the zeolite, and the nature of the alkali atom, (and possible counterions), and dopant concentration, all modify these optical properties.

In the zeolites X, Y and A the large number of potential absorption sites for the M ion along with issues regarding Al siting make comparison of theory and experiment difficult; however alkali atoms can also be absorbed in sodalite systems, where issues of Si/Al ratios do not cloud experiment and theory with additional uncertainty, and so provide us with an excellent starting point for theory. The sodalite systems have been extensively investigated and thus provide a useful database for theoretical investigations. In what follows two limits will be considered. One in which only small concentrations of alkali metals are introduced into the zeolite, such that the ratio of β cage to excess alkali is around 50:1, and the second case where this ratio is much closer to 1. In the first example we study the optical properties of isolated electrons solvated within a sodalite cage, while in the second example we address the perhaps more profound issue of the extended electronic structure of such compounds, and specifically whether we will be able to construct new classes of microporous conductor and semi-conductor materials.

From a theoretical standpoint, the low concentration limit is an excellent starting point for addressing the issue of the electrostatic fields experienced by an electron in the cage of a zeolite, and allows us to infer the likely magnitude of the charges on the framework atoms. Assuming that a reasonable electrostatic potential can be constructed for these systems it is then possible to theoretically address the issue of high dopant concentrations and issues such as nonmetal-metal transitions. From the viewpoint of chemistry itself however, it is perhaps instructive to consider the electron solvation properties of these compounds as a special class of charge transfer process. An understanding of the process of electron solvation therefore opens up a more exciting future, in which one attempts to use zeolites as a medium for the controlled charge transfer chemistry.

3.2. Alkali metal clusters in the low dopant concentration limit

3.2.1. Cathodochromic systems

At first it may seem surprising that in a section entitled alkali metal clusters in zeolites, that we would discuss exposure of zeolites to high energy electron beams. Nonetheless, there are historical reasons for doing this, and it will become evident why this section exists as we go on. Zeolites can be made cathodochromic by mounting a 'fired' form of the polycrystalline zeolite upon the screen of a CRT. The material is then exposed to a 5-20keV electron beam, with a screen current density in the range 0.01-100mC/cm^2, for periods ranging from several seconds to several minutes. In the anion sodalites, with stoichiometric formula Na$_4$[AlSiO4]$_3$X {X = Cl, Br, I}, one sees, after even very short exposure times, the manifestation of color. This effect is termed *cathodochromism*. This effect is by no means restricted to electron beams, and the first reports of this property were noted by Kasai as long ago as 1965[82] when he considered the effects of γ irradiation of Na-Y. On irradiation the sample turned pink, and subsequent E.S.R. investigations of the pink solid revealed that the color center could be attributed to the ionic cluster. Most of the subsequent research in this specific field was conducted by engineers. The observation that one could bleach these compounds by further exposure to the UV radiation led to interest in these cathodochromic compounds for application in both display and read/write storage devices.

The recognition of a possible commercial application led to extensive investigations of the sodalite family of zeolites. Marshall, Forrester and MacLaughlan[83] and Bolwijn, Schipper and van Doorn[84] illustrated that the wavelength of maximum absorption was linearly related to the lattice constant of the host. Careful choice of host therefore allows one to produce compounds with colors that span the spectrum. The exact mechanism for electron trapping remains somewhat elusive, and current thought has it that the electrons, once injected into the conduction band of the zeolite, get trapped at p-type defect sites of the type {Na$_4$(AlSiO$_4$)$_3$}$^{+1}$ Such a theory could adequately describe the formation of color centers in sodalite, Na-X and Na-Y. In the case of zeolite A other defect sites may also play a role since additional Na$_2^+$ and Na$_2^{+3}$ clusters have been identified within the α-cages[85].

3.2.2 Alkali metal doping of zeolites

One year after Kasai had shown that alkali metal clusters of the type could be created through exposure of NaY to γ-radiation, Shomaker and co-workers[86] illustrated that dehydrated Na-Y zeolite, when exposed to Na vapor, produced the same red compound reported by Kasai. This type of reaction was by no means restricted to the Na/Na-Y system and, in general, clusters of the type M$_n^{(n-1)+}$ (M = Li, Na, K, Cs) can be prepared by exposure of appropriately cation exchanged Z-X, Z-Y, Z-A or sodalite to alkali metal vapor. Interestingly, this is not the only technique by which one can dope the zeolite to produce alkali metal clusters. Mixing the zeolite with the corresponding azide[87] or thermal decomposition of an alcoholic mixture of the zeolite and the azide[88] produce the same results. Additionally one can reduce the zeolite cations by exposure of the zeolite to solvated electron mixtures, including Eu/primary amine solutions[89,90] Li/primary amine solutions[91], organolithium reagents[92] or alkali metals in THF.[93]

Within a short period of time following the work of Shomaker similar phenomena were found in other zeolite derivatives. Barrer and Cole[94] exposed 'dry' sodalite (6:0:0)[95], $Na_6[AlSiO_4]_6$ to Na vapor to obtain fully loaded (8:0:0) "black" sodalite. The synthesis temperature for sodium inclusion is about 350° C. At about 400° C, thermogravimetric analysis showed loss of one sodium atom per sodalite cage. The stability of "black" sodalite is remarkable. Once the metallic sodium was in the sodalite cage no significant change in color was observed with 48 hour soxhlet extraction with methanol or 12 hour extraction with water. Longer extraction with water caused the sodalite to become progressively violet, red, pink and ultimately white with a corresponding increase in alkalinity of the water. This ability to form and store reactive species or enhance the lifetime of organic molecules by inorganic packaging is an important attribute for potential technological applications. Again E.S.R. and E.S.E. studies indicated that the color center was an electron delocalized over a Na_4^{3+} center .

While the cathodochromic studies of the late 60's and early 70's demonstrated that the optical properties varied according to the nature of the anion, these synthetic pathways led workers in the 80's to consider the effect of the cation upon the optical properties of the solvated electron in particular in the larger Na-X, Na-Y and Na-A systems. This work supplied some quite interesting findings. E.S.R. investigations by the group of Edwards et al.[96] demonstrated that two distinct species (the Na and K clusters) could be accommodated within the sodalite β cages of zeolite Y, and that no evidence for mixed ionic clusters was found. K was not the only ion that can replace Na in the Na-Y zeolite. Breuer and co-workers[97] further prepared Rb^+ and Cs^+ exchanged Na-Y in addition to K^+ exchanged Na-Y, and demonstrated that by varying the nature of the cation it was possible to produce compounds that spanned the visible spectrum from deep red (in the case of Na-Y), to deep blue (K-Y). Equally profound changes in the optical properties of the solvated electron are observed when one changes the nature of the host framework, thus while Na doping of Na-Y leads to red solid, Na doping of Na-X results in a blue complex attributable to the supercluster[98]. In addition to these color centers, a further three have been identified, K_3^{2+} [99], Na_5^{4+} [100], and Na_3^{2+} [101].

Additional investigations by Geismar, Westphal and co-workers[102] demonstrated that color centers could also be prepared in the halogen sodalites, $Na_8[AlSiO_4]_6$ 2X (where X represents a halogen) by exposing halogen sodalite to Na vapor. It is perhaps surprising that a halogen sodalite cage could uptake an alkali atom, where all six ring windows are taken by the Na ions. The answer appears to be that the synthetic route chosen leads to solid solution in which 10% of the cages are of $Na_3[AlSiO_4]_3$ stoichiometry[103,104], and it is migration of the alkali atoms into the impurity cages that leads to the appearance of color centers. As with Na-Y zeolites the authors found the characteristic E.S.R. signal regardless of the alkali vapor used in the experiment while electron spin echo measurements revealed a hyperfine splitting attributable to the nearest neighbor Na ions. Fourier transform e.s.e.e.m spectra show modulations due to the 24 neighboring sodium atoms at a distance of 5.7 Å in adjacent cages[103]. From the separation of the hyperfine lines the percentage of "atomic character," which is a measure of the occupancy of the metal cation ns orbital, is estimated to be 41% for Na_4^{3+} and 80% for K_4^{3+} in sodalite. The increased occupancy for the potassium cluster is attributed to the fact that this cluster is more compressed within the cage[104]. Furthermore they were able to demonstrate that these compounds exhibited the same linear dependence of the wavelength of maximum absorption on the lattice constant of the host. Thus we are forced to conclude that the *color center must be the same irrespective of whether the sample is prepared cathodochromically or by alkali metal doping.*

3.2.3. Theoretical studies of electron solvation in the low dopant limit

While a large number of experimental investigations of such compounds have been conducted, little in the way of theoretical investigations have been reported. Harrison and Edwards proposed a simple particle-in-a-box model for the solvated electron[104] This, as we shall show, is a gross simplification of the problem. First, more sophisticated treatments demonstrate that, upon excitation, the electron is able to pass through the hexagonal windows into the neighboring β-cages. Second, as both experimental and theoretical data suggest, electrons in neighboring cages are coupled through electron exchange, which in turn requires that the electron is not confined to a single cage with infinitely high walls. Third, pseudopotential calculations show that, above all, the size of the Na_4^{3+} cluster (i.e. the distance of the sodium ions from the center of the cage) determines the frequency of absorption, which is only indirectly related to the size of the unit cell.

Of course the complexity of the problem demands consideration of a simplified model. The main reason for this lies in the fact that to perform calculations at the *ab initio* level would demand calculations involving several hundred atoms and this is beyond present technology. Model calculations have therefore concentrated on considering the problem of an electron solvated within the Coulombic field supplied by the partially ionic zeolite lattice. The interaction potential 'seen' by the electron is therefore described by a series of pseudopotentials - which in the early work were taken to be of Shaw form (truncated Coulomb potentials) where the charges are those assigned by Mulliken analysis of model *ab initio* calculations[61,73]. At this level Haug, Srdanov, Stucky and Metiu[105] considered the effect of different charge and truncation radius schemes, and the influence of Na ion tetrahedral radii upon the absorption spectrum. These model calculations indicated that the spectrum was critically sensitive to the tetrahedral radius of the Na cluster - thus realistic calculations would require energy minimisation of the Na positions within the lattice. In addition however it was found that the calculations were quite sensitive to the truncation radii of the framework atoms and so a systematic procedure for their assignment would be crucial.

These and other issues prompted Blake, Srdanov, Stucky and Metiu[106] to reconsider the issue of the framework pseudopotentials. First it was felt that the original reasons for choice of the charge scheme used in those calculations may have been suspect, and was more an artifact of the truncation radii chosen and not the charge that brought the theory and experiment into apparent accord. Secondly it was by no means apparent that the tetrahedral dimensions chosen previously on Na doped sodalite[105] corresponded to the minimum energy structure. This being so it would be a mistake to assign likely framework charges on the strength of the absorption spectrum calculations with the erronous nuclear configuration. While the charges used were not at variance with MNDO/2 calculations it is well-known that such calculations over emphasize charge separation in such systems[58]. In accord with Monnier et al.[107], a more sophisticated model was developed whereby the charge distributions at each nuclear site of the framework were taken to be Gaussian. The width of the charge distribution was set to the ionic crystal radius for the composite ions of the framework[108], and the charge assigned again from higher level *ab initio* calculations at the STO 3G level[58]. With this model it is possible to construct the exact Coulombic field by solving Poisson's equation. It can be shown that the ionic crystal radius is related to the truncation radius in the Shaw potential, and so this model arrives at a systematic method for the assignment of the terms in the pseudopotential. In order to assess the performance of such potentials it was necessary to minimize the Na positions. For this purpose a Born-Mayer force field was derived for the interaction of the Na ions with the framework. It was found necessary to abandon the formal charge interaction potentials of Catlow and co-workers[109] in describing the Na framework force field in favor of a force field that utilizes the same ionic charges seen by the electron. The exponential repulsion term for the Na-O interaction was adjusted to ensure that the force field generates the correct perfect lattice structures in the absence of the electron. These potentials were used to model the absorption spectrum for the Na-doped sodalite (6:0:0) and the chloro-sodalites both containing low concentration of Na_4^{3+} clusters.

Sodalite	Shell 1 (Å)	Shell 2 (Å)
Cl	2.65 (2.73)	4.97
Br	2.83 (2.88)	4.88
I	2.95 (3.12)	4.71
6.0.0	3.32 (3.71)	4.14

TABLE 1. Radial displacement of the Na ions from cage center. The values given in parentheses correspond to the corresponding values in the perfect crystalline host.

For this one-electron problem one can solve the Schrödinger equation exactly within the time domain (see Haug et al[105]). The ground electronic state configuration was generated using either imaginary time propagation or a grid-based Lanczos method married with steepest descent for the nuclei (see Blake et al.[106]). In each system considered it was found that the Na tetrahedron of the impurity cage contracts and the electron is indeed localized within the central cage in agreement with ESR data. The next nearest neighbor positions are in agreement with the calculations of Smeulders et al[102] to within 3pm for the chloro-sodalite system (**Table 1**).

Sodalite	**$h\nu_{max}$ experimental(eV)**	**$h\nu_{max}$ calculated(eV)**
Cl	2.48	2.42
Br	2.25	2.25
I	2.06	2.12
6.0.0	2.00	1.90

TABLE 2. A comparison of the calculated and observed frequencies of maximal absorption (in eV). The experimental data is taken from the papers of Bolwijn et al.[84] for the anion sodalites, and of Srdanov, Haug, Metiu and Stucky for the Na-doped dry sodalite[110].

The absorption spectrum was generated through use of standard wavepacket propagation techniques, where the electron is coupled explicitly to the electric field through the dipole moment operator. This approach leads to the frequency dependent absorption cross-section in the first order perturbation theory limit (e.g. **Figure 1**). In each case the framework was taken to be that of the host lattice hence the calculations are strictly considered in the infinite dilution limit in which no appreciable distortion of the lattice can occur. In the case of (6:0:0) the structure elucidated by Felsche, Luger and Baerlocher was used[111] while structures of Hassan and Grundy were used for the chloro-sodalite[112] and Beagley, Henderson and Taylor for the bromo and iodo-sodalites[113]. The results obtained using this simple model are in excellent agreement with experiment (**Table 2, Figure 1 and Figure 2**).

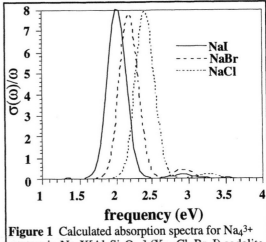

Figure 1 Calculated absorption spectra for Na_4^{3+} centers in $Na_4X[Al_3Si_3O_{12}]$ (X = Cl, Br, I) sodalite structures.

In **Figure 2** we compare theory and experiment for the Na doped (6:0:0) sodalite. The experimental spectrum was obtained using a high vacuum (10^{-8} torr) apparatus for metal vapor deposition. In this way it was possible to diffuse sodium atoms into the "dry" $Na_6(AlSiO_4)_6$ sodalite cages and form a series of samples containing different concentrations of Na_4^{3+} (see reference 110). The absolute absorbance of the sample shown here that contains one Na_4^{3+} cluster per 50 empty cages shows a prominent peak centered at 628 nm. In the theoretical spectrum the

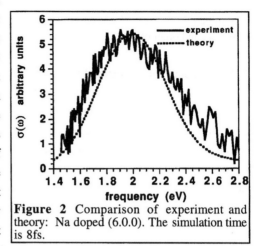

Figure 2 Comparison of experiment and theory: Na doped (6.0.0). The simulation time is 8fs.

calculation is performed over a simulation time of 10 fs, and it is this that leads to the broadened signal. The physics for this broadening is found to be attributable to the strong coupling of electron motion with the Na ions in both the central and surrounding cages and is thus primarily a homogeneous broadening effect[114]. Clearly both theory and experiment are in satisfactory agreement at this level of treatment.

In **Figure 3** we demonstrate how the wavelength of maximum optical absorption exhibits a linear dependence upon the *host* lattice constant. The experimental data are taken from the cathodochromic studies of Bolwijn et al[84], and the agreement between theory and experiment forced us to conclude that theory supports the view that cathodochromism in the anion sodalites is attributable to isolated Na_4^{3+} clusters embedded within a

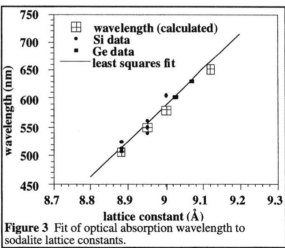

Figure 3 Fit of optical absorption wavelength to sodalite lattice constants.

perfect anion sodalite lattice. The electron leads in all cases to a contraction of the tetrahedral dimensions of the metal cluster. This linear relationship has perhaps overemphasized the role of the host lattice on the optical absorption maximum, and has led to particle in a box models where the box dimensions are those of the lattice. Results from these pseudopotential calculations indicate that it is really the dimensions of the ionic cluster that are paramount in determining the maximal frequency of absorption, and the lattice dimensions indirectly affect this as a second order effect.

3.3. High dopant concentration limit: metal or non-metal?

In one area, however the grail has remained elusive. From the above it is apparent that zeolites X, Y and A with their large supercages allow for a large number of ionic alkali metal clusters within their cavities - can they play host to bulk metallic behavior ? If so it may be possible by varying the size of these clusters to examine the issue of quantum size effects and the onset of metallic character. In the early 80's several candidates for metallic behavior were found. Edwards discovered that prolonged exposure of Na-Y to Na, K and Rb vapors led to the formation of dark solids, whose single line ESR spectra were attributed to the presence of metallic clusters within the cavities of the zeolite[104]. Assignments were made by comparison to E.S.R. measurements of alkali metal clusters in paraffin mixtures. In both cases the g-values were the same of both ESR signals were singlet in structure. However, doubts regarding this assignment were evident.

Similar results were found by Srdanov, Monnier, and Stucky[115] for the Na doping of the (6:0:0) sodalite, which again turns black on prolonged exposure. The ESR signal is that of a singlet, but the Reitveld analysis of the sodalite reveals the existence of just one extra sodium atom per cage. The ESR signal in this case is clearly not attributable to metallic clusters yet exhibits all ESR properties that led to the original assignment of the metallic clusters in Na-Y and Na-X. Anderson and Edwards[116] recently revisited this topic, and concluded that the experimental findings are not consistent with a metal particle model. Several important findings were given in support of this assertion. First, the absence of the metallic Knight shifted peak at around 1130ppm in the ^{23}Na NMR spectrum precludes the presence of large metal particles in the zeolite. Second, the temperature dependence of the line-width of the troublesome ESR feature is also not consistent with conduction electrons. Finally the ESR signal collapses to a singlet at concentrations at which the probability that 2 adjacent cages contain clusters. The explanation afforded by Anderson and Edwards that loss of the hyperfine structure occurs because of the coupling of the electrons in neighboring cages through an exchange interaction is extremely plausible. Their assertion is further reinforced by the fact that such a model leads to a barely resolvable hyperfine splitting with just 2 interacting centers.

Given these conclusions it will be necessary to re-evaluate some of the more recent work on metallic clusters in particular that of Xu and Kevan[117] who considered alkali metal inclusion within the cation exchanged Na-X, through controlled decomposition of the metal azides. In this work they interpreted the singlet ESR signals as attributable to the metallic clusters citing the early Harrison and Edwards paper[104] in assignment of the metallic cluster peaks which is most likely to be incorrect.

The only theoretical study to be reported until now on the fully doped Na sodalite is that of Monnier, Srdanov, Stucky and Metiu[107]. They considered the black phase sodalite characterized by Srdanov et al.[115] calculating the one-electron Bloch states using grid-based wavepacket methods. Several issues were addressed in this paper, in particular, is this phase a conductor or an insulator? This system would perhaps be expected to be an insulator in the ground state since the unit cell contains a total of two electrons, being isoelectronic with the alkali earths. In the latter case however a curve crossing ensures of the ground state with the conduction band ensures the metallic character of the alkali earths. In contrast to earlier work the pseudopotentials were taken as $V(r) = Q_i erf(r/rc)/r$ which is the Coulombic field originating from a Gaussian charge distribution. The 'truncation radius', rc, were set to be those used by Haug et al.[105], with the exception of the Si which was too small leading to electron density upon the framework. The charge scheme used to describe the framework was one of those

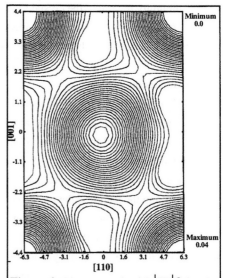

Figure 4 A contour plot of $|\Psi|^2$ for the ground electronic state of an electron within the unit cell of the 8.0.0.2e-sodalite. Plotted is a projection along [110] and [001] axes. The minimum value is 0.0 Bohr⁻³ while the maximum is 0.04 Bohr⁻³ with a contour interval of 0.001 Bohr⁻³.

considered by Haug et al[105] and deemed to be in best agreement with experiment. The band structure calculated for this system could be fitted very well with the tight binding model, with a band gap of 2eV at the Γ point. With the inclusion of electron-electron interaction into this Bloch treatment of the band structure it seemed likely that the solid would be a Mott insulator in the ground state, thus forming a kind of Wigner lattice[118] with electrons being tied to lattice sites (in this case the cage centers). Unfortunately this is not in agreement with experiment where the band gap is nearer to 0.8 eV[115]. Furthermore it appears that a tight binding scenario is not consistent with the singlet ESR spectrum observed for this system.

More recent work with the improved potentials of Blake, Srdanov, Stucky and Metiu[119] which assign a smaller ionicity to the sodalite predict a direct band gap of ~1eV, in much better agreement with the experiment. Furthermore we find that in the ground electronic state the

138

electron is appreciably delocalised throughout the lattice (**Figure 4**) in accordance with the ESR experiments. Further it suggests that introduction of low concentrations of electron acceptors within the sodalite could result in a conducting solid. Thus preliminary work suggests that *the band structure is extremely sensitive to the charge distribution of the sodalite*. In addition, there is no ambiguity regarding the crystal structure of the "black " sodalite[115] so such studies may lead to our best estimates for the charge distribution and potential surface continuum of the sodalite cage and, ultimately, of aluminum silicate zeolite in general.

3.4. Conclusion

The alkali clusters afford us with an excellent example of how the zeolitic host can modify both the physical and chemical properties of sorbates. We have shown how the alkali atoms are autoionised to form new compounds unknown in any other system. and that these compounds have different charge transfer energetics. These compounds are colored, with useful cathodochromic properties that have actually led to applications in visual display units. By controlling the nature of the zeolite we can achieve control over the optical properties of the sorbate, and by varying the concentrations of the dopant we can modify the electronic properties of the solid, to the extent that we may be able to synthesize new conductors and semiconductors. In addition, by marrying theory with experiment we have been able to show that such compounds are sensitive measures of the electrostatic field supplied by the zeolite framework. The modelling of electron solvation has shown that the optical properties of the electron are very sensitive to charge distributions and provide us with an independent means to assess charge schemes. These in turn can be introduced into classical force fields important in the simulation of sorption of molecules in all zeolites.

These systems, which perhaps represent the simplest of charge transfer process to occur in zeolites also give us insight into the nature of more complicated charge transfer processes. Theoretically the pseudopotential used in describing the interaction of an electron with the sodalite can be extended to more elaborate quantal subsystems embedded in a zeolite cage, and so opens up the future to more elaborate studies of the effects of zeolites upon the electronic properties of sorbates.

ACKNOWLEDGMENTS. This work was partially supported by the Office of Naval Research and by the NSF Science and Technology Center for Quantized Electronic Structures, Grant No. DMR 91-20007.

4. REFERENCES

1. K. Moller, T. Bein, N. Herron, W. Mahler, J. Macdougall and G. D Stucky, Mol. Cryst. Liq. Cryst. **181**, 305 (1990).

2. Y.A. Alekseev, V.N. Bogomolov, T.B. Zhukova, V.P. Petranovskii, S.G. Romanov and S.V. Kholodkevich, Izv. Akad. Nauk SSSR, Ser. Fiz. **50(3)**, 418 (1986).

3. M.A. Fox and T.L. Pettit, Langmuir **5(4)**, 1056 (1989).

4. Y. Wang and N.Herron, J. Phys. Chem. **91(2)**, 257 (1987); Y. Wang and N.Herron, J. Phys. Chem. **92(17)** ,4988 (1988).

5. Y. Nozue, Z. K. Tang and T. Goto, Solid State Commun. **73(8)**, 531 (1990).

6. S. Miyazaki and H. Yoneyama, Denki Kagaku oyobi Kogyo Butsuri Kagaku **58(1)**, 37 (1990).

7. O. Terasaki, K. Yamazaki, J. M. Thomas, T. Ohsuna, D. Watanabe, J. V. Sanders and J. C. Barry, Nature (London) **330**, 58(1987); O. Terasaki, K. Yamazaki, J. M. Thomas, T. Ohsuna, D. Watanabe, J. V. Sanders and J. C. Barry, J. Solid State Chem. **77**, 72 (1988).

8. G. A. Ozin, S. Ozkar and R. A. Prokopowicz, Acc. Chem. Res. **25**, 553 (1992); A. Stein, M. Meszaros, P. M. Macdonald, G. A. Ozin and G. D. Stucky, Adv. Mater. **3**, 306 (1991); G. A. Ozin, Adv. Mater. **4**, 612 (1992).

9. M. Wark, G. Schulz-Ekloff, N. I. Jaeger and W. Lutz, Mater. Res. Soc. Symp. Proc. **233**, 133 (1991); G. Schulz-Ekloff, Stud. Surf. Sci. Catal. **69**, 65 (1991).

10. X. Liu and J. K. Thomas, Langmuir **5**, 58 (1989); R. D. Stramel, T. Nakamura and J. K. Thomas, J. Chem. Soc. Faraday Trans. 1, **84**, 1287 (1988).

11. G. D. Stucky and J. E. MacDougall, Science **247**, 669 (1990); G. D. Stucky, Mater. Res. Soc. Symp. Proc. **206**, 507 (1991); G. D. Stucky, Prog. Inorg. Chem. **40**, 99 (1992).

12. A. I. Ekimov and A. A. Onushchenko, Sov. Phys. Semicond **16**, 775 (1982).

13. Al. L. Efros and A. L. Efros, Sov. Phys. Semicond. **16**, 772 (1982).

14. See, for example, the tutorial review of E. Hanamura, Opt. Quant. Electronics **21**, 441 (1989).

15. P. Horan and W. Blau, Z. Phys. D: At., Mol. Clusters **12**, 501 (1989).

16. P. Roussignol, D. Ricard, C. Flytzanis and N. Neuroth, Proc. SPIE-Int. Soc. Opt. Eng. **1017**, 20 (1989).

17. F. Henneberger, U. Woggon, J. Puls and C. Spiegelberg, Appl. Phys. B, **46**, 19 (1988).

18. R. C. Powell, R. J. Reeves, M.G. Jani, M. S. Petrovic, A. Suchocki and E. G. Behrens, Proc. SPIE-Int. Soc. Opt. Eng. **1105**, 136 (1989).

19. A. P. Alivisatos, A. L. Harris, N. J. Levinos, M. L. Steigerwald and L. E. Brus, J. Chem. Phys. **89**, 4001 (1988).

20. G. D. Stucky, Naval Research Reviews **43**, 28 (1991).

21. G. W. Bryant, Phys. Rev. Lett. **59**, 1140 (1987); Y. Kayanuma, Phys.Rev. B **44**, 13085 (1991).

22. For a comprehensive review see M. L. Steigerwald and L. E. Brus, Annu. Rev. Mater. Sci. **19**, 471 (1989).

23. C. B. Murray, D. J. Norris and M. G. Bawendi, J. Am. Chem. Soc. **115**, 8706 (1993).

24. N. Herron, J. C. Calabrese, W. E. Farneth and Y. Wang, Science **259**, 1426 (1993).

25. L. T. Canham, Appl. Phys. Lett. **57**, 1046 (1990).

26. M. S. Brandt et al., Solid State Commun. **81**, 307 (1992).

27. Z. Y. Xu, M. Gal and M. Gross, Appl. Phys. Lett. **60**, 137 (1992).

28. S. M. Prokes et al., Phys. Rew. B **45**, 13788 (1992).

29. X. Wang et al., Phys. Rev. Lett. **71**, 1265 (1993).

30. W. P. Wilson, P. F. Szajowski, and L. E. Brus, Science **262**, 1242 (1993).

31. R. Bedard, L. D. Vail, S. T. Wilson and E. M. Flanigen, U.S. patent 4,880,761, 1989; R. L. Bedard, S. T. Wilson, L. D. Vail, J. M. Bennett and E. M. Flanigen, "Studies in Surface Science and Catalysis", in *Zeolites: Facts, Figures, Future*, P.A. Jacobs and R. A. Van Santen, eds., Elesevier, V. 49, Part A, 375 (1989).

32. J. B. Parise, J. Chem. Soc. Chem. Commun.**22**, 1553 (1990); J. B. Parise, Science **251** (1991) and **293** (1991); J. B. Parise and Y. Ko, Chem. Mater.**4**, 1446 (1992).

33. C. L. Bowes and G. A. Ozin, Mater. Res. Soc. Symp. Proc. **286**, 93 (1993).

34. J. Liao, C. Varotsis and M. G. Kanatzidis, Inorg. Chem. **32**, 2453 (1993).

35. O. M. Yaghi, Z. Sun, D. A. Richardson and T. L. Groy, J. Am. Chem. Soc. **116**, 807 (1994).

36. L. E. Brus, J. Chem. Phys. **80**, 4403 (1984); **90**, 2555 (1986).

37. D. Leonard, M. Krishnamurthy, C. M. Reaves, S. P. DenBaars and P. M. Petroff, Appl. Phys. Lett. (in press).

38. A. Gossard, private communication.

39. Y. Wang and N. Herron, J.Phys. Chem. **92**, 4988 (1988).

40. N. Heron, Y. Wang, M. M. Eddy, G. D. Stucky, D. E. Cox, K. Moller, and T. Bein , J. Am. Chem. Soc. **111**, 530 (1989).

41. Y. Wang, N. Herron and G. D. Stucky (unpublished results).

42. N. Herron, Y. Wang, H. Eckert, J. Am. Chem. Soc. **112**, 1322 (1990).

43. A. Jentys, R. W. Grimes, J. D. Gale, and C. R. A. Catlow, J. Phys. Chem., in press.

44. X. Liu and J. K. Thomas, Langmuir **5**, 58 (1989).

45. M. Wark, G. Schulz-Ekloff, and N.I. Jaeger, Catalysis Today **8**, 467 (1991).

46. S. Miyazaki and H. Yoneyama, Denki Kagaku **58**, 37 (1990).

47. G. D. Stucky and J. E. MacDugal, Science **247**, 669 (1991).

48. G. A. Ozin, R. A. Prokopowicz and S. Ozkar, J. Am. Chem. Soc. **114**, 8953 (1992).

49. S. D. Cox, Ph.D. Thesis, University of California, Santa Barbara, 1989.

50. J. E. MacDougall, H. Eckert, G. D. Stucky, N. Herron, Y. Wang, K. Moller, T. Bein, and D. Cox, J. Am. Chem. Soc. **111**, 8006 (1989).

51. M. Steele, A. J. Holmes, and J. A. Ozin, Proceedings from the Ninth International Zeolite Conference, II, 185 (1992).

52. T. Sun and K. Seff, J. Phys. Chem. **97**, 7719 (1993).

53. M. Wark, H. J. Schwenn, G. Shulz-Ekloff, and N. Jaeger, Ber. Bunsen-Ges. Phys. Chem. **96,** 1727 (1992).

54. K. Moller, T. Bein, N. Herron, W. Mahler, J. MacDougall, and G. D. Stucky, Mol. Cryst. Liq. Cryst. **181**, 305 (1990).

55. O. Teraskai, K. Yamazaki, J. M. Thomas, T. Ohsuna, D. Watanabe, J. V. Sanders, and J. C. Barr, Nature **330**, 6143 (1987); O. Teraskai, K. Yamazaki, J. M. Thomas, T. Ohsuna, D. Watanabe, J. V. Sanders, and J. C. Barr, J. Sol. State Chem. **77**, 72 (1988); O. Terasaki, Z. K.Tang, Y. Nozue, and T. Goto, Materials Research Society Symposium Proceedings **233**, 139 (1991).

56. Y. Nozue, T. Kodaira, O. Terasaki, K. Yamazaki, T. Goto, D. Watanabe, and J. M. Thomas, J. Phys. Condens. Matter **2**, 5209 (1990); Y. Nozue, Z. K. Tang and T. Goto, Solid State Commun. **73**, 531 (1990).

57. S. Ozkar, G. A. Ozin, K. Moller, and T. Bein, J. Am. Chem. Soc. **112**, 9575 (1990); G. A. Ozin, M. D. Baker, J. Godbe, and C. J. Gil, *J.* Phys. Chem. **93**, 2899 (1989); G. A. Ozin, S. Ozkar and P. Macdonald, J. Phys. Chem. **94**, 6939 (1990); G. A. Ozin, S. Kirkby, M. Meszaros, S. Ozkar, A. Stein, and G. D. Stucky, in *Materials for Nonlinear Optics*, S. Marder, J. Sohn and G. D. Stucky, eds., ACS Symp. Ser. **455** (1991), 554.

58. J. Sauer, Chem. Rev. **89**, 199 (1989).

59. R. Vetrival, C. R. A. Catlow and E. A. Colbourn, "Innovation in Zeolite Materials Science", in *Studies in Surface Science and Catalysis,* P. J. Grobet et al., eds., Elsevier, (1988), 309.

60. L. Uytterhoeven, W. J. Mortier and P. Geerlings, J. Phys. Chem. Solids **50**, 479 (1989).

61. K. A. Van Genechten, W. J. Mortier and P. J. Geerlings, J.Chem. Phys. **86**, 5063.

62. J. B. Nicholas, A. J. Hopfinger, F. R. Trouw, and L. E. Iton, J. Am. Chem. Soc. **113**, 4792 (1991).

63. J. D. Gale, A. K. Cheetham, R. A. Jackson, R. C. Richard, A. Catlow, and J. M. Thomas, Adv. Mater. **2**, 487 (1990).

64. A. K. Nowak, C. J. J. DenOuden, S. D. Pickett, B. Smit, A. K. Cheetham, M. F. M. Post, and J. M. Thomas, J. Phys. Chem. **95**, 848 (1991).

65. R. L. June, A. T. Bell and D. N. Theodorou, J. Phys. Chem. **94**, 8232 (1990).

66. R. A. van Santen, D. P. De Bruyn, C. J. J. DenOuden, and B. Smit, in *Studies in Surface Science and Catalysis: Introduction to Zeolite Science and Practice*, H. van Bekkum, E. M. Flanigen and J. C. Jansen, eds., Elsevier, New York (1991), **58**, 317.

67. P. Demontis, G. B. Suffritti, S. Quartieri, E. S. Fois, and A. J. Gamba, J. Phys. Chem. **92**, 867 (1988).

68. R. Dovesi, C. Pisani, C. Roetti, M. Causa, and V. R. Saunders, *CRYSTAL88: An ab Initio All -Electron LCAO Hartree-Fock Program for Periodic Systems*; QCPE Program Nr. 577, Quantum Chemistry Program Exchange, Indiana University, Bloomington IN, 1988.

69. J. Sauer and D. Deininger, Zeolites **2**, 114 (1982).

70. R. T. Sanderson, *Chemical Bonds and Bond Energy*, Academic Pres, New York, 1976, p. 138.

71. R. J. Hill, M. Newton and G. V. Gibbs, J. Solid State Chem. **47**, 185 (1983).

72. S. Grigoras and T. H. Lane, J. Comp. Chem. **9**, 25 (1988).

73. J. Sauer, P. Hobza and Z. Zahradnik, J. Phys. Chem. **89**, 3318 (1980).

74. V. Bosacek, D. Freude, T. Froehlich, H. Pfeifer, and H. Schmiedel, J. Colloid Interface Sci. **85**, 502 (1982).

75. T. Masuda, K. Tsutsumi, and H. Takahashi, J. Colloid Interface Sci. **77**, 238 (1980).

76. E. CohendeLara and Y. Delaval, J. Chem. Soc. Faraday Trans. 2 **74**, 790 (1978).

77. D. Denney, V. M. Mastikhin, S. Namba, and J. Turkevich, J. Phys. Chem. **82**, 1752 (1978).

78. K. Tsutsumi and H. Takahashi, J. Phys. Chem. **76**, 110 (1972).

79. E. Dempsey, in *Molecular Sieves*, Society of Chemical Industry, London, 1968, p. 293.

80. P. H. Kasai and R. J. Bishop, Jr., in *Zeolite Chemistry and Catalysis*, J. A. Rabo, ed., American Chemical Society, 1976, pp. 350-391.

81. V. Ramamurthy, J. V. Caspar and D. R. Corbin, J. Am. Chem. Soc. **113**, 594 (1991).

82. P.H. Kasai, J. Chem. Soc. **43**, 3322 (1965).

83. M. J. Taylor, D. J. Marshall, P. A. Forrester and S. D. MacLaughlan, Radio Electron. Eng. **40,** 17 (1970).

84. P. T. Bolwijn, D. J. Schipper and C. Z. van Doorn, J. Appl. Phys. **43**, 132 (1972).

85. K. Iu, X. Liu and J. K. Thomas, J. Phys. Chem. **97**, 88165 (1993).

86. J. A. Rabo, C. L. Angell, P. H. Kasai and V. Shomaker, Discuss. Faraday Soc. **11**, 328 (1966); J. A. Rabo and P.H. Kasai, Progr. Solid State Chem. **9**, 1 (1975).

87. B. Xu, X. Chen and L. Kevan, J. Chem. Soc. Faraday Soc. **87**, 3157 (1991).

88. L. R. M. Matens, P. J. Grobet and P. A. Jacobs, Nature **315**, 568 (1985).

89. S. L. Suib, R. P. Zerger, G. D. Stucky, R. M. Emberson, P. G. Debrunner, and L.E. Iton, Inorg. Chem. **19**, 1858 (1980).

90. S. L. Suib, R. P. Zerger, G. D. Stucky, T. I. Morrison, and G. K. Shenoy, J. Chem. Phys. **80**, 2203 (1984).

91. P. A. Anderson, D. Barr and P. P. Edwards, Angew. Chem. Int. Ed. Engl. **30**, 568 (1991).

92. K. B. Yoon and J. K. Kochi, J. Chem. Soc. Chem. Commun., 510 (1988).

93. Y. S. Park, Y. S. Lee and K. B. Kochi, J. Am. Chem. Soc. **115**, 12220 (1993).

94. R. M. Barrer and J. F. Cole, J. Phys. Chem. Solids **29**, 1755 (1968).

95. The notation refers to the number of extra-framework cations:cage anions: water per unit cell of the sodalite structure.

96. P. P. Edwards , M. R. Harrison, J. Klinowski, S. Ramdas, J. M. Thomas, D.C. Johnson and C. J. Page, J. Chem. Soc., Chem Commun. 982 (1984)

97. R. E. H. Breuer, E. de Boer and G. Geismar, Zeolites **9**, 336 (1989).

98. J. A. Rabo, C. L. Angell, P. H. Kasi and V. Shomaker, Transactions of the Faraday Society **41**, 328 (1966).

99. P. A. Anderson Ph.D. thesis, Cambridge University, 1990; P. A. Anderson, R. J. Singer, and P. P. Edwards, J. Chem. Soc. Chem Comm. 914 (1991); T. Sun and K. Seff, J. Phys. Chem. **97**, 5213 (1993).

100. P. A. Anderson and P. P. Edwards, J. Chem. Soc. Chem Comm., 915 (1991).

101. P. A. Anderson, D. Barr, P. P. Edwards, Angew. Chem. Int. Ed. **30**, 1501 (1991).

102. U. Westphal and G. Geismar, Z. Anorg. Allg. Chem. **508**, 165 (1984); J.B.A.F. Smeulders, M.A. Hefni, A.A.K. Klaassen, E. De Boer, U. Westphal and G. Geismar, Zeolites **7** , 347 (1987); U. Westphal and G. Geismar, Z. Anorg. Allg. Chem. **508**, 165 (1984); G. Geismar and U. Westphal, Chem. Ztg. **111**, 2772 (1987).

103. R. E. H. Breuer, E. DeBoer and G. Geismar, Zeolites **9**, 336 (1989).

104. M. R. Harrison, P. P. Edwards, J. Klinowski, J. M. Thomas, D. C. Johnson, and C. J. Page, J. Sol. State. Chem. **54**, 330 (1984).

105. K. Haug, V. I. Srdanov, G. D. Stucky and H. Metiu, J. Chem. Phys., **96**, 3495 (1992)

106. N. P. Blake, V. I. Srdanov, G. D. Stucky and H. Metiu, in preparation.

107. A. Monnier, V. I. Srdanov, G. D. Stucky , and H. Metiu, J. Chem. Phys., in press.

108. R. D. Shannon, Acta. Cryst.A **32**, 751 (1976).

109. G. Ooms, R.A. van Santen, J. J. den Oudar, R.A. Jackson and C.R.A. Catlow, J. Phys. Chem. **92**, 4462 (1988).

110. V. I. Srdanov, K. Haug, H. Metiu and G. D. Stucky, J. Phys. Chem. **96**, 9039 (1992).

111. J. Felsche, S. Luger and Ch. Baerlocher , Zeolites **6**, 367 (1986).

112. I. Hassan and H. D. Grundy, Acta. Cryst. B **40**, 6 (1984).

113. B. Beagley, C. M. B. Henderson and D. Taylor, Miner. Mag. **46**, 459 (1982).

114. N.P. Blake and H. Metiu, J.Chem. Phys., to be submitted.

115. V. I. Srdanov, A. Monnier and G. D. Stucky, to be submitted.

116. P. A. Anderson and P. P. Edwards, J. Am. Chem. Soc.**114**, 10606 (1992).

117. B. Xu and L. Kevan, J. Phys. Chem. **96**, 242 (1992).

118. E. Wigner, Trans. Faraday Soc. **34**, 678 (1938).

119. N. P. Blake, V. I. Srdanov, G. D. Stucky, and H. Metiu, to be submitted.

J.C. Jansen, M. Stöcker, H.G. Karge and J. Weitkamp (Eds.)
Advanced Zeolite Science and Applications
Studies in Surface Science and Catalysis, Vol. 85
© 1994 Elsevier Science B.V. All rights reserved.

Nonlinear optical effects of dye-loaded molecular sieves

Günter Schulz-Ekloff

Institut für Angewandte und Physikalische Chemie
Universität Bremen, D-28334 Bremen, Germany

Molecular sieves represent suitable host systems for the accomodation of polarizable molecules since their inorganic framework guarantees stability against chemical attack by excited dyes forming radicalic triplet states. Moreover, the incorporated chromophores exhibit increased resistance towards photobleaching. The applicability of dye-loaded molecular sieves for optical second harmonic generation, based on the organized arrangement of molecular dipoles, and for persistent spectral hole burning, revealing a peculiar stability towards hole broadening by spectral diffusion, is summarized.

1. INTRODUCTION

In information technology, a photonic or optical material is any material with which light interacts for the purposes of information generation, transmission, detection, conversion, display, storage or processing. Recent advance has transformed the field of optical communications from a visionary dream to a commercial reality in just over a decade. Substantial progress has been made in fields like optical phase conjugation, two-wave mixing and energy exchange, erasable optical memories, phase transition materials, flat panel displays or optical window materials. Optical bistability provides the key to all-photonic switching and logic. The phenomena of spectral hole burning and photorefractive effects represent examples [1].

A challenge of great significance is the extension to three dimensions of the kind of control in materials fabrication that photolithography or vapor deposition provides in one and two dimensions. Molecular sieve materials represent host systems where, in principle, photoactive guests can be accomodated in the channels and cages with translational symmetry on a nanometer scale. Guest species, like clusters of metals and semiconductors, can form superlattices which could couple to minibands and exhibit novel optical or optoelectronic properties. The host-guest inclusion chemistry for mineralic molecular sieves is in an early stage of development. Silver clusters in zeolites belong to the most extensively studied mineralic host-guest system, where clusters of well-defined size are formed, exhibiting peculiar optical properties and being considered as quantum dots weakly interacting with each other [2-7]. Generally, high metal loadings of zeolites having low Si/Al ratio result in formation of randomly distributed metal particles larger than the zeolite void dimensions [8-11]. Quantum size effects are found for the optical properties of zeolite-enclosed semiconductors clusters [12-17], where again local lattice fragmentation occurs for high loadings [18-22]. Besides the blue shifts with decreasing cluster size, which are expected from quantum theory [23], unusual fluorescence spectra occur [22,24] which might be explained by peculiar interactions between zeolite host and cluster guest, i.e. as matrix effect on quantum-sized particles [25-27].

In the following, nonlinear optical properties of dye-loaded molecular sieves will be discussed focussing (1) on a peculiar effect of the second-order susceptibility, i.e., the generation of the second optical harmonic or frequency doubling, and (2) on a special effect of the third-order susceptibility, i.e., an optical bistability giving rise to a high-density optical information storage by persistent spectral hole burning.

2. FREQUENCY DOUBLING

2.1 Principle of Frequency Doubling

The electromagnetic field of a laser beam impinging on a material of high polarizability generates a nonlinear polarization. In the presence of an external field E, the microscopic polarization p induced at an atom or a molecule is given by

$$p = \alpha E + \beta E^2 + \gamma E^3 + \dots \tag{1}$$

where α, β and γ are the linear, the quadratic and the cubic polarizability tensors. Similarly, the macroscopic polarization induced in the bulk media can be expanded in a power series

$$p = \aleph^{(1)}E + \aleph^{(2)}E^2 + \aleph^{(3)}E^3 + \dots \tag{2}$$

where $\aleph^{(1)}$, $\aleph^{(2)}$ and $\aleph^{(3)}$ are the first-, second- and third-order susceptibility tensors obtained by summarizing over all microscopic subunits s, e.g., $\aleph^{(1)} = \Sigma\alpha(s)$. The higher orders for the electric field terms, i.e., the nonlinearity of the polarization, can be considered to result from the motion of the electron oscillator under the influence of the nonlinear forces (Fig. 1). For the case that the field strength E can be represented by a sinus wave, the linear and the quadratic terms of the polarization of an oscillating dipole yield

$$p = \aleph^{(1)}E_o\sin \omega t + \aleph^{(2)}E_o{}^2\sin^2\omega t \qquad (3)$$

From simple trigonometric rearrangement follows

$$p = \aleph^{(1)}E_o\sin \omega t + \tfrac{1}{2}\aleph^{(2)}E_o{}^2 - \tfrac{1}{2}\aleph^{(2)}E_o{}^2\cos(2 \omega t) \qquad (4)$$

This means that the polarization wave contains the overtone frequency 2ω and that the oscillating dipole emits electromagnetic radiation of this frequency. A treatment of the anharmonic oscillator model for nonlinear optical effects is given elsewhere [28a].

   If frequency doubling is considered to result from the interaction of two photons polarized parallel to each other [29], then the the frequency doubling results from the conservation of impulse, i.e., $\hbar\, 2\pi/\lambda_1 + \hbar\, 2\pi/\lambda_2 = \hbar\, 2\pi/\lambda_3$, and for $\lambda_1 = \lambda_2$ the vector addition of the two photons gives $2\, \hbar 2\pi/\lambda_1 = (\hbar 2\pi)/(\lambda_1/2)$. In case that the hyperpolarizable molecule, crystal or supramolecular structure exhibits inversion symmetry, then the dipole second harmonic fields average to zero [29].

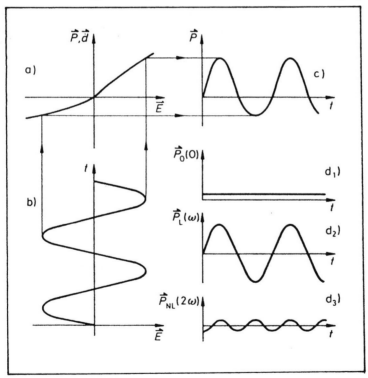

Figure 1. On the electron having the nonlinear characteristic (a) a sinus type field strength of frequency ω operates (b). The anharmonic oscillation of the electron results in a corresponding polarization (c). The Fourier resolution of the anharmonic oscillation yields the nonoscillating term (second term in eq. (4)) (d_1), the basic wave (d_2) and the second harmonic (d_3) (from [28b]).

For the selection of suitable chromophores, exhibiting large first hyperpolarizabilities ß and, thus, large effects in frequency doubling, the Electric-Field-Induced Second-Harmonic Generation (EFISHG) technique is applied, usually. Since this method gives only the vector part of ß, a new technique, i.e., the Hyper-Rayleigh Scattering (HRS) has been proposed recently to overcome the shortcomings of EFISHG [30,31]. The HRS technique makes a complete tensor analysis of the microscopic first-order polarizability available and enables the study of ionic species or of molecules with octopolar charge distribution. Octopolar molecules have distinct advantages over dipolar molecules for aplication in nonlinear optics, since they combine favorable NLO properties with a strict cancellation of all vector-like properties [32].

2.2 Organized Dye/Mineral Composites

Minerals as host materials for an organized anchoring of dye molecules offer various advantages as compared to polymers, like higher thermal, mechanical or chemical stability. The regular crystal structure of minerals offers the preconditions for an organized arrangement of dipole molecules resulting in the addition of the molecular effects to a macroscopic bulk effect. A variety of preliminary results have been published, demonstrating the high innovative potential of mineral-anchored dyes or polarizable molecules for optical processing, especially, in the case of clay-type minerals.

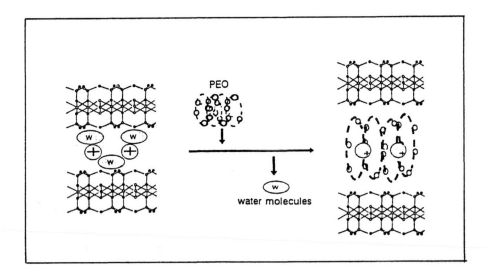

Figure 2. Representation of polyethylenoxide (PEO) intercalation in a 2:1 charged phyllosilicate showing the replacement of the water shell of the cations by helical PEO chains (from [40]).

Clay minerals exhibiting plate-like structures and optical transparency, e.g., hectorite, laponite or montmorillonite, are ideal model hosts for the preparation of composites with nonlinear optical properties [33]. The polyanionic nature of the clays enables the intercalation of cationic dyes, e.g., crystal violet or phthalocyanines [34], by ion exchange in the interlamellar space. Qualitative conclusions on the orientation of dye molecules were drawn by diffuse reflectance spectroscopy [35,36]. The intercalation of π-electron molecules in the layered structure of phosphates and phosphonates of tetravalent metals has been demonstrated [37,38]. It has been shown that the water shell around the cations of the homoionic 2:1 charged phyllosilicates of the smectite group can be substituted by polyoxyethylene units in a helical conformation (Fig. 2) [39]. Further, the

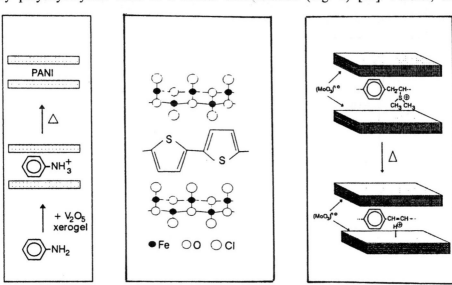

Figure 3 (left). Illustration of the intercalation of aniline in vanadium pentoxide xerogel and polymerization by thermal treatment (from [40]).

Figure 4 (center). Idealized structural representation of polythiophene intercalated into FeOCl (from [40]).

Figure 5 (right). Illustration of poly(p-phenylene-vinylene) generation in the interlayer space of MoO_3 (from [40]).

intercalations of aniline and subsequent formation of polyaniline in montmorillonites or layered phosphates and xerogels were described (Fig. 3) [40]. Intercalations of polymers have also been reported for layered structures of FeOCl (Fig. 4), MoO_3 (Fig. 5) and MoS_2 [40]. These examples demonstrate that a wide variety of combined materials can be prepared exhibiting a marked anisotropy.

2.3 Frequency Doubling by Dye/Molecular Sieve Composites

Composite materials of chromophores and crystalline molecular sieves offer several advantages for the application in nonlinear optics, i.e., tunability of the materials by variation of their components, oriented arrangement of the chromophores in the channels of molecular sieves by steric constraints, thermal resistance of the mineral host or chemical stability towards excited molecules. The pore system of the molecular sieve causes the incorporated dye molecules to form highly organized arrangements. This results in a variety of optical effects, e.g., (i) the light absorption becomes strongly anisotropic and the loaded molecular sieve act as polarizer [41], (ii) the birefringence of molecular sieve crystals is controlled by dye incorporation, and (iii) such materials can exhibit large nonlinear optical susceptibilities of the second order, which results in optical second harmonic generation being observed for these materials (Fig. 6) [42-46].

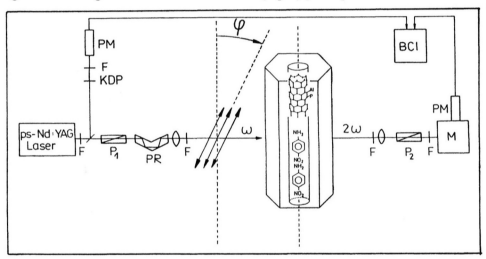

Figure 6. Dye-loaded $AlPO_4$-5 crystals in a ps-Nd:YAG laser apparatus for optical frequency doubling (second harmonic generation). φ denotes the angle between the axis of the $AlPO_4$-5 crystal and the polarization plane of the incident laser beam. F, filter. P1, P2 polarizers. M, monochromator. PM, photomultiplier. BCI, gated boxcar integrator. KDP, potassium dihydrogenphosphate crystal. PR, polarization rotator (fresnel rhombs). ω and 2ω denote the frequencies of the incident light and the second harmonic wave, resp. (from [44, 45]).

The existence of a second harmonic generation (SHG) effect for molecular sieves with one-dimensional channel structures ($AlPO_4$-5, $AlPO_4$-11, VPI-5) containing p-nitroaniline (PNA) was discovered by Cox et al. on powder samples [42-43]. Although PNA molecules have a large second-order hyperpolarizability, crystals of PNA do not exhibit a macroscopic effect since they exhibit an antiparallel arrangement of PNA chains resulting in a compensation of the microscopic dipoles and a gross macroscopic dipole of zero.

For the pearl-string-like arrangement of PNA molecules in molecular sieve channels (Fig. 6) the molecular dipole moments are expected to superimpose giving a macroscopic hyperpolarization. The PNA molecules exhibit head-to-tail arrangements with intermolecular hydrogen bonds resulting in a pseudo-quinonoid electron distribution [47]. The straight arrangement of the PNA molecule chains in molecular sieves is concluded from polarization dependences in Raman scattering [47]. Pyroelectric studies have shown that the highly organized structure of the incorporated molecules is a result of a special uptake mechanism, i.e., an entering of the molecules in head-to-tail arrangement from both sides of the one-dimensional channels [48]. Polarization dependent frequency doubling was for the first time demonstrated for PNA in large single crystals of $AlPO_4$-5 or ZSM-5 using a ps-Nd:YAG laser [44,45]. From single crystal SHG studies the different tensor components of the SHG can be identified [46] which is the basis for phase matching as the condition for any technical application of the SHG based on dye-loaded molecular sieve crystals. From the appearance of frequency doubling effect it is evident (1) that the molecules enter the one-dimensional channels from both ends of a crystal, (2) that the domains with dipole chains having parallel orientation are larger than the wavelength of the incident laser light (0.532 µm), and (3) that the second harmonic order light from different domains does not interfere. For the occurence of the SHG effect in the bimodal and bidirectional MFI framework of the ZSM-5 crystals it is proposed that only the PNA molecules aligned by the straight channels of ZSM-5 contribute to the frequency doubling.

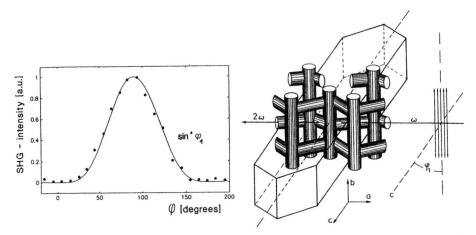

Figure 7. SHG intensity for PNA moleculaes in a large ZSM-5 crystal as a function of the angle ϕ between the c-axis of the ZSM-5 crystal and the plane of polarization of the incident light (from [44]).

For PNA in $AlPO_4$-5 or ZSM-5 the frequency-doubled radiation is polarized parallel to the dipole chains, i.e., along the length direction of the hexagonal prisms ([001]) or the straight channels ([010]), respectively. However, for (dimethylamino)benzonitrile (DMABN) two components of the SHG have been found [46], i.e., besides the expected effect parallel to the dipole chain, a second component perpendicular to the chain is observed (Fig. 8). It is discussed that the reason for the different tensor elements are the close-lying lowest excited states in DMABN having transition dipole matrix elements perpendicular to each other. As a consequence, mixing of waves of different polarizations is observed enabling potential technical applications for phase-matching [46].

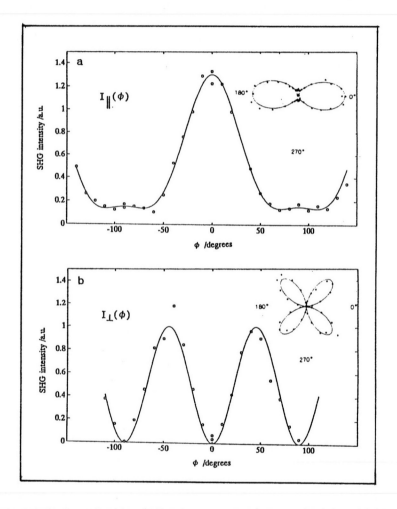

Figure 8. Intensity of the SH radiation polarized parallelly (a) and perpendicularly (b) to the crystal length axis (z-axis) for DMABN in $AlPO_4$-5 as a function of the polarization direction ϕ of the incident light. The measured values are well fitted by $I_{\parallel}(\phi) = a^2 \cos^4\phi + b^2 \sin^4\phi$ and $I_\perp(\phi) = (\cos\phi \sin\phi)^2$, respectively. Inset: Polar plot of the same measurements. Polar angle: ϕ, radius: $I(\phi)$ (from [46]).

2.4 Future Aspects

The void structures of molecular sieves offer a unique possibility to impose molecular arrangements which are unknown for the hyperpolarizable molecules in their crystalline bulk states. Attempts to produce extended conjugated aromatic systems in the void structures of molecular sieves are at the beginning [49-51]. Results in this area would inspire the development of new materials with strong optical nonlinearities. The discovery of mesoporous molecular sieves [52] broadens the arsenal of host structures for the incorporation of hyperpolarizable molecules. It will be exciting to learn, wether the liquid crystal structure of the detergents, used as templates for the hydrothermal synthesis of mesoporous molecular sieves, can contribute to the organized arrangement and incorporation of chromophores having high hyperpolarizabilities.

3. HIGH DENSITY OPTICAL DATA STORAGE

3.1 Principle of Persistent Spectral Hole-Burning

The term persistent spectral hole-burning (PSHB) comprises photochemical hole-burning (PHB) as well as nonphotochemical hole-burning (NPHB) [53]. Both

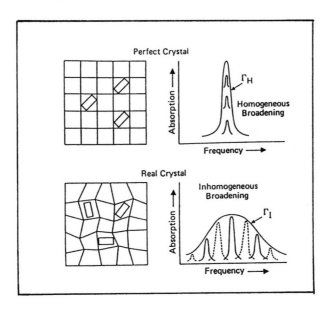

Figure 9. Illustration of homogeneous and inhomogeneous broadening of absorption bandwidths (from [53]).

hole-burning mechanisms result in persistent holes created in the absorption band of molecularly dispersed photoreactive guests in host matrices at low temperatures. The former is based on a photochemical change or rearrangement of guest-molecules whereas the latter is explained with the existence of so-called two-level systems (TLS) generated by the interaction with the host matrix. Based on the first observation of the effect [54], its application for high-density frequency domain optical storage [55] as well as for high-resolution spectroscopy [56] was proposed.

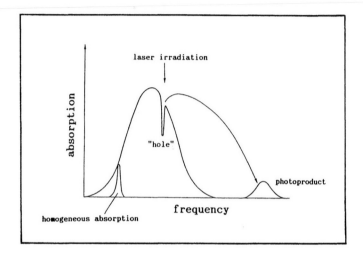

Figure 10. Representation of hole formation (from [53]).

The optical transition energy in an assembly of photosensitive molecules is not uniform due to the spatial nonuniformity of the microenvironment and the crystal field around the photosensitive molecules. This leads to the inhomogeneous broadening of optical spectra (Fig. 9). The PSHB technique creates a hole as a replica of a homogeneous line in an inhomogeneously broadened absorption band (Fig. 10). PHB can occur due to a variety of photochemical mechanisms, like proton tautomerization (Fig. 11), rearrangement of hydrogen bonding between dihydroxyanthraquinone (DAQ) (Fig. 12) or photoinduced electron transfer. The NPHB mechanism is based on the assumption of two dye-host arrangements separated by an energy barrier which can be crossed by the photoexcitation.

The PSHB process is affected by a coupling of the electronic transitions to the phonons, i.e., the collective vibrations of the host. This results in additional emissions in the luminescence spectrum (Fig. 13), i.e., the band created by the pure electronic excitation, called zero-phonon line, is generally accompanied by a phonon wing. This means, that holes burnt by the PSHB process can be accompanied by phonon side holes. This effect is largely suppressed if (1) a weak coupling between host and guest exists or (2) the phonon modes exhibit high frequencies.

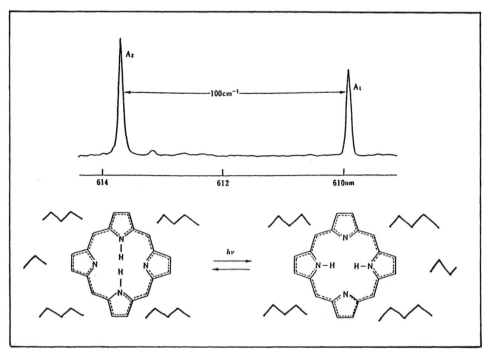

Figure 11. The two tautomers of porphyrins caused by the difference in microenvironment (from [53]).

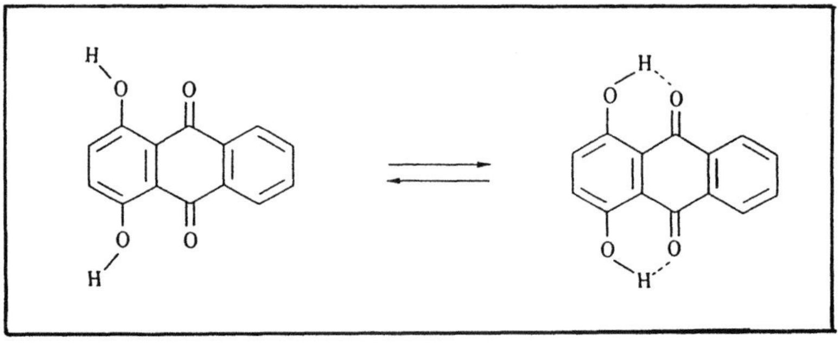

Figure 12. Hydrogen bond rearrangement of DAQ (from [53]).

156

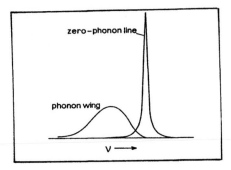

zero-phonon line

phonon wing

ν ⟶

Figure 13. Representation of the shape of a vibronic band in the luminescence spectrum of a guest molecule in a solid matrix.

3.2 Preparation of Dye-loaded Molecular Sieves
3.2.1 Ion Exchange of Cationic Dyes

The polyanionic nature of zeolites in combination with the high mobility of the charge compensating cations enables the introduction of cationic dye molecules via ion exchange from aqueous solutions. It is obvious, that only those dye molecules will enter which fit into the openings of the internal voids, i.e., channel openings or cage windows. In dependence on the dye molecule size, extended exchange periods, i.e., weeks or months, can be expected [59,60]. Additionally, the strong interaction of the π-electron systems with the coulomb centers may give rise to accumulations of the dyes at the external surface and/or at the pore openings, thus, impeding the exchange process. Complete adsorption isotherms have been determined for the exchange of thionine into zeolite only, up to now (Fig. 14) [61]. Further parameters, influencing the rate and extent of dye ion exchange, are the Si/Al ratio and the solvent type, e.g., water or ethanol [62].

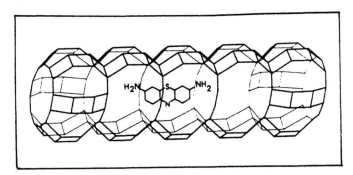

H_2N ⟶ S ⟶ NH_2
N

Figure 14. Schematic view of a thionine molecule in the main channel of a zeolite L (from [61]).

3.2.2 Incorporation of Neutral Dyes

Neutral dyes can be incorporated into dehydrated molecular sieves via gas phase absorption, if the chromophore can be evaporated and deposited without decomposition. Dye molecules larger than the windows of a three-dimensional cage system, like the faujasites, can be synthesized from precursors, resulting in stably anchored chromophores, like e.g. phthalocyanines (Fig. 15), which cannot be removed by extraction with a solvent [63-65]. A recent review on zeolite-encaged chromophores of metal complexes used in catalysis and photoprocesses is published by DeVos et al. [66].

Neutral dyes can also be incorporated by impregnation from solution. This process is again influenced by various parameters, like Si/Al ratio or type of solvent [62].

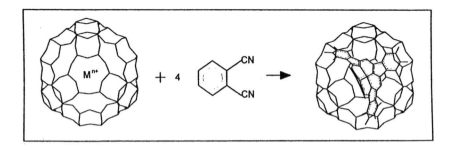

Figure 15. Synthesis of phthalocyanines in the zeolite Y supercage (from [66]).

3.2.3 Crystallization Inclusion

Dye molecules have been added to batches for the hydrothermal crystallization of zeolites resulting in new phases [67,68]. This effect has been denoted as a structure directing one. The inclusion of dissolved chromophores into the matrix of molecular sieves during the crystallization is achieved [59, 69-72], but depends on prerequisites which have to be met.

Firstly, the dye must be stable under the conditions of the molecular sieve synthesis. Especially, the high pH values, which are valid for the preparation of zeolites with low Si/Al ratios, are harmful for many dyes. For example, the dye methylene blue (MB) can be partially demethylated by a base-catalyzed selfoxidation or disproportionation and hydrolysis [73] as formulated in Fig. 16. The product of total demethylation, i.e. thionine, can be further degraded to the imine as expressed in Fig. 17. The problem of the relatively slow degradation can be overcome largely, if rapid zeolite syntheses are applied.

Secondly, as in the case of loading by ion exchange, only dyes with positive charge will be incorporated to a considerable amount, i.e., with yields in the range of 0.1 - 0.2 dye molecules per unit cell [59,74].

Thirdly, the incorporation of chromophores in monomolecular dispersion requires a solubility in the aqueous solutions of the synthesis mixtures. However, it

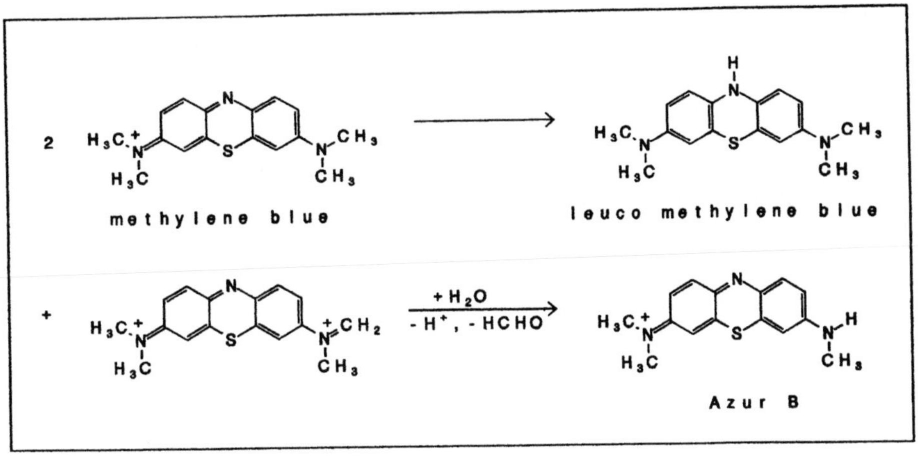

Figure 16. Mechanism of base-catalyzed demethylation of methylene blue via self-oxidation (from [73]).

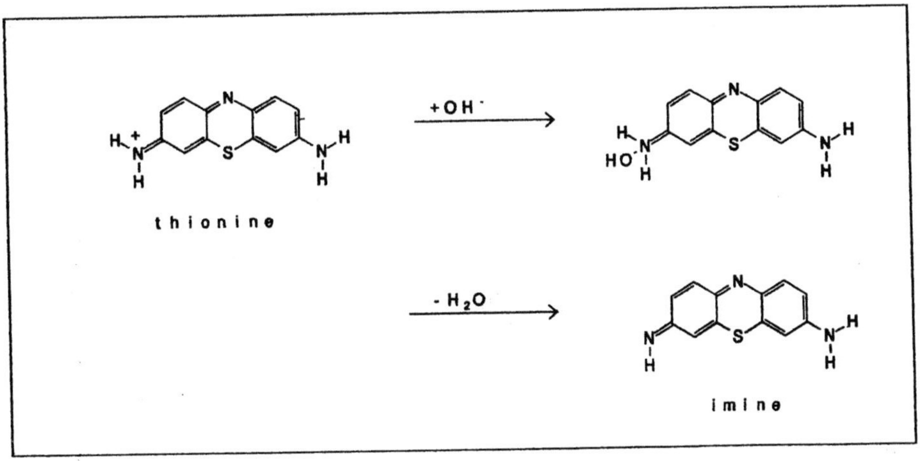

Figure 17. Mechanism of imine formation from phenothiazine dyes at high pH values (from [73]).

is demonstrated that the neutral thioindigo, which is insoluble in water, could be enclosed during hydrothermal crystallization of zeolite NaX due to its solubility in tri-ethanolamine used as templating additive in the faujasite synthesis [75]. In this case up to 0.4 molecules per unit cell were incorporated.

Relatively large amounts of chromophores, i.e., loadings up to 0.1 molecules per unit cell, are incorporated by crystallization inclusion into AlPO$_4$-5 molecular sieve exhibiting much smaller unit cell dimensions than the faujasites [76]. Due to the favorable pH regions for the hydrothermal synthesis of the aluminophosphate family molecular sieves, ranging around 3-7 [77], a large number of cationic dyes can be enclosed, since any base-catalyzed degradation does not take place. Furthermore, the acidic pH of the synthesis mixtures enables the enclosure of a large number of neutral dyes, requiring simply amine substituents. Then, the ammonium cations formed in an acidic synthesis batch are sufficiently soluble and are readily enclosed [72].

3.3 Characterization of Guest/Host Systems
3.3.1 Location on Lattice Sites
   The location of dye molecules on lattice sites of the structure of molecular sieves can be revealed from the analysis of X-ray powder diffractograms by means of

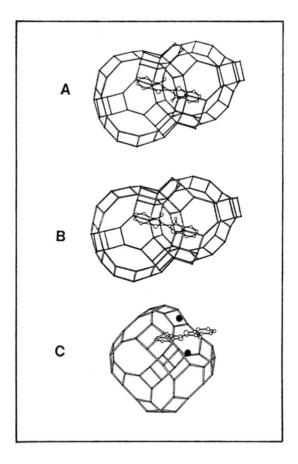

Figure 18. Location of trans-(A,C) and cis-(B) thioindigo in the center of a 12-ring window between two faujasite supercages. Black dots (C) mark the sodium cation positions (S_{II*} sites) for mildly dried samples (from [75]).

Rietveld refinement in combination with electron density maps. An example is given, as follows, for thioindigo (cf. Fig. 24) incorporated in faujasite NaX by

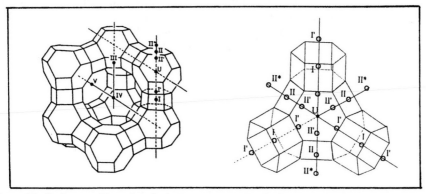

Figure 19. Cation positions in a faujasite structure: In the hexagonal prism (I), in the sodalite unit facing I(I'), in the supercage at the 6-ring centers (II), in the sodalite unit facing II(II'), in the supercage (II*), in the supercage at the center of the 4-ring ladder (III), in the center of the supercage (IV), and in the supercage at the center of the 12-ring (V).

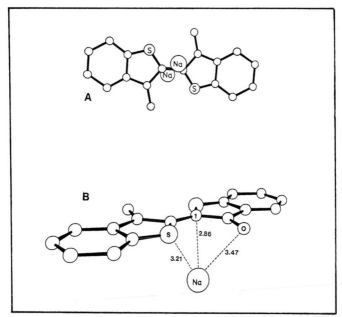

Figure 20. Positions (A) and distances (B) of sodium ions in the neighborhood of trans-thioindigo (from [75]).

crystallization inclusion [75]. The thioindigo (TI) molecule is located in the center of the 12-ring window connecting two neighboring supercages (Fig. 18). This arrangement enables the closest approach of the donating S- and O-atoms of the dye molecule to an accepting cation position in a supercage, i.e., $S_{II}(S_{II}*)$ or S_{III} or S_V site (Fig. 19). In dependence on the thermal treatment of the sample for the removal of the solvent ethanol used for Soxhlet extraction the cations in closest neighborhood of the chromophore occupy different positions. Mild drying at room temperature results in Na cations on S_{II} sites being closest to the TI (Fig. 18 A-C). Drying at 350 K (2h) exhibits a preferential occupation of sites close to $S_{II}*$ by Na^+ (Fig. 20). Finally, thorough drying (350 K, 12 h) results in Na^+ on sites close to S_{III}. In this position Na^+ forms a member of a 6-ring, i.e., exhibits a coordination like a chelate (Fig. 21).

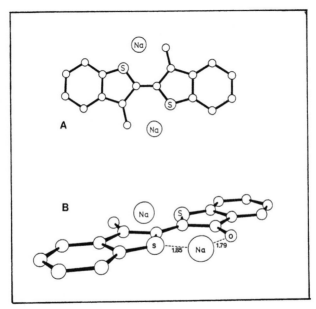

Figure 21. Positions and distances of sodium cations in the thioindigo chelat complex (from [75]).

Faujasite-incorporated methylene blue exhibits different sites for the location of the dye molecules in dependence on the type of incorporation. Ion exchange results in a preferential location close to the center of the supercage, whereas a favored occupation of the 12-ring windows is observed for the dye molecules encaged by crystallization inclusion [60,75]. Obviously, ion exchange results preferentially in the substitution of very mobile Na^+ close to the center of the supercage, whereas the dye is attached to cations which are essential for the zeolite structure if it is incorporated during crystallization [60].

3.3.2 Location in Mesopores

Dye molecules can be incorporated in the matrix of $AlPO_4$-5 (Fig. 22) by crystallization inclusion, even if the size of the chromophore exceeds the channel diameter of the molecular sieve. For example, the porphyrin type chromophores (cf.

162

Fig. 26) tetrakis[N-methyl-3-(pyridyloxy)]phthalocyanine-zinc (ZnTpyP) or tetrakis[N-ethyl-2,3-pyridino]tetraazaporphyrin-zinc (ZnTaP) have been incorporated with average loadings of 1 dye molecule per 10^3 unit cells [72,78].

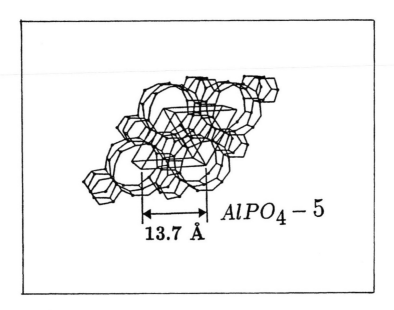

Figure 22. Model of the AlPO$_4$-5 framework. The expansion of the hexagonal unit cell is represented by the two trigonal prisms (from [76]).

Since the channel diameter of AlPO$_4$-5, i.e., ca. 0.73 nm, amounts to less than one half of the molecule size (1.5 - 2.5 nm), an incorporation into defect sites of the molecular sieve lattice has to be taken into account. This assumption is proved from adsorption isotherms, taken after burn-off of the dye, indicating the presence of mesopores [72].

3.3.3 Optical Properties

The absorption of NaX-encaged TI in the visible region for the $S_0 \rightarrow S_1$ excitation occurs at 530 - 560 nm (Fig. 23 B), i.e., is in the range valid for the solid TI in the thermodynamically stable trans conformer state [79-81]. The band is broader (FWHM {full width of half maximum} = 90 nm) than in solution (FWHM = 65 nm) and poorly resolved. The band of TI adsorbed at the external surface of the NaX crystals (Fig. 23 C) is even broader (FWHM = 120 nm), and shifted bathochromically, indicating a stronger interaction in this state, presumably, towards silanol groups.

The fluorescence spectrum of NaX-encaged TI (Fig. 24 A) exhibits a maximum at approx. 625 nm, which is red-shifted by 82 nm in comparison to TI

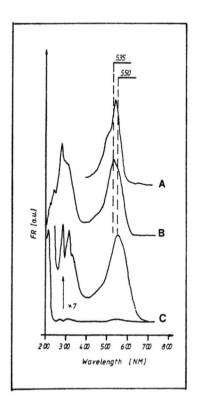

Figure 23. UV/VIS absorption spectra of thioindigo in toluene (A), encaged in NaX (B) and adsorbed on the external surface of zeolite crystals (C). The transmission spectrum (A) is represented by the absorbance and the diffuse reflectance spectra (B,C) are processed by the Kubelka-Munk function (from [75]).

dissolved in toluene. This large shift indicates a relatively strong polarization of the environment during the excitation of the TI molecule. The rapid relaxation of the host polarization lowers the energy of the excited state of the guest, thus, decreasing the energy of the emitted radiation. The readily polarizable environment is represented by the sodium cations. The strong coupling of the electronic transitions to the phonons is supported by the absence of the zero phonon lines and of any fine structure in the fluorescence spectrum, which are generally observed for dissolved TI [79-81]. The fluorescence spectrum of TI adsorbed at the external surface of zeolite crystals (Fig. 24 B) exhibits a less extended red shift of 51 nm, supporting the assumption given above, that another type of interaction is valid for the externally adsorbed TI.

Water molecules, which are always present in molecular sieves, have an influence on the optical spectra. For example, the fluorescence spectra of NaY-encaged thionine strongly deviate in either the hydrated or the dehydrated state (Fig. 25) [82]. As expected, the dehydrated form shows (1) a stronger red shift, presumably, due to stronger coulomb interaction with the host lattice in the absence of shielding water molecules, and (2) an essentially featureless spectrum.

The fluorescence spectrum of ZnTpyP and ZnTaP exhibit no difference in the hydrated or in the dehydrated state (Fig. 26). This might be referred either to the

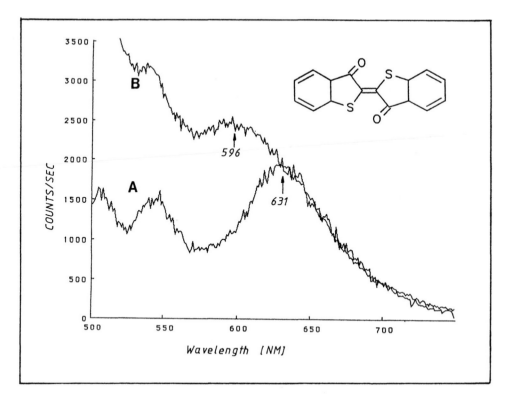

Figure 24. Fluorescence spectra (excitation wavelength: 450 nm) of the thioindigo encaged in NaX (A) and adsorbed on the external surface of NaX (B) (from [75]).

absence of mobile metal cations or to the location in mesopores [72,78]. The maximum of the spectrum for ZnTpyP is more in the red region than that for ZnTaP, pointing to a stronger interaction of the former, i.e., a lower energy of the excited state [78].

Chromophores incorporated in molecular sieves exhibit enhanced stabilities towards photobleaching, as is demonstrated for thioindigo in faujasite NaX [75], methylene blue in NaY [60], or porphyrins in AlPO$_4$-5 [72]. It is assumed that singlet oxygen, which is produced by photoexcitation, will be quenched largely by the molecular sieve host, thus, suppressing the attack of the dye molecules by this highly reactive $^1O_2$ molecule.

3.4 Hole-Burning in Mineral-Enclosed Dyes
4.3.1 Low-Temperature Hole-Burning

Low-temperature hole-burning experiments with zinc porphyrins, i.e., ZnTaP and ZnTpyP, which are enclosed in $AlPO_4$-5, are carried out with average loadings around 5 x 10^{-7} mol/g [72,78]. The dye contents could be determined with high precision based on zinc concentration measurements by atomic absorption

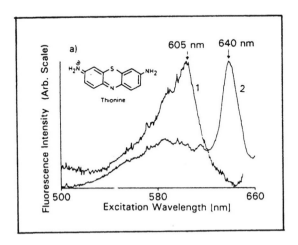

Figure 25. Fluorescence excitation spectra of thionine/ NaY samples. Liquid helium temperature (T = 1.3 K) spectra of a hydrated (1) and a dehydrated (2) sample (from [82]).

spectroscopy. The fluorescence excitation spectra are used for analysis of the burnt holes to avoid the problems of light scattering which impede the use of UV/VIS absorption spectra. The applied dye loadings represent optimal values, i.e., overcome detection problems related either to too small or to too large dye contents, where quenching processes reduce the fluorescence intensities in the latter case. For the removal of the template triethylamine the samples are heated up (5 K min^{-1}) to ca. 400 K in a vacuum (10^{-3} Pa) and subsequently rehydrated at room temperature over saturated aqueous KCl solution (83% relative humidity). The sample cuvettes were held in a Cryovac helium bath/flow cryostat (1.3 - 300 K). Spectral holes were burnt using an autoscan laser (CR 899-29) supplied with DCM dye and pumped by an argon ion laser (Innova 200). The dye laser line width was ≤ 1 MHz. The spectra were recorded in fluorescence excitation mode using a photomultiplier (CRC C 31034 A) and a single-photon counting system (SI). Appropriate cut-off filters were used to select fluorescence light only. High resolution spectra were recorded scanning the autoscan laser. Deep-saturated holes were burnt at 20 K and 685.0 nm (ZnTpyP) or 655.0 nm (ZnTaP) with 1-15 $mWcm^{-2}$ for 1800 s. The hole was read out at temperatures between 2 and 50 K.

In samples with solvent-filled pores, Lorentzian-shaped holes could easily be

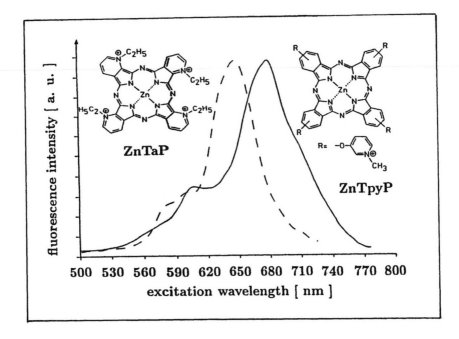

Figure 26. Fluorescence excitation spectra of ZnTaP (dashed line) and ZnTpyP (solid line) in hydrated AlPO$_4$-5. Both dyes consist of up to four structural isomers, as is usual for tetra-substituted phthalocyanine derivatives. Hole-burning experiments were carried out near the maxima of 650 and 680 nm (from [78]).

burnt in both porphyrin/AlPO$_4$-5 systems at temperaturs between 2 and 20 K. In order to detect the photoproduct, deep and saturated holes were burnt at 20 K (Fig. 27). A photochemical hole-burning mechanism is ruled out on the basis of the positions of the product bands [83,84], i.e., they are too close to the zero phonon hole (ZPH) and do not depend on the type of solvent, e.g., water or chloroform. Electron transfer from higher triplet states as found in other matrices is not efficient as could be demonstrated in two-color hole-burning experiments [78]. Hole erasure by temperature excursion was comparable to the behavior well-known for bulk glasses. However, hole broadening by spectral diffusion was found to be significantly reduced in AlPO$_4$-5 samples in comparison to ethanol glass.

3.4.2 High-Temperature Hole-Burning

Spectral holes were burnt up to 85 K in hydrated porphyrin/AlPO$_4$-5 samples (Fig. 28) [78]. The found features fit well into those of currently known high-temperature dye systems, where a photochemical burning mechanism is claimed. Obviously, the porphyrin/AlPO$_4$-5 systems are exceptions where a non-

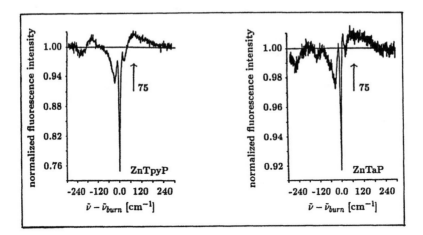

Figure 27. Deep saturated holes of phthalocyanine-zinc derivatives in hydrated AlPO$_4$-5. (a) Chromophore = ZnTpyP, hole burnt at 20 K and 685.0 nm with 1.5 mW/cm^2 for 1800 s. At both sides of the ZPH the PSB can clearly be seen (28 cm^{-1}). On the high energy side of the hole, the maximum of the product band is marked by an arrow at 75 cm^{-1}. The dip at -220 cm^{-1} results from a vibronic pseudo side-hole. Its product band lies 75 cm^{-1} to higher energy at -145 cm^{-1}. (b) Chromophore = ZnTaP, hole burnt at 20 K and 655.0 nm with 1.0 mW/cm^2 for 1800 s. At both sides of the ZPH the PSB can clearly be seen (30 cm^{-1}). On the high energy side of the hole, the maximum of the product band is marked by an arrow at 75 cm^{-1}. Vibronic pseudo side-holes are at -250 and -130 cm^{-1} (from [78]).

photochemical mechanism has to be assumed. Presumably, the chromophores couple to extrinsic two-level sytems with high barriers enabling the formation of stable product states. The high barriers are provided by the water in the pores of the mineral host, i.e., the thin water shells surrounding the dye molecules. Besides this NPHB mechanism at elevated temperature, the porphyrin /AlPO$_4$-5/ solvent samples exhibit a number of other interesting features like drastically reduced hole broadening by spectral diffusion and phonon frequencies higher than in bulk glasses.

3.4.3 Dynamics of Encaged Chromophores

A deep hole burnt in a hydrated thionine/faujasite sample (Fig. 29 a) [82] reveals strong phonon and pseudo-phonon side bands at 25 and 35 cm^{-1}. A similarly deep hole in the dehydrated sample (Fig. 29 b) exhibits much weaker electron-phonon coupling to a mode of higher energy (ca. 50 - 70 cm^{-1}).

A peculiar effect is observed, if a fluorescence spectrum is recorded after narrow-band excitation of the dehydrated sample (Fig. 29 c). One finds a

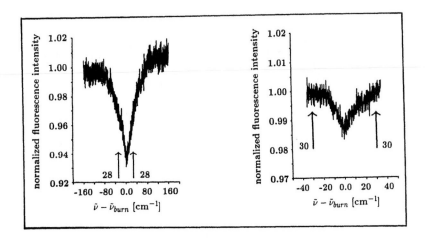

Figure 28. (a) Shallow spectral hole of ZnTaP in hydrated AlPO$_4$-5 burnt at 80 K and 655.0 nm with 13 mW/cm^2 for 100 s. Saturation effects and sideband enhancement (marked by arrows) are still absent and allow reasonable detection of the hole width, which is 19 cm^{-1} in this case. (b) Deep spectral hole of ZnTpyP in hydrated AlPO$_4$-5 burnt at 77 K and 685.0 nm with 10 mW/cm^2 for 1000 s. The hole width is 67 cm^{-1}. Saturation effects and PSB enhancement (marked by arrows) dominate the line shape of the hole (from [78]).

broad band spectral increase around the exciting wavelength, and in addition, the appearance of a narrow band (ca. 5 cm^{-1}) on top of this broad band right at the excitation wavelength, i.e., one finds an antihole at the burning wavelength. Multiple antiholes can be burnt right next to each other by changing the excitation wavelength. If one increases the burning time, these antiholes disappear and holes are detected at the excitation wavelength.

The low temperature optical spectroscopy of the thionine/NaY system indicates that the geometry of and the microenvironment in the zeolite cage determine the dynamics of the chromophore guest. The characteristics of thionine in the hydrated zeolite are similar to those of chromophores in glassy systems. This suggests that the water molecules in the faujasite pores interact strongly with the dye and form a kind of solvent matrix around the chromophore. This water matrix supplies the degrees of freedom responsible for the strong electron-phonon coupling as well as those necessary for hole burning.

The broad band spectral increase and antihole formation for the dehydrated system is explained by the existence of two different spectral forms of the thionine in the faujasite cages. The incident light causes a change of the population of the two forms. This assumption is supported by the appearance of different fluorescence intensities in dependence on reversible thermal treatment cycles [82].

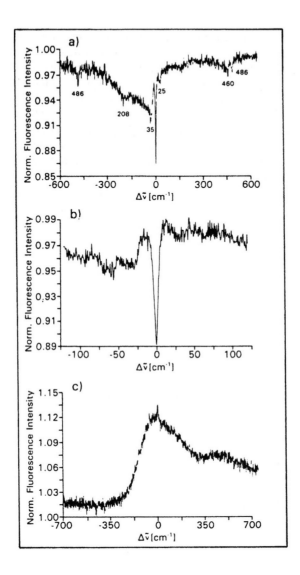

Figure 29. a: Hole spectrum of hydrated thionine/NaY sample at T = 1.3 K (burning wavelength λ = 605.05 nm, laser intensity P = 1 mW/cm^2, burning time τ = 3600 s). b: Hole spectrum of dehydrated thionine/NaY sample at T = 1.3 K (burning wavelength λ = 644.95 nm, laser intensity P = 5 mW/cm^2, burning time τ = 1000 s). c: Spectral change after excitation of a dehydrated thionine/NaY sample at T = 1.3 K (excitation wavelength λ = 630.06 nm, laser intensity P = 5 mW/cm^2, excitation time τ = 100 s). Note the extra narrow band increase on top of the broad band feature (from [82]).

3.5 Future Aspects

PSHB will find increasing application in the very high-resolution spectroscopy for research on composite materials, since one can obtain information about (i) solid-state photochemistry at very low temperatures, (ii) dephasing of electronic transitions in glassy systems, (iii) low-energy excitation modes of host matrices (iv) or the dynamics of molecules in a microenvironment [57].

The application for optical memory systems might depend on progresses in the nondestructive readout and data addressing [85]. Stable systems are reported to be based on photon-gated photochemical hole-burning, i.e., two-step photoionization [86], two-step photodecompositions [87] or two-step photoinduced donor-acceptor electron transfer reactions [89-90]. It can be summarized that PSHB is still in a state of fundamental research with respect to practical application in high-density data storage, but represents a powerful method for the elucidation of photoprocesses in advanced materials.

4. OUTLOOK

The basic advantage of a mineral host as compared to a polymer matrix for chromophores exhibiting NLO properties is obvious, since the former cannot be attacked chemically by excited dye molecules. The application of crystalline molecular sieves is essential for SHG applications. In addition to this function to disperse and stabilize guest atoms or molecules, the guests will be also oriented by the pore system of a molecular sieve. Fig. 30 (right) shows that dipolar guest molecules can be very well aligned by the straight pores forming dipole chains of parallel orientation. To have for intensity reasons a macroscopic ensemble of oriented molecules, it is possible to orient the host structure containing the guests as shown in Figure 30.

The application of molecular sieves as host materials requires the solution of light scattering problems. Different routes are possible to overcome this shortcoming, i.e., application of (1) large molecular sieve single crystals as well as assemblies of single crystals, e.g., for SHG (cf. Fig. 30) or (2) nanocrystals much smaller than the wavelength of the incident light, e.g., for optical data storage.

Several attempts are reported to deposit aligned dye molecules at fused silica surfaces for second harmonic generation, since scattering problems are largely avoided for these systems [91-94]. Very prospective seems to be the usage of silica glasses prepared by sol-gel processes. The applicability of dye-doped silica glasses for the study of nonlinear optical effects based on the third-order susceptibility has been published [95,96].

Thus, the conclusion of Stucky and MacDougall is still valid: The combination of molecular sieve quantum confinement and porous glass chemistry could lead to the development of practical NLO devices. With further study of both systems (crystalline and glass), one can expect new levels of understanding and applications of the fascinating regime of the solid state [97].

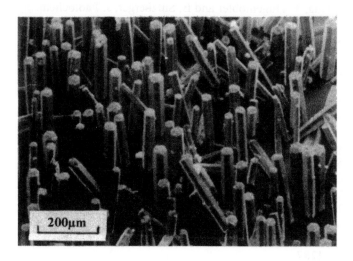 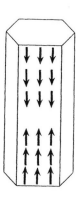

Figure 30. Alignment of large $AlPO_4$-5 crystals by an electric field of the strength $E=3$ kVcm^{-1}. Aligned crystals are fixed by water glass. The orientation of guest molecules by the unimodal and unidirectional channels where they form parallel dipole chains is shown schematically (from [45]).

ACKNOWLEDGEMENT

I acknowledge gratefully the critical reviews and important improvements by Drs. J. Caro, F. Marlow and F.W. Deeg.

REFERENCES

1.   Materials, Science and Engineering for the 1990s (National Research Council, ed.), National Academy of Sciences, Washington 1990.
2.   Y. Kim and K. Seff, J. Phys. Chem. 82 (1978) 1307.
3.   P.A. Jacobs and H.K. Beyer, J. Phys. Chem. 83 (1979) 1174.
4.   D. Hermerschmidt and R. Haul, Ber. Bunsenges. Phys. Chem. 84 (1980) 902.
5.   a) L.R. Gellens, W.J. Mortier, R.A. Schoonheydt and J.B. Uytterhoeven, J. Phys. Chem. 85 (1981) 2783.
        b)L.R. Gellens, W.J. Mortier, R. Lissilour and A. LeBeuze, J. Phys. Chem. 86 (1982) 2503.

6. a)G. Calzaferri, S. Hug, T. Hugentobler and B. Sulzberger, J. Photochem. 26 (1984) 109.
    b) M. Brändle and G. Calzaferri, Research on Chemical Intermediates (1994).

7. A. Stein, G.A. Ozin, P.M. Macdonald, G.D. Stucky and R. Jelinek, J. Am. Chem. Soc. 114 (1992) 5171.

8. D. Exner, N.I. Jaeger, R. Nowak, G. Schulz-Ekloff and P.L. Ryder, in: Proc. 6th Int. Zeolite Conf. (D. Olson and A. Bisio, eds.), Butterworths, Guildford 1984, p. 387.

9. D. Exner, N. Jaeger, A. Kleine and G. Schulz-Ekloff, J. Chem. Soc. Faraday Trans. 1,84 (1988) 4097.

10. G. Schulz-Ekloff and N.I. Jaeger, Catalysis Today 3 (1988) 459.

11. A. Tonscheidt, P.L. Ryder, N.I. Jaeger and G. Schulz-Ekloff, Surface Science 281 (1993) 51.

12. Y. Wang and N. Herron, J. Phys. Chem. 91 (1987) 257.

13. R.D. Strahmel, T. Nakamura and J.K. Thomas, J. Chem. Soc. Faraday Trans. 1, 84 (1988) 1287.

14. N. Herron, Y. Wang, M.M. Eddy, G.D. Stucky, D.E. Cox, K. Möller and T. Bein, J. Am. Chem. Soc. 111 (1989) 530.

15. X. Lui and J.K. Thomas, Langmuir 5 (1989) 58.

16. M.A. Fox and T.L. Pettit, Langmuir 5 (1989) 1056.

17. K. Möller, M.M. Eddy, G.D. Stucky, N. Herron and T. Bein, J. Am. Chem. Soc. 111 (1989) 2564.

18. M. Wark, G. Schulz-Ekloff and N.I. Jaeger, Catalysis Today 8 (1991) 467.

19. M. Wark, G. Schulz-Ekloff, N.I. Jaeger and A. Zukal, Stud. Surf. Sci. Catal. 69 (1991) 189.

20. M. Wark, G. Schulz-Ekloff, N.I. Jaeger and W. Lutz, Mat. Res. Soc. Symp. Proc. 233 (1991) 133.

21. M. Wark, H.-J. Schwenn, G. Schulz-Ekloff and N.I. Jaeger, Ber. Bunsenges. Phys. Chem. 96 (1992) 1727.

22. O.P. Tkachenko, E.S. Shpiro, M. Wark, G. Schulz-Ekloff and N.I. Jaeger, J. Chem. Soc. Faraday Trans. 89 (1993) 3987.

23. L. Brus, J. Phys. Chem. 90 (1986) 2555.

24. Y. Wang and N. Herron, J. Phys. Chem. 92 (1988) 4988.

25. L. Spanhel, M. Haase, H. Weller and A. Henglein, J. Am. Chem. Soc. 109 (1987) 5649.

26. M. O'Neill, J. Marahn and G. McLendon, J. Phys. Chem. 94 (1990) 4356.

27. A. Eychmüller, A. Hässelbarth, L. Katsikas and H. Weller, Ber. Bunsenges. Phys. Chem. 95 (1991) 79.

28 a) P.N. Prasad and D.J. Williams, "Introduction to Nonlinear-Optical Effects in Molecules and Polymers", Wiley, New York 1991.
    b) Lehrbuch der Experimentalphysik (Bergmann/Schaefer), Bd. III (Optik), 8. Auflage, S. 925, Walter de Gruyter, Berlin 1987.

29. F.A. Hopf and G.I. Stegeman, Applied Classical Electrodynamics, Vol. II: Nonlinear Optics, Wiley, New York 1986.

30. K. Clays and A. Persoons, Phys. Rev. Lett. 66 (1991) 2980.

31. K. Clays and A. Persoons, Rev. Sci. Instrum. 63 (1992) 3285.

32. J. Zyss, Nonlinear Opt. 1 (1991) 3.

33. H. van Olphen and J.J. Fripiat, Data Handbook for Clay Materials and Nonmetallic Minerals, Pergamon Press, Oxford 1979.

34. B.K.G. Theng, The Chemistry of Clay-Organic Reactions, Hilger, London 1974.

35. J. Cenens and R.A. Schoonheydt, Clays and Clay Min. 36 (1988) 214.

36. R.A. Schoonheydt and L. Heughebaert, Clay Min. 27 (1992) 91.

37. G. Alberti and U. Costantino, in: Intercalates of Zirconium Phosphates and Phosphonates (J.L. Atwood and J.E.D. Davies, eds.) Oxford University Press, 1991, p. 136.

38. U. Costantino, M. Casciola, S. Chieli and A. Peraio, Solid State Ionics 46 (1991) 53.

39. E. Ruiz-Hitzky and P. Aranda, Adv. Mater. 2 (1990) 545.

40. E. Ruiz-Hitzky, Adv. Mater. 5 (1993) 334; and references therein.

41. F. Marlow and J. Caro, Zeolites 12 (1992) 433.

42. a) S.D. Cox, T.E. Gier, G.D. Stucky and J. Bierlein, J. Am. Chem. Soc. 110 (1988) 2986.
b) S.D. Cox, T.E. Gier and G.D. Stucky, Chem. Mater. 2 (1990) 609.

43. S.D. Cox, T.E. Gier and G.D. Stucky, Solid State Ionics 32/33 (1990) 514.

44. L. Werner, J. Caro, G. Finger and J. Kornatowski, Zeolites 12 (1992) 658.

45. J. Caro, G. Finger, J. Kornatowski, J. Richter-Mendau, L. Werner and B. Zibrowius, Adv. Mater. 4 (1992) 273.

46. F. Marlow, J. Caro, L. Werner, J. Kornatowski and S. Dähne, J. Phys. Chem. 97 (1993) 1286.

47. F. Marlow, W. Hill, J. Caro and G. Finger, J. Raman Spectrosc. 24 (1993) 603.

48. F. Marlow, M. Wübbenhorst and J. Caro, J. Phys. Chem., submitted.

49. P. Enzel and T. Bein, J. Phys. Chem. 93 (1989) 6270.

50. P. Enzel and T. Bein, J. Chem. Soc., Chem. Commun. (1989) 1326.

51. P. Enzel and T. Bein, Chem. Mater. 4 (1992) 819.

52. C.T. Kresge, M.E. Leonowicz, W.J. Roth, J.C. Vartuli and J.S. Beck, Nature 359 (1992) 10.

53. K. Horie and A. Furusawa, in: Progress in Photochemistry and Photophysics (J.F. Rabek, ed.), CRC Press, Boca Raton 1990, vol. V, chapter 2; and references therein.

54. a) . Kharlamov, R.I. Personov and I.A. Byskovskaya, Opt. Commun. 12 (1974) 191.
b) A.A. Gorokhovskii, R.K. Kaarli and L. Rebane, JETP Lett. 20 (1974) 216.

55. G. Castro, D. Haarer, R.M. MacFarlane and H.P. Trommsdorff, U.S. Pat. 410 1976 (1978).

56. J. Friedrich, H. Scheer, B. Zickendraht-Wendelstadt and D. Haarer, J. Chem. Phys. 74 (1981) 2260.

174

57.   W.E. Moerner, Ed., Persistent Spectral Hole-Burning: Science and Applications, Springer, Berlin 1988.

58.   S. Voelker and R.M. MacFarlane, IBM J. Res. Dev. 23 (1979) 547.

59.   R. Hoppe, G. Schulz-Ekloff, D. Wöhrle, M. Ehrl and C. Bräuchle, Stud. Surf. Sci. Catal. 69 (1991) 199.

60.   R. Hoppe, G. Schulz-Ekloff, D. Wöhrle, C. Kirschhock and H. Fuess, Stud. Surf. Sci. Catal. in press.

61.   G. Calzaferri and N. Gfeller, J. Phys. Chem. 96 (1992) 3428.

62   R. Hoppe, Dissertation, Bremen 1992.

63.   B.V. Romanovsky, Proc. 8th Intern. Congress on Catalysis (DECHEMA, ed.), Verlag Chemie, Weinheim 1984, vol. IV, p. 657.

64.   G. Meyer, D. Wöhrle, M. Mohl and G. Schulz-Ekloff, Zeolites 4 (1984) 30.

65.   R.F. Parton, L. Uytterhoeven and P.A. Jacobs, Stud. Surf. Sci. Catal. 59 (1991) 395.

66.   D.E. DeVos, F. Thibault-Starzyk, P.P. Knops-Gerrits, R.F. Parton and P.A. Jacobs, Makromol. Chem. (1994), in press.

67.   T.V. Whittam, Brit. Pat. 1453 115 (1974).

68.   M.J. Desmond, F.A. Pesa and J.K. Currie, US Pat. 4582693 (1986).

69.   S. Kowalak and K.J. Balkus Jr., Collect. Czech. Chem. Commun. 57 (1992) 774.

70.   S. Wohlrab, R. Hoppe, G. Schulz-Ekloff and D. Wöhrle, Zeolites 12 (1992) 862.

71.   K.J. Balkus Jr., C.D. Hargis and S. Kowalak, ACS Symp. Ser. 499 (1992) 347.

72.   O. Franke, A. Sobbi, G. Schulz-Ekloff and D. Wöhrle, submitted.

73.   F.C. Schaefer and W.D. Zimmermann, Nature 220 (1968) 66.

74.   R. Hoppe, G. Schulz-Ekloff, D. Wöhrle, E.S. Shpiro and O.P. Tkachenko, Zeolites 13 (1993) 222.

75.   R. Hoppe, G. Schulz-Ekloff, D. Wöhrle, C. Kirschhock and H. Fuess, Langmuir, accepted.

76.   W.M. Meier and D.H. Olson, Atlas of Zeolite Structure Types, Butterworths, London 1988.

77.   S.T. Wilson, B.M. Lok, C.A. Messina, T.R. Cannan and E.M. Flanigan, J. Am. Chem. Soc. 104 (1982) 1146.

78.   M. Ehrl, F.W. Deeg, C. Bräuchle, O. Franke, A. Sobbi, G. Schulz-Ekloff and D. Wöhrle, J. Phys. Chem. 98 (1994) 47.

79.   D. Schulte-Frohlinde, H. Herrmann and G.M. Wyman, Z. Phys. Chem. (Frankfurt) 101 (1976) 115.

80.   T. Karstens, K. Kobs and R. Memming, Ber. Bunsenges. Phys. Chem. 83 (1979) 504.

81.   A. Corval and H.P. Trommsdorf, J. Phys. Chem. 91 (1987) 1317.

82.   F.W. Deeg, M. Ehrl, C. Bräuchle, R. Hoppe, G. Schulz-Ekloff and D. Wöhrle, J. Luminescence 53 (1992) 219.

83.   I.J. Lee, J.M. Hayes and G.J. Small, J. Chem. Phys. 91 (1989) 3463.

84.   L. Shu and G. J. Small, J. Opt. Soc. Am. B 9 (1992) 724.

85. R. Ao, S. Jahn, M. Kümmel, R. Weiner and D. Haarer, Jpn. J. Appl. Phys. 26 (1992) 227.
86. R.M. MacFarlane, J. Luminescence 38 (1987) 20.
87. S. Machida, K. Horie and T. Yamashita, Appl. Phys. Lett. 60 (1992) 286.
88. H.W.H. Lee, M. Gehrtz, E.E. Marinero and W.E. Moerner, Chem. Phys. Lett. 118 (1985) 611.
89. T.P. Carter, C. Bräuchle, V.Y. Lee, M. Manavi and W.E. Moerner, J. Phys. Chem. 91 (1987) 3998.
90. E.I. Alshits, B.M. Kharlamov and R.I. Personov, Opt. Spectrosc. 65 (1988) 173 and 326.
91. T.F. Heinz, C.K. Chen, D. Ricard and Y.R. Shen, Phys. Rev. Lett. 48 (1982) 478.
92. N.E. van Wyck, E.W. Koenig, J.D. Byers and W.M. Hetherington, Chem. Phys. Lett. 122 (1985) 153.
93. L. Werner, W. Hill. F. Marlow, A. Glismann and O. Hertz, Thin Solid Films 205 (1991) 58.
94. D.A. Higgins, S.K. Byerly, M.B. Abrams and R.M. Corn, J. Phys. Chem. 95 (1991) 6984.
95. J.I. Zink, B. Dunn, R.B. Kaner, E.T. Knobbe and J. McKiernan, ACS Symp. Ser. 455 (1991) 541.
96. M. Nakamura, H. Nasu and K. Kamiya, J. Noncrystall. Solids 135 (1991) 1.
97. G.D Stucky and J.E. MacDougall, Science 247 (1990) 669.

J.C. Jansen, M. Stöcker, H.G. Karge and J. Weitkamp (Eds.)
Advanced Zeolite Science and Applications
Studies in Surface Science and Catalysis, Vol. 85
© 1994 Elsevier Science B.V. All rights reserved.

METAL IONS ASSOCIATED TO THE MOLECULAR SIEVE FRAMEWORK: POSSIBLE CATALYTIC OXIDATION SITES

G. Bellussi[a] and M.S. Rigutto[b]

[a]*ENIRICERCHE S.p.A., Via F. Maritano 26, 20097 San Donato, Milano, Italy*
[b]*Delft University of Technology, Julianalaan 136, 2628 BL Delft, The Netherlands*

CONTENTS

1. INTRODUCTION

The terms "isomorphous substitution" and "framework species" are often used to describe the situation of the catalytic agent in certain molecular sieves containing transition metals. Since there are few catalytic reactions which do not involve coordinative interactions between one of the reactants and the catalytic center, the usage of such terms implies that these framework species should have some coordination chemistry. Nevertheless, the role of molecular sieve frameworks in catalytic reactions is often viewed as a static one, which is perhaps not surprising in view of the fact that studies on isomorphous replacements in zeolitic frameworks have traditionally emphasized structural and physical aspects [1,2]. Until the late seventies, exchangeable cations [3] and other extraframework species have been the primary focus of catalyst researchers. Since then, several examples of coordination of framework atoms by sorbed molecules have been reported, such as reversible hydration of framework aluminium in $AlPO_4$'s [4] and zeolites [5] and Lewis acidity of framework aluminium in ZSM-5 [6]. But in fact, the discovery of titanium silicalite – 1 and its wide catalytic potential by Taramasso et al. [7,8] and other ENI workers [9,10] already had triggered interest in this area, since it was soon realized that some rather specific coordination chemistry of lattice titanium ions was responsible for the unique catalytic properties of this material (A detailed treatment of this case is given in chapter 2).

In this paper, we will try to deal with the catalytic properties of isolated transition metal ions connected to the framework of a molecular sieve. More specifically, we will focus on the materials that are active catalysts in oxidation reactions, which include titanium, vanadium and chromium silicalites, as well as cobalt-, vanadium-, and chromium-substituted AlPO's. The metals in this list are known to catalyze a host of different oxidation reactions in which different mechanisms are operating, some with electron transfer steps and autoxidation chains, others with heterolytic or homolytic oxo or peroxo oxygen transfer steps [11]. Although the mechanisms have little in common, they are all likely to involve ligand exchange and non-tetrahedral coordination states, perhaps accompanied by reversible framework bond lysis [12].

When trying to understand what can be specific for catalysis by molecular sieves containing transition metal framework sites, one needs to address the following topics:

i.   The local structure of the site. The case where the metal occupies one of the regular tetrahedral framework sites (and is tetrahedrally coordinated) is called *isomorphous framework substitution*.

ii.   The interaction of the site with reactants, products and solvents: its coordination chemistry. In one extreme view, it is dominated by the requirements of an only slightly flexible molecular sieve framework. In another extreme view, a flexible framework adapts to the coordination chemistry of the ion.

iii. Specific   properties   of   the   surrounding   framework,   such   as   its hydrophilicity/hydrophobicity characteristics, or its shape selective properties.

The first two items describe properties of the site which are expected to depend on the metal in question, so we will discuss them in the appropriate sections; one can mention one property which different framework-connected metal ions are supposed to have in common, which is their monomeric nature. Concerning the last item, it is possible to make a few general remarks before treating specific catalysts.

Shape selectivity effects are well known and often well understood in acid catalysis [13], but clear examples in oxidation reactions are rare. Since oxidations are mostly irreversible, and usually not accompanied by reactions that can effect the interconversion of isomeric products, only *reactant selectivity* or *restricted transition state selectivity* can be of practical importance. Slow diffusion of, for instance, some undesired isomer, will not lead to an improved selectivity, but only to low reaction rates.

Specific sorbent – or solvent – properties of zeolites can be of great significance. The pronounced hydrophobicity of titanium silicalites is undoubtedly one of their key properties, and it allows them to be used as catalysts for selective oxidations with aqueous hydrogen peroxide. Selective sorption by molecular sieves can also give rise to unexpected solvent effects. Although it is a well-established phenomenon [14], we feel that its importance in liquid-phase catalytic reactions is sometimes underestimated.

2. TITANIUM-CONTAINING SILICALITES

Among the metal-containing molecular sieves, titanium-containing silica-based materials are the most intensively studied ones. The reason for this is that one member of this family, titanium silicalite – 1 (usually denoted as TS – 1), selectively catalyzes a broad range of oxidations with hydrogen peroxide [9,10]. Its catalytic properties are of great interest, both commercially and scientifically. Mechanistic aspects of the oxidation reactions catalyzed by TS – 1 will be discussed in section 2.2. Before that is done, a brief review of the literature on the synthesis of titanium-containing silicalites is given (section 2.1). Finally, an effort is made to draw some conclusions concerning the structure of the titanium sites in these materials, and their ability to react with various molecules including hydrogen peroxide (section 2.3).

Most of this chapter is devoted to TS – 1, which has the MFI structure, and its closest relatives, which are also compositional variants of known high-silica zeolites and that contain only a small amount of titanium. As we will see in section 2.3., there is now sufficient evidence for tetrahedral framework siting of titanium in TS – 1, substituting for silicon, but the situation is less clear for other materials. A few titanium-rich molecular sieves are known; they have unique structures composed of infinite chains of titanium octahedra connected to silicon tetrahedra and actually constitute a separate family of materials [15]. We only mention them once here, since they seem to bear little relevance to catalysis. A likely reason for their reported poor catalytic performance in oxidations [16] is the simple fact that framework titanium ions are complete shielded by silicon tetrahedra and hence inaccessible from the pores.

2.1. Synthesis

Tetravalent titanium usually assumes an octahedral coordination, and has a strong tendency to polymerize in aqueous systems, usually resulting in precipitation and eventual formation of one of the very stable forms of titanium dioxide. One could therefore expect the synthesis of materials containing isolated tetrahedral titanium sites to be rather difficult, and this expectation proves to be true: some specific features of the synthesis of TS – 1 are now understood to a certain extent, and it is known to involve a number of conditions which are highly critical. The synthesis of zeolitic materials with other structures (MEL, BEA, ZSM-48), containing titanium in framework positions, has been described, but with the exception of [Ti]-BEA [17], these materials have not yet been deeply investigated.

Synthesis of TS – 1

In the patents of Taramasso et al. [7] two methods are described for synthesizing TS – 1. Both are specific variations on general recipes for the preparation of silicalite – 1, and both use the same template, the tetrapropylammonium cation. The methods differ mainly in the way in which the precursor mixture for crystallization is prepared: following the first method, it is obtained by controlled hydrolysis of tetraethyltitanate and tetraethylsilicate; we shall label this method the "mixed alkoxide" method. Following the second method, the precursor is obtained by mixing colloidal silica and a solution containing tetrapropylammonium hydroxide, hydrolyzed titanium alkoxide, and hydrogen peroxide. (The peroxide forms complexes with titanate species, thus surpressing their tendency to polymerize. Padovan et al. [18] describe an alternative procedure without peroxide.) We shall label this second method the "dissolved titanium" method. The composition of the mixtures and the conditions that can be used to crystallize TS – 1 from them are roughly the same [8].

Many workers have experienced that the preparation of TS – 1 *via* the mixed alkoxide method is highly critical. Van der Pol et al. [19] showed, that deviations from the optimum mixing temperature, the hydrolysis temperature and the speed of hydrolysis easily leads to unwanted precipitate formation in the early stage of gel preparation. It might be of importance to note that the tetrapropylammonium ion seems to act like a phase transfer catalyst during the hydrolysis of the alkoxide mixture, and allows it to proceed smoothly. Less reactive titanium sources, such as the tetrabutoxide [20], have been used in an attempt to make the preparation procedure less critical, but the effects of this approach have been strongly questioned [21]. Kraushaar [22] proposed a simple model, according to which, prior to hydrolysis, all titanium should be present as single titanium alkoxide units in mixed oligomers with the silicon alkoxide. Acceptor properties of the titanium alkoxides towards dioxane [23] suggest that in $Si(OEt)_4$, titanium ethoxide could form species like $Ti(OEt)_4[(\mu^2-OEt)Si(OEt)_3]_2$ or $Ti(OEt)_4(\mu^2-OEt)_2Si(OEt)_2$, under conditions where other alkoxides are either unassociated or form oligomers containing $Ti(\mu-OR)Ti$ units. Unassociated monomers or oligomers would then give rise to aggregated oxidic phases upon hydrolysis. As isolated titanium species are expected to be metastable relative to such phases, the formation of the latter can be considered irreversible an should be prevented.

Another possible pitfall in TS – 1 synthesis is the presence of alkali metal ions in the synthesis mixture. Several authors [9,24,25] have pointed out that a small amount of sodium or potassium, which can for instance originate from a commercial tetrapropyl-ammonium hydroxide solution, suffices to prevent the insertion of titanium into the molecular sieve framework. Instead, titanium species aggregate to form an amorphous phase, and ultimately anatase [25]. The role of alkali metal ions in the synthesis of high-silica zeolites is obscure [26]; they seem to catalyze the equilibration of silicate species, presumably by promoting hydrolysis and recondensation reactions of $Si-O-Si$ bonds, stabilizing terminal $Si-O^-$ groups by complexation [27]. A similar reasoning might apply to $Si-O-Ti$ bonds. A very different effect is caused by the addition of salts of trivalent metals like Al^{3+}, Ga^{3+} or Fe^{3+} to the synthesis mixture, which also surpresses Ti incorporation to a certain extent [25], but does not cause segregation of titanium. As a consequence, it is possible to synthesize doubly substituted analogues of TS – 1 (see Fig. 1), like e.g. [Ti,Al] – MFI [25,28], which can act as bifunctional catalysts. Further study of such competitive crystallizations might give clues about the gel chemistry involved in titanium incorporation.

The use of fluoride ions in TS – 1 synthesis was first reported by Guth and coworkers [29] and Qiu et al. [30]. In an effort to synthesize TS – 1 in the presence of fluoride, Kooyman et al. [31] observed anatase impurities. In addition, lower activities and selectivities were found in phenol hydroxylation and alkane oxygenation, compared to a conventional TS – 1 sample. Recent EXAFS work has confirmed that the use of fluoride leads to the formation of octahedral, extraframework titanium [32].

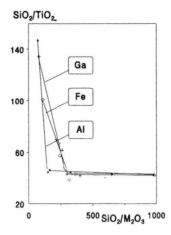

Figure 1. Competition between trivalent ions and titanium in the crystallization of [M^{III},Ti]-MFI [20] from syntheses mixtures with constant Si/Ti ratios and varying Si/M^{III} ratios.

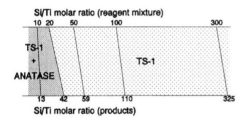

Figure 2. Influence of the gel composition on Ti contents in TS-1 [8].

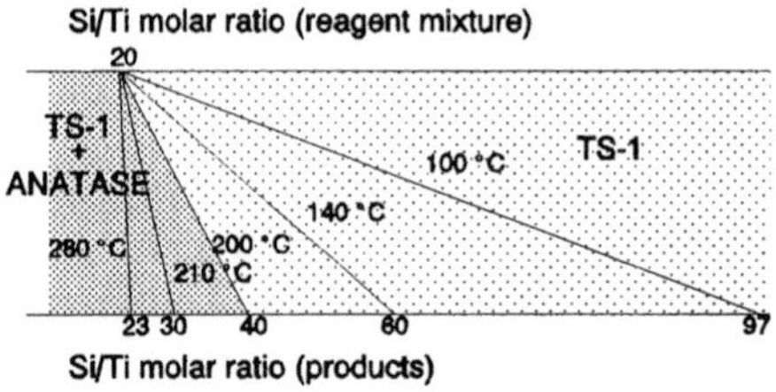

Figure 3. Influence of the crystallization temperature on Ti contents in TS-1 [8].

Crystallization of TS‑1 can be performed in a fairly wide temperature range, as can be seen in Fig. 2. The results shown were obtained with the mixed alkoxide method. The effectiveness of titanium incorporation increases with temperature, up to the point where anatase formation starts to occur, at 200 °C. At a fixed temperature, the titanium content of the crystalline product is always lower than that of the precursor mixture, as is shown in Fig. 3. This means that the solution is enriched in titanium during synthesis, and that unnecessary long crystallization times ultimately lead to the deposition extraframework titanium phases [18].

Perego et al. pointed out [33], that titanium could be incorporated in the MFI framework up to a Ti/(Ti+Si) molar ratio of 0.025. This apparent compositional ceiling is probably not a thermodynamic one and TS‑1 is already a metastable phase, so it might be possible to break through it. For instance, TiO_2–SiO_2 glasses can contain titanium substituting for silicon in tetrahedral sites up to a Ti/(Si+Ti) ratio of 0.07 – 0.09 [34]. In fact, Thangaraj et al. [20a] claimed the preparation of TS–1 with much higher titanium contents using an alternative synthesis method (*vide supra*), and found the upper limit of the Ti/(Ti+Si) ratio to be 0.10. However, Millini et al. [21] observed that the XRD data reported in that paper did not agree with the results of Perego et al. [33], and concluded from a detailed re-investigation by Rietveld analysis that in the materials of Thangaraj, a large fraction of the titanium was present as extraframework material. Further discrepancies in ^{29}Si MAS NMR and IR spectra [20,33] confirm this conclusion. The XRD data of van der Pol et al. [19], who has claimed an upper limit for Ti/(Ti+Si) of 0.04, also disagree with the data of Perego et al [33], although to a lesser extent.

Finally it should be mentioned that Tuel et al. [35] reported the synthesis of TS-1 from alkali-free gels containing TBA/TEA and TPA/TEA mixtures, which indicates that some variation in the template is apparently possible.

Synthesis of TS–2, [Ti]–ZSM–48 and [Ti]–beta

As silicalite–1 and silicalite–2 are structurally similar [36], and as they can be prepared in similar ways using somewhat different templates, it is perhaps not surprising that a titanium-containing equivalent of silicalite–2, denoted as TS–2, can be synthesized [37,38]. Bellussi et al. [37] also reported the synthesis of MFI-MEL structural intergrowths containing titanium, which they labeled TS–3. In fact, materials classified as having the MEL structure are in reality often disordered intergrowths of MEL and MFI [36]. Reddy et al. [38] reported the possibility to synthesize TS–2 with rather high titanium contents using the method of Thangaraj et al. [20] (*vide supra*). The synthesis of TS–2 is not as well studied as that of TS–1, but one can expect it to respond in similar ways to the parameters discussed in the above.

The synthesis of [Ti]–ZSM–48 has recently been reported by Serrano et al. [39]. A variant of the dissolved titanium method, with added hydrogen peroxide, was used to prepare materials with Ti/(Ti+Si) ratios up to 0.037. The synthesis of zeolite beta containing titanium was reported by Camblor et al. [40]. Here, again the dissolved titanium method was used, but without added peroxide. The addition of aluminium to the synthesis mixture was found to be necessary for crystallization of beta, yielding products with Al/(Al+Si) ratios of 0.010 or higher. This apparently does not surpress titanium incorporation as strongly as it does in the case of TS–1 synthesis [25].

The synthesis of a titanium-containing porous silica resembling MCM-41 was very recently described by Corma et al [41]. X-Ray powder data suggest that the material is a lamellar phase rather than a hexagonal phase [42]. A more thorough characterization of this material is required.

Conclusive remarks

Although the synthesis of TS–1 has been studied rather intensively in the past decade, many aspects of the specific titanium chemistry involved are still unclear. The stabilizing interactions that allow the formation of neutral, tetrahedral titanium species –even in the presence of water– are worth identifying in their own right, but they might give some important clues about the catalytic properties of the tetrahedral complex as well. The deleterious effect on titanium incorporation of mobilizing agents or conditions, such as fluoride ions, alkali metal cations or high temperatures, and the competition by trivalent metals, need a better understanding. This would be of great help in the development of synthesis methods for other titanium-substituted materials.

2.2. Catalytic reactions

The scope of TS–1 catalyzed reactions and the advantages of TS–1 over other catalysts have been reviewed extensively [9,10,33,43], so we will limit our treatment to the data and trends that are of interest for mechanistic considerations. In all reactions mentioned below, hydrogen peroxide was used as the oxidant.

Hydroxylation of aromatics
The hydroxylation of phenol to catechol and hydroquinone is currently the most important application of TS – 1, and a number of studies has been devoted to this reaction [9,10,24,31,43-46]. It is often used as a test reaction, because of its sensitivity to the quality of the catalyst: the presence of extraframework titanium significantly lowers the reaction selectivity by catalyzing the decomposition of hydrogen peroxide [9] and the formation of quinones and coupling products [47]. Acetone is the solvent of choice, but methanol or water give similar results. The use of acetonitrile gives somewhat lower rates, and concomitantly hydrogen peroxide decomposition becomes a relatively important side reaction [48]. The reaction is probably diffusion-limited as observed rates strongly depend on crystal size [46]. The selectivity for *para*-hydroxylation is higher with TS – 1 than with Brønsted acids or Fenton-type catalysts [9] which is commonly interpreted as a shape selectivity effect. It also indicates that, for the TS – 1 catalyzed reaction, coordination of phenol to the site is not part of the reaction sequence, in contrast to some other cases [49,50]. This in turn might be one of the reasons why further oxidation of dihydroxybenzenes to quinones is relatively slow on TS – 1.

TS – 1 also catalyzes the hydroxylation of benzene itself and other aromatics that are more electron-rich than benzene [33,43,51]. Chlorobenzene, nitrobenzene, benzoate esters and other substrates bearing electron-withdrawing substituents do not react. Shape selectivity manifests itself in the fact that substrates carrying an otherwise activating bulky alkyl group, e.g. *i*-propyl- or *t*-butylbenzene, are also unreactive. Only a few other cases of aromatics hydroxylation involving group 4 - 6 metals and hydroperoxides have been reported: relevant examples are the electrophilic, heterolytic hydroxylation of naphtols by $Ti(O^iPr)_4$/*t*-butylhydroperoxide or $[MoO(O_2)_2] \cdot py \cdot HMPT$ [52], and the hydroxylation of benzene and benzene derivatives by the vanadium peroxo picolinate complex $VO(O_2)(Pic) \cdot 2H_2O$ [53]. According to Mimoun et al. and Bonchio et al. [53] oxidations by the latter reagent involve homolysis of the $O-O$ bond and a radical chain pathway. Many other examples of the reaction involving homolytic [49] or heterolytic [50] oxygen transfer are known. For the case of TS – 1, both pathways can be envisaged. Reactivity trends and isomer ratios could lead to the conclusion that the reaction involves an heterolytic, electrophilic attack on the aromatic ring [49,50,53], but similar results could also be obtained through a pathway involving an electrophilic radical attack. On the other hand, benzylic oxidation, usually a homolytic reaction involving radicals, prevails over hydroxylation of the nucleus when ethylbenzene is the substrate [43,51]. This seems to imply that the same species can perform (or initiate) homolytic as well as heterolytic

184

oxygen transfer. C–H bond activation at saturated carbon atoms is discussed further on in this section.

Epoxidation

TS–1 is a particularly active catalyst for the epoxidation of alkenes [43], and allows this reaction to be carried out at temperatures as low as 0 °C [54,55]. Even electron-poor alkenes, such as allyl chloride and allyl ethers, are epoxidized with high yields and selectivities. Epoxidation occurs with retention of configuration [56], evidencing a heterolytic mechanism. Relative reactivities among alkenes depend mainly on the nucleophilicity of the double bond, but steric factors sometimes dominate [54], as in the case of cyclohexene, which reacts two orders of magnitude slower than 1-hexene [43]. Relative reactivities of some lower alkenes are shown in Fig. 4. In contrast, homogeneously catalyzed epoxidation rates are mainly governed by electronic factors, while steric factors only have a minor influence [57]. Restricted transition state selectivity as well as differences in sorption strength and diffusion rates are thought to cause these differences. The similar reactivities of allyl alcohol and allyl ethers and esters observed with TS–1 [43], indicates that the mechanism does not involve substrate coordination [58]. It would be interesting to confirm this by measuring *threo/erythro*-selectivities for suitable substrates.

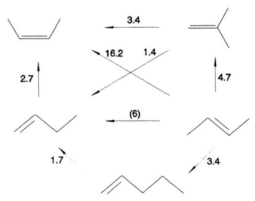

Figure 4. Relative reactivities of lower alkenes in epoxidation catalyzed by TS-1 [54]. The arrows point to the more reactive alkene.

Allylic oxidation is usually not observed [43], although recent work by Tatsumi et al. [59] on the oxidation of allylic alcohols shows that it can sometimes compete with epoxidation, especially for the bulkier substrates. The authors attributed this effect to a lack of reactivity of the double bond due to its decreased ability to approach the titanium site. Alkene coordination was invoked by Huybrechts et al. [94] to explain the inhibition of *n*-hexane oxidation by *n*-octene. This view was countered by Clerici et al. [54] who did not observe such inhibition with a pure catalyst.

With TS–1, epoxidation rates depend on the solvent used [54,55] in a way that is different from the solvent effects observed for homogeneously catalyzed epoxidations [60].

The use of methanol or ethanol gives higher rates than the use of aprotic solvents: e.g., in methanol, which is the solvent of choice, the rate is an order of magnitude higher than in acetonitrile. The addition of fluorides or alkali metal acetates inhibits the reaction [54]. Clerici et al. [54] and Bellussi et al. [61] proposed a mechanism that provides an explanation for both the solvent and the inhibitory effects. This mechanism contains a proposal for the active complex and will therefore be discussed in Section 2.3.

Recently Camblor et al. [40] reported the use of [Ti,Al]−beta as an epoxidation catalyst. The reportedly unavoidable presence of aluminium in these materials causes almost complete solvolysis of the initially formed epoxides. The authors showed that turnover frequencies in the epoxidation of 1-hexene were an order of magnitude lower for [Ti,Al]−beta than for TS−1, which was interpreted to be an intrinsic difference in site activities [62]. Identification of the cause of such a difference is important, as it could shed light on the origin of the high activity of the TS−1 site and on possible structure-acitivity relationships.

Ammoximation of cyclohexanone and related reactions

Another reaction catalyzed by TS−1 is the ammoximation of ketones. The application to cyclohexanone is commercially important: its reaction with ammonia and hydrogen peroxide in *t*-butanol gives cyclohexanone oxime with high selectivity (>95 % at 80 - 90 % conversion) [63]. Amorphous titanium-on-silica with a high titanium loading (9.8 wt %) [64] and phosphotungstic acid [65] also catalyze this reaction, but they do so with somewhat lower selectivities and less efficient use of hydrogen peroxide. Two different reaction sequences have been proposed (see Fig. 5). The first one involves (uncatalyzed) intermediate formation of the imine, followed by an oxygenation step: it was proposed by Thangaraj et al. [66] and Reddy et al. [67] to explain the by-products formed in the ammoximation reaction. Interaction of cyclohexanone with TS−1 and formation of imine upon addition of ammonia, as shown by infrared spectroscopy, was interpreted by Tvaruzkova et al. [68] as evidence for this reaction sequence. In addition, imine oxygenation by hydrogen peroxide is a known reaction and is catalyzed by molybdenum and tungsten complexes [11]. However, Zecchina et al. [69] made the following observations which disfavour this first sequence: (i) bulky substrates that are not sorbed by the catalyst, like cyclododecanone and 4-*t*-butylcyclohexanone, are nevertheless converted, reacting at relative rates equal to those observed for oximation with hydroxylamine; (ii) TS−1 catalyzes the formation of hydroxylamine from ammonia with good selectivity; (iii) UV-Vis spectra of TS-1 contacted with H_2O_2/NH_3 indicate that hydroxylamine is formed on titanium sites. The second reaction sequence, which involves catalytic formation of hydroxylamine (see Fig. 5), is therefore more likely.

TS−1 also catalyzes the oxidation of dialkylamines to the corresponding hydroxylamines [70], and of sulfides to sulfoxides [71]. The homogeneous analogues of both reactions are well studied, and are known to involve a nucleophilic attack of the substrate on a peroxo oxygen, followed by O−O bond heterolysis [72].

Oxidation of alcohols

TS−1 catalyzes the oxidation of both primary and secondary alcohols at moderate temperatures [43,73]. With primary alcohols, high selectivities to aldehyde can be obtained at < 30 % conversion of the alcohol. Most secondary alcohols react much faster than the

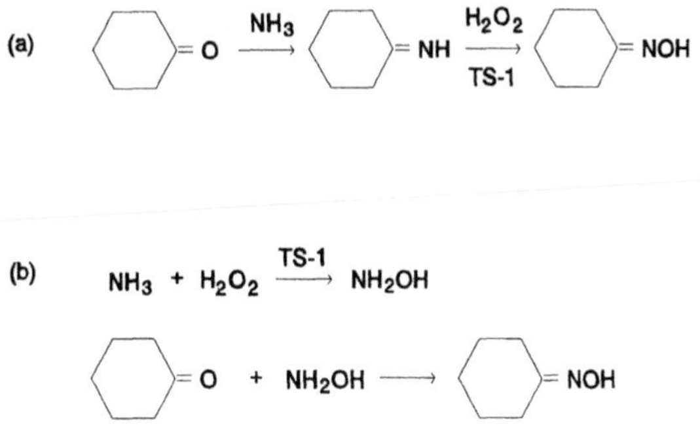

Figure 5. Possible reaction pathways for the ammoximation of cyclohexanone.

primary alcohols. Shape-selective effects are also observed in this reaction: a remarkable one is the difference in reactivity between 2-pentanol and 3-pentanol, the latter reacting approximately ten times slower. Some acid catalysis occurs and leads to the formation of side products (acetals, esters). This reaction has not received as much attention as some of the other reactions catalyzed by TS–1, and little is known about its mechanism. A radical process can be envisaged as well as a two-electron process; the relatively high selectivities to aldehydes favour the latter. It is interesting to note that alcohol oxidation by neutral or anionic tungsten or molybdenum peroxo complexes is probably a two-electron process [74,75]. The molybdenum picolinate n-oxido complex $MoO(O_2)_2(pico)^-$ is especially selective to aldehydes [74]. It is unclear to which extent heterolytic alcohol oxidation by peroxo complexes can still be considered an "electrophilic" reaction [76]; it is commonly said that less electrophilic reagents are needed to effect alcohol oxidation compared to alkene epoxidation.

Oxygenation of alkanes

Examples of alkane oxygenation where good selectivities to alcohols and ketones are obtained at appreciable conversions of the substrate are rare. Huybrechts [77] and Clerici [78] independently discovered the TS–1 catalyzed oxidation of alkanes to secondary alcohols and ketones. Under the somewhat different conditions used, somewhat different results were obtained; a comparison is made in Table 1. The most important difference is the fact that Huybrechts et al. [77] observed a statistical distribution of oxygenate isomers, whereas Clerici [78] found a preference for the 2-position, which he interpreted as a steric effect. It is not unlikely that under the severe conditions used by Huybrechts, the participation of radicals and molecular oxygen in the process becomes more important.

Table 1.   *n*-Hexane oxidation catalyzed by TS-1.

Reference	[77]	[78]
Solvent	Acetone (two-phase)	Methanol (one-phase)
Temperature	100 °C	55 °C
H_2O_2 conversion	> 90 %	98 %
Sel. based on converted H_2O_2	70 %	91 %
(2-ol+2-one)/(3-ol+3one)	1.1	2.3

Clerici observed that in methanol at 55 °C, addition of 2,6-di-*t*-butyl-4-methylphenol, carbon tetrachloride, chloroform or dichloromethane did not influence reaction rates or product distributions. Free-radical *chain* mechanisms can therefore be considered less likely under these mild conditions. Evidence for the electrophilic nature of the process was obtained from the oxidation of 1-chlorohexane and methyl heptanoate: in both cases oxygenation occured preferentially on the remote secondary C−H bonds [77].

The fact that TS−1 is able to oxidize alkanes in the presence of an oxidizable alcohol such as methanol is remarkable. It could be due to an intrinsic property of the active complex, or perhaps to selective sorption properties of the silicalite [14,43]. Solvent effects similar to those observed for TS−1 catalyzed epoxidation are also found in this reaction, although there are a few differences. Methanol gives the best selectivities, but in *t*-butanol the reaction is faster and selectivities are only slightly lower. On the other hand, epoxidation in *t*-butanol is much slower than in methanol [54]. This difference has been explained in terms of steric hindrance of the site active in olefin epoxidation, formed upon the interaction of a titanium hydroperoxo complex with a solvent molecule [54]. Another common feature of epoxidation and alkane oxygenation is the fact that they are both inhibited by fluorides and alkali metal salts, especially the acetates [78].

It is as to yet unclear what the mechanism of the oxygen insertion reaction is, and whether it consists of consecutive one-electron steps or a single two-electron step. A vanadium peroxo complex is known that can effect alkane oxygenation, but it is different from TS−1 in the sense that it also oxygenates primary C−H bonds and operates *via* a radical chain mechanism [74]. A few strongly electrophilic species, such as ozone or the $H_3O_2^+$ ion in superacidic media [79], can effect heterolytic oxygen insertion into C−H bonds. Such reactions are supposed to involve formation of a three-center two-electron bond by attack on a σ-bond ("oxenoid" insertion). Reactions involving a hydrogen abstraction step followed by a fast recombination of radicals in a so called *solvent cage* are also known [80]. Such solvent cage reactions are also not affected by radical chain inhibitors or chloroalkanes. Clerici [78] prefers a solvent cage mechanism on the basis of the large isotope effect observed for TS−1 ($k_H/k_D = 4.1$).

2.3. Structure and reactivity of the titanium site

Influences on the surrounding lattice

The original identification of titanium in TS–1 as a true tetrahedral framework atom had been based on indirect evidence, obtained from infrared spectroscopy, X-ray powder diffraction, EDX microprobe analysis and ^{29}Si MAS NMR measurements [33]. It was shown that the unit cell volume of calcined TS–1 increases linearly with titanium content. Provided that the neither the site nor the lattice are strongly distorted, the unit cell volume V can be expected to depend on the degree of substitution x = Ti/(Ti+Si) following the equation:

$$V(x) = V(0) - V(0) \cdot [1 - (d_{Ti-O}/d_{Si-O})^3] \cdot x \qquad \text{(Eq. 1)}$$

Where V(x) is the observed unit cell volume, V(0) the unit cell volume of pure silicalite and d is an average bond length. This is known as *Vegard's law* [81], and it is usually obeyed when substitution is both isomorphous and isovalent. Fitting of X-ray data of TS–1 with Eq. 1 (assuming a value of 1.59 Å for d_{Si-O} [82]) gives a titanium-oxygen bond length of 1.78 Å, which closely agrees with bond lengths found in other compounds containing tetrahedral titanium in oxidic environments [83]. Linear dependence of unit cell expansion on titanium content was also observed for other titanium-containing materials. In an effort to treat the phenomenon in a quantitative fashion, we have summarized some of the data from literature in Table 2. Data for [Ti]-beta is omitted since this material contains aluminium, which complicates the interpretation.

Table 2. Summary of XRD data for titanium-substituted materials.

Material [ref]	R[a]	hypothetical d_{Ti-O} (Å)
TS–1 [33]	0.395	1.78
TS–1 [20a]	0.113	1.65
TS-2 [20b]	0.308	1.74
[Ti]-ZSM-48 [39]	0.230	1.70
idem	0.337[b]	1.75[b]

[a]R according to $V(x) = V(0)(1+Rx)$, $R = (d_{Ti-O}/d_{Si-O})^3 - 1$; $d_{Si-O} = 1.59$ Å (for symbols, see Eq. 1).
[b]Values obtained when the point corresponding to the highest Ti content (x=0.037) is ignored (ref. 39).

The shortest known average Ti–O bond length for titanium in a tetrahedral environment is 1.764 Å [83]. Apart from the first entry, the bond lengths calculated from XRD data therefore seem to be out of range. If Eq. 1 is valid for all the above materials, one possible way to explain the data is to assume that these materials contain some extraframework titanium [21]. In any case, it seems advisable to excercise some caution in interpreting unit cell expansion data, and to aquire independent evidence to ascertain the purity of new molecular sieves materials containing titanium.

An observation which is difficult to understand is the decrease in the temperature of transition to orthorhombic symmetry that is caused by incorporation of titanium [86]. Such a decrease has also been observed for substitution of silicon by boron and aluminium [87], whereas substitution by germanium effects a strong *increase* [88]. Little can be said about the cause of these phenomena.

Small distortions of the lattice effected by the presence of framework titanium sites are also apparent from ^{29}Si NMR spectra of a series of TS–1 samples: the well-resolved features of the silicalite spectrum already become blurred at low titanium contents [33].

Direct methods for spectroscopic probing of the site geometry

The most easily accessible method for direct probing of titanium(IV) environments in solids is diffuse reflectance UV-Vis spectroscopy, although charge transfer spectra are obtained, which are somewhat difficult to interpret. Pure, as-synthesized as well as calcined, dehydrated samples of TS–1 show a single absorption maximum around 48,000 cm^{-1}. By comparison of spectra with those obtained from model compounds, Boccuti et al. [84] were able to assign the absorption to tetrahedral titanium(IV) species. In addition, the use of UV-Vis spectroscopy was shown to allow the detection of octahedral (supposedly extraframework) titanium impurities with good sensitivity [85].

Recent EXAFS and XANES measurements have confirmed the tetrahedral geometry of titanium sites in TS–1 [32,83,89]. The strongest evidence comes from the average Ti–O bond lengths, which were determined to be 1.83 ± 0.02 Å [32], 1.80 ± 0.01 Å [83] and 1.81 ± 0.01 Å [89], respectively (Note that these values are consistent with the above interpretation of the XRD data). In addition, Pei et al. [83] were able to conclude from their data that a second shell of oxygen and/or silicon atoms was present within a radial distance range of 3.1 - 3.8 Å, which is consistent with the environment of a framework site. Previous EXAFS and XANES data indicated the presence of multiple, five- and six-coordinated species in TS–1 [90]. Pei et al. [83] and Bordiga et al. [89] showed that this could have been due to the presence of extraframework material, while site hydration might also have played a role [32,89]. (We will discuss site hydration separately).

Trong On et al. [91] also found a tetrahedral site in TS–1 using XANES and EXAFS, but they proposed a structure with tetrahedra sharing edges rather than corners, based on the presence of a maximum at 2.1 Å in the Fourier-transformed data. The authors explained this maximum by assuming a second coordination shell of two silicon atoms at 2.1 Å. However, Pei et al. [83] showed that the maximum is an artefact. In addition, recent calculations by Jentys et al. [92] indicate that an edge-sharing tetrahedral species is not stable.

The 960 cm^{-1} band and its interpretation

As-synthesized and calcined, dehydrated samples of TS–1 show a characteristic band at approximately 960 cm^{-1} in their infrared [33] and Raman spectra [93]. The infrared feature is shared by TS-2, [Ti]-ZSM-48 and [Ti]-beta (For these materials, no Raman data in the 800 - 1100 cm^{-1} range have been published). The interpretation of this band has been a matter of debate: originally, it had been attributed to a stretching vibration of the silicon-oxygen bond in Si–OTi bridges [33,84], but later titanyl groups (ν Ti=O) [9,94], and silanol groups (ν Si–O) [95], which are expected to resonate in the same frequency region, have also been invoked. In addition to what has been mentioned earlier in this section, there is substantial evidence for the original assignment which can be summarized as follows:

- The position of the 960 cm^{-1} band is not changed by treatments with D_2O [61]. In contrast, a clear secondary isotope effect was found for Si–O vibrations in silanols on silica [96].
- A titanium-oxygen double bond would give rise to a charge transfer band in the visible region (25,000 - 30,000 cm^{-1}): such a band is not observed. Instead, UV-Vis absorption can be adequately explained as originating from tetrahedral titanium [84].
- EXAFS results show that the range of titanium-oxygen bond lenghts is very narrow [83], which renders the presence of a titanium-oxygen double bond short enough to resonate at 960 cm^{-1} unlikely.
- No exchange with $^{18}O_2$ occurs, even at elevated temperatures [84].

A detailed treatment of the Si–OTi vibrational mode and of some of the arguments listed here was given by Boccuti et al. [84].

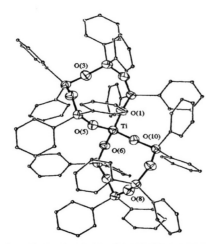

Figure 6. The tetrahedral titanium site in the siloxide Ti[O(OSiPh$_2$)$_4$]$_2$ [97].

In summary, we think that for TS–1, the model according to which titanium is occupying tetrahedral framework positions, substituting for silicon, is the only model that can consistently explain all the present data. We could expect the site to look very much like the central tetrahedral titanium atom in the siloxide whose structure is displayed in Fig. 6 [97]. The question then remains to be answered: what can be the reactivity of such a tetrahedral framework atom, and what are the interactions that give it its catalytic properties ?

On the reactivity of the titanium site

Boccuti et al. [84] studied the interaction of protic molecules with TS – 1 using infrared and DR UV-Vis spectroscopy. Adsorption of water, ammonia or methanol was shown to broaden and shift the 960 cm^{-1} band somewhat and to lower its intensity. Adsorbed water also causes a shift of the UV-Vis adsorption maximum from 48,000 to 42,000 cm^{-1} which indicates an increase in the coordination number of titanium. Absorbed ammonia causes an additional resonance at 38,000 cm^{-1}, which is ascribed to a NH$_3$ → Ti(IV) ligand to metal charge transfer [89]. The changes in infrared and UV-Vis spectra are reversed on outgassing. When TS – 1 is contacted with labeled water and subsequently degassed, a shift of the 960 cm^{-1} band to 952 cm^{-1} (H$_2$^{17}O) or 937 cm^{-1} (H$_2$^{18}O) is observed [61]. In addition, H$_2$^{17}O treatment of TS – 1 causes the appearance of a narrow signal at 360 ppm in its ^{17}O MAS NMR spectrum. In view of the observed shift of the 960 cm^{-1} band, it is likely that this signal arises from a Si – ^{17}O – Ti bridging oxygen. Taken together, these results show that expansion of the titanium coordination shell is accompanied to some degree by hydrolysis and recondensation reactions of Si – O – Ti bonds (see Fig. 8).

The use of XANES and EXAFS spectroscopy [89] allowed detailed analysis of the interaction between the titanium site and adsorbed ammonia: the coordination number of titanium was shown to increase from four to six, while the average titanium-ligand bond length increased from 1.81 Å to 1.84 Å, accompanied by an increase in bond length variation. Outgassing at 400 K largely reversed these changes. Similar results were observed by Lopez et al. [32] for water adsorption on TS – 1 using XANES, EXAFS, ^{17}O MAS NMR and ^{47,49}Ti NMR. ^{47,49}Ti NMR signals were only observed for hydrated samples, which might have been due to the averaging effect of water exchange.

The above results can be summarized in the following reaction scheme:

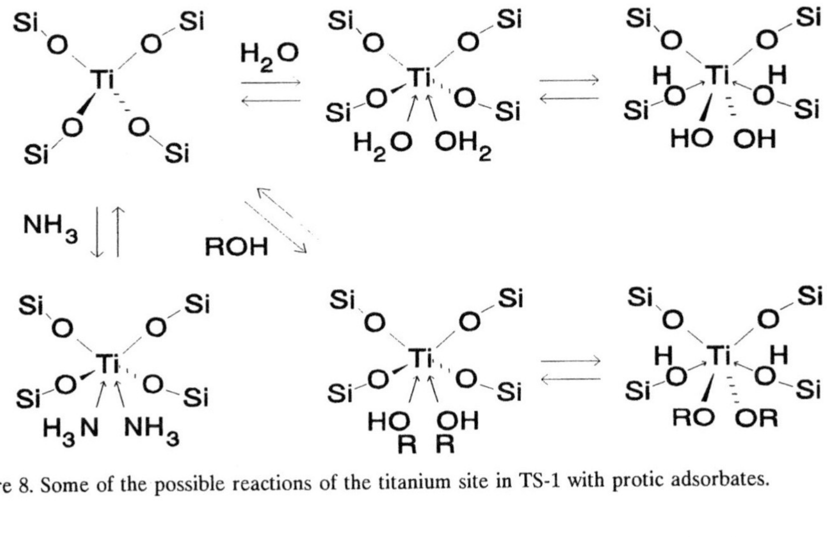

Figure 8. Some of the possible reactions of the titanium site in TS-1 with protic adsorbates.

A remarkable aspect of the site's coordination chemistry, and an important one in terms of catalysis, is its resistance to more extensive hydrolysis. In contrast, isolated titanium species on amorphous silica are hydrolized by ambient moisture [93]. Tetrakissiloxides of titanium are particularly stable to hydrolysis [98], which is not merely a kinetic effect. For example, the compound depicted in Fig. 6 is synthesized from $Ti(OR)_4$ and $Ph_2Si(OH)_2$ with simultaneous formation of water; its stability is probably due to $O \rightarrow Ti$ π-bonding [97]. The pronounced hydrophobic character of TS–1 should provide additional kinetic stability. It is reasonable to assume that the primary driving force for hydrolysis of the titanium site is the expansion of its coordination shell (rather than the formation of Ti–OH bonds), and that further hydrolysis cannot take place unless Ti–O–Ti bridges are formed.

Of more direct interest with respect to catalysis is the interaction of the site with hydrogen peroxide. A limitation in studying surface (hydro-) peroxo complexes of TS-1 is their low stability at ordinary temperatures, which makes them less suitable for X-ray spectroscopy or solid state NMR studies. Using DR UV-Vis spectroscopy, Geobaldo et al. [99] observed a new charge transfer band at 26,000 cm^{-1} when TS–1 was contacted with aqueous hydrogen peroxide. A CT band at 26,000 cm^{-1} with the same shape was also observed for the $[TiF_5(O_2)]^{3-}$ ion, and it was assigned to a $O_2^{2-} \rightarrow Ti^{4+}$ ligand to metal transfer. Huybrechts [24] observed a broadening and shift of the infrared 960 cm^{-1} band (*vide supra*) and the simultaneous appearance of a weak band at 880 cm^{-1} upon contacting TS–1 with hydrogen peroxide. These results allow one to conclude that hydrogen peroxide is directly coordinating to the titanium site forming a peroxo or hydroperoxo surface complex which is probably octahedral. The binding mode of the (hydro-) peroxo anion is less obvious. In the $[TiF_5(O_2)]^{3-}$ ion, as in most stabe titanium peroxo complexes, the peroxo species is bonded side-on, and the spectral similarities suggest side-on bonding for the case of TS–1 as well. However, the TS–1 complex is a reactive species, and it might have a geometry which is different from known titanium peroxo compounds, which are rather inert [100]. The bonding mode will be discussed in more detail later in this section.

When hydrogen peroxide is present, TS–1 acts as a Brønsted acid, catalyzing the hydrolysis or alcoholysis of epoxides [61]. This finding suggests that the complex contains a hydroperoxo rather than a peroxo ligand [101], or alternatively that the presence of a peroxo ligand enhances the acidity of another protic ligand [102]. Moreover, the rate of these acid-catalyzed reactions is solvent dependent (in much the same way as the epoxidation reaction, see Section 2.2), which suggests involvement of a solvent molecule in the complex. In a following study, Clerici et al. [103] were able to obtain stable alkali metal containing peroxo complexes of TS–1 by treatment with alkaline hydrogen peroxide solutions (these can be considered the ion-exchanged forms of the above Brønsted-acidic complex). The complexes show an additional charge transfer band at 33,000 cm^{-1} in the UV-Vis spectrum, and a weak infrared band at 866 cm^{-1}. They do not act as oxidizing agents unless they are reacidified. The authors concluded that the complex probably contained a peroxo ion bonded side-on to a framework titanium species (see Fig. 9).

The above findings and considerations led to the following reaction scheme:

The acidic nature of the TS–1/H_2O_2 is probably due to stabilization of the conjugate anion which can form a five membered cyclic structure, i.e. structures (**2**) and (**4**) in the above scheme.

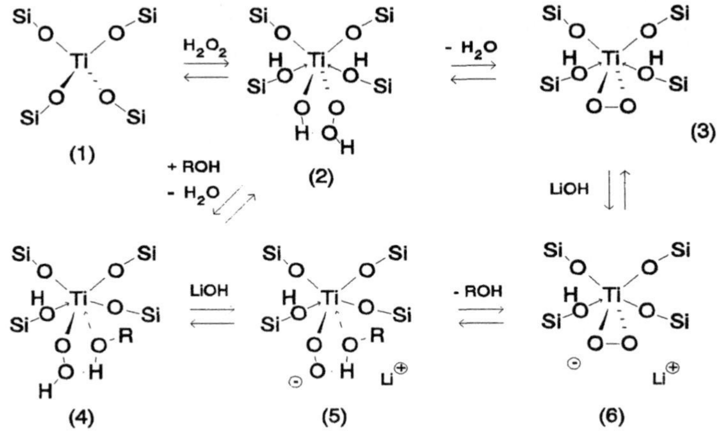

Figure 9. The interactions of a titanium site in TS-1 with hydrogen peroxide.

About the acitive complex, and the mechanism of electrophilic oxygenations catalyzed by TS−1

Discussions on mechanism in heterogeneous catalysis often rely on analogies with homogeneous systems, which are usually easier to characterize, and allow one to do kinetic studies in a straightforward way. In the case of TS−1, it is difficult to find suitable homogeneous analogues, since soluble titanium complexes do not catalyze oxidations with hydrogen peroxide. The related titanium-on-silica system, which catalyzes the epoxidation of alkenes with alkyl hydroperoxides [104], does not offer many clues, since it has hardly been studied in any detail.

To simplify the discussion, let us focus on the TS − 1 catalyzed epoxidation reaction, which is in several aspects similar to its homogeneously catalyzed counterpart (see Section 2.2). Looking for clues, one is confronted with the many unsolved problems that even surround the homogeneous case [60,105,106]. Nevertheless, consensus exists on the following points that can be relevant to our case:

i.   Complexes that catalyze the heterolytic epoxidation of simple alkenes with hydrogen peroxide or alkyl hydroperoxides are almost always Lewis acids.

ii.   Complexes containing a transition metal with an empty d-shell (e.g. Mo(VI), Ti(IV), Re(VII) in CH_3ReO_3) are much more effective than other Lewis acids.

iii. In all known peroxo complexes of the group 4 - 7 transition metals, the peroxo moiety is bonded side-on. An explanation for *ii.* has been based on a frontier-orbital model of the reaction between a peroxo complex and an alkene [106]. According to this model, smooth oxygen transfer is possible because a simultaneous 1,2-migration of the metal occurs (see Fig. 10), so that disruption of metal-oxygen bonding is avoided.

iv.   Three-membered rings (Fig. 10) are not the only structures from which oxygen can be transferred smoothly: organic peracids epoxidize *via* a somewhat different mechanism involving a five-membered ring (see Fig. 11). Note that also in this case, no bonds are disrupted during the transfer reaction.

v. Metal-catalyzed epoxidation and epoxidation by peracids experience similar solvent effects: polar solvents enhance the reaction rate, except for strongly coordinating or protic solvents which slow down or completely inhibit the reaction. With the exception of TS – 1, this also applies to homogeneous and heterogeneous catalysis by titanium.

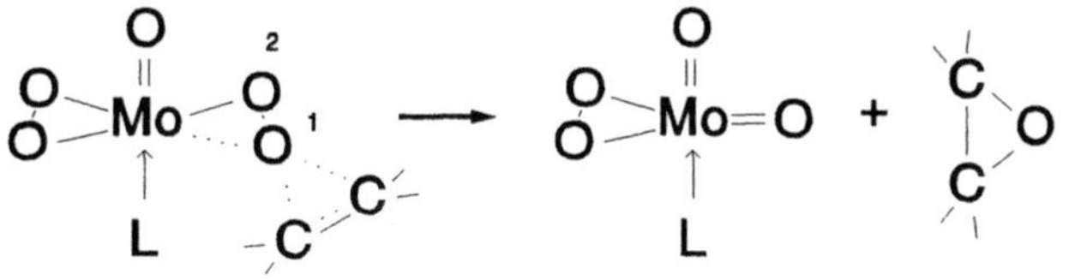

Figure 10. The mechanism of alkene epoxidation by MoO(O$_2$)•HMPT involves a 1,2-shift of the metal [60].

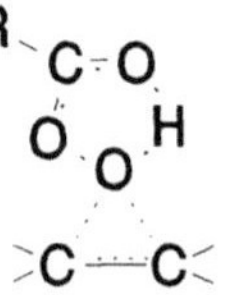

Figure 11. The "butterfly" transition state of epoxidation of alkenes with organic peracids.

In addition, a proposal for the mechanism of epoxidation catalyzed by TS – 1 has to account for the following observations:

vi. None of the known mechanisms *per se* can adequately explain the solvent effects observed for TS – 1, and in particular the beneficial influence of methanol and ethanol (see Section 2.2).

vii. A number of stable titanium peroxo complexes is known that contain a side-on bonded peroxo ligand (η^2-O$_2$), but they are more stable and less reactive than hydrogen peroxide itself, and none of them acts as an oxygen transfer agent. In one known case, such an unreactive Ti(η^2-O$_2$) unit can even be converted into a Ti(OH)(OOR) unit which *does* effect epoxidation of alkenes [107].

viii. The active species possesses Brønsted acidity and ion exchange properties (see also Section 2.3).

Oxygen transfer from a Ti(η^2-O$_2$) species was proposed by Huybrechts et al. [94] and Notari et al. [10]. It does not explain the solvent effects (*vi*), but an explanation for solvent effects in terms of selective sorption, as suggested by Romano et al. [43], cannot be ruled out on the basis of existing data. More importantly, it does not deal with consideration *vii*.

Taking the above facts and considerations into account, Clerici et al. [54] formulated a mechanism for TS – 1 catalyzed epoxidation which involves a hydroperoxo rather than a peroxo species, and coordination of an alcohol molecule to the site:

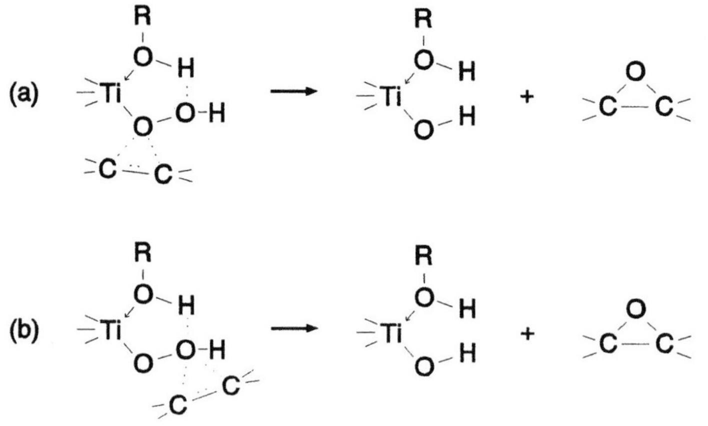

Figure 12. Mechanisms of oxygen transfer in the epoxidation of alkenes catalyzed by TS-1 as proposed by Clerici et al [54].

Note that pathway (a) also involves a metal 1,2-shift. Coordination of the alcoholic solvent is expected to become increasingly difficult with increasing bulkiness, which would explain the rate decrease in the series: methanol > ethanol > t-butanol. In aprotic solvents, water can take over the role of the alcohol, but is apparently less effective. The five-membered ring thus formed should then allow for smooth oxygen transfer, analogous to the mechanism of oxygen transfer from peracids (*vide supra*). It can also account for the acidity of the TS–1/H_2O_2 complex.

According to this description, the catalytic action of the framework titanium(IV) ion is essentially that of a neutral, oxophilic Lewis acid. The question whether or not a specific chemical property of titanium, such as its d^0 configuration, contributes to the oxygen transfer step has to remain unanswered at this moment. Some plausible hypotheses have been sketched for the oxygen transfer steps in other reactions catalyzed by TS–1 (we briefly discussed some possibilities in Section 2.2), but as to yet they lack sufficient experimental support.

3. VANADIUM-CONTAINING SILICALITES

The encouraging results obtained in selective oxidations by the use of TS-1 gave rise to a growing interest in the isomorphous substitution of transition metals other than Ti in silicalites. The well known catalytic properties of supported vanadium oxides for selective oxidations and ammoxidations of organic compounds [108], as well as some similarities between vanadium and titanium in their catalytic peroxo chemistry have prompted a number of studies on the possibility to insert vanadium in the framework of crystalline microporous silicas and aluminosilicates. In this chapter we will briefly review the synthesis and characterization of V-containing silicalites having the MFI or the MEL structure. Some other V-containing materials (ZSM-48 and NCL-1) are also briefly discussed. Finally, we will try to describe some aspects of the catalytic chemistry of V-containing silicalites.

3.1 Synthesis

For a better understanding of the problems related to the synthesis of V-containing silicalites, it is useful to briefly recall some of the relevant aqueous chemistry [109]. The most common oxidation states of vanadium are III, IV and V. In basic solutions ($pH > 9$), vanadium(III) and vanadium(IV) species are both easily oxidized by air to give vanadate(V) anions. In fluoride-containing solutions, vanadium forms octahedral air-stable anions in oxidation states IV and V, predominantly of the type $VOF_n(H_2O)_m^{p-}$. All of these air-stable anions are highly soluble and do not seem to interact readily with silicate species [110].

The lower oxidation states seem to provide better starting points for synthesis as their oxides are less acidic and a better interaction with silica is expected. Little is known about their aqueous chemistry in basic solutions: vanadium(IV) can form vanadate ions (most of them not well-characterized) whose composition and degree of polymerization strongly depend on pH and concentration. The main species in dilute, basic ($pH > 11$) solutions is reportedly monomeric $VO(OH)_3^-$ [109b]. The aqueous chemistry of vanadium(III) in basic solutions is very poorly documented.

The known coordination chemistry of vanadium does not seem to justify the expectation that vanadium ions will be able to occupy tetrahedral lattice positions in a silica. A truly stable and more or less symmetrically bonding tetrahedral coordination is only found for vanadium(V) in vanadates of metals that form basic or amphoteric oxides. On silica, tetrahedral monooxo or dioxo surface complexes are formed [111], but the interaction is weak and the species are easily hydrolyzed [112]. In this respect, vanadium(V) chemistry is similar to that of phosphorus(V).

The usual coordination geometry of vanadium(IV) is square pyramidal or distorted octahedral, and compounds containing tetrahedral vanadium(IV) are rare. In most compounds of vanadium(IV), the vanadyl(IV) unit (formally VO^{2+}) is present: one of the $V-O$ bonds is very short and has essentially the character of a double bond. Tetrahedral coordination is also rarely found for vanadium(III), one possibly relevant example being the VCl_4^- ion that occurs discretely in some salts [109].

Synthesis of V-containing MFI and MEL type zeolites.

A summary of the results of hydrothermal syntheses of V-containing MFI- and MEL-type silicalites that have been reported in literature is given in Table 3. Because of the widely different gel compositions and crystallization conditions used by different authors, it is difficult to distinguish trends in these results. We will nevertheless try to make some general remarks.

The importance of the choice of vanadium source was pointed out by Miyamoto et al. [114]; they reported that it was possible to prepare V-containing silicalites starting from V(III) or V(IV) salts. The use of NH_4VO_3 as the vanadium source resulted in an MFI-type material which did not contain vanadium. Later, this result was partly confirmed by other authors, who obtained V-containing silicalites using V(V) salts but with a content of vanadium much lower than in the cases where V(III) or V(IV) salts were used. The use

Table 3: Syntheses of V-containing MFI and MEL type zeolites selected from the literature.

N*	Reagent Mixture				Crystal. solid obtained			Ref.
	Source of V	Templ. agent[1]	$Si/V^{(2)}$	$M^+/Si^{(2,3)}$	$Si/V^{(2)}$	$Si/Al^{(2)}$	Crystal. phase	
1	VCl_3	HMDA	53	0.5	63	120	MFI	113
2	$VOSO_4$	TPA-Br	(4)	(4)	174	123	MFI	114
3	VCl_3	TPA-Br	(4)	(4)	178	132	MFI	114
4	NH_4VO_3	TPA-Br	42	(4)	-	106	MFI	114
5	VCl_3	HMDA	(4)	(4)	21.7	198	$MFI^{(7)}$	115
6	VCl_3	HMDA	(4)	(4)	21.6	455	$ZSM-48^{(7)}$	115
7	VCl_3	TBA-OH	(4)	(4)	30.8	669	$MEL^{(7)}$	115
8	$VO(C_2O_4)$	TPA-Br	5	~0.5	42	-	MFI	116
9	$VOSO_4$	TPA-OH	50	(5)	98	770	MFI	117
10	M^+VO_3	TPA-OH	(4)	(4)	~85	-	MFI	118
11	VCl_3	TPA-OH	31	-	50	-	MFI	119
12	VCl_3	TPA-OH	31	0.3	28	-	$MFI^{(7)}$	119
13	VCl_3	TPA-OH	15	-	58	-	MFI	120
14	VCl_3	TPA-OH	15	-	118	-	MFI	120
15	$VOSO_4$	TBA-OH	20	-	41	-	MEL	121
16	$VOSO_4$	TBA-OH	40	-	78	-	MEL	121
17	NH_4VO_3	TBA-OH	40	-	182	-	MEL	121
18	NH_4VO_3	TPA-Br	40	$1.0^{(6)}$	152	-	MFI	122

(1) HMDA = hexamethylenediamine, TPA = tetrapropylammonium, TBA = tetrabutylammonium; (2) Molar ratios; (3) M^+ = alkali cation; (4) present but it is not possible to determine in which amount; (5) 0.11 equivalents of NH_3 were used; (6) M^+ = NH_4^+ and the synthesis is performed in the presence of F^- at pH = 7.9; (7) a transformation to crystobalite occurs upon calcination in air at 550°C.

of vanadium(IV) salts in combination with fluoride at neutral pH also leads to very low vanadium contents [123].

Some authors [115,119] reported that vanadium-containing zeolites and silicalites that had been prepared using VCl_3, underwent a structural collapse leading to the formation of crystobalite upon calcination in air at 550°C; the same was not observed when the samples were calcined at 550°C in H_2 atmosphere. Such behaviour was in fact only observed when sodium ions were present in the starting reaction mixture [119]. The structural collapse has been tentatively attributed to the oxidation of V(III) or V(IV) species very well dispersed in the zeolite matrix. Since air had not been excluded in any of the reported preparations using V(III), the vanadium source was, at least partially, oxidized to V(IV) and V(V) already during the preparation of the precursor gel or the subsequent hydrothermal treatment [114,119].

The use of vanadyl(IV) sulfate as the vanadium source in the absence of sodium was found to lead to thermally stable materials with more reproducible vanadium contents in the final product [117] compared to the use of V(III) sources [119], provided that the oxidation of vanadium(IV) during gel preparation or crystallization was avoided. This approach was shown to be applicable to the synthesis of MEL-type materials as well [121]. In contrast to the compositional ceiling of Si/V≈90 found for MFI-type materials [117], Hari Prasad Rao et al. [121] reported that the V content of the MEL-type materials increased linearly with the V content of the precursor gel when the latter was in the range $20 < Si/V < 160$; the vanadium uptake was reported to be around 50%. On the basis of semiquantitative ESR spectra, the same authors concluded that the V(IV) content of MEL type silicalite decreased when the Si/TBAOH molar ratio in the precursor gel was chosen higher than 5.0, and that during crystallization, the V(IV) content of the solids increased with an increasing degree of crystallinity.

Synthesis of other vanadium-containing zeolites.

The synthesis of a V-containing ZSM-48 was reported by Habersberger et al. [115] from a gel containing VCl_3 as source of vanadium, hexamethylene diamine as the structure-directing agent and sodium hydroxide (see Table 3). The crystalline material thus obtained was transformed into cristobalite upon calcination in air. Recently Tuel et al. [124] described the preparation of a V-containing ZSM-48, starting from $VOSO_4$, 1,8-diaminooctane, fumed silica and water. The zeolite was stable towards calcination in air, confirming that as for the V-containing MFI silicalite, the presence of sodium ions is responsible for the instability to thermal treatment in an oxidative atmosphere [119]. Again, the Si/V molar ratio in the crystalline solid was about double that in the precursor gel.

The first example of a large pore V-containing zeolite was reported by Reddy et al. [125], who described the synthesis of a V-containing NCL-1 silicalite from a reaction mixture containing fumed silica, $VOSO_4$, NaOH and hexamethylenebis(triethylammonium) bromide. The authors claim NCL-1 to be a new molecular sieve and have classified it as a large pore material on the basis of its sorption properties [126].

V. Dubanska [127] reported the synthesis of vanadium-containing low-silica zeolites through hydrothermal digestion of vanadium alumino-silicate glasses in solutions of sodium and potassium carbonate and sodium hydroxide. Analcime, cancrinite, phillipsite, sodalite and other zeolites were obtained, but the state of vanadium in the solid samples was not investigated.

3.2. Characterization of vanadium species

Most characterization studies on vanadium-containing molecular sieves have focussed on V(IV) or V(V) species. No effort has been described to characterize vanadium(III) species in the materials prepared from vanadium(III) salts. Centi et al. [120] reported that MFI-type materials prepared from vanadium(III) mainly contain vanadium(IV) and vanadium(V). They found at least three different species, one of which was an undefined vanadium-rich oxidic phase. Washing an air-calcined sample with ammonium acetate at room temperature was reported to remove most (>80%) of the vanadium. These and other complications arising from the use of vanadium(III) salts (see also section 3.1) do not seem to make them very suitable precursors.

Most of the V-containing zeolites described in the literature are reported to contain predominantly vanadium(IV) after crystallization, even if the materials had not been prepared from a vanadium(IV) salt. ESR spectra of these V(IV)-containing samples show anisotropic hyperfine splitting caused by the ^{51}V nucleus with or without the presence of an underlying broad singlet. The singlet is thought to arise from vanadium-rich phases or oligomeric species, while the hyperfine pattern originates from isolated ions. A proper analysis of the spectra should include pattern simulation and estimation of the contribution of the singlet. A quantitative approach is preferred, because antiferromagnetic interactions can make vanadium(IV) oligomers difficult to detect.

Linewidths in the hyperfine pattern are usually not reported, although they can be used to calculate a lower limit for the average spin-spin distance in a sample [128]. For example, a linewidth of 20 G would correspond to a spin-spin distance of at least 12 Å, which would imply for a sample having a Si/V ratio of 100 that the vanadium is rather well distributed over the crystals, and not concentrated in certain regions such as grain boundaries or occluded phases. Using electron probe microanalysis, Rigutto and van Bekkum [117] showed that in large crystals of vanadium-containing MFI, the metal is distributed more or less homogeneously. In fact, concentration depressions even coincided with grain boundaries in intergrown crystals.

In Table 4, the ESR parameters are listed for V(IV) species in several V-containing zeolites. The values reported for g and hyperfine coupling tensors generally agree with the presence of approximate C_{4v} symmetry (or, less likely, D_{4h}). In one case a low value of $A_{//}$ (154 G) was ascribed to tetrahedral V(IV), presumably in C_{3v} symmetry [120].

Table 4: ESR parameters of samples prepared by direct synthesis, obtained from X–band spectra taken at room temperature.

Sample	V-source in initial gel	(1)	$A_{//}$(G)	$g_{//}$	A⊥ (G)	g⊥	(2)	Ref.
2 (3)	Tab. 3	H_2	196	1.957	91	2.015	-	114
3 (3)	Tab. 3	H_2	191	1.927	89	1.994	-	114
8 (3)	Tab. 3	A.S.	186	1.949	72	1.990	sq.pyr.	116
9 (3)	Tab. 3	A.S.	199	1.933	74	1.991	sq.pyr.	117
10 (3)	Tab. 3	A.S.	192.5	1.939	87.8	1.961	sq.pyr	118
11 (3)	Tab. 3	A.S.	191	1.95	55	2.00	sq.pyr.	119
12 (3)	Tab. 3	A.S.	191	1.93	89	1.97	sq.pyr.	119
13 (3)	Tab. 3	air$^{(4)}$	203		80		d.octa.	120
14 (3)	Tab. 3	$H_2^{(4)}$	154	1.911	68	1.963	d.tetra.	120
15 (3)	Tab. 3	A.S.	185	1.932	72	1.981	-	121
18 (3)	Tab. 3	H_2	189	1.925	84	1.997	-	122
ZSM-48	$VOSO_4$	A.S.	188	1.936	70	1.99	-	124
NCL-1	$VOSO_4$	A.S.	197	1.929	72	1.973	-	125

(1) Synthesis post-treatment: A.S. = as synthesized, air = calcined in air, H_2 = calcined in air then reduced in H_2; (2) Proposed coordination of the V(IV) species; (3) Reference number of sample listed in Tab. 3; (4) spectrum recorded at 77K.

In spite of small but somewhat puzzling differences among g tensors and A parameters for the C_{4v} symmetrical species, which might indicate some variation of local symmetry and covalent character, the values of the ESR parameters fall within the range reported for square pyramidal vanadyl complexes in an oxygen environment [129]. (It does not seem possible to distinguish the vanadium species in vanadium-containing silicalites from vanadyl ions in exchangeable positions [130] on the basis of ESR spectra alone). Obviously, a vanadyl species cannot be located at a regular tetrahedral framework site, and other possible situations of vanadium, e.g. as a *framework satellite*, have to be considered. We will come back to this later.

Several papers report the possibility to oxidize and reduce the vanadium species reversibly [114,116,117,121-124]. Upon calcination in air, V(IV) is oxidized to V(V); the former species is restored (judged from ESR spectra) by thermal treatment in hydrogen or a hydrocarbon reductant. Although nonquantitative, these results were also taken to evidence the lack of mobility of vanadium species undergoing red/ox cycles at high temperatures. Immobility, albeit on a micrometer scale, was also evidenced by micropobe analysis [117].

Table 5: ^{51}V-NMR chemical shifts for vanadium-containg zeolites and reference samples.

Sample	(1)	Pretreat. (2)	δ_{max} (3) (ppm)	δ_{iso} (4) (ppm)	(5)	Ref.
V-MFI	8	A.S.	-	-680	tetrah.	116a
V-MFI	9	C.	-500	-500	tetrah.	117
V-MFI	10	C.	-	-557	tetrah.	118
V-MFI	11	C.	-385	-	octah.	119
V-MFI	12	A.S.	-570	-570	tetrah.	119
V-MFI	13	C.	-480	-	tetrah.	120
V-MEL	16	C.	-	-573	tetrah.	134
V-ZSM-48	-	C.	-530	$-575^{(6)}$	tetrah.	124

(1) References number of samples listed in Tab. 3; (2) A.S.= as synthesized, C. = calcined in air; (3) location of the maximum absorption intensity in the static spectrum; (4) location of the center band in MAS spectra; not corrected for second order quadrupolar shifts (5) attribution of the vanadium coordination made in the refs.; (6) value obtained from simulation of the powder pattern. All shifts are relative to $VOCl_3$.

The local symmetry of the V(V) species present in calcined samples has been investigated by ^{51}V NMR spectroscopy; Table 5 lists some results reported in the literature. The many ^{51}V NMR spectra of reference vanadium compounds that have been published in the last few years [111,131] have greatly facilitated the interpretation of the usually poorly resolved spectra obtained from solid catalysts. From the data shown in Table 5, it appears that in most calcined vanadium-containing silicalites, V(V) is present in tetrahedral coordination. In some cases, the MAS spectra display poorly resolved spinning sideband patterns due to chemical shift dispersion; in all of these materials, multiple tetrahedral vanadium(V) species are present. Very narrow MAS signals are observed for some other materials [117,124,134], probably due to reaction of the site with water; a fast hydrolysis/condensation equilibrium is expected to give rise to anisotropy averaging. Recent data show that in completely dehydrated V-MFI samples the main species is a tetrahedral monomer with low symmetry (C_{2v} or lower: $\sigma_{11} = -370$ ppm, $\sigma_{22} = -560$ ppm, $\sigma_{33} = -950$ ppm) [132]. ^{51}V spectra also showed that the species reacts more or less readily with ambient water, depending on the hydrophobicity of the molecular sieve matrix.

Several authors observed a new band at about 960 cm^{-1} in the IR [118,119,121,124,125] or Raman [132] spectra of calcined V-containing silicalites that was not present in the corresponding silicalite prepared in the absence of vanadium. As was discussed in chapter 2 of this paper, in TS-1 a IR or Raman band at about 960 cm^{-1} may best be attributed to an asymmetric stretching mode of framework SiO$_4$ tetrahedra bound to another ion (*in casu* Ti) through an oxygen bridge [84]. By analogy, the IR band at 960 cm^{-1} has been taken as evidence for the presence of V-O-Si bonds in vanadium-containing silicalites [121,124]. A direct correlation between vanadium content and unit cell parameters has been reported for V-containing MEL [121] and V-containing ZSM-48 [124], while for MFI-type materials such a correlation was absent [117,118]. For V-containing MEL type silicalite in particular, the relative intensity of 960 cm^{-1} IR band and the unit cell volume were reported to increase linearly with the vanadium content [121b].

Adsorption of ammonia, pyridine and acetonitrile revealed the presence of Lewis and weak Brønsted sites on V-MFI [120] and V-MEL [134] that were not observed in the corresponding silicalites. In agreement with the presence of weak Brønsted acidity, an ion exhange capacity was observed for V-containing MFI-type silicalites [117].

On the basis of the results mentioned above, hypotheses on the environment of V(IV) and V(V) were given by some authors. Figure 13 depicts the local vanadium environments proposed respectively by Rigutto et al. [117] and Centi et al. [120]. (In view of the conclusion by Centi et al. [120] that the vanadium(V) site is tetrahedral, it is not clear whether or not the dotted line in Fig. 13(b) should be interpreted as a coordinative bond). Prasad Rao et al. [121b] proposed a slightly different structure for the site in oxidation state IV without the terminal oxo ligand, but preferred the vanadyl(IV) structure in a later paper [121a]. For comparison, the structure of the vanado(IV)silicate anion chain in the mineral haradaite is shown also in Fig. 13.

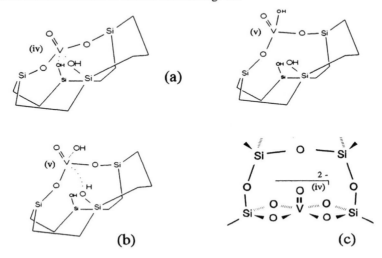

Figure 13. Structure of vanadium sites in silicalites as proposed in (a) ref. 117, (b) ref. 120; (c) structure of a part of the anion chain in haradaite, Sr$_2$[(VO)$_2$Si$_4$O$_{12}$] [133].

Conclusive remarks

The results reported in literature suggest that it is possible to prepare vanadium-containing zeolites and silicalites in which isolated vanadium species are probably connected to the framework at defect sites forming so-called *framework satellites*. There is no tendency for vanadium to adopt a symmetrically binding tetrahedral coordination, and no evidence is found in the literature for *isomorphous* substitution. One can only speculate about the nature of the defect sites to which vanadium binds. The structures (a) and (b) shown in Fig. 13 depict a lattice vacancy as the binding site. Although silicalites are often found to contain lattice defects, the presence of vacancies is a matter of controversy [135]; one nevertheless could imagine their formation to be induced by the presence of vanadium. Dessau et al. [135a] proposed a different structure for the defects in silicalites, consisting of a hydrolyzed Si–O–Si bond in a four ring; such a site could also bind a vanadyl ion. The structure of the vanadium site in the mineral haradaite (see Fig. 13(c)) provides us with an indication of the coordinative requirements of vanadium(IV) in a similar environment. The satellite species in vanadium-containing silicalites is accesible to molecules that can enter the micropores, and can apparently change its coordination and oxidation state reversibly. The oxidized vanadium(V) species is both a weak Lewis acid and a weak Brønsted acid.

The formation of vanadium-rich impurities during crystallization and the related thermal instability of the molecular sieve matrix can best be prevented by using a vanadium(IV) salt as the precursor while avoiding the presence of alkali metal ions [117,121] and oxygen [117].

3.3 Catalytic reactions

Vanadium containing zeolites have been tested as oxidation catalysts in the presence of different kinds of oxidants: O_2 and N_2O have been used in gas phase reactions [116,136], and H_2O_2 in liquid phase reactions [134,137,138].

Reactions performed in gas phase.

In the oxidative dehydrogenation of lower alkanes, a V-containing-silicalite showed a lower activity but an higher selectivity towards alkenes [127,136a] and aromatics [116] compared to H-ZSM-5. In the oxidative dehydrogenation of propane, Zatorski et al. [136a] observed a very high selectivity to propene with either O_2 or N_2O, using a catalyst with a very low vanadium content (see Table 6). The observed behaviour for V-silicalite and other modified silicalites was at least partially attributed to the presence of defect sites but no hypothesis was given on the reaction mechanism. On the basis of an accurate physical-chemical characterization of the same catalyst, Bellussi et al. [136b] later suggested the following reasons for its remarkable selectivity:

i	the presence of isolated tetrahedral V(V) as the sole site of oxygen activation;
ii.	a limited tendency for oxygen insertion or alkene activation at allylic positions due to the presence of stable V-O-Si bonds containing non-electrophilic bridging oxygen ions, and a stable coordination of vanadium;
iii.	the absence of strong Brønsted acid sites.

Table 6: Oxidative dehydrogenation of propane [136a][(1)]

Sample	Si/M	Oxidant	T (°C)	Conv (%)	Selectivity (%)				
					C_1+C_2	C_3H_6	C_4+C_5	arom.	COx
H-ZSM-5	254	O_2	450	18.5	5	45	5	19	26
Fe-MFI	250	O_2	450	20.1	3	40	2	25	30
B-MFI	36	O_2	450	5.6	2	74	-	-	24
V-MFI	270	O_2	450	12.5	2	73	-	-	25
H-ZSM-5	254	N_2O	350	18.9	4	39	5	32	20
B-MFI	36	N_2O	350	2.1	3	84	-	-	13
V-MFI	270	N_2O	350	12.5	1	95	-	-	4

(1) Performed in a plug flow reactor, space velocity 564 h[-1], feed: 2.3% propane, 1.65% O_2 (or 9.6% N_2O), balance He, atmospheric pressure, activities measured after 1h time on stream.

In methanol oxidation, vanadium-containing MFI-type silicalites prepared from Ludox and $Si(OEt)_4$ and a 1 % vanadium-on-silica catalyst were found to show similar selectivities to formaldehyde (> 90%) [132]. Higher activities were found for the silicalites, which could be explained on the basis of different sorption strengths. Intrinsic rate constants were found to differ by a factor of two at most.

Reactions performed in the liquid phase.

Most of research activity on oxidations with H_2O_2 in the presence of V-containing silicalite have been performed by the research group of the National Chemical Laboratory in Pune (India) [134,137]. The catalyst used was a V-containing MEL-type silicalite (see Table 3, entries 15 and 16). The results reported in paraffins oxidations and phenol hydroxylation are comparable to that of a reference TS-2 and lower with respect to the best results reported in the literature for TS-1 (Tables 7 and 8). Results similar to those obtained with V-MEL have been reported for V-containing ZSM-48 [124].

Table 7. Oxyfunctionalization of hexane catalyzed by titanium- and vanadium-containing silicalites.

Catalyst [(1)]	Conv. (%)	H_2O_2 conv. (%)	H_2O_2 sel. (%)[(2)]	Product selectivity (%)							Ref.
				1-ol	2-ol	3-ol	1-al	2-one	3-one	other	
V-MEL [(3)]	14.6	100	57.1	3.7	9.2	8.2	7.2	26.3	25.0	21.4	137b
TS-2 [(3)]	15.9	100	58.6	-	19.1	17.1	-	23.7	23.0	16.6	137b
TS-1 [(4)]	27.6	98	86.0	-	32.1	25.9	-	39.8	2.0	-	78

(1) V-MEL, Si/V = 79; TS-2, Si/Ti = 77; TS-1, Si/Ti = 46; (2) H_2O_2 used for monofunctional product formation; Reaction conditions: (3) cat = 0.1 g, n-hexane = 5 g, n-hexane/H_2O_2 (mol) = 3, solvent (acetonitrile) = 20 g, T = 373 K, reaction duration = 8 h; (4) cat = 0.856 g, n-hexane = 2.6 g, n-hexane/H_2O_2 (mol) = 2.19, solvent (methanol) = 50 g; T = 328 K, reaction duration = 1h.

V-MFI also catalyzes phenol hydroxylation and toluene oxidation with hydrogen peroxide giving product distributions similar to those reported for V-MEL, but it does with low hydrogen peroxide efficiencies (< 20 %) [139]. It is difficult to explain these

Table 8. Phenol hydroxylation catalyzed by titanium- and vanadium-containing silicalites.

Catalyst [1]	Conv. mol (%)	H_2O_2 conv.(%)	H_2O_2 [2] sel.(%)	Product Distribution (%)			Ref.
				pBQ[3]	HQ[4]	Catechol	
V-MEL [5]	24.3	100	55.7	3.9	52	44.1	137b
TS-2 [6]	20.8	100	70.0	0.9	49.8	49.3	137b
TS-1 [7]	27.0	100	82.0	1.0	50.0	49.0	44

(1) V-MEL, Si/V = 79; TS-2, Si/Ti = 29; TS-1, Si/Ti = 34; (2) H_2O_2 used for monofunctional product formation; (3) p-benzoquinone; (4) hydroquinone; (5) cat. = 0.1g, phenol = 1g, phenol/H_2O_2 (mol) = 3, T = 333K, reaction duration = 8h, solvent = water; (6) cat = 0.1g, phenol = 1g, phenol H_2O_2 (mol) = 3, T = 342K, reaction duration = 29h, solvent = acetone; (7) cat = 0.72g, phenol 20.7g, phenol/H_2O_2(mol) = 3.8, T = 373K; reaction duration = 1h, solvent = water/acetone.

differences in view of (a) the fact that TS-1 and TS-2 behave similarly in catalytic experiments (see section 2.2) and (b) the fact that the vanadium sites in both materials have similar spectroscopic signatures (see section 3.2).

The main reported differences in the behaviour of Ti-containing silicalites and V-containing silicalites are the higher activity of the latter in the oxyfunzionalization of primary carbon atoms in alkanes and alkylaromatics (toluene) [137,138], indicating a more pronounced radical character of the active site. Following a proposal by Mimoun et al. [53] for the mechanism of hydroxylation by oxoperoxovanadium(V)picolinate complexes, Ramaswamy et al. [137b] assumed the involvement of vanadium(IV) superoxo complexes in these reactions for the case of V-MEL as well.

4. OTHER TRANSITION METAL-CONTAINING SILICALITES

Ferrisilicate analogs of many different high-silica zeolites have been described and framework siting of iron(III) ions in these materials is well-established [140]. Framework iron(III) species in silicalites do however not seem to have redox properties.

Some interesting results in the catalytic oxidative cleavage of unsaturated compounds to aldehydes with hydrogen peroxide have recently been reported for chromium-containing MFI-type silicalites prepared from chromium(III) salts [141,142]. E.g., oxidation of methyl methacrylate with non-alkaline hydrogen peroxide was reported to give methyl pyruvate in 80% selectivity at 95% conversion [141]. In homogeneous systems, chromium is not a practical catalyst because it efficiently decomposes hydrogen peroxide [143]. The chromium species in these silicalites have not been characterized; Sheldon et al. [142] claim isomorphous framework substitution by chromium, but evidence is not provided. Tetrahedral coordination of chromium(III) is very rare and the reported green colour of as-synthesized materials [142] suggests octahedral coordination. For MFI-type silicalites prepared from $Si(OEt)_4$, $Cr(NO_3)_3$ and alkali-free Pr_4NOH, low chromium incorporation efficiencies were observed [144]; no effect on unit cell parameters was found, which led the authors to conclude that the chromium species is not a framework species. Spoto et al. [145] prepared a chromium-containing silicalite by grafting of chromic acid onto internal silanol groups at 973 K followed by reduction, and concluded from infrared studies that a framework-connected chromium(II) species is formed which readily interacts with NO and CO.

5. TRANSITION METAL-CONTAINING ALUMINOPHOSPHATES

Szostak [146] already remarked that in general, $AlPO_4$'s seem more flexible in their ability to incorporate low levels of foreign ions into their structures than silica-based materials. This generalization does not seem to apply to transition metals, for which there are only three well-established cases of isomorphous substitution for each class of materials: iron, cobalt and zinc for aluminophosphates, and titanium, iron and zinc for silica-based molecular sieves. There is however reason to believe that aluminophosphate frameworks are more tolerant towards non-tetrahedral ions. Some structures are known to contain non-tetrahedral aluminium bearing hydroxide ions (*e.g.* $AlPO_4$-17 [147]) or water molecules (*e.g.* VPI-5 [148]). Several studies indicate that reversible framework hydration of calcined $AlPO_4$'s, producing octahedral aluminium, is a general phenomenon [4,147]. In this chapter, one case of isomorphous substitution (CoAPO) and two supposed cases of non-isomorphous substitution (VAPO and CrAPO) are treated.

5.1. CoAPO-5, CoAPO-11 and other cobalt-containing $AlPO_4$'s

Low levels of cobalt(II) can be incorporated into the framework of $AlPO_4$-5 [149] and $AlPO_4$-11 [151,152], substituting for aluminium on tetrahedral framework sites, as evidenced by (quantitative) diffuse reflectance UV-Vis and near-infrared spectra (DREAS), ESR and ^{31}P NMR measurements. There is quite some disagreement among the authors cited here about the maximum attainable content of framework cobalt, but cobalt incorporation up to a level of $Co/(Co+Al+P)=0.005$ (molar ratio) is apparently feasible. According to Schoonheydt et al. [149a], octahedral (*i.e.* assumingly non-framework) cobalt(II) is very difficult to detect by DREAS when much more strongly absorbing tetrahedral species are also present. Still lower maximum contents of framework cobalt have been reported for CoAPO-16 and CoAPO-34 [150,151]. In view of the very high framework cobalt content of CoAPO-50 ($Co/(Co+Al+P)=0.19$) these apparent compositional ceilings are somewhat difficult to explain [153].

Upon calcination in air or oxygen, both tetrahedral cobalt(III) and strongly distorted tetrahedral cobalt(II) species are formed in CoAPO's [149,150]. The oxidized species are reduced back to tetrahedral cobalt(II) by treatment with hydrogen at 773 K [149] or even with a weak one-electron donor like methanol at room temperature [150,151]. Infrared studies indicate that tetrahedral framework cobalt(II) ions in CoAPO's (in the proton-form) are capable of coordinating acetonitrile [154].

CoAPO-11 and CoAPO-5 have both been used as catalysts in the autoxidation of p-cresol to p-hydroxybenzaldehyde in methanolic sodium hydroxide [155]. A selectivity of 90 % at 90 % conversion was claimed. It would be interesting to investigate the role of substrate-cobalt surface complexes in this reaction. The formation of these should not be too difficult, as framework aluminium is also easily hydrated in calcined AlPO-5 and AlPO-11 [4]. In a further examination of p-cresol oxidation, Peeters et al. [156] found leaching of cobalt from CoAPO's and questioned whether CoAPO acts as a truly heterogeneous catalyst. Very recently, Lin et al. [157] reported CoAPO−5 to be an active and moderately selective catalyst for the autoxidation of cyclohexane in acetic acid. Selectivities to adipic acid up to 45 % were reported at conversions in the range 30−40%. The cobalt(III) ion is a very strong oxidant; its regeneration from cobalt(II) in the above reactions is believed to be effected by homolytic reduction of the hydroperoxide intermediates formed in the autoxidation cycle [157], as in the homogenous case.

5.2. VAPO−5 and VAPO−11

The synthesis of pure VAPO−5 using V_2O_5 as the vanadium source has been reported by Montes et al. [158] and Jhung et al. [159]. In both cases, vanadium was largely reduced during synthesis, and the crystalline product was shown by ESR to contain mainly well-dispersed monomeric vanadyl(IV) species, which were oxidized to vanadium(V) upon calcination at 773 K. Using quantitative ESR measurements, Jhung et al [159] showed that up to 80% of the vanadium(V) could be reduced to monomeric vanadyl(IV) species with ESR parameters nearly identical with those of the species present in the as-synthesized materials, by treatments with one-electron donors like xylene or trimethylbenzene at 373 K. The presence of well-dispersed, immobile vanadium species was taken to evidence framework siting of the materials. Both Jhung et al. [158] and Montes et al. [159] made the assumption that vanadium(IV) subtituted for phosphorus(V), mainly on the basis of elemental analyses and a generalization of the predictions by Flanigen [160] on the incorporation of tetravalent and pentavalent elements. Rigutto et al. [161] found compositions consistent with substitution for aluminium rather than phosphorus for VAPO−5 synthesized in the presence of fluoride using vanadyl(IV) sulphate as the vanadium source. In addition, ^{51}V NMR was used to show that at low vanadium contents, the dominant species in calcined materials is a square pyramidal or distorted octahedral vanadyl(V) species, consistent with a phosphate environment but difficult to explain when a second coordination shell of aluminium ions is assumed.

VAPO−5 was found to catalyze oxidations with *tert*-butyl hydroperoxide (TBHP), such as the epoxidation of allylic alcohols (>95% selectivity at 50% conversion of the substrate) and benzylic oxidations [161]. Similar results were obtained with VAPO−11 [162]. Based on ^{51}V NMR studies on the interaction of vanadium species in VAPO−5 with TBHP and some other sorbates [161] a model was proposed for the catalytic site (see Fig. 14).

5.3. Chromium-containing AlPO$_4$'s

Although several papers report the synthesis of chromium-containing AlPO$_4$'s [163], little effort has been made to characterize these materials. Helliwell et al. [164] concluded from a single crystal study of a chromium-containing aluminophosphate with the GaPO−14 structure that chromium was occupying 6% of the octahedral framework sites (GaPO-14 contains tetrahedral, five-coordinated and octahedral gallium sites [165]). Non-tetrahedral framework-connected ions (cf. vanadyl species in VAPO's) might also occur in other chromium-containing AlPO's. Diffuse reflectance UV-Vis spectra seem to indicate that octahedral chromium(III) is present in as-synthesized CrAPO-5 which is

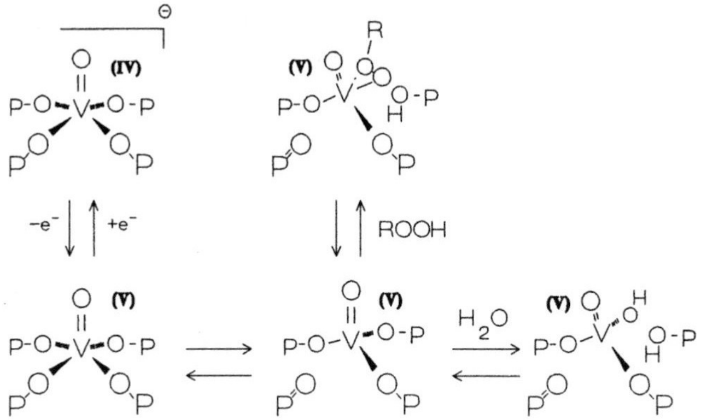

Figure 14. Structure of the vanadium site in VAPO-5 proposed in ref. 161 and its possible interaction with sorbates.

oxidized to tetrahedral chromium(VI) upon calcination [163b].

Very recently, Chen et al. [142,166] reported chromium-containing AlPO–5 to be an active and selective catalyst in the decomposition of hydroperoxides towards ketones (e.g. 86% selectivity to cyclohexanone from cyclohexyl hydroperoxide at 87% conversion), benzylic oxidations and the (aut-)oxidation of secondary alcohols with TBHP and oxygen. Chromium was not leached from the catalyst, which was found to be recyclable. These results justify further characterization studies.

6. CONCLUSIVE REMARKS

The majority of studies on the chemistry of framework transition metal sites in molecular sieves has been devoted to titanium silicalite-1. These studies show that a tetrahedral framework ion can have considerable freedom in reversibly coordinating small molecules, and that the view of a rigid framework dominating the coordination chemistry of the transition metal ion is too simple. The resistance of the titanium site to extensive hydrolysis however imposes restraints on the accesibility of the site; as a consequence, the site is apparently sufficiently electron-deficient to be a very active catalyst for oxygen transfer from hydrogen peroxide. Together with those properties, site separation and selective sorption by the molecular sieve matrix can explain several features of the catalytic chemistry of titanium silicalites. A better insight will perhaps also allow us to predict to some extent whether it is possible to synthesize catalysts based on other molecular sieves or containing other transition metals.

Futhermore, we have seen several examples of molecular sieve materials containing non-tetrahedral transition metal species chemically bound to their frameworks. Such materials are less-well investigated than (supposedly) isomorphously substituted materials. There is no obvious reason for this; in fact, most transition metals do not fulfill the requirements for isomorphous substitution, but might nevertheless yield interesting catalysts when anchored to the framework of a molecular sieve.

References

1. R.M. Barrer, in D.H. Olson and A. Bisio (Eds.), Proc. 6th Int. Zeol. Conf., Butterworths, 1983, 870.
2. For a recent review see: J.C. Vedrine, Stud. Surf. Sci. Catal., **69**, (1991), 25.
3. We include protons in this category for the sake of argument. It is of course possible to view protonation, deprotonation, alkylation etc. of anionic sites in the zeolite lattice as coordinative interactions of a trivalent framework ion.
4. R. Meinhold, N. Tapp, J. Chem. Soc., Chem. Commun., **1990**, 219.
5. E. Bourgeat-Lami, P. Massiani, F. Di Renzo, P. Espiau, F. Fajula, T. Des Courières, Appl. Catal., **71**, (1991), 139.
6. A.S. Medin, V.Yu Borovkov, V.B. Kazansky, A.G. Pelmetschikov, G.M. Zhidorimov, Zeolites, **10**, (1990), 668.
7. M. Taramasso, G. Perego, B. Notari, U.S. Patent No. 4,410,501 (1983), to Snamprogetti S.p.A; M. Taramasso, G. Manara, V. Fattore, B. Notari, U.S. Patent No. 4,666,692 (1987), to Snamprogetti S.p.A.
8. G. Bellussi, V. Fattore, Stud. Surf. Sci. Catal., **69**, (1991), 79.
9. B. Notari, Stud. Surf. Sci. Catal., **37**, (1987), 413 and refs. cited therein.
10. B. Notari, Stud. Surf. Sci. Catal., **60**, (1991), 343 and refs. cited therein.
11. R.A. Sheldon, J.K. Kochi, Metal-Catalyzed Oxidations of Organic Compounds, Academic Press, New York, 1981.
12. Some highly relevant model studies have been performed on transition metal siloxides. See: F.J. Feher, J.F. Walzer, Inorg. Chem., **30**, (1991), 1689, and refs. cited therein.
13. S.M. Csicsery, Zeolites, **4**, (1984), 202.
14. R.M. Dessau in W.H. Frank (Ed.), "Adsorption and Ion Exchange with Synthetic Zeolites", ACS Symp. Ser., **135**, (1980), 123.
15. D.M. Chapman, A.L. Roe, Zeolites, **10**, (1990), 730; S.M. Kuznicki, K.A. Trush, WO Patent No. 91/18833, (1991), to Engelhardt Corp.
16. P.J. Kooyman, unpublished results.
17. T. Blasco, M.A. Camblor, A. Corma, J. Pérez-Pariente, J. Am. Chem. Soc., **115**, (1993), 11806.
18. M. Padovan, F. Genoni, G. Leofanti, G. Petrini, G. Trezza, A. Zecchina, Stud. Surf. Sci. Catal., **63**, (1991), 431.
19. A.J.H.P. van der Pol, J.C. van Hooff, Appl. Catal. A, **92**, (1992), 93.
20. (a) A. Thangaraj, R. Kumar, S.P. Mirajkar, P. Ratnasamy, J. Catal., **130**, (1991), 1; (b) A. Thangaraj, M.J. Eapen, S. Sivasanker, P. Ratnasamy, Zeolites, **12**, 943.
21. R. Millini, E. Previde Massara, G. Perego, G. Bellussi, J. Catal., **137**, (1990), 503.
22. B. Kraushaar, Ph.D. Thesis, Eindhoven Univ. of Technology, 1989.
23. C.G. Barraclough, R.L. Martin, G. Winter, J. Chem. Soc., **1964**, 758.
24. D.R.C. Huybrechts, I. Vaesen, H.X. Li, P.A. Jacobs, Catal. Lett., **8**, (1991), 237.
25. G. Bellussi, A. Carati, M.G. Clerici, A. Esposito, Stud. Surf. Sci. Catal., **63**, (1991), 421.
26. M. Goepper, H.-X. Li, M.E. Davis, J. Chem. Soc., Chem. Commun., **1992**, 1665.
27. R.K. Iler, "The Chemistry of Silica", John Wiley, New York, 1979, p. 150.
28. A. Thangaraj, R. Kumar, S. Sivasanker, Zeolites, **12**, (1992), 135.
29. J.L. Guth, H. Kessler, J.M. Higel, J.M. Lamblin, J. Patarin, A. Seive, J.M. Chezeau, R. Wey, in M.L. Occelli, M.E. Robson (Eds.), "Zeolite Synthesis", ACS Symp. Ser., **398**, (1989), 176; J.M. Popa, J.L. Guth, H. Kessler, Eur. Patent Appl. No. 292,363, (1988).
30. Qiu Shilun, Pang Wenqin, Yao Shangqing, Stud. Surf. Sci. Catal., **49A**, (1989), 133.
31. P.J. Kooyman, J.C. Jansen, H. van Bekkum, in R. von Ballmoos, J.B. Higgins, M.M.J. Treacy (Eds.), Proc. 9th Int. Zeolite Conf., Butterworth-Heinemann, Boston, 1992, p. 505.
32. A. Lopez, M.H. Tuilier, J.L. Guth, L. Delmotte, J.M. Popa, J. Solid State Chem., **102**, (1993), 480.
33. G. Perego, G. Bellussi, C. Corno, M. Taramasso, F. Buonuomo, A. Esposito, in Y. Murakami, A. Iijima, J.W. Ward (Eds.), "Proceedings of the 7th International Conference on Zeolites", Kodansha Elsevier, Amsterdam, 1987, p 129.

210

34.   R.B. Greegor, F.W. Lytle, D.R. Sandstrom, J. Wong, P. Schultz, J. Non-Crystalline Solids, **55**, (1983), 27; F. Lytle, R.B. Greegor, in F.L. Galeener, D.L. Griscom, M.J. Weber, "Defects in Glasses", Mater. Res. Soc. Symp. Proc., **61**, (1985), 259.
35.   A. Tuel, Y. Ben Taarit, Microporous Mater., **1**, (1993), 179; A. Tuel, Y. Ben Taarit, C. Naccache, submitted for publication in Zeolites.
36.   G.T. Kokotailo, P. Chu, S.L. Lawton, W.M. Meier, Nature, **275**, (1978), 119; G. Perego, M. Cesari, G. Allegra, J. Appl. Cryst., 17 (1984), 403; G. Perego, G. Bellussi, A. Carati, R. Millini, V. Fattore, in "Zeolite Synthesis", M.L. Occelli, H.E. Robson (eds.), ACS Symp. Ser., **398**, (1989), 360.
37.   G. Bellussi, A. Carati, M.G. Clerici, A. Esposito, R. Millini, F. Buonuomo, Bel. Patent No. 1,001,038, (1989), to Eniricerche S.p.A. and Enichem Synthesis S.p.A.
38.   J.S. Reddy, R. Kumar, Zeolites, **12**, (1992), 95.
39.   D.P. Serrano, H.-X. Li, M.E. Davis, J. Chem. Soc., Chem. Commun., **1992**, 745.
40.   M.A. Camblor, A. Corma, J. Pérez-Pariente, Zeolites, **13**, (1993), 82.
41.   A. Corma, M.T. Navarro, J. Pérez-Pariente, J. Chem. Soc., Chem. Commun., **1994**, 147.
42.   A. Monnier, F. Schüth, Q. Huo, D. Kumar, D. Margolese, R. Maxwell, G. Stucky, M. Krishnamurty, P. Petroff, A. Firouzi, M. Janicke, B. Chmelka, Science, **261**, (1993), 1299.
43.   U. Romano, A. Esposito, F. Maspero, C. Neri, M.G. Clerici, Stud. Surf. Sci. Catal., **55**, (1990), 33; U. Romano, A. Esposito, F. Maspero, C. Neri, M.G. Clerici, Chim. Ind., **72**, (1990), 610.
44.   B. Kraushaar-Czarnetzki, J.H.C. van Hooff, Catal. Lett., **2**, (1989), 43; P.J. Kooyman, P. van der Waal, P.A.J. Verdaasdonk, J.C. Jansen, H. van Bekkum, Catal. Lett., **13**, (1992), 229.
45.   A. Tuel, S. Moussa-Khouzami, Y. Ben Taarit, C. Naccache, J. Mol. Catal., **68**, (1991), 45; J.A. Martens, Ph.L. Buskens, P.A. Jacobs, A.J.H.P. van der Pol, J.H.C. van Hooff, C. Ferrini, H.W. Kouwenhoven, P.J. Kooyman, H. van Bekkum, Appl. Catal. A, **99**, (1993), 71.
46.   A.J.H.P. van der Pol, A.J. Verduyn, J.H.C. van Hooff, Appl. Catal. A, **92**, (1992), 113.
47.   M. Constantini, E. Garcin, M. Gubelmann, J.-M. Popa, Eur. Patent Appl. 385,882, (1990), to Rhône Poulenc Chimie.
48.   A.J.H.P. van der Pol, A.J. Verduyn, J.H.C. van Hooff, submitted for publication in Appl. Catal.
49.   E.A. Kharakanov, S.Y. Narin, A.G. Dedov, Appl. Organomet. Chem., **5**, (1991), 445.
50.   R.O.C. Norman, R. Taylor, "Electrophilic Substitution in Benzenoid Compounds", Elsevier, Amsterdam, 1965, p. 110.
51.   A. Thangaraj, R. Kumar, P. Ratnasamy, Appl. Catal., **57**, (1990), L1.
52.   K. Krohn, K. Brüggmann, D. Döring, P.G. Jones, Chem. Ber., **125**, (1992), 2439.
53.   H. Mimoun, L. Saussine, E. Daire, M. Postel, J. Fischer, R. Weiss, J. Am. Chem. Soc., **105**, (1983), 3101; M. Bonchio, V. Conte, F. Coppa, F. Di Furia, G. Modena, Stud. Surf. Sci. Catal., **66**, (1991), 497.
54.   M.G. Clerici, P. Ingallina, J. Catal., **140**, (1993), 71.
55.   M.G. Clerici, G. Bellussi, U. Romano, J. Catal., **129**, (1991), 159.
56.   M.G. Clerici, G. Bellussi, Eur. Patent Appl. 315,248, (1992).
57.   J. Sobczak, J.J. Ziolkowski, J. Mol. Catal., **13**, (1981), 11.
58.   T. Itoh, K. Jitsukawa, K. Kaneda, S. Teranishi, J. Am. Chem. Soc., **101**, (1979), 159.
59.   T. Tatsumi, M. Yako, M. Nakamura, Y. Yuhara, H. Tominaga, J. Mol. Catal., **78**, (1993), L41.
60.   K.A. Jørgensen, Chem. Rev., **89**, (1989), 431.
61.   G. Bellussi, A. Carati, M.G. Clerici, G. Maddinelli, R. Millini, J. Catal., **133**, (1992), 220.
62.   M.A. Camblor, A. Corma, A. Martinez, J. Pérez-Pariente, S. Valencia in V. Cortés Corberán, S. Vic Bellón (Eds.), Preprints of the 2nd World Congress on New Developments in Selective Oxidation, Benalmádena, Spain, 1993, paper G.4.
63.   P. Roffia, G. Leofanti, A. Cesana, M. Mantegazza, M. Padova, G. Petrini, S. Tonti, P. Gervasutti, R. Varagnolo, Chim. Ind., **72**, (1990), 598.
64.   P. Roffia, G. Leofanti, A. Cesana, M. Mantegazza, M. Padova, G. Petrini, S. Tonti, P. Gervasutti, Stud. Surf. Sci. Catal., **55**, (1990), 43.
65.   S. Tsuda, Chem. Econ. Eng. Rev., **1970**, 39.
66.   A. Thangaraj, S. Sivasanker, P. Ratnasamy, J. Catal., **131**, (1991), 394.
67.   J. Sudhakar Reddy, S. Sivasanker, P. Ratnasamy, J. Mol. Catal., **69**, (1992), 383.
68.   Z. Tvaruzkova, M. Petras, K. Habersberger, P. Jiru, Catal. Lett., **13**, (1992), 117.

69. A. Zecchina, G. Spoto, S. Bordiga, F. Geobaldo, G. Petrini, G. Leofanti, M. Padovan, M. Mantegazza, P. Roffia, Stud. Surf. Sci. Catal., 75, (1992), 719.

70. S. Tonti, P. Roffia, A. Cesana, M.A. Mantegazza, M. Padovan, Eur. Patent Appl. 314,147, (1989), to Montedipe S.p.A.

71. R.S. Reddy, J.S. Reddy, R. Kumar, P. Kumar, J. Chem. Soc., Chem. Commun., 1992, 84.

72. See e.g. O. Bortolini, F. Di Furia, P. Scrimin, G. Modena, J. Mol. Catal., 7, (1980), 59 and refs. cited therein.

73. F. Maspero, U. Romano, J. Catal., in press.

74. V. Conte, F. Di Furia, in G. Strukul (Ed.), "Catalytic Oxidations with Hydrogen Peroxide as Oxidant", Kluwer, Dordrecht, 1993, p. 223.

75. S.E. Jacobson, D.A. Muccigrosso, F. Mares, J. Org. Chem., 44, (1979), 921.

76. For an explanation of the electrophilic nature of metal peroxo complexes see R.D. Bach, G.J. Wolber, B.A. Coddens, J. Am. Chem. Soc., 106, (1984), 6098, and refs. cited therein.

77. D.R.C. Huybrechts, L. de Bruyker, P.A. Jacobs, Nature, 345, (1990), 240.

78. M.G. Clerici, Appl. Catal., 68, (1991), 249.

79. G. Olah, Acc. Chem. Res., 20, (1987), 422.

80. See e.g. M. Faraj, C.L. Hill, J. Chem. Soc., Chem. Commun., 1987, 1487.

81. To be more precise, it is a (very close) linear approximation of Vegard's law. It should be noted that, to the knowledge of the authors, it has not been verified extensively for cases where one of the pure end members does not exist. See for example: L.V. Azároff, "Introduction to Solids", McGraw-Hill, New York, 1960, p. 290.

82. H. van Koningsveld, J.C. Jansen, H. van Bekkum, Zeolites, 10, (1990), 235.

83. S. Pei, G.W. Zajac, J.A. Kaduk, J. Faber, B.I. Boyanov, D. Duck, D. Fazzini, T.I. Morrisson, D.S. Yang, Catal. Lett., 21, (1993), 333 and refs. cited therein.

84. M.R. Boccuti, K.M. Rao, A. Zecchina, G. Leofanti, G. Petrini, Stud. Surf. Sci. Catal., 48, (1989), 133.

85. A. Zecchina, G. Spoto, S. Bordiga, A. Ferrero, G. Petrini, G. Leofanti, M. Padovan, Stud. Surf. Sci. Catal., 69, (1991), 251.

86. A. Tuel, Y. Ben Taarit, J. Chem. Soc., Chem. Commun., 1992, 1578.

87. D.G. Hay, H. Jaeger, J. Chem. Soc., Chem. Commun, 1984, 1433.

88. Z. Gabelica, J.L. Guth, Stud. Surf. Sci. Catal., 49A, (1989), 421.

89. S. Bordiga, F. Boscherini, F. Buffa, S. Coluccia, F. Genoni, C. Lamberti, G. Lefanti, L. Marchese, G. Petrini, G. Vlaic, A. Zecchina, J. Phys. Chem., accepted for publication.

90. P. Behrens, J. Felsche, S. Vetter, G. Shultz-Ekloff, N.I. Jaeger, W. Niemann, J. Chem. Soc., Chem. Commun., 1991, 678; E. Schultz, C. Ferrini, R. Prins, Catal. Lett., 14, (1992), 221.

91. D. Trong On, A. Bittar, A. Sayari, S. Kaliaguine, L. Bonneviot, Catal. Lett., 16, (1992), 85.

92. A. Jentys, C.R.A. Catlow, Catal. Lett., 22, (1993), 251.

93. G. Deo, A.M. Turek, I.E. Wachs, D.R.C. Huybrechts, P.A. Jacobs, Zeolites, 13, (1993), 365.

94. D.R.C. Huybrechts, Ph.L. Buskens, P.A. Jacobs, J. Mol. Catal., 71, (1992), 129.

95. M.A. Camblor, A. Corma, J. Pérez-Pariente, J. Chem. Soc., Chem. Commun., 1993, 557.

96. F. Boccuzzi, S. Coluccia, G. Ghiotti, C. Morterra, A. Zecchina, J. Phys. Chem., 82, (1978), 1298.

97. M.B. Hursthouse, A. Hossain, Polyhedron, 3, (1984), 95; V.A. Zeitler, C.A. Brown, J. Am. Chem. Soc., 79, (1959), 4618.

98. F. Schindler, H. Schmidbaur, Angew. Chem., 79, (1967), 697.

99. F. Geobaldo, S. Bordiga, A. Zecchina, E. Giamello, G. Leofanti, G. Petrini, Catal. Lett., 16, (1992), 109.

100. H. Mimoun, M. Postel, F. Casabianca, J. Fischer, A. Mitschler, Inorg. Chem., 21, (1982), 1303.

101. A.F. Ghiron, R.C. Thompson, Inorg. Chem., 29, (199), 4457.

102. J.D. Lydon, L.M. Schwane, R.C. Thompson, Inorg. Chem., 26, (1987), 2606.

103. M.G. Clerici, P. Ingallina, R. Millini, in R. von Ballmoos, J.B. Higgins, M.M.J. Treacy (Eds.), Proc. 9th Int. Zeolite Conf., Butterworth-Heinemann, Boston, 1992, p. 445.

104. R.A. Sheldon, J. Mol. Catal., 7, (1980), 358.

105. H. Mimoun, Angew. Chem., 94, (1982), 750.

106. R.D. Bach, G.J. Wolber, B.A. Coddens, J. Am. Chem. Soc., 106, (1984), 6098, and refs. cited therein.

107. H.J. Ledon, F. Varescon, Inorg. Chem., 23, (1984), 2735.

212

108. (a) G. Centi, F. Trifirò, J.R. Ebner, V. Franchetti, Chem. Rev., **28**, (1989), 400; (b) G.C. Bond, A.J. Sarkany, G.D. Parfitt, J. Catal., **57**, (1979), 476; (c) G.C. Bond, P. König, J. Catal., **77**, (1982), 309; P.J. Gellings, Catalysis, **7**, (1985), 105.

109. (a) R.J.H. Clark, in J.C. Bailar et al. (eds.), "Comprehensive Inorganic Chemistry", Pergamon, Oxford, 1973, p. 491; (b) L.V. Boas, J.C. Pessoa in G. Wilkinson (ed.), "Comprehensive Coordination Chemistry", Pergamon Press, Oxford, Vol. 3, 453; (c) A. Butler in D. Chasteen (ed.), "Vanadium in Biological Systems", Kluwer Academic, Dordrecht, 1990, p. 25.

110. R.K. Iler, "The Chemistry of Silica", John Wiley, New York, 1979, p. 665.

111. N. Das, H. Eckert, H. Hu, I.E. Wachs, J. Walzer, F. Feher, J. Phys. Chem., **97**, (1993), 8240.

112. G. Deo, I.E. Wachs, J. Phys. Chem., **95**, (1991), 5889.

113. L. Marosi, J. Stabenow, M. Schwarzman, DE Pat. 2,831,631, (1980).

114. A. Miyamoto, D. Medhanavyn, T. Inui, Appl. Catal., **28**, (1986), 89.

115. K. Habersberger, P. Jiru, Z. Tvaruzkova, G. Centi, F. Trifirò, React. Kintet, Lett., **39**, (1989), 95.

116. (a) P. Fejes, J. Halasz, I. Kiricsi, Z. Kele, I. Hannus, C. Fernandez, J.B. Nagy, A. Rockenbauer, Gy. Schöbel, in "New Frontiers in Catalysis", L. Guczi et al. (eds.), Akademiai Kiado, Budapest, 1993, part A, 421; (b) P. Fejes, I. Marsi, I. Kiricsi, J. Halasz, I. Hannus, A. Rockenbauer, Gy. Tasi, L. Korecz, Gy. Schöbel, Stud. Surf. Sci. Catal., **69**, (1991), 173.

117. M.S. Rigutto, H. Van Bekkum, Appl. Catal., **68**, (1991), L1.

118. J. Kornatowski, M. Sychev, V. Goncharuk, W.H. Baur, Stud. Surf. Sci. Catal., **65**, (1991), 581.

119. G. Bellussi, G. Maddinelli, A. Carati, A. Gervasini, R. Millini, in R. Von Ballmoos et al. (eds.), Proc. 9th Int. Zeol. Conf., Butterworth-Heinemann, Boston, 1992, Vol.I, 207.

120. G. Centi, S. Perathoner, F. Trifirò, A. Aboukais, C.F. Aissi, M. Guelton, J. Phys. Chem., **96**, (1992), 2617.

121. (a) P.R. Hari Prasad Rao, R. Kumar, A.V. Ramaswamy, P. Ratnasamy, Zeolites, **13**, (1993), 663, (b) P.R. Hari Prasad Rao, A.V. Ramaswamy, P. Ratnasamy, J. Catal., **137**, (1992), 225.

122. S.B. Hong, C.G. Kim, Y.S. Uh, Y.K. Park, S.I. Woo, Korean J. Chem. Eng., **9**, (1992), 16.

123. M.S. Rigutto, unpublished work.

124. A. Tuel, Y. Ben Taarit, Appl. Catal. A. General, **102**, (1993), 201.

125. K.R. Reddy, A.V. Ramaswamy, P. Ratnasamy, J. Chem. Soc., Chem. Commun., **1992**, 1613.

126. R. Kumar, K.R. Reddy, A. Raj, P. Ratnasamy, Proc. 9th Int. Zeol. Conference, R. Von Ballmoos et al. (eds.), Butterworth-Heinemann Publ., Montreal 1992, Vol. 1, 189.

127. V. Dubanska, CERAMICS-Silikaty, **36**, (1992), 31.

128. H. Takahashi, M. Shiotany, H. Kobayashi, J. Sohma, J. Catal., **14**, (1969), 134.

129. S.S. Eaton, G.R. Eaton, in D. Chasteen (ed.), "Vanadium in Biological Systems", Kluwer Academic, Dordrecht, 1990, p. 199.

130. G. Martini, M.F. Ottaviani, G.L. Seravalli, J. Phys. Chem., **79**, (1975), 1716.

131. See e.g.: H. Eckert, I.E. Wachs, J. Phys. Chem., **93**, (1989), 6796 and refs. cited therein.

132. M.S. Rigutto, H. van Bekkum, G. Deo, N. Arora, I.E. Wachs, H. Eckert, manuscript in preparation.

133. F. Liebau, "Structural Chemistry of Silicates", Springer Verlag, Berlin, 1985, p. 195.

134. P.R. Hari Prasad Rao, A.A. Belhekar, S.H. Egde, A.V. Ramaswamy, P. Ratnasamy, J. Catal., **141**, (1993), 595.

135. (a) R.M. Dessau, K.D. Schmitt, G.T. Kerr, G.L. Woolery, L.B. Alemany, J. Catal., **104**, 484; (b) R.M. Dessau, K.D. Schmitt, G.T. Kerr, G.L. Woolery, L.B. Alemany, J. Catal., **109**, 472; (c) K. Yamagishi, S. Namba, T. Yashima, **95**, (1991), 872.

136. (a) L.W. Zatorski, G. Centi, J. Lopez Nieto, F. Trifirò, G. Bellussi, V. Fattore, Stud. Surf. Sci. Catal., **49B**, (1989), 1243; (b) G. Bellussi, G. Centi, S. Perathoner, F. Trifirò, in S. Oyama, W. Hightower (eds.), "Catalytic Selective Oxidation", ACS Symp. Ser., **523**, 1993, p. 281.

137. (a) P. Ratnasamy, R. Kumar, Catal. Lett., **22**, (1993), 227; (b) A.V. Ramaswamy, S. Sivasanker, Catal. Lett., **22**, (1993), 239; (c) P.R. Hari Prasad Rao, K.R. Reddy, A.V. Ramaswamy, P. Ratnasamy, Stud. Surf. Sci. Catal., **78**, (1993), 385; (d) P.R. Hari Prasad Rao, A.V. Ramaswamy, Appl. Catal. A. General, **93**, (1993), 123; (e) P.R. Hari Prasad Rao, A.V. Ramaswamy, J. Chem. Soc., Chem. Commun., (1992), 1245.

138. C. Marchal, A. Tuel, Y. Ben Taarit, Stud. Surf. Sci. Catal., **78**, (1993), 447.

139.	M.S. Rigutto, H. van Bekkum, Recent Research Reports of the International Symposium "Zeolite Chemistry and Catalysis", Prague, 1991, abstract no. 2.
140.	P. Ratnasamy, R. Kumar, Catal. Today, **8**, (1991), 329.
141.	M. Kawai, T. Kyora, Japan. Pats. 0356,439 and 0358,954, (1991), to Mitsui Toatsu Chemicals.
142.	R.A. Sheldon, J.D. Chen, J. Dakka, E. Neeleman, Preprints of the IInd World Congress on Selective Oxidation, Belmaneda, Spain, paper G.1.
143.	J. Muzart, Chem. Rev., **92**, (1992), 113.
144.	U. Cornaro, P. Jiru, Z. Tvaruzkova, K. Habersberger, Stud. Surf. Sci. Catal., **69**, (1991), 165.
145.	G. Spoto, S. Bordinga, E. Garrone, G. Ghiotti, A. Zecchina, J. Mol. Catal., **74**, (1992), 175.
146.	R. Szostak, "Molecular Sieves; Principles of Synthesis and Identification", Van Nostrand Reinhold, New York, 1989.
147.	C.S. Blackwell, R.L. Patton, J. Phys Chem., **88**, (1984), 6135.
148.	L.B. McCusker, C. Baerlocher, E. Jahn, M. Bülow, Zeolites, **11**, (1991), 308;
149.	(a) R.A. Schoonheydt, R. de Vos, J. Pelgrims, H. Leeman, Stud. Surf. Sci. Catal., **49 A**, (1989), 559; (b) C. Montes, M.E. Davis, B. Murray, M. Narayana, J. Phys. Chem., **94**, (1990), 6425.
150.	L.E. Iton, I. Choi, J.A. Desjardins, V.A. Maroni, Zeolites, **9**, (1989), 535.
151.	B. Kraushaar-Czarnetzki, W.G.M. Hoogervorst, R.R. Andrea, C.A. Emeis, W.H.J. Stork, J. Chem. Soc., Faraday Trans., **87**, (1991), 891.
152.	M.P.J. Peeters, L.J.M. van de Ven, J.W. de Haan, J.H.C. van Hooff, Colloids Surf., **72**, (1993), 87.
153.	J.M. Benett, B.K. Marcus, Stud. Surf. Sci. Catal., **37**, (1988), 269.
154.	J. Jänchen, M. Peeters, J. van Wolput, J. Wolthuizen, J. van Hooff, U. Lohse, submitted for publication.
155.	J. Dakka, R.A. Sheldon, Netherl. Pat. 9,200,968, (1992).
156.	M.P.J. Peeters, M. Busio, P. Leijten, J.H.C. van Hooff, submitted for publication in Appl. Catal.
157	S.S. Lin, H.S. Weng, Appl. Catal. A, **105**, (1993), 289.
158.	C. Montes, M.E. Davis, B. Murray, M. Narayana, J. Phys. Chem., **94**, (1990), 6431.
159.	S.H. Jhung, Y.S. Uh, H. Chon, Appl. Catal., **62**, (1990), 61.
160.	E.M. Flanigen, R.L. Patton, S.T. Wilson, Stud. Surf. Sci. Catal., **37**, (1987), 13.
161.	M.S. Rigutto, H. van Bekkum, J. Mol. Catal, **81**, (1993), 77; M.S. Rigutto, H. van Bekkum, Extended Abstracts of the 9th Int. Zeolite Conf., Montreal, 1992, abstract no. RP54.
162.	M.J. Haanepen, J.H.C. van Hooff, DGMK Tagungsbericht, **9204**, (1992), 227.
163.	(a) X. Yu, P.J. Maddox, J.M. Thomas, Polyhedron, **8**, (1989), 819; B.Z. Wan, K. Huang, T.C. Yang, C.T. Tai, J. Chin. Inst. Chem. Eng., **22**, (1991), 17; (c) S. Hočevar, J. Batista, J. Kaučič, J. Catal., **139**, (1993), 351.
164.	M. Helliwell, V. Kaučič, G.M.T. Cheetham, M.M. Harding, B.M. Kariuki, P.J. Rizkallah, Extended Abstracts of the 9th Int. Zeolite Conf., Montreal, 1992, abstract no. RP203.
165.	J.B. Parise, Acta. Crys. C, **42**, (1986), 670.
166.	J.D. Chen, J. Dakka, E. Neeleman, R.A. Sheldon, J. Chem. Soc., Chem. Commun., **1993**, 1379; J.D. Chen, J. Dakka, R.A. Sheldon, Appl. Catal. A, **108**, (1994), L1.

J.C. Jansen, M. Stöcker, H.G. Karge and J. Weitkamp (Eds.)
Advanced Zeolite Science and Applications
Studies in Surface Science and Catalysis, Vol. 85
© 1994 Elsevier Science B.V. All rights reserved.

PREPARATION OF COATINGS OF MOLECULAR SIEVE CRYSTALS FOR CATALYSIS AND SEPARATION

J.C. Jansen[a], D. Kashchiev[b] and A. Erdem-Senatalar[c]

[a] Laboratory of Organic Chemistry and Catalysis, Delft University of Technology, Julianalaan, 136, 2628 BL, Delft, The Netherlands

[b] Institute of Physical Chemistry, Bulgarian Academy of Sciences, ul. Akad. G. Bonchev 11, Sofia 1113, Bulgaria

[c] Department of Chemical Engineering, Istanbul Technical University, 80626 Maslak-Istanbul, Turkey

1. INTRODUCTION

The development of coatings of molecular sieves in general, but of zeolites in particular, is rapidly gaining interest.

Two types of molecular sieve coatings can be distinguished which will further be referred to as films and layers.

A film is defined as a continuous solid phase of microporous crystals oriented in a parallel mode on the support. The thickness of the film is smaller than a few micrometer.

A layer consists of a (dis)continuous solid phase of microporous crystals more or less disorderly oriented on a support. The thickness of the layer lies between five and fifty micrometer.

The specific properties of a coating are:

In general:

improved adsorption capacity

reduced diffusion times

216

Reactant flow along the coating:

low pressure drop in preshaped catalysts

Reactant flow through the coating:

(catalytic) membrane

Reactant flow within the coating:

optimal heat transfer in metal modules

maximal response in sensor and electrochemical devices

A number of potential applications are discussed in Chapters 4, 5 and 15 as well as the references cited therein.

The preparation of coatings, mainly deposits of polycrystalline zeolite, is well known [1-6]. In general, the phases formed are layers, often showing twinned, intergrown and via amorphous silica phases connected crystals resulting in reduced diffusion properties.

In three studies a continuous layer of crystals of the MFI-type, see Figure 1, was prepared [7-9]. The behaviour of the zeolite layer in a membrane configuration could be tested this way. The results are discussed in Chapter 15.

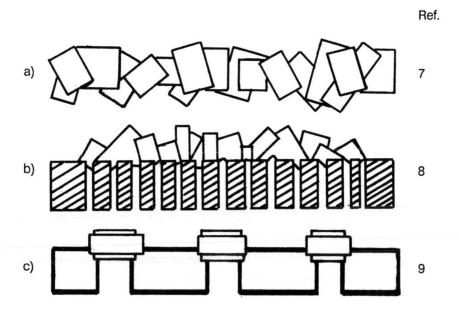

Figure 1. Different layers of MFI-type zeolite in a membrane configuration. a) and b) unsupported and porous metal supported silicalite-1 crystals. c) twinned crystals of silicalite-1 on porous metal support.

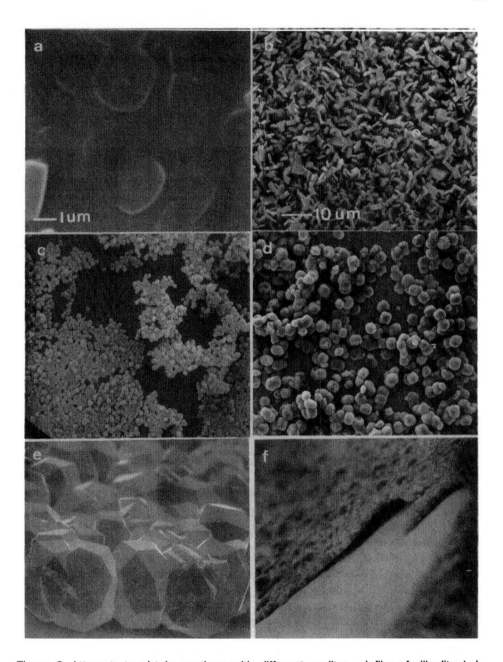

Figure 2. Attempts to obtain coatings with different zeolites. a) film of silicalite-1, b-direction perpendicular to support, b) ZSM-5 layer, c-direction perpendicular to support, c) areas of gmelinite crystal layers, d) stacks of mordenite needles parallel to support, e) analcime crystals developed from layer of gel spheres on support and f) continuous mono-layer of NaA crystals in random orientation, and not bonded, to support.

In a recent publication the formation of a coating of oriented silicalite-1 crystals with a thickness of 2 μm was reported [10]. Preliminary results indicate that the preparation of films of silicalite-1 crystals with a thickness of less than .2 μm are possible [41]. In both cases the crystal orientation and thus the pore direction were characterized by scanning electron microscopy (SEM) and X ray diffraction. The XRD-pattern and an atomic force microscope (AFM)-view of a film are given in Figures 5 and 6 d, respectively. The major advantage of such a film is: a well defined pore system; with unique diameter, short length, small length distribution and no blocking. Apart from the silicalite-1 films, discontinuous layers of Cu-ZSM-5 were prepared as well. The crystals were then oriented perpendicular to the support surface, actually indicating the ability to control synthesis of coatings of MFI-type zeolite parallel as well as axial on support [10]. In this chapter the theory on thin coating nucleation and crystallization will be presented. The characteristics of the films and layers of zeolites silicalite-1, ZSM-5, Gmelenite and NaA are related to the precipitation conditions. The possibilities to prepare films of zeolites, see Figure 2, the nucleation, crystal orientation and crystal growth history are reported.

2. THEORY

Some general theoretical results obtained in the investigation of the kinetics of nucleation and growth of thin coatings will be outlined briefly. In many cases of deposition of such films from vapours, for example, the results were found to agree with the experiment. The general theory of film growth is therefore to be expected to be a valuable guide also in the case of zeolite films. That is why, in this section first some general results are presented and then used to obtain dependencies directly applicable to zeolite films which, typically, are grown from solutions or gels. For a more thorough examination of the general theory of thin film growth the reader is referred to a number of review articles [11-21] and the references cited therein.

2.1. Growth mode

The mode by which a thin film grows during the initial stages of its formation is of great importance for its structure and properties. Most frequently observed are [14, 17-19, 21] the layer (or Frank-van der Merwe), the island (or Volmer-Weber), the layer-plus-island (or Stranski-Krastanov) and the continuous (normal of liquid-like) modes of growth. The substrate (or the support) is covered either by spreading of only one or a few monatomic layers (MLs) or by nucleation and growth of separate island-like crystallites of multiatomic height when the film growth is the layer or the island one, respectively. The Stranski-Krastanov growth is a combination of layer and island growth: the building up of one or several MLs is followed by formation of island-like crystallites on top of them. Continuous growth occurs when the substrate is again covered by simultaneous filling of several monolayers, but this filling is a random rather than nucleation-mediated process. Films grown by the continuous mode tend to be of lower degree of crystallinity and even amorphous, while those grown by the island mode may develop grain boundaries between the separate crystallites and become polycrystalline. The layer mode of growth is evidently the most favourable one for growing thin films with single crystal structure. The question therefore arises what is the reason for the different growth modes and whether it is possible to change them by varying the experimental conditions.

Bauer [22] has shown that the simplest answer to this question can be obtained on thermodynamic grounds. Indeed, before the formation of the film we have a single substrate/solution (or gel) interface with specific free energy σ_s (J/m^2). After the formation of a continuous, uniformly thick film the interfaces are two, the film/solution (or gel) and the substrate/film ones, with specific surface free energies σ_f (J/m^2) and σ_{sf} (J/m^2), respectively. All other conditions being the same, the energy change associated with the formation of the film is therefore $\Delta\sigma = \sigma_f + \sigma_{sf} - \sigma_s$.

Obviously, if $\Delta\sigma \leq 0$, the formation of the film as a continuous layer is energetically favourable and this is the thermodynamic criterion for layer growth [22]. Conversely, when $\Delta\sigma > 0$, the system is in a lower energy state if the film is broken into separate parts (island) and this, thermodynamically, is the criterion for island growth [22]. The $\Delta\sigma \leq 0$ condition implies that binding between a molecule of the film and a molecule of the substrate stronger than or equal to the binding between two molecules of the film and corresponds to the condition for complete wetting of substrates by liquid films. The opposite is true when $\Delta\sigma > 0$ which is also the condition for incomplete wetting: a film-substrate molecular pair is then more weakly bound than two molecules of the film. The conclusion from the above consideration is that for a given film/substrate system (i.e. $\Delta\sigma$) thermodynamics favour either the layer of the island mode of growth.

As pointed out by Venables and Price [14], however, the film growth mode is essentially a kinetic phenomenon and can be changed by varying the growth conditions. The existence of a critical supersaturation (corresponding, e.g., to a critical temperature or solution concentration) for growth mode transition has indeed been observed in laboratory [23-28] and computer [29] experiments. It is clear, therefore, that the modes of thin film growth and the transitions between them can be analyzed comprehensively solely with the aid of kinetic considerations. The theory of polylayer growth of thin solid films proposed by Kashchiev [30] provides a framework for such an analysis.

2.2. Theory of polylayer growth

In the theory of polylayer growth of thin films [30] the modes of film growth are described from a unified point of view and the condition for transition between them is obtained in a general form. The theoretical model is as follows. The initially bare, atomically smooth substrate is free of screw dislocations. It can, however, possess different nucleation-active sites due, for example, to adsorbed molecules, edge dislocations, etc. The building up of the film occurs via simultaneous filling of a number of layers (not necessarily of ML thickness), each of them being deposited only on top of the layer underneath. For a given film/substrate system, the number of simultaneously filled layers and, thereby, the film growth mode is entirely determined by the growth conditions.

The kinetics of overall filling of a given film layer are characterized quantitatively by a time constant θ (s) which is a phenomenological parameter. This quantity can be determined only by means of concrete model considerations and this is the way the theory accounts for the kinetic peculiarities of the case under particular study. In some simple cases of layer filling θ reads [29-31]

$$\theta = (3/\pi J v^2)^{1/3} \tag{1}$$

$$\theta = (1/\pi N v^2)^{1/2} \tag{2}$$

$$\theta = 1/(\omega^+ - \omega^-) \tag{3}$$

Eq. (1) is applicable to layer filling by progressive nucleation (PN) and lateral growth of disk-shaped clusters which appear and spread with time-independent nucleation rate J ($m^{-2}s^{-1}$) and spreading velocity v (m/s). Eq. (2) gives θ when a layer is filled by instantaneous nucleation (IN) of N (m^{-2}) disk-shaped clusters which spread radially as in the PN case. Eq. (3) is valid for continuous (or liquid-like) layer filling which occurs without nucleation simply by random incorporation of molecules into the layer with attachment and detachment frequencies ω^+ (s^{-1}) and ω^- (s^{-1}), respectively.

Given the time constants of filling of the successive layers, it is possible to determine a number of important film characteristics: the film growth rate and mean thickness at all stages of deposition [30], the shape factor f of the individual crystallites on the substrate [32], the mean film thickness h (m) at the moment of 99% covered substrate [30], etc. As f is a convenient measure of the ratio between the crystallite mean height and the radius of the crystallite base [32] and since h can be interpreted as the mean thickness at which the growing film reaches continuity [31], both f and h are very informative about the film growth mode.

Most simply, the film growth can be analyzed upon assuming that the influence of the substrate is felt only within the first film layer (since this layer is in immediate contact with the substrate) and that the molecules of the second, third, ect. layers are deposited as if onto a substrate of their own bulk phase. In this case it becomes possible to express f and h only through the time constants of filling of the first and the second layers, θ_1 and θ_2, respectively. It then follows that [32]

$$f = d/4\Gamma v_b \theta_2 \tag{4}$$

for a crystallite increasing steadily its height and the radius of its base and that [30]

$$h = h^* + \psi d\theta_1/\theta_2 \tag{5}$$

Here Γ and ψ are numerical factors close to unity, d (m) is the molecular diameter, v_b (m/s) is the constant spreading velocity of the crystallite base, and h^* (m) is the height of the crystal nucleus on the substrate. The latter determines the thickness of the first film layer and is given by the thermodynamic Gibbs-Thomson formula [30]

$$h^* = 2d\alpha_m\Delta\sigma/\Delta\mu, \qquad (0 < \Delta\mu < 2\alpha_m\Delta\sigma) \tag{6}$$

$$h^* = d, \qquad (\Delta\mu \geq \alpha_m\Delta\sigma) \tag{7}$$

in which α_m (m^2) is the area of a molecule in the nucleus, and $\Delta\mu$ (J) is the super-saturation defined as the difference between the chemical potentials of the molecules in the parent phase (e.g., solution or gel) and in the film. According to eq. (7), the first film layer is of ML height only when the supersaturation is high enough.

Since f = 1/2 corresponds to hemispherical crystallites [32], eq. (4) shows that needle-like crystallites will grow on the substrate when G $>>$ v_b (G = $d/\Gamma\theta_2$ is approximately the crystallite growth rate normally to the substrate [30, 32]). Conversely, if G $<<$ v_b, the crystallites will be very flat, platelet-like. Thus f $>>$ 1 and f $<<$ 1 are indicative for island and layer mode of growth, respectively. The same indication comes from eq. (5): h $>>$ h^* when θ_1 $>>$ θ_2 (slow filling of the first layer and, hence, island growth) and h $\approx$ h^* if θ_1 $<<$ θ_2 (quickly filled first layer, i.e. layer growth). These conclusions are supported also by Monte Carlo simulation data for f [33] and h [29].

2.3. Nucleation rate

Most generally, the stationary rate of nucleation on a substrate is given by the expression [34]

$$J = z\omega^* n_o \exp(-W^*/kT). \tag{8}$$

Here $z \approx 0.01$ to 1 is the so-called Zeldovich factor, ω^* (s^{-1}) is the attachment frequency of molecules to the nucleus, n_o (m^{-2}) is the density of substrate sites on which a nucleus can be formed, W^* (J) is the nucleation work, k is the Boltzmann constant and T is the absolute temperature. If N_α (m^{-2}) nucleation-active sites are present on the substrate, $n_o = N_\alpha$, and in the absence of such sites $n_o = 1/\alpha_m \approx 10^{19}$ m^{-2}. The nucleation work depends on $\Delta\sigma$ and $\Delta\mu$ according to [17, 30, 34]

$$W^* = \beta' v^2_m \sigma^3_{ef}/\Delta\mu^2, \qquad (0 < \Delta\mu < 2\alpha_m\Delta\mu) \tag{9}$$

$$W^* = \beta'' \alpha_m \chi^2/(\Delta\mu - \alpha_m\Delta\sigma), \qquad (\Delta\mu \geq 2\alpha_m\Delta\sigma) \tag{10}$$

where β' and β'' are shape factors (e.g., $\beta' = 4\pi$ and β'' for disk-shaped nuclei), v_m (m^3) is the volume of a molecule in the nucleus, σ_{ef} (J/m^2) is an effective specific peripheral energy of the nucleus. The nucleation in the two $\Delta\mu$ ranges specified by eqs. (9) and (10) is often called three-dimensional (3D) and two-dimensional (2D), respectively, because while at lower supersaturations the nucleus height is $\Delta\mu$-dependent, at higher supersaturations it is fixed and equal to the molecular diameter (see eqs. (6) and (7)).

For nucleation of crystals from solutions $\Delta\mu$ is given by [34, 35]

$$\Delta\mu = kT \ln(C/C_e) \tag{11}$$

where C (m^{-3}) is the actual concentration of film molecules in the solution, and C_e (m^{-3}) is the respective equilibrium concentration (also called solubility) at which the film neither grows nor dissolves. In many cases C_e obeys the Arrhenius-type temperature dependence

$$C_e = C_o e^{-\lambda/kT} \tag{12}$$

where λ (J) is the heat of dissolution, and $C_o(m^{-3})$ is a practically T-independent concentration. The attachment frequency ω^* depends on the concrete mechanism of transport of molecules to the nucleus surface and/or their incorporation into the nucleus, but, typically, it is of the form [34, 35]

$$\omega^* = \omega Ce^{-E/kT}. \tag{13}$$

Here the frequency factor ω (m^3/s) is approximately constant with respect to C and T, and E (J) is the activation energy for molecular transport across the nucleus/solution interface.

Combining eqs. (8)-(11) and (13) thus yields

$$J = KCe^{-E/kT}\exp(-A/T^3\ln^2 S), \qquad (o < T\ln S < 2\tau) \tag{14}$$

$$J = KCe^{-E/kT}\exp[-B/(T^2\ln S-\tau)], \qquad (T\ln S \geq 2\tau). \tag{15}$$

These equations give the dependence of J on C and T for, respectively, 3D and 2D nucleation on a substrate. In them the temperature-dependent supersaturation ratio $S > 1$ and the virtually C,T-independent kinetic factor K (m/s) and thermodynamic parameters A (K^3), B (K^2) and τ (K) are given by

$$S = C/C_e = (C/C_o)e^{-\lambda/kT} \tag{16}$$

$$K = z\omega n_o \tag{17}$$

$$A = \beta' v_m^2 \sigma_{ef}^3/k^3 \tag{18}$$

$$B = \beta'' \alpha_m \chi^2/k^2 \tag{19}$$

$$\tau = \alpha_m \Delta\sigma/k. \tag{20}$$

It should be noted that eqs. (14) and (15) can also be used to describe nucleation in concentrated solutions and gels if everywhere in them concentrations are replaced by activities.

2.4. Spreading velocity

The spreading velocity of the clusters within a given film layer is expressed as (e.g., [36])

$$v = d(\omega^+ - \omega^-). \tag{21}$$

For a molecularly rough cluster periphery, ω^- is approximately equal to ω^+ at $C = C_e$. Since ω^+ is of the form of ω^* from eq. (13), we then have

$$\omega^+ = \omega C e^{-E/kT} \tag{22}$$

$$\omega^- = \omega C_e e^{-E/kT} \tag{23}$$

and eq. (21) leads to the following C,T-dependence of v [35]

$$v = d\omega(C - C_e)e^{-E/kT}. \tag{24}$$

It is worth noting that, numerically, ω and E in this formula may in some cases be different from ω and E in eqs. (14), (15) and (17). Also, eq. (24) gives the spreading velocity v_b of the crystallite base in the case of molecularly rough crystallite periphery.

2.5. Number of nuclei

The number N of nuclei formed instantaneously in the first film layer is not easily obtainable in a general form, because it depends on factors of different nature. Physically, N can be related to the maximum (or saturation) number of nuclei formed in the layer during its filling [31]. Since IN occurs usually in the presence of strongly nucleation-active sites in the layer, it may be expected that in many cases

$$N = N_\alpha \tag{25}$$

It should be noted, however, that ingestion of the active sites by nucleation-forbidden zones formed around the laterally growing clusters or by the clusters themselves [37, 38] can considerably reduce N below N_α and make it depend also on J and v.

2.6. Crystallite shape factor

We can now use the results given above for finding the C,T-dependence of the shape factor f of zeolite crystallites on a substrate. Jansen et al. [39] have shown that such crystallites can be nucleated on solid substrates in contact with a solution or at a gel/solution interface. In the latter case the zeolite crystallites grow into the gel so that the solution can be regarded formally as a liquid substrate. Also, both the nucleation and the growth of the crystallites take place in the presence of template molecules.

We limit the consideration to a crystallite which has a molecularly rough periphery of its base and grows unperturbed by neighbouring crystallites and examine the cases of filling of the crystallite upper MLs by the PN or by the continuous mechanism. Using eqs. (1), (3) and (24) to express θ_2 and v_b in eq. (4) and recalling eqs. (15), (22) and (23), we find that

$$f = [(S_2 - 1)/4(S_1 - 1)]\exp(-B/3T^2\ln S_2) \tag{26}$$

for the PN mechanism and that

$$f = [(S_2 - 1)/4(S_1 - 1)] \tag{27}$$

for the continuous mechanism. In deriving these equations it is assumed that ω and E in eq. (15) are equal to those in eq. (24) and it is taken into account that $\Gamma \approx 1$, $(\pi z/3^{1/3} \approx 1$ and $(C_2-C_e)^{2/3}C_2^{1/3} \approx (C_2 - C_e)$, and that $\Delta\sigma = 0$ and $n_o = 1/\alpha_m \approx 1/d^2$ (the nucleation in the second and next layers of the crystallite) and that $\Delta\sigma = 0$ (the nucleation in the crystallite second and next layers is on own substrate). Since during growth the concentration closer to the substrate is lower, the crystallite upper layers are filled at higher supersaturation ratio $S_2 = C_2/C_e$ than the supersaturation ratio $S_1 = C_1/C_e$ for the first layer, C_2 and C_1 being the respective concentrations away and at the substrate ($C_2 \geq C_1$).

As already noted, smaller f-values correspond to growth of flatter crystallites, i.e. to the layer mode of growth. Examining eqs. (26) and (27), we conclude that such growth is favoured (i) by the PN mechanism (because of the exponential factor), (ii) by lower concentrations when the PN mechanism is operative (then $S_1,S_2 \to 1$ and f from eq. (27) tends to zero), (iii) by temperatures decreasing the product $T\ln[C_2/C_e(T)] = \Delta\mu_2/k$ (again when the PN mechanism is operative), and (iv) by the absence of a concentration gradient normally to the substrate (then $S_1 = S_2$ and the absolute upper limit of f is 1/4 which corresponds to nearly hemispherical crystallites). It must be emphasized that since these conclusions are drawn by means of concrete mechanisms of spreading of the crystallite base and of filling of the crystallite upper

MLs, some of them may not be valid for other mechanisms.

2.7. Film thickness at the moment of covering the substrate

The C,T-dependence of the mean thickness h of a zeolite film at the moment when the substrate becomes 99% covered can be determined from eq. (5) provided the filling mechanisms of the first and the next film layers are specified. In the experiments of Jansen et al. [39] the nucleation of the zeolite crystallites on the substrate occurs in the presence of template molecules which seem to play the role of nucleation-active sites. That is why we shall consider filling of the first film layer by either the PN or the IN mechanism combined with filling of the second, third, etc. film layers by the PN mechanism. In these two cases, using eqs. (1) and (2) to express θ_1 and θ_2 in eq. (5) and allowing for eqs. (6), (7), (14)-(17) and (24), we find that

$$h/d = 2\tau/T\ln S_1 + (\alpha_m N_\alpha)^{-1/3}[(S_2 - 1)/(S_1 - 1)]$$

$$\times \exp(A/3T^3\ln^2 S_1 - B/3T^2\ln S_2), \qquad (0 < T\ln S_1 < 2\tau) \qquad (28)$$

$$h/d = 1 + (\alpha_m N_\alpha)^{-1/3}[(S_2 - 1)/(S_1 - 1)]$$

$$\times \exp[B/3(T^2\ln S_1 - \tau T) - B/3T^2\ln S_2], \qquad (T\ln S_1 \geq 2\tau) \qquad (29)$$

for PN filling of both the first and the second (and the next) film layers and that

$$h/d = 2\tau/T\ln S_1 + (\alpha_m N_\alpha)^{-1/2}[(S_2 - 1)/S_1 - 1)]$$

$$\times \exp(-B/3T^2\ln S_2), \qquad (0 < T\ln S_1 > 2\tau) \qquad (30)$$

$$h/d = 1 + (\alpha_m N_\alpha)^{-1/3}[(S_2 - 1)/(S_1 - 1)]$$

$$\times \exp(-B/3T^2\ln S_2), \qquad (T\ln S_1 \geq 2\tau) \qquad (31)$$

for IN filling of the first and PN filling of the second (and the next) layers of the film. As in the previous section, ω and E in eqs. (14) and (15) are assumed to be equal to those in eq. (24) and it is taken into account that $\psi \approx 1$, $(z/3\pi^{1/2})^{1/3} \approx 1$, and $(C - C_e)^{2/3}C^{1/3} \approx (C - C_e)$, that $N_0 = N_\alpha$ and $N = N_\alpha$ for the first film layer (nucleation of active sites provided, for instance, by template molecules of density N_α which are adsorbed on the substrate) and that $\Delta\sigma = 0$ and $n_0 = 1/\alpha_m \approx 1/d^2$ for the second and the next film layers (nucleation on own substrate).

Eqs. (28)-(31) represent the sought C,T-dependence of h for the considered mechanisms of nucleation-mediated filling of the of the successive layers of the film. Eqs. (28) and (29) show that when the first layer is filled by the PN mechanism, h decreases with increasing concentration and for $S_1, S_2 \to \infty$ reaches the limiting value of $h = (1 + \alpha_m N_\alpha^{-1/3})d$. When the first film layer is filled by the IN mechanism, however, according to eqs. (30) and (31), h is a different function of the concentration: with increasing S_1 and S_2 it first decreases to a minimum and then increases up to the limiting value of $h = (1 + \alpha_m N_\alpha^{-1/2})d$. This increase of h with the concentration is analogous to that of the crystallite shape factor f from eq. (26) and the similarity between eqs. (26) and (31) is physically understandable in view of the fact that they describe the same mechanism of layer filling. As to the temperature dependence of h, eqs. (28)-(31) predict again diametrically different behaviour when the first film layer is filled by the two different nucleation mechanisms. Lower h-values can be obtained by changing the temperature so that the products $T\ln[C_1/C_e(T)] = \Delta\mu_1/k$ and $T\ln[C_2/C_e(T)] = \Delta\mu_2/k$ are increased or decreased in the PN and the IN cases, respectively. Looking at eqs. (28)-(31), we see also that during both the PN and the IN filling, lower mean film thickness is favoured by eliminating concentration gradients normally to the substrate (then $S_1 = S_2$) and/or by higher density N_α of nucleation-active sites on the substrate. Concerning the $h(N_\alpha)$ behaviour, the IN mechanism is more effective than the PN one: $h \propto N_\alpha^{-1/2}$ and $h \propto N_\alpha^{-1/3}$, respectively. Again, it must be stressed that the above conclusions pertain to particular mechanisms of layer filling and some of them may not have a general validity.

The inference is therefore that studying experimentally the h(C), h(T) and $h(N_\alpha)$ dependencies in various cases of growth of thin zeolite films makes it possible to discriminate between the PN and the IN mechanism of filling of the first film layer. Also, the above conclusions concerning the behaviour of h may be a useful guide in finding the optimal conditions for growing thin zeolite films by the layer mode of growth which, as already noted, is characterized by lower values of h/d down to 1 ML.

3. EXPERIMENTS

A simple zeolite crystallization i.e. of silicalite-1 was performed to test if the above theory is applicable. In addition low Si/Al ratio zeolite crystallization is compared with the silicalite-1 formation.

228

3.1 Supports

The supports used for the crystallization were mainly metals and ceramics as given in Table 1.

Table 1, Supports applied in the crystallization of zeolites

Metals			Composites	Single crystals
aluminum	copper	Au	cordierite	quartz
silicon	nickel	Pt	mullite	rutile
titanium	st. steel		mica	sapphire

In most cases dissolution of the metal support is expected under the zeolite synthesis conditions. Crystal coatings are formed, however, before dissolution takes place. Apart from Au, Pt and mica zeolites could be anchored on the support in particular cases. Before using the support, of which the geometry can be a plate, gauze, wire or module a thorough cleaning procedure is necessary in order to achieve ultimate wetting of the surface. A most adequate cleaning procedure is the following. Thirty minutes boiling in xylene, subsequently fifteen minutes at 80 °C in a mixture of 1 H_2O_2 (30%), 1 NH_4OH (25%) and 5 H_2O parts by volume. Next, storage after sonification before use in distilled, deionized water. The supports were generally positioned in the upper part of the synthesis mixture with the help of a simple holder of teflon mounted on the inner side of the cover of the autoclave, see Figure 3.

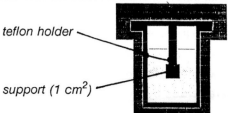

teflon holder

support (1 cm^2)

Figure 3. Holder of teflon in the cover of the autoclave. This holder configuration was chosen in order to avoid convection, influencing nucleation, as much as possible.

3.2. Silicalite-1-type

In the case of MFI, a film as well as a layer type coating of 1 cm^2 can be prepared of which the synthesis procedure has been reported elswhere [10].
In the course of an attempt to prepare a film of zeolite it became clear that crystals of silicalite-1 with smooth rounded top crystal faces in the c-direction and a crystal

thickness of less than 2 μm were the most promising building blocks to comprise such a coating. The crystals could be oriented parallel with the b-direction, and thus the straight channels of the pore system, perpendicular to the support surface.

The crystal growth history of such crystals has been extracted from an earlier investigation into the development of MFI crystals of this form [39]. For reasons of clarity this history will be summarized first.

According to an *in situ* observation of a synthesis, without supports, large gel spheres are developed during the synthesis. Infrared - and elemental analysis indicated that no tetrapropylammonium ions (TPA) were present in the gel phase. Therefore the crystallization started at the interface of gel and liquid phase while proceeding *into* the gel sphere as schematically given in Overview 1. Crystals are formed with smooth rounded top faces apparently caused by kinetic roughening under the high supersaturation prevailing.

As the MFI-crystals on support show smooth rounded top faces and as a gel layer was observed on support preceding the crystallization it is concluded that a similar growth process occurred. Thus the silicalite-1 crystals nucleated at the interface of gel phase formed on support and the synthesis liquid. Crystal growth is then relatively fast for those nuclei oriented with the fastest growth directions parallel to the interface of both nutrient pools, being TPA in the synthesis liquid and SiO_2 in the gel phase. It therefore looks like as from the start crystallites are already more or less parallel oriented to the support surface. In the course of the crystallization *into* the gel phase, as this is the highest concentration of SiO_2 nutrients, a TPA nutrient pool is gradually formed between the growing crystal and the diminishing gel phase, probably caused by the difference in density of the gel phase and the crystal, see Overview 1. As soon as the crystal touches the support surface it is attracted with the ac-plane, the largest plane of the crystal, by electrical and surface tension forces. The crystal then becomes laterally oriented to the support surface. As shown in Figure 6 a, based on comparable forces crystals sometimes show mutual attraction leading to parallel orientation. Subsequently, the crystal can be chemically bonded to the support by condensation reactions of the terminal OH-groups of the crystal and the support as well.

Although the crystal size distribution is small, see Figure 6 a,b and c, the nucleation occurs in a progressive mode, see Figure 4 a, with crystals of different crystal size. All crystals grow, however, in the beginning very fast in a very high local concentration of nutrients, the gel phase. As soon as the gel phase is consumed the crystal growth rate reduces drastically. Therefore, as shown in Figure 4 b, crystals can start at different nucleation times and pass different crystal growth times, however, showing rather the same size at the end of the crystallization time.

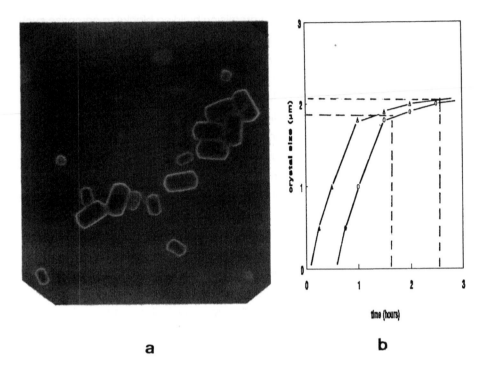

a   b

Figure 4. Progressive nucleation of silicalite-1 on support. a) crystals of different sizes, b) nucleation and crystal growth times of two species resulting in almost the same crystal size.

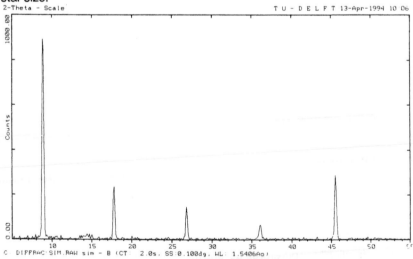

Figure 5. XRD-pattern of a film of silicalite-1 shown in figure 6 d.

Overview 1, comparable events of MFI crystallization in free gel spheres and supported gel film.

Synthesis mixture	SiO_2, NaOH, TPABr, H_2O	SiO_2, TPAOH, H_2O
Temperature (K)	443	436
Time (h)	72 - 120	2 - 4

View on **gel sphere** **gel film**

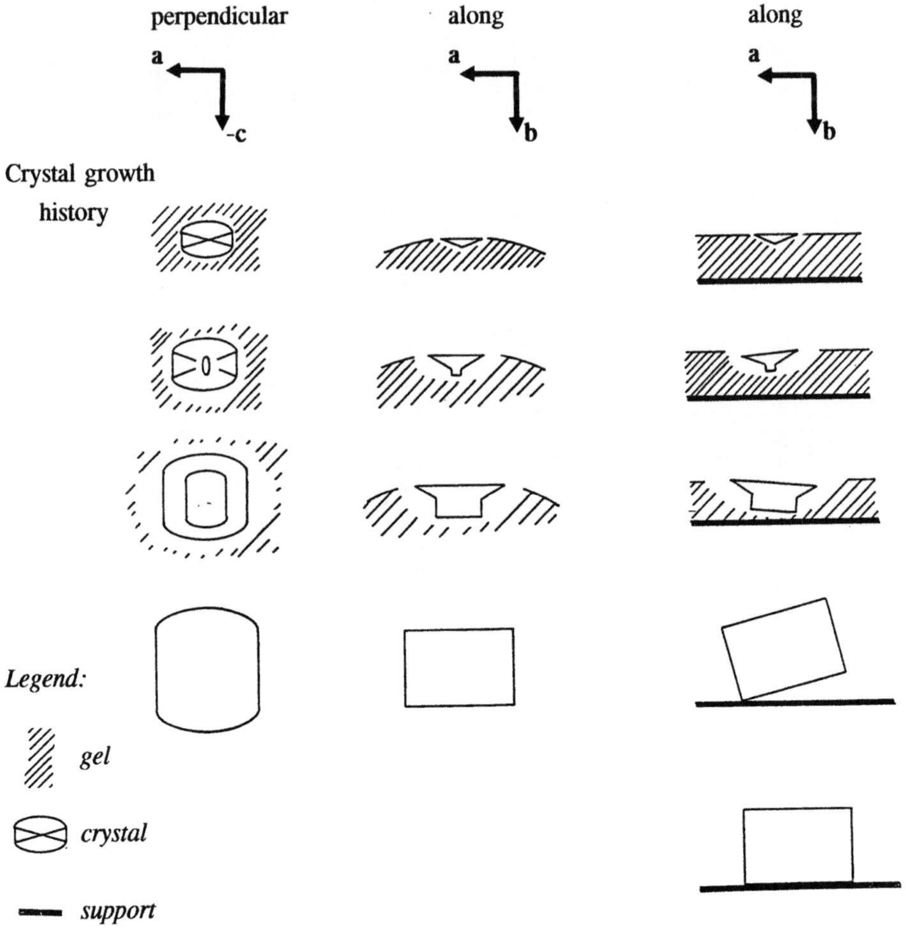

perpendicular along along

a ← a ← a ←

↓ -c ↓ b ↓ b

Crystal growth history

Legend:

gel

crystal

support

| Single crystal size (μm) | 50 - 500 | .0 - 2 |

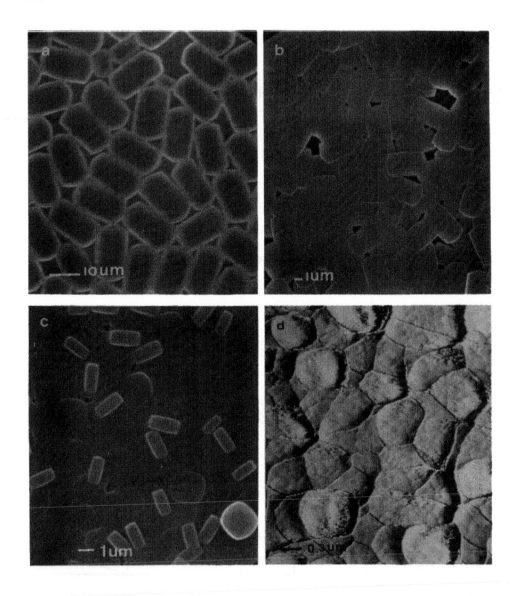

Figure 6. Improvement of silicalite-1 coating to obtain a film. a) and b) silicalite-1 crystals parallel to support, c) smallest crystals of which the form is still recognized, d) film of still oriented, see figure 5, crystallites of silicalite-1.

According to a recent publication thin coatings of smaller crystals can be prepared [41].

Based on a higher concentration of TPA the number of crystals on support increased resulting in smaller crystal dimensions. The thickness of this coating is about 1 um while the orientation of the crystals parallel to the support is maintained according to XRD analysis and SEM, see Figure 6 c.

In the case of higher concentrations a coating was obtained of crystals of .3 μm. This film of crystals is homogeneously covering a 1 cm^2 support as far as light microscopy can show. The XRD diffractogram, see Figure 5, shows only reflections with oko indices indicating the film is comprised of parallel oriented crystallites with the b-direction perpendicular to the support. AFM depicts that the film is continuous while the common form and shape of the MFI-crystals is not recognized. The crystallites without showing any crystal faces have grown together in an irregular pattern due to the very high concentration of the available nutrient pool. In all three cases discussed zeolite crystalline material nucleated and developed in a gel phase on a certain distance, often less than the crystal thickness, from the support surface. In combination with a flat, chemically potential active surface this is an important condition to prepare an oriented layer of microporous phase, connected to the support.

As shown in Figure 6 a-d the length/ width ratio of the crystals approaches unity while the size decreases, clearly resulting in a better two dimensional packing. From the derivation of the shape factor f given by equation (26) for the PN-mechanism it is concluded that the smaller f-values correspond to the growth of flatter crystallites and that such growth is favoured by lower concentrations. It is indeed found here that flatter crystals are developed at low TPA concentrations. Closing the space between the flat relatively large crystals seems not feasable as the nutrient concentration is too low for further growth of the already excisting relatively large crystals. Apparently very small crystals are needed.

In general the experiment shows that smaller the crystal size and the more spherical the shape the smaller the intercrystalline space, see Figure 7. Nucleation experiments reveal that only increasing the TPA concentration results in smaller sized, a larger number and a larger f-factor of crystals [41]. The relation given by expression (8), *vide supra,* seems applicable here. The nucleation rate J depends exponentially on the square of the supersaturation. This supersaturation is exclusively determined by the TPA concentration since the concentration of the other building unit, SiO_2, remains virtually constant in the dense gel phase.

Larger f-values thus a more spherical shape of the smaller crystals, see Table 2, upon increasing TPA concentration is difficult to interprete in the complicated crystal growth system, see Overview 1. From the liquid substrate, as defined by Kashchiev, *vide supra*, the first crystal layer at S1 becomes depleted of SiO_2, while the upper layer(s) at S2 can still grow in a SiO_2-rich environment, however, with an axial growth rate which merely depends on the transport of TPA molecules into the gel phase. On the other hand the S1/ S2 ratio and thus the influence of the SiO_2 nutrient can hardly be predicted in this system.

Table 2
Width/height/length, the a/b/c ratio, respectively of silicalite-1 crystals in μm.

crystal	decreasing size (μm)					
width (a)	70	5	2	1	1	.5
height (b)	13	1	.75	.5	.45	.3
length (c)	150	12	4	2	1	.5

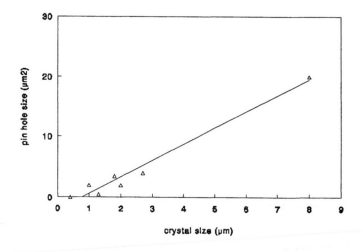

Figure 7. The intercrystalline area as a function of the crystal size.

As shown in Figure 6 d a film could be made. As, however, the density of the crystal is larger than the gel density the SiO_2 nutrient for this crystallization must have not only be supplied by the gel but the solution as well. This is the first indication that at very high supersaturation better confirmation regarding the theory is possible as both ingredients for the crystallization are obtained from the same nutrient pool.

Work on converging theory and the experimental conditions in order to obtain possible confirmation is in progress.

Although the crystal growth history is comparable the crystals formed in the gel spheres are a few hundred micrometers in size while the crystals on the support are less than one micrometer. This is apparently due to the number of nuclei and thus crystals formed. The attachment of TPA on the gel sphere surface is strongly inhibited by the presence of Na^+ present in the synthesis mixture. In the thin coating formulation no Na^+-source was applied.

Thus the more frequent interaction of TPA with the gel phase resulted in a larger number of crystals.

While the ac crystal face of the individual crystallites on the top side of the coating show a slight deviation in orientation (4°) the bottom side reflects the same flatness as the support surface. According to HREM analysis no additional interface is observed [41]. The number of terminal groups in the ac-plane of the MFI-crystals is 12 OH/nm^2 and in the natural oxide layer of the silicon wafer 4-6 OH/nm^2. Frequent chemical bonding via condensation reactions is possible resulting in strong connection of the crystal film to the support. Mutual attachment of the crystals results in a continuous phase without grain boundaries as concluded from HREM observations [41]. The coating formed this way can be used at temperatures up to 600 °C. Heating as well as cooling does not affect the stability of the crystals on the substrate. The thermal expansion coefficient of supports like SiO_2 and stainless steel are 110 - and 193 x $10^{-7}/°C$ while those of the crystals are, in a temperature range between 20 and 500 °C, in the a-direction 207 -, b-direction -222 - and in the c-direction 5 x $10^{-7}/°C$, respectively [42]. The crystal thermal expansion coefficients between 20 and 500 °C are overall values including the calcination step and phase transition which both result in relatively large changes in the cell dimensions, see Table 3. Despite the possible chemical bonding of the crystals on support and the differences in occurring stresses based on the expansion coefficients no cracks or loss of crystals is observed after calcination at 550 °C. Frequently heating and cooling does not affect the quality of the coating based on the following. Free silicalite-1 crystals show upon cooling a phase transition at about 80 °C. Actually, the orthorhombic single crystal changes into an aggregate of monoclinic twin domains. The deviation of a from 90° (a= 90.5°) is ascribed to a mutual shift of successive (010) pentasil layers along the c-direction. By applying an appropriate mechanical force during the symmetry change a nearly single crystal in the monoclinic form could be achieved. According to a recent publication

[43] oriented crystals in a coating indeed do not follow the orthorhombic/monoclinic phase transition upon cooling after calcination. Therefore after calcination there is hardly any cell expansion or reduction as a function of temperature as witnessed by the data in Table 2. Thus in the case of a continuous coating unique passage through the pores of zeolite can be achieved in a large temperature range. It looks like this is the reason that maximal four p-xylene molecules per unit cell can be adsorbed in supported crystals [43] instead of eight molecules as found in a structure analysis and adsorption measurements of free crystals as well [44].

In the case of the original concentration of TPA and SiO_2 in the presence of Na^+ separate gel particles are observed on the surface of the support. Crystals with smooth curved top faces are developed in a random orientation, partly grown a thin gel phase and separate from and gel particles on the surface as shown in Figure 8 a.

3.3. ZSM-5 Type

Direct nucleation on the support surface is accomplished at concentrations .5 and .25 of the original TPA concentration. As no gel layer was observed on the support preceding the crystallization the nucleation started at the oxidic surface of the silicon wafer support. The random orientation of the nuclei formed, after two hours, is reflected in the final crystal orientations see Figure 8 b. The elongated prismatic forms clearly reveal the crystal growth at relatively low supersaturation. Continuation of the crystallization results in a layer of mainly axially oriented crystals of which the thickness is about 10 μm, as shown in Figure 2 b. In the case of crystallization on stainless steel gauze $NaAlO_2$ was added to the synthesis mixture. The elongated prismatic form developed with the c-direction axially oriented to the wire surface of the gauze. This

Table 3

Lattice parameters of as-synthesized, calcined orthorhombic and calcined monoclinic silicalite-1 single crystals, respectively

Structure	T (K)	a (Å)	b (Å)	c (Å)	α (°)
TPA-Silicalite-1 orthorhombic	293	20.022(2)	19.899(2)	13.383(1)	
----Silicalite-1 orthorhombic	350	20.078(6)	19.894(7)	13.372(3)	
----Silicalite-1 monoclinic	293	20.107(2)	19.879(2)	13.369(1)	90.5

Figure 8. Random orientation of silicalite-1 crystals on support. a) crystals with smooth curved top faces developed from a discontinuous gel phase. b) direct nucleation on support.

composite can be washed thoroughly and calcined at 550 °C without loss or cracking of the crystalline material. Subsequently the material can be exchanged in order to obtain either a M-ZSM-5 or H-ZSM-5 catalyst for low pressure drop reactor systems. Further details regarding the applications are discussed in Chapter 15.

3.4. Gmelinite

Recently, the preparation of self-supporting films of gmelinite were reported [6]. The films with a thickness of one crystal show two dimensional order, i.e. the c-axis of all crystals is parallel and perpendicular to the support. DABCO seems to be an excellent structure-directing agent for the crystallization of mordenite [46], ZSM-12 [47] and gmelinite as well. The diameter of the molecule is about 6.1 Å and the length 8.7 Å. These values are comparable with the gmelinite pore dimensions. The channels have a diameter of 7 Å while the repetition unit is 10 Å. The use of DABCO as template in trials to prepare a film of gmelinite is two-fold. Firstly, electrostatic interactions of the DABCO with the support result in a consistent start and orientation of the individual crystallites which have to grow together. Secondly, the formation and intergrowth with chabazite is avoided as well as stacking faults blocking the channels in the c-direction.

In principle the film can be removed from the substrate. One of the synthesis methods discussed in the paper was used to prepare films in order to have a comparison with the MFI-film preparation. A synthesis mixture of the following composition in molar ratio was used.

SiO_2	NaOH	DAB*-4Br	$NaAlO_2$	H_2O
Aerosil 200 Degussa	p.a. Baker	*vide infra*	a.g. Riedel-deHahn	distilled
7.5	8.0	.83	1.0	197.3

* 1,4-diazobicyclo[2,2,2]butane.

Figure 9. Gmelinite crystals on support. Although not extended it looks like a monolayer can be formed.

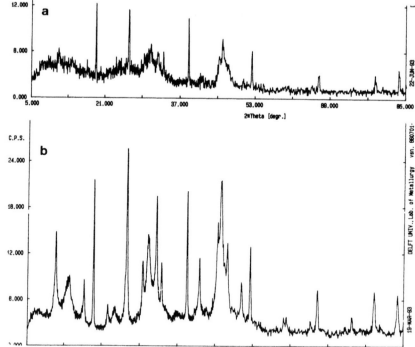

Figure 10. XRD-patterns of gmelinite. a) on support, showing hko and ool reflections, thus indicating at least two crystal orientations, b) free crystals in random orientation.

The DAB-4Br was synthesized by the Menschutkin-reaction according to the description of alamone and Snider [48]. The molecular weight was estimated from the intrinsic viscosities derived from GPC measurements. The calculated molecular weights were: M_n = 24000 and M_w = 40000. This corresponds to about 100 units and thus to a length of the extended polymer chain of about 0.1 μm.

The film preparation was as follows: The aluminate solution was added to the silica/DABCO solution under vigorous stirring. In a 35 ml teflon lined autoclave 30 g of the mixture was kept at 90 °C for 4 days. The supports used were silicon wafer, teflon, aluminium, polyimide (Kapton; DuPont) and polystyrene. The supports were positioned in the upper part of the synthesis mixture in an identical way as discussed earlier, see Figure 3. Layers of gmelinite, with a thickness of 1 um, although not extended, but with areas which show the possible preparation of a continuous phase, see Figure 9, were achieved on the silicon support. According to ^{13}C-NMR and FTIR the DABCO is present in the crystals on the support. TG measurements confirm the presence of organic material in side the zeolite and show that the water content is about 10 wt %. The crystal layers show an XRD powder pattern which differs substantially from the powder pattern of the free gmelinite crystals, see Figure 10. The pattern of the oriented crystal layer indeed shows a strong reduction in the peak intensity of the 203 reflection. The 002 as well as the 110 and 220 reflections are present with the highest intensity indicating that the material is oriented still in mainly two positions. As the length of the DABCO is 0.1 μm frequently short order orientation resulted in combination with potentially stacking faults on a longer range partially orienting monolayers of crystals.

3.5. NaA

The formation of a zeolite A crystal coating is in contrast to a silicalite-1 coating prepared without a structure directing agent. The gmelinite synthesis is the most complicated with sodium as well as DABCO present. In the silicalite-1 synthesis without sodium and tetrapropylammonium ions a precursor of silica gel in a continuous phase is present interfacing TPA in solution. The precursor phase is comprised in the case of zeolite A of an aggregate of sodiumaluminosilicate bodies.

Zeolite A is applied a.o. in gas- and n-iso alkane separations and water adsorption systems.

A continuous monolayer of zeolite A crystals in a membrane configuration is interesting as an absolute separation tool.

A monolayer of zeolite A crystals on the inner wall of a GC-column as a stationary phase might result in promising separation properties.

Furthermore, the large adsorption capacity of water makes zeolite A grown on metal support suitable as a relatively efficient tool in heat-pump engineering [48].

Although the formation of zeolite A layers on support has been presented [49] the crystal growth history and investigations into the possibilities of the layer formation of especially one crystal layer have not been discussed.

Therefore the study of the preparation of a coating of zeolite A on support was started, recently.

Zeolite A crystallization experiments were carried out to understand whether a particular gel phase must be formed or excluded on support in order to prepare a monolayer of zeolite A crystals either from a precursor phase or directly on the support surface, respectively. In addition the effect of support properties on the crystallization, like continuity of the crystal layer, possible orientation of crystals and number of initial nucleation points were studied.

The type of support, concentration of synthesis mixture, temperature as well as time in experiments resulting in monocrystal orientation of zeolite A on monocrystal support are given in Table 4.

Table 4

Synthesis mixtures, temperature and time for monocrystal formation of zeolite A on monocrystal support

Support Single crystal Metal oxide	Molar oxide ratio				T	Time	Remarks
	SiO_2	Al_2O_3	H_2O	Na_2O	^0C	(h)	
Sapphire(001)	1.0	0.2	200	10	80,100	1 - 3	wafer1cm^2
Rutile(001)	1.0	0.2	100	10	80,100	1 - 3	wafer1cm^2
Rutile	1.0	0.2	100	10	80,100	1 - 5	single crystal
Corund	1.0	0.2	100	10	65,80,100	2 - 19	single crystal
Quartz	1.0	0.2	100	10	65,80,100	.5 - 27	single crystal

For the crystallization of zeolite A polyethylene bottles were, after a cleaning procedure, filled with 25 ml of synthesis mixture based on a solution of sodium hydroxide (AR grade of Merck), sodium silicate (AR grade of BDH) and sodium aluminate (AR grade Riedel deHahn).

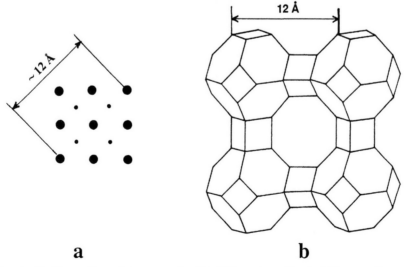

a   b

Figure 11. Repeating distances of rutile 001 and zeolite A 100 as well.

The aluminate solution was added under vigorous stirring to the silicate solution. Subsequently, the mixture was brought into a polyethylene bottle of 50 ml. The supports were precleaned with a standard technique, *vide supra,* and placed vertically in a holder in the upper part of the liquid phase. The bottle was then placed in a preheated hot air oven.

The choice of the monocrystal supports used was based on chemical stability in the synthesis mixture as well as possible attachment on terminal groups of the zeolite nuclei and developing crystals. In the case of the 001 rutile, the possibility of matching of lattice dimensions of the monocrystal support and the monocrystal of zeolite A, see Figure 11 was investigated.

In a particular series of experiments of which the formulation is given in Table 4 the formation of a zeolite A crystal layer was followed in time at 65 °C on quartz single crystals.

The results on a sapphire as well as a corund single crystal, both of interest for zeolite A membrane development on Al_2O_3 tubes, were similar. In general the nucleation and crystal growth started earlier on the substrate than the crystal formation in the bulk solution. Figure 2 a shows the initial crop of crystals, which look like hemispheres, some already larger than 1 um in diameter, after 2.5 hours of reaction time. The smaller crystals which are barely visible are about .1-.2 μm in size. On a similar surface, see Figure 12 b, after 3.5 hours a progress in crystallization is observed. The zeolite A morphology is recognized for the larger crystals on the support surface after 5 hours, as depicted in Figure 12 c. The corner and edges are still rounded due to the prevailing high concentration in the vicinity. Apparently, the initial rate of nucleation is strongly dependent on the typical surface of the single crystal support exposed, as can be seen from the markedly different crystal populations on the different faces of the quartz crystal, see Figure 12 d. After 7 hours the crystals become more facetted, indicating a lower concentration of the nutrient, see Figure 13 a. Probably based on the way the experiment is performed the small crystallites shown here were nucleated and formed upon cooling, however, not in a progressive but in an explosive event. A bimodal crystal size distribution is the result therefore after terminating the crystallization. Apparently nucleation starts at the support surface as the small crystals as well as the large ones are not oriented. As depicted in Figure 13 b the large crystals do not exceed 4 μm in size. In Figure 13 c relatively high populated surfaces are shown after 9 and 11 hours, respectively. A truly continuous layer is not obtained until much later. Figure 13 d shows such a layer formed after 27 hours at 65 °C and after 11 hours at 70 °C. The layer has a thickness of about 4.5 μm.

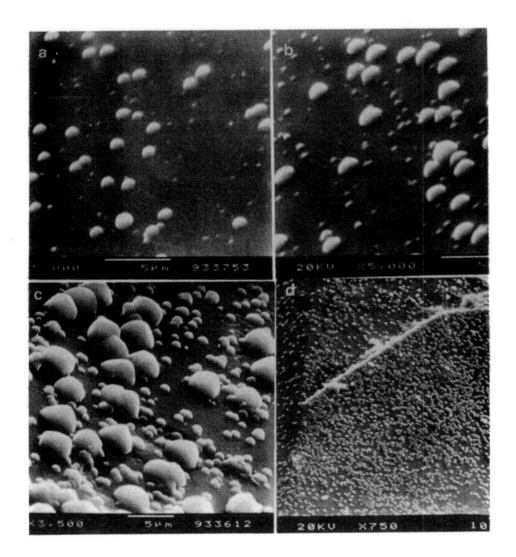

Figure 12. The crystal growth history of zeolite A on a single crystal of quartz. a) first observation shows hemispheres in a bimodal distribution, the smaller crystallites are probably born upon cooling the supersaturated mixture. b) illustrating the progress in crystallization. c) orientation of the crystals becomes visible. d) on different quartz surfaces, different populations are observed. Although a chemical factor is suggested, convection of the liquid phase might also have an influence here.

244

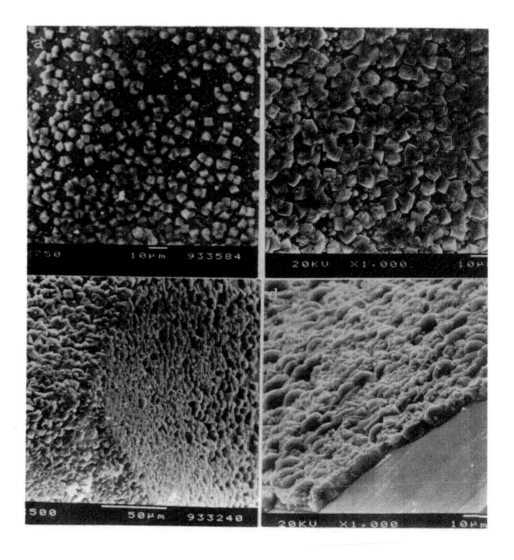

Figure 13. Continuation of events of figure 12 on the crystal growth history of zeolite A on a single crystal of quartz. a) crystallization after 7 hours. The crystal thickness is 4 μm. b) almost closed layer of crystals. c) discontinuous layer of crystals with still a bimodal size distribution on three crystal faces of quartz. d) continuous layer on support of quartz.

Although the solution was clear a thin gel layer was observed on the support surface, probably initiated by a dissolution effect of the support material in the case of quartz. It is supposed that this gel layer is build up by aluminosilicate bodies in the sodium form. The heterogeneous nucleation thus started on/close to the interface of the slightly dissolving support and the gel particles resulting in a random crystal orientation. A surprising effect is the relatively small crystal size distribution of the larger crystals. The thickness of the crystal layer is not exceeding about 7 μm throughout the synthesis time, apparently due to a drop in the local nutrient concentration at the moment the precursor phase is exhausted. The crystal growth rate is then substantially decreased. A phenomenon discussed already for silicalite-1.

Comparable results were obtained on the surface of rutile, which is considered to be a support more inert than quartz and sapphire.

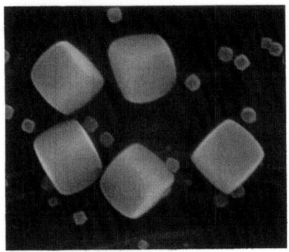

Figure 14. Depicting the start of the crystallization of single crystals zeolite A on the 001 plane of a single crystal wafer of rutile. A bimodal crystal size distribution is shown.

A discontinuous mono-layer of crystals could be prepared on different supports, shown in Figure 14 for a rutile 001 mono crystal support at 80 °C. The bimodal crystal size distribution of zeolite A crystals indicates again two events of nucleation. The first nucleation being relatively poor started on the support surface in a diluted synthesis mixture resulting in relatively large crystals with smooth rounded edges and corners. Most of the large crystals are randomly mainly edge-oriented on the surface, indicating that despite lattice matching factors no orientation relation occurred. After termination of the synthesis, an explosion of nucleation took place and small crystals are formed all showing no alignment. One crystal face, however, is parallel oriented to the support surface indicating the nucleation event did not occur close to the surface but in the gel phase probably because this support is not dissloving. Based on these observations it is concluded that the remaining gel phase on the support surface initiated the explosion of small crystal formation upon cooling. As soon as the growing crystallites touch the support surface they become oriented by electrical and surface tension forces. A comparable crystal growth history as described for the cubic shaped crystals of the MFI-type, *vide supra.*

The preliminary results from the above experiments show that a mono layer of zeolite A crystals oriented in one direction on support at least needs a thin gel phase, which is not thicker than necessary for the first crystal layer.

Relatively thin ($< 5\ \mu m$) continuous monolayers of randomly oriented crystals of zeolite A, on essentially flat surfaces can be prepared. As the pore system is three dimensional optimal adsorption properties are still expected especially in thin layers.

Table 5
Poly-crystallization of zeolite NaA on amorphous support

Support	Molar oxide ratio				T	Time	Remarks
	SiO_2	Al_2O_3	H_2O	Na_2O	^0C	(h)	
Metal oxide							
Nickel	1.0	0.4	50	10.0	80	1.5, 2.0, 2.5	1 cm^2
Copper	1.0	0.4	50	10.0	80	2.0	
Stainl. st.	1.0	0.4	50	10.0	80	2.0	gauze
Silicon	1.0	0.4	100	5.0	80	1.0, 1.5, 2.0	dissolved
Titanium	1.0	0.4	50		80	1.0, 1.5, 2.0	
Aluminum	1.0	0.4	50		80	2.0	
Ceramic							
Cordierite	1.0	0.4	40		80	6.0	monolith

Crystallization experiments have been carried out to obtain thicker layers of crystals of zeolite A.

In this case the concentrations were chosen much higher than in the thin monolayer crystallization as given in Table 5.

Different metal supports were tested. The natural oxide layer of the metal being an excellent interface to initiate and anchor the zeolite crystal layer in a way as discussed for the MFI-type crystals as well.

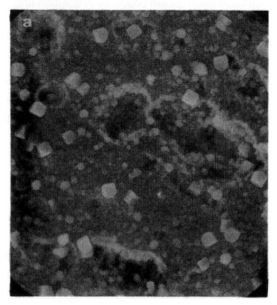

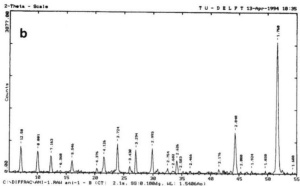

Figure 15. a) Multi crystal layer of zeolite A, thickness about 5 μm, grown on nickel support. The larger crystals are supposed to be co-precipitated from solution in the course of the termination of the synthesis. b) The XRD-powder diffractogram of this layer on Ni-support does not indicate any crystal orientation.

Silicon and Cordierite have just been chosen for reasons of comparison.

Apart from silicon on all supports crystals of zeolite A were observed. In these crystallization experiments the platelets were positioned in the lower part of the synthesis mixture. The best results in terms of a massive relatively thick layer of 5 μm was observed on the bottom side of the platelet, see Figure 15. In contrast the top side showed a highly corrugated layer of different thickness apparently caused by uncontrolled settlement of gel particles.

The larger crystals have been grown probably in solution and were co-precipitated on the developed layer of small crystals. Based on this phenomenon again a bimodal crystal size distribution is observed.

In contrast to the silicalite-1 coating the NaA layer is often not bonded to the support surface as, for example, shown in Figure 2 f. The mechanism suggested for silicalite-1, based on approach of the OH groups of the crystal as well as the support is in the case of NaA obstructed by the Na-ions. Further investigations are carried out.

4. CONCLUSIONS

general

The theory of film nucleation/ crystallization is a promising tool to refine experiments concerning film formation.

Crystal orientation could only be obtained if a gel phase was present and the support surface is flat on a nanometer scale.

Film development highly depends on the chemical composition and crystal shape and size of the zeolite used.

in particular

A continuous film of silicalite-1 over 1 cm^2 can be prepared.

A discontinuous layer of ZSM-5 on larger support areas can be made.

Possible bonding of the zeolites with high Si/Al ratio is of a chemical nature.

The potentiality of a gmelinite coating [6] has been confirmed.

A continuous monolayer of NaA crystals has been achieved.

5. REFERENCES

1. I.M. Lachman, US Patent 4,800,187 (1989).
2. H. Suzuki, US Patent 4,699,892 (1987).
3. G.D. Stucky and J.E. MacDougall, Science, 247 (1990) 669.
4. G.A. Ozin, A. Kuperman and A. Stein, Angew. Chem., 101 (1989) 373.
5. S.P. Davis, E.V.R. Borgstedt and S.L. Suib, Chem. Mater., 2 (1990) 712.
6. M.W. Anderson, K.S. Pachis, J. Shi and S.W. Carr, J. Mater. Chem., 2 (1992) 255.
7. J.G. Tsikoyiannis and W.O. Haag, Zeolites, 12 (1992) 126.
8. E.R. Geus, H. van Bekkum, W.J.W. Bakker and J.A. Moulijn, Microporous Mat.,
9. J. Caro, P. Kölsch, E. Lieske, M. Noack and D. Venzke, Congress Abstracts, New Directions in Separation Technology, Noordwijkerhout, The Netherlands, 1993, 6.1 (1993) 131.
10. J.C. Jansen, W. Nugroho and H. van Bekkum, in: "Proceedings of the 9th International Zeolite Conference, Montreal 1992", Eds. R. von Ballmoos et al., Butterworth-Heinemann, U.S.A., 1993, p. 247.
11. G. Zinsmeister, Kristall Technik, 5 (1970) 207.
12. B. Lewis, Thin Solid Films, 7 (1971) 179.
13. M.J. Stowell, J. Crystal Growth, 24/25 (1974) 45.
14. J.A. Venables and G.L. Price, in: "Epitaxial Growth", Ed. J.W. Matthews, Academic Press, New York, 1975, p. 381.
15. R. Kern, G. Le Lay and J.J. Metois, in: "Current Topics in Materials Science", Vol. 3, Ed. E. Kaldis, North-Holland, Amsterdam, 1979, p. 131.
16. B. Lewis and J.C. Anderson, "Nucleation and Growth of Thin Films", Academic Press, London, 1978.
17. S. Stoyanov and D. Kashchiev, in: "Current Topics in Materials Science", Vol. 7, Ed. E. Kaldis, North-Holland, Amsterdam, 1981, p. 69.
18. R.W. Vook, Intern. Metals Revs., 27 (1982) 209.
19. J.A. Venables, G.D.T. Spiller and M. Hanbucken, Rept. Progr. Phys., 47 (1984) 399.
20. L. Ickert and H.G. Schneider, in: "Advances in Epitaxy and Endotaxy", Eds. H.G. Schneider, V. Ruth and T. Kormany, Elsevier, Amsterdam, 1990, p. 227.
21. M. Zinke-Allmang, L.C. Feldman and M.H. Grabow, Surf. Sci. Repts., 16 (1992) 377.
22. E. Bauer, Z. Krist., 110 (1958) 372; 395.

250

23. W.A. Jesser and J.W. Matthews, Philos. Mag., 15 (1967) 1097; 17 (1968) 461; 595.
24. D.G. Lord and M. Prutton, Thin Solid Films, 21 (1974) 341.
25. C.T. Horng and R.W. Vook, J. Vac. Sci. Technol., 11 (1974) 140.
26. R.W. Vook, C.T. Horng and J.E. Macur, J. Crystal Growth, 31 (1975) 353.
27. U. Gradmann, W. Kummerle and P. Tillmanns, Thin Solid Films, 34 (1976) 249.
28. U. Gradmann and P. Tillmanns, Phys. Stat. Sol. (a), 44 (1977) 539.
29. D. Kashchiev, J.P. van der Eerden and C. van Leeuwen, J. Crystal Growth, 40 (1977) 47.
30. D. Kashchiev, J. Crystal Growth, 40 (1977) 29.
31. D. Kashchiev, Thin Solid Films, 55 (1978) 399.
32. D. Kashchiev, J. Crystal Growth, 67 (1984) 559.
33. A. Trayanov and D. Kashchiev, J. Crystal Growth, 78 (1986) 399.
34. A.C. Zettlemoyer (Ed.), "Nucleation", Dekker, New York, 1969.
35. A.E. Nielsen, "Kinetics of Precipitation", Pergamon, Oxford, 1964.
36. J.D. Weeks and G.H. Gilmer, Adv. Chem. Phys., 40 (1979) 157.
37. I. Markov and D. Kashchiev, J. Crystal Growth, 13/14 (1972) 131; 16 (1972) 170; Thin Solid Films, 15 (1973) 181.
38. W. Obretenov, Electrochim. Acta, 33 (1988) 487.
39. J.C. Jansen, C.W.R. Engelen and H. van Bekkum, in: Zeolite Synthesis, M.L. Occelli and H.E. Robson (Eds.), ACS, Washington DC, Vol. 398 (1989) 257.
40. J.C. Jansen and G.M. van Rosmalen, J. Cryst. Growth, 128 (1993) 1150.
41. J.H. Koegler, H.W. Zandbergen, J.L.N. Harteveld, M. Nieuwenhuizen, J.C. Jansen and H. van Bekkum, accepted for publication in the Proceedings of the 10th IZC.
42. E.R. Geus, PhD Thesis, Delft, 1993.
43. N. van der Puil, J.C. Jansen and H. van Bekkum, accepted for publication in the Proceedings of the 10th IZC.
44. H. van Koningsveld, F. Tuinstra, H. van Bekkum and J.C. Jansen, Acta Cryst., B45 (1989) 423.
45. R.H. Daniels, G.T. Kerr and L.D. Rollmann, J. Am. Chem. Soc., 100 (1978) 3097.
46. M.E. Davis and C.S. Saldarriaga, J. Chem. Soc., Chem. Commun., 1988, 920.
47. J.C. Salamone and B. Snider, J. Polym. Sci., Part A-1, 8 (1970) 3495.
48. G. Gacciola and G. Restuccia, Heat Recovery Systems 8 CHP, accepted for publication.
49. G.J. Myatt, P.M. Budd, C. Price and S.W. Carr, J. Mater. Chem., 2 (10) (1992) 1103.

J.C. Jansen, M. Stöcker, H.G. Karge and J. Weitkamp (Eds.)
Advanced Zeolite Science and Applications
Studies in Surface Science and Catalysis, Vol. 85
© 1994 Elsevier Science B.V. All rights reserved.

THE CATALYTIC SITE FROM A CHEMICAL POINT OF VIEW

V.B.Kazansky

Zelinsky Institute of Organic Chemistry, Moscow B-334, Russia.

1. INTRODUCTION

The practical importance of catalysis by zeolites for refining chemistry, such as cracking, isomerization, alkylation of paraffins, etc. is well known. All these reactions represent the examples of heterogeneous acid catalysis, and the study of their mechanism is of interest for the development of general theory of catalysis.

Another attractive feature of catalysis by zeolites is that the crystal structure of these materials and the nature of their acid active sites have been studied in detail. Such sites are represented by the so called bridging or structural hydroxyl groups compensating the excessive negative charge of tetrahedrally coordinated aluminum atoms in the framework:

$$
\begin{array}{c}
\text{H} \\
| \\
\text{H–O} \quad \text{O} \quad \text{O–H} \\
\text{H–O–Si} \quad \text{Al–O–H} \qquad\qquad (1)\\
\text{H–O} \quad\quad \text{O–H}
\end{array}
$$

The main factors affecting their acid strength are the Si-O-Al bond angle and the amount of aluminum atoms in first coordination sphere [1-3].

The mechanism of hydrocarbon conversion on Broensted acid sites of zeolites was first formulated almost 30 years ago in a way similar to homogeneous acid catalysis. Since that time it remained practically unchanged. For instance, in the case of catalytic transformations of olefins, it is generally believed that the active intermediates are represented by adsorbed carbenium ions resulting from protonation of the double bonds in the initial molecules:

$$
\begin{array}{c}
\text{H} \quad\quad \text{H} \\
\backslash \quad / \\
\text{C=C} \quad + \quad \overset{+}{\text{H}}_{ads} \quad \longrightarrow \quad \text{H–}\overset{\text{H}}{\text{C}}\text{–}\overset{+}{\text{C}}_{ads} \qquad (2)\\
/ \quad \backslash \\
\text{H} \quad\quad \text{H}
\end{array}
$$

The subsequent reactions of these species, such as skeletal isomerization, cracking, hydride transfer, double bond shift, etc. result in final products, composition of which is governed by the well known rules of carbenium ion chemistry.

Protonation of paraffins represents another type of active intermediates - adsorbed nonclassical carbonium ions with penta-coordinated carbon atoms:

$$R\text{--}\overset{\overset{\displaystyle H}{|}}{\underset{\underset{\displaystyle H}{|}}{C}}\text{--}H \;+\; H_{ads}^{+} \;\longrightarrow\; R\text{--}\overset{\overset{\displaystyle H}{|}}{\underset{\underset{\displaystyle H}{} \quad \underset{\displaystyle H}{}}{C}}\overset{+}{\text{--}}H_{ads} \qquad (3)$$

Most typical reactions of these species involve cracking and dehydrogenation and also yield carbenium ions which further react in a similar way as those resulting from the protonation of olefins.

Of course, such schemes don't represent a real reaction mechanism but can pretend only for a rational description of the final reaction products. Indeed, they can not even properly explain the difference between homogeneous and heterogeneous acid catalysis, since the solvation of active intermediates or their interaction with the surface of zeolites is not depicted by the above structures. Moreover, they can not explain many of other important details of heterogeneous acid catalysis. For instance, it is well known that the migration of a double bond in olefins requires much weaker acid sites than those necessary for skeletal isomerization or cracking. This difference can not be explained by equations (2) and (3), since a heterogeneous catalyst itself is even not considered explicitly. In other words, such schemes don't give a real description of heterogeneous acid catalysis on the molecular level and therefore are out of date.

In this connection, the aim of the present paper is to present and to discuss more realistic mechanisms of acid catalysis by zeolites which take into account the interaction of active intermediates with the surface and describe the catalytic reactions on a molecular level with respect to the real geometry and other properties of surface acid active sites.

2. A COVALENT CHARACTER OF ACID HYDROXYL GROUPS IN ZEOLITES.

Usually acid hydrogen forms of zeolites are prepared from their sodium forms either by direct exchange of cations with protons or by exchange with ammonium ions which then decompose at higher temperatures. Therefore the resulting hydroxyl groups similar to the sodium ions in the starting zeolites are intuitively believed to be of an ionic nature like the adsorbed protons in the above equations (2) and (3). However, such a conclusion is absolutely incorrect, since there are strong evidences in favor of covalent character of these hydroxyls. Among them the most convincing arguments were obtained by IR diffuse reflectance spectroscopy.

This technique allows the measurements of overtones and a combination of bending plus stretching OH frequencies. Then, in - plain and out-of-plain OH bending frequencies can be obtained as a difference between corresponding combination and stretching modes. Some of the bending frequencies obtained in this way are collected in Table 1.

Table 1.
Characteristic frequencies of bridging OH groups in hydrogen forms of different zeolites.

Zeolite	Type of hydroxyl groups	Frequencies cm^{-1}		
		Fundamental stretching	In-plain bending	Out-of-plain bending
HZSM-5	bridging OH	3610	1050	290
HM	bridging OH	3610	1050	290
HY	bridging OH	3655	1050	?
HY	bridging OH	3645	1020	280
HX	bridging OH	3660	995	?
NaMgM	terminal MgOH	3620	960	-
NaCaM	terminal CaOH	3600	930	-

The difference of in-plain and out-of-plain bending frequencies is quite remarkable. It certainly indicates the planar structure of Si-OH-Al fragments with sp^2 hybridization of the bridging oxygen orbitals. This is also in agreement with the results of cluster quantum - chemical calculations which confirm the planar structure of Si-OH-Al fragments with Si-O-H and Al-O-H bond angles of about 120^0. In addition, the Mulliken charges of protons are rather low thereby confirming the covalent nature of the OH bonds.

However, the most convincing argument in favour of covalent character of hydroxyl groups in zeolites follows from reconstruction of the bottom parts of their potential curves. This was done in our papers [4-7] within the formalism of Morse potential:

$$U(r) = D_0 \left[1 - e^{-\beta(r-r_0)} \right]^2 = \frac{\omega_e}{4X_e} \left[1 - e^{-\beta(r-r_0)} \right]^2 \quad (4)$$

254

This function was originally introduced for diatomic molecules, but also nicely describes the shape of O-H potential in surface hydroxyl groups, since their vibrations are highly characteristic and practically don't interact with those of the oxide crystal lattice.

Expression (4) includes two parameters: the harmonic frequency at the bottom of the potential well ω_0 and the anharmonicity coefficient X_e. To find both of them, the experimental frequencies of at least two vibrational transitions are required. They can be easily detected from IR diffuse reflectance spectra which allow measurements in the near infrared region. Then, both the shape of the potential curve of the OH bond and its dissociation energy D_0 can be reconstructed from the spectral data. Of course, if several vibrational transitions are known, the result of such reconstruction is more precise and reliable.

Table 2 summarizes the positions of experimentally observed fundamental frequencies and overtones for OH groups in hydrogen forms of zeolites and on the surface of some other oxides with basic (MgO) or weakly acidic (SiO_2) properties. The OH dissociation energies calculated from spectral data are presented in the last column.

Table 2. ·

The dissociation energies of the surface hydroxyl groups with different acid-base properties calculated from spectral data [8].

Oxide or catalyst	Frequencies of fundamental vibrations and overtones, cm^{-1}			D_0 kcal/mol
MgO	3750	7350	10760	140
SiO_2	3749	7326	10735	124
$AlPO_4$	3679	7147	-	124
HY	3645	7130	10460	126
HM	3611	7065	10360	125

The obtained D_0 values are rather low and surprisingly close to each other. This definitely points to the homolytic dissociation of the surface hydroxyl groups (compare for instance the calculated bond strengths ca. 125 kcal/mol with the energy of homolytic dissociation of water 118 kcal/mol). It is also remarkable that hydroxyls with very different acid-base properties have almost the same homolytic dissociation energies and the shapes of the corresponding potential curves. Thus, the ground states of hydroxyl groups are rather insensitive to their acid-base properties. This shows that the bridging groups in zeolites are of covalent nature and that the naive belief, according to which these materials are sometimes considered as solid electrolytes with almost free protons incorporated in their cages is certainly only a very rough simplification.

On the other hand, the energies of heterolytic dissociation of surface hydroxyls are quite different from those of homolytic dissociation given in Table 2. The former are much higher and closely related with the acid-base properties.

To illustrate this, let us at first split the surface O-H bond homolytically, then ionize the hydrogen atom and transfer the released electron to the surface O$^-$. This would result in heterolytic dissociation with the following total energy:

$$D_{het} = D_0 + U_H - \phi \tag{5}$$

Since the hydrogen atom ionization potential U_H (13.6 eV) is much higher than the electron affinity ϕ of surface O$^-$ (only ca 6-7 eV), this expression results in D_{het} value of about 11-12 eV (250-280 kcal/mol). This well agrees both with the results of quantum - chemical calculations [3,9] and with the spectroscopically estimated values [10,11] and is, at least, by 6-7 eV higher than the energy of homolytic dissociation.

It is also clear from expression (5) that the proton splitting energy is directly related to the acid-base properties of the surface hydroxyl groups. Indeed, the higher is the electron affinity of O$^-$ surface holes the lower is proton abstraction energy D_{het}. Thus, the acid - base properties should be correlated both with electronegativity of solids and with corresponding proton abstraction energies. This is consistent with the generally accepted approach, when the energy of proton abstraction from the surface of zeolites is used as a measure of the strength of surface Broensted acid sites [10,11].

On the other hand, the difference in the frequencies of O-H stretching vibrations for the ground states of the groups with different acid-base properties or the corresponding NMR chemical shifts are too small to be used as a measure of the acid strength. Therefore, any attempt to determine this charactristic of the surface OH groups in a zeolite from their stretching frequencies or from any other properties of their ground state will fail.

The only proper way to distinguish hydroxyl groups of different acid strengths is to use the energy of their heterolytic dissociation. Unfortunately, it can not be directly measured by any physical methods, but can be estimated only in an indirect way from the response of surface hydroxyls to their interaction with adsorbed bases. This is actually a traditional, most commonly used approach when the low - frequency shifts of OH stretching vibrations resulting from hydrogen bonding with adsorbed molecules, the heats of ammonia adsorption, the positions of ammponia thermodesorption peaks, or the adsorption of Hammett indicators are used for this purpose.

3. CATALYTIC TRANSFORMATIONS OF OLEFINS

3.1. Nature of adsorbed aliphatic carbenium ions

Altghough adsorbed aliphatic carbenium ions were postulated as active intermediates already in the very first works devoted to catalysis by zeolites, their existence was never proved by any direct physical method. These species were directly detected by NMR only in homogeneous solutions of a "magic acid" [12]. However, these experiments were carried out in absolutely unhydrous conditions at low temperatures and their results could not be directly transferred to catalysis by zeolites. Thus, the information on the nature and properties of carbenium ions adsorbed in zeolites is purely speculative and based only on the composition of final reaction products or on the kinetics of corresponding reactions.

Among such kind of data the study of the double bond shift in 1-butene on partially decationated faujusites is of the most interest [13]. The number of authors concluded that this reaction proceeds via adsorbed butyl cations resulting from protonation of 1-butene. These intermediates likely represent the unstable short - lived excited complexes. The following reaction pathway with the activation barrier E of 15-17 kcal/mol was also suggested [14]:

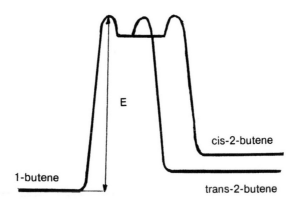

Fig. 1. The energy profile of 1-butene isomerization according to [13].

However, these conclusions put forward the following two important questions:

i What is the relation between ionic species represented by adsorbed carbocations and isomeric covalent alkoxides which can be also formed by protonation of adsorbed olefins?

ii. What is the possibility for the highly excited ionic species to exist on the surface of a zeolite as real active intermediates without fast deactivation or decomposition into more stable final products?

The first of these problems was discussed in our paper [15] in frames of electronegativity equalisation theory modified by Pearson for the analysis of ion character of chemical bonds [16].

According to this theory, the adsorption of carbenium ion on the surface of a zeolite should result in a electronegativity equalisation. This is accompanied by a partial electron density transfer from surface basic oxygen to adsorbed species and by a decrease of their positive charge that makes the adsorption bond essentially covalent:

$$R^+ \ + \ O_{surf} \ \longrightarrow \ \overset{(1-\delta)+}{R_{ads}} \ \overset{\delta+}{O_{surf}} \qquad (6)$$

As a first approximation, the extent of such charge transfer can be estimated from the following expression:

$$\delta \ = \ \frac{\chi_{R^+} \ - \ \chi_{Osurf}}{2 \, (\eta_{R}{}^+ \ + \ \eta_{Osurf})} \qquad (7)$$

where χ and η are the electronegativities and hardnesses of a carbenium ion and a surface basic oxygen, respectively

It follows from the eq. (7) that the nature of adsorption bond depends on the Lewis acid strength of a cation and on the basicity of a surface oxygen which are directly related with the electronegativity. For instance, the bonding of adsorbed species with lattice oxygen of an X zeolite should be more covalent than that in a case of a high silica zeolite, where the negative charge of oxygen of the zeolite crystal lattice is lower and the electronegativity is higher. Therefore, according to the electronegativity equalisation principle, the ground states of adsorbed carbenium ions, similar to the surface hydroxyls already discussed above, should be always more or less covalent.

In our papers [17,18] this conclusion was confirmed by semi-empirical cluster quantum chemical calculations using a CNDO method with full optimization of geometry of adsorbed species. A fragment of a high silica zeolite with adsorbed methyl, ethyl, or iso-propyl carbenium ions denoted as R was modelled by the following cluster:

$$\begin{matrix} & & \overset{\delta+}{R} & & \\ H{-}O & & O\overset{\delta-}{} & O{-}H & \\ H{-}O{-}Si & & Al{-}O{-}H & & \\ H{-}O & & & O{-}H & \end{matrix}$$

(8)

The obtained results confirmed that the ground states of all the above - mentioned adsorbed species are essentially covalent. This follows from their geometry with the short covalent C-O bond length, from tetrehedral O-C-H bond angles and from the relatively low Mulliken positive charges of the alkyl fragments presented in Table 3.

Table 3.

The results of semi-empirical quantum - chemical calculations for carbenium ions adsorbed on the surface of a high silica zeolite [17].

Alkyl fragment	$\overset{+}{CH_3}$	$\overset{+}{C_2H_5}$	$\overset{+}{C_3H_7}$
qR	+0.32	+0.33	+0.33
qO	-0.42	-0.41	-0.41
qC	+0.18	+0.13	+0.07
$\angle$ C-O-H	115.1	112.0	110.1
r_{C-O} A	1.461	1.474	1.466
D_{het} eV	7.25	6.05	4.93

The insigniffican increase in the C-O distance and in the positive charge of alkyl fragments in this series is quite surprising. This certainly indicates a small difference in the ionic character of the ground states of the adsorbed species. At first glance, this looks like a contradiction with the general belief that the ionic properties of iso-propyl group are expressed much stronger than those of methyl group. However, this question is absolutely clarified if one takes into account the energy of heterolytic alkyl - fragment splitting (Table 3, last line). When going from methyl to iso-propyl group, this energy decreases by more than 2 eV. Thus, similar to the above case of surface hydroxyl groups, the main difference between the adsorbed species is associated not with their covalent ground states but with the energies of their heterolytic abstraction from the surface. In other words, the chemical properties of adsorbed methyl, ethyl and iso-propyl species similar to those of surface hydroxyl groups are determined by positions of the electronically excited heterolytic terms.

In reality, in the course of heterogeneous catalytic transformations of hydrocarbons, the complete heterolytic dissociation of C-O bonds in surface esters never occurs. Instead, similar to reactions of proton transfer, the stretching of C-O bonds originating from their vibrational excitation makes the alkyl fragments more ionic. Their positive charge increases, and they become more flat resembling in their geometry and electronic structure free aliphatic carbocations. This is a particular case of a general rule formulated by us earlier in [19], according to which the stretched chemical bonds are usually more polar than the corresponding covalent ground states. Thus, the real meaning of the more ionic character of iso-propyl fragment in comparison with methoxy group consists in easier vibrational polarization of the corresponding C-O bonds resulting in easier formation of an excited unstable ion pair with a carbenium -ion - like positively charged alkyl fragment.

This idea was recently developed in more detail in [20], where nonempirical quantum - chemical calculations were performed for the proton transfer to adsorbed ethylene molecule in a high silica zeolite and for the reverse reaction of surface ethoxy - group decomposition. The geometry of both adsorbed ethylene and the resulting alkoxide was fully optimized with the help of gradient procedure of a "Gaussian 80" program. Acid groups of a high silica zeolites were modelled by a simplest $HO(H)Al(OH)_3$ cluster.

The obtained results demonstrated that ethylene is at first adsorbed as a π-complex of Fig. 2 (a). Its formation energy is 6.90 kcal/mol. Practically no change in the ethylene geometry was found upon adsorption, except of only a very small (+0.031e) positive charging of the adsorbed molecule.

In accordance with the above discussion, the most stable protonated structure is a covalent ethoxy group of Fig. 2 (c). It has a relatively low positive charge of the ethyl fragment (+0.384 e) and a geometry that is typical of covalent organic compounds. Indeed, the calculated C-O bond length is close to the mean length of this bond in alcohols or esters. The obtained C-C distance of 1.553Å and the nearly tetrahedral O-C-H and O-C-C valence angles in the alkyl fragment also confirm its covalent character. Finally, the calculated heat of zeolite surface ethoxylation (11 kcal/mol) is consistent with the thermochemical data on the heat of decomposition of covalent sulfuric acid ester [21].

The transformation of adsorbed ethylene into the surface ethoxide proceeds via a transition state of Fig. 2 (b) with an activation energy barrier of 15.4 kcal/mol. The geometry and electronic structure of such an activated complex resemble those of a classical form of ethyl carbocation. This follows from a marked enhancement in the positive charge of the ethyl group from +0.384 to +0.563 e and from change in the the the new C-C bond length that becomes intermediate between double and single C-C bonds. In addition, an essential flattening of methylene fragment occurs with O-C-H valence angles now close to 90^0. The marked deprotonation tendency of ethyl group is also clearly revealed in the formation of a hydrogen bond between one of the protons of CH_3 group with the neighbouring basic oxygen of the zeolite surface.

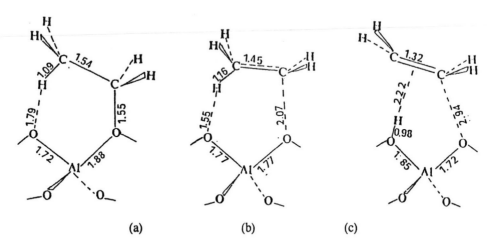

(a)   (b)   (c)

Fig. 2. The results of nonempirical quantum - chemical calculation of ethoxylation of bridging OH groups in high silica zeolites [20].

The reaction coordinate of ethoxylation is very complicated. It includes a stretching of the O-H bond in the acidic surface hydroxyl group and of C-C bond in the adsorbed molecule as well as a change of the C-C-H valence angles in methyl and methylene fragments and a shift of the adsorbed ethylene molecule towards the neighbouring surface basic site. The activation energy of this elementary step is low enough only if all of these parameters are changing simultaneously. Thus, the proton transfer to an adsorbed ethylene molecule proceeds via a concerted mechanism which provides maximum energy compensation of the bonds to be broken by those to be formed.

The decomposition of the surface alkoxy structure follows the opposite pathway. The corresponding reaction coordinate includes at the beginning the stretching of the C-O bond in the surface ester resulting in the considerable increase in its polarity which is maximal in the carbenium-ion-like activated complex.

The further heterolytic dissociation is extremely energetically unfavourable. Therefore instead of abstraction of ethyl carbocation from the zeolite surface, a deprotonation of the activated complex takes place. This requires to overcome an activation energy barrier of only about 20 kcal/mol that is much lower than the homolytic dissociation energy of about 90 kcal/mol or the heterolytic dissociation energy of about 150 kcal/mol (Table 3, Fig. 3). Thus, the reaction coordinate of the surface ethyl - fragment decomposition represents the combination of the C-O bond stretching in the surface ester and the deprotonation of the resulting carbenium - ion - like transition state.

Figure 3 demonstrates the energy diagram for the surface hydroxyl - group ethoxylation and for the reverse reaction of its decomposition. It differs from the above Fig. 1 in two respects. Firstly, the adsorbed carbenium ion is a transition state rather than a short - lived excited complex. Secondly, instead of adsorbed ethylene, the most stable surface intermediate is a covalent surface alkoxide.

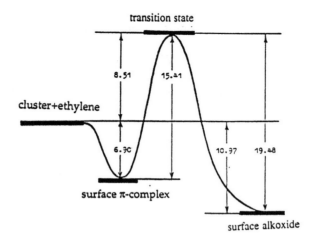

Fig. 3. The energy profile of bridging OH - group ethoxylation in a high silica zeolite.[20]

3.2. ^{13}C NMR MAS study of protonated intermediates

The above conclusion about the covalent bonding of adsorbed aliphatic carbenium ions with the surface of zeolites was supported by a ^{13}C NMR study of adsorbed olefins and alkohols. Due to very large chemical shifts in aliphatic carbenium ions of about 330 ppm, this technique is most suitable for their detection. However, until recently, the line broadening caused by low mobility of adsorbed species limited the NMR application to only purely homogeneous systems. The situation radically changed when MAS NMR was introduced to chemical study as a rutine technique.

At the moment the only attempt no apply this technique for the study of olefin adsorption is oligomerization of propene on HY zeolite [22,23]. The first of these works is of a more qualitative nature, whereas in the second paper the experiments were carried out with ^{13}C enriched samples selectively labelled in CH$_3$, CH or CH$_2$ groups in a broad temperature interval with the use of cross polarization.

Propene was shown to be highly mobile in the zeolite cavities at temperatures far below the onset of chemical reactivity. It was also concluded that adsorbed secondary or tertiary carbocations either do not exist or are so transient that they are not detected by NMR even at temperatures as low as 163 K. Only the long - lived stable alkyl substituted cyclopentadienyl carbocations were detected at room temperature. However, these species do not form until all of the propene is consumed and hence do not play a significant role in the oligomerization. On the other hand, in accordance with the above results of quantum - chemical calculations, the most important long lived intermediates are not the adsorbed aliphatic carbenium ions but alkoxy species having a 87 ppm ^{13}C chemical shift on carbon atoms bonded with oxygen of the zeolite framework.

Another example, where ^{13}C NMR MAS was used to study of protonated active intermediates, is adsorption of alcohols. They are also believed to produce aliphatic carbenium ions by dehydration of adsorbed species when interacting with acid protons:

$$H^+_{ads} \quad + \quad ROH \quad ----- \quad R^+_{ads} \quad + \quad H_2O \tag{9}$$

For instance, in [24] adsorption of 2-methyl-2-propanol, $(CH_3)_3{}^{13}COH$, on a HZSM-5 zeolite was studied in the broad temperature range from 10 to 373 K. No resolvable spectral features could be distinguished in the noise in the region expected for carbenium ions even though the final products were similar to those following olefins adsorption and were consistent with formation of the carbenium ions. This also implies that the steady- state concentration of carbenium ions during tert-butanol decomposition on the HZSM-5 zeolite is very low or that lifetime of these species is very short.

On the other hand, the observed ^{13}C NMR spectrum showed two prominent features of formation of adsorbed tert-butyl alkoxydes. Indeed, the line at 77 ppm was assigned to the tertiary carbon atom bonded to the surface oxygen of the zeolite framework, and another line at 29 ppm was ascribed to aliphatic carbons in their methyl groups or in the reaction products. After heating to 373 K, the further oligomerization of hydrocarbon fragments occurs resulting in the appearance of additional lines in this region with somewhat smaller chemical shifts.

All these conclusions were essentially confirmed in the later publication which was also devoted to the study of tert-butyl alcohol dehydration selectively labelled with ^{13}C on a HZSM-5 zeolite [25]. The simultaneous measurements of ^{13}C MAS and deuterium solid state NMR were carried out. Similar to [24], no signals from adsorbed tertiary butyl carbenium ions were detected. Instead only those from ^{13}C tertiary carbon atoms in the surface tertial-butyl-silyl ether or corresponding oligomeric-silyl ether with chemical shift of 86 ppm together with the lines from oligomeric aliphatic final products were observed. The silyl ether intermediates were stable within the temperature range of 296-373 K and decomposed at 448 K to produce additional amount of butene oligomers. Another important feature of the reaction was the scrambling of ^{13}C label among the reaction products.The authors also supposed that dehydration proceeds through invisible tert-butyl carbenium ions and postulated the equilibria between the initial tert-butyl alcohol, tert-butyl cation, and butene that is formed from the intermediate carbocation.

Some further details of the acid - catalysed dehydration on HZSM-5 were clarified by the study of less reactive iso-butanol labelled with a ^{13}C isotope in the CH_2 group [26,27]. In this case, the chemical shift of secondary carbon atom of iso-butyl sylil ether was equal to 73 ppm, and this intermediate was stable below 398 K, whereas no oligomerization products were formed. Within the temperature range of 398-423 K, the adsorbed alkoxy species gradually decomposed to produce at first butene dimer, probably 2,5-dimethyl-1-hexene, and then other butene dimers and oligomers. At 423 K the scrambling of the selectively labelled carbon in the initial dimeric product over various positions in the carbon skeleton took place. Although no NMR signals from iso-butyl carbenium ions were observed, this was explained, as in the case of tert-butanol, by the formation of excited carbenium - ion intermediates similar to the above structure (8):

$$
\begin{array}{ccc}
\begin{array}{c}
CH_3 \quad CH_3 \\
\diagdown CH \diagup \\
| \\
^*CH_2 \\
| \\
O \\
\diagup \diagdown \\
Si \quad Al
\end{array}
&
\begin{array}{c}
CH_3 \quad CH_3 \\
\diagdown CH \diagup \\
| \\
^*CH_2^+ \\
O^- \\
\diagup \diagdown \\
Si \quad Al
\end{array}
&
\begin{array}{c}
^*CH_3 \quad CH_3 \\
\diagdown CH \diagup \\
| \\
CH_2^+ \\
O^- \\
\diagup \diagdown \\
Si \quad Al
\end{array}
\end{array}
\qquad (10)
$$

The skeletal isomerization of the butyl fragment and subsequent oligomerization was also explained through excited carbenium-ion-like species.

3.3. The real meaning of carbenium ion mechanisms

When discussing the acid - catalyzed transformations of olefins, the following main reactions are usually considered: double bond isomerization, cracking via β-scission of adsorbed carbenium ions, the opposite reaction of oligomerization, skeletal isomerization, and, finally, hydride transfer between iso-paraffins and adsorbed carbenium ions. Below we will discuss the more realistic mechanisms of these reactions from the standpoint of the above - developed ideas, according to which adsorbed carbenium ions represent unstable excited ion pairs or transition states.

a) Double - bond isomerization.

The mechanism of this reaction was already discussed above in section 3.1., when we considered the decomposition of surface alkoxy groups. It consists in strething of the C-O bond in surface alkoxide resulting in the formation of an excited ion pair representing an unstable carbenium-ion-like intermediate. One of the possible ways of its stabilization is deprotonation. If proton is splitted off from the CH_2 group next to the positively charged carbon, the double bond shift in the final olefin molecule takes place:

$$CH_3-CH-CH-CH_3$$

$$+ \quad CH_3-CH=CH-CH_3 \quad (11)$$

This is also a concerted mechanism similar to the decomposition of surface alkoxides. The reaction coordinate involves mainly the C-O bond stretching followed by the proton transfer from the resulting carbenium-ion-like transition state to the surface oxygen and by the transformation of the single C-C bond into the double bond. Thus, the activation energy of the double - bond shift should be equal to that of the corresponding surface alkoxide decomposition. This is in agreement both with the above - calculated activation energy of this elementary step and with that of 1-butene isomerization of ca. 15-20 kcal/mol [13].

b) Cracking.

Cracking of adsorbed carbenium ions which proceeds via β-scission and results in desorption of lighter olefins from the surface is another way to stabilize the excited ion pairs. Here, the neutralization of the positively charged carbon atom occurs due to the formation of double bond. The positive charge migrates to the splitted alkyl fragment, which then forms a new surface alkoxy group, when interacting with the neighbouring basic oxygen:

$$CH_3-CH--CH_2\text{-}R \quad CH_3-CH=CH_2 \quad R$$

$$\longrightarrow \quad C_3H_6 \quad + \quad (12)$$

The corresponding reaction coordinate is extremely complicated. It includes stretching of the C-O and C-R bonds, compression of the $CH\text{-}CH_2$ bond when transforming into the double bond, the bonding of the splitted alkyl fragment R with the neighbouring basic oxygen, and desorption of propylene from the surface.

This is another example of a rather complicated concerted reaction mechanism which makes possible the compensation of the dissociating bonds by the newly formed ones. This allows to minimize the activation barrier which would be much higher, if the reaction involves formation of almost free carbenium ions followed by their subsequent cracking into the final products. An attractive feature of this mechanism is a logical explanation of the β-scission rule by the simultaneous dissociation of the C-R bond and by the formation of new double bond with the neutralisation of the positive charge in the excited carbenium ion. Finally, this mechanism naturally explains the easier cracking of branched paraffins, where the newly formed branched carbenium ions are more stable and therefore are formed easier.

c). Oligomerization of olefins.

This reaction is the reverse of cracking. It also starts with the formation of surface alkoxide which then may be excited into an unstable ion pair. If this intermediate reacts during its lifetime with another olefin molecule, a larger alkyl fragment is formed. After decomposition, it converts into a higher olefin as the final product of the reaction.

Since the excitation of a tertiary surface alkyl fragment into an ion pair occurs easier than that of secondary and primary alkoxides, the rate of oligomerization of branched iso-olefins is higher than that of primary or secondary olefins.

d) Skeletal isomerization of olefins.

Skeletal isomerization of olefins is known to require very strong acid sites similar to those necessary for cracking. This probably indicates the common features of the mechanisms for both reactions.

Indeed, the initial step of isomerization is likely the same as that of cracking represented by reaction (12). However, if, instead of adsorption on neighbouring basic oxygen of the active site, the splitted smaller carbenium ion attacks the double bond of the olefin, and the central proton of the olefin is transferred back to the catalyst surface, this may result in skeletal isomerization:

$$
\begin{array}{ccc}
\underset{\underset{\substack{|\\ \overset{|}{H}\quad \overset{+}{R}\\ \underset{Al}{\diagup O\ \ O^-}\\ \diagup O\quad O}}{}}{CH_3\text{-}CH\text{=}CH_2}
& \longrightarrow &
\underset{\underset{\substack{|\quad|\\ \overset{|}{H}\quad R\\ \underset{Al}{\diagup O\ \ O^-}\\ \diagup O\quad O}}{}}{CH_3\text{-}C\text{=}CH_2}
\end{array}
\longrightarrow
\underset{\underset{R}{|}}{CH_3\text{-}C\text{=}CH_2}
\;+\;
\underset{\underset{\substack{\underset{Al}{\diagup O\ \ O}\\ \diagup O\quad O}}{}}{\overset{H}{\overset{|}{\ }}}
\quad (13)
$$

This mechanism naturally explains a close relation between cracking and skeletal isomerization. In addition, it predicts the importance of the splitted carbenium ion interaction with neighbouring basic site in case of cracking and with olefin for skeletal isomerization. This might be related to the different geometry of the active sites or to the different strength of their basic oxygen atoms.

e) Hydride - ion transfer.

This reaction represents one more way to neutralize the positive charge of the excited carbenium-ion-like fragment in the surface ion pair by the hydride - ion transfer from iso-paraffin:

$$(14)$$

This is also a concerted mechanism involving a combination of $O-R_1$ bond stretching in the surface ester, heterolytic dissociation of the C-H bond in iso-paraffin, and formation of a new C-O bond with the neighbouring basic oxygen of the active site. In addition, the newly formed H_2C-R fragment exhibits the Walden inversion.

The common feature of all the above mechanisms is the bifunctional nature of the active sites. Their Broensted acid part protonates the initial molecules, while the interaction with basic oxygen of the zeolite framework stabilizes the resulting carbenium ions as the covalent surface alkoxides. Another way to stabilize the excited ion pairs with carbenium-ion-like alkyl fragments is the splitting off protons (reaction 11) or carbenium ions (reaction 12).

The combination of proton transfer to adsorbed molecules with simultaneous deprotonation of transition states by neighbouring basic oxygen results in concerted mechanisms with low activation energy barriers. This is an important feature of catalysis by acid - base pairs which has to be considered instead of a simple protonation according to schemes (2,3).

4. CATALYTIC TRANSFORMATIONS OF PARAFFINS AND THE NATURE OF ADSORBED NONCLASSICAL CARBONIUM IONS

Adsorbed carbenium ions play also the role of active intermediates in cracking and isomerization of paraffins, which are traditionally considered as chain reactions. The chain initiation occurs through the interaction of bridging hydroxyls with olefins similar to the reaction already discussed above. This is then followed by the cracking of adsorbed carbenium ions and by the chain propagating through hydrogen transfer in the rate determining step:

$$H^+_{ads} + olefin \longrightarrow R^+_{ads}$$

$$R^+_{ads} \longrightarrow R_1^+{}_{ads} + olefin \qquad (15)$$

$$R_1^+{}_{ads} + RH \longrightarrow R_1H + R^+_{ads} \quad etc.$$

A similar chain mechanism is commonly accepted for isomerization of paraffins with the only difference that skeletal isomerization of carbenium ions takes place instead of cracking:

$$R^+_{ads} \longrightarrow i\text{-}R^+_{ads}$$

$$\qquad (16)$$

$$i\text{-}R^+_{ads} + RH \longrightarrow i\text{-}RH + R^+_{ads} \quad etc.$$

The adsorbed carbenium ions involved in both of these reactions are certainly similar to those already discussed above. Therefore, they also don't really exist as relatively stable adsorbed active intermediates but rather represent the corresponding transition states.

It was also recently shown that at low partial pressure of olefins and high temperatures, when the concentration of adsorbed carbenium ions is low, the mechanism of chain initiation is different. It involves adsorbed nonclassical carbonium ions that are formed by direct protonation of paraffins [28]. Consequent dehydration or cracking of these species also result in formation of adsorbed carbenium ions which react further according to the already discussed above schemes:

$$
H^+_{ads} + R\!-\!\overset{\displaystyle R}{\underset{\displaystyle H}{C}}\!-\!CH_3 \longrightarrow R\!-\!\overset{\displaystyle R}{\underset{\displaystyle H\ \ H}{C}}\!-\!CH_3
\begin{cases}
H_2 + R\!-\!\overset{\displaystyle R}{C}\!-\!R_{ads} \quad (a)\\[2pt]
\qquad\qquad + \\[6pt]
CH_4 + R\!-\!\overset{\displaystyle R}{\underset{\displaystyle H}{C^+}}{}_{ads} \quad (b)
\end{cases}
\qquad (17)
$$

The most convincing evidence of this mechanism was obtained for cracking of light paraffins at very low conversions. For instance, at 450-500 C, iso-butane conversion over a HZSM-5 zeolite in a steady - state flow reactor was less than 1 %. The main products were methane and propylene formed in equal amounts with apparent activation energy of 57 kcal/mol [29]. This indicates that cracking proceeds through carbenium ion intermediate according to scheme (17,b). On the other hand, the activation energy for n-butane cracking was much lover (only about 35-40 kcal/mol), and the rates of hydrogen + butenes, methane + propylene, and ethane + ethylene formation were quite comparable [30].

A similar conclusion that the chain initiation occurs by direct protonation of C-H and C-C bonds was also made for i-butane ant n-pentane cracking on faujusites. Cracking of i-butyl carbonium ions results both in the splitting of molecular hydrogen and methane, whereas penthyl carbonium ions split off methane. The resulting carbenium ions propagate chain reactions mainly by isomerization followed by hydrogen transfer. The estimated chain length varied from 3 to 15 depending on the reaction conditions.

Similar to carbenium ions, such adsobed carbonium ions certainly also represent highly excited intermediates or transition states. This follows from high activation energies of protolytic cracking of paraffins on zeolites according to reactions (11) and is consistent with by about 1.5 eV lower proton affinity of paraffins in comparison with olefins (Table 4).

Table 4.

the hydrocarbon	Proton affinity, eV.
CH_4	5.1
C_2H_6	5.6
C_3H_8	6.1
C_2H_4	6.9
C_4H_8	7.8
C_3H_6	7.9

The reduced tendency of proton transfer to adsorbed paraffins is also supported by spectral data (smaller low frequency shifts of stretching vibrations caused by interaction of hydroxyl groups with paraffins in comparison with olefins). Finally, this concliusion was also confirmed by our quantum - chemical analysis of protolytic cracking on a high silica zeolite performed in [32].

The "ab initio" quantum chemical calculations were carried out in this paper using a "Gaussian 80" program and 3-21 basis set. The geometry of adsorbed complexes was optimized by the gradient technique, whereas the transition state of cracking was found by minimizing the norm of gradient. Both procedures required rather long computations. Therefore only the simplest case of ethane interaction with the simplest H_2O $Al(OH)H_2$ cluster was considered. It is certainly too small for quantitative description of the proton transfer or cracking energetics. For instance, the proton - abstraction energies calculated for this cluster with a 3-21 basis set is 15.38 eV. This is considerably higher than the values of 12-13.5 eV estimated from spectral data [10,11] or those calculated for larger clusters with a more sophisticated basis set [9]. Therefore the following results should be considered only as a first preliminary attempt of qualitative quantum-chemical analysis, whereas a real quantitative treatment of protolytic cracking requires more accurate calculations.

The geometry of the corresponding activated complex and the Mulliken charge distribution are shown in Fig. 4 (a) and (b), respectively:

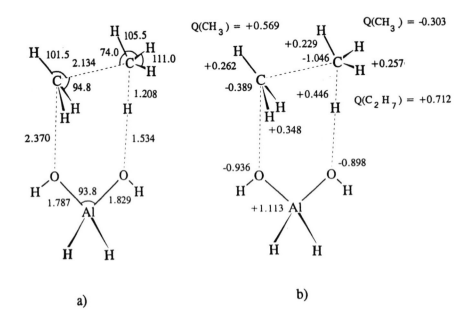

a)

b)

Fig. 4. The transition state of protolytic cracking of ethane on a high silica zeolite.

Such a transition state should be considered as a bipoint interaction of ethane with the bridging acid hydroxyl group and neighbouring basic oxygen of the active site. This results in both polarization and positive charging of the adsorbed molecule due to the partial proton transfer from the surface of the zeolite. Indeed, the net positive charge of the hydrocarbon fragment in such complex is rather high (+0.712e), whereas the positive charge of the left methyl group is only +0.297e in comparison with +0.569 e for the right metyl group. The considerable increase of the C-C bond length is also quite remarkable indicating dissociation of ethane to methane and positively charged methyl group. The latter resembles a free methyl carbenium ion not only due to a high positive charge, but also because it becomes considerably more flat than initial ethane molecule.

Thus, the reaction coordinate of protolytic cracking represents the concerted stretching of the O-H bond in surface acid hydroxyl group combined with the heterolytic splitting of the C-C bond in protonated ethane molecule. This results in formation of methane and adsorption of splitted methyl carbenium ion on the neighbouring surface oxygen of the active site as a methoxy group:

$$(16)$$

The calculated activation energy of this reaction is equal to 93.4 kcal/mol while its ethalpy is only 2.5 kcal/mol.

The first of these figures is certainly strongly overestimated. Indeed, If proton transfer is a limiting step of cracking, then the calculated energy should be close to the experimentally measured activation energy of cracking. The latter is much lower, and as indicated above, changes for different butane isomers within 35-60 kcal/mol.

However, the following points should be taken into account when comparing experimental and calculated figures:

i. The protonation of ethane is certainly more difficult than that of butane.
ii. The cluster used in our calculations is too small. This results by ca 1.5-2 eV higher proton - abstraction energy than that estimated from spectral data or obtained from more precise calculations.
iii. The effective activation energies reported in the literature represent the difference between real activation energies and the heats of hydrocarbon adsorption. Therefore the real activation energy of cracking should be igher.
iiii. The 3-21 basis set used in our calculations can not pretend for quantitative description of the interaction energy.

Due to all these reasons our calculations strongly overestimate the endothermicity of proton transfer. Therefore, the real discrepancy between the calculated and the experimentally measured activation energies is not as large as one could imagine at the first glance. Anyway, our results clearly demonstrate that adsorbed nonclassical carbonium ions even in a more extent than adsorbed carbenium ions represent highly excited active intermediates which could not be considered as really existing surface compounds. Such short - lived surface complexes likely arise from the interaction of thermally excited at elevated temperatures surface OH groups with adsorbed paraffins or from the collisions of gaseous molecules with such OH groups.

CONCLUSION

It follows from the above discussion that the ground states of the surface acid OH groups in hydrogen forms of zeolites are essentially covalent, whereas the main difference between the strongly and weakly acidic groups consists in the positions of their electronically excited heterolytic terms. In a similar way, the interaction of acid OH groups with olefins instead of yielding adsorbed carbenium ions or ion pairs gives rise to the formation of covalent surface alkoxides which are much less ionic than it is usually believed. These conclusions are supported by quantum chemical calculations and by IR and MAS NMR data.

The reason for the covalent character of the surface protons and adsorbed alkyl fragments consists in their interaction with the surface basic oxygens of the zeolite framework which neutralizes the positive charges of these species. Therefore, adsorbed carbenium ions represent not really existing surface intermediates of acid catalysis but transition states or electronically excited unstable ion pairs resulting from excitation of the covalent ground states of surface alkoxy groups.

The similar conclusion is even more true for adsorbed carbonium ions which are formed by direct protonation of paraffins. Due to about 1-2 eV lower proton affinity, these species also represent highly excited short - lived complexes resulting from the bipoint interaction of paraffins with the acid hydroxyl groups and neighbouring basic oxygen atoms of the active sites.

The above ideas clearly demonstrate the bifunctional nature of acid active sites in heterogeneous acid catalysis. Their Broensted acid part protonates the adsorbed molecules, while the interaction with the neighbouring basic oxygen converts the resulting excited protonated species into more stable covalent intermediates. Since the simultaneous interaction of adsorbed molecules with acid - base pairs strongly depends on their geometry, such dual nature of active sites may also result in their structural inhomogeneity.

Another important feature of heterogeneous acid catalysis is the significance of concerted mechanisms involving very complicated reaction coordinates. This provides reaction pathways with the low activation barriers due to most complete compensation of the energy required for the rupture of the reacting chemical bonds in the initial coumpounds by the energy of newly formed chemical bonds in the final products. Such concerted mechanisms in some sense resemble the well known SN2 mechanism that is effective for many acid - catalyzed reactions in organic chemistry.

All these ideas should be certainly regarded not as a criticism of the classical carbenium - ion conception but rather as an attempt of its further development and modernization.

REFERENCES

1. V.B.Kazansky. "Structure and Reactivity of Modified Zeolites". Ed. by
   P.Jacobs, P.Jiru and V.Kazansky. Elsevier Sci. Pub., Amsterdam 1984, P. 61.

2. I.N.Senchenya, V.B.Kazansky, S.Beran. J. Phys. Chem. 1986, V.90, P.4857.
3. G.J.Kramer, R.A.van Santen. J. Am.Chem.Soc. 1993, V. 115, P. 2887.
4. V.B.Kazansky. Kinetika i Kataliz (Russ.) 1977, V. 18, P. 966.
5. V.B.Kazansky, A.M.Gritskov, V.M.Andreev, G.M.Zhidomirov. J. Molec. Catal. 1978, V. 3, P.135.
6. V.B.Kazansky. Kinetika i Kataliz (Russ.) 1980, V. 21, P. 128.
7. V.B.Kazansky. Acc. Chem. Res. 1991, V. 24, P. 379.
8. V.B.Kazansky. Kinetika i Kataliz (Russ) 1980, V. 21, P. 159.
9. H.V.Brand, L.A.Kurtiss, L.Elton. J. Phys. Chem 1992, V. 96, P. 7725.
10. J.Datka, M.Boczar. J. Catal 1988, V.114, P. 368.
11. L.Kubelkova, S.Beran, J.A.Lercher. Zeolites 1989, V. 9, P. 539.
12. H.A.Klinovski, S.Berger, P.Braun. "^{13}C-NMR Spectroscopy". Tholone Verlag, Stuttgart 1985.
13. P.A.Jacobs. "Carboniogenic Activity of Zeolites".Elsevier Sci. Pub. Co., 1977, Amsterdam,. Oxford, N.Y.
14. P.A.Jacobs, L.J.Declerck, L.J.Vandamme, J.B.Uytterhoeven. J.Chem.Soc.Far. Trans. I, 1975, V. 71, P. 1545
15. V.B.Kazansky, W.J.Mortier, B.G.Baekelandt, J.L.Lievens. J. Molec. Catal. 1993, V. 83, P. 135.
16. R.G.Pearson. Inorg. Chem 1988, V.27, P. 734.
17. I.N.Senchenya, V.B.Kazansky. Kinetika i Katalis (Russ) 1987, V. 28, P. 448.
18. V.B.Kazansky, I.N.Senchenya. J. Catal. 1989, V. 119, P.108.
19. V.B.Kazansky. In Proc. of the 6th Int. Congr. on Catalysis, Imperial College, London 1976, Royal Society of Chemistry, Lethworth, Herts 1977.
20. V.B.Kazansky, I.N.Senchenya. Catal. Lett. 1991, V. 8, P. 317.
21. "Thermochemistry of Organic and Metalloorganic Compounds". Ed. by J.D.Cox, G.L.Pilcher, Academic Press, N.Y. 1979, P. 390.
22. M.Zardhoohi, J.F.Haw, J.H.Lunsford. J. Am. Chem. Soc. 1987, V. 109, P. 5278.
23.J.F.Haw, B.R.Richardson, I.S.Oshiro, N.D.Lazo, J.A.Speed. J. Am. Chem. Soc. 1989, V. 111, P. 2052.
24. M.T.Aranson, R.J.Gorte, W.E.Farneth, D.J.White. J. Am. Chem. Soc. 1989, V. 111, P. 840.
25. A.G.Stepanov, K.I.Zamaraev, J.M.Thomas. Catal. Lett. 1992, V. 13, P. 407.
26. A.G.Stepanov, V.N.Romannikov, K.I.Zamaraev. Catal. Lett. 1992, V. 13, P. 395.
27. A.G.Stepanov, K.I.Zamaraev. Catal. Lett. 1992, V. 19, P. 153.
28. W.O.Haag, R.M.Dessau. Proc. 6th Int. Congr. on Catalysis, Berlin 1984, V. 2, P. 305.
29. C.Stefanidis, B.C.Gates, W.O.Haag. J.Molec. Catal.1991, V.67, P. 363.
30. H. Krannila, W.O.Haag, B.C.Gates. J. Catal. 1992, V. 135, P. 115.
31. P.V.Shertugde, G.Marcelin, G.A.Still, W.K.Hall. J. Catal. 1992, V. 136, P.446.
32. V.B.Kazansky. I.N.Senchenya, M.Frash, R.A.van Santen. Catal. Lett., in press.

J.C. Jansen, M. Stöcker, H.G. Karge and J. Weitkamp (Eds.)
Advanced Zeolite Science and Applications
Studies in Surface Science and Catalysis, Vol. 85
© 1994 Elsevier Science B.V. All rights reserved.

Theory of Brønsted Acidity in Zeolites

R.A. van Santen

Schuit Institute of Catalysis, Eindhoven University of Technology,
P.O. Box 513, 5600 MB Eindhoven, The Netherlands.

1. ABSTRACT

The nature of the chemical bond of protons in a zeolite is analysed on the basis of theoretical and spectroscopic results. Of interest is the dependence on zeolite structure as well as composition. The zeolitic OH bond is mainly covalent. Proton attachment to the zeolite lattice causes a weakening of neighbouring Si-O and Al-O bonds. The effective increase in volume of the bridging oxygen atom causes a local deformation, that changes the strength of the lattice-chemical bonds over a few bond distances. Proton concentration effects as well as lattice-composition effects can be understood on the basis of the lattice-relaxation model. The energetics of proton transfer is controlled by the need to stabilize the resulting Zwitter-ion. The positive charge on the cation becomes stabilized by contact with basic lattice-oxygen atoms.

2. Introduction

How the reactivity of the acidic protons attached to the framework-oxygen atoms in the micropores of zeolites depends on composition, location and zeolite framework is of obvious importance to the application of zeolites in acid catalyzed reactions. Whereas significant progress has been made to determine this relationship, it is still not completely understood. One of the reasons is the complexity of the catalytic reaction cycle, that is composed of different reaction steps. Proton transfer between lattice and substrate is only one of them and not necessarily always the rate limiting step. Secondly the acidity of the proton, as probed in the catalytic event depends on its state before reaction as well as the stability of the "Zwitter-ion" state, generated upon proton transfer to the reacting molecule. The latter may vary dependent on substrate and, as we will see, depends also strongly on the interaction between ion and zeolite wall.

Here we will discuss recent results of theoretical as well as spectroscopic studies probing the details of the chemical bond of the proton with the zeolite and the response of the proton with reacting basic molecules.

Protons in zeolites have been extensively studied using many different physical techniques, especially infrared spectroscopy, NMR spectroscopy and neutron scattering. Because of their well-defined nature zeolites belong to the exceptional class of catalysts that allow for a definitive structural characterization of its catalytically active site. Computational chemical techniques require such detailed knowledge. The use of ab-initio calculations has contributed in an essential way to an understanding of the chemical features that control zeolite acidity. An important theoretical question we will address is the validity of the cluster approach to approximate the protonic site. A second important theoretical approach is the use of lattice-energy calculations and lattice-structure simulations based on force fields determined from accurate quantum-chemical calculations.

In the following two sections we will highlight our present understanding of the nature of the chemical bond of the proton. In a subsequent section we will present the results of theoretical studies on proton transfer to adsorbed basic molecules. It will appear that two properties of the solid state are very important to zeolite acidity:
- electrostatics, the positive charge developing on the proton has to be compensated by a negative charge generating on the zeolite lattice.
- lattice relaxation, changes in chemical bonding affect the forces acting on the lattice atoms involved.

In a final section the implications of the physical chemistry of proton transfer to acid catalysis will be shortly outlined.

3. The proton Bond

The zeolite lattice consists of a three dimensional network of tetrahedra connecting four valent or three valent metal cations such as Si or Al, each having four oxygen atoms as neighbours. One oxygen atom has two metal cations as nearest neighbours (figure 1). When all lattice ions are Si, the zeolite lattice has the composition SiO_2 and is a polymorph of quartz. Brønsted acidic sites are generated when silicon, which has a formal valency of four, is replaced by a metal atom with a lower valency. Most common is the replacement of silicon by aluminum with a formal valence of three. A proton is attached to the oxygen atom connecting a neighbouring silicon and aluminum atom, resulting in a chemically stable situation. Note that now the oxygen atom becomes three coordinated.

Quantum-chemical calculations have convincingly shown that the SiO, AlO as well as the OH bonds have considerable covalency, resulting in a relatively weak OH bond. The "onium" type coordination of oxygen is the fundamental reason for the high acidity of the attached proton. Cluster calculations on H or OH terminated rings in figure 2 give illustrations of such geometries providing detailed geometric information on the acidic site [1]. The results of ab-initio cluster calculations on protonated rings formed from four tetrahedra are shown. As we will see crystallographically different sites in a zeolite show small differences. One of the most interesting features is the long Al-O bond compared to the Si-O bond (0.188 nm and 0.172 nm respectively). The acidic site can also be seen as a silanol group promoted by a the Lewis acidic group. Whereas the OH bondlength changes little, the heterolytic bond energy changes from approximately 21.8 eV [2] to 13.7 eV [3].

Kazansky [4] has analyzed the OH interaction-potential by an analysis of the overtones of the OH stretch frequencies measured in the near infrared spectroscopic range.

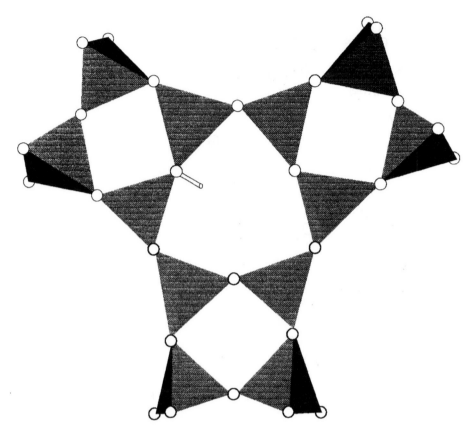

Figure 1. Connectivity of T-atoms. The T-atoms are in the centre of the tetrahedra formed by four oxygen atoms: when one silicon atom is replaced by an aluminum atom and a neighbouring oxygen atom becomes protonated.

He found that the covalent OH interaction-energy varies little with acidity. The constant homolytic bondstrength of OH indicates that the factor that contributes most to the strong acidity is mainly the increased stabilization of the negative charge left on the lattice with deprotonation.

1H MAS NMR spectra enable the experimental determination of the Al-H distance [5]. For zeolite Y values of 0.237 nm and 0.248 nm have been found for protons located in the supercages and sodalite cages respectively.

The cluster calculations indicate a large change in the equilibrium geometry of the ditetrahedral cluster upon deprotonation (compare figures 2a and 2b, atom-position 8). As is shown in figure 2b, the AlO and SiO distances shorten. These changes are a consequence of covalent bonding and the Bond Order Conservation principle [6]. Since no electrons of the bridging oxygen atom are involved any more in bonding with the OH bond, they become available to additional bonding for AlO and SiO. The result is an increase in the SiO and AlO bondstrengths.

276

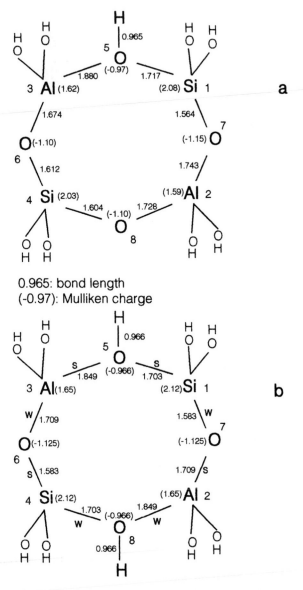

0.965: bond length
(-0.97): Mulliken charge

0.966: bond length
(-0.966): Mulliken charge
bond strengthened (S) or weakened (W) by second proton

Figure 2. 165 Four-ring cluster containing two silicon and two aluminum atoms and one (a) or two (b) acidic protons [1].

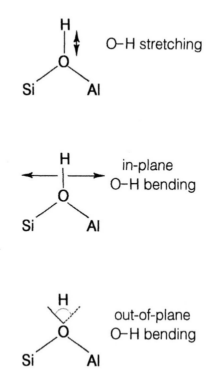

Figure 3. The three vibrational modes of the Brønsted acidic group.

Ab-initio quantum-chemical calculations also provide information on the vibrational modes of the OH group. Infrared spectroscopy and inelastic neutrons scattering [7] can be used to compare predictions with experiments. The three vibrational modes of the Brønsted acidic group are sketched in figure 3. Typical values of the OH stretch frequencies of the Brønsted acidic protons vary between 3650 and 3550 cm^{-1}, to be compared with the silanol frequency at 3745 cm^{-1}. The relatively small decrease in stretch frequency corresponds to a slightly weakened OH force constant of the acidic protons. The in plane bending mode of the Brønsted acidic protons has a frequency of approximately 1050 cm^{-1} [7-10], the out-of plane bending mode of approximately 400 cm^{-1} [7,9,10]. Indirect measurement of the bending modes are available from infrared combination bands [7,8]. Direct measurements have been done using inelastic neutron scattering spectroscopy [7,10]. Experimental and theoretical values [9] are in close agreement.

Experimental indications of local structural relaxation upon deprotonation have been provided by infrared spectroscopic studies of the lattice-modes of zeolite Y, comparing the spectra of NH$_4$-Y with the H-Y form [11]. Figure 4a shows the corresponding infrared

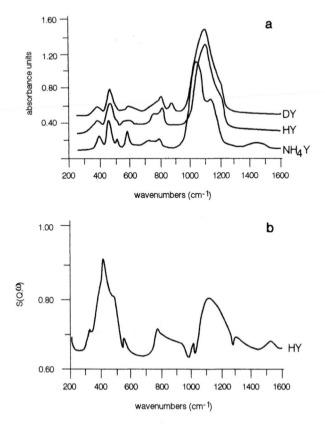

Figure 4. Infrared active lattice modes of NH₄Y, HY, DY (a) [11,12], and inelastic neutron scattering spectrum of HY (b) [7].

spectra. Also shown are the frequencies of the bending OH modes as measured by inelastic neutron scattering (figure 4b).

Protonation of the zeolite lattice results in significant changes of the lattice vibrational modes. One observes an upwards shift around 750 and 1100 cm^{-1}, decreases in intensity around 600 cm^{-1} and smaller changes at 400 cm^{-1}. Figure 4a shows also the infrared spectrum of deuterated zeolite Y. New peaks arise near 860 cm^{-1}[12]. The shifts around 750 and 1100 cm^{-1} and the losses in intensity below 600 cm^{-1} remain. Clearly in the region from 700 to 1100 cm^{-1} there is a strong coupling between the in-plane OH bending mode and the Si-O-Al stretching modes, that are resonant in energy. The existence of this coupling between the OH bending modes and the lattice modes is concluded from results from inelastic neutron scattering experiments [10]. Here, due to the coupling, the intensity of the lattice modes of H-Y zeolites is increased compared to those for Na-Y zeolites. The intensity of the modes between 600 - 300 cm^{-1} change very similarly for the protonated as well as deuterated system. The modes are mainly combinations of tetrahedral bending modes as well as Si-O-Al symmetric stretching modes, sensitive to deformation of the lattice [13]. These changes are due to the deformation of the lattice upon protonation.

Whereas the results of the ab-initio cluster calculations indicate the possibility of structural deformation upon protonation, embedding of the cluster in the three dimensional zeolite lattice is needed in order to evaluate to which degree it actually will occur in the solid. This can be conveniently done using energy minimization schemes applicable to the solid state. The equilibrium atomic geometries can be computed using atomic force fields [14]. Such force fields can be developed based on a fit of the parameters of classical interatomic potentials to potential energy surfaces computed quantum-chemically for small clusters [15]. One concludes from such studies that the changes in local chemical bonding due to proton attachment also lead to deformation of the lattice. However, the deformations are very local and disappear at a few atomic distances from the proton location [16]. This is illustrated in figure 5 for proton attachment to the O(1) atom of the double-six ring of faujasite. The geometry optimized structures of the protonated and non-protonated sites are compared. Atom positions before and after proton attachment are shown. One observes the longer SiO and AlO distances of the protonated system and the rapid decline of the bondlength changes at a larger distance from the proton.

Sites generated upon protonation require more space than the proton-free site. In zeolites atom distances of the lattice atoms vary for crystallographically different positions. One may expect to find a relation between bond angles and bondlengths around a site before protonation and the protonation energy. Structural and compositional effects will be discussed in the next section. The main conclusion of this section is that experimentally as well as theoretically local structural relaxation can be considered well established.

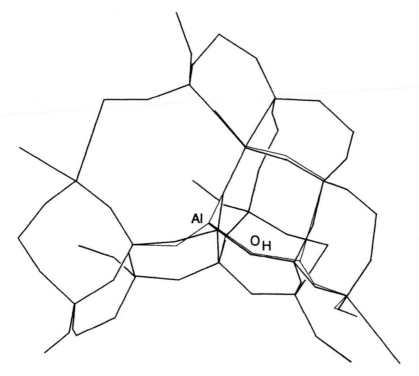

Figure 5. Lattice relaxation due to protonation of the double-six ring in the Faujasite lattice [16].

4. Proton affinity differences.

When one compares the stretch frequency of the OH bond in a zeolite in the same structure but varying the aluminum content, small changes in the infrared frequency are observed. Figure 6 shows the change in the infrared frequencies of high-frequency OH stretching modes in zeolite Y as a function of increasing Si/Al ratio [17]. Whereas at a ratio of Si/Al = 2.4, one observes only the silanol and two Brønsted acidic hydroxyl groups for zeolite Y (the high-frequency and low-frequency bands), upon an increase of the Si/Al ratio infrared HF OH peaks at lower frequencies appear, indicative of a stronger Brønsted acidity. This is in agreement with the shift to lower field of the corresponding lines in the 1H MAS NMR spectra (see figure 6).

In a zeolite according to the Löwenstein rule, aluminum ions cannot occupy neighbouring tetrahedra. This exclusion principle derives from the negative charge build up around the two tetrahedra leading to repulsive effects. At a low Si/Al ratio each [Si-O-Al] unit is connected to six tetrahedra of which a maximum of three is occupied by Al.

In the faujasite lattice positions are equivalent, but the four oxygen atoms forming a tetrahedron are non-equivalent (figure 7). Of these four oxygen atoms, protons are

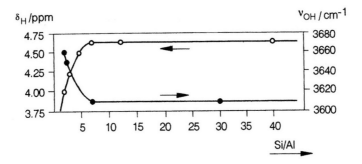

Figure 6. Infrared high-frequency OH stretch frequency (OH) as a function of the Si/Al ratio [17] and chemical shift (H) of the corresponding line in the 1H MAS NMR spectrum as a function of the Si/Al ratio.

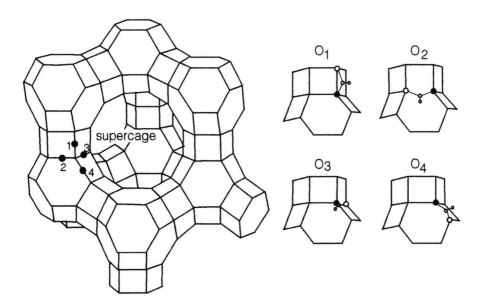

Figure 7. The four different oxygen atoms in the Faujasite lattice[21].

Table 1.
Total energies (in atomic units) and proton affinities (in eV) of 4-ring clusters. Complete geometry optimizations were performed at the STO-3G level; the 3-21G results refer to the STO3G-optimized geometry with an additional optimization of the proton position. Results marked by an asterisk refer to the abstraction of one proton only; the result in parentheses is influenced by a different configuration of the terminal hydrogen atoms [1]

		A:STO3G optimized geometry		B:3-21G STO3G geometry		C:3-21G optimized geometry	
	ring	E(a.u.)	PA(eV)	E(a.u.)	PA(eV)	E(a.u.)	PA(eV)
1	SiSiSiSi	-203.318558		-2048.522833			
2	Si^{H}SiSiSi	-203.779993	12.56	-2048.839850	8.63		
3	AlSiSiSi	-201.331399		-2001.777284		-2001.850643	
4	Al^{H}SiSiSi	-202.002739	18.27	-2002.279709	13.67	-2002.374317	14.25
5	AlSi^{H}SiSi	-201.925292	16.16	-2002.208117	11.72	-2002.309532	12.49
6	Al^{H}SiSiSi^H					-2002.749718	10.22*
7	AlSiAlSi					-2002.988098	
8	AlAlSiSi	-199.161940		-1954.885995			
9	Al^{H}SiAlSi	-200.024318	21.98	-1955.542784	16.77	-1955.626118	17.36
10	Al^{H}Si^{H}AlSi	-200.633693	16.58*	-1955.994844	12.29*	(-1956.120927)	13.46*)
11	Al^{H}SiAl^{H}Si	-200.726481	19.10*	-1956.067643	14.27*	-1956.163893	14.63*
12	AlSi^{H}Al^{H}Si	-200.683936	17.94*	-1956.024463	13.11*	-1956.1256167	13.58*

attached most strongly to O(1) and O(3) sites, giving rise to a HF (3650-3610 cm^{-1}) and LF (3550-3570 cm^{-1}) OH infrared absorption peak. The reasons for this will be discussed below. With a decrease of aluminum content, the probability for a [Si-OH-Al] unit to be surrounded by three tetrahedra with aluminum decreases. With a statistical distribution of aluminum over the tetrahedra (constrained by Löwenstein's rule) now different embedding possibilities of the [Si-OH-Al] unit exist. The frequency changes for the HF OH depend only on changes of composition in tetrahedra in the first coordination shell of the [Si-OH-Al] unit and converge to the same value at Si/Al>>10.

A decreasing heterolytic proton bond dissociation energy with increasing Si/Al ratio is also found from ab-initio calculations on clusters of varying Si/Al content. Figure 8 shows rings of four tetrahedra, terminated by end-on OH groups with varying Si/Al ratio. As follows from the corresponding table 1 the deprotonation energy is smallest for the tetrahedral four ring with one aluminum atom and the protons prefer attachment to an oxygen atom also coordinated to an aluminum atom. When the aluminum content increases, the deprotonation energy increases for a cluster with 2 protons as well as the negatively charged cluster, simulating compensation by cations. In figure 2 the corresponding geometries and charges are shown.

Low-aluminum zeolitic clusters

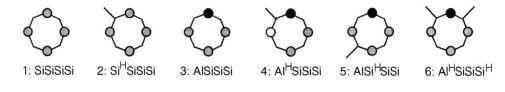

1: SiSiSiSi   2: Si^{H}SiSiSi   3: AlSiSiSi   4: Al^{H}SiSiSi   5: AlSi^{H}SiSi   6: Al^{H}SiSiSi^H

High-aluminum zeolitic clusters

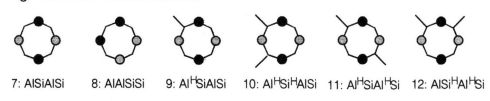

7: AlSiAlSi   8: AlAlSiSi   9: Al^{H}SiAlSi   10: Al^{H}Si^{H}AlSi   11: Al^{H}SiAl^{H}Si   12: AlSi^{H}Al^{H}Si

Figure 8. Four-ring clusters with different Si/Al ratios and different proton concentrations[1], see also table 1.

Experimentally the Brønsted acidity of zeolite Y appears to be a strong function of proton content. In figure 9 calorimetric measurements are presented of the ammonia adsorption on zeolite Y with three different compositions [18,19]. The interaction with zeolite Y (Si/Al = 2.4) is compared at 100% proton exchange and 85% proton exchange. One notes a significant increase in the initial heat of adsorption due to protonation of ammonia. The same increase in heat of adsorption is found for the dealuminated material.

Adding a proton to position O(8) results in a weakening of the O(5) - H bond, (enhancing the intrinsic Brønsted acid strength) and a strengthening of the neighbouring Si(3) - O(5) and Al(1) - O(5) bonds. This can be deduced from the bondlength differences in figure 2. Whereas proton attachment to position O(8) lengthens the Si(2) - O(8) and Al(4) - O(8) bonds, the next Si(2) - O(7) and Al(4) - O(6) bonds become stronger. As a function of distance to the proton on O(8) these changes alternate causing the O(5) - H bond to weaken. This again is a consequence of covalent bonding and the Bond Order Conservation principle. When one of the bonds to an atom weakens, the other will become stronger. The changes in atomic charges are very small hence one concludes that proton-strength differences are dominated by changes in covalent bondstrength.

For the LF OH peaks in zeolite Y a different behaviour upon dealumination is observed. With increasing Si/Al ratio the maximum for this band is shifted to higher frequencies. This behaviour cannot be explained by the small clusters discussed here, since the acidic protons, which are responsible for this infrared band, interact with two close oxygen neighbours at a distance of 0.26 nm [20].

To relate changes in OH bondstrength with composition to changes in lattice covalent bonding is one way of interpreting consequences of changes in electron density. As discussed elsewhere analysis in terms of HOMO-LUMO interactions leads to very similar conclusions [6]. It is important to realize that the changes in acidity are ascribed to electron-density differences and not to changes in atom charges.

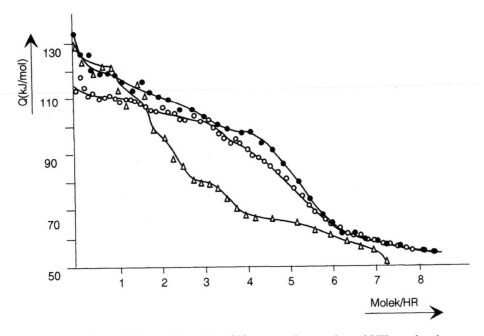

Figure 9. Differential heat of sorption (Q) versus the number of NH_3 molecules per supercage, measured at 423 K on Y type zeolites [19]: o NaY(85), Si/Al=2.4; • HY(100), Si/Al=2.4; △ DY-3, Si/Al=4.1.

Using force-fields, differences in proton affinity have been theoretically analyzed for two low aluminum content zeolite frameworks: zeolite Y which is a low density zeolite and silicalite of high density [1,21]. In zeolite Y the four oxygen-atom positions are non-equivalent. In silicalite 12 atom-positions are non-equivalent and 48 different oxygen atom positions are in principle possible. The deprotonation energy is calculated for fully relaxed bridging oxygen sites in the presence and absence of an added proton. These studies have demonstrated the importance of the zeolite lattice structure for the intrinsic Brønsted acidity of lattice protons. It is found that the deprotonation energies vary within an interval of 20 kcal/mole for different sites. For the MFI structure the amount of structural relaxation and corresponding energy gain is found to vary very little for the different sites, whereas there are large differences for the protonation energies. The parameter of the fully relaxed SiO_2 polymorph that correlates strongly with the OH interaction strength at a particular site is the SiO bondlength (figure 10). The larger the average SiO bondlength the stronger the OH bond. This is in line with the previously discussed bulkiness of the [Si-OH-Al] group compared to the Si-O-Si unit. The strongest Brønsted acidic sites are found at sites that are spatially most constrained.

According to the results presented, the dominant effect controlling the OH bondstrength is the proton concentration (see table 1). When the proton concentration is low, the in-

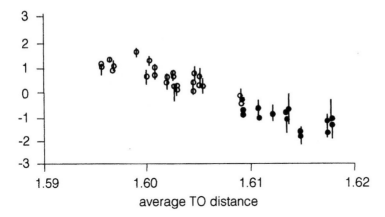

Figure 10. Relation between average T-O distance, O-O distance and T-O-T angle in the relaxed all-silica structure of MFI, and the (averaged) proton energy of the corresponding oxygen site [1].

trinsic Brønsted acidity per proton increases with decreasing Al-concentration. Structural diffences have a comparable effect as variation in Si/Al ratio.

5. Proton-weak base interaction

An easily accessible probe of the intrinsic acidity of the zeolitic proton is the change in the infrared adsorption spectrum, when the proton interacts with a weakly interacting base. This is illustrated in figure 11 that shows the difference spectrum of the undisturbed P-OH group of a CoAPO II zeolitic material and that interacting with basic CD_3CN. The OH peak is found to be shifted to approximately 3100 cm^{-1}[22]. The spectral adsorption is significantly broadened and increased in intensity. The lower peak maximum is due to the weakening of the OH bond by interaction with the nitrile group:

$$\text{O-H} \diagup\!\!\!\!\diagdown \text{N}\equiv\text{C-CD}_3$$

The lone pair on nitrogen pushes electron density away from the proton, this results in an increase of its positive charge. It is reflected in the high absorption intensity that is proportional to the square of the induced OH dipole moment. The broadening of the peak is characteristic of hydrogen bonding and stems from anharmonic coupling to the motion of the $N\equiv C\text{-}CD_3$ molecule against of the proton. The features around 2300 cm^{-1} are due to the perturbed CN group and indicate the presence of Lewis as well as Brønsted acid sites[23].

Whereas the interaction of acetonitrile weakens the OH stretching modes, it shifts the

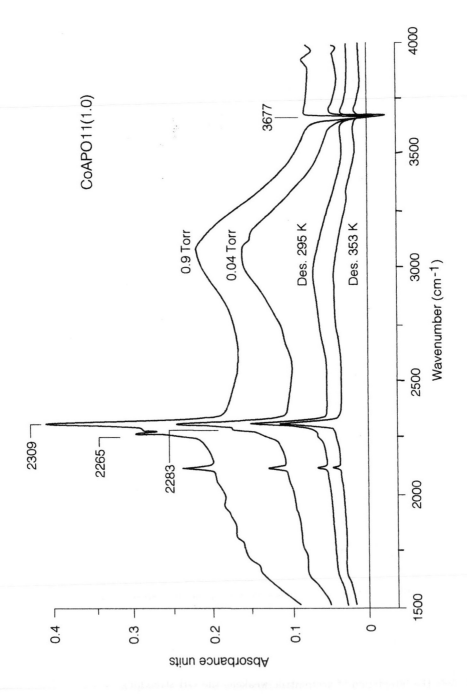

Figure 11. Difference Ft-IR spectra of CoAPO II by adsorption with CD_3CN.

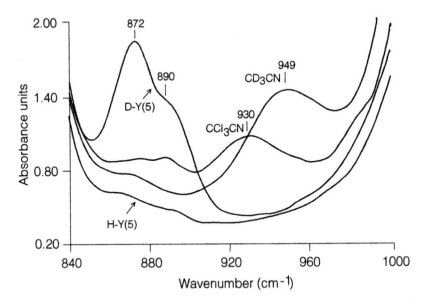

Figure 12a. Comparison of in-plane OD bending modes of DY (Si/Al=5) in the absence and presence of interacting CCl_3CN and CD_3CN.

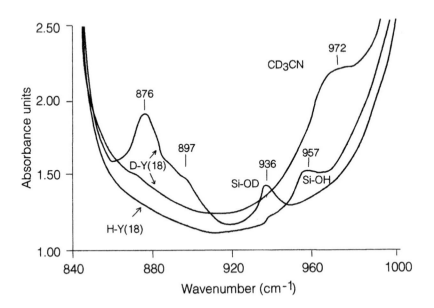

Figure 12b. In-plane OD bending modes of deuterated zeolite Y (Si/Al=18) in the absence and presence of CD_3CN. Also the spectrum of non-deuterated HY is shown.

frequency the OH bending modes upwards. This is due to the induced limited motion of the bending mode. Using deuterated zeolites, these upward shifts have been experimentally demonstrated[24]. In figures 12 the non-disturbed and disturbed OD bending modes of DY are shown. In the absence of base a weak shoulder is observed ≈ 25 cm^{-1} higher than the main peak at 872 cm^{-1}. The shoulder corresponds to the hydrogen bonded LF stretching peak (O(3)). Interaction with Cl$_3$CN gives a smaller upwards shift than CD$_3$CN. This is because Cl$_3$CN is a weaker base. The interaction with the basic adsorbates gives an upwards shift with 60-80 cm^{-1}. This is much more than the 25 cm^{-1} shift of the LF proton, due to interaction with other lattice oxygen atoms. Comparison of figures 12a and 12b illustrates the effect of the difference in intrinsic proton acidity. With a higher Si/Al ratio the proton is more weakly bonded and the upwards shift upon interaction with the basic molecule larger. The spectrum of the shifted zeolitic OH stretching frequency in contact with CD$_3$CN is more complex than that of the POH group. The spectra are shown in figure 13[23]. Instead of one single shifted peak as shown in figure 11, now a double peak appears in the spectrum. The average position of the shifted intensity, summed over the two peaks is approximately 2700 cm^{-1}. The stronger intrinsic acidity of the bridging OH group, compared to the POH group is the cause of the much larger shift of the bridging OH group. The double peak is due to large vibrational coupling effects that arise when the features of the shifted OH bond overlap with other protonic modes. Such a mode is the overtone of the upwards shifted in-plane OH bending modes. The corresponding fundamental frequency will have a value of approximately 1300 cm^{-1} and hence its overtone a frequency of 2600 cm^{-1}. The resulting Fermi resonance[23] causes the dip in the spectrum. It appears that the interaction with many basic molecules shows this doubly peaked structure in its spectrum. It can be considered a signature of hydrogen bonding in a zeolite. When H$_2$O or methanol interacts with protons the double peaked feature also appears, indicating that hydronium[25] or methoxonium[24] formation does not occur for these molecules in the ground state. In the next section we will discuss the proton-transfer reaction.

6. Proton-transfer

In the acid-base reaction:

$$ZH + B \rightleftharpoons Z^- + HB^+$$

charge is separated. In solution the cost of charge separation is overcome by solvation of the ions generated by solvent molecules. The dielectric constant of a zeolite is very small (~ 4), the energy of charge separation now has to be counteracted by the interaction between the positively charged protonated base and the negative charge of the zeolite wall.

Cluster-calculations on the protonation of NH$_3$ provide an interesting illustration of the importance of Zwitter-ion stabilization [26,27]. Very accurate ab-initio calculations of the NH$_3$-[Si-OH-Al] dimer interaction with the ammonia nitrogen atom coordinating to the bridging OH site show that the OH distance remains 0.098 nm to be compared with the long HN distance of 0.277 nm (figure 14a). So proton transfer does not occur. The

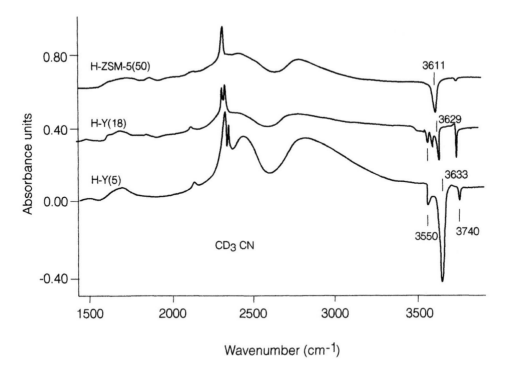

Figure 13. Difference spectra of H-ZSM-5 (Si/Al=50), H-Y (Si/Al=18) and H-Y (Si/Al=5) interacting with CD_3CN.

interaction between ammonia and OH shifts the OH frequency downwards by 380 cm^{-1} indicating a weakening of the OH bond. The computed interaction energy is 55 kJ/mole. The experimentally measured heat of adsorption of NH_3 is between 110 kJ/mole and 160 kJ/mole. Coordination as shown in figures 14b and 14c makes Zwitter-ion formation possible [27]. The NH_4^+ ion coordinates with two or three of its hydrogen atoms to the negatively charged oxygen atoms surrounding the Al atom. The computed interaction energies are very similar and equal to 114 kJ/mole. For the bidentate and tridentate bonded NH_4^+-ions, the infrared spectrum can be computed. Each coordination type is characterized by a different spectrum. Comparison with experimentally measured spectra enables an assignment of the two different coordination types (see table 2). Also the frequency of the external vibration of NH_4^+ versus zeolite wall can be computed. For the bidentate and tridentate bonded NH_4^+-ions the obtained frequencies vary from 250 to 350 cm^{-1} [28]. These low frequencies compare very well with typical values of NH_4^+ (80-200 cm^{-1}) found for zeolite Y [29]. The absence of hydrogen bonding is also evident from the non-appearance of the double-peaked infrared adsorption band around 2600 cm^{-1}.

Comparison of the geometry of the aluminum-tetrahedral cluster before NH_3 interaction (figure 13d) and the interacting clusters (figure 13b and 13c) show that the geometry around the tetrahedra changes when the proton is transferred. In a rigid lattice this would be prevented and as a consequence no proton transfer would occur. This is what is found theorically if no geometric relaxation is allowed when computing proton transfer. So proton-transfer requires a flexible lattice, that allows for local deformations.

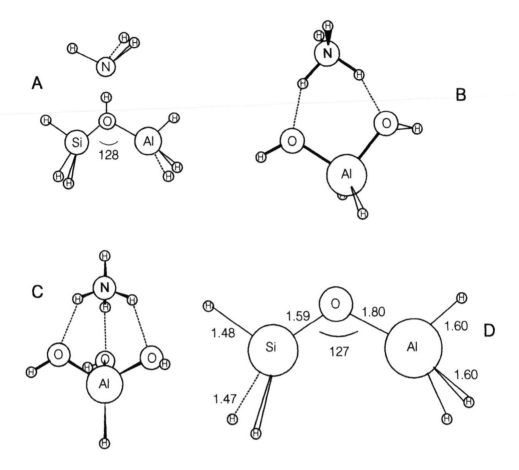

Figure 14. Clusters showing the interaction of NH_3 with the Brønsted site [27].

The equal interaction energies of NH_4^+ coordinated bidentate or tridentate implies that the NH_4^+ ion moves in a very flat potential. Neutron scattering energy loss data show that NH_4^+ is nearly freely rotating in the supercages of the Y-zeolite, but has a slightly more hindered rotation when in the smaller sodalite cage [30].

Whereas in water the basicity of ammonia is larger than that of pyridine, temperature programmed desorption experiments show always a higher temperature of desorption for adsorbed pyridine than ammonia [31,32]. The low basicity of pyridine compared to ammonia in water stems from the high heat of hydration of ammonia compared to pyridine. The gas phase protonation energy of pyridine, however, is 10.8 eV, whereas

Table 2.
The frequencies of the experimental and calculated infrared spectra given in cm^{-1}. Experimental values are for NH$_4^+$ in: Mordenite (MOR), Faujasite (FAU), Beta (BET) and Erionite (ERI). The other spectra are calculated ones. 2H and 3H are the notation for the doubly and triply bonded NH$_4^+$. H is the structure in which the proton is attached to the zeolite, the peak given here thus is the OH stretching. COAD is a singly bonded NH$_4^+$ with one NH$_3$ coadsorbed (N-H stretching of the proton pointing towards the zeolite is shifted to 2123 cm^{-1}, but is not included in the table because it is mixed with a Si-H stretching).
(I) means that these peaks have a reasonable intensity, the other peaks have negligible intensity but are included for completeness [28]

MOR	FAU	BET	ERI	2H	3H	H	COAD
2780	2800	2970	2840	2623(I)	3103(I)	3142(I)	3153(I)
2930	3040		3068	2740(I)	3141(I)	3360	3363
3180	3270	3200	3260	3418	3478(I)	3478	3401
3400	3360	3460	3384	3495		3483	3473
							3476
							3483

that of NH$_3$ is 9.9 eV [33]. The difference relates to the higher s-character of the pyridine lone pair compared to that of ammonia. The absence of hydration and the vacuum-type environment of the zeolite is responsible for the stronger basicity of pyridine in the zeolite environment.

Other quantum-chemical cluster calculations on the zeolite proton-polar molecule interaction show similar results.

As discussed in the previous section protonation of H$_2$O and CH$_3$OH does not occur. However interaction with the zeolitic proton is very strong and a proton of water and the proton of methanol is strongly interacting with a basic lattice oxygen atom. [24,25]. The ground state is very close to the protonated state. This agrees with the rapid proton-deuterium exchange reaction of deuterated water or methanol with zeolitic protons. The transition-state for the protonation of methanol is sketched in figure 14. Protonated methanol becomes coordinated as bidentate. This can be considered Brønsted acid-Lewis base assisted protonation. The Brønsted acid is the proton, the Lewis base is the other oxygen atom around aluminum to which the methanol proton attaches (figure 15). This can be considered Brønsted acid-Lewis base assisted protonation. The Brønsted acid is the proton, the Lewis base is the other oxygen atom around aluminum to which the methanol proton attaches (figure 14).

These results have an important consequence for the transition state of reactions that proceed via carbenium-ions or carbonium-ion intermediates.

Carbenium-ion formation from ethylene has been studied quantum chemically [35,36]. Ethylene can interact in two ways with the proton. It can form initially a π-complex without proton transfer. In this state interaction is weak. When the proton becomes attached to ethylene (the transition state), the positive charge that develops at the other end of the carbenium-ion becomes stabilized by interaction with the negative charge on another atom around aluminum. This is the Brønsted acid-Lewis base assisted proton

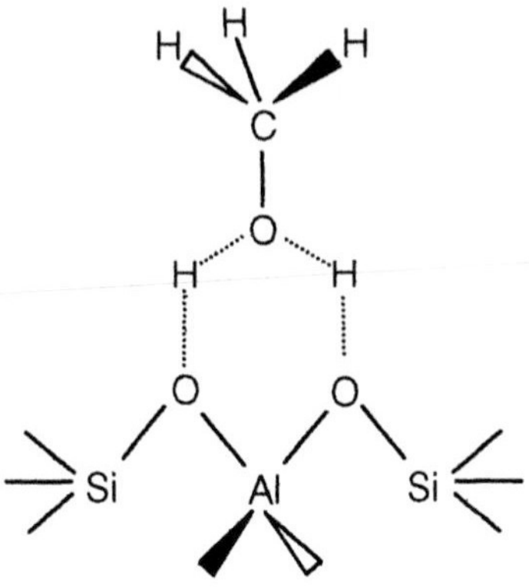

Figure 15. Transition state in the protonation of methanol [34].

transfer mechanism. In the case of ethylene this leads to formation of a stable O-C bond, the σ-complex or surface alkoxide. The formation of σ-complexes has been confirmed by NMR [37,38].

Recently we demonstrated that also for methane, hydrogen-deuterium exchange in a zeolite proceeds by a carbonium-ion state which is not a stable intermediate, but a transition state [39]. The penta-coordinated carbon atom is coordinated bidentate to the two basic oxygen atoms around an $[AlO_4]$ center. The experimentally measured activation energy for proton-deuterium exchange of 120 kJ/mole has such a low value due to the stabilizing zwitter-ionic nature of the carbenium-ion transition state.

7. Concluding remarks

In the previous section the quantum chemistry of the interaction of adsorbates with Brønsted acidic zeolite sites has been discussed. The relative stabilities depend on the deprotonation energy as well as the stabilization of the organic substrate by the negative charge generated on lattice oxygen atoms. The deprotonation energy is sensitive to lattice relaxation and depends on local geometric constraints due to long range structure of the zeolites as well as zeolite composition.

Acid catalysis is controlled by the chemistry of the protonation site as we discussed it here, as well as the interaction of reactant and products molecules with the stabilizing zeolite.

8. References

*. Based in part on: R.A. van Santen, G.J. Kramer, W.P.J.H. Jacobs, In: "Elementary Reaction Steps in Heterogeneous Catalysis", Eds. R.W. Joyner, R.A. van Santen, NATO-ASI 1993, p. 398.

1. G.J. Kramer and R.A. van Santen, J. Am. Chem. Soc, 115 (1993) 2887.

2. J. Sauer and W. Schirmer, In: "Innovation in Zeolite Materials Science (Studies in Surface Science and Catalysis 37), Eds. P.J. Grobet, W.J. Mortier, E.F. Vansant and G. Schulz-Eklhoff, Plenum, New York, 1986, p.323.

3. J. Sauer, J. Phys. Chem. 91 (1987) 2315.

4. V.B. Kazansky, Acc. Chem. Res. 24 (1991) 379.

5. D. Fenzke, M. Hunger, H. Pfeifer and J. Magn-Reson, 95 (1991) 477.

6. R.A. van Santen, Theoretical Heterogeneous Catalysis, World Scientific (Singapore) 1991, p.204, 320.

7. W.P.J.H. Jacobs, H. Jobic, J.H.M.C. van Wolput and R.A. van Santen, Zeolites 12 (1992) 315.

8. L.M. Kustov, V. Yu. Borovkov and V.B. Kazansky, J. Catal. 72 (1981) 149.

9. J. Sauer, J. Mol. Catal. 54 (1989) 312.

10. H. Jobic, J. Catal. 131 (1991) 289.

11. W.P.J.H. Jacobs, A.J.M. de Man, J.H.M.C. van Wolput and R.A. van Santen, Proceedings, 9th Int. Zeol. Conf., July 5-10, 1992, Montreal;
W.P.J.H. Jacobs, J.H.M.C. van Wolput, R.A. van Santen, Chem. Phys. Lett. 210 (1993) 32.

12. W.P.J.H. Jacobs, H. Jobic, J.H.M.C. van Wolput and R.A. van Santen, in Zeolites, to appear.

13. R.A. van Santen and D.L. Vogel, Adv. Solid. State Chem. 1 (1989) 151.

14. C.R.A. Catlow and W.C. Mackrodt, Computer Simulations of Solids, Lecture Notes in Physics 66, Springer, Berlin, 1982.

15. G.J. Kramer, N.P. Farragher, B.H.W. van Beest and R.A. van Santen, Phys. Rev. B 43 (1991) 5068.

16. G.J. Kramer, A.J.M. de Man and R.A. van Santen, J. Am. Chem. Soc. 113 (1991) 6435.

17. D. Freude, M. Hunger, H. Pfeifer and W. Schwieger, Chem. Phys. Lett. 128 (1986) 62.

18. U. Lohse, B. Parlitz, V. Parlitz and V. Patzelova, J. Phys. Chem. 93 (1089) 3677.

19. U. Lohse and J. Jänchen, unpubl. results.

20. M. Czjzek, H. Jobic, A.N. Fitch and T. Vogt, J. Phys. Chem. 96 (1992) 1535.

21. K.P. Schröder, J. Sauer, M. Leslie, C.R.A. Catlow and J. Thomas, Chem. Phys. Lett. 188 (1992) 320.

22. M. Peeters, Thesis Eindhoven 1993.

23. A.G. Pelmenchikov, R.A. van Santen, J. Jänchen, E. Meyer, J. Phys. Chem. 97 (1993) 11071.

24. A.G. Pelmenchikov, R.A. van Santen, J.H.M.C. van Wolput, to be published.

25. A.G. Pelmenchikov, R.A. van Santen, J. Phys. Chem. 97 (1993) 10678.

26. E.H. Teunissen, F.B. Duijneveldt and R.A. van Santen, J. Phys. Chem. 96 (1992) 366.

27. E.H. Teunissen, R.A. van Santen, A.P.J. Jansen and F.B. van Duijneveldt, J. Phys. Chem. 97 (1993) 203.

28. E.H. Teunissen, W.P.J.H. Jacobs, A.P.J. Jansen and R.A. van Santen, Proceedings, 9th Int. Zeol. Conf., July 5-10, 1992, Montreal.
29. G.A. Ozin, M.D. Baker, J. Godber and C.J. Gil, J. Phys. Chem. 93 (1989) 2899.
30. W.P.J.H. Jacobs, Thesis Eindhoven 1993.
31. H.G. Karge and V. Dondur, J. Phys. Chem. 94 (1990) 765.
32. H.G. Karge, V. Dondur and J. Weitkamp, J. Phys. Chem. 95 (1991) 283.
33. W.J. Hehre, L. Radow, P.V.R. Schleyer and J.A. Pople, In: "Ab Initio Molecular Orbital Theory", John Wiley & Sohns, (1986).
34. J. Sauer, C. Kölmel, F. Haase and R. Ahlrichs, Proceedings, 9th Int. Zeol. Conf., July 5-10, 1992, Montreal.
35. V.B. Kazansky and I.N. Senchenya, J. Catal. 119 (1989) 108.
36. I.N. Senchenya and V.B. Kazansky, Catal. Lett. 8 (1991) 317.
37. M.T. Aronson, R.J. Gorte, W.E. Farneth and D. White, J. Am. Chem. Soc. 111 (1989) 840
38. J.F. Haw, B.R. Richardson, I.S. Oshiro, N.D. Lazo and J.A. Speed, J. Am. Chem. Soc. 111 (1989) 2052.
39. G.J. Kramer, R.A. van Santen, C.A. Emeis, A.K. Novak, Nature 363 (1993) 529.

J.C. Jansen, M. Stöcker, H.G. Karge and J. Weitkamp (Eds.)
Advanced Zeolite Science and Applications
Studies in Surface Science and Catalysis, Vol. 85
© 1994 Elsevier Science B.V. All rights reserved.

Analysis of the guest-molecule host-framework interaction in zeolites with NMR-spectroscopy and X-ray diffraction

Hermann Gies, Inst. f. Mineralogie, Ruhr-Universität Bochum, Deutschland

1 Guest-host system in zeolite science

Stoichiometry in the classical sense of Proust is inappropriate for zeolites. The wide range of chemical composition of zeolite 'host frameworks' and the large variety of template and sorbate molecules makes it difficult to describe zeolites in chemical formula units. Zeolites are not "non-stoichiometric" compounds because this term is used for materials with ions in different oxidation states and, therefore, varying amounts of charge compensating counter ions. Zeolites are composite materials on the atomic scale with a well defined host framework and adjustable amounts of 'guest-molecules'. The composition of the host framework might be considered as solid solution series of T-atoms of different nature tetrahedrally coordinated by oxygen. Zeolites are classified in structure types describing the topology of the TO_2-host framework in a three letter code regardless of its composition or manufacture (1). On the other side, a wide variety of material names stand for one zeolite host structure type having a specific composition or composition range. However, there is no criterion classifying the guest molecules and/or assigning it to a material or structure type. The "non-framework constituent" might be a cation, an anion, both hydrated or dehydrated, a water molecule, or a neutral organic or inorganic molecule. In this context templates are those non-framework constituents which are used as structure stabilising species for the synthesis of zeolites. However, usually metal cations are excluded when one talks about templates. Sorbates are those guest molecules which are loaded onto a host framework after calcination of the synthesis product. Water is a sorbate, however, is most often not considered to play an important role during synthesis as template. Looking at zeolites as host-guest systems the host framework is the tetrahedral network TO_2 built from $[SiO_4]$-, $[AlO_4]$-, $[PO_4]$-, etc. units. The guest molecules on the other side include all non framework constituents present inside the host's pore system. There has been some effort in the past to classify cations according to the zeolite host framework they preferentially form (2). Obviously, the result was not very convincing. This article will try again to analyse the guest molecules and, in particular, the organic templating species with respect to their interaction with the silicate host framework during synthesis and inside the cavity of the intact solid. It will focus on high silica and all silica materials which have been selected from a vast variety of examples for their obvious dependence of formation on template molecules.

2 Guest host analysis in the porosil system and in porosil-like materials

Porosils (3) have the composition $SiO_2{}^*M$ with silica forming the neutral host framework and M as the templating guest molecule. The porosil system is much simpler than the general zeolite system. Since the composition of the host framework is SiO_2 only, no charge compensating cations and hydrating water molecules are required. The template molecule M is the only compositional variable and thus determines the nature of the host system formed. So far there are 16 different porosil structure types with cage-like or channel-like pore systems (4) which are a good basis for a general analysis of the influence of the template molecule given (Table 1). This porosils have been synthesized with more than 100 different template molecules allowing for their systematic classification (Table 2). In addition to the type of template molecule M, the synthesis parameters temperature and pressure of synthesis, pH of the starting solution, and the concentration ratio template/silica have been investigated systematically and included in the evaluation.

Table 1
Summary of porosil structure types which have been obtained using templates as structure directing agents. The structure type which are approved by the structure commission have a 3-letter code and are contained in the atlas of zeolite structure types (1) to which is referred for more information.

POROSILS

Zeosils	Frame (T-atoms p1000A^3)	Code	Clathrasils	Frame (T-atoms p1000A^3)	Code
RUB-3	17.3	a)	RUB-3	17.3	a)
Silica-SSZ-24	17.5	AFI	Silica-sodalite	17.4	SOD
Decadodecasil 3R *,*PD* 17.6		DDR	Decadodecasil 3R *,*PD* 17.6		DDR
Decadodecasil 3H *,*PD* 17.6		b)	Decadodecasil 3H *,*PD* 17.6		b)
Silica-ZSM-5	17.9	MFI	Octadecasil	17.6	AST
Silica-ZSM-11	17.9	MEL	Sigma-2	17.8	SGT
Silica-ZSM-12	19.4	MTW	Dodecasil 1H *PD*	18.4	DOH
Silica-theta-1	19.7	TON	Dodecasil 3C *PD*	18.6	MTN
Silica-ZSM-48	19.9	c)	Melanophlogite	19.0	MEP
			Nonasil	19.3	NON

*clathrasil in as synthesized form, zeosil in calcined form; PD: silica framework composed of layers of pentagonal dodecahedra; a) Recent Research Reports, Garmisch, 1994; b) Z. Kristallogr. 174(1986) 64; c) Zeolites 5 (1985) 355

Table 2
a) Summary of guest molecules used for the synthesis of clathrasils.

Skeletal atoms guest / Host structure	1	2	3	4	5	6	7	Code
dodecasils 3C	Kr Xe							0/80
melanophlogites	Kr Xe	CH_3NH_2	CO_2, N_2O					0/160
dodecasils 3C		CH_3NH_2	$C_2H_5NH_2$ $(CH_3)_2NH$	$(CH_3)_2CHNH_2$ $(CH_3)_3N$ $i\text{-}C_3H_7SH$	$(CH_3)_3CHNH_2$ $(C_2H_5)NH$ $CH_3CH(NH_2)CH_2NH_2$ $(CH_3)_2NC_2H_5$ $CH_3CH(NH_2)C_2H_5$	$(CH_3)_3CCH_2NH_2$	SF_6	0/240
nonasils					$CH_3CH(NH_2)C_2H_5$ $CH_2(NH_2)C_3H_7$	$CH_3CH(NH_2)C_3H_7$		0/290
dodecasils 1H						$CH_3CH(NH_2)C_3H_7$		0/390
silica-sodalites				$HOCH_2\text{-}CH_2OH$				0/

Table 2
b) Summary of guest molecules used for the synthesis of zeosils.

Host structure	Guest molecule
s-ferrierite	$(HO)_3B + H_2N\text{\textasciitilde}NH_2$. $(H_2N\text{\textasciitilde})_2NH + (OH)_3B$
s-ZSM-5	
s-ZSM-11	
s-beta	$+ H_3BO_3$

Host structure		Guest molecules	Code number
10MR pores	s-theta-1	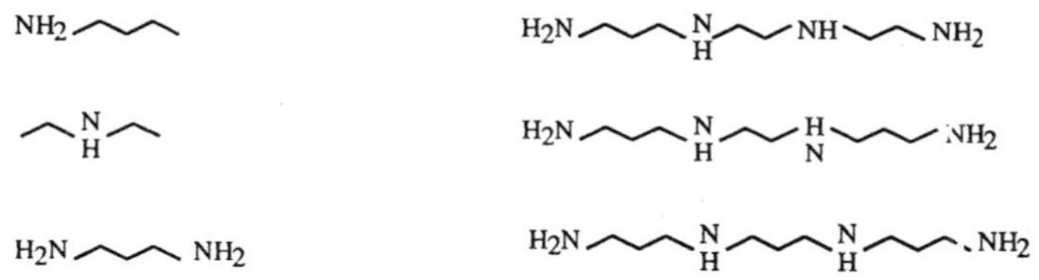	1/10
	s-ZSM-48	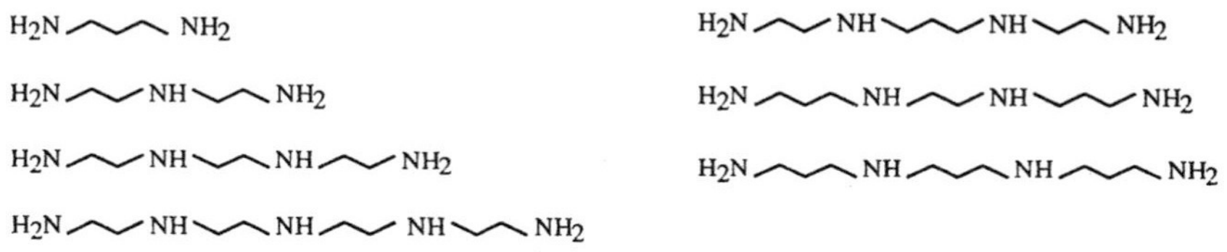	
12MR pores	s-ZSM-12	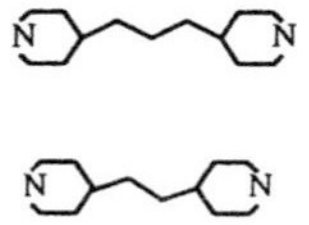	1/12

Table 2
c) Summary of cyclic guest molecules used for the synthesis of clathrasils

Table with columns:

skeletal atoms (guest) / host structure	5 monocyclic	6 monocyclic	7 monocyclic	8 monocyclic	8 bicyclic	9 monocyclic	9 bicyclic	10 monocyclic	10 bicyclic	code number
silica sodalite		(structure)								0/130
dodecasils 3C	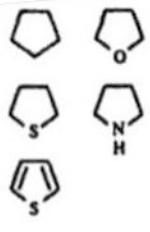	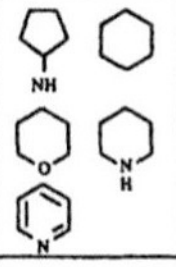 (structures)			(structure)					0/250
octadecasil					(structure)					0/280
nonasils		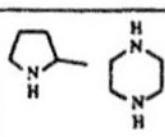 (structures)	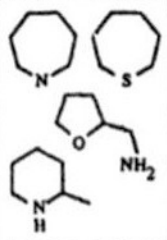 (structures)	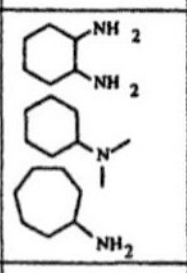 (structures)		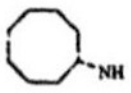	(structure)			0/290
deca-dodecasils 3R					(structure)		(structure)			0/350
levyne					(structure)		(structure)			0/370
sigma-2					(structure)		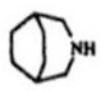 (structure)			0/390
dodecasils 1H		 (structure)	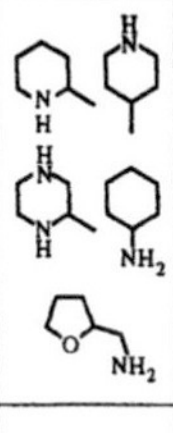 (structures)	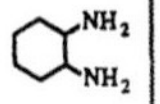 (structure)	(structure)	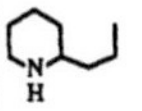 (structure)	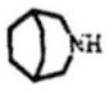 (structure)			0/430
ZSM-12				(structure)			 (structure)		 (structure)	
RUB-4							(structure)			

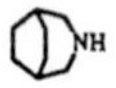

skeletal atoms / guest → host structure ↓	monocyclic	11 bicyclic	tricyclic	12 tricyclic						code number
silica sodalite										0/130
dodecasils 3C										0/250
octadecasil										0/280
nonasils										0/290
deca-dodecasils 3R										0/350
levyne										0/370
sigma-2										0/390
dodecasils 1H										0/430
ZSM-12										
RUB-4										

2.1 Classification of template molecules in porosils

The overview of the entire system SiO_2/M only gives a good idea of the interaction between host and guest on the molecular scale, however, a more detailed analysis of the interaction between host structure and guest molecules is required. It is evident that amongst the template molecules the chemical class of organic amines is by far the most important. Presumably, the ambiphilic character of amines facilitates the formation of crystal seeds and supports the crystal growth of the porosil host framework. As can be seen from an idealized structure plot amines could interact with their hydrophobic tail with the completed silica surface of a half cavity formed on the growth surface of a crystal, which is hydrophobic too (Figure 1). The hydrophilic functional group at the same time might

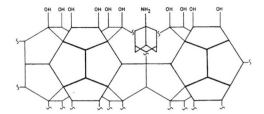

Figure 1. Schematic view of the growth surface of dodecasil 1H showing the template molecule 1-aminoadamantane docked in the half-cavity of the large 20-hedron. The hydrophobic tail of the template interacts with the hydrophobic SiO_2 surface whereas the hydrophilic aminogroup built up hydrogen bridges to the silanol groups at the surface and the solvent.

interact with the silanol groups of the silica surface and the solvent water forming hydrogen bridges. Thus, the strongest possible interactions between host and guest are formed at the interface and lead to an effective fixation of the template at the surface at least much stronger than it would be possible for other molecules comparable in size and shape. This also explains why hydrocarbons, alcohols, or thiols are much less efficient as templates compared to amines if molecules of similar size and shape are used. In a series of experiments with substituted adamantanes, 1-amino-adamantane has proved to be by far the most effective template molecule for the synthesis of the clathrasil dodecasil 1H (DOH) (Table 3).

Table 3
Series of substituted adamantanes used template molecules similar in size and
shape arranged in decreasing activity as template for the synthesis of DOH.

$R = NH_2 > SH > OH > Br \approx CH_3$

As can be seen from Table 2, for all porosil structure types there are several
guest molecules which have been used successfully as templates for the synthesis
of one particular structure type (4). These template molecules occupy preferentially
one specific type of cage in clathrasils or one specific type of channel in zeosils.
The volume of the cage and the pore opening of the channel, respectively, are,
therefore, the critical quantities classifying the guest molecule for a specific
structure type. In Table 4 a scheme for the classification of guest molecules in the
porosil system is given.

There is a peculiarity in the case of dodecasil 3C. There are two types of cage
in the host framework, the $[5^{12}]$-cages and the $[5^{12}6^4]$-cages which are
considerably larger. The chemical formula per unit cell is 136 SiO_2 *$16M^{12}$ *$8M^{16}$.
The ratio of the cages is $2[5^{12}]:1[5^{12}6^4]$. In the case of the noble gases Kr and Xe
as templates M, the small $[5^{12}]$ is the cage preferentially housing the structure
stabilising guest molecule. From single crystal structure analysis of the dodecasil
3C-Xe (5) the occupation of the $[5^{12}]$-cages with Xe is slightly above 50%
and is slightly more than in the larger 16-hedra. The content per unit cell,
136 SiO_2*8 Xe^{12}*4 Xe^{16}, clearly shows that the host guest interaction in the
pentagonal dodecahedron predominantly contributes to the lattice energy of the
material and that the $[5^{12}]$ is the cage for the structure stabilising template. For
those templates which can fill out the much larger volume of the $[5^{12}6^4]$ the 16-
hedron become the cage housing the structure controlling guest species. In this
materials the small pentagonal dodecahedron is occupied by so called help guests,
usually atmospheric gases present during synthesis. Structure analysis and mass
spectroscopy shows that most of the cages are filled with help guests and that

argon and nitrogen are preferentially incorporated whereas oxygen can not be detected (6).

Table 4
Classification of the guest molecules in groups according to the type of cavity they occupied. The cavity is the pore containing the structure stabilising guest molecule.

groups of guest species	cage volume [Å³]	structure type	code
guest species favouring $[5^{12}]$	80	MTN	0/80
guest species favouring $[4^{6}6^{8}]$	130	SOD	0/130
guest species favouring $[5^{12}6^{2}]$	160	MEP	0/160
guest species favouring $[5^{12}6^{4}]$	250	MTN	0/250
guest species favouring $[4^{6}6^{12}]$	280	AST	0/280
guest species favouring $[5^{8}6^{12}]$	290	NON	0/290
guest species favouring $[4^{4}5^{4}6^{2}]$	300	"RUB-3"	0/300
guest species favouring $[4^{35}12^{6}18^{3}]$	350	DDR	0/350
guest species favouring $[5^{12}6^{8}]$	390	DOH	0/390
guest species favouring $[5^{18}6^{2}8^{3}]$	540	"DD3H"	0/540

	channel diameter [Å]		
guest species favouring 10-MR	5.5	"ZSM-48" TON	1/10MR
guest species favouring 12-MR	6	MTW	1/12MR

The examples given above should be useful for the synthesis of new host guest systems. For a particular porosil structure type one should be able to choose a templating molecule according to the criteria given above and adjust the synthesis parameters to the specific requirements of the guest molecule. In the search for a new host structure type one should look out for new templates different in size, shape, but of good thermal and chemical stability and, so, design the new zeolite framework.

Table 2 also shows that one guest molecule can act as template for different porosil structure types under different synthesis conditions (4). The most sensitive synthesis parameter is the synthesis temperature. Templates which fit the cages tightly at lower synthesis temperature become to large at higher synthesis temperatures and, therefore, stabilise structure types with larger cages. This might be explained with an increase of the volume of the guest molecule through thermal expansion and, at the same time, a decrease of the volume of the cage because of the increased thermal motion of the oxygen atoms representing the inner surface of the cage. A typical example for this is the template quinuclidine. At low synthesis temperature dodecasil 3C is formed with quinuclidine residing in the $[5^{12}6^4]$-cage. The template exactly fits the cage in size and shape. At higher synthesis temperature dodecasil 1H is formed with quinuclidine in the $[5^{12}6^8]$-cage. More examples are summarized in Table 5.

Table 5
Compilation of guest molecules which stabilise different clathrasil structure types at different temperatures of synthesis. The main product is listed first, the following phases are listed with decreasing yield. Synthesis have been performed in silica tubes and confirm the kinetically controlled mechanism of crystal growth.

Temperature/ Guest molecule	160° C	170° C	180°C	190° C	200° C	220° C
Krypton	MTN MEP	MEP MTN				
Methylamine	MEP	MEP MTN	MEP MTN	MTN	MTN	MTN
2-Butylamine		NON	MTN			
2-Azabicyclononane	DDR SGT RUB-4	SGT DOH RUB-4				
2-Aminobutane	NON MTN	NON MTN	NON MTN	NON MTN		
Diaminocyclohexane	NON DOH	NON DOH	NON DOH	NON DOH		DOH
Quinuclidine					MTN	DOH

Another parameter taking influence on the structure type formed is the concentration of the templating species in the synthesis mixture. This correlates with the cage density (7) in clathrasils giving the concentration of cages of a specific type per 1000Å^3. For several clathrasil structure types the volume of the cage housing the structure stabilising guest molecule is very close. The cage density, however, is different and also the framework density. The higher the cage density is the lower the framework density. Table 6 gives a compilation of this specific property for all clathrasil structure types. Although the predominant contribution to the lattice energy of clathrasils comes from the silica host framework it is possible to form different structure types depending on the concentration of the specific guest molecule. At low concentrations the form with low cage density and higher framework density is built, however, increasing the concentration of the guest while keeping all other variables constant the structure type with high cage density but lower framework density is formed. In the case of 1-aminoadamantane dodecasil 1H with a cage density of 0.5 is formed at low concentration template whereas DDR with a cage density of 1.5 is formed at high concentration. Table 7 gives a summary of more guest molecules which act concentration sensitive.

The templating species for the channel like pores in zeosils with 1-dimensional pore systems are unbranched chain-like molecules as can be seen from Table 2. A template hydrocarbon chain is ideally accommodated in a 10-membered ring (10MR) channel. The structure refinement of silica ZSM-22 in the as synthesized form (8) shows the close correspondence of the pore geometry and the template size and shape (Figure 2).

Table 6

a) Compilation of zeosil structure types. Information of the composition per unit cell, crystallographic data, and the pore geometry is given.

Structure type	Composition per unit cell	Lattice parameters topological symmetry space group	Type, direction of channel, dimensionality of pore system
RUB-3	$24[SiO_2] \cdot 2M^{16}$	$a_0=14.0$Å $b_0=13.7$Å $c_0=7.4$Å ß$=102.5°$ C2/m	8-MR pore in [010] straight channel, 1D pore system
Silica-SSZ-24, AFI	$24[SiO_2] \cdot M^{12-MR}$	$a_0=13.7$Å $c_0=8.4$Å P6/mcc	12-MR pore in [001], straight channel, 1D pore system
Decadodecasil 3H	$120[SiO_2] \cdot 6M^{10}$ $9M^{12}1M^{15}4M^{19}$ $1M^{23}$	$a_0=13.9$Å $c_0=40.9$Å P3m	8-MR pores in [100] and [010], zigzag channels, 2D pore system
Decadodecasil 3R, DDR	$120[SiO_2] \cdot 6M^{10}$ $9M^{12}6M^{19}$	$a_0=13.9$Å $c_0=$Å R$\bar{3}$m	8-MR pores in [100] and [010], zigzag channels, 2D pore system
Silica-ZSM-5, MFI	$96[SiO_2] \cdot M^{10-MR}$ $M^{10'-MR}$	$a_0=20.1$Å $b_0=19.9$Å $c_0=13.4$Å Pnma	10-MR pores in [100] sinusoidal, in [010] straight channels, 3D pore system
Silica-ZSM-11, MEL	$96[SiO_2] \cdot M^{10-MR}$	$a_0=20.1$Å $c_0=13.4$Å I$\bar{4}$m2	10-MR pores in [100] and [010], straight channels, 3D pore system
Silica-ZSM-12, MTW	$56[SiO_2] \cdot M^{12-MR}$	$a_0=24.9$Å $b_0=5.0$Å $c_0=24.3$Å ß$=107.7°$ C2/c	12-MR pores in [010] straight channel 1D pore system
Silica-theta-1, TON	$24[SiO_2] \cdot M^{10-MR}$	$a_0=13.8$Å $b_0=17.4$Å $c_0=5.0$Å Cmc2$_1$	10-MR pores in [001], zigzag channel, 1D pore system
Silica-ZSM-48	$48[SiO_2] \cdot M^{10-MR}$	$a_0=14.2$Å $b_0=20.1$Å $c_0=8.4$Å Cmma	10-MR pores in [001], stright channel, 1D pore system

Table 6

b) Compilation of the clathrasil structure types. Information on the composition per unit, crystallographic parameters, and cage types is given.

Structure type	Composition per unit cell	Lattice parameters maximum topological symmetry space group	Number, type and symmetry of the cages	Fundamental cage
RUB-3	$24[SiO_2] \cdot 2ABN$	$a_0=14.0$Å, $b_0=13.7$Å, $c_0=7.4$Å, $\beta=102.5°$, C2/m	$2[4^5 4^6 2]$ 2/m; $2[4^6 5^4 6^6]$ 2/m	$[4^4 5^4 6^2]$
Silica-Sodalite	$12[SiO_2] \cdot 2M^{14}$	$a_0=8.8$Å, Im3m	$2[4^6 6^8]$ m3m	$[4^6 6^8]$
Decadodecasil 3H	$120[SiO_2] \cdot 6M^{10} 9M^{12} 1M^{15} 4M^{19} 1M^{23}$	$a_0=13.9$Å, $c_0=40.9$Å, P3m	$6[4^3 5^6 6^1]$ 3m; $9[5^{12}]$ m3; $1[4^6 5^6 8^3]$ 62m; $4[4^3 5^{12} 6^3 8^1]$ 3m; $1[5^{18} 6^{28} 3]$ 6/mmm	$[4^3 5^6 6^1]$ + $[5^{12}]$
Decadodecasil 3R	$120[SiO_2] \cdot 6M^{10} 9M^{12} 6M^{19}$	$a_0=13.9$Å, $c_0=40.9$Å, R$\overline{3}$m	$6[4^3 5^6 6^1]$ 3m; $9[5^{12}]$ m3; $6[4^3 5^{12} 6^{18} 3]$ 3m	$[4^3 5^6 6^1]$ + $[5^{12}]$
Octadecasil	$20[SiO_2] \cdot 2M^6 2M^{18}$	$a_0=9.2$Å, $c_0=13.4$Å, I4/m	$2[4^6]$ 4/mmm; $2[4^6 6^{12}]$ 4/mmm	$[4^6 6^{12}]$
Sigma-2	$64[SiO_2] \cdot 8M^9 4M^{20}$	$a_0=10.2$Å, $c_0=34.4$Å, I41/amd	$8[4^3 5^6]$ 2mm; $4[5^{12} 6^8]$(SGT) $\overline{4}$	$[4^3 5^6]$
Dodecasil 1H	$34[SiO_2] \cdot 3M^{12} 2M^{12'} 1M^{20}$	$a_0=13.8$Å, $c_0=11.2$Å, P6/mmm	$3[5^{12}]$ m3; $2[4^3 5^6 6^3]$ 62m; $1[5^{12} 6^8]$ 6/mmm	$[5^{12}]$
Dodecasil 3C	$132[SiO_2] \cdot 16M^{12} 8M^{16}$	$a_0=19.4$Å, Fd3m	$16[5^{12}]$ m3; $8[5^{12} 6^4]$ m3	$[5^{12}]$
Melanophlogite	$46[SiO_2] \cdot 2M^{12} 6M^{14}$	$a_0=13.4$Å, Pm3n	$2[5^{12}]$ m3; $6[5^{12} 6^2]$ $\overline{4}$ m2	$[5^{12} 6^2]$
Nonasil	$88[SiO_2] \cdot 8M^8 8M^9 4M^{20}$	$a_0=22.2$Å, $b_0=15.1$Å, $c_0=13.6$Å, Fmmm	$8[5^4 6^1]$ 222; $8[4^1 5^8]$ mm2; $4[5^8 6^{12}]$ mmm	$[4^1 5^8]$

Table 7
Compilation of some guest molecules which stabilise different clathrasil structure types dependent on their concentration or partial pressure. The number given below the structure code is the cage density indicating the correlation between syntheses variable and crystal structure.

| Guest molecule | Host framework stabilised with cage density pressure (P) and concentration (C) dependency | | | | |
	Low P,C				High P,C
krypton (P)	MEP 2.5		P ──>		MTN 2.2 + 1.1
methylamine (C)	MTN 1.1		C ──>		MEP 2.5
1-aminoadamantane (C)	DOH 0.5	c ──>	Deca- dodecasil 3H 0.6 + 0.15	c ──>	DDR 0.9

At lower temperatures of synthesis silica ZSM-48 is formed with this type of template whereas silica ZSM-22 is favoured at higher temperature of synthesis. Since there is no structure refinement for ZSM-48 so far it is difficult to discuss reasons for different templating behaviour. It is most likely that the diameter of the pore of ZSM-48 becomes to narrow at higher temperatures for the hydrocarbon and the slightly wider channel of ZSM-22 is stabilised. Increasing the diameter of the template molecule leads to the formation of a 1-dimensional 12MR channel system. This has been shown for only two template molecules and silica ZSM-12 as zeosil structure type so far but seems to be conclusive (Table 2).

Intersecting channel systems have been obtained with branched template molecules (cf. Table 2). Using trialkyl amines as templates silica ZSM-5 with an intersecting 10MR channel system has been obtained. The guest molecule here resides most likely at the intersection of the two types of channel as can be shown

310

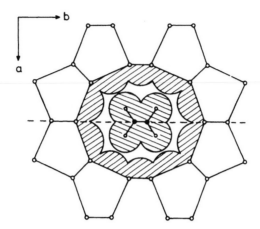

Figure 2. View parallel to the channel axis of silica-ZSM-22. The drawing demonstrates the close fit of the organic template inside the pore with the silica framework if the van der Waals radii are taken into account.

in a structure refinement of a high silica analogue (Figure 3, (9)). There is no example so far for the logic continuation of the series of templates in order to produce silica end members of intersecting 12MR channel systems. However, it is believed that using the appropriate template molecule this synthesis goal can be achieved and even the formation of still wider pore systems should be possible.

There is one clathrasil structure type which has been synthesized only with ion pairs as templates so far, octadecasil (10). Octadecasil has been obtained in the presence of quinuclidinium fluoride and tetramethyammonium fluoride as templating medium. The two types of cage building the structure are occupied with one template species each, F^- in the $[4^6]$-cage and quinuclidinium in the $[4^6 6^8]$-cage. The ion pair is effectively separated as can be visualized by a van der Waal model of the $[4^6]$-cage including the fluoride ion (Figure 4). This unit represents a quasi dense packed section of the structure which might be essential for the stability of low density materials. Ion pairs as templates have been detected in other high silica porosils synthesized with the fluoride method (9, 11, 12) with the fluorine ion always residing in small cavities.

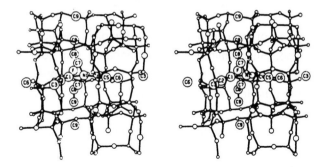

Figure 3. Representation of the result of the structure refinement of as synthesized silicalite 1 including the template tetrapropylammonium ion residing at the intersection of the two 10MR channels.

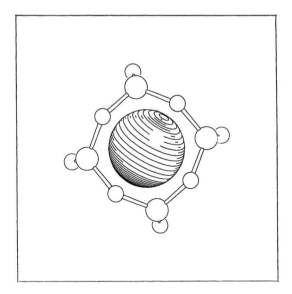

Figure 4. Projection of the double 4-ring in octadecasil including the fluoride ion shown as van der Waal sphere. The close fit of the ion inside the cage is obvious.

2.2 Templates in high silica porous systems

In the presence of trivalent T-atoms the template molecules usually form the same structure type. Since amines in their protonated form compensate the charge introduced by the substitution of silicon by e.g. Al, the host guest relationship is still maintained although the interaction energy is ionic in its character. However, there is a small number of template molecules which form certain porosil structure types only in the presence of the trivalent T-atoms and, on the other side, this structure type is only formed with Si partially substituted by Al or B. Examples are given in Table 8.

Table 8
Summary of high silica porosils. The phases shown here have only been obtained in the presence of organic templating molecules, however, only when Si in the framework is partially replaced by trivalent T-atoms.

HIGH-SILICA POROSILS

Zeosile-Type	Frame (T-atoms p1000A^3)	Code	Clathrasile-Type	Frame (T-atoms p 1000A^3)	Code
Beta	15.1	BEA*	RUB-13	16.6	___a)
Levyne	15.2	LEV	RUB-10	18.0	___b)
RUB-13	16.6	___a)			
Mordenite	17.2	MOR			
ZSM-23	17.7	MTT			

*a) Recent research reports, Garmisch, 1994; b) J. Appl. Crist., submitted for publication, 1993 (14).

Tetramethylamonium (TMA) is used as template for a number of porous materials. In the silica system the degradation product trimethylamin formed under synthesis conditions gives dodecasil 3C (MTN). However, in the presence of small amounts of B as trivalent substituents for Si the new porous structure RUB-10 is formed. RUB-10 is the boron analogue of the alumosilicate Nu-1 of which the structure has not been determined yet (13). The structure analysis of RUB-10 clearly shows that TMA is the template in a cage-like void and compensates the charge introduced by B on the host framework (14). RUB-10 is only formed with a well defined number of B-atoms on T-sites which corresponds to the number of cationic template molecules TMA in the cages. In the absence of B RUB-10 has not been obtained yet. Similar findings have been observed for the alumosilicate Nu-1 and the Ga- and Fe-silicate derivatives (15).

Quinuclidine (Q) and azabicyclononane (ABN) both stabilise in the presence of boron the high silica form of levine and DDR, SGT is only formed with Q. In addition, the porosil phases MTN, MTW, and DOH with Q and SGT and DOH with ABN have been obtained in the all silica form (7). Whereas DDR and SGT are known in the all silica form with 1-aminoadamantane as template high silica levine only crystallizes in the presence of B. Again, templates are quinuclidine and azabicyclononane in their protonated form and should compensate for the negative charge of the framework. It is interesting to note that the framework structure of levyne is composed of even membered rings only which is typical for silicate zeolites whereas the all silica phases stabilised with the same template molecules contain a high number of 5-membered ring as usually found in high silica and all silica zeolite phases. With 1-aminoadamantane as template the levine structure type has also been obtained only in the presence of Al as substituent for Si on T-site (16). Here the corresponding all silica phases are the same as for the above mentioned templates Q and ABN, namely DOH, DDR, and SGT.

Zeolite beta is synthesized according to the patent literature in the presence of tetraethylammonium cation as templating agent and Al as charge compensating T-atom in the silicate host framework. Using 4,4'-trimethylendipiperidine as template a high silica form of beta is formed in the presence of boron (17). In the absence of boron silica ZSM-12 crystallizes. In direct synthesis beta has not been obtained yet in the silica form.

2.3 X-ray diffraction and NMR analysis of the host-guest interaction in porosils and high silica porous systems

In order to proof the effect of the template molecule and to analyse the host-guest interaction non-destructive analysis techniques have been applied such as diffraction and NMR. The analysis and localization of the template molecule as structure stabilising agent, however, only makes sense if silica end-members are investigated. Otherwise, ionic interaction might overrule the templating effect and might lead to wrong conclusions.

Solid state NMR and X-ray diffraction are complementary to each other. NMR probes the local environment of the nucleus investigated, X-ray diffraction experiments yield information on the periodicity of structural motives. Table 9 gives an overview on host-guest systems where NMR and X-ray diffraction have been applied for the structure analysis together and where only the combined use of the techniques lead to the solution of the problem. Besides X-ray diffraction electron- (18) and neutron diffraction (19) are valuable techniques which have been used for the structural characterization and should be considered. X-ray diffraction analyses, however, are by far more abundant and the results of some structure refinements will be discussed in the following.

Table 9
Compilation of examples of structure analyses where X-ray diffraction and NMR spectroscopy have been combined for the analysis of the as synthesized or loaded material.

Method System	X-ray	NMR	Literature
Silica-ZSM-5	high resolution powder X-ray diffraction	^{29}Si MAS NMR	(20)
Silica-ZSM-23	high resolution powder X-ray diffraction	^{29}Si MAS NMR	(21)
RUB-10	low resolution powder X-ray diffraction	^{29}Si MAS NMR ^{11}B MAS NMR ^{13}C MAS NMR	(14)
DOH	single crystal structure analysis	^{29}Si MAS NMR ^{11}B MAS NMR	(22)
SGT	single crystal structure analysis	^{29}Si MAS NMR ^{11}B MAS NMR	(23)

X-ray diffraction shows the long range order of the material. Since templates often are dynamically disordered and symmetry requirements for their detailed analysis are not fulfilled - the symmetry of the guest molecule should be at least a subgroup of the symmetry of the pore-centre- a random orientation of the guest inside the pore is found. The analysis of the electron density distribution gives an approximation of the time and space behaviour of the guest molecule but makes it impossible to identify it with certainty. In the case where the group/subgroup symmetry relationship between the pore and the guest molecule allows for an interpretation of the electron density inside the pore a very detailed analysis of the geometrical arrangement and kinetic behaviour of the guest molecule is possible.

NMR probes the local environment of the nuclei excited. Since the chemical shift of the carbon guest atoms is close to the chemical shift of the carbon atoms of the free molecule identification of the clathrated species is easily possible. However, the chemical shift of the atoms of a template molecule inside the pore also reflects the space restrictions of the host framework and is correlated with the void size (24). The chemical shift of ^{13}C in TMA for example has be used to estimate the volume of the cages in zeolites. Beside the qualitative analysis of the guest molecule in magic angle spinning experiments, a spectrum of the static sample gives also a rough estimate of the dynamic disorder of the template in the pore within the limits of the NMR time scale.

2.3.1 Guest molecules analysed in diffraction experiments

The analysis of X-ray diffraction data of host guest systems has been pioneered by Jeffrey and co-workers in studies on clathrate hydrates in the 60th (25). In a series of single crystal structure analyses they have shown how valuable information can be obtained from accurate measurements and careful data processing. Since the contribution of the guest molecule to the scattered X-ray intensity is weak because of low atomic weight of the constituents and the severe disorder of the molecule, the symmetry analysis of the intensity data set is dominated by information on the symmetry of the host framework. In most cases the superstructure which would include the ordering behaviour of the guest molecule can not be detected and an averaged structure is refined. However, if the symmetry of the guest molecule is equal or in a subgroup supergroup relationships with the site symmetry of the cage or channel the evaluation of the electron density distribution inside the cages gives valuable information on the arrangement of the guest molecule inside the cage.

The 1-aminoadamantane/porosil system is an example where symmetry requirements of host cage and guest molecule are fulfilled, at least in two structure types. In the clathrasil dodecasil 1H, DOH, the large cage has site symmetry 6/mmm (Table 6). Neglecting the influence of the hydrogen atoms the symmetry of the guest molecule is 3m which is a subgroup of 6/mmm. This allows to interpret the residual electron density inside the cage including the template's symmetry and geometry. The model of the template obtained is an average of 4 distinct but symmetrically equivalent molecule orientations (Figure 5). The delocalised electrondensity distribution of the guest inside the cage indicates the high dynamic or static disorder of the molecule. Because of the prolate shape of the cage and the guest molecule the disorder is most likely caused by a rotation of the molecule about the 3-fold axis. Simulating a van der Waals model of the template molecule inside the cage using the coordinates of the guest molecule obtained from the structure refinement the close geometric relation of the size and shape of the guest species

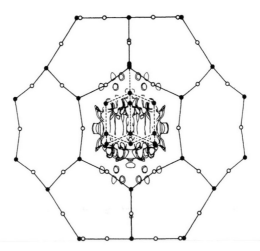

Figure 5. Residual electron density distribution as obtained after the refinement of the silica framework of DOH in a single crystal structure refinement. The comparision to the ball and stick model of the 1-aminoadamantane template shows the correspondance of the experimental result and the template model. The calculation confirmed the integrity of the guest molecule.

and the size and shape of the cavity becomes obvious (Figure 6). Since the guest molecule and the host framework are electroneutral only van der Waals interactions between host and guest are effective. An estimate of the interaction energy calculated with the BIOSYM package of structure simulation programs with the template azabycyclononane (ABN) shows that the interactions are very weak indeed indicating that the formation of the materials is kinetically controlled and supporting the growth mechanism discussed before (Table 10, (26)).

Table 10
Nonbonded energy of sorption sites in the large cavities of DOH, DDR and SGT for the template molecule ABN in the perfect solid

	DOH	SGT	DDR
nonbonded energy [kcal]	- 13.21	- 12.14	-10.50

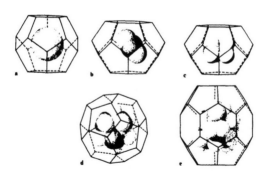

Figure 6. Van der Waals models of the guest species Xe, CO_2, CH_3NH_2, piperidine, and 1-aminoadamantane inside the respective cages [5^{12}], [$5^{12}6^2$], [$5^{12}6^2$], [$5^{12}6^4$], and [$5^{12}6^8$].

The same template 1-aminoadamantane is also a structure stabilising agent in deca-dodecasil 3R (DDR). The cage containing the guest molecule and the guest molecule itself have both site symmetry 3m allowing for a precise analysis of the guest molecule's geometrical orientation inside the cavity. The cage is also prolate and does not posses a mirror plane perpendicular to the 3-fold axis. Therefore, the analysis of the X-ray data set gives information on the orientation of the functional group of the template inside the cavity which might reveal information on the mechanism of crystal growth. The refinement of the two orientations yielded a preference of 90% of the amino group pointing towards the 6MR of the cavity. The rotational disorder of the template showed a preference for a close contact of the template with the silica frame (eclipse position) compared to the staggered alternative as can be seen from the residual electron density distribution inside the cage (Figure 7, (27)). 1-aminoadamantane is also a template for the formation of the clathrasil sigma-2 (SGT) (16). In this example the site symmetry of the cage and the symmetry of the template do not correspond. The information on the host guest interaction which can be obtained from an X-ray analysis only describes an averaged electron cloud smeared out on a sphere-like shell obeying the site symmetry of the cage. For the structure refinement the geometry of the template was taken into account restricting C-C-distances to theoretical values. This, however does not confirm the templates integrity without doubt. Only complementary spectroscopic characterization would prove the nature of the template and confirm its structure stabilising nature.

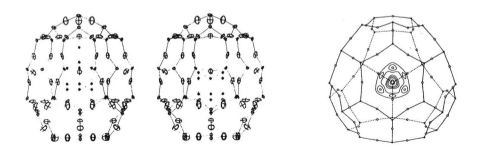

Figure 7. View of the large 19-hedral cage of DDR showing the two orientations of the templateing molecule (left) and a section through the electron density distribution perpendicular to the 3-fold axis of the cage. The model which is placed on atomic positions obtained from the refinement demonstrates the close relation in size and shape of template and cavity. The delocalized electron density indicates disorder of the guest, however, with higher probability for the eclipse orientation.

There are 3 structure analyses of porosil structure types with quinuclidine (Q) as template (Table 2). The symmetry analysis of the system Q/DDR and Q/LEV shows that the site symmetry of the respective cages and the symmetry of the template is equal and ideally suited for an analysis of the host guest interaction in a diffraction experiment. The site symmetry of the cage would even enable to localize the nitrogen site in the template molecule. The residual electron density found inside the cage clearly shows the template molecule. It's orientation, however, is not well defined but averaged in time or space. The delocalized electron density maxima indicate high orientational disorder about the 3-fold axis of the molecule and the fact that nitrogen could not be localized signifies that there is no difference in the "up" and "down" orientation of the molecule in the cages. This is supported by the refinement of the silicate framework. The charge compensating B-atoms are replacing Si statistically on all symmetrically equivalent tetrahedral sites.

As an example for chain-type guest molecules as templates in porosiles analysed in the refinement of a crystal structure the system diethylamine/silica ZSM-22 will be discussed (Table 2 (8)). The 10MR-pore is one-dimensional and shows a typical zigzag course of a zeolite channel. The crystallographic periodicity is 5.05Å due to the 2er chain of [SiO$_4$]-units. This is slightly smaller than the periodicity of the carbon chain of the template molecule which is approximately 6Å. Thus the periodicity of the template molecule is incomensurate with the periodicity of the host framework. Because of the poor scattering power of the template in X-ray experiments, the template-substructure has not been detected yet and, therefore, the composite structure of the host-guest system has not been analysed yet. In a classical diffraction experiment the residual electron density maxima inside the channel clearly show up representing the averaged arrangement of the template molecule in the host framework channel. The view perpendicular onto the channel shows that the template molecule in its extended arrangement which fits the channel-like pore very well building van der Waal contacts (Figure 2).

The Rietveld analysis of a high resolution synchrotron X-ray data set, ^{29}Si 2D solid state MAS NMR, and electron diffraction were applied for the characterization of silica-ZSM-23. Here, NH$_4^-$ and F$^-$ incoorporated through ion exchange occupy the 10MR channel. The schematic representation of the electron density maxima within the channel shows that cation and anion alternate and follow the zig-zag course of the pore.

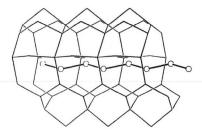

Figure 8. Topology of the channel of silica-ZSM-23 including the zig-zag chain of the guest ions. The electron density maxima were interpreted as NH_4^+ and F^- ions sorbed during the washing process.

The examples given above demonstrate that the guest molecule really act as geometrical templates for the pores of the silica or silicate host-frameworks during synthesis determining their size, shape, and dimensionality. Although the X-ray analysis looks at the 'frozen in' templates residing inside a perfect cage it is obvious that nucleation and crystal growth only would proceed after the sorption and fixation of the template molecule at the growth surface. The calculation of template/host-framework interaction in perfect cages does not elucidate the process of formation or growth but only allows to estimate the contribution of the van der Waals energy to the total lattice energy of the crystal. It would be a great help for understanding of the mechanism of nucleation and growth if one would experimentally determine or numerically simulate the host-guest interaction in the nucleation state and at the growth surface.

The structural analysis of sorbate molecules in pure silica and high silica zeolites has been neglected so far. There is a series of low resolution X-ray studies of Mentzen et al. (28) analysing the location and orientation of the sorbate molecules in ZSM-5. A number of neutron studies has been performed with the aim to describe the position of the sorbate including the hydrogen atoms precisely (cf. 19). Finally few high resolution synchrotron X-ray Rietvelt analysis have successfully described the fitting of the guest species within the pores of the host framework. In general, the information of the analysis is the location of the guest molecule which are in close contact to the zeolite host framework. The limiting condition of the diffraction experiment, however, is the symmetry constraint analysis which also applies to the guest molecule and the averaging nature in time and space of the diffracted intensity data.

2.3.2 Guest molecules analysed in NMR experiments

The short range order probed in NMR spectroscopy is complementary to the information on the long range periodicity given in diffraction experiments. The technique has mainly been used for the non destructive analysis of the template molecules (e.g. 29). The spectra of the occluded template molecules is very similar to the spectra of the pure compounds registrated in solution. The host-guest interaction, therefore, must be very weak and in the van der Waals range which does not affect the geometry of the first coordination sphere of the carbon atoms in template molecules. In Figure 9 the spectrum of 1-aminoadamantane as template in DDR is shown. Although the cross polarization experiment distorts the relative intensities of the signals their chemical shifts are very close to those of the pure compound. Similar is true for all templates analysed so far.

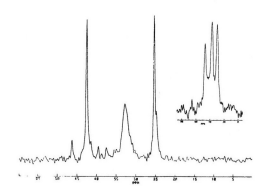

Figure 9. ^{13}C NMR spectra of 1-aminoadamantane inside the 19-hedral cavity of DDR. The MAS-spectrum confirms the nature of the template. The insert shows the spectrum of the static sample. Although there is considerable line broadening compared with the MAS spectrum, the experiment confirms the high mobility of the guest molecule.

The spectra usually show whether there is a single template molecule or a mixture of different templates in the pores of a porosil. The spectrum in Figure 9 shows an example where ethylenediamine has been detected in the pores of a clathrasil together with the template molecule 1-aminoadamantane. An interesting example is the incorporation of methane in MTN synthesized in the presence of ethlyenediamine where methane occupies the $[5^{12}]$-cage ($\approx$-3ppm) as well as the $[5^{12}6^4]$-cage in different quantities. The chemical shift of the two methane signals is significantly different and their relative intensity ratio can be used to calculate the relativeoccupation of the respective cages with the guest species (Figure 10).

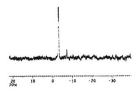

Figure 10. ^{13}C spectrum of CH_4 in the different cages of MTN. The integral intensity of the NMR signals allows the calculation of the relative occupancies of the two cages.

A simple experiment gives a good estimate of the dynamic disorder on the NMR time scale of the template molecule inside the cavity of the porosil. Performing the solid state NMR experiment with the spinner at standstill the line width of the resonance signal is narrow if the template molecule is mobile. A typical example is given in Figure 9 where the insert shows the spectrum of 1-aminoadamantane of a static sample. Although the signal line width is broader than in the corresponding MAS spectrum only dynamic disorder of the template inside the cage would explain such narrow lines. This also explains the X-ray diffraction experiment where the delocalized electron density maxima in the cage indicated disorder of the template, without differentiating, however, between its dynamic or static nature. For all templates investigated so far similar results have been obtained showing that the template molecule is intact and despite the close geometrical relationship of host and guest the very weak interaction energy does not pin down the template but allows for dynamic disorder.

Most NMR experiments carried out on templates were ^{13}C experiments. Since the chemical shift is very sensitive for changes in the first coordination sphere the NMR experiment also probes the charges on the amino group. In the case where B or Al substitute for Si on T-sites in the host framework the ^{13}C spectrum of the template also shows whether or not the template is incorporated in its protonated form. The examples of the high silica porosils discussed above all contained organoammonium cations as charge compensating species. The NMR-analysis might be quantified in combination with classical chemical analysis of the trivalent substituent proving the charge compensating character of the templating species. In Figure 11 the ^{13}C NMR spectrum of pyrrolidine in NON clearly shows the downfield shift of the signal of the carbon atom close to the protonated nitrogen. The insert on Figure 11 shows the static experiment and proves that the cationic template is also dynamically disorderd.

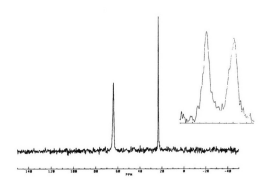

Figure 11. ^{13}C NMR spectrum of pyrrolidine in boron containing nonasil. The spectrum clearly shows that the templating species was the pyrrolidinium cation compensating the charge of the framework introduced through the substitution of Si with B. The insert shows again that the template molecule is highly mobile inside the cage.

An interesting NMR experiment has been carried out with a series of sorbate molecules in order to differentiate between the influence of size and shape, dipolar- and polarizationproperties of the guest molecule on the host-guest interaction (30). P-xylene, p-chlorotoluene, and p-dichlorobenzene were loaded on silica ZSM-5 in amounts that the phase transition to the orthorhombic high form was just terminated at room temperature (2μl/250 mg sample). In ^{29}Si MAS NMR spectra the influence of the different guest species which are very similar in size and shape on the silica host framework were investigated. The comparison of the 3 spectra showed that only minor changes are observed (Figure 12). Therefore it was concluded that size and shape of the sorbate are the dominating factors which take influence on the host framework and that electronic interaction between host and guest can not be judged from NMR experiments.

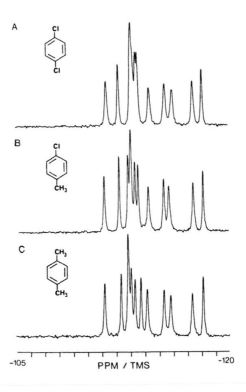

Figure 12: ^{29}Si MAS NMR spectra of the orthorhombic high symmetry form of silica-ZSM-5 which was loaded with 3 different sorbate molecules. the close similarity of the spectra demonstrates that it is the guest geometry which influences the host framework and not electronic interaction between guest and host.

3 Conclusion

Host guest interaction in porosils and high silica porosils is the key phenomenon in the formation of this class of compounds. Pore size, shape and geometry is correlated with the nature of the templating species. The large number of experiments allows to classify the template molecules and to work out rules for the synthesis of in particular porosils. It is most likely that seed formation in the presence of templating agents is the structure determining step in crystallisation. From the crystal morphology a mechanism for the interaction of host and guest on the growth surface is proposed taking into account the ambiphilic nature of the most active templating species, the amines, and the ambiphilic character of the growth surface with terminating hydrophilic silanol groups and perfectly formed hydrophobic halve spheres of the cavities.

The analysis of the host-guest interaction in the final product discribes a different situation and focuses on the close correspondence of geometrical features of the template and of the cavity. In combination with NMR experiments it is shown that the templates are in close contact with the inner surface of the porosil, however, the molecule still is dynamically disordered. It has been shown in an experiment, that the influence of the geometry of the guest molecule is most important for the status of the host framework and that electronic interaction can be neglected.

Acknowledgement: The author thanks M. Hochgräfe, A. Grünewald-Lüke, and B. Marler for information prior to publication. Technical assistance of S. Vortmann during the preparation of the manuscript is gratefully appreciated.

References

1. W.M. Meier and D.H. Olson, Atlas of Zeolite Structure Types, Butterworth-Heinemann,London, 1992.

2. R.M. Barrer, Hydrothermal Chemistry of Zeolites, Academic Press, London, pp. 157.

326

3. a) R.M. Barrer, in New Developments in Zeolite Science and Technology,. Y. Murrakami, A. Iijima and J.W. Ward, (ed.), Elsevier, Amsterdam, 1986, p. 6.

   b) F. Liebau et al., Zeolites 6 (1986) 373.

4. a) H. Gies, in Inclusion Compuonds 5, eds. J.L. Atwood, J.E.D. Davies, D.D. MacNicol, Oxford University Press, 1991, p. 1.

   b) H. Gies, B. Marler, Zeolites 12 (1992) 42.

5. a) H. Gies et al., Z. Kristallogr. Suppl. 7 (1993) 57.

   b) B. Marler et al., Acta Cryst. A49 (1993) C260.

6. a) H. Gies et al., N. Jb. Miner. Mh. 3 (1982) 119.

   b) H. Gies et al., Angew. Chem 94/3 (1982) 214.

7. A. Grünewald-Lüke, H. Gies, Microp. Mat., in press (1994).

8. B. Marler, Zeolites 6 (1986) 393.

9. G.D. Price et al., J. Am. Chem. Soc. 104 (1982) 5971.

10. P. Coullet et al., J. Solid State Inorg. Chem. 28 (1991) 345.

11. J.-L. Guth et al., in Proceedings from the 9th International Zeolite-conference, R. von Ballmoos, J.B. Higgins, M.M.J. Treacy, (ed.), Butterworth-Heinemann, Boston, 1992, p. 215.

12. L. Delmotte et al., Zeolites 10 (1990) 778.

13. a) T.V. Wittum, B.Youll, US Patent 4060590 (1977).

    b) M. Taramaso et al., in Proceedings of the 5th International Conference on Zeolites, L.V. Rees (ed.), Heyden, London, 1980.

    c) U. Oberhagemann et al., in Proceedings of the 10th International Zeolites Conference, in press.

14. H. Gies, J. Rius, J. Appl. Cryst. submitted 1993.

15. G. Bellussi et al., Zeolites 10 (1990) 642.

16. A. Stewart et al., Innovation in Zeolite Material Science, P.J. Grobet et al,. (ed.), Elsevier, Amsterdam (1987) 57.

17.   B. Marler et al., in Proceedings of the 9th International Zeoliteconference, R. von Ballmoos, J.B. Higgins, M.M.J. Treacy, (ed.), Butterworth-Heinemann, Boston, 1992, p. 425.

18. a) C.A. Bursill et al., Nature (1980) 111.

    b) O. Terasaki, Electron Microscopy 2 EUREM 92 (1992) 615.

19. a) R. Goyal et al., J. Chem. Soc., Chem. Commun. 1990 (1990) 1152.

    b) J. Newsam, J. Phys. Chem. 93 (1989) 7689.

20.   H. Gies et al., J. Phys. Chem. Solids 52 (1991) 1235.

21.   B. Marler et al., J. Appl. Cryst. 26 (1993) 636.

22.   H. Gies, J. Incl. Phenomena 4 (1986) 85.

23.   A. Grünewald-Lüke and H. Gies, Z. Kristallogr. Suppl. 5 (1992) 88.

24.   S. Huyashi et al., Chem. Phys. Let. 113 (1984) 368.

25.   G.A. Jeffrey, Inclusion Compunds 1, J.L. Atwood, J.E.D. Davies, D.D. McNicol, (ed.), Academic Press, London, 1984, p. 135.

26.   M. Hochgräfe and H. Gies, in "Fifth German Workshop on Zeolites", J. Kärger (ed.), Universität Leipzig, Leipzig, 1993, p. PT I.

27.   H. Gies, Z. Kristallogr. 175 (1986) 93.

28. a) B.F. Mentzen and F. Bosselet, C.R. Acad. Sci. Paris 309 (1989) 539.

    b) B.F. Mentzen, Comptes Rendus 303 (1986) 681.

29. a) G. Boxhoorn et al., J. Chem. Soc., Chem. Commun. 1982 (1982) 264.

    b) G. Boxhoorn et al., in Proceedings of the 6th International Zeoliteconference, D. Olson, A. Bisio, (ed.), Butterworth, London, 1984, 694.

30.   C.A. Fyfe et al., Can. J. Chem. 66 (1988) 1942.

J.C. Jansen, M. Stöcker, H.G. Karge and J. Weitkamp (Eds.)
Advanced Zeolite Science and Applications
Studies in Surface Science and Catalysis, Vol. 85
© 1994 Elsevier Science B.V. All rights reserved.

329

The preparation and potential applications of ultra-large pore molecular sieves : A review

J. L. Casci

ICI Katalco, RT&E Department, P O Box 1, Billingham, Cleveland TS23 1LB, UK

This paper describes the preparation and potential applications of ultra-large pore molecular sieves; materials with entry ports greater than 1nm. While this review attempts to cover a wide spectrum of materials, including zeotypes and pillared layered solids, it will concentrate on patent and open literature reports concerning the, recently discovered, mesoporous molecular sieves typified by M41S. Data on the preparation and structure of these mesoporous materials is reviewed and conclusions drawn as to their true crystallinity. In describing their method of formation an analogy is drawn with layer lattice materials. General methods of characterising ultra-large pore materials are discussed and comments made on the usefulness of generating theoretical frameworks.

1. DEFINITION, NOMENCLATURE AND SCOPE

From the IUPAC definition [1] microporous materials are those which have pores < 2nm in diameter, while those porous solids with channel diameters in the range 2 to 50nm are classed as mesoporous.

Molecular sieve zeotypes are an important subset of inorganic microporous solids. Historically these materials are classified as small-, medium- or large-pore depending on whether the channels giving entry to a particular structure are made up from <= 8 T-atoms, 10 T-atoms or 12 T-atoms respectively. From the above it would seem reasonable to conclude that zeotypes with pore openings bounded by more than 12 T-atoms could be described as ultra-large pore materials. This description, however, extends from conventional microporous zeotypes to inorganic mesoporous materials. Figure 1 schematically illustrates materials with pore sizes from 0.3 to 10nm, highlighting material types and the definitions of small-, medium-, large- and ultra-large pore materials used in this review.

This paper will attempt to review ultra-large pore molecular sieves, both zeotypes and ordered mesoporous materials; porous solids with pores > 1nm.

2. INTRODUCTION

The term "Molecular Sieve" is used, quite simply, to describe species which can discriminate between molecules solely on the basis of size. This ability is based on such porous solids having pore dimensions close to the kinetic diameter of many common molecules.

Materials, other than zeolites which exhibit this property, include some carbons (charcoals), clays, silicas and aluminas.

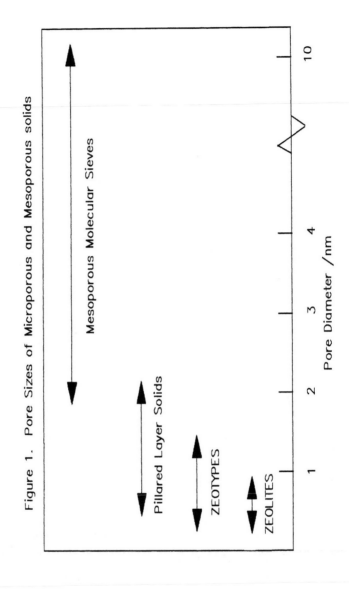

Figure 1. Pore Sizes of Microporous and Mesoporous solids

It has long been recognised that among these materials, molecular sieve zeotypes are unique; only they have ordered 3-D, crystal structures. This gives rise to a much narrower pore size distribution than can be achieved by other sorbents - although very significant advances have been made in recent years.

Although zeolites have found widespread use as ion-exchangers and sorbents there can be little doubt that the great interest in, and development of, these materials has been fuelled by their use as catalysts particularly in Fluid Catalytic Cracking (FCC) and other Petroleum Refining Processes [2]. The use of faujasite based catalysts in FCC is, to a large extent, due to this material having a 3-D channel system based on 12-ring windows [3]. The channel dimensions (0.74nm) allow the access and egress of a number of the common aliphatic and aromatic hydrocarbons found in petroleum fractions. The importance of FCC in the development of zeolites cannot be overstated. The requirements of this process has had a very significant influence on much of the zeolite research carried out in the last 30 years. It is interesting to note that two reviews of this area [2,4] carried out some 10 years apart both predicted that a major focus of future research, in FCC, would be on octane number enhancement and the (preparation and) use of very large pore materials to allow processing of ever heavier petroleum fractions.

Smaller ring zeolites, principally MFI type materials, are finding increasing use as additives in FCC catalysts to allow the generation of high octane gasoline, taking advantage of their shape selectivity in cracking linear (as opposed to branched) hydrocarbons [5]. The search continues, however, for very large pore materials to allow "bottom of the barrel" processing - a search which was given new impetus by the discovery of the 18-ring VFI structure [6], first reported in 1989.

While new large, or ultra-large, pore molecular sieve zeotypes are likely to be of interest for the other traditional uses, ion-exchange and sorptive/separation, it seems probable that research into their preparation will be driven by the historic catalytic applications and the new emerging areas associated with potential electronic or optical properties [7-10].

From the early beginnings of international meetings on molecular sieves there was a move to provide a proper definition of materials that could called zeolites [11]. Concerns over definition were again raised in the early 1980s [12] partly as a response to the wide range of aluminophosphate molecular sieves first reported by workers at Union Carbide [13]. Rigid definitions have now been relaxed and currently, to be included in the "Atlas of Zeolite Structure Types" all that is required is that a framework has to be composed of atoms which have a tetrahedral coordination (other than the oxygen bridges) and have less than about 21 T-atoms per 1000 Å3 [3] - a definition which allows the inclusion of materials with a variety of framework composition and, indeed, interrupted frameworks.

This paper will attempt to review work on the preparation, characterisation and applications (particularly catalytic applications) of ultra-large pore molecular sieves. In addition to summarising recent work on zeolites and related zeotypes it will also provide some information on Pillared Clays (PILCs) and attempt to review the available literature (patent and open literature) on the recently discovered mesoporous sorbents such as Mobil's MCM-41 [14].

3. CHARACTERISATION

While the pore size of molecular sieve zeolites can be tailored by a number of techniques, for example ion-exchange, it is set principally by the framework topolgy (effectively the number of T-atoms or more correctly the number of oxygen atoms which make up the pore opening) [15]. However, the shape of the opening is also important; there is ample evidence for a variety of configurations for windows containing the same number of T-atoms: circular, elliptical and even "tear drop" shaped pores [3].

Structure solution of zeotypes is often a far from trivial exercise partly because most synthetic materials are formed with crystals too small for single crystal studies even using synchrotron radiation sources. In addition a number of frameworks are formed as intergrowths further complicating the issue. Hence most structure solution work relies heavily on a multi-technique approach, involving electron microscopy based techniques, powder diffraction and model building.

It is not the purpose of this paper to review such techniques these will be discussed in greater detail in other parts of this volume. However, the characterisation of ultra-large pore materials does pose some additional problems compared to conventional microporous materials and it is these problems which this section attempts to address.

In the multi-technique approach to structure solution the starting point is usually electron diffraction, to provide both the symmetry of the material and (crude) unit-cell parameters. While lattice imaging can provide information on channel configuration and the presence of intergrowths and x-ray powder diffraction used to refine unit-cell dimensions the next major step is the starting model for the structure. One piece of experimental information which can assist in the development of this model is sorption: the molecular sieving effect which occurs when the uptake of sorbates of varying molecular dimensions is examined can give an indication of window size [15,16]. While newer modelling approaches to structure solution such as simulated annealing [17] and the APS formalism [18] are less reliant on such input, the results of sorption remain a valuable tool in the characterisation of novel zeotypes.

For non-zeolitic molecular sieves in which the entry apertures are not defined by a certain number of T-atoms sorption is of crucial importance.

For true ultra-large pore materials, regardless of their structure type, the problem is one of how to characterise the pore opening using conventional molecular sieving experiments which rely on having a range of sorbate molecules of differing kinetic diameters. The problem was highlighted in the original report on VPI-5 [6] in which the adsorbtion of perfluorotributylamine (kinetic diameter approximately 1.05nm) was interpreted as showing the presence of 18 T-atom ring windows. For materials with even larger windows the problem of probe molecule selection becomes even more complicated.

Paradoxically, for these ultra-large pore materials it is useful to obtain adsorption isotherms for small inert molecules such as Argon or Nitrogen. The data from the adsorption isotherm can then be treated to give an indication of channel dimensions. For materials with channel dimensions greater than about 5nm the Kelvin equation can be applied - in which the adsorption is treated as being capillary condensation. For smaller diameter pores this approach is not valid (hence the use of probe molecules described above). For pores up to about 2.5nm a potentially

Figure 2   Pore Size Information on a Titania Pillared Clay
            (Data supplied by Coulter Electronics Ltd)

Figure 2a   Adsorption Isotherm

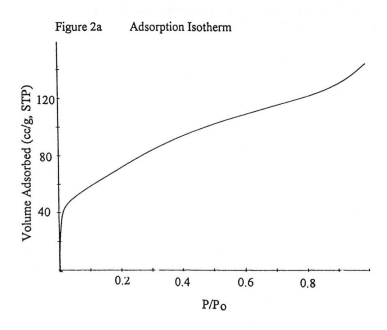

Figure 2b   Horvath-Kawazoe Plot

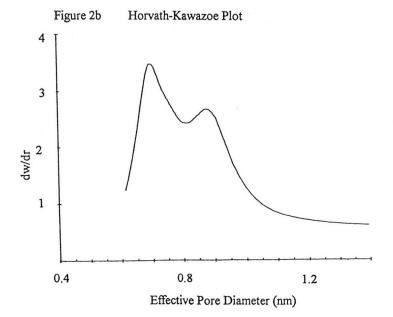

valuable technique is that developed by Horvath and Kawazoe [19] which it has been suggested [20] can be extended to pores up to about 6nm.

Using the Lennard-Jones functions and Gibbs free energy of adsorption, the Horvath and Kawazoe technique uses an expression that correlates the effective pore diameter to the adsorption isotherm. The expression is:

$W/W_o = f(l-d_a)$

where W = volume of gas adsorbed.

      Wo = maximum volume of gas adsorbed.

      $l-d_a$ is the effective pore diameter.

A Horvath-Kawazoe type plot for a titania-pillared monmorillonite clay is shown in Figure 2 (supplied by Coulter Electronics Ltd, UK). Effectively, the figure shows the conventional sorption isotherm and pore size plot illustrating the effective window dimension for the material. Interestingly molecular sieve zeotypes are excellent materials with which to calibrate any instrument.

The potential of this technique is evident particularly for ultra-large pore materials.

4. MATERIALS

4.1 Zeotypes

Of the 85 frameworks described in the "Atlas of Zeolite Structure Types" only three can be considered as ultra-large pore according to the definition used in this review (> 12 T-atom channels). To the three zeotypes described in the Atlas, -CLO, VFI and AET, a fourth can be added JDF-20 [21-23].

Table 1 summarises aspects of these structure types, in particular, details of the framework composition, (maximum) ring size and template(s) used in their synthesis. Most striking is the fact that none of these materials are traditional aluminosilicate zeolites; all contain phosphorous and aluminium except Cloverite which is a gallophosphate. This last material is also somewhat unusual in that its synthesis was facilitated by the use of fluoride rather than hydroxide as the mineralising species.

The AET structure, which contains 14 T-atom windows, was first reported in 1982 [24] yet its structure was only solved some time later [25], after indeed the first reports on VFI [27,28] and as a consequence the AET structure has been somewhat ignored. In reality, however, this material was the first ultra-large pore molecular sieve zeotype: > 12 T-atom windows. One can only speculate on the impact this would have had on molecular sieve research had the structure been solved earlier.

There have been about 100 reports describing various aspects of VFI. A large proportion of these have centred on the controversy of whether VPI-5 is actually the same phase as H1 described by D'Yvoire [34] over 25 years before the first reports on VPI-5. The current status of the various arguments has been summarised recently [35], with the opposing views being that they are the same material [29], similar but different materials [36] or that without a pure sample of H1 there is insufficient information on which to pass a judgement [37]. This controversy will not be dwelt on in this review.

Table 1
Ultra-Large Pore Molecular Sieve Zeotypes

Material	Structure Code	Ring Size	Framework Composition	Template	Reference
AlPO4-8	AET	14	[Al,P]	DPA(a)	24,25,26
VPI-5	VFI	18	[Al,P] -(c) [Co,Al,P]	TBA(b), DPA TBA, DPA	27,28 29 30
Cloverite	-CLO	20	[Ga,P]	Quinuclidinium (+ F-)	31,32,33
JDF-20	-	20	[Al,P]	Triethylamine (+ glycol)	21,22,23

(a) = n-Dipropylamine (b) = Tetrabutylammonium (c) = Inorganic

The 20 ring structures cloverite and JDF-20 are both incomplete frameworks with hydroxyl groups protruding into the channels.

The fact that none of these materials is an aluminosilicate has already been commented upon yet it is particularly important when aspects of their preparation is discussed. Both AET and VFI are prepared from conventional template types, materials used previously in the preparation of aluminosilicate zeolites (tending to yield MFI or MEL structure types [38]), although VFI can also be prepared "template-free" [29]. Hence it can be argued the significant feature of their synthesis is the use of the aluminophosphate system. This is also true for JDF-20, while cloverite is formed from the gallophophste system. These latter materials, however, also employ unusual reaction conditions: cloverite is crystallised using fluoride as the mineraliser while JDF-20 requires the use of an organic solvent (glycol).

The significance of the framework composition will be returned to later when hypothetical molecular sieve frameworks are discussed. The use of non-aqueous solvents and fluoride mineraliser could well have great significance for novel (ultra-large pore) zeotype synthesis. Of these, the use of fluoride may well be the most important, since it should allow investigation (and re-investigation) of templates which undergo significant Hoffmann degradation under the more usual high pH conditions.

4.2 PLS

The second group of materials that can be described as ultra-large pore molecular sieves are the Pillared Layered Solids (PLS). Historically, the preparation of such materials was first demonstrated by Barrer and MacLeod in 1955 when they replaced the alkali and alkaline-earth cations in the smectite clay montmorillonite by a quaternary ammonium compound. The resulting material becoming known as PILCs: Pillared Inter Layer Clays. The use of organic cations to pillar or cross-link lamellar solids has been reviewed recently [39]. Simply inserting organic cations between the lamellar regions does, however, restrict the thermal stability of the resulting materials, a significant disadvantage to their use as catalysts. A notable advance was the use of oxyhydroxyaluminium cations [40] as the pillaring agent. Such materials not only provide enhanced thermal, and hydrothermal, stability but can also provide catalytic centres.

As with molecular sieve zeolites the last decade has seen a dramatic increase in the number, and type, of PLS type materials which have been prepared. This diversity encompasses both the type of layered material, 2-D substrate or host compound, and the pillaring species.

While the initial focus was on layered silicates other host materials are metalhydrogen phosphates, metal oxyhalides and layered double hydroxides. The majority, and arguably the most significant work has been carried out on:

(a) Smectite Clays
(b) Phosphates and Phosphonates such as Zirconium Phosphate.
(c) Layered double hydroxides such Hydrotalcite.

Similarly, pillaring species have evolved from organic and oxyhydroxyaluminium cations to now include pillars based on zirconium, chromium, silicon, titanium, iron and (transition) metal complexes.

A number of excellent reviews have appeared recently describing the preparation, characterisation and use of PLS [41-45].

In spite of the structural diversity the basic principle underlying the preparation of these Pillared Layer Solids are very similar and is shown schematically in Figure 3. Although this figure shows the idealised case it does illustrate some of the major issues in the preparation of pillared solids: the size ("height") of the pillaring species and its distribution in the host compound.

The choice of the pillaring compound is motivated by two factors: the need to "cement" the layers of the host compound together providing high mechanical and thermal stability and to provide "active" centres for the ultimate use whether that be as a catalyst, sorbent or electronic or sensing device.

Irrespective of the chemical nature of the pillar its physical dimensions and distribution within the host will have a major impact on the ultimate utility. As Figure 3 demonstrates it is possible to generate a non-porous intercalate if the pillaring species are, in effect, in lateral contact filling the inter-layer space. Alternatively, if the pillars are distributed too thinly, or are too large, the pillared structure can be unstable. The ideal situation, therefore, is for the pillar height to be of the same order as the lateral separation between pillars. Such a situation also provides a near uniform pore size distribution.

Figure 3. Idealised Schematic of the Preparation of Pillared Layer Solids.

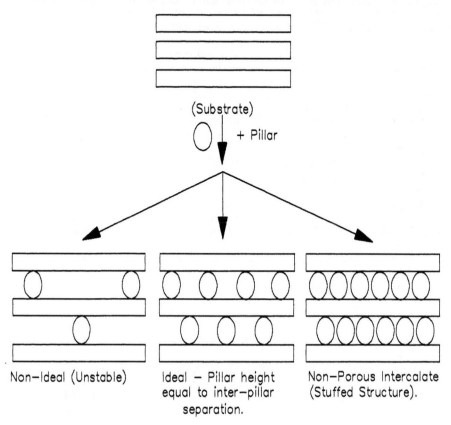

Non-Ideal (Unstable)

Ideal — Pillar height
equal to inter-pillar
separation.

Non-Porous Intercalate
(Stuffed Structure).

Figure 4. Schematic of a Pillared Solid Illustrating Faulting.

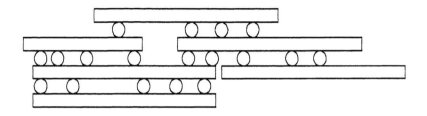

A more realistic view of pillared solids, allows for the non-parallel stacking of layers, for the layers not being of uniform length and for the presence of stacking disorders [46,47]. Combinations of these, coupled, with the variations in the lateral separations of pillars, can give rise to mesoporous regions, illustrated in Figure 4.

The three principle families of lamellar substrate, mentioned above, are very different chemically. Such differences can be simply exemplified by comparison of cationic clays with the layered double hydroxides. Both have charged sheets, but the difference in sign of this charge means that the interlayer species in the clays are cations whereas for the double hydroxides the species are anions. In both cases metal complexes can be used as pillaring species: cations (such as the aluminum Keggin ion) for the clays and anionic materials (such as vanadate species) for the double hydroxides.

While the Zirconium Phosphates can be pillared by similar cations to those used for the smectite clays, since these materials do not swell in water it is necessary to "artificially" swell the layered phosphate precursor using alkylamines [46,48]. Pre-doping of alkylamines is also a favoured tool to restrict the number of cations intercalated and hence not only prevent the formation of a non-porous material (see Figure 3) but produce a more ordered pillaring; hence a narrower pore size distribution.

The diversification into the variety of compositions described above has been prompted by three main driving forces: to improve the thermal (and hydrothermal) stability, improve (narrow) the pore size distribution and produce mesoporous materials. Employing various combinations of precursor and pillar types, major advances have been made in all three areas. Perhaps the most exciting is the preparation of meso-porous materials ("Supergallery" Pillared Materials) by a sol-gel technique one aspect of which is direct intercalation of metal oxide sols (DIMOS) [42].

Another technique employed [49] is the pillaring of a montmorillonite clay by Imogolite, an unusual naturally occurring tubular aluminosilicate with an internal diameter of about 0.8nm. While the Imogolite-pillared structure may have limited future it serves as an excellent introduction to the next type of mesoporous solids.

4.3 Mesoporous Molecular Sieves

True mesoporous solids, with narrow pore size distributions, have been reported by two groups using preparative techniques which, ostensibly, are very different but which the author believes may actually have some common features. The similarities and differences between the two sets of materials, themselves, will be returned to later.

The two groups reporting these ultra large pore materials were from Mobil and a group from Toyota and Waseda University. Already there have numerous comments and reports [50-54] on the Mobil work but by comparison the reports from the group in Japan have been ignored. This probably reflects the very large number of patents and papers - see later - from Mobil and to the superior phase characterisation from that group.

The first reports on ultra large pore molecular sieves of this type, however, came from the group in Japan describing the intercalation of long chain (typically C16) alkyltrimethylammonium cations in the layered silicate Kanemite [55,56] followed by calcination to remove the organic. It was claimed that during this process the silicate layers condensed to form a 3-D structure with "nano-scale pores". Pore size distribution studies, based on nitrogen adsorption isotherms, suggested a narrow range of pore sizes centred on about 3nm. Powder x-ray diffraction was largely uninformative. It did, however, indicate that the Kanemite structure was "lost" on treatment with the quaternary ammonium cation to be replaced by a much simpler pattern containing, essentially, one feature: an intense line at low angle. 29Si solid state NMR provided evidence for the condensation of the Q3 species to generate a large proportion of Q4 species. Examination of the acid sites in aluminium containing samples, by IR and TPD, was reported to show that the material had characteristics similar to an amorphous silica-alumina.

The first published report on the Mobil materials was most probably the issue of a PCT Application [57] in August 1991. Since that time there have been a large number of patents [58-77] and reports [14,78] on these materials, principally the M41S family. To simplify matters, apart from the PCT Application mentioned above, this review will restrict itself to US Patents only.

Tables 2 and 3 attempt to summarise Mobil (US) Patents, with those relating to Preparation, Characterisation and Modification being contained in Table 2 while Table 3 contains details of patents concerned with Applications or Use. Obviously there is considerable overlap between various areas with, for example, patents concerned with a particular use also describing the method of preparation. In the tables an attempt has been made to categorise each patent according to (the author's view of) its main technical feature. Two dates are also given in the tables; filing and publication. Although not strictly accurate, the former was used to set the patents in chronological order, while the publication date sets when the information was available to the public.

The nomenclature used by the Mobil group indicates that they regard their material as not being a single phase but a family of structures; termed M41S [78]. There are a number of members of this family including both hexagonal and cubic phases (see later). The hexagonal phase is given the notation MCM-41, yet even this material is not a single phase but has a number of members which differ, principally, in the size of the pore openings. For this reason, at least, in the open literature reports, Beck and co-workers employ the term MCM-41(X) to describe individual members, where X represents the approximate pore size in angstroms.

The preparation of M41S materials is accomplished using compositions and procedures similar to those used for the preparation of templated zeotypes. There are however a number of modifications which allow the formation of specific members of the MCM-41 family. To describe these it is helpful to consider the "simplest" or most basic preparation. This involves the reaction of a system containing:

$$SiO_2 - Al_2O_3 - (Na_2O) - TAA_2O - RTMA_2O - H_2O$$

where TAA is a symmetrical tetralkylammonium species with a C1-C3 chain and RTMA is a long chain alkyltrimethylammonium cation. Sodium oxide is shown in parenthesis since its inclusion would appear to be optional.

A single preparation, of the above general type, is repeated in a number of publications (reference 14, reference 78 (as method A) and in numerous patents, for example reference 60, example 4). This specific reaction mixture has the following (molar) composition:

$$33.2 \ SiO_2 - Al_2O_3 - 0.23 \ Na_2O - 5.2 \ TMA_2O - 6.1 \ CTMA_2O - 780 \ H_2O$$

Table 2
Mobil Patents on Mesoporous Materials : Preparation and Modification

Ref	Patent Number	Filing Date (a)	Publication Date	Type	Comment/Summary (b)
58	US 5102643	25.01.90	07.04.92	Preparation	Synthesis and Characterisation
59	US 5057296	10.12.90	15.10.91	Preparation	Use of auxilliary organics
60	US 5108725	10.12.90	28.04.92	Preparation	Importance of TEM in determining purity
61	US 5098684	10.12.90	24.03.92	Preparation	See US 5108725 except use of TMB in addition to other "templates"
62	US 5110572	24.06.91	05.05.92	Preparation	Use of organoaluminosilane source to produce product of higher Al content
63	US 5112589	25.06.91	12.05.92	Preparation	Addition of mineral acids to improve synthesis - product has higher surface area
64	US 5156828	18.07.91	20.10.92	Modification	Use of steaming
65	US 5145816	20.06.91	08.09.92	Modification	Functionalising product to reduce pore size
66	US 5156829	20.06.91	20.10.92	Modification	Functionalising to improve (hydro) thermal stability
67	US 5104515	24.06.91	14.04.92	Modification	Treatment of product with basic solution to remove impurities
68	US 5246689	24.07.91	21.09.93	Preparation	Synthesis and Characterisation

(a) Date is given as Day, Month, Year   (b) Author's view on the key technical feature(s) of the patents

Table 3
Mobil Patents on Mesoporous Materials : Applications

Ref	Patent Number	Filing Date (a)	Publication Date	Type	Comment/Summary (b)
69	US 5105051	14.04.92	14.04.92	Catalysis	Cr impregnated MCM-41 for olefin oligomerisation (C10) - see also US 5200058
70	US 5134241	21.06.91	28.07.92	Catalysis	Olefin (C3) oligomerisation
71	US 5134242	21.06.91	28.07.92	Catalysis	Olefin (C3) oligomerisation
72	US 5134243	21.06.91	28.07.92	Catalysis	Olefin (C3) oligomerisation
73	US 5191144	07.10.91	02.03.93	Catalysis	Olefin (C5-C12) Cracking and Isomerisation
74	US 5200058	08.06.92	06.04.93	Catalysis	Hydrocarbon conversion - C10 oligomerisation over functionalised material (see US 5145816)
75	US 5232580	21.06.91	03.08.93	Catalysis	Dealkylation and Naphtha Cracking
76	US 5220101	22.07.92	15.06.93	Sorption	Functionalising product to alter pore size and generate hydrophobicity
77	US 5143707	24.07.91	01.09.92	Catalysis	Selective Catalytic Reduction of NOx

(a) Date is given as Day, Month, Year (b) Author's view on the key technical feature(s) of the patents where TMA is tetramethylammonium and CTMA is cetyltrimethylammonium or hexadecyltrimethylammonium. The alumina source is a pseudo-boehmite and a combined silica source (tetramethylammoniumsilicate solution and precipitated silica) is employed. The sodium is present only as an impurity. The resulting mixture is reacted, without agitation, at 150°C for 24 hours. The product is isolated by filtration, washed and dried in air at ambient. Such a preparation yields MCM-41(40).

A range of raw material sources can be used in the standard composition. In addition to the pseudo-boehmite, sodium aluminate (reference 58 example 6) and aluminium sulphate (reference 58 example 3) can be employed. For silica, both quaternary ammonium and sodium silicate solutions are exemplified (reference 78 synthesis B) in addition to precipitated silica and colloidal silica (reference 58 example 9).

While the reaction mixture compositions given above indicate that the long chain surfactant is added as the free base it is actually added as a mixed halide/hydroxide solution - prepared by partial exchange of the halide using an anion exchange resin.

In terms of "framework" composition the MCM-41 materials can be prepared from reaction mixtures containing a wide range of silica/alumina ratios from about 25/1 to those in which alumina is present only as an impurity in the other raw material sources. The use of a combined silica-alumina source, a dialkoxyaluminotrialkoxysilane [62], is claimed to allow the preparation of M41S materials with higher aluminium contents: silica/alumina ratios extended to below 10/1.

To prepare M41S materials with different pore sizes essentially three techniques can be used:

(i)   Alter the surfactant chain length [78].
(ii)   Add an auxiliary organic [59,61,78]
(iii) Post-synthetic treatment to reduce the pore size [65].

The post-synthetic treatment will be described later when modifications are discussed.

The basic preparation described above employs a C16 chain length surfactant as the "primary template". Use of shorter chain lengths can affect the pore size. Using data supplied by Beck et al [78] Figure 5 has been constructed. This plots the M41S product pore size (determined from the argon sorption isotherm) against surfactant chain length. It can be seen that there is a, reasonably, linear correlation between surfactant length and pore size for both siliceous and aluminosilicate M41S materials. This plot also suggests that it is possible to select and prepare a particular pore size by judicious choice of surfactant.

In "standard" M41S preparations a short chain tetralkylammonium species is added, as the hydroxide, (TAAOH) in addition the primary surfactant template. The precise role of the TAAOH is not clear. It is possible, however, that M41S requires moderate to high alkalinities but low alkali metal contents thus the TAA may be added purely as a source of base. From the data available it would seem that the nature of the TAA (tetramethyl, tetraethyl or tetrapropyl) makes little difference to the M41S species formed; this is rather surprising when one considers that these can be strongly directing species in zeolite syntheses.

Figure 5. M41S Pore Size against Surfactant Chain Length.

Plot uses data from Beck et al [78].

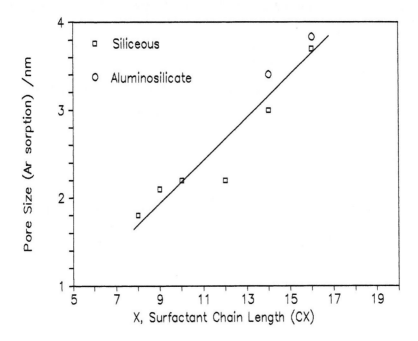

It follows from the above that the auxiliary organic [59,61,78] added to syntheses to expand the M41S pore size is not a tetralkylammonium compound. In fact the auxiliary organic is a neutral species. While 1,3,5-trimethylbenzene (TMB) is frequently used in examples a wide range of organics have been evaluated [59] including substituted aromatics, paraffins and alcohols. Most were reported as having some effect although there was little reported success with alcohols - most failing to crystallise.

Using data supplied [78], and making some assumptions about the sodium aluminate composition, it is possible to estimate the reaction mixture used by Beck and co-workers to demonstrate the effect of an auxiliary organic, TMB. The reaction mixture studied by the Mobil group had the (approximate) molar composition:

$$31 \; SiO_2 - Al_2O_3 - 0.01 \; Na_2O - 2.5 \; TMA_2O - 10 \; CTMA_2O - 750 \; H_2O$$

To this were added 0, 7, 10 and 25 moles of TMB. The pore size of the resulting product (from argon sorption studies) increases with increasing TMB content as can be seen from Figure 6. As with surfactant chain length the apparent linear correlation should allow selection of pore size. The above series was, however, over a fairly limited range of auxiliary/surfactant ratios. Beck and co-workers report [78] that at significantly higher levels of TMB even larger pore MCM-41 materials are prepared but at the expense of pore size distribution and a more "irregular" pore arrangement.

The above techniques illustrate that the pore size of MCM-41 can be altered by synthesis conditions allowing the preparation of materials over a wide range of pore sizes.

The final method of altering pore size involves a post-synthesis treatment: the reaction or "functionalisation" of internal silanol groups. A wide variety of materials are claimed [65] and a number exemplified including silanes, aluminiumalkoxides and combined Si-Al compounds such as a dialkoxyaluminotrialkoxysilane - the same materials that are used as raw materials in some syntheses [62]. Boron- and phosphorous-containing compounds have also been used [65]. Only for the silane modified materials was there any evidence presented for pore size reduction - based on pore size measurements before and after treatment [65]. The only data presented for the functionalisation by other species (Al, P and B) was elemental analysis and/or total pore volume reductions [65]. Such data, of course, is not definitive; it merely shows that the species has been occluded: pore blockage could account for the reduction in pore volume.

The functionalisation however is a recurrent theme in the Mobil patents not only is it credited with pore size reduction but also with adding catalytic centres [66] and, most significantly, with improving the thermal and hydrothermal stability [66].

Two further patents describe "modifications" to the standard synthesis composition. These detail the addition of mineral acids [63] and a method of isolating or purifying the final product which involves washing with a basic solution [67].

The advantage of adding a mineral acid is claimed to be in providing material with a higher benzene adsorption and/or higher surface area [63]. The examples, and comparative examples, given, however, are unconvincing. While two products are described which have higher benzene capacities than the comparative material there

Figure 6. Plot of Pore Size against moles TMB in reaction mixture.

Reaction mixture composition estimated from [78]:

31SiO2–Al2O3–0.01Na2O–2.4TMA2O–10CTMA2O–750H2O + XTMB

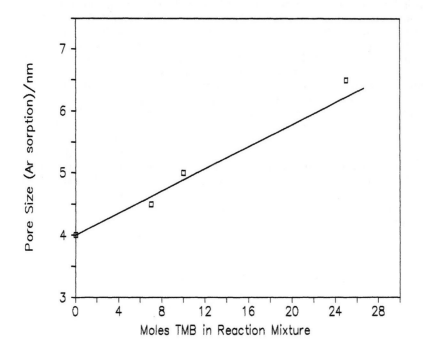

is no direct comparison with materials prepared by similar methods (except for the addition of the acid). Thus while the addition of acid may indeed have a beneficial effect there is insufficient data to allow any effect to be discerned.

Washing MCM-41 products contaminated with other crystalline phases with dilute sodium carbonate solution at modest temperatures (about 80°C) apparently results in the dissolution of the contaminant and the generation of a purer product. Treatment is carried out on an "as synthesised, washed product" [67]: it can be presumed that the surfactant stabilises the MCM-41 structure itself against dissolution. In the examples presented the "additional" phase is described as having a "broad line" close to the high intensity line at low angle for MCM-41. No other description of the impurity is presented nor is there any suggestion of the contaminant level. Interestingly, there is no difference in total surface area before and after treatment and little difference in the benzene capacity. This is a somewhat puzzling result and suggests that the impurity is present at low concentrations - not the impression given by the x-ray diffractograms presented [67].

With the exception of the effect of surfactant/silica ratio, which will be reviewed below, most of the key synthesis parameters have now been discussed. The exceptions are reaction temperature, time and agitation. Patents [58] suggest that crystallisation temperatures of between 50 and 250°C can be used with corresponding times being 14 days and 5 minutes and that reactions can be carried out with stirring or under static conditions. This last point would seem to be confirmed by patent examples where there are approximately equal numbers of preparations with and without agitation. Reaction temperature/time would not seem to be straightforward for these materials. The lack of examples of identical compositions carried out at different temperatures makes detailed comment difficult. The vast majority of syntheses are carried out between 95 and 150°C but reaction times do not seem to follow usual patterns: it is as common for a 95°C reaction to be carried out for 16 hours as a reaction at 150°C. To illustrate this, Table 4 has been constructed from examples 1 and 2 of one of the earlier patents [58].

Table 4
MCM-41 Syntheses illustrating different temperature times.
(Taken from reference 58. Both preparations were static.)

Example [58]	Composition						Temp °C	Time hours
	SiO_2	Al_2O_3	Na_2O	TMA_2Oa	$CTMA_2O$b	H_2O		
1	60	0.18	0.58	5.6	3.3	576	95	16
2	60	0.22	0.72	45.7	8.0	1371	150	16

a = tetramethylammonium. b = hexadecyltrimethylammonium.

Both preparations were reported to be carried out without agitation. The reaction mixtures have similar silica/alumina ratios and contain comparable levels of sodium. Major differences are in the water contents and most importantly levels of TMA and CTMA; although the concentration of surfactant in both examples is almost identical. One might imagine that example 2, with the higher (TMA) base content, would crystallise more rapidly yet both were reported to have been terminated after 16 hours. Even allowing for the fact that this was the shut-down time rather than a true measure of the crystallisation time itself this is an unusual result, repeated in numerous examples. It may be that the M41S materials are very stable to reaction over-run and that it crystallises very rapidly even at low temperatures. This would seem to be confirmed by a recent publication [92] in which Monnier et al, report that at 75°C the "hexagonal mesophase" was detected (from xrd) after 20 minutes and attained full crystallinity after 10 hours.

In the introduction to the Mobil work on M41S it was noted that both hexagonal and cubic materials had been prepared. All that has been described above is based on the hexagonal form and this reflects the balance in all publications from the Mobil group. The only real key to the differences in the preparative methods lies in a description [78] of the effect of the surfactant/silica ratio. At ratios less than 1, the hexagonal phase dominates while above 1 "... a cubic phase can be produced" [78]. Interestingly, at even higher surfactant/silica ratios other materials are prepared. These are reported as being being lamellar in nature with poor thermal stability. In addition there was evidence for the formation of the cubic octamer, a Q^3 species:

$$8 \ (C_{16}H_{33}(CH_3)3N^+) \ [Si_8O_{20}]^{8-}$$

Mechanistically, the M41S materials are reported to be formed by a Liquid Crystal Templating (LCT) mechanism [78]. Schematically, this has been represented in Figure 7 using information and diagrams from Beck et al [14,78]. The key feature in the Mobil proposal is the formation of liquid crystal micelles which are cylindrical in shape. It is these cylinders rather than the individual molecules which are the templates, hence the term: LCT. The Mobil group propose that either these "micellar templates" form and are then surrounded by silicate species or that it may be the addition of the silicate which results in the ordering of the "encased surfactant micelles". To this proposal one other feature can be added: the presence of highly condensed silicate species such as the cubic octamer. This may well be a result of the high TMAOH concentrations in syntheses since TMA is known to favour the formation of the silicate "cube" [79].

From the above it could be argued that the key feature of M41S syntheses is the addition of a surfactant to a silicate solution containing predominantly condensed silicate species. The work by Kuroda et al [55,56] in which they effectively delaminate Kanemite in the presence of a surfactant may well produce, in effect, a very similar situation: micellar cylinders and condensed silicate species. Interestingly, Monnier et al, noted that an intermediate phase in the formation of MCM-41 type materials was a layered material which could be isolated. This material could then be converted [92] by hydrothermal treatment to the "hexagonal mesostructure".

Only by carrying out detailed characterisation of M41S and the material produced by the group from Japan could any similarity be determined. Unfortunately, there is not much characterising information available in the published work on the Kanemite derived material.

In contrast a great deal of characterising information is available on M41S materials, information presented in both the open literature reports [14,78] and patents, with the majority of patents providing an identical overview of this material's characteristics. This review will summarise information from both sources.

Figure 7. Schematic Representation of M41S Formation.
(Mechanism according to Beck et al [14,78])

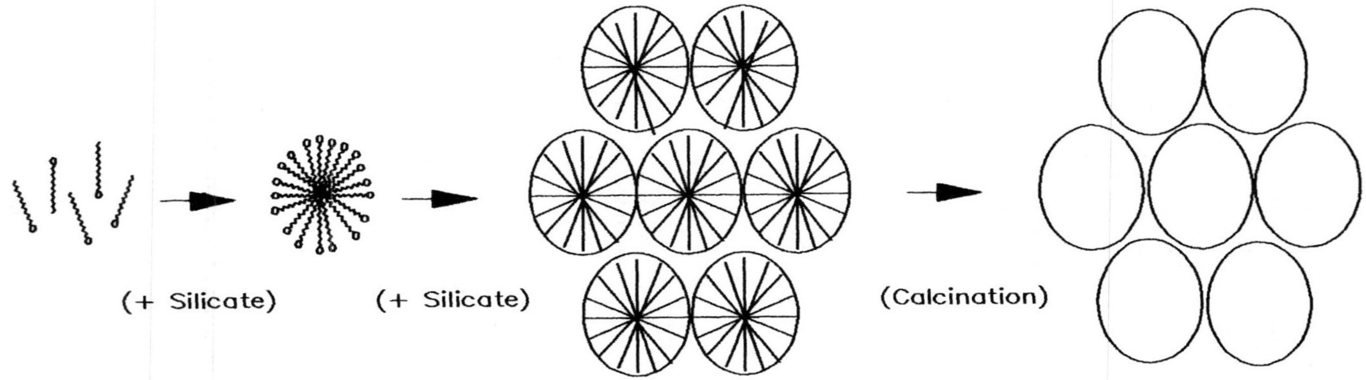

(+ Silicate)   (+ Silicate)   (Calcination)

In terms of its composition, in its as-synthesised form, M41S is described as:
$$rRM_{n/q} (W_a X_b Y_c Z_d O_h)$$

where: R = surfactant template
    M = alkali metal such as sodium
    W = divalent first row transition metal eg Mn, Co or Fe
    X = trivalent element eg Al, B, Ga or Fe
    Y = tetravalent element eg Si or Ge
    Z = pentavalent element eg P

In the patents published to date the majority of the M41S materials are either silicates or aluminosilicates although phosphorous containing materials are exemplified.

M41S materials are described as having a characteristic powder x-ray diffraction pattern, which contain " ..at least one peak". Both patents and papers show diffractograms with a single intense line at very low angle (usually less than two degrees two theta). Other reflections are virtually absent and certainly there is no significant intensity by ten degrees two-theta. The diffraction patterns of the MCM-41 materials can be indexed on a hexagonal unit cell but only hk0 reflections are observed [78]. It has been suggested [53] that the wall structure of these materials is amorphous and this is not inconsistent with data from the Mobil group [78] for a calculated pattern based on cylindrical shells.

The regularity of the pore structure is illustrated by lattice images which show [60,78] the honeycomb like structure. In addition to lattice images two techniques are used to give an estimate of channel dimensions: Ar sorption (Horvath-Kawazoe technique [19]) and unit cell dimensions from xrd. As Beck et al note [78] the difference between these two values provides an estimate of wall thickness. Table 5 has been constructed from information from the Mobil group [78] and shows unit cell dimensions, pore size and "wall thickness" for a series of siliceous MCM-41 materials prepared using surfactants of different chain lengths.

These figures for wall thickness have, obviously, to be treated with some caution, this is especially true for the values obtained for the larger pore size materials. However, up to pore sizes of about 2.2 nm the values for wall thickness are fairly constant at about 1.1 nm. The physical significance of this value may be gauged by consideration of work on zeolite mesh generation [18] and the natural material imogolite [80]. A formalism, the APS formalism, for describing zeolite structures and the "2-D" meshes from which the structures can be constructed, identifies certain key repeat distances associated with particular symmetry operations which result in certain "layer" configurations. One of these characteristic repeat distances is 1 nm. While this repeat is usually associated with a particular corrugated sheet it is interesting to note how similar it is to the values for wall thickness in Table 5.

The natural mineral, imogolite, has some similiarities with the M41S materials, its wall thickness is reported [42] to be about 0.9 nm a value sufficiently close to that shown in Table 5 to suggest that there may be similarities in structure.

Table 5
Wall Thickness of MCM-41 materials.
(Calculated from Unit Cell and Pore Size data [78])

Surfactant Chain Length $[RN(CH_3)_3]^+$ R	a_0 /nm from xrd [78]	Pore Size/nm Ar sorption [78]	Wall Thickness /nm (a)
8	3.1	1.8	1.3
9	3.2	2.1	1.1
10	3.3	2.2	1.1
12	3.3	2.2	1.1
14	3.8	3.0	0.8
16	4.0	3.7	0.3

(a) Wall Thickness = a_0 - Pore Size .

Information on the wall structure, itself, is available from Si-NMR studies [78] which suggest that it resembles amorphous silica, in addition, deconvolution suggested that up to 40% of the Si atoms (in as prepared materials) were associated with silanol groups.

In many ways the definitive characterising information on any molecular sieve is sorption. MCM-41 displays very high capacities for organic molecules such as benzene; the capacity increasing with pore diameter. Pore size information from argon physisorption is presented in numerous patents (for example) [61]. Both the mesoporous nature and narrow pore size distribution of a M41S material has been confirmed in a recent publication [81]. The sample, prepared at Mainz University had an average pore diameter of 3.2 nm with a range of 3.0-3.7 nm.

While M41S type materials, with uniform, mesoporous channels, may be expected to find application as sorbents there is little demonstration of utility in the published reports. Sorption separation is claimed over modified forms [76]. The modification can be used to generate hydrophobicity or reduce the pore size as discussed previously. Exemplification is restricted to the demonstration of the removal of a hydrocarbon from water and the suggestion that it can be used to separate adamantane and its bulky dimers and trimers.

Table 3 indicates the majority of Mobil patents, describing applications, are concerned with catalytic uses. Catalytic applications demonstrated are dominated by those associated with refining, in particular, olefin oligomerisation [69-74]. Most of these used the "acid" form of MCM-41 but two patents described modified materials: Cr-impregnated [69] and a functionalised material (to reduce pore size) [74]. A range of olefins are used from C3 to C10. The Cr-MCM-41 [69] is compared to a "Chromium on silica", the MCM-41 catalyst is shown to have superior activity - but only one catalyst of each type was tested (at two temperatures). The same basic Cr-MCM-41 is also described in a separate patent [74].

Aside from oligomerisation (and isomerisation) of olefins, there are descriptions of MCM-41 in cracking (Naphtha [75] and C5-C12 olefins [73]) and dealkylation reactions [75]. Indeed the dealkylation of tri-tertiarybutylbenzene (TBB) would seem to be standard test reaction of the Mobil group used, presumably, because of the large kinetic diameter of TBB.

The use of MCM-41 in a quite different reaction, NOx reduction , is also described [77]. This patent also illustrates one other feature of many of these application patents: of the 21 examples only one is concerned with the catalysis itself the remaining examples describe preparation and characterisation. In NOx reduction the catalyst, Ti,V-MCM-41 was compared with both a commercial SCR catalyst (Süd Chemie, AG) and a Ti,V-Silica. Data is presented which shows that the MCM-41 catalyst displays a higher NOx conversion than the silica based catalyst but is less active than the commercial material (V, W, Ti Oxides). However, no information is presented about the effect of time-on-line.

One other description of a catalytic use for an MCM-41 derived material is available [54]. This describes the preparation of a titanosilicate MCM-41 and its use in selective oxidation. The material was prepared by direct synthesis and characterising information presented included xrd and IR and UV spectra. Interestingly, the material had a diffraction pattern which was consistent with those reported by Mobil and was very similar to that reported by Kuroda et al [56] again suggesting that a true comparison of these phases may be worthwhile. Oxidation of hex-1-ene was reported by Corma et al using hydrogen peroxide and of norbornene using tertiarybutylhydroperoxide. Comparison with zeolites Beta and ZSM-5 show that the MCM-41 has superior activity in oxidising both larger molecules and using organic hydroperoxides.

4.4 Other Materials

One other specific class of materials will be mentioned here, porous, 3-D structure types based on frameworks containing both octahedral and tetrahedral atoms. Two examples of this type of material are the natural mineral cacoxenite and the novel, synthetic materials, ETS, from Engelhard Corporation.

The ETS materials are a (potential) family of structures based on Si, Ti and O [82-84]. Currently two members have been described, ETS-4 and ETS-10, both of which are microporous.

The mineral cacoxenite has attracted recent interest [85,86] since it has 1.4 nm unidimensional pores and exhibits sorptive properties similar to many molecular sieve structures. The framework is based on Fe,Al octahedra and Al,P tetrahedra.

Materials of this general type hold particular promise for the generation of mesoporous molecular sieves. As with molecular sieve zeotypes it should be possible to model and predict hypothetical framework types - a useful tool both as an aid to structure solution and potentially to allowing framework design.

5. THEORETICAL FRAMEWORKS

The development of physical models was (is) an integral part of structure solution. Theoretical zeolite structures almost certainly had their origins in this same model building, indeed Breck [15] describes the deliberate generation of theoretical frameworks in the early 1960s.

Much of the current work on theoretical framework generation continues to be based around structure solution. This is typified by the work of Smith and co-workers [87] who have as their objective the generation of all known topologies so that a "library" of powder patterns are available which can be matched to the patterns of new, as yet unknown phases. This approach has already been successful, as an example, it has been noted that the VFI framework actually corresponds to one of the Smith theoretical nets: 81(1) [88]. Other groups with a similar objective have used a different approach to framework generation [89,90].

Such approaches, however, provide much more than aids to structure solution; they provide insights into how such frameworks are constructed. An excellent example of this is the work by Meier [91] who noted that very open networks must contain large numbers of small ring constituents (3 and 4 rings). He also noted that Lovdarite, a beryllosilicate, contains 3-rings [3]. Thus Be-containing syntheses may well be a useful starting point in the generation of ultra large pore zeotypes.

While none of the current (ultra) large pore zeotypes contain Be this concept of attempting to link compositional (and therefore synthesis) features with desired structures is likely to have tremendous potential.

6. APPLICATIONS: CONVENTIONAL AND EMERGING

The traditional catalytic applications of zeolites have been in the refining and petrochemical sectors. Not surprisingly therefore the initial applications summarised above have been, largely, in these areas. To take full advantage of these new material types it is likely that two main types of process will be investigated: firstly, processes in which the mesoporous material functions purely as an ordered, regular support and secondly those which will involve large molecules, taking full advantage of the mesoporous nature of the pores and carrying out the equivalent of "shape selective" catalysis.

For future sorption/separation applications the situation is somewhat more straightforward: true molecular sieving should now be possible on a range of "biologically active" molecules offering unique opportunities for separation.

Perhaps the most exciting prospects, however, are those involving the new, emerging areas of electronic, optical or sensing devices. As this review should has shown a number of material types are now available covering a wide range of compositions and pore sizes. What now remains is for the required degree of control over their formation to be developed to produce crystals of the required size and purity.

ACKNOWLEDGEMENTS

The author gratefully acknowledges Coulter Electronic Ltd, UK who supplied the Horvath-Kawazoe plot and Dr J.S. Beck, Dr D. Jones and Dr N.E. Thompson for supply of some references and ICI C&P Ltd. for permission to publish.

REFERENCES

1. IUPAC Manual of Symbols and Terminology, Pure Appl. Chem., 31, 1978, 578.
2. A. Corma, Proc. VIIIth Int. Zeolite Conf., Eds. P.A. Jacobs and R.A. van Santen, Elsevier, Amsterdam, 1989, 49.
3. W.M. Meier and D.H. Olson, "Atlas of Zeolite Structure Types", 3rd Revised Edition, 1992, Butterworth-Heinemann.
4. D.E.W. Vaughan, "The Properties and Applications of Zeolites", 1980, 294, R.P Townsend Editor, The Chemical Society.
5. P.H. Schipper, F.G. Dwyer, P.T Sparell, S. Mizrahi and J.A. Herbst, ACS Symp. Ser. 375, 1988, 64, M.L. Occelli Editor.
6. M.E. Davis, C. Montes and J.M. Garces, ACS Symp. Ser., 398, 1989, 291.
7. G.A. Ozin, A. Kuperman and A. Stein, Angew. Chemie. (intl. edition), 28, 1989, 359.
8. G.D. Stucky and J.E. MacDougall, Science, 247, 1989, 669.
9. G.A. Ozin, Adv. Mater., 10, 1992, 612.
10. M.W. Anderson, J. Shi, D.A. Leigh, A.E. Moody, F.A. Wade, B. Hamilton and S.W. Carr, J. Chem. Soc., Chem. Comm., 1993, 533.
11. W.M. Meier, "Molecular Sieves", 1968, Society of Chemical Industry (London), 10.
12. L.V.C. Rees, Nature, 296, 1982, 491.
13. S.T. Wilson, B.M. Lok, C.A. Messina, T.R. Cannan and E.M. Flanigen, J. Am. Chem. Soc., 104, 1982, 1146.
14. C.T. Kresge, M.E. Leonowicz, W.J. Roth, J.C. Vartuli and J.S. Beck, Nature, 359, 1992, 710.
15. D.W. Breck, "Zeolite Molecular Sieves", 1974, J. Wiley and Sons, New York.
16. I.D. Harrison, H.F. Leach and D.A. Whan, Proc. VIth Int. Zeolite Conference, Butterworth, 1984, 479.
17. M.W. Deem and J.M. Newsam, Nature, 342, 1989, 260.
18. M.D. Shannon, Proc. IXth Int. Zeolite Conference, Eds. R. von Ballmoos et al, Butterworth-Heinemann, 1993, 389.
19. G. Horvath and K. Kawazoe, J. Chem. Eng. Jpn., 16, 1983, 470.
20. J.S. Beck, J.C. Vartuli, W.J. Roth, M.E. Leonowicz, C.T. Kresge, K.D. Schmitt, C. T-W. Chu, D.H. Olson, E.W. Sheppard, S.B. McCullen, J.B. Higgins and J.L. Schlenker, J. Am. Chem. Soc., 114, 1992, 10834.
21. Q. Huo, R. Xu, S. Li, Z. Ma, J.M. Thomas, R.H. Jones, A.M. Chippindale, J. Chem. Soc., Chem. Commun., 1992, 875.
22. R.H. Jones, J.M. Thomas, J. Chen, R. Xu, Q. Huo, S. Li, Z. Ma and A.M. Chippindale, J. Solid State Chem., 102, 1993, 5605.
23. Q. Huo, R. Xu, S. Li, Y. Xu, Z. Ma, Y. Yue and L. Li, Proc. IXth Int. Zeolite Conference, Eds. R. von Ballmoos et al, Butterworth-Heinemann, 1993, 279.
24. S.T. Wilson, B.M. Lok and E.M. Flanigen, U.S. Patent 4310440, 1982.
25. R.M. Dessau, J.L. Schlenker and J.B. Higgins, Zeolites, 10, 1990, 522.
26. J.W. Richardson, Jr. and E.T.C. Vogt, Zeolites, 12, 1992, 13.
27. M.E. Davis, C. Saldarriaga, C. Montes, J. Garces and C. Crowder, Nature, 331, 1988, 698.
28. M.E. Davis, C. Montes and J.M. Garces, ACS 398, Eds. M.L. Occelli and H.E. Robson, 1989, 291.

354

29. B. Duncan, R. Szostak, K. Sorby and K. Ulan, Catal. Lett., 7, 1990, 367.
30. M.E. Davis, C. Montes, P.E. Hathaway and J.M. Garces, Proc. VIIIth Int. Zeolite Conference, Eds. P.A. Jacobs and R.A. van Santen, Elsevier, 1989, 199.
31. M. Estermann, L.B. McCusker, Ch. Baerlocher, A. Merrouche and H. Kessler, Nature, 352, 1991, 320.
32. A. Merrouche, J. Patarin, H. Kessler, M. Soulard, L. Delmotte, J.L. Guth and J.F. Jolly, Zeolites, 12(3), 1992, 226.
33. J.L Guth, H. Kessler, P. Caullet, J. Hazm, A. Merrouche and J. Patarin, Proc. IXth Int. Zeolite Conference, Eds. R. von Ballmoos et al, Butterworth-Heinemann, 1993, 215.
34. F. D'Yvoire, Bull. Chim. Soc. Fr., 1961, 1762.
35. K.C. Franklin, Ph.D. Thesis, University of Edinburgh, 1993.
36. J.O. Perez, N.K. McGuire and A. Clearfield, Catal. Lett., 8, 1991, 145.
37. H-X. Li and M.E. Davis, J. Chem. Soc. Farady Trans., 89(6), 1993, 957.
38. B.M. Lok, T.R. Cannan and C.A. Messina, Zeolites, 3, 1983, 282.
39. M.L. Occelli, P.S. Iyer and J.V. Sanders, Proc. VIIIth Int. Zeolite Conf., Eds. P.A. Jacobs and R.A. van Santen, Elsevier, Amsterdam, 1989, 469.
40. G.W. Brindley and R.E. Sempels, Clay Minerals, 12, 1977, 229.
41. T.J. Pinnavaia and H. Kim, NATO ASI Series Vol. 352, Eds. E.G. Derouane et al., Kluwer Academic Publishers, 1992, 79.
42. T.J. Pinnavaia, T. Kwon and S.K. Yun, NATO ASI Series Vol. 352, Eds. E.G. Deruoane et al., Kluwer Academic Publishers, 1992, 91.
43. "Pillared Layer Structures", Ed. I.V. Mitchell, Elsevier, 1990.
44. R. Burch (Ed.), Catalysis Today, 2, 1988.
45. "Multifuctional Mesoporous Inorganic Solids", NATO ASI Series Vol. 400, Ed. C.A.C. Sequeira and M.J. Hudson, Kluwer Academic Publishers, 1993.
46. A. Clearfield, "Multifuctional Mesoporous Inorganic Solids", NATO ASI Series Vol. 400, Ed. C.A.C. Sequeira and M.J. Hudson, Kluwer Academic Publishers, 1993, 169.
47. T.J. Pinnavaia, "Preparation and Properties of Pillared and Delaminated Clay Catalysts", Ed. B.L. Shapiro, Texas Univ. Press, 1984.
48. A. Clearfield and R.M. Tindwa, Inorg. Nucl. Chem. Lett., 15, 1979, 251.
49. I.J. Johnson, T.A. Werpy and T.J. Pinnavaia, J. Amer. Chem. Soc., 110, 1988, 8545.
50. M.E. Davis, Nature, 364, 1993, 391.
51. Editorial comment, Science, 259, 1993, 891.
52. Catalytica Highlights, 19(3), 1993.
53. P. Behrens, Adv. Mater., 5(2), 1993, 127.
54. A. Corma, M.T. Navarro and J. Perez Pariente, J. Chem. Soc. Chem. Commun., 1994, 147.
55. T. Yanagisawa, T. Shimizu, K. Kuroda and C. Kato, Bull. Che. Soc. Japan, 63, 1990, 988.
56. S. Inagaki, Y. Fukushima, A. Okada, T. Kurauchi, K. Kuroda and C. Kato, Proc. IXth Int. Zeolite Conference, Eds. R. von Ballmoos et al, Butterworth-Heinemann, 1993, 305.
57. J.S. Beck, C.T-W. Chu, I.D. Johnson, C.T. Kresge, M.E. Leonowicz, W.J. Roth and J.C. Vartuli, WO 91/11390, 1991.

58. C.T Kresge, M.E. Leonowicz, W.J. Roth and J.C Vartuli, US Patent 5102643, 1992.
59. J.S. Beck, US Patent 5057296, 1991.
60. J.S. Beck, C.T-W. Chu, I.D. Johnson, C.T. Kresge, M.A. Leonowicz, W.J. Roth and J.C. Vartuli, US Patent 5108725, 1992.
61. C.T. Kresge, M.E. Leonowicz, W.J. Roth and J.C. Vartuli, US Patent 5098684, 1992.
62. D.C. Calabro, K.D. Schmitt and J.C. Vartuli, US Patent 5110572, 1992.
63. I.D. Johnson and J.P. McWilliams, US Patent 5112589, 1992.
64. T.J. Degnan, I.D. Johnson and K.M. Keville, US Patent 5156828, 1992.
65. J.S. Beck, D.C. Calabro, S.B. McCullen, B.P. Pelrine, K.D. Schmitt and J.C. Vartuli, US Patent 5145816, 1992.
66. S.B. McCullen and J.C. Vartuli, US Patent 5156829, 1992.
67. C.T-W. Chu and C.T. Kresge, US Patent 5104525, 1992.
68. J.S. Beck, W.S. Borghard, C.T. Kresge, M.E. Leonowicz, W.J. Roth and J.C. Vartuli, US Patent 5246689, 1993.
69. B.P. Pelrine, K.D. Schmitt and J.C. Vartuli, US Patent 5105051, 1992.
70. Q.N. Le, R.T. Thomson and G.H. Yokomizo, US Patent 5134241, 1992.
71. Q.N. Le, R.T. Thomson and G.H. Yokomizo, US Patent 5134242, 1992.
72. N.A. Bhore, Q.N. Le, and G.H. Yokomizo, US Patent 5134243, 1992.
73. Q.N. Le and R.T. Thomson, US Patent 5191144, 1993.
74. J.S. Beck, D.C. Calabro, S.B. McCullen, B.P. Pelrine, K.D. Schmitt and J.C. Vartuli, US Patent 5200058, 1993.
75. Q.N. Le and R.T. Thomson, US Patent, 5232580, 1993.
76. J.S. Beck, D.C. Calabro, S.B. McCullen, B.P. Pelrine, K.D. Schmitt and J.C. Vartuli, US Patent 5220101, 1993.
77. J.S. Beck, R.F. Socha, D.S. Shihabi and J.C. Vartuli, US Patent 5143707, 1992.
78. J.S. Beck, J.C. Vartuli, W.J. Roth, M.E. Leonowicz, C.T. Kresge, K.D. Schmitt, C.T-W. Chu, D.H. Olson, E.W. Sheppard, S.B. McCullen, J.B. Higgins and J.L Schlenker, J. Am. Chem. Soc., 114, 1992, 10834.
79. D. Hoebbel and W. Wieker, Z. Anorg. Allg., 384, 1971, 43.
80. P.D. Cradwick, V.C. Farmer, J.D. Russel, C.R. Mason, K. Wada, N. Yoshinaga, Nat. Phys. Sci., 240, 1972, 187.
81. P.J. Branton, P.G. Hall and K.S.W. Sing, J. Chem. Soc., Chem. Commun., 1993, 1257.
82. S.M. Kuznicki, US Patent 4853202, 1989.
83. S.M. Kuznicki, US Patent 4938939, 1990.
84. S.M. Kuznicki, K.A. Thrush, F.M. Allen, S.M. Levine, M.M. Hamil, D.T. Hayhurst and M. Mansour, "Molecular Sieves, Volume 1", Ed. M.L. Occelli and H.E. Robson, Van Norstrand Reinhold, 1992.
85. P.B. Moore and J. Shen, Nature, 306, 1983, 356.
86. R. Szostak, R. Kuvadia, J. Brown and T.L. Thomas, Proc. VIIIth Int. Zeolite Conf., Eds. P.A. Jacobs and R.A. van Santen, Elsevier, Amsterdam, 1989, 439.
87. J.V. Smith, Proc. VIIIth Int. Zeolite Conf., Eds. P.A. Jacobs and R.A. van Santen, Elsevier, Amsterdam, 1989, 29.
88. J.W. Richardson, Jr., J.V. Smith and J.J. Pluth, J. Phys. Chem., 93, 1989, 8212.

89. D.E. Akporiaye and G.D. Price, Zeolites, 9, 1989, 23.
90. I.G. Wood and G.D. Price, Zeolites, 12, 1992, 320.
91. W.M. Meier, Proc. VIIth Int. Zeolite Conf., 1986, Eds. Y. Murakami et al, Elsevier, 13.
92. A. Monnier, F. Schuth, Q. Huo, D. Kumar, D. Margolese, R.S. Maxwell, G.D. Stucky, M. Krishnamurty, P. Petroff, A. Firouzi, M. Janicke and B.F. Chmelka, Science, 261, 1993, 1299.

J.C. Jansen, M. Stöcker, H.G. Karge and J. Weitkamp (Eds.)
Advanced Zeolite Science and Applications
Studies in Surface Science and Catalysis, Vol. 85
© 1994 Elsevier Science B.V. All rights reserved.

Advances in the *in situ* ^{13}C MAS NMR characterization of zeolite catalyzed hydrocarbon reactions

I.I. Ivanova* and E.G. Derouane

Facultés Universitaires N. D.de la Paix, Laboratoire de Catalyse, 61, Rue de Bruxelles, B-5000 Namur, Belgium

The different techniques (flame sealed glass NMR cells, fitting MAS rotors or home-made probes; CAVERN apparatus designed for sealing and unsealing a MAS rotor on a vacuum line) for realization of *in situ* ^{13}C MAS NMR measurements are reviewed. The application of *in situ* MAS NMR to unravel mechanisms of zeolite catalysed reactions is illustrated by examples taken from our studies of alkylation, isomerization, cracking and aromatization reactions.

Benzene alkylation with propylene was studied over H-ZSM-11 with and without excess of benzene at 293 K. Propene 1-^{13}C and propene 2-^{13}C were used to follow the fate of propene carbon atoms during alkylation.

Cumene - n-propylbenzene isomerization was conducted under excess benzene over H-ZSM-11 at 473 K. To follow the fate of different carbon atoms during the reaction, cumenes labeled with ^{13}C-isotopes either on α- or on ß-positions of the alkyl chain or in the aromatic ring have been synthesized.

^{13}C MAS NMR has been performed in situ to investigate the early stages in the conversion of propane to aromatics on Ga-containing H-ZSM-5 catalysts. Propane 2-^{13}C was used as labelled reactant. The effect of the total and partial pressure of propane and the reaction temperature on the mechanism of propane conversion was studied. The proposed mechanism was tested by addition of coadsorbates (H_2 and C_6H_6).

n-Hexane 1-^{13}C conversion was followed in situ to investigate the mechanism of n-hexane isomerization, aromatization and cracking over Pt/KL.

The state, the mobility and the interaction of adsorbed species with surface active sites and the identification of reactions intermediates were inferred from in situ ^{13}C MAS NMR experiments. The testing of the reaction mechanisms was carried out by the addition of coadsorbates and the study of the reaction kinetics. Reaction mechanisms are proposed.

---

* On leave from Moscow State University, Russia

1. INTRODUCTION

Heterogeneous catalytic reactions are usually followed by analyzing the desorbed reactants and products. ^{13}C NMR spestroscopy enables one to study *in situ* catalytic transformations. In addition, it has the advantage of differentiating between different carbon atoms and of identifying the specific behavior of the adsorbed state.

High resolution ^{13}C NMR spectra of systems presenting a catalytic interest can be obtained in two different modes. The first one, which is the approach used over the last twenty years, consists in studies of reacting adsorbates, looking into the product formation and reaction kinetics under static conditions. A satisfactory resolution is achieved in this case by considering a statistical number of reactant monolayers such that the molecule will retain sufficient mobility in the adsorbed phase. The results obtained using this mode were reviewed and discussed several times [1-6].

The second, more recent mode is referred to as high resolution magic-angle spinning (MAS); the sample is spun with a relatively high frequency (3 - 10 kHz, or even higher) at an angle of 54° 44' 8.2" relative to the external magnetic field. This procedure averages out chemical shift anisotropies and dipolar interactions between nuclei and leads to sharp NMR lines from strongly adsorbed species in which molecular motion is reduced.

In situ ^{13}C MAS NMR has been demonstrated to be a superior technique for the investigation of mechanisms of zeolite catalysed reactions. During the last five years reactions such as methanol-to-gasoline conversion [7-15], olefin oligomerization [16-19], alcohol dehydration and oligomerization [19-24], alkylation of aromatics [25,26], alkanes isomerization [27], aromatization [27-30] and cracking [27-32], conversion of alkylaromatics [25-33] etc. were studied using this method.

The aim of this paper is to review different approaches for realization of *in situ* ^{13}C MAS NMR experiments and to illustrate the use of this techniques by examples of our recent results.

Controlled - atmosphere magic-angle spinning NMR experimental techniques

Studies of the catalytic reactions in the adsorbed phase require conditions of controlled atmosphere, which become especially challenging if the adsorbate is only weakly adsorbed or catalyst-adsorbate samples are air or moisture sensitive. The NMR cells must thus be carefully sealed. On the other hand, to achieve a sufficiently high rotation speed the NMR cells must be well balanced.

There are at least three general approaches suiting these requirements. The first approach makes use of flame sealed glass ampules containing a catalyst and the adsorbate, and fitting precisely into MAS rotors [34-37]. While Andrews-Beams rotors [34-36] were easily designed to hold sealed NMR tubes (Figure 1), this was not the case, until recently, with high-speed double-bearing MAS rotors which must be used when high angular stability is essential. Highly symmetrical NMR cells are required in the latter case, because any asymmetry makes it impossible to obtain high rotation speeds and can also lead to explosion of the sample ampule in the rotor due to centrifugal forces. A simple design of an NMR tube with constriction (Figure 2), in which a powdered sample can be contained and admitted to an adsorbate, allowed to achieve such highly symmetric sealing [37].

Figure 1. NMR cell designed to fit Andrew-Beams rotors.

This design was later on used for the commercial production of inserts for NMR rotors of different types, which were successfully applied during the last five years by several research groups for *in situ* MAS NMR measurements [7-9, 17, 21-30, 33].

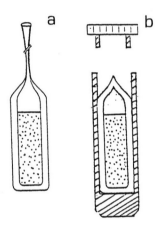

Figure 2. NMR tube insert incorporating sample (a). Sealed NMR cell fitting double-bearing MAS NMR rotor (b).

In the second approach glass ampules fit directly into home-made probes [36-40]. Sealed NMR cells are contained in a plastic disk playing the role of rotor in the probe (Figure 3). This disk is placed at some distance from the radiofrequency coil and consequently the magnetic nuclei of its material do not contribute appreciably to the NMR spectra of the sample under study. Highly symmetric sealing is not necessary in this case.

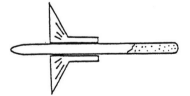

Figure 3. NMR cell fitting directly into home-made probes.

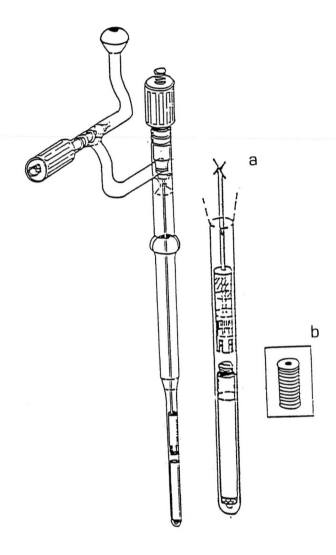

Figure 4. Second-generation CAVERN apparatus: left diagram shows the CAVERN apparatus prior to capping; right diagram shows an enlarged view of the apparatus lower section after capping (a) and of Kel-F cap used (b).

The third approach developed by the research group of Haw [14-16,41] makes use of rotor caps with 10 to 15 deformable ridges that create an air-tight seal when the cap is driven into the rotor [42]. This sealing method works well for temperatures up to 523 K, but glass ampules are generally required for higher temperatures, especially if higher pressures are generated. CAVERN apparatus were designed for sealing and unsealing a MAS rotor on a vacuum line. There are three generations of CAVERN apparatus already reported in literature [14, 16, 41]. They differ in design, mode of catalyst activation and reactant adsorption onto the catalyst. Figure 4 shows a drawing of the second-generation CAVERN apparatus, which is the most reliable and the easiest to fabricate.

While sealed NMR cells of the first and the second types are desirable when high temperature reactions are investigated or when long-term sample storage is required, the CAVERN design shows undoubtable advantages when the experiments must be carried out at criogenic temperature or when the adsorption of the additional reagents in course of the reaction is neaded.

2. The study of catalytic reactions

The application of *in situ* MAS NMR to unravel mechanisms of zeolite catalysed reactions will be illustrated by experiments excerpted from our studies of alkylation, isomerization, cracking and aromatization reactions over zeolite catalysts [25-30,33].

2.1. Experimental.

In situ ^{13}C MAS NMR measurements were carried out on CXP-200 or MSL-400 Bruker spectrometers operating at 50.3 and 100.6 MHz, respectively. Quantitative conditions were achieved using high-power gated proton-decoupling with suppressed NOE effect (90° pulse, recycling delay = 4 s). Spinning rate was 3 kHz. Some non-^{1}H-decoupled spectra were recorded to identify reaction products and intermediates. Cross-polarization experiments were performed to distinguish between species with different mobilities.

Powdered catalyst samples (0.07±0.03 g) were packed into NMR tubes (Wilmad, 5.6 mm o.d. with constrictions) fitting exactly into the double-bearing Bruker zirconia rotors. The catalysts were evacuated to a pressure of $6 \cdot 10^{-6}$ torr after heating for 8 h at 573 K and cooled to 298 K before adsorption. Liquid reactants were dosed gravimetrically, while gaseous reactants were dosed volumetrically. After introduction of the reactants, the NMR cells were maintained at 77 K to ensure quantitative adsorption and carefully sealed to achieve proper balance and high spinning rates in the MAS NMR probe.

In a typical *in situ* experiment, the sealed NMR cell is rapidly heated to a selected temperature and maintained at this temperature for various duration of time. The MAS NMR spectrum is recorded at 293 K after quenching of the sample cell. After collection of the NMR data, the NMR cell is returned to reaction conditions and heated for progressively longer periods of time.

Conversion of reactant r at time t was determined as:

$X_{r,t} = (1 - I_{r,t}/I_{r,0}) \cdot 100$ [%], where

$I_{r,t}$ is integral intensity of reactant r resonance in the NMR spectrum after heating for t min; $I_{r,0}$ is integral intensity of reactant r resonance in the initial NMR spectrum.

Yield of product p at time t was determined as:

$Y_{p,t} = I_{p,t}/I_{r,0} \cdot 100$ [%], where

$I_{p,t}$ is integral intensity of product p resonance in the NMR spectrum after heating for t min.

2.2. An *in situ* ^{13}C MAS NMR study of benzene isopropylation over H-ZSM-11

Figure 5 shows the *in situ* ^{13}C MAS NMR spectra observed immediately after adsorption of propene 2 - ^{13}C (a) or propene 1-^{13}C (b) with excess of benzene over H-ZSM-5 at 298 K. The weak resonance at 128 ppm in both spectra corresponds to aromatic carbons of benzene and cumene (natural abundance). NMR lines observed in the aliphatic regions were assigned to cumene α-^{13}C and cumene ß-^{13}C, respectively.

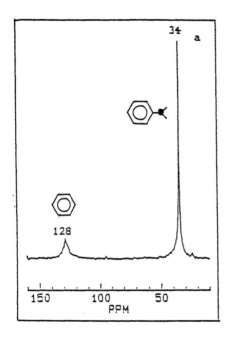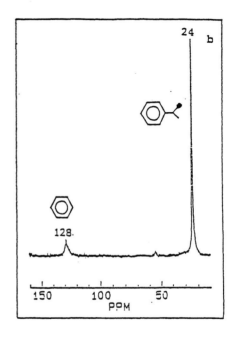

Figure 5. ^{13}C MAS NMR spectra observed immediately after adsorption of 1 molec./u.c. of propene 2 - ^{13}C (a) or propene 1-^{13}C (b) and 9 molec./u.c. of benzene over H-ZSM-11 at 298 K.

No resonance from reactant propene is observed, indicating rapid and complete conversion of propene in cumene at 298 K upon contact with benzene. This is consistent with recent reports by Haw et al. [16] showing that propene is very mobile and reactive on zeolite catalysts even at temperatures below 273 K.

The expanded aliphatic regions of the spectra are presented in Fig. 6. Interestingly, cumene α-^{13}C has only one signal, whereas the NMR line corresponding to cumene ß-^{13}C is split into two lines at 24.2 and 23.7 ppm, respectively. These lines have a different linewidth and approximately the same integrated intensities estimated from the decomposed spectrum B (Fig. 6b). Splitting of the resonance, corresponding to cumene ß-^{13}C is probably due to different mobility of the two methyl groups in the adsorbed cumene. The broader line at 23.7 ppm corresponds to a less mobile methyl group. Thus, one of the isopropyl methyl groups interacts more favourably with the H-ZSM-11 channel walls.

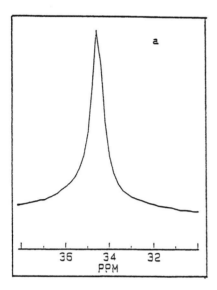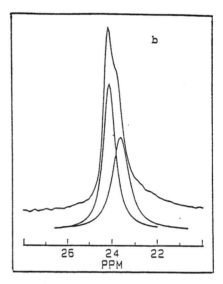

Figure 6. The expanded aliphatic regions of spectra presented in Figure 5.

It is noteworthy that no broad lines in the range of 14 - 40 ppm corresponding to oligomerization products were detected when excess benzene was applied. It provides an indication for suppression of the competitive oligomerization of propene in the presence of excess benzene. Benzene acts like a trap for the active species obtained from propene.

In contrast, when the molar ratio of the reacting propene 1-^{13}C and benzene was equal to 1 additional two broad resonances at ca. 30 and 14 ppm were detected (Figure 7). These NMR lines were assigned to propene oligomers. Interestingly, after heating the sample to 413 K a new broad resonance line at ca. 40 ppm appeared, suggesting isomerization and ^{13}C scrambling in oligomers. Further heating led to oligomer cracking and benzene alkylation with the fragments formed. As a result weak resonances corresponding to cumene α-^{13}C and sec-butylbenzene were observed (Fig. 7).

These experimental observations can be accounted for by the mechanism presented in Scheme 1.

According to classical concepts [43], applied later to zeolites [44], alkylation occurs via olefin protonation, formation of carbenium ion intermediate, and subsequent electrophilic attack on the aromatic π-system to form a benzenium cation which can re-aromatize by proton loss. The formation of a carbenium ion is believed to be the rate-limiting step.

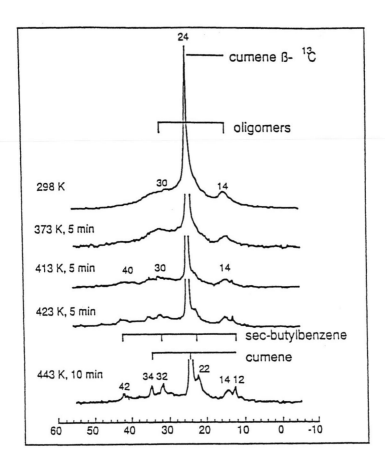

Figure 7. ^{13}C MAS NMR spectra observed after reaction of 9 molec./u.c. of propene 1 - ^{13}C and 9 molec./u.c. of benzene over H-ZSM-11 at progressively increasing temperatures.

Our results on tracing the fate of 1-^{13}C and 2-^{13}C carbon atoms of propene in the course of alkylation are fully in line with the classical carbenium ion reaction mechanism as shown in Scheme 1. However, the nature of the rate-limiting step could be questioned. Indeed, extremely rapid alkylation at 298 K in the adsorbed phase is in conflict with significantly higher temperatures ($\approx$ 473 K) required when the reaction is conducted in the liquid phase [45], suggesting that desorption and diffusion of product might be a rate-limiting step at temperatures below 473 K. Rapid catalyst deactivation at 473 K [45] confirms this.

The second pathway of propene conversion (Scheme 1) suggested by our *in situ* experiment (Fig. 7) explains high concentrations of sec-butylbenzene observed during benzene alkylation with propene in flow conditions at comperatively low temperatures [45, 46].

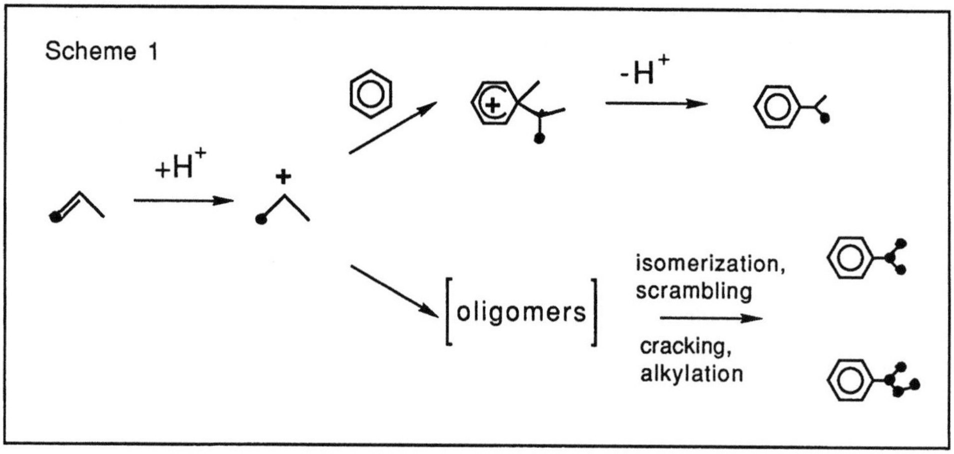

Scheme 1

(•) - indicates ^{13}C labelled carbon atom.

The classical alkylation mechanism is based on carbenium ion intermediates. However, recent investigations [16, 18, 46] question the existence of a free carbenium ion at least in the case of non-branched, light olefins. Alkoxy species formed from protonated alkenes and zeolite framework oxygen anions adjacent to aluminium sites were suggested to be important intermediates in oligomerization and bond shift reactions. A concerted mechanism including subsequent formation and decomposition of surface alkoxy groups was proposed [47]. Since the transition state of this mechanism has a cationic character it obeys all rules of the carbenium ion formalism.

2.3. On the mechanism of cumene - n-propylbenzene isomerization over H-ZSM-11

The mechanisms proposed for cumene (IPB) - n-propylbenzene (NPB) isomerization over acidic catalysts [29-34] suggest that both inter- and intramolecular pathways for different types of catalysts and reaction conditions may occur.

Neitzescu et al. [49, 50] reported an intramolecular mechanism for cumene - NPB isomerization over AlCl3 at 373 K. In their opinion, the isomerization of the side chain occurs via hydride abstraction resulting in the formation of a benzylic ion, which then rearranges into a primary cation, the latter undergoes rearrangement into n-propylbenzene secondary cation by methyl shift.

A similar mechanism was proposed by Douglas and Roberts [51] for the same type of catalyst and reaction conditions. In this proposal, a phenonium ion intermediate is postulated.

The latter mechanism was suggested to be preferred for NPB formation in course of cumene conversion at 635 - 775 K over different types of zeolite catalysts

[52, 53]. The remarkable tenfold increase in NPB selectivity over H-ZSM-5 compared to HY, LaY and amorphous Si/Al was reported by Fukase and Wojciechowski [53]. The authors explained this by the relative enhancement of monomolecular isomerization versus other bimolecular reaction pathways constrained by steric effects in the small pore diameter H-ZSM-5.

Monomolecular isomerization is, however, in conflict with the results of Beyer and Bordely [54] for cumene - NPB isomerization over H-ZSM-5 catalysts at comparatively low temperatures (500 - 600 K). No conversion was detected with pure cumene as a feed, while addition of benzene or toluene provided considerable amounts of n-propylbenzene and propyltoluene, respectively. Hence, the conclusion was that the side chain isomerization over H-ZSM-5 proceeds via intermolecular alkyl transfer.

The question whether isomerization occurs inter- or intramolecularily is best answered by *in situ* tracing either cumene or benzene labelled in the aromatic ring [33] as shown in Table 1. Indeed, starting for example with labelled cumene and unlabelled benzene, one should end up with labelled benzene if the reaction is intermolecular and with labelled NPB if the reaction is intramolecular. Analogously, when starting with labelled benzene and unlabelled cumene, labelled or unlabelled NPB will be observed depending on reaction mechanism. The ability of ^{13}C MAS NMR to distiguish between aromatic carbons of benzene, cumene, and NPB allows one to discriminate unambigously between these two mechanisms.

Table 1
Effect of the isomerization mechanism and of the initial ^{13}C label position in the aromatic rings of cumene and benzene, on the ^{13}C label distribution in resulting NPB and benzene.

(•) - indicates ^{13}C labelled carbon atom.

In the first experiment, cumene labelled in the ring as well as in the alkyl chain was used to follow the fate of the aromatic ring atoms and to follow simultaneously the cumene conversion to NPB and other products. Aromatic regions of the ^{13}C MAS NMR spectra are shown in Figure 8a. The spectrum of initial unheated sample contains 3 lines: at 149 ppm, corresponding to ^{13}C-1 in cumene, at 128.5 ppm, ascribed to ^{13}C-3,5 in cumene, and natural abundant ^{13}C in benzene and at 126.5

ppm - to ^{13}C-2,4,6 in cumene. After heating for 60 min, the conversion of cumene to NPB, estimated from the aliphatic part of the spectrum, was more than 60%. The concentration of the other by-products was comparatively low. The line at 149 ppm, corresponding to ^{13}C-1 in cumene disappeared and the line at 142.5 ppm corresponding to ^{13}C-1 in NPB was not observed.

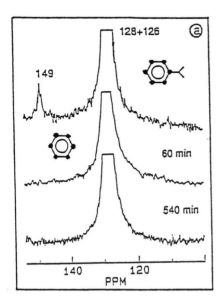 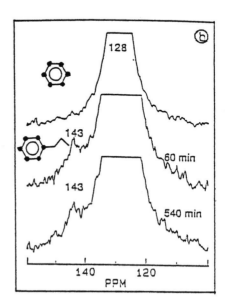

Figure 8. ^{13}C MAS NMR spectra of labelled cumene and unlabelled benzene (a) and labelled benzene and unlabelled cumene (b) before and after reaction over H-ZSM-11 at 473 K. Benzene/cumene=2

The complete disappearance of the ^{13}C-1 resonance of cumene and the absence of NPB labelled in the aromatic ring demonstrates the intermolecular character of the mechanism of the cumene - NPB isomerization .

Similarly, when using labelled benzene and unlabelled cumene (Fig. 8b) a benzene - NPB aromatic ring label transfer was evidenced by the appearance of the resonance at 143 ppm. This confirms unambiguously the intermolecular character of cumene isomerization.

Consequently, transalkylation mechanisms should be considered for the cumene - NPB isomerization in the presence of benzene.

Transalkylation of alkylaromatics can proceed by both monomolecular $S_{N}1$ and bimolecular $S_{N}2$ mechanisms [55, 56].

The $S_{N}1$ mechanism implies isopropyl group cleavage, followed by the formation of a 2-propenium ion. To form NPB by this mechanism, 2-propenium ions must further rearrange either to unstable 1-propenium cations or to the cycloproponium ions with intermediate stability [57, 58]. Both of them, in principle, may give NPB by alkylation of benzene as shown in Scheme 2 (1).

Scheme 2

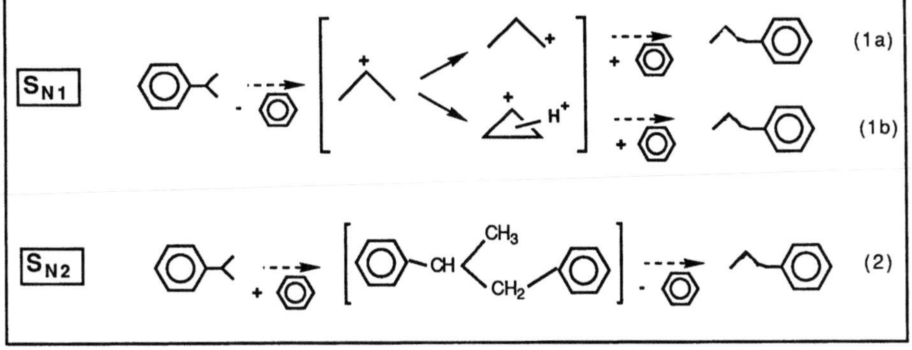

Formation of NPB by the S_{N2} mechanism will need a 1,2-diphenyl-1-methylethane intermediate (Scheme 2 (2)). 1,2-Diphenyl-1-methylethane formation also requires unstable primary cation as an intermediate. However, this route may be the lower energy process due to resonance stabilization of the carbocation intermediate.

^{13}C MAS NMR experiments with ^{13}C label in the alkyl chain may help to discriminate between the (1b) and two other mechanisms as shown in Table 2, since only (1b) mechanism leads to complete scrambling of ^{13}C in the alkyl chain. Mechanisms (1a) and (2) lead to similar label distributions in the alkyl chain.

Results obtained in experiments with cumene differently labelled in alkyl chain (Fig.9) suggest that the isomerization may proceed by any of the latter two mechanisms.

Table 2.
Effect of the isomerization mechanism and of the initial ^{13}C label position in the aliphatic chain of cumene on the ^{13}C label distribution in resulting NPB.

Reagent / Mechanism	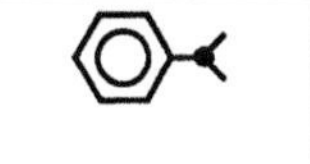	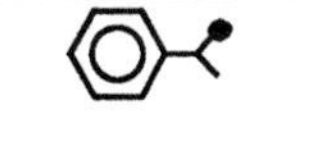
S_{N1}(1a)	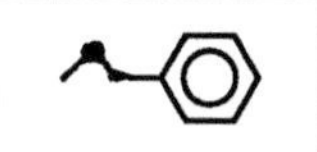	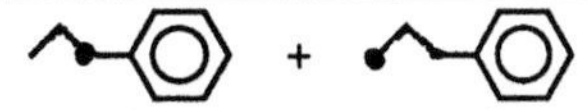 1 : 1
S_{N1}(1b)	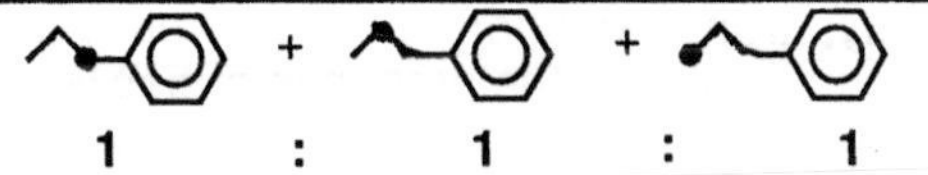 1 : 1 : 1	
S_{N2}(2)	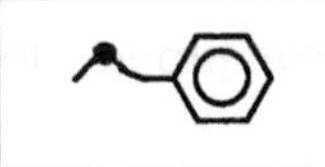	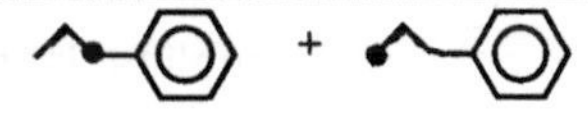 1 : 1

(•) indicates the position of ^{13}C label.

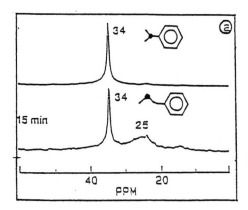 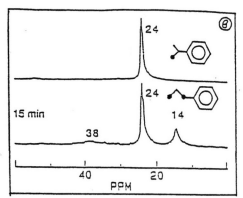

Figure 9. ^{13}C MAS NMR spectra observed before and after reaction of cumene α-13C (a) or cumene ß-13C (b) with benzene on H-ZSM-11 at 473 K. Benzene/cumene=8.

The lines corresponding to cumene are narrower compared to those of NPB. It indicates a faster motion of the isopropyl group and a stronger adsorption of the n-propyl group in comparison with the isopropyl chain. This might be due to a better accomodation of the straight alkyl chain of NPB along the wall of the zeolite channel. The narrowing of the lines corresponding to the alkyl chain of NPB from α- to γ-carbon atoms indicates an increasing sequential mobility in that order.

To discriminate further between bimolecular S_{N2} (2) and monomolecular S_{N1} (1b) mechanisms the effect of pressure on the reaction rate was studied. The pressure was varied by changing the volume of the NMR cell, the amounts of catalyst and reagents being the same.

The aliphatic regions of the ^{13}C MAS NMR spectra observed after 15 min reaction on samples with different partial pressures of cumene α-13C and benzene can be compared in Figure 10. Spectra are qualitatively similar. However, the rate of isomerization increases with the pressure of reagents. It indicates that the rate-limitting step of isomerization must be bimolecular. The S_{N1} (1b) mechanism was, therefore, excluded by default.

On the basis of the above observations an intermolecular S_{N2} pathway involving diphenylmethylethane as an intermediate could be proposed (Scheme 2).

Despite the fact that the supposed intermediates look rather bulky, its cross section is close to that of cumene, and therefore this mechanism could occur in the channels of H-ZSM-11.

To differentiate between reactions taking place on the external surface and inside the zeolite channels during alkylation over H-ZSM-11, the catalysts external surface passivation by silica deposition was achieved [59]. The observed NMR spectra were qualitatively similar to those obtained on non-modified H-ZSM-11. Thus, isomerization can occur inside the channels of H-ZSM-11. This result is in good agreement with a recent report [60] stating that the isomerization of paracymene into n-propyltoluene takes place inside the channels of MFI zeolites, contrarily to Fraenkel and Levy's conclusions [61].

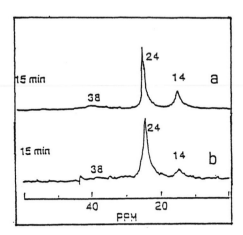

Figure 10. ^{13}C MAS NMR spectra observed after reaction of cumene α-13C and benzene on H-ZSM-11 at 473 K. Benzene/cumene=8. NMR cell volumes are of 0.24 cm^3 (a) and 0.59 cm^3(b).

2.4. Mechanistic studies of initial stages in propane aromatization over Ga-modified H-ZSM-5 catalysts

Dehydrocyclodimerization converts light parafins to aromatics, as was disclosed by Csicsery in the early seventies [62]. This reaction is known to occur on pure acid catalysts (e.g., H-MFI) or on bifunctional catalysts having a dehydrogenating and an acid function (e.g., Ga/H-MFI).

Most of the research efforts dealt with Ga-modified H-MFI catalysts as evidenced by two recent reviews [63, 64]. According to the overall conventional aromatization pathway propane is first dehydrogenated to propene which is then di- or oligomerized, resulting in a pool of olefins. These olefins are then further dehydrogenated to naphthenes, aromatics, and eventually to undesired coke. Discrepancies exist in the literature between the various detailed mechanistic proposals.

According to Ono et al. [64, 65] the initial step in the activation of propane involves acidic sites only and proceeds via the formation of carbonium ions as described by Haag and Dessau for hexane activation over H-ZSM-5 [66]. In contrast, Mériaudeau et al. favour a route involving dehydrogenation of propane on Ga species [67, 68] as also proposed by Gnep et al. [69] and reviewed recently [63].

On the basis of *in situ* ^{13}C NMR data we have recently demonstrated that the initial activation of propane on Ga/H-MFI is bifunctional [28].

The reaction of propane on Ga/H-MFI begins at 573 K. Typical spectra observed before and after 5 min reaction of propane 2-^{13}C are shown on Fig. 11. The initial spectrum shows the only resonance at ca. 17 ppm, corresponding to the initially labelled methylene group of propane. After 5 min reaction the line at ca. 16

ppm corresponding to a methyl group in propane appeared. We have confirmed this assignement by acquiring the spectrum without proton decoupling to find that the signal splits into a quadruplet [28]. No other significant signals were observed in the spectrum.

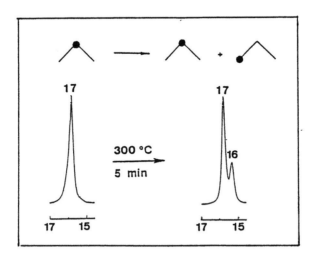

Figure 11. ^{13}C MAS NMR spectra observed before and after reaction of propane 2-^{13}C (4 molec./u.c.) over Ga/H-MFI catalyst at 573 K.

Scrambling of the 13C label in propane, in the absence of any large amount of other reaction products, can only occur via cyclic intermediate.

Three *in situ* NMR experiments on Ga/H-MFI, H-MFI and Ga/Silicalite were performed to investigate the complementary synergy between gallium species and acidic sites in the early stage of propane 2-^{13}C activation. Reactivity of these catalysts in propane activation is compared in Figure 12. Extremlely low activity of Ga/Silicalite compared to Ga/H-MFI proves that Brønsted acidity is necessary to activate propane at 573 K on Ga-containing H-MFI catalysts.

On the other hand faster ^{13}C scrambling observed on Ga/H-MFI than on H-MFI and the induction period (15 min) in propane activation over H-MFI (Figure 12) lead us to conclude that gallium species also play a vital role in propane activation. This conclusion is supported by our *in situ* NMR experiments in the presence of hydrogen [30]. As shown in Figure 13 hydrogen has an inhibition effect in the initial stages of propane activation. This effect might be due to concurrent adsorption of H_2 on the Ga_2O_3 species or even due to reduction of Ga_2O_3 to Ga_2O. The reduction of Ga_2O_3 in Ga_2O was already reported in literature[70-72].

As both Ga sites and acidic sites have a complementary function to each other during initial ^{13}C scrambling in propane, we propose a bifunctional reaction step (BREST) mechanism for propane activation (Scheme 3).

The active site responsible for the propane activation consists of an (Ga^{+3}, O^{-2}) ion pair closely associated with an acidic Brønsted site. The acidic site is provided by the bridge OH of the zeolite. The (Ga^{+3}, O^{-2}) ion pair involves either ion-exchanged gallium (Ga_{ionex}) and an oxygen anion from the zeolite lattice, or dispersed extra-framework Ga species (Ga_D) which may contain oxygen if they are (hydroxy)-cations or neutral species, or may also need zeolite oxygen anions otherwise.

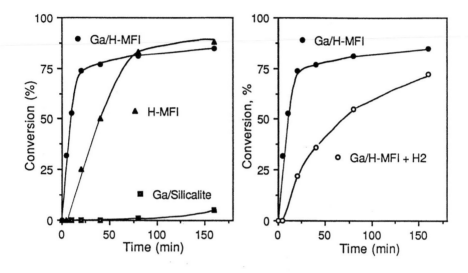

Figure 12. Variations of propane 2-^{13}C conversion as a function of reaction time on Ga/H-MFI, H-MFI and Ga/Silicalite at 573 K.

Figure 13. The effect of hydrogen on propane 2-^{13}C conversion over Ga/H-MFI at 573 K.

Scheme 3

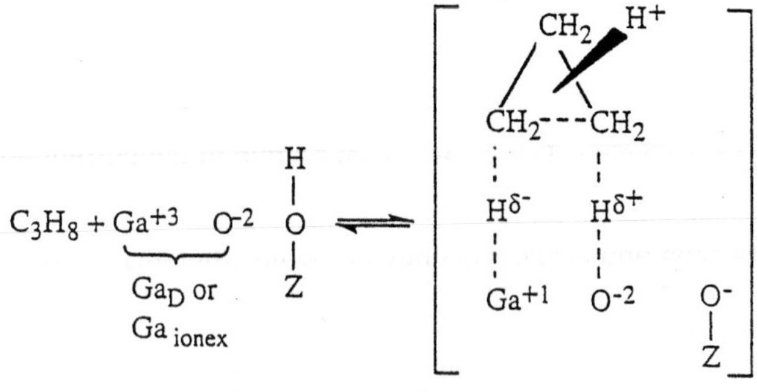

It is proposed that propane interacts heterolytically with the (Ga^{+3}, O^{-2}) ion pair via a positive and a negative charged hydrogen atoms, and that it is converted into a pseudo cyclopropane entity which can be protonated by a neighbouring Brønsted site.

The heterolytic activation/dissociation of propane as $(C_3H_7^+, H^-)$ on an oxidic Ga-component was already claimed previously [68, 73]. However, in these reports it was proposed that the activation of propane involves Ga sites only and the existence of a cyclic intermediate was not identified. The formation of the protonated pseudocyclopropane (PPCP) is the key step for the initial ^{13}C scrambling in propane.

Upon chemosorption of propane, Ga^{+3} in the active site is formally reduced to Ga (1) and Ga is present as dispersed Ga hydride (Scheme 3). This model provides an explanation for the recent Ga K-edge EXAFS data [72] claiming that active Ga species in Ga/H-MFI are highly dispersed, probably as Ga hydride coordinated to basic oxygens in the zeolite channels.

It was shown that PPCP intermediate can evolve in different ways depending on reaction temperature, propane coverage and total pressure in the NMR cell [28, 29].

A typical spectrum observed after 1 hour of propane $2\text{-}^{13}C$ reaction at 573 K is shown on Fig. 14. The main lines correspond to propane $2\text{-}^{13}C$ (C_{32}), propane $1\text{-}^{13}C$ (C_{31}), ethane ^{13}C (C_2), methane ^{13}C (C_1), n-butane $1\text{-}^{13}C$ (nC_{41}), n-butane $2\text{-}^{13}C$ (nC_{42}), i-butane $1\text{-}^{13}C$, and i-butane $2\text{-}^{13}C$. Two last resonances are not resolved and will be further considered together as labelled i-butane (iC_4). A more detailed description of these line assignements is given elsewhere [28].

^{13}C carbon balances indicate that in some experiments up to 20% of ^{13}C labels is not observed in the NMR spectra. The average H/C ratio estimated for non-observed products [29] was 2.3 after 3 hours of reaction. Thus, non-observed products are probably long chain oligomers or polymers. Their resonances can be broadened beyond the detection limit. The appearance of weak resonances at 28 and 45 ppm, corresponding to $(C_3)_n$ - polymers observed in some experiments supports this supposition.

It was suggested that PPCP intermediate can evolve in at least four different ways, as illustrated on Scheme 4.

(1)   It can reorganize (by rupture along 1) to yield a $C_3H_9^+$ carbocation bound to the zeolite framework. Deprotonation of proponium ion recovers the zeolite acid site and initial propane, explaining ^{13}C scrambling.

(2)   A $C_3H_7^+$ carbenium ion bound to the zeolite framework and dihydrogen are obtained if rupture occurs along 1 and 2. The $C_3H_7^+$ carbenium ion may further decompose into propene or oligomerize and in the event, restore the original zeolite Brønsted site.

(3)   A third route involves rupture along 1 and 3. Methane and a $C_2H_5^+$ carbenium ion bound to the zeolite framework are formed. $C_2H_5^+$ may restore the Brønsted acid site if it decomposes to ethene or oligomerizes. This mechanism pathway along with pathway (2) account for non-observed species, supposed to be oligomers. It also explains how methane, observed even at very low conversions, is a primary reaction product [29, 63].

(4)   Finally, PPCP intermediate can also decompose into ethane and CH_3^+ carbocation bound to the zeolite framework if rupture occurs along 1 and 4.

374

CH₃⁺ carbocations formed can react further with olefins and alkanes. For example, reaction of CH_3^+ (stabilized by basic zeolite oxygen) with propane (activated on the Ga site) may yield n-butane or i-butane. These products are indeed observed in the intial reaction stages.

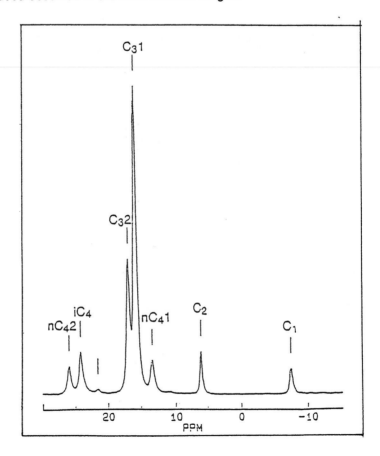

Figure 14. ^{13}C MAS NMR spectrum observed after 80 min of propane-2 ^{13}C reaction over Ga/H-MFI catalyst at 573 K.

The PPCP intermediate model agrees with the activation of propane by dissociative adsorption on Ga species [68, 69, 73], the recently disclosed role of gallium as hydrogen "porthole" [74], and the negative reaction order with respect to hydrogen when extra framework Ga species are present [69]. It also rationalises several earlier mechanistic proposals claiming either the propane activation by acidic sites only [64, 65] or the dehydrogenation of propane on Ga-species [63,67,68].

Scheme 4

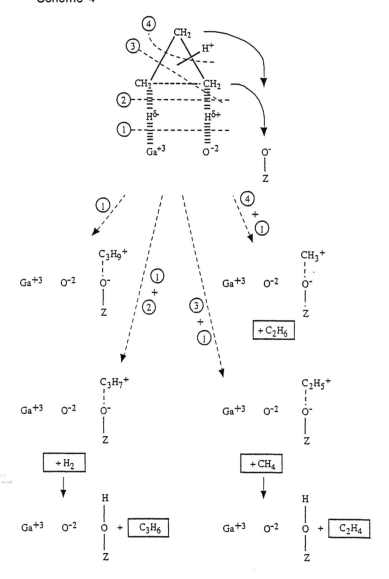

The effect of pressure on propane conversion can be evaluated from Figure 15 [29]. Different pressures of propane were achieved by varying the amounts of propane 2-^{13}C loaded in the NMR cell.

The chemical shifts and linewidths of the observed resonances depend strongly on propane coverages and are controlled by an exchange process between chemisorbed, physisorbed and gaseous species.

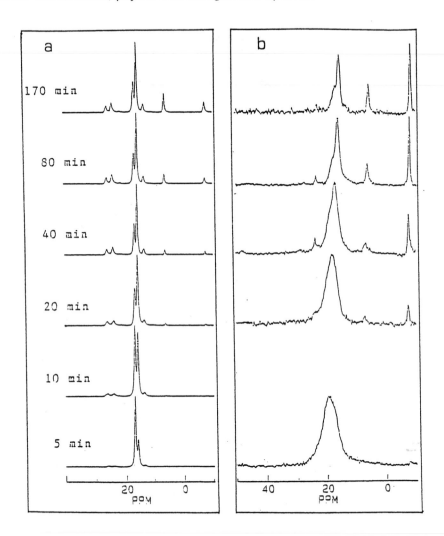

Figure 15. ^{13}C MAS NMR spectra observed for propane-2 ^{13}C reaction over Ga/H-MFI at 573 K as a function of reaction time: a) 4 molec/u.c. loaded; b) 0.5 molec/u.c. loaded.

According to the exchange model [4, 75], the measured resonance is given by a weighted average of the different contributions arising from the molecules adsorbed on sites of stronger and weaker energies. It depends in general on surface coverage and on the rate of molecular exchange between the different sites.

Low surface coverage (Fig. 15b) results in a higher population of strong sites, giving broad resonance shifted relative to the position of propane in solution. With increasing number of adsorbed molecules (Fig. 15a) the adsorption sites of weaker energies become populated and the resulting resonance is comparatively narrow and only slightly shifted.

High pressure of propane enhance bimolecular primary formation of butanes via BREST mechanism (Scheme 4 (4)). Butanes are formed by addition of a CH_3^+ carbenium ion (resulting from PPCP decomposition) to propane activated on the Ga site. This mechanism explains primary butane formation and the observation of both butane isomers at high pressures.

Low pressure shifts reaction equilibria towards fragmentation leading to dihydrogen, methane and olefins, the latter are immediately converted to oligomers and polymers. Routes (2) and (3) hence become significant pathways for PPCP decomposition. Secondary butane formation observed at low pressures cannot be accounted for by route (4).

To account for the low pressure butane formation we proposed [29] another mechanistic pathway. It involves the formation of polymeric hydrocarbon chain intermediates by classical cationic polymerization of primary $C_2H_5^+$ or $C_3H_7^+$ carbenium ions formed upon decomposition of the PPCP intermediate. These polymeric intermediates are known to give preferencially i-alkanes [76] by cracking, in our case i-butane. Ethane can be formed also by this way. This pathway explains the selective formation of i-butane at low pressure.

The model presented in Scheme 5 rationalizes further all the above observations. Propane first converts to PPCP via the BREST mechanism. It then decomposes in different ways depending on total pressure. At high pressure, ethane and CH_3^+ carbenium ion are formed. CH_3^+ reacts further with propane to yield butanes. At low pressure PPCP decomposes into methane, dihydrogen, and $C_2H_5^+$ and $C_3H_7^+$ carbenium fragments. These fragments then polymerize and the polymers eventually crack to yield i-butane and ethane as a secondary products.

Scheme 5

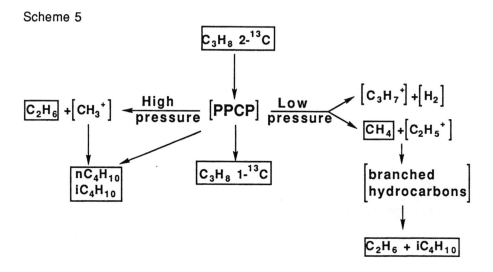

The propane 2-^{13}C reaction over Ga/H-MFI at higher temperatures is shown in Figure 16. Heating the sample up to 673 K results in the formation of small amounts of aromatic products (mainly, benzene, toluene and xylenes) evidenced by the set of weak resonances observed in the 120 - 140 ppm range of the spectra. The lines belonging to alkyl chains of alkylaromatics could also be identified in the range 10 - 30 ppm. Increasing of the temperature up to 723 -753 K results in the formation of some condensed aromatics, steming from increasing formation of the side bands.

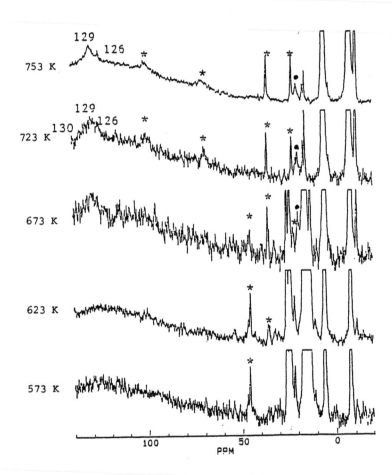

Figure 16. ^{13}C MAS NMR spectra observed for propane-2 ^{13}C reaction over Ga/H-MFI (4 molec/u.c.) as a function of progressively rising temperature. Each spectrum was recorded after 5 min heating. * denoted spinning side bands; • denotes resonances assigned to alkyl chain carbons in aromatics.

The mechanism presented in Scheme 5 enables us to suggest two alternative pathways for hydrocarbon chain grow leading to aromatic products: the first and the conventional one, operating at low pressures, is based on carbenium-ion oligomerization. The second and new one, operating at high pressures, is based on alkylation of alkanes with carbenium-like species yielding cabonium ions which decompose giving higher alkane and restoring acid sites.

2.5. In situ ^{13}C NMR study of the n-hexane conversion on Pt/KL
Recent reports [77-89] demonstrated a new generation of more active and

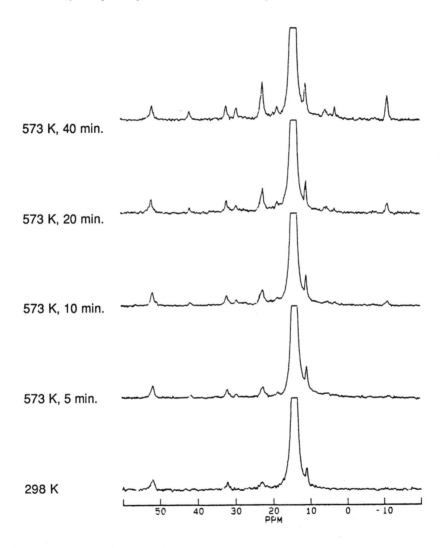

Figure 17. Aliphatic regions of the ^{13}C MAS NMR spectra observed after reaction of n-hexane 1-^{13}C over Pt/KL catalyst at 573 K as a function of reaction time.

selective reforming catalysts for light naphtha consisting of Pt supported on non-acidic materials such as KL zeolite (Pt/KL) [77-85] and aluminium-stabilized magnesium oxide (Pt/Mg(Al)O) [86-89].

Figures 17, 18 show ^{13}C MAS NMR spectra observed before and after reaction of n-hexane 1-^{13}C over Pt/KL catalyst at 573 K and 653 K, respectively.

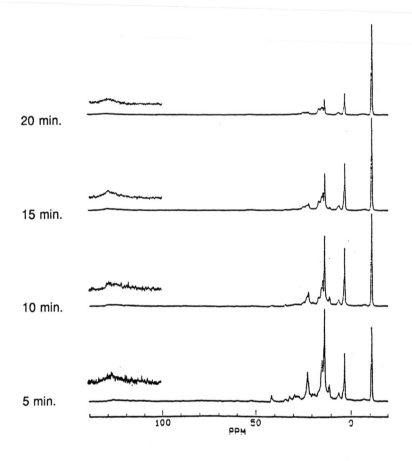

Figure 18. ^{13}C MAS NMR spectra observed after reaction of n-hexane 1-^{13}C over Pt/KL catalyst at 653 K as a function of reaction time.

The attribution of the resonance lines observed in the spectra was based on the comparison of their chemical shifts with solution [90] and gaseous phase [5] data, and the examining of multiplet patterns in the absence of proton decoupling. When a given signal could not be assigned unambiguously to a single molecular species using the above criteria, the direct adsorption of the corresponding compound in separate experiments was performed to confirm the assignement.

Table 7
Assignements of the resonances observed after conversion of n-hexane 1-^{13}C over Pt/KL and Pt/Mg(Al)O catalysts.

Assignement [a]	Experimental δ (ppm) [b]		Splitting [c]
	Pt / Mg(Al)O	Pt / KL	
Reactant :			
(13,7s)	13,8	14,1	4
(22,8s)	23,2	23,0	3
(31,9s)	32,3	32,1	3
non-identified	-	11,0	(4)
Isomerisation products :			
(22,4s)	22,2	22,4	4
(27,6s)	28,3	28,2	(2)
(41,6s)	41,8	41,7	3
(20,5s)	21,0	20,8	3
(14,0s)	13,8	14,1	4
(11,1s)	11,0	11,1	(4)
(29,1s)	29,5	29,8	(3)
(36,5s)	36,7	-	2
(18,4s)	18,4	18,5	(4)
(19,1s)	19,1	19,3	(4)
(33,9s)	34,1	34,2	(2)
Dehydrocyclization products :			
(21,4s)	20,3 *	-	(4)
(35,8s)	34,9 *	-	(3)
(27,6s)	27,4 *	27,2	3

Table 7 (continued)

Assignement [a]	Experimental δ (ppm) [b]		Splitting [c]
	Pt/Mg(Al)O	Pt/KL	
Aromatization products :			
⬡ (128,5$_s$)	130,3	110 - 135	-
Cracking products :			
∿ (13,7$_s$)	13,8 (∿)	14,1 (∿)	4
∿ (22,6$_s$)	22,7	22,6	-
∿ (34,6$_s$)	34,8 (⃝)	34,2 (><)	(3)
⋏ (21,9$_s$)	21,7	21,8	-
⋏ (29,7$_s$)	30,2	29,7	-
⋏ (31,7$_s$)	31,9	31,5	-
⋏ (11,4$_s$)	11,0 (⋏)	11,0 (⋏ imp)	(4)
∿ (13,2$_s$)	13,2	13,3	(4)
∿ (24,9$_s$)	25,3	24,8	(3)
⋏ (24,6$_s$)	24,3 (⋏)	24,1 (⋏)	-
⋏ (23,3$_s$)	23,9 (⋏)	23,4 (⋏)	-
⌒ (16,1$_s$; 15,4$_g$)	15,6	15,0	4
⌒ (16,3$_s$; 16,1$_g$)	16,2 *	16,5	4
CH_3-CH_3 (6,5$_s$; 3,2$_g$)	3,2 ; 5,1	3,2 ; 5,1	4
CH_4 (-2,3$_s$; -11$_g$)	-11,0 ; -9,7	-10,9 ; -5,5	5

a   Chemical shifts reported in literature are given in brackets: (s) - solution data
    [90], (g) - gaseous phase data [5].
b   Other assignements of the corresponding lines are given in brackets.
c   Splitting was not determined if the intensity of the corresponding line was low,
    or if the line was significantly broadened. Splittings which are not identified
    unambiguously because of multiple patterns overlapping are given in brackets.
•   indicates ^{13}C labelled carbon atoms;
*   indicates that the NMR line assignement was confirmed in independent
    reference experiment.

Table 7 lists the observed resonances, their splittings in non-[1]H-decoupled conditions, their assignements, and the corresponding chemical shifts (referred to TMS) in gaseous state and in solution.

The initial spectrum of n-hexane 1-^{13}C adsorbed on Pt/KL (Fig. 17) possesses an intensive resonance at 14.1 ppm which corresponds to the labeled methyl group in n-hexane and three weak resonances at ca. 11.0 ppm, 23.0 and 32.1 ppm attributed to an impurity presenting originaly in the reactant used and to natural-abundant methylene carbons, respectively (Table 1).

The reaction begins at 573 K after 5 min heating (Fig. 17). However, noticeable amounts of products appear only after 40 min heating at 573 K. The NMR lines corresponding to aromatic compounds appear at 653 K.

Comparison of the ^{13}C MAS NMR spectra and chemical shifts of the resonances observed over Pt/KL with those observed over Pt/Mg(Al)O catalyst [26, 27] (Table 7) pointed to the following differences:

* The resonance lines are in general broader on Pt/KL than those on Pt/Mg(Al)O, due to restricted mobility of the adsorbed species within the zeolite channels.

* The resonances corresponding to hexanes and pentanes methyl groups on Pt/KL are shifted down-field with respect to those observed on Pt/Mg(Al)O, due to their higher distortion upon the adsorption on these catalyst. This observation is in line with end-on or terminal adsorption to the Pt surface in unidimensional channels of zeolite L [80, 84].

* No methylcyclopentane was observed over Pt/KL catalyst, indicating preferential 1/6 ring closure. This conclusion supports the confinement model [81].

The comparison of observed products with those usually obtained in conventional flow conditions (Table 8) shows that the most important discrepancy is related to the absence of olefinic products. Neither hexenes, nor olefinic products of cracking and dehydrocyclization are formed to any appreciable extent under our experimental conditions, indicatinig that high pressures induced in the NMR cells at high temperatures shift the reaction equilibria [26, 29], favouring bimolecular hydrogenation with respect to dehydrogenation and lead to a different product distribution.

At 573 K the main reaction pathway is n-hexane isomerization. Appearance of the ^{13}C label in all the positions of n-hexane and methylpentanes suggest that isomerization proceeds via both possible mechanisms [91] as illustrated in Scheme 6. The determination of the ratios between differently labelled isomers allowed us to estimate the contribution of cyclic and bond-shift mechanisms. The estimated contribution of bond-shift mechanism on Pt/KL was higher than that determined on Pt/Mg(Al)O [27], indicating that the steric constraints within KL zeolite prevent the formation of a methylcyclopentane intermediate, as indeed observed in NMR experiments.

At 653 K cracking is the major reaction pathway (Fig. 18). In agreement with earlier reports [80, 84, 85] ethane and methane were preferentially observed, indicating favorization of terminal cracking in KL constrains.

Table 8
Comparison of n-hexane conversion products obtained in the experiments using *in situ* ^{13}C MAS NMR and continuous flow microreactor / GC techniques

Product group	NMR experiments; 653 K	Flow / GC experiments; 750 K
Isomerization	2-methylpentane 3-methylpentane 2,2-dimethylbutane	2-methylpentane 3-methylpentane
Hydrocracking and cracking	methane ethane propane butanes pentanes cyclopentane	methane ethane, ethylene propane, propenes butanes, butenes pentanes, pentenes cyclopentane, cyclopentene
Dehydrocyclization	methylcyclopentane cyclohexane	methylcyclopentane cyclohexene, cyclohexadiene
Dehydrogenation	-	linear and branched hexenes
Aromatization	benzene methyl and ethyl substituted benzene	benzene

It has been proposed in several contributions [87, 88, 92] that aromatization on Pt/KL catalyst occurs via partially dehydrogenated linear C_6 intermediates. Our experiments performed *in situ* show that another pathway leading to benzene formation operates parallely on Pt/Kl (Scheme 7). The formation of aromatics in NMR experiments is most probably due to 1/6 cyclization reported for nobel metal supported catalysts [91, 93]. This conclusion is based on the observation of significant concentrations of cyclic hydrocarbons at low and intermediate conversions of n-hexane (Table 8) and on the monotonous decrease of cyclic hydrocarbons yields with increasing benzene yeilds.

Scheme 6

Cyclic mechanism:

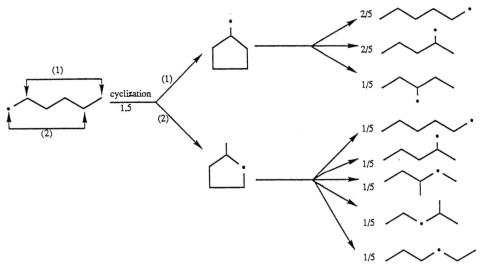

Bond-shift mechanism:

- Methyl shift :

- Ethyl shift :

- Propyl shift :

- Interconversion:

Scheme 7

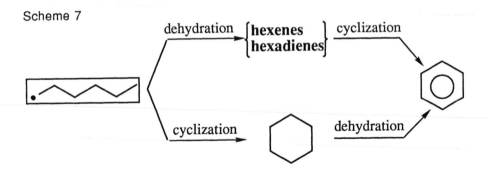

3 CONCLUSIONS

The potential of the *in situ* 13C MAS NMR technique for studying hydrocarbon reactions on various zeolites has been fully demonstrated:

* The state, mobility and interaction of adsorbed species with surface active sites and identification of surface active sites can be inferred from the usual chemical parameters, i.e. chemical shifts, linewidths and relaxation times.

* The direct observation and the indirect identification of reaction intermediates using specifically labeled compounds can be performed.

* As an analytical tool, 13C MAS NMR spectroscopy can also be used to determine concentrations of reactants and products as a function of time and hence can serve for the study of the reaction kinetics.

* The testing of the reaction mechanisms can be carried out by addition of coadsorbates (H_2, O_2, H_2O, C_6H_6 etc.)

* Finally, the direct observation of the shape selectivity, confinement effects in molecular-sieve catalysts and catalyst deactivation proved to be possible.

    As a result one can arrive at the proposition of a reaction mechanism.

    As an illustration the mechanistic studies of benzene isopropylation, cumene isomerization, propane aromatization and hexane conversion are presented.

(i) Benzene isopropylation proceeds in the adsorbed phase of H-ZSM-11 at temperatures lower than 298 K yielding cumene with 100% conversion and selectivity when the benzene/propene ratio is higher than 3. In contrast, when the molar ratio of the reacting propene and benzene is equal to 1 alkylation is accompanied by oligomerization. Oligomers further crack and benzene is alkylated with the fragments formed. The major product of this pathway is sec-butylbenzene. The results on tracing the fate of 1-^{13}C and 2-^{13}C carbon atoms of propene during alkylation are fully in line with the classical carbenium ion reaction mechanism. Desorption and/or diffusion of product is a rate-limiting step at temperatures below 473 K.

(ii) n-Propylbenzene is formed inside H-ZSM-11 channels by intermolecular S_{N2} transalkylation between cumene and benzene. The probable reaction intermediate is 1,2-phenyl-1-methylethane.

(iii) The activation of propane occurs on bifunctional catalytic sites of Ga/H-MFI catalyst consisting of a (Ga^{3+}, O^{2-}) ion pair and a Brønsted acidic site. The intermediates are pseudo-protonated cyclopropane species (PPCP) which can decompose in various ways depending on the propane pressure. High total pressure favours bimolecular reaction, leading to primary formation of butanes. Low pressure shifts the reaction equilibria towards fragmentation of PPCP and the primary formation of methane, dihydrogen and carbenium fragments which further polymerize and crack to give i-butane and ethane.

(iv) Isomerizaton of n-hexane over Pt/KL occurs via both cyclic and bond-shift mechanisms. Cyclohexane is a possible intermediate for n-hexane aromatization. Terminal cracking leading to ethane and methane occurs preferentially at high temperatures.

Acknowledgments

I.I. Ivanova thanks the Belgian Program on Interuniversity Attraction Poles (PAI), Haldor Topsøe A/S and Laboratoire de Catalyse (FUNDP) for a research postdoctoral position. The authors thank G. Daelen for technical assistance.

REFERENCES

1. H. Pfeifer, Physics Reports (Section C), 26 (1976) 293.
2. J. Tabony, Progr. Nucl. Magn. Reson. 14 (1980) 1.
3. E.G. Derouane and J. B.Nagy, ACS Symposium Series 348, In: Catalytic materials: Relationship between structure and reactivity, T.E. White et al. (eds.), San Francisko, 1983, p. 101
4. G. Engelhardt, and D. Michel, "High-Resolution Solid-State NMR of Silicates and Zeolites", John Wiley & Sons, Chichester · New York · Brisbane · Toronto · Singapore, 1987, p. 379..
5. J. B.Nagy, G. Engelhardt and D. Michel, Advances in Colloid and Interface Science 23 (1985) 67.
6. J.-Ph. Ansermet, C.P. Slichter and J.H. Sinfelt, Progr. in NMR Spectroscopy, 22 (1990) 401.
7. M.W. Anderson and J. Klinowski, Nature, 339 (1989) 200.
8. M.W. Anderson and J. Klinowski, J. Am. Chem. Soc., 112 (1990) 10.
9. M.W. Anderson and J. Klinowski, Chem. Phys. Lett., 172 (1991) 275.
10. E.J. Munson and J.F. Haw, J. Am. Chem. Soc., 113 (1991) 6303.
11. E.J. Munson, N.D. Lazo, M.E. Moellenhoff and J.F. Haw, J. Am. Chem. Soc., 113 (1991) 2783.
12. E.J. Munson, A.A. Kheir, N.D. Lazo and J.F. Haw, J. Phys. Chem., 96 (1992) 7740.
13. W. Kolodziejski and J. Klinowski, Appl. Catal., 81 (1992) 133.
14. E.J. Munson, D.B. Ferguson, A.A. Kheir, N.D. Lazo and J.F. Haw, J. Catal., 136 (1992) 504.
15. J.F. Haw, Spec. Publ. - R. Soc. Chem., 114 (1992) 1.
16. J.F. Haw, B.R. Richardson, I.S. Oshiro, N.D. Lazo and J.A. Speed, J. Am. Chem. Soc., 111 (1989) 2052.
17 K.P. Datema, A.K. Nowak, J. van Braam. Houckegeest and A.F.H. Wielers, Catalysis Letters, 11 (1991) 267.
18. B.R. Richardson, N.D. Lazo, P.D. Schettler, J.L. White and J.F. Haw, J. Am. Chem. Soc.,112 (1990) 2886.

19. N.D. Lazo, B.R. Richardson, P.D. Schettler, J.L. White, E.J. Munson and J.F. Haw, J. Phys. Chem.,95 (1991) 9420.
20. M.T. Aronson, R.J. Gorte, W. E. Farneth and D. White, J. Am. Chem. Soc., 111 (1989) 840.
21. A.G. Stepanov, K.I. Zamaraev and J.M. Thomas, Catal. Lett. 13 (1992) 407.
22. A.G. Stepanov, V.N. Romannikov and K.I. Zamaraev, Catal. Lett. 13 (1992) 395.
23. A.G. Stepanov and K.I. Zamaraev, Catal. Lett. 19 (1993) 153.
24. A.G. Stepanov, V.N. Zudin and K.I. Zamaraev, Solid state NMR, 2 (1993) 89.
25. I.I. Ivanova, D. Brunel, J. B.Nagy and E.G. Derouane, submited to J. Mol. Cat.
26. I.I. Ivanova, A. Pasau-Claerbout, M. Seirvert, N. Blom and E.G. Derouane, submitted to J. Catal.
27. I.I. Ivanova, M. Seirvert, A. Pasau-Claerbout, N. Blom and E.G. Derouane, submitted to J. Catal.
28. E.G. Derouane, S.B. Abdul Hamid, I.I. Ivanova, N. Blom and P.-E. Højlund-Nielsen, J. Mol. Cat., 86 (1994) 371.
29. I.I. Ivanova, N. Blom, S.B. Abdul Hamid and E.G. Derouane, Recueil des Travaux Chimiques des Pays-Bas, in press.
30. I.I. Ivanova, N. Blom and E.G. Derouane, to be published.
31. J.L. White, N.D. Lazo, B.R. Richardson and J.F. Haw, J. Catal, 125 (1990) 260.
32. F.G. Oliver, E.J. Munson and J.F. Haw, J. Phys. Chem., 96 (1992) 8106.
33. I.I. Ivanova, D. Brunel, G. Daelen, J. B.Nagy and E.G. Derouane, Stud. Surf. Sci. Catal., 78 (1993) 587.
34. R.E. Taylor, L.M. Ryan, P. Tindall and B.C. Gerstein, J. Chem. Phys., 73 (1980) 5500.
35. H. Pfeifer, D. Freude and M. Hunger, Zeolites 5 (1985) 274.
36. V.M. Mastikhin, I.L. Mudrakovsky and A.V. Nosov, Progr. Nucl. Magn. Reson. Spectrosc., 23 (1991) 259.
37. T.A. Carpenter, J. Klinowski, D.T.B. Tennakoon, C.J. Smith and D.C. Edwards, J. Magn. Reson., 68 (1986) 561.
38. I.D. Gay, J. Magn. Reson., 58 (1984) 413.
39. F. Rachdi, J. Reichenbach, L. Firlej, P. Bernier, M. Ribet, R. Aznar, G. Zimmer, M. Helmil and M. Mehring, Solid State Com., 87 (1993) 547.
40 W. Buckerman, L.-C. de Ménorval, F. Figueras and F. Fajula, to be published.
41. E.J. Munson, D.K. Murray and J.F. Haw, J. Catal. 1441 (1993) 733.
42. J.F. Haw and J.A. Speed, J. Magn. Reson. 78 (1988) 344.
43. G.A. Olah (ed.), Friedal-Crafts and Related Reactions, vol. 2, Willey, New York, 1964.
44. P.B. Venuto and P.S. Landis, Adv. Catal., 18 (1968) 259.
45. W.W. Kaeding and R.F. Holland, J. Catal., 109 (1988) 212.
46. I.I. Ivanova, I.A. Zen'kovich, T.A. Chemleva, K.V. Topchieva, N.F. Meged' and L.A. Lapkina, Neftekhimija, 24 (1984) 805.
47. V.B. Kazansky and I.N. Senchenya, Catal. Lett., 8 (1991) 317.
48. V.B. Kazansky, Sov. Chem. Rev., 1 (1988) 1109.
49. C.D. Neitzescu, Experimenta, 16 (1960) 332.
50. C.P. Neitzescu, I. Necsoiu, A. Geatz and M. Zolman, Ber., 92 (1959) 10.
51. J.E. Douglas and R.M. Roberts, Chem. Ind. (1959) 926.
52. D. Best and B.W. Wojciechowski, J. Catal., 47 (1977) 11.
53 S. Fukase and B.W. Wojciechowski, J. Catal., 109 (1988) 180.

54. H.K. Beyer and G. Borbely, New Developments in Zeolite Science and Technology, Proceedings of the 7th International Zeolite Conference, Y. Murakami, A. Iijima and J.W. Ward, Eds., Elsevier, New York, 1986, p. 867.
55. P.A. Jacobs, Carboniogenic Activity of Zeolites, Elsevier, New York, 1977.
56. M.L. Poutsma, ACS Monograph, 171 (1971) 431.
57. L. Random, J.A. Pople, V. Buss and P.V.R. Schleyer, J. Amer. Chem. Soc., 93 (1971) 1813.
58. J. Sommer and J. Bukala, Acc. Chem. Res., 26 (1993) 370.
59. O.O. Parenago, O.E. Lebedeva, I.I. Ivanova, N. Elizondo, L.E. Latisheva, S.A. Skornikova, V.V. Chenets and E.V. Lunina, Kinet. and Catal., 34 (1993) 162.
60. P.A. Parikh, N. Subrahmanyam, Y.S. Bhat and A. B. Halgeri, Appl. Catal., 90 (1992) 1.
61. D. Fraenkel and M. Levy, J. Catal., 118 (1989) 10.
62. S.M. Csicsery, J. Catal, 17 (1970) 207; ibid. 216; ibid. 315; ibid. 323.
63. M. Guisnet, N.S. Gnep and F. Alario, Appl. Catal. 89 (1992) 1.
64. Y. Ono, Catal. Rev. Sci. Eng., 34 (1992) 179.
65. M. Shibata, H. Kitagawa, Y. Sendoda, and Y. Ono, "New Developments in Zeolite Science and Technology", Y. Murakami, A. Iijima, J.W. Ward, eds.; Kodansha-Elsevier, Tokyo, 1986, p. 717.
66. W.O. Haag and R.M. Dessau, in Proc.8th Int. Congr. Catal., Berlin, Vol. 2, Dechema, Frankfurt; 1984; p. 305.
67. P. Mériaudeau and C. Naccache, J. Mol. Cat. 59 (1990) 431.
68. P. Mériaudeau, G. Sapaly and C. Naccache, J. Mol. Cat. 81 (1993) 293.
69. N.S. Gnep, J.Y. Doyemet, A.M. Seco, F. Ribeiro and M. Guisnet, Appl. Catal., 35 (1987) 93.
70. J.L. Price and V. Kanazirev, J. Mol. Cat., 66 (1991) 115.
71. J.L. Price and V. Kanazirev, J. Catal., 126 (1990) 267.
72. G.D. Meitzner, E. Iglesia, J.E. Baumgartner and E.S. Huang, J. Catal., 140 (1993) 12.
73. C.R. Buckles, G.H. Hutchings, and C.D. Williams, Catal. Letters, 11 (1991) 89.
74. E. Iglesia, J.E. Baumgartner, and G.L. Price, J. Catal., 134 (1992) 549.
75. A. Michael, W. Meiler, D. Michel, H. Pfeifer, D. Hoppach and J. Delmau, J. Chem. Soc., Faraday Trans. 1, 82 (1986) 3053.
76. M. Daage and F. Fajula, J. Catal. 81 (1983) 394.
77. J.R. Bernard, "Proc. 5th Int. Zeolite Conf." (L.V. Rees, Ed.), p.686. Heyden, London, 1980.
78. T.R. Hughes, W.C. Buss, P.W. Tamm and R.L. Jacobson, Stud. Surf. Sci. Catal., 28 (1986) 725.
79. C. Besoukhanova, J. Guidot, D. Barthomeuf, M. Breysse and J.R. Bernard, J. Chem. Soc. Faraday Trans. I, 77 (1981) 1595.
80. S.J. Tauster and J.J. Steger, Mater. Res. Soc. Proc., 111 (1988) 419.
81. E.G. Derouane and D. Vanderveken, Appl. Catal., 45 (1988) L15.
82. G. Larsen and G.L. Haller, Catal. Lett., 3 (1989) 103.
83. I. Manninger, X.L. Lu, P. Tetenyi and Z. Paal, Appl. Catal., 51 (1989) L7.
84. S.J. Tauster and J.J. Steger, J. Catal, 125 (1990) 387.
85. G.S. Lane, F.S. Modica and J.T. Miller, J. Catal., 129 (1991) 145.
86. R.J. Davis and E.G. Derouane, Nature, 349 (1991) 313.
87. R.J. Davis and E.G. Derouane, J. Catal., 132 (1991) 269.
88. E.G. Derouane, V. Jullien-Lardot, R.J. Davis, N. Blom and P.E. Højlund-Nielsen, Stud. Surf. Sci. Catal., 75 (1993) 1031.

89. E.G. Derouane, V. Jullien-Lardot, A. Pasau-Claerbout, N.J. Blom and P.E. Højlund-Nielsen, North American Catalysis Meeting (1993).

90. E. Breitmaier and W. Voelter, "Carbon-13 NMR Spectroscopy", V.C.H., Weinheim, 1987.

91. F.G. Gault, Mechanisms of skeletal isomerization of hydrocarbons on metals, Academic Press, New-York, 1981.

92. D.S. Lafyatis, G.F. Froment, A. Pasau-Claerbout and E. G. Derouane, J. Catal., in press.

93. B.C. Gates, J.R. Katzer and G.C.A. Schuit, Chemistry of Catalytic Processes, Mc Graw-Hill, New-York, 1979.

J.C. Jansen, M. Stöcker, H.G. Karge and J. Weitkamp (Eds.)
Advanced Zeolite Science and Applications
Studies in Surface Science and Catalysis, Vol. 85
© 1994 Elsevier Science B.V. All rights reserved.

Practical Aspects of Powder Diffraction Data Analysis

Christian Baerlocher and Lynne B. McCusker

Crystallography
ETH-Zentrum, CH-8092 Zurich, Switzerland

1. INTRODUCTION

Zeolites are notoriously difficult to synthesize in a form suitable for single crystal analysis. As a result, zeolites and powder diffraction analysis have been intimately associated with one another from the beginning of modern zeolite science. Because of their complexity, zeolites have proved to be a challenge for the powder method and have thereby helped to promote the progress of the method itself. Although powder diffraction instrumentation and methodology have made substantial progress during the last decades, many zeolite problems are still pushing the limits of the method. In fact, they have gained the reputation of being worst-case test examples for new developments in powder diffraction data analysis.

The principal limitation of powder data is the fact that the 3-dimensional diffraction data are projected onto one dimension with a corresponding loss of information. Single crystal refinements are usually overdetermined (ratio between observations and variables) by a factor of about 10. For zeolite refinements with powder data, this ratio often comes dangerously close to 1. In the former case, it is not disastrous if some of the observations are in error, or if the model used has some deficiencies. The overdetermination will help to pinpoint any shortcomings. With powder data, however, such errors can easily lead to seriously wrong results, and a person working with powder data must always be aware of this fact. The sections which follow have been written with this concern in mind.

It is not and cannot be the purpose of this paper to give a full account of all the problems encountered in powder diffraction. Rather, we hope to provide a brief overview of the powder structure analysis techniques pertinent to zeolite structures. Some practical aspects of the techniques which are not normally discussed in the literature, but which we consider to be important, are also covered. We have limited ourselves to two main topics: data collection and Rietveld analysis. In the first part, the various methods for powder data acquisition are briefly described, some guidelines given, and a few pitfalls highlighted. The Rietveld refinement section concentrates on several of the problems presented by a zeolite structure refinement and the strategies used to solve them. To illustrate the technique, an actual refinement is described step by step.

For the interested reader, there are a number of good texts available on structure analysis with powder data. A very recent book, *The Rietveld Method* [1], gives an excellent introduction to and a detailed treatment of various aspects of Rietveld refinement. The classic book, *X-ray Diffraction Procedures for Polycrystalline and Amorphous Materials*, by Klug and Alexander [2] is still a valuable source on the underlying principles of powder diffraction and for specific details regarding the technique of data acquisition. Finally, a volume of *Reviews in Mineralogy*, entitled *Modern Powder Diffraction*, edited by Bish and Post [3] covers some of the more recent developments in the field, and is also highly recommended.

This paper is intended to help both newcomers to the field and experienced zeolite scientists from other areas to read the structural literature with an appreciation of the difficulties inherent to powder diffraction (i.e. with appropriate caution). It is hoped that all readers will gain a better understanding of the concepts involved in a Rietveld refinement, and that they will perceive the power of powder diffraction data analysis, while recognizing its limits.

2. DATA COLLECTION

Powder data are, by their very nature, inferior to single crystal data. It is therefore of prime importance, that the data are collected in such a way, that the damage done by the fact that powder data have to be used is limited as far as possible. Within reason, it pays to spend time on optimizing the sample crystallization and preparation, and on collecting data carefully.

2.1. Film Cameras and Powder diffractometers

Powder diffraction patterns are frequently obtained with film cameras. The most common ones are the Debye-Scherrer powder camera, which is the simplest camera type, and the Guinier focusing camera, which yields high resolution data. Their main advantages are that they are simple to use, need very little sample, and are low cost (provided one already has a dark room facility). Film is a reasonably good storage medium, but it is normally difficult to duplicate film accurately for inclusion in a document or to analyze the data with a computer.

However, with the availability of relatively cheap high-resolution flatbed scanners, which can also be operated in transmission mode, new possibilities arise. One does not require an expensive film scanner or densitometer to digitize the powder pattern. Proffen and Hradil [4] have written a program, which analyzes film data measured with a simple scanning device connected to a PC. Nonetheless, the problem of the nonlinear response of the film at higher intensities remains. To obtain accurate intensities from a film, a gray scale calibration has to be done, but most people do not bother. The reason for this is simple. The diffractometer is a very practical alternative.

In a diffractometer, the film is replaced by an X-ray detector system, usually a proportional counter or a scintillation detector. Increasingly, so-called solid-state detectors, which have a better energy resolution, are being used. No filters are needed with such a detector, because it can discriminate the K_β from the K_α radiation. It generally reduces the background (by discriminating some of the incoherent scattering) and eliminates fluorescence radiation, without attenuating the radiation of primary interest – the K_α-radiation. The overall gain is therefore higher intensity and a better peak/background ratio, and this allows faster data collection. Even higher speeds in data acquisition are achieved with position-sensitive detectors (PSD's), and for this reason, they are used increasingly in X-ray diffraction. PSD's can record data over a range of 2θ values simultaneously (between 5 and 120 $°2\theta$ at a time, depending on the design of the detector). The angular resolution (apparent step size) varies, but can be as low as .01 $°2\theta$.

2.2. Diffractometer geometries

There are several different designs for diffractometers, all having their specific advantages, and, of course, disadvantages. Diffractometers can be operated in reflection or in transmission mode. The former has been the more popular one, but diffractometers with a transmission geometry are catching up fast, thanks to improved designs and the use of PSD's. The following is a brief discussion of the merits of the two principle operating modes.

2.2.1. Reflection mode

Most commercially available diffractometers employ the Bragg-Brentano parafocusing geometry depicted in Figure 1. It uses a flat plate sample holder, which lies tangent to the focusing circle. With this arrangement, the divergent beam from the anode or the monochromator is 'focused' into the receiving slit in front of the detector, and thus yields sharp, narrow peaks. By choosing a smaller slit, one can improve the resolution to some extent, but at the expense of intensity. The principle advantages of the reflection mode are:

i) a large sample area can be irradiated, and the high intensity thereby produced allows fast scan rates,

ii) no adsorption correction has to be made, if an 'infinitely' thick sample is used (see below).

On the negative side, one has to mention the fact that a larger sample volume is required, and that the preparation of a flat sample tends to increase preferred orientation (see section 2.6). Preferred orientation seriously affects the diffracted intensities, but has no influence on the diffraction angle, and therefore presents no problem for lattice constant determination. For a structure determination and/or a Rietveld analysis, however, accurate intensities are mandatory, and the data collection for such purposes requires special care. There are a few points (in addition to the preferred orientation problem) which have to be considered.

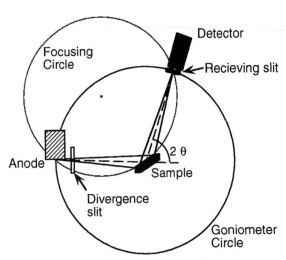

Figure 1. Bragg-Brentano parafocusing geometry

Fixed divergence slit setting: During a scan, the sample rotates with half the angular velocity of the detector arm, i.e. the angle between the primary beam and the sample surface is θ while that of the detector arm and the primary beam is 2θ. Because the angle between the sample surface and the beam changes, the illuminated area of the sample changes as well (see Figure 1). It is frequently overlooked, that the illuminated area can become larger than the sample at low angle. This results in incorrect (too low) intensities for the low angle peaks, since not all of the primary beam is utilized for the diffraction.

To correct for this error, smaller divergent slits have to be used for the low 2θ regions. By decreasing the slit size, one ensures, that the beam never goes outside the sample area, but at the same time, one produces a series of 2θ ranges which have been measured with different incident beam intensities. These ranges have to be scaled to one another, and the appropriate scale factors can be determined in several ways. Two possibilities are mentioned here.

First, one can choose the ranges so that they overlap, and so that the overlapping region contains at least one reasonably strong peak. This peak can then be used to calculate the scale factor between the adjacent ranges. Since it is sometimes difficult to find a suitable peak between the ranges, we normally use a second method. We have modified our data collection program so that it will periodically measure a standard peak. This peak has to be in the range of the largest divergent beam slit. The program will determine its integrated intensity by scanning over the peak and measuring the background on both sides of the peak. Thus, the same peak is measured with all the different slit settings and the program can automatically scale the different ranges to the same incident beam intensity.

As a side benefit, the standard peak is also a control on the experiment. Since the peak is usually measured several times in each range, any large deviation within a range will point to sample deterioration or some failure of the diffractometer (X-ray tube or detector). If a sample changer is available, the 'standard peak' can also be measured on a different sample (e.g. the silicon standard) by changing back and forth between the two samples. This is especially useful, when no suitable standard peak can be found in the pattern.

Variable divergence slit: To avoid the problem of having to change the slits manually and to calculate scale factors, most diffractometers are now equipped with a variable divergence slit. This slit is coupled to the movement of the diffractometer and ensures, that the maximum sample area is irradiated. Thus, the variable slit system maximizes the intensity one can obtain from a sample. This is especially useful at higher angles, where the diffraction peaks tend to be small.

Since the intensity of the primary beam changes constantly with this arrangement, the diffraction patterns can not be compared directly with measurements taken with a constant slit. They need to be corrected by multiplying the intensity of each step with sin θ. However, this correction is only approximately valid, and a variable slit should *not* be used if accurate intensity measurements are required.

Constant sample area vs. constant sample volume: The intensity scattered from a sample is proportional to the irradiated sample volume. The irradiated volume depends on the penetration depth of the X-ray beam and this is a function of the adsorption of the X-ray beam by the sample. Figure 2a depicts the situation at the sample. To simplify the arguments, the X-ray beam is assumed to be parallel. Since the divergence of the beam normally lies between 0.5° and 2° this approximation is not a serious one. Nonetheless, it can be shown, that the result is independent of the divergence of the beam.

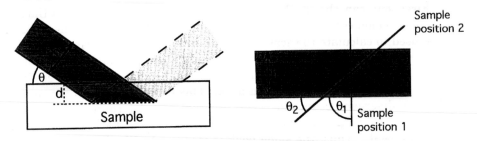

Figure 2. Effect of sample transparency and diffraction angle on the irradiated volume (see text).

Let us assume that the beam travels a distance t into the sample until it is attenuated below say 1%. The thickness d of the irradiated volume is then $t \cdot \sin \theta$. In a constant slit system, the irradiated sample surface changes with the glancing angle θ (Figure 2b). For $\theta = 90°$, the area irradiated is $h \cdot l$, where h is the height of the beam at the sample and l the length along the axis of the diffractometer. In this case d would be equal to t. For any other value of θ, the area is larger by a factor of $1/\sin \theta$ (see Figure 2b). Since the dimension l does not change, the two effects cancel out exactly, and the exposed sample volume remains constant for the constant slit arrangement. Of course, this is only true if the sample thickness is larger than $d = t \cdot \sin \theta_{max}$ (i.e. the sample is 'infinitely' thick). To be precise, the sample thickness must be only $d/2$, since the scattered beam has to travel the same distance out of the sample.

With a variable slit system, the height h changes continuously with $\sin \theta$ and thus the irradiated volume changes too. This can only be corrected exactly, if the intensity distribution of the primary beam along h is constant or has been exactly determined. For small divergence, the distribution may be constant, but for larger openings, this is certainly not the case and an error is introduced when the simple $\sin \theta$ correction mentioned above is applied.

A third possibility is sometimes recommended. This is a thin film of the sample on a low background sample holder (e.g. a quartz plate, which is cut at such an angle that no reflections are possible). Here, a constant area also means constant volume, but since the sample is not 'infinitely' thick the path length in the sample depends on the angle θ. Hence one has to correct for adsorption.

2.2.2. Transmission mode.

Diffractometers operated in transmission mode have appeared in the laboratories only in recent years. This is probably mainly due to the fact that reliable commercial PSD's have only just become available. To reduce the absorption effects, very little sample material can (or has to) be used, either in a thin capillary (0.5mm or smaller) or as a thin film. The resulting low intensity, and hence long counting time, can be compensated for by the use of a PSD. Normally, a focusing geometry is employed (Guinier-type diffractometer or some variation thereof) resulting in very good resolution of the powder patterns (see Figure 3). For capillaries, the resolution depends directly on the diameter of the capillary.

The principle advantages of the transmission mode are:

i)   a small amount of sample can be used,
ii) when working with capillaries, the environment can be controlled easily (dry atmosphere, sorption complex under its own vapor pressure, etc.),
iii) preferred orientation is reduced in capillaries, and
iv) very rapid measurements can be made, with the use of a curved position sensitive detector (e.g. in combination with a furnace)

The disadvantage is that samples with a high X-ray absorption factor will be difficult to measure, but zeolites generally contain only lighter atoms, and can be investigated easily using samples in capillaries.

2.3. Resolution.

The resolution of a powder pattern is an important issue in zeolite crystallography. Zeolites have comparatively large unit cells and although the symmetry is usually high, they still have a large number of partially overlapping reflections. The better resolved these reflections are, the higher is the information content of a pattern and, consequently, the more reliable are the structural data obtained in the analysis.

The full width at half maximum (FWHM) of the peaks in a pattern is given in $^{\circ}2\theta$, and depends on two factors: the instrument and the sample. Modern laboratory instruments can be tuned to give fairly high resolution as can be seen in Figure 3. In this figure, the instrumental function of the FWHM for a number of experimental setups is plotted. Often, laboratory instruments are built for high intensity, i.e. high scanning speeds, rather than for high resolution. For these instruments, the minimum FWHM can be as high as 0.15 or 0.20 $^{\circ}2\theta$, compared with the 0.05 $^{\circ}2\theta$ shown in Figure 3 for 'high resolution' laboratory instruments.

In this context, sample broadening has to be considered briefly here. The two main reasons for peak broadening by the sample are crystallite size and stress/strain. These effects can be distinguished in a powder diffraction pattern, and can give valuable information about the sample. For zeolites, the

398

size effect is probably the more common and therefore the more important (or annoying) one.

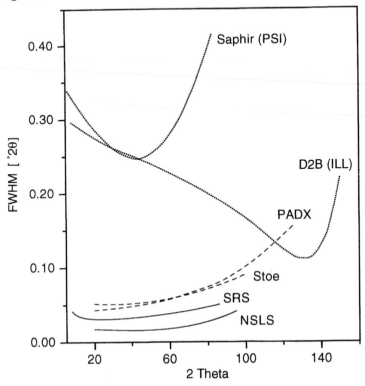

Figure 3. Full width at half maximum as a function of 2θ for several diffractometers. The solid lines are synchrotron, the dashed lines laboratory and the dotted lines neutron instruments. All are in transmission mode with the exception of PADX, which has a Bragg-Brentano geometry.

The simplest relationship between peak broadening and the size of a crystallite is given by the Scherrer equation

$$\beta_{cs} = \frac{K \lambda}{L \cos(\theta)} \tag{1}$$

where β_{cs} is the part of the peak width broadening (in radians) due to the crystal size, K is a constant (approx. 1.0), λ is the wavelength, and L is the length of the crystal perpendicular to the lattice plane responsible for the reflection at the Bragg angle θ. Strictly speaking, the total FWHM is a convolution of the instrumental and the sample broadening and β_{cs} must be obtained by a deconvolution process. However, for a quick estimation, β_{cs} can be calculated by subtracting the instrumental contribution from the total FWHM. Very often, zeolite crystals have only one short coherent dimension (e.g. resulting from stacking faults) and consequently have peak broadening only along the reciprocal axis parallel to this short dimension. Thus, broad and sharp peaks can occur in a powder pattern side by side, and this requires special software additions in a pattern fitting or Rietveld program.

2.4. 2θ calibration

In a carefully aligned diffractometer, the 2θ values can be calibrated with a standard sample to within .003 °2θ over the whole 2θ range. However, by mounting a new sample, the conditions are changed, and the 2θ values are usually off by slightly more than that. The main error comes from the sample height (which depends on the sample preparation) and the transparency of the sample. Since the latter changes the sample thickness d and thus the mean height of the 'reflection surface' in the sample, both effects result in a(n) (apparent) sample displacement away from the focusing circle. As will be discussed in chapter 3, good 2θ values are important in a profile fit, and for indexing a pattern, they are mandatory. For an accurate measurement, an internal standard (usually silicon) has to be used (i.e. mixed with the sample) to avoid the 2θ discrepancy caused by changing the sample.

In the case of zeolites, the 2θ calibration at low angles is critical. Since the first peak in the Si diffraction pattern occurs at 28.4 °2θ with CuKα radiation, another standard is needed for this region. Unfortunately, well characterized materials with only a few, but very well defined, reflections at low angle are difficult to find. In a zeolite laboratory, however, it should not be difficult to find a good zeolite A sample (first peak at 7.2 °2θ with CuKα radiation), which can be mixed with the Si and the sample under investigation. In this way, the lattice parameter of zeolite A can be freshly determined using the Si standard, and the zeolite A lines can then be used in turn to calibrate the low angle region.

2.5. Indexing

Indexing a powder pattern is of course fairly easy when the lattice parameters are already known. It is much trickier for a new, as yet unknown and never indexed, phase. Several computer programs, which use different algorithms for indexing powder patterns, have gained popularity over the years [5,6,7,8]. All these programs work better if accurate positions for the powder lines are determined, since whatever the algorithm used, they all have to compare these positions with one another.

It goes without saying, that an internal standard must be used when one is indexing a pattern. The low angle peaks are the most crucial ones, since they are used to start the indexing process. For zeolites, with their large unit cells, these peaks pose special problems (see last section).

An indexing procedure may fail for several reasons. For example,

i)   if the 2θ values are not calibrated well enough,

ii) if the first line or lines were omitted because they were thought to be poorly determined,

iii) if the limits in the program are set too low (To limit the search, a program defines values for a maximum cell volume and/or maximum lengths of cell edges. These limits are often too low for zeolites. If you are lucky, the program may find a smaller subcell – perhaps with unindexed lines.),

iv) if the problem is not suited to the strengths of the particular algorithm used (another program with a different algorithm may work), or

v)   there is an impurity present.

To evaluate the success of indexing a pattern, the figure of merit M(20) introduced by de Wolff [9] is normally calculated. With good laboratory data, M(20)'s of 30 to 40 can be obtained. However, using synchrotron data (see section 2.7.1), figures of merit of 200 and more can be obtained relatively easily. Of course, with values like this, one can be confident of the results.

2.6. Preferred orientation

It is well known, that in a sample consisting of platelets, the platelets will arrange themselves more or less parallel to the surface of the flat sample holder. The intensities of the reflections with hkl planes parallel to this surface will be much stronger (since there are many more planes available for reflection) than the intensities of those reflections from planes perpendicular to the sample surface.

This preferred orientation, or texture as it is sometimes called, is probably the biggest headache for a person interested in powder structure analysis. All too often it is ignored, in the hope that it may not be too bad. In many cases this hope is justified. It may be annoying that zeolites do not crystallize well, but it is also an advantage for powder diffraction, since zeolites tend to form multiple twins and/or to grow in larger agglomerates. This makes them or their agglomerates more spherical, which, of course, increases the chances of a random orientation of the individual crystallites.

Unfortunately, zeolite chemists have learned to improve their syntheses to make more perfect crystals. Especially the $AlPO_4$'s and SAPO's seem to crystallize more easily, and form larger crystals and fewer agglomerates. It is surprising how quickly a sample exhibits preferred orientation. Figure 4 shows a micrograph of an AlPO$_4$-34 (CHA) preparation [10]. It has nice rhombohedral crystals, which do not look like platelets at all, and one would not immediately suspect this sample to be prone to texture effects. Nonetheless, the sample has preferred orientation, and this can be seen easily by comparing the two powder patterns in Figure 5. Curve (a) was measured in reflection mode with a flat plate sample, whereas curve (b) was obtained in transmission mode using a capillary.

Figure 4. Scanning electron micrograph of AlPO$_4$-CHA

From this example two things can be learned. First, preferred orientation occurs more readily than one might expect, and second, preferred orientation can be spotted easily by measuring the sample in both reflection and the transmission modes. This is because the orientation of the primary beam differs by 90° between the two modes. If no transmission diffractometer is available, it can also be measured with a film camera (preferably a Guinier type camera).

What can be done about preferred

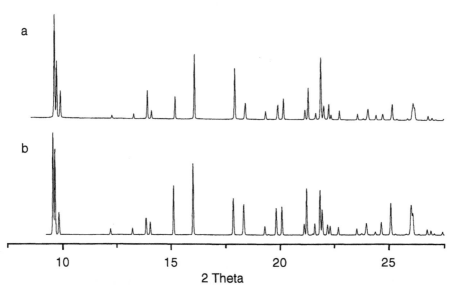

Figure 5. Powder diffraction data collected on AlPO$_4$-CHA (a) in reflection mode, and (b) in transmission mode.

orientation? The simplest measure usually taken is to rotate the sample about its axis. In the case of a flat sample this increases the number of crystallites which come into reflection position and therefore increases the random orientation, but it does nothing to improve the situation where platelets or needles lie flat on the sample holder.

Is there anything one can do in that situation? Apart from the possibility of asking the synthesis person to make a preparation with a little more synthesis gel left in the sample, there are a few other measures one can take. If there are indeed some larger lumps of aggregates of zeolite bound together by some unreacted gel in the sample, no attempt to grind the sample and thus free the individual crystals should be made. Loading a flat sample holder from the back or from the side is often recommended, since this minimizes the preferred orientation caused by the pressure exerted while preparing the smooth surface of a sample. In our opinion, the best alternative is to use a capillary. The sample is not pressed into the capillary, and while it rotates in the beam, the crystallites can tumble around and reorient themselves to some degree. With synchrotron radiation, where preferred orientation effects are even more of a problem (see below), capillary measurements are now the rule for intensity measurements.

The third possibility is to correct for the preferred orientation in the profile refinement itself. Although it is always better to avoid the problem experimentally, preferred orientation can be calculated and refined as long as the effect is not too pronounced. So far only simple models, which use just one parameter to correct for what in reality is a fairly complex effect, have been widely incorporated into refinement programs. Better models are needed, and have in fact been developed, for instance by Ahtee, Nurmela & Suortti [11] and Popa [12].

2.6. Step size and counting time

Customarily, powder data are collected in stepscan mode with a step size of ca. 0.02 $°2\theta$ for a given time per step. The time spent per step is usually chosen to give 'reasonable' counting statistics. Just what 'reasonable' means is, to some extent, a matter of debate. The counting statistics of a pattern can be expressed in the so-called expected R-value,

$$R_{exp} = \left[(N - P) / \sum_{i}^{N} w_i y_i (obs)^2 \right]^{1/2} \tag{2}$$

where N is the number of steps measured, P the number of refinable parameters, w_i the statistical weight ($w_i = 1/y_i(obs)$) and $y_i(obs)$ the observed count rate at step i. Normally an R_{exp} of about 10% is considered to be sufficient (see section 3.12).

In a whole pattern refinement, the intensity measured at step i is considered to be an observation, so decreasing the step size increases the number of observations. However, the structural information is contained only in the integrated intensities of the Bragg peaks, and increasing the number of steps measured across a peak, will add nothing to the accuracy of the structural parameters determined from such data. In fact it will only falsify the estimated standard deviations as they are calculated by most refinement programs and give an incorrect impression of high accuracy (see section 3.13).

But what is the correct step size and counting time? Is there a way to evaluate this in advance? Hill and Madsen [13] have made a very valuable study of the effect of the step width and counting time per step on the results of a Rietveld structure refinement. Their main arguments and conclusions are summarized in section 3.13. In short, they use a statistic test at the end of the refinement to decide what the correct step size should have been. From a statistics point of view, it turns out that one needs to collect the data with a step size corresponding to approximately the minimum FWHM in the pattern (however, see the discussion in section 3.13). We recommend using a step size that produces approximately 5 points in the upper half of a narrow peak in the pattern. The counting time should be such, that not more than a few thousand counts are accumulated as a maximum step intensity.

2.7. Radiation Sources

2.7.1. Synchrotron radiation

The most convenient radiation source is the sealed X-ray tube in the home laboratory. It is easily accessible, it is relatively cheap and it can give very good results. The alternative, of course, is one of the dedicated *synchrotron radiation sources*, which now offer reliable, high quality X-ray beams on an almost routine basis. Their main advantages are normally listed as

i)   high intensity,
ii) high collimation (small divergence of the beam), and
iii) tunable wavelength.

What do these properties mean for the powder method? Can they be utilized, or is synchrotron radiation just a fanciful (and expensive) new toy? Certainly, in some cases, it is used without justification. This is partly because researchers are still learning what can be done with this new source.

However, the powder method can indeed benefit from all three advantages listed above.

Let us start with the high intensity. Contrary to what one might expect, this does not necessarily mean fast scan rates. It could be exploited in this way, but more often, the higher intensity is sacrificed for high resolution, and the actual data collection time is then comparable to that needed for a laboratory instrument. However, some preliminary experiments have demonstrated that data collection times in the range of ms or below can be achieved. The high intensity allows the use of an analyzer crystal in place of a simple receiving slit [14]. This analyzer crystal acts as a slit with a very narrow angular acceptance, accepting only the reflected parallel beam. A similar effect can be achieved by using a high-resolution parallel-slit system (i.e. a long Soller slit) [15] in front of the detector.

The high collimation, i.e. the parallel nature of the beam, is the key property for the high resolution obtained with synchrotron radiation (see Figures 3 and 6). A parallel beam also means that no focusing geometry is necessary, and therefore an experiment cannot be 'out of focus'. Thus the problems discussed in section 2.4 with sample height displacement and the aberrations caused by the fact that a flat rather than a curved sample is used in the Bragg-Brentano geometry simply do not arise! If an analyzer crystal or a long Soller slit system is used, no additional 2θ-corrections have to be made, and this is a real hit when it comes to indexing a powder pattern or refining

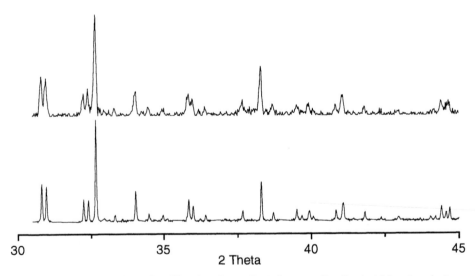

Figure 6. Comparison of powder diffraction data collected on a well-calibrated laboratory instrument (top) and at a synchrotron facility (bottom).

lattice constants. It also means that the size of a capillary has no effect on the resolution, its dimension is only limited by the X-ray absorption in the sample. High resolution is of great benefit in a Rietveld refinement, since it increases the number of observable peaks in a complex pattern. This is obvious in the comparison of patterns offered in Figure 6. On the other hand, the high resolution is of no use for most cubic structures or for phases with small unit cells, since the peaks are usually already well separated with the resolution obtained with modern laboratory instruments. Similarly, for phases which exhibit strong line broadening, the trip to a synchrotron facility may be a waste of time and money. Even a high resolution instrument cannot reduce the line broadening caused by the sample itself.

The free choice (tunability) of the wavelength allows one to select a wavelength close to an adsorption edge of an element. At the adsorption edge, the scattering power of an atom changes by a few electrons, and by comparing a near-edge with an off-edge measurement, one can highlight the contributions from that element. With this technique, Pickering et. al. [16] could show that Fe in synthetic zeolite L is indeed located in the tetrahedral sites of the framework and that it is distributed equally in the two crystallographic T-positions. However, the method can only be used for elements heavier than Cr, since the lighter elements require too short a wavelength, and there must be a sufficient amount of the anomalous scatterer in the material.

2.7.2. Neutron radiation

The principle advantages of neutrons are probably well known. The scattering power or scattering cross section does not vary that much from element to element (or more precisely, from isotope to isotope), although it can also be negative. This makes it possible to determine the positions of light elements and heavy ones in the same structure with equal precision. This fact can be exploited in zeolite structure analysis to determine hydrogen positions or to locate sorbed organic species. The other important property of neutrons, their magnetic moment, which allows magnetic structure to be determined, may prove to be important in the future, when or if it becomes possible to make zeolites with magnetic properties.

There are two types of sources for neutron radiation: a nuclear reactor (used, for instance, at the Institute Laue-Langevin (ILL) in Grenoble) or a spallation source (e.g. ISIS at the Rutherford and Appleton Laboratory in Didcot, UK). In the latter, the neutrons are obtained by bombarding a heavy metal target with a proton beam. In both cases, thermal neutrons are used and the wavelength is proportional to the velocity of the neutron. Since the spallation source has a natural time structure, the time-of-flight technique

(TOF) can be used to great advantage. In this technique, the sample is irradiated with a 'white beam' (all wavelengths) and the wavelength used for a particular reflection is simply determined by measuring the time it takes the neutron to arrive at the detector.

The resolution of instruments at neutron facilities is generally inferior to that of X-ray instruments (see Figure 3). However, the high-resolution time-of-flight instrument HRPD at the Rutherford Laboratory competes well with X-ray instruments, and for the low d-spacing region of the powder pattern, offers a resolution approaching that of instruments at synchrotron facilities.

3. RIETVELD REFINEMENT

In a single-crystal structure refinement, the measured intensities of a set of (usually several hundred) reflections are compared with those calculated for a structural model. The differences between the two are then minimized in a least-squares procedure by adjusting (refining) the atomic parameters in the model. In a powder diffraction experiment, the 3-dimensional single-crystal data (e.g. each reflection is defined by the three angles 2θ, ω and χ) are condensed into one dimension (e.g. defined only by 2θ), because *all* crystal orientations are present in a polycrystalline sample.

Since all reflections with similar 2θ values diffract simultaneously, only the sum of the intensities (i.e. not the individual contributions) can be measured. This loss of information in a powder diffraction experiment is the root of many of the difficulties that arise in the analysis of powder data. The 3-dimensional data can be reconstructed from the 1-dimensional pattern if it is simple enough and contains no exactly overlapping reflections (i.e. reflections unrelated by symmetry, but with identical 2θ values). Unfortunately, this is rarely the case.

However, the integrated intensities of the peaks can still be measured, and in the early refinements with powder data, these were used in single-crystal programs modified to take into account the reflection overlap. This integrated intensity approach proved to be satisfactory for simple structures, but not for the more complex ones with large unit cells and/or many atoms (e.g. for zeolites). As the number of reflections increases, the pattern becomes more and more difficult to decompose into discrete entities (i.e. well-defined groups of reflections), and the loss of information becomes even more severe.

3.1. The Theory

An alternative structure refinement procedure for neutron powder diffraction data that circumvented the reflection overlap problem was published in 1969 by Hugo Rietveld [17]. Instead of comparing the observed and calculated intensities of individual reflections or groups of reflections, he compared each step of the observed and calculated powder diffraction *patterns*, and used these differences to refine the atomic parameters of the model. The function that is minimized can be expressed as

$$S_y = \sum_i w_i \big[y_i(obs) - y_i(calc) \big]^2 \tag{3}$$

The calculated intensity at step i, $y_i(calc)$, is calculated according to the formula

$$y_i(calc) = s \sum_{hkl} L_{hkl} M_{hkl} |F_{hkl}|^2 P_{hkl} \phi (2\theta_i - 2\theta_{hkl}) + b_i \tag{4}$$

where s is the scale factor, hkl the Miller indices, L_{hkl} the Lorentz-polarization factor, M_{hkl} the reflection multiplicity, F_{hkl} the structure factor, P_{hkl} the preferred orientation correction function, ϕ the peak profile function (which is itself dependent on FWHM, peak asymmetry and shape parameter(s)), and b_i the background count at step i.

The intensities of the reflections ($\propto |F_{hkl}|^2$) are calculated from the atomic parameters in the model. To generate the powder diffraction pattern, Rietveld had to introduce additional "profile" parameters to describe the shape of the diffraction maxima (ϕ). These parameters are also adjusted in the refinement until the form in the calculated pattern mimics that in the observed one. Rietveld used a Gaussian curve to describe the peak shape of each reflection, and allowed the width of the peak to vary as a function of diffraction angle. The peaks in an X-ray diffraction pattern require a more complex peak shape function, and this was implemented later by Malmros and Thomas [18] and by Young et al. [19] in the late '70's. Since then, the Rietveld or whole-pattern refinement method has become the standard structure refinement technique for all powder data.

For more theoretical information on Rietveld refinement, the reader is again referred to the book *The Rietveld Method* edited by Young [1]. The following sections of this chapter focus on the practical aspects of the Rietveld refinement of zeolite structures.

3.2. Steps in the structure refinement process

The steps involved in a structure refinement with powder data can be summarized as follows:

1) Preparation of the powder diffraction data
2) Determination of the peak shape function

3) Evaluation of the starting values for the profile parameters
4) Selection of the space group
5) Refinement of the profile parameters
6) Addition of geometric restraints
7) Scaling of the calculated pattern to the observed data
8) Generation of a difference electron density map
9) Interpretation of the difference electron density map
10) Repetition of steps 8 and 9 until the structural model is complete
11) Rietveld refinement of the structural and profile parameters

Experience over the years has served to identify the critical steps, to highlight the common problems, and to evolve a general refinement strategy. To allow the discussion of these points to take on a more concrete form, a relatively straightforward example of a typical Rietveld refinement of a zeolite structure has been selected for detailed examination in the subsequent pages. Details relevant to this example are included at the end of each section, and to allow the reader to recognize them more easily, a different typeface has been used.

NU-3: The material chosen is the high silica zeolite NU-3 (zeolite structure type **LEV**) synthesized in the presence of 1-aminoadamantane, and the aim of the structural investigation is to establish the location of the template and to examine the details of the framework geometry [20]. The refinement is described in terms of the Rietveld refinement program package XRS-82 [21], but the concepts can be transferred quite easily to other programs.

Powder diffraction data were collected using the diffractometer on Station 9.1 at the synchrotron facility in Daresbury, U.K. [22] Details of the data collection are summarized in Table 1. The data were normalized for the gradual loss of X-ray beam intensity as the electron beam in the storage ring decayed, and the resulting raw data is shown in Figure 7.

Table 1. Powder diffraction data collection for NU-3

Synchrotron facility	SRS, Daresbury, U.K.
Diffractometer geometry	Debye-Scherrer
	0.5mm glass capillary
	Simple slit system
	no analyzer crystal
Monochromator	Channel-cut Si 111
Wavelength	1.5388 Å
Step size	0.01 °2θ
Time per step	8-28 °2θ: 10 s
	28-92 °2θ: 15 s
2θ range	8-92 °2θ

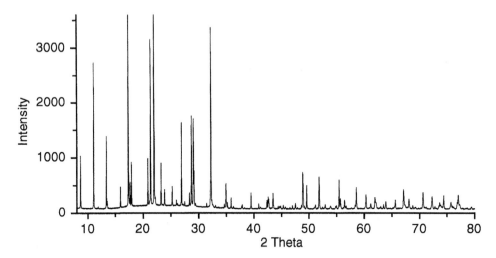

Figure 7. Observed powder diffraction pattern for NU-3. The intensity axis has been cut at 1/2 of the maximum peak height to show more detail at high angles.

3.3. Background

In XRS-82, the first step in a Rietveld refinement is the estimation and subtraction of the background from the raw data. In general, the background decreases rapidly as the detector moves away from the direct beam at low angles, and then becomes almost constant, decreasing only very slowly, at higher angles. In the case of zeolites, there is usually an additional "hump" centered around 25-30 °2θ (for a wavelength of 1.5 Å) resulting from amorphous scattering (the angle corresponds to 3.2 Å, i.e. a typical T-T distance). If the sample is measured in a glass capillary, the scattering from the glass will enhance this "hump".

The anchor points selected for the background interpolation need to be sufficiently close together in regions of high curvature, so that the resulting background curve is smooth. At low angles they are easily estimated in the regions between the peaks. However, the true background level at higher angles is difficult to judge, because of the high density of reflections in this region, and one or more re-estimations are usually necessary during the course of the refinement. The extra time spent in doing a careful background correction at this point is generally rewarded by a more successful profile parameter refinement later.

Some Rietveld programs use a polynomial to describe the background, and include the coefficients as additional profile parameters in the refinement process. This approach has the advantage that the background can be adjusted as the refinement progresses, and the standard deviations of the final parameters are better estimated. The potential difficulty is that the background may not be well-modelled by the function used.

NU-3: The initial background estimation for NU-3 is shown in Figure 8.

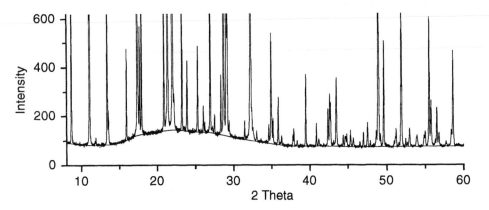

Figure 8. Initial background correction for NU-3. The intensity axis has been cut at 1/10 of the maximum peak height to focus on the shape of the background.

3.4. Peak Shape Function

It is obvious that an accurate description of the shapes of the peaks in the powder diffraction pattern is critical to the success of a Rietveld refinement. It is perhaps less obvious that the realization of this description is non-trivial. The peak shapes are a function of both the sample and the diffractometer (see section 2.3), and they vary as a function of 2θ (see Figure 3). In other words, every diffraction pattern requires a different peak shape. Whether one uses one of the analytical peak shape functions [23] or a tabulated experimental one [24], the observed peaks must be well described.

In XRS-82, a peak in the measured powder pattern is used to calculate a peak shape function, and this is then stored in tabulated form with a symmetric part and an asymmetric part (learned or experimental peak shape function). The fit obtained for a strongly asymmetric peak and the corresponding symmetric and asymmetric parts of the function are shown in Figure 9. When this type of function is used to describe the other reflections in the pattern, both the full width at half maximum (FWHM) and the relative contributions of the symmetric and the asymmetric part are varied (see section 3.5).

Ideally, the peak selected to calculate the experimental peak shape function should be a single reflection (or several exactly overlapping

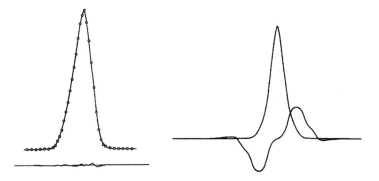

Figure 9. (a) The fit of an experimental peak shape function for a strongly asymmetric peak. (b) The symmetric and asymmetric parts (not to scale) of the function shown in (a).

observed in the data, should have a significant asymmetric part. If a single reflection is not available, a peak from a similar substance measured on the same instrument can be substituted. The intensity requirement simply ensures that the peak shape is statistically well defined. If a symmetric (or almost symmetric) reflection is selected, the asymmetric part of the function will be zero (or almost zero), and then the asymmetric reflections in the pattern cannot be properly described.

Another potential source of error for the peak shape description is the range selected for the peak (expressed in number of FWHM's). A peak is generally considered to be "finished" when the intensity is less than 0.1% of the maximum intensity, and the range should be chosen accordingly. This value may have to be adjusted during the course of the refinement.

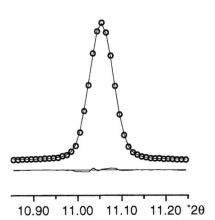

10.90 11.00 11.10 11.20 °2θ

Figure 10. Observed (circles), calculated (line) and difference (bottom) profiles for the experimental peak shape function used in the NU-3 refinement.

NU-3: The fit for the relatively symmetric peak used for the NU-3 refinement (very little asymmetry is observed in the whole pattern) is shown in Figure 10. The symmetric peak shape is an advantage of the simple diffractometer geometry used on Station 9.1 at the Daresbury facility combined with the parallel nature of the synchrotron beam. The range for the peak was calculated to be 9 FWHM's.

3.5. Profile parameters

Profile parameters describe the positions and shapes of the peaks in the calculated powder diffraction pattern. The positions are determined primarily by the lattice parameters, since these determine the 2θ values of the reflections, but a small contribution is made by the instrumental error in the 2θ scale. In XRS-82, there are two zero correction options. One is an arbitrary function:

$$2\theta(\text{corr}) = 2\theta(\text{obs}) + r + s \cdot 2\theta(\text{obs}) + t \cdot 2\theta(\text{obs})^2 \tag{5}$$

where $2\theta(\text{corr})$ is the corrected 2θ value, $2\theta(\text{obs})$ the observed one, and r, s and t are refinable parameters. The other accounts for the displacement of a capillary (in x and y) from the true axis of the 2θ arm of the diffractometer:

$$2\theta(\text{corr}) = 2\theta(\text{obs}) + x \cdot \sin \theta - y \cdot \cos \theta \tag{6}$$

Starting values for the lattice parameters are usually obtained from an indexing program (see section 2.5), and those for the 2θ zero point come from the calibration of the diffractometer with a standard material such as Si (see section 2.4). Not surprisingly, the two sets of parameters are highly correlated. By collecting a second data set on the sample, this time mixed with a standard material, the 2θ scale for the sample can be well calibrated using the standard lines, and then accurate lattice parameters for the sample can be refined. These lattice parameters can then be used with confidence in the refinement with the data collected on the sample without an internal standard, and any misfit in 2θ in that refinement can be adjusted by refining the zero correction parameters.

The other profile parameters deal with the shape of the peaks as a function of 2θ. In general, the peak width increases with 2θ, but with certain geometries a minimum is observed (see Figure 3). The change in the peak width is generally described using the formula Caglioti et al. derived for neutron diffraction [25]

$$\text{FWHM}^2 = u \cdot \tan^2\theta + v \cdot \tan\theta + w \tag{7}$$

where FWHM is the full width at half maximum of the peak at diffraction angle 2θ, and u, v and w are refinable parameters. In XRS-82, a numerical function derived from the FWHM's of the well-defined peaks in the pattern is used in place of equation 7. These values are tabulated, and then an FWHM is assigned to each reflection in the pattern by interpolation. They can then be adjusted singly or collectively, and are not constrained to fit an equation. The advantage of this rather pragmatic approach is that index dependent variations in the FWHM's (e.g. caused by an extreme crystallite morphology or stacking faults) can be modelled more easily.

The asymmetry of the peaks in a powder diffraction pattern tends to decrease as the diffraction angle increases. In XRS-82, this change is simulated in the calculated pattern by varying the relative contribution of the asymmetric part of the peak shape function (see section 3.4). As in the case of the FWHM's, the asymmetric contribution is estimated for well defined peaks in the pattern, interpolated, and then a peak asymmetry assigned to each reflection in the pattern. These values can also be refined singly or collectively and are not constrained to fit an equation.

NU-3: The zero correction for the instrument was accounted for in the observed data (i.e. the 2θ values for each step were corrected) and no further correction for the calculated pattern was needed. The indexing program TREOR [5] indicated that the system was hexagonal with $a = 13.221$ and $c = 22.279$ Å. Initial estimates of the FWHM's ranged from 0.059 to 0.20 $^\circ 2\theta$, and the peak asymmetry contribution from 0.012 to 0.000.

3.6. Space Group Selection

In a single crystal experiment, the symmetry of the crystal structure is usually established by examining the diffraction pattern for systematically absent reflections (for centering, glide planes and screw axes) and for intensity symmetry (e.g. mirror planes or rotation axes). In a powder diffraction experiment, this approach is complicated by the overlap of reflections. Generally, only a few low angle reflections can be used to detect systematic absences, so the result is rarely definitive. Intensity symmetry cannot be observed. Consequently, several space groups are usually consistent with the powder pattern.

Fortunately, there are other techniques available that are sensitive to symmetry. Electron diffraction from small sections of the crystallites can be very useful in this respect. Given the framework topology (i.e. the connectivity of the T-atoms), certain space groups can be eliminated from consideration because either the connectivity or the geometry of the framework atoms cannot be satisfied. The latter can be tested with the distance least squares program DLS-76 [26] (see section 3.8). If the DLS R-value is not below 0.005 after refinement of the geometry, the model is unlikely to be correct. For certain zeolites, NMR can also contribute symmetry information. For example, in a high silica zeolite, ^{29}Si MAS NMR can indicate the number of Si sites and their relative occupancies, and this is extremely useful in the space group determination step. Fyfe et al. have been particularly active in this area [27].

It is likely that a choice of several space groups still remains. One of these is selected for the refinement (usually the highest symmetry), but the others should not be forgotten. If the refinement does not proceed satisfactorily, another symmetry may yield better results.

NU-3: Only reflections with -h+k+l=3n were observed. This indicated the presence of rhombohedral centering, and the possibilities were thereby limited to seven space groups. A c-glide in the [110] direction could be ruled out because significant violations of the required systematic absences were apparent. In this way, two more space groups could be eliminated, leaving only R3, R̄3, R32, R3m, and R̄3m. The **LEV** topology has the symmetry R̄3m, so this was chosen as the starting space group for the refinement. The other four are all subgroups of R̄3m (full symbol: R̄32/m) and therefore remain possible candidates.

3.7. Initial refinement of profile parameters

At this point, the approximate lattice parameters and the indices of the reflections allowed by the symmetry are known, so the profile parameters can be refined. This is best done at this stage without a structural model. Instead of refining the positions of atoms in a model to obtain a better fit of the reflection intensities, the intensities themselves are allowed to vary. In this way, the peak shapes and the positions of the reflections can be refined even before the structural model is complete (or even known). With a partial model (e.g. the atoms of the framework, but not those of the template), the intensities in the calculated pattern will not be correct, so a refinement of the profile parameters using the structural model is difficult. The refined intensities are not important for the Rietveld refinement (though they could be important if a reflection intensity extraction is needed in order to attempt an *ab initio* structure solution), but accurate profile parameters are. This method of intensity refinement can also be used to obtain accurate lattice parameters for a material without doing a full fledged Rietveld refinement of the structure.

The order in which the profile parameters are refined can be critical. Our experience has shown that the 2θ values must be close to correct before a sensible FWHM and peak asymmetry refinement is possible. Otherwise, the FWHM refinement tries to compensate for the 2θ mismatch, and this results in peaks in the calculated pattern that are too broad. To refine the lattice parameters, the FWHM's are set to be slightly too small. In this way, the refinement is forced to concentrate on fitting the maxima of the peaks. Once the 2θ values are correct, the FWHM and peak asymmetry parameters can be refined. Incorrect profile parameters yield characteristic difference profiles, so the examination of a suitably enlarged profile plot can be very useful in deciding which parameters need to be refined. Three examples are illustrated in Figure 11. The effect of an attempt to fit an asymmetric peak with a symmetric peak shape function is shown in (a), the symmetric difference typical of a FWHM discrepancy in (b), and the effect of a 0.005° error in 2θ in (c). The integrated intensities of all three calculated peaks are the same.

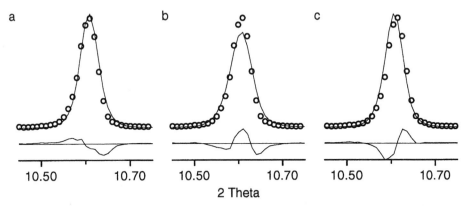

Figure 11. The observed (circles), calculated (line) and difference (bottom) profiles for (a) a peak asymmetry that is too small, (b) a FWHM that is too large, (c) a 2θ mismatch.

NU-3: In XRS-82, the module EXTRACT [28] is used to refine the profile parameters without a structural model. The resulting peak asymmetries and FWHM's are added to the reflection data, and accurate lattice parameters (and/or zero correction parameters) are added to the crystal data. For NU-3, the refined lattice parameters are $a = 13.2251(1)$ and $c = 22.2916(1)$ Å, the peak asymmetries did not change, and the FWHM's now range from 0.048 to 0.164 °2θ.

3.8. Geometric Restraints

3.8.1. Distance least squares refinement

Since the interatomic distances and angles in zeolite frameworks are well established, this information can be used to optimize the geometry of a framework structure for which only the connectivity and symmetry are known. This is possible because there are more independent distances and angles in a framework structure than there are positional parameters x,y,z. The program DLS-76 [26] minimizes the differences between the values prescribed for the distances and angles and those calculated from the approximate structural model by adjusting the coordinates of the atoms in the model in a least-squares procedure. The function that is minimized can be expressed as

$$S_R = \sum_n w_n \left[R_n(obs) - R_n(calc) \right]^2 \tag{8}$$

where $R_n(obs)$ is an expected distance or angle, $R_n(calc)$ is the value calculated from the model, and w_n is the weight. In this way, a geometrically sensible starting model for the refinement can be obtained.

NU-3: Approximate coordinates for the framework atoms in the asymmetric unit of NU-3 are given in Table 2, and the **LEV** framework topology is shown in Figure 12.

Table 2. Approximate coordinates for the framework atoms of NU-3.
($R\bar{3}m$, hexagonal axes)

Atom	x	y	z	U (Å^2)	Position	Point Symmetry
Si(1)	0.235	-0.001	0.067	0.01	x,y,z	1
Si(2)	1/3	-.092	1/6	0.01	1/3,y,1/6	2
O(1)	0.270	0	0	0.02	x,0,0	2
O(2)	0.099	-.099	0.074	0.02	x,-x,z	m
O(3)	0.252	0.126	0.083	0.02	2x,x,z	m
O(4)	0.447	-.106	0.157	0.02	x,2x-1,z	m
O(5)	0.313	-0.31	0.110	0.02	x,y,z	1

The connectivity of the atoms can be described schematically:

```
        Si(1')                              Si(2')
          |                                   |
        O(1)                                O(4)
          |                                   |
Si(2) – O(5) – Si(1) – O(2) – Si(1")   Si(1*) – O(5') – Si(2) – O(5) – Si(1)
          |                                   |
        O(3)                                O(4')
          |                                   |
       Si(1''')                            Si(2*)
```

where the apostrophe (') indicates a symmetry equivalent atom, and the asterisk (*) a symmetry equivalent atom that is not needed to describe the geometry but has been included to complete the diagram. For example, if Si(1) has the coordinates x,y,z, then Si(1') has the coordinates -y,-x,z (i.e. related by a mirror plane), Si(1") x,x-y,z, and Si(1''') x-y,-y,-z.

Si(1) lies in a general position and Si(2) on a twofold axis, so there are four symmetrically independent Si-O bonds for Si(1) and two for Si(2). Similarly, there are six symmetrically independent tetrahedral angles for Si(1) and four for Si(2) (of course, in each case, one of these is redundant). Finally, there is one Si-O-Si angle at each of the five O-atoms. The complete list of geometric information for the NU-3 framework is given in Table 3. The weight w_n used is 1/esd^2, where the esd is the standard deviation estimated for this distance or angle.

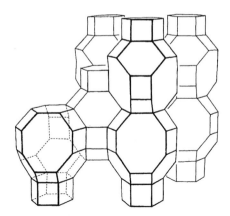

*Figure 12. The **LEV** framework topology. Each node represents a Si atom. For clarity, bridging oxygens have been omitted.*

Table 3. Distance and angles and their prescribed values for NU-3.

Distance or Angle	R(obs)	esd
Si(1) - O(1)	1.61 Å	0.01 Å
Si(1) - O(2)	1.61 Å	0.01 Å
Si(1) - O(3)	1.61 Å	0.01 Å
Si(1) - O(5)	1.61 Å	0.01 Å
Si(2) - O(4)	1.61 Å	0.01 Å
Si(2) - O(5)	1.61 Å	0.01 Å
O(1) - Si(1) - O(2)	109.5°	1°
O(1) - Si(1) - O(3)	109.5°	1°
O(1) - Si(1) - O(5)	109.5°	1°
O(2) - Si(1) - O(3)	109.5°	1°
O(2) - Si(1) - O(5)	109.5°	1°
O(3) - Si(1) - O(5)	109.5°	1°
O(4) - Si(2) - O(5)	109.5°	1°
O(4) - Si(2) - O(4')	109.5°	1°
O(4) - Si(2) - O(5')	109.5°	1°
O(5) - Si(2) - O(5')	109.5°	1°
Si(1) - O(1) - Si(1')	145°	8°
Si(1) - O(2) - Si(1")	145°	8°
Si(1) - O(3) - Si(1''')	145°	8°
Si(2) - O(4) - Si(2')	145°	8°
Si(1) - O(5) - Si(2)	145°	8°

Thus, there are a total of 21 pieces of geometric "data" for NU-3, and 14 coordinates to be refined. A geometric refinement is usually done at this stage to obtain idealized starting coordinates for the framework atoms.

3.8.2. Geometric restraints in a Rietveld refinement

As pointed out earlier, a Rietveld refinement is hampered by the loss of information inherent to a powder diffraction experiment. To supplement the powder diffraction data, geometric data such as that used in DLS can be added as additional "observations" [29]. In both Rietveld refinement and DLS, atomic coordinates are refined, and this means that both X-ray and geometric data can be used simultaneously in a structure refinement. The function to be minimized is a simple combination of equations 3 and 8 to give

$$S = S_y + c_w S_R \qquad (9)$$

where the common weight factor c_w allows the relative weight of the two functions to be regulated.

The use of geometric restraints not only increases the number of observations, thereby allowing more parameters to be refined, it keeps the geometry sensible. Especially in the early stages of refinement, when a

418

geometry sensible. Especially in the early stages of refinement, when a structural model is incomplete, interatomic distances can easily become unreasonably long or short, but if the refinement includes these additional geometric restraints, such false minima can be avoided.

3.9. Scaling the calculated pattern

The calculated pattern has to be scaled to the observed data. For zeolites, the low angle reflection intensities are particularly sensitive to the presence or absence of non-framework species, whereas the high angle reflection intensities are determined primarily by the atoms of the framework (see Figure 13). Therefore, when the partial model consists of only the atoms of the framework, the high angle reflections are best suited for the scaling of the two patterns to one another, even though the resulting fit at low angles will look

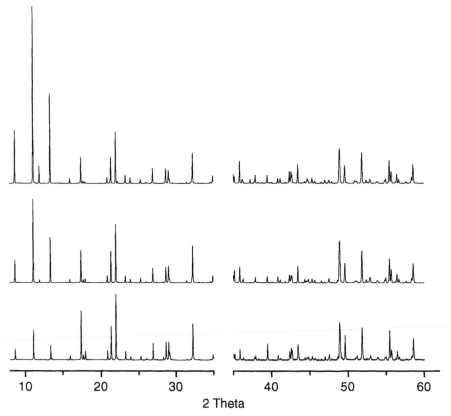

Figure 13. Powder diffraction patterns for NU-3 (a) calculated using the framework atoms only, (b) calculated using the framework atoms and half a template, and (c) the observed data. To facilitate comparison, the scale for the higher angle data has been increased by a factor of five.

disastrous. The intensity differences at low angles are real, and reflect the incompleteness of the model.

3.10. Location of non-framework atoms

Everything is now set up for the Rietveld refinement, but the structural model is incomplete. The atoms of the template molecule are missing. In order to locate these, a single-crystal method for generating an electron density map can be used. Of course, the problem of reflection overlap still exists, but with a partial structural model, the distribution of the intensities of reflections overlapping in the powder pattern can be estimated by assuming that the distribution is the same as that of the intensities calculated for the partial model. Such a partitioning of the overlapping reflections is easily done in a Rietveld program. The more complete the model, the more valid is this approximation.

With the resulting *pseudo* single crystal data set (a list of the *hkl*'s and their corresponding intensities), an approximate electron density map (albeit model biased) can be generated via a Fourier transform (using phases calculated from the partial model). In general, maps calculated from powder data are more diffuse than those calculated from single crystal data, but they are still quite usable. By subtracting the electron density calculated for the model from that calculated using the "observed" intensities, a so-called difference Fourier map (or difference electron density map), which highlights the electron density not accounted for in the model, is generated..

For optimum results, it is important that the scaling described in the previous section be correct. The differences between the intensities of the observed pattern and those of the calculated one give rise to the peaks in the difference electron density map. If the whole pattern rather than just the high angle data is used to determine the scale factor, these differences will be too low at low angle and too high at high angle, and the resulting difference electron density map will be much less definitive.

The interpretation of the difference electron density map is the next problem. Non-framework species rarely obey the high symmetry of the framework, so for example, an atom near, but not on, a mirror plane of the framework will appear as half an atom in each of two symmetrically equivalent positions. Since the two positions are too close to one another for simultaneous occupation, the maximum occupancy of the position is 0.5. This problem of partial occupancy, or disorder, is a common one in zeolite structures. It always complicates, and can even prevent, a successful structure analysis.

NU-3: The calculated pattern was scaled as shown in Figure 13a, and a difference electron density map generated. Four distinct peaks in the asymmetric unit appeared (at 0,0,0.197; 0.061,0.122,0.249; 0.062,0.124,0.318; and 0.122,0.061,0.337, see Figures 14 and 15). With the application of the space group symmetry, these positions describe an adamantane molecule. Addition of these peaks as C-atoms in the structural model improved the profile fit considerably. A second difference map, generated using the improved model, revealed at weak peak at 0.120, 0.240, 0.340, which was 1.41 Å from C(3). This position was interpreted as an N-atom, and added to the model. Although 1-aminoadamantane has only one NH_2 group, the $R\bar{3}m$ symmetry produces three. The electron density at this position is weak simply because only 1/3 of the positions are occupied. The orientation of the 1-aminoadamantane template molecule in the NU-3 framework is shown in Figure 15. With the

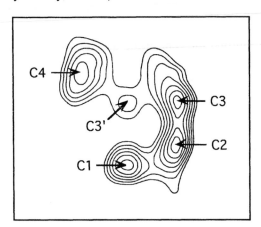

Figure 14. A section of the first difference electron density map for NU-3 showing the four C-atom positions. See Figure 15 for the orientation of the adamantane molecule.

addition of the N-atom position (with a population parameter of 1/3), the profile fit improved slightly (Figure 16a-c) and the structural model could be considered complete. Since the geometry of 1-aminoadamantane is well-defined, geometric

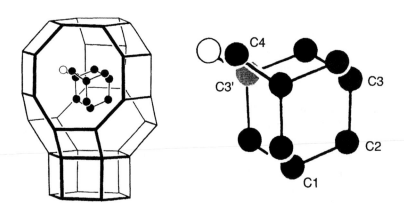

Figure 15. The orientation of the 1-aminoadamantane template molecule in the NU-3 framework. The positions shown in the electron density map in Figure 14 have been highlighted in the enlargement on the right. (The molecule is rotated slightly from the plane shown in Figure 14.)

restraints were also added for its atoms.

3.11. Rietveld refinement strategy

With a complete structural model, the actual Rietveld refinement of the structural parameters can begin. Because the minima of the least squares residual function are much shallower with powder data than they are with single crystal data, the refinement needs constant monitoring. A typical zeolite structure refinement requires several hundred cycles, and this is usually done in sets of two to five cycles at a time. To monitor the progress of a refinement the two most useful pieces of information are the nature of the parameter shifts (i.e. are they converging?) and the profile fit. The latter is best seen in a plot of the observed and calculated patterns, but can also be followed numerically with a reliability factor or R-value (see section 3.12).

It is difficult to cover all the details of a full refinement, but an approximate strategy for zeolite structures can be described. First the positions of the framework atoms can be refined with a high weight on the geometric restraints (e.g. $c_w \approx 10$). The weight for the restraints can then be reduced, until the interatomic distances and angles start to deviate too strongly from ideal values. Then the non-framework atom positions can be refined. If a molecule with a known geometry is involved, geometric restraints can also be used here. If that refinement converges, all atomic positions in the model can be refined simultaneously.

At this point, the refinement of the somewhat trickier parameters can be attempted. The scale, the population parameters and the thermal parameters are highly correlated with one another, and are more sensitive to the background correction than are the positional parameters. The population parameters for the non-framework species might be refined and the occupancies fixed at chemically sensible values if they are less than one. Refinement of the thermal parameters (framework atoms first) can be attempted, but this is notoriously difficult with X-ray powder data. It is perhaps prudent to constrain the thermal parameters for similar atoms (e.g. all framework Si atoms, all framework O atoms, all non-framework organic atoms, all Na^+ ions, etc.) to be equal, and thereby reduce the number of thermal parameters required. The profile parameters can be further refined at any time (now using the model to generate the intensities) if the plot indicates that it is necessary (see section 3.7 and Figure 10).

NU-3: For NU-3, the strategy described above works quite well. Thermal parameters for the Si-atoms, the O-atoms, and the non-framework atoms, were set arbitrarily at 0.01, 0.02 and 0.03 $Å^2$, respectively. Both the framework and the non-framework atoms refined satisfactorily, and the common weight factor c_w for the restraints could be reduced to 1.0. Refinement of the population parameters of the non-framework atoms resulted in occupancies above 1.0. Of course, this is not possible, but it is caused by the presence of H-atoms not accounted for in the model. Although a single H-atom has only one electron, the H-atoms in a CH_2 group account for 1/4 of the scattering power, and in 1-aminoadamantane they account for 20% of the scattering power. This was first approximated by increasing the population parameters for the C-atoms appropriately (Figure 16d), but then positions for the hydrogens were calculated and added to the model (their positions were not refined).

The background correction was improved slightly, the peak shape function was recalculated using a larger 2θ range (peak range increased to 12 FWHM), and then thermal parameters for each of the framework atoms and a single one for the C- and N-atoms were refined. At this point, the geometric restraints on the framework atoms (but not on the organic ones) could be removed without the geometry becoming unreasonable. Simultaneous refinement of all the structural parameters then converged (Figure 16e). A difference electron density map was generated to check that the model was, in fact, complete, and was found to be featureless. Profile plots at various stages of the refinement are shown in Figure 16.

3.12. R-values

Although a profile plot is the best way of following and guiding a Rietveld refinement, the fit of the calculated pattern to the observed data can also be given numerically. This is usually done in terms of reliability factors or R-values. The weighted profile R-value, R_{wp}, is defined as

$$R_{wp} = \left\{ \sum_i w_i[y_i(obs) - y_i(calc)]^2 \, / \, \sum_i w_i[y_i(obs)]^2 \right\}^{1/2} \tag{10}$$

where $y_i(obs)$ is the observed intensity at step i, $y_i(calc)$ the calculated intensity, and w_i the weight. The expression in the numerator is the value that is minimized during the refinement. The final R_{wp} should approach the statistically expected R-value, R_{exp}, which reflects the quality of the data (see section 2.6). The ratio between the two

$$\chi^2 = R_{wp} / R_{exp} \tag{11}$$

is also quoted quite often in the literature.

An R-value similar to that reported for single crystal refinements, based on the agreement between the structure factors, F_{hkl}, can also be calculated by distributing the intensities of the overlapping reflections according to the structural model. It is, of course, biased towards the structural model. That is

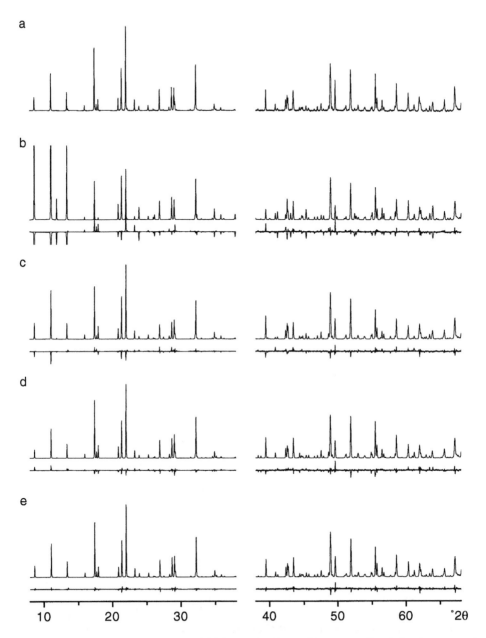

Figure 16. (a) Observed data for NU-3. (b)-(e) Calculated and difference profiles during the course of the Rietveld refinement. (b) Initial model with framework atoms only, (c) C and N positions from difference Fourier added, (d) H positions simulated with increased population parameters, all positions refined with restraints, (e) Final model, all parameters refined. The scale for the second half of the profiles has been increased by a factor of 5.

defined as

$$R_F = \sum_{hkl} |F_{hkl}(obs) - F_{hkl}(calc)| / \sum_{hkl} |F_{hkl}(obs)| \tag{12}$$

This quantity is not actively used in the refinement, but should decrease as the refinement progresses.

These R-values are useful indicators for the evaluation of a refinement, but they should not be overinterpreted. The most important criteria for judging the quality of a Rietveld refinement are (1) the fit of the calculated pattern to the observed data, and (2) the chemical sense of the structure. The former can be evaluated on the basis of the final profile plot, and the latter on a careful examination of the final atomic parameters. Interatomic distances (both bonding and non-bonding) should be reasonable, bond angles sensible, and population parameters consistent with the chemical composition of the material.

NU-3: The R-values at various stages of the refinement described in the previous pages are listed in Table 4. They correspond to the profiles shown in Figure 16.

Table 4. R-values during the Rietveld refinement of NU-3

Model	R_F	R_{wp}	R_{exp}
Framework only	0.178	2.353	0.100
C & N positions from Fourier added	0.102	0.192	0.100
H atoms added, non-H positions refined	0.100	0.187	0.101
Final model, improved profile parameters	0.060	0.103	0.103

3.13. Standard deviations of the final structural parameters

The estimated standard deviation (esd) calculated in a least-squares program is

$$\sigma = \left\{ A_{jj}^{-1} \sum_i^N w_i \left[y_i(obs) - y_i(calc) \right]^2 / (N - P) \right\}^{1/2} \tag{13}$$

where A_{jj}^{-1} is the diagonal element of the inverse matrix for the jth parameter. From this equation, it is clear that the esd's will become smaller as the number of observations N (i.e. the number of steps) becomes larger. Furthermore, increasing the counting time and thereby the counts per step will decrease the variance of $y_i(obs)$.

Both these effects can, and often do, lead to artificially low esd's in a Rietveld refinement. This has been a major criticism of the Rietveld method

[30]. The esd's of a Rietveld analysis are 'wrong' if there is significant correlation between residuals across a Bragg peak and/or if the errors in the fit are dominated by errors in the model rather than by counting statistics (the basis of the weighting scheme). To test for serial correlation between adjacent steps in a powder pattern, Hill and Flack [31] proposed the use of the Durbin-Watson d statistic. They used it in a weighted form as

$$d = \sum_{n=2}^{N} \left(w_i \Delta_i - w_{i-1} \Delta_{i-1} \right)^2 / \sum_{n=2}^{N} \left(w_i \Delta_i \right)^2 \tag{14}$$

where $\Delta_i = y_i(\text{obs}) - y_i(\text{calc})$. The value of d should be close to 2. Using this statistic, Hill and Flack could show that a step size corresponding to approximately the minimum FWHM in the pattern should be chosen. Hill and Madson [13] recommend that smaller step sizes be used in the data collection itself, but that in the course of the refinement, a d-statistic test be used to decide how many of the steps should be skipped.

Since it may be difficult to accept the above findings, a practical example is illustrated in Figure 17. The second part of the powder pattern of VPI-5 is reproduced. The data were collected at the synchrotron radiation source in Daresbury, UK with a step size of .01 °2θ [32]. The FWHM's range from 0.043 to 0.069 °2θ. In (a), the pattern is plotted as measured, and in (b), only every fifth step (i.e step size of 0.05 °2θ), corresponding to approximately the half width of

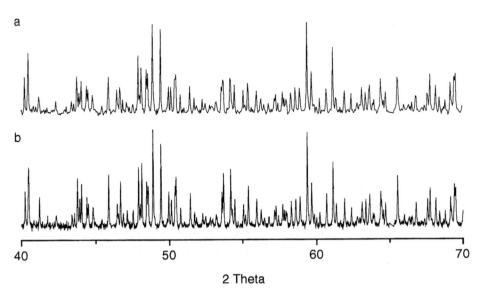

2 Theta

Figure 17. Powder diffraction data for VPI-5 (a) with a step of 0.01 °2θ, and (b) with a step size of 0.05 °2θ.

the peaks, is plotted. It is quite surprising to see how well the structure of the pattern is preserved despite the large step size, and it is therefore understandable that all the necessary information for a structure refinement is preserved in the thinned-out pattern. This is only true, of course, if the peak shape, the half width, the asymmetry and the peak position are already known exactly. In this case, one or two measurements across a peak are indeed sufficient to determine the integrated intensity of that peak, but a smaller step size is needed to determine the profile parameters.

NU-3: To obtain the correct standard deviations for the structural parameters, the final cycles of refinement were performed using every seventh data point. This yielded a Durbin-Watson *d* of 2.07.

4. CONCLUSION

A number of interesting aspects of zeolite powder diffraction data analysis could not be covered in this paper. In particular, approaches to the determination of zeolite structures using the information in a powder diffraction pattern have not been discussed. Of course, this is an important and exciting area of zeolite structural analysis, which is still undergoing rapid development. Many new ideas and computer programs are currently being explored. A brief report on this topic can be found in a paper by one of us (LBM) in the proceedings of the main conference.

Powder diffraction data analysis is a powerful tool in the elucidation of zeolite structures, be it the determination of new structures, the location of template molecules or the investigation of sorbed species inside the zeolite channels or cages. Reitveld refinement has now progressed to the point where zeolite structures with up to 200 structural parameters can be refined successfully. It is now taken for granted that the cations of an as-synthesized or of an ion-exchanged zeolite or zeolite-like material can be located with no problem using powder data.

However, not all of the desperately needed structural information can be delivered. Regardless of the progress made in X-ray and neutron diffraction, this will always be the case. This thirst for more is positive, since it continues to provide the stimulus for the search for new pathways to solve specific problems. Perhaps an indirect impression of this stimulating environment has been conveyed in this paper. It is hoped that the reader has gained a better understanding of the importance of small details in powder diffraction data analysis, and an appreciation of the potential *and* of the limitations of the method.

5. ACKNOWLEDGEMENTS

We thank Prof. H. Kessler for providing the SEM micrograph of $AlPO_4$-CHA. This work was supported in part by the Swiss National Science Foundation.

REFERENCES

1 *The Rietveld Method*, R.A. Young (ed.), Oxford University Press, Oxford, 1993.
2 H.P. Klug and L.E. Alexander, *X-ray Diffraction Procedures for Polycrystalline and Amorphous Materials*, 2nd ed., John Wiley & Sons, New York, 1974.
3 *Modern Powder Diffraction*, D.L. Bish & J.E. Post (ed), Reviews in Mineralogy, **20** (1989).
4 T. Proffen & K. Hradil, *Z. Kristallogr. Suppl.* **7** (1993), 155.
5 P.-E. Werner, L. Eriksson & M. Westdahl, *J. Appl. Crystallogr.*, **18** (1985) 367-370.
6 J.W. Visser, *J. Appl. Crystallogr.*, **2** (1969) 89-95.
7 A. Boultif & D. Louër, *J. Appl. Crystallogr.*, **24** (1991) 987-993.
8 D. Taupin, *J. Appl. Crystallogr.*, **22** (1989) 455-459.
9 P.M. De Wolff, *J. Appl. Crystallogr.*, **1** (1968) 108-113
10 H. Kessler in *Synthesis, Characterization and Novel Applications of Molecular Sieve Materials*, R.L. Bedard, T. Bein, M.E. Davis, J. Garces, V.R. Maroni & G.D. Stucky (eds.), Materials Research Society, Pittsburg, PA, 1991, pp. 47-55.
11 M. Ahtee, M. Nurmela & P. Suortti, *J. Appl. Crystallogr.*, **22** (1989) 261-268.
12 N.C. Popa, *J. Appl. Crystallogr.*, **25** (1992) 611-616.
13 R.J. Hill & I.C. Madsen, *Powder Diffraction*, **2** (1987) 146-162.
14 D.E. Cox in *Handbook on Synchrotron Radiation*, G. Brown & D.E. Moncton (eds.), Elsevier Science Publishers, Amsterdam, 1991, pp. 155-199.
15 M. Hart, R.J. Cernik, W. Parrish & H. Toraya, *J. Appl. Crystallogr.*, **23** (1990) 286-291.
16 I.J. Pickering, D.E.W Vaughan, K.G. Strohmaier, G.N. Geroge & G.H. Via, *Proceedings 9th Intl. Zeolite Conf. Montreal 1992*, R. von Ballmoos, J.B. Higgins & M.M.J. Treacy (ed), Butterworth-Heinemann, Boston, (1993), pp. 595-600.
17 H.M. Rietveld, *J. Appl. Crystallogr.*, **2** (1969) 65-71.
18 G. Malmros & J.O. Thomas, *J. Appl. Crystallogr.*, **10** (1977) 7-11.
19 R.A. Young, P.E. Mackie & R.B. von Dreele, *J. Appl. Crystallogr.*, **10** (1977) 262-269.
20 L.B. McCusker, *Mat. Sci. Forum*, **133-136** (1993) 423-434.
21 Ch. Baerlocher, *XRS-82. The X-ray Rietveld System*, ETH, Zurich, 1982.

428

22 P. Pattison, R.J. Cernik & S.M. Clark, *Rev. Sci. Instrum.*, **60** (1988) 2376-2379.

23 R.A. Young & D.B. Wiles, *J. Appl. Crystallogr.*, **15** (1982) 430-438.

24 A. Hepp & Ch. Baerlocher, *Austral. J. Physics*, **41** (1988) 229-236.

25 G. Caglioti, A. Paoletti & F.P. Ricci, *Nucl. Instrum.*, **3** (1958) 223-228.

26 Ch. Baerlocher, A. Hepp & W.M. Meier, *DLS-76. A distance least squares refinement program*, ETH, Zurich, 1977.

27 C.A. Fyfe, Y. Feng, H. Grondey, G.T. Kokotailo & H. Gies, *Chem. Rev.*, **91** (1991) 1525-1543.

28 Ch. Baerlocher, *EXTRACT. A Fortran program for the extraction of integrated intensities from a powder pattern*, ETH, Zurich, 1992.

29 Ch. Baerlocher in *The Rietveld Method*, R.A. Young (ed.), Oxford University Press, Oxford, 1993, pp. 186-196.

30 M. Sakata & M.J. Cooper, *J. Appl. Crystallogr.*, **12** (1979) 554-563.

31 R.J. Hill & H.D. Flack, *J. Appl. Crystallogr.*, **20** (1987) 356-361.

32 L.B. McCusker, Ch. Baerlocher, E. Jahn & M. Bülow, *Zeolites*, **11** (1991) 308-313.

J.C. Jansen, M. Stöcker, H.G. Karge and J. Weitkamp (Eds.)
Advanced Zeolite Science and Applications
Studies in Surface Science and Catalysis, Vol. 85
© 1994 Elsevier Science B.V. All rights reserved.

429

Review on recent NMR Results

Michael Stöcker

SINTEF SI, Department of Hydrocarbon Process Chemistry,
P.O.Box 124 Blindern, N-0314 Oslo, Norway

Dedicated to Mikkel

Contents

1.   INTRODUCTION

The objective of this review is to survey recent solid-state NMR results of zeolites and $AlPO_4$ molecular sieves, especially with respect to new techniques, methods and pulse sequences.

The impact of solid-state NMR as a powerful analytical tool for studies of zeolites and $AlPO_4$ molecular sieves has been dramatic during the last years. Since the pioneering work of Lippmaa, Engelhardt and coworkers (1, 2) in the beginning of the 1980 decade, dealing for the first time with high-resolution solid-state ^{29}Si NMR of zeolites, tremendous progress has been made, aiming towards enhanced resolution, sensitivity and improved multinuclear capabilities. Solid-state NMR spectroscopy is nowadays a well established technique for characterization of zeolites and related materials with respect to structure elucidation, catalytic behaviour and mobility properties.

Solid-state NMR provides some advantages compared to X-ray diffraction (XRD), since amorphous materials as well as crystalline materials can be studied. In addition, silicon and aliminium have almost the same scatterings factors, which means no discrimination between those two nuclei can be achieved by XRD. While X-ray diffraction provides information about the long-range ordering and periodicities, the NMR techniques allows investigations on the short-range ordering (local environment) and structure. Powder XRD reveals only limited information with respect to zeolite lattice structures and only single crystal XRD could be used to elucidate zeolite structures. As known, zeolite crystals are usually to small for single crystal XRD investigations. However, XRD is a very complementary technique to solid-state NMR, and the use of both methods is still a powerful combination in connection with zeolite and $AlPO_4$ molecular sieve characterization.

The analytical potential of NMR spectroscopy in studies of liquids is well accepted. The measured frequency of a NMR active isotop is very sensitive to the electronic and structural environment of that nucleus. The observed fine structure pattern gives rise to the high-resolution spectrum. The potential of high-resolution solid-state NMR has been known for a long time. However, the challenge has always been to overcome the problems in connection with recording solid-state NMR spectra with sufficient resolution. The main difference in the NMR characteristics of liquids and solids is that distinct nuclear spin interactions like chemical shift anisotropy, dipolar and quadrupolar interactions, which lead to excessive line broadening, are averaged in liquids due to the fast thermal/molecular motions of molecules, but are operative in the rigid lattice of solids (the molecules are less mobile) (3). As a consequence, the fine structure is lost since broad lines are obtained, hiding the essential information of analytical character. In addition, long T_1-values are controlling the relaxation of nuclei in solids, due to the lack of translation- and rotation motions.

However, during the last years, advanced techniques, methods and pulse sequences were developed, allowing the recording of high-resolution solid-state NMR spectra which line pattern show a resolution comparable to that of liquid state spectra. Todays high-resolution solid-state NMR equipment allow the study of solids in a routine manner and provide information of analytical quality.

A large number of authors have been working in this field and contributed to the amount of results and literature available. Therefore, it is a difficult task to review and evaluate the entire field of solid-state NMR related to zeolites and $AlPO_4$ molecular sieves. However, several books (4, 5) and review articles (6-14) are available, presenting investigations on high-resolution solid-state NMR of zeolites and related material in an excellent way, and I will refer to those books/papers whenever appropriate. Furthermore, I would like to focus on the **latest** developments (during the last four years) and advice the reader to consult the addressed books/reviews with respect to **former** studies.

In this chapter a brief overview will be presented regarding the general aspects of solid-state NMR spectroscopy with emphasis on the experimental techniques to achieve spectra with the fine structure quality of high resolution. This part will be kept on a level acceptable for experienced zeolite scientists without requirering all the fundamental knowledge of the general and/or specific solid-state NMR theory. The following survey will deal with the information available from high-resolution solid-state NMR and its analytical application to investigations of zeolites and AlPO$_4$ molecular sieves. The author hopes that the reader at the end of this chapter will have some understanding of the present and future potential of solid-state NMR spectroscopy within this field.

2. HIGH RESOLUTION SOLID STATE NMR - GENERAL ASPECTS

The broad lines recorded for solid-state NMR spectra are due to the different behaviour of nuclear spin interactions in solids compared to liquids. These interactions are averaged to zero or to the isotropic values in liquids by the fast molecular motions, whereas the fixed (and different) orientations (with respect to the external magnetic field B$_o$) of the local environments of NMR active isotops in the rigid lattice of a solid cause line broadenings. Figure 1 clearly demonstrates the difference between liquid and solid state NMR spectra recorded for ethanol (^{1}H NMR) (5). The recorded broad NMR line patterns are superpositions of resonances from randomly oriented individual nuclei due to a random distribution of different orientations, since zeolites or AlPO$_4$ molecular sieves usually are microcrystalline powders. Table 1 summarizes the nuclear spin interactions and their behaviour in liquids versus solids (15), whereas Table 2 shows the NMR properties of some relevant nuclei investigated in connection with NMR of zeolites and AlPO$_4$ molecular sieves.

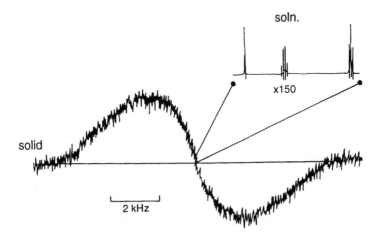

Fig. 1. Proton NMR spectrum of solid ethanol (77 K) presented as the derivative with (inset) the high-resolution NMR spectrum with the scale expension indicated (Reproduced by permission of The Royal Society, London).

Table 1
NMR Nuclear Spin Interactions

Interaction	Behaviour	Liquid phase	Solid state	Range [a] (solid state)
Zeeman interaction	Linear with H_0, at high H_0: high population difference	Enery level splitting $\varpi_0 = \gamma \cdot H_0$ (same effect in both phases)		50 MHz
Dipol-dipol interaction	depending on γ and distance $(1/r^3)$, independent on H_0	averaged to zero	dominant, short T_2, broad lines	$\approx$ 15 kHz
Chemical shift anisotropy (CSA)	linear with H_0 ("Chemical diagnostic")	isotropic values	anisotropic values	up to 10 kHz
Spin-spin coupling (scalar coupling)	independent on H_0	small values	yes, but not dominant	$\approx$ 200 Hz
Quadrupolar interaction $(I > 1/2)$	independent on H_0, depending on $(3 \cos^2 \Theta - 1)$	averaged to zero	dominant	1 - 15 MHz

In addition to those NMR nuclear spin interactions we have to deal with long T_1 relaxation times due to the lack of translations and rotations in the solid state. The T_1 relaxation times usually control the repetition times of NMR experiments.
[a] from reference 15.

2.1. Line broadening phenomena

As already mentioned, the lack of molecular motion in solids gives rise to broad resonances and the received spectral pattern consists of overlapped lines, hiding the valuable analytical information available from the isotropic chemical shifts. In principle, there are three line broadening mechanisms, described in the following (13).

a) Dipolar Interactions

Besides the magnetic field causing the Zeeman splitting, there are usually additional magnetic interactions between the magnetic moments of the observed nucleus and those located in the neighbour environment. The strength of these so-called **dipolar couplings** depends on the magnitude of the magnetic moments of the interacting neighbour nuclei, the distance (decrease with the internuclear distance, $1/r^3$) and the orientation of the internuclear vector with respect to the external field B_0. This interaction is independent of the applied

Table 2
Properties of NMR active nuclei related to zeolites and $AlPO_4$ molecular sieves

NMR active nucleus	Spin quantum number	NMR frequency (MHz)[a]	Natural abundance (%)	Absolute sensitivity[b]
^1H	1/2	400	99.98	1.00
^7Li	3/2	155.45	92.58	0.27
^{11}B	3/2	128.34	80.42	0.13
^{13}C	1/2	100.58	1.10	$1.76 \cdot 10^{-4}$
^{17}O	5/2	54.23	0.04	$1.08 \cdot 10^{-5}$
^{23}Na	3/2	105.81	100	$9.25 \cdot 10^{-2}$
^{27}Al	5/2	104.23	100	0.21
^{29}Si	1/2	79.46	4.7	$3.69 \cdot 10^{-4}$
^{31}P	1/2	161.92	100	$6.63 \cdot 10^{-2}$
^{51}V	7/2	105.15	99.8	0.38
^{69}Ga	3/2	96.01	60.4	$4.17 \cdot 10^{-2}$
^{71}Ga	3/2	121.98	39.6	$5.62 \cdot 10^{-2}$
^{129}Xe	1/2	110.64	26.44	$5.60 \cdot 10^{-3}$
^{133}Cs	7/2	52.47	100	$4.74 \cdot 10^{-2}$

[a] at $H_O = 9.395$ Tesla.
[b] product of relative sensitivity and natural abundance.

magnetic field (11, 13). In liquids, fast molecular motion averages the dipolar couplings to zero. However, in a solid no such effect occurs, and broad resonance lines are obtained. The shape of the line is broad and featureless, and the line width ($\Delta \upsilon_{1/2}$) can be up to many kHz (see Figure 1). There are two types of dipolar couplings: the **homonuclear dipolar couplings** (for example ^1H-^1H), which means interaction between the spins of like nuclei, and the **heteronuclear dipolar couplings** (for example ^{29}Si-^1H), in which the interaction occurs between the nuclear species under observation and spins of different nuclei.

Dipolar interactions with protons are usually very dominating, whereas homonuclear interactions of low natural abundant nuclei, like ^{29}Si and ^{13}C, can usually be neglected due to strong internuclear distance dependence.

b) Chemical Shift Anisotropy (CSA)
In addition, the external magnetic field induces magnetic moments due to electron circulation in connection with chemical bonds. The observed chemical shift for a given nucleus depends on the orientation of the molecule and of the chemical bond containing the nucleus relative to the magnetic field B_O. In a microcrystalline powder, this chemical bond will have a distribution of orientations relative to the external magnetic field, which leads to a distribution of chemical shifts. The resulting spectrum will have the shape of a powder pattern (cf. Figure 2) and this phenomenon is called **chemical shift anisotropy (CSA)** (13). The CSA is determined by the symmetry of the electronic charge distribution around the

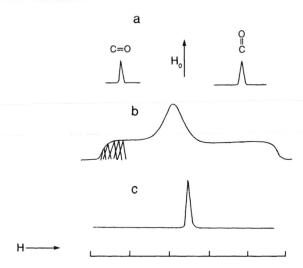

Fig. 2. Schematic representation of the ^{13}C NMR absorption of a carbonyl functionality:
a) Single crystal with two different orientations b) Polycrystalline sample (contributions
from the random distribution of orientations, chemical shift anisotropy, CSA) c) In solution
(random motion of the molecules yields the isotropic average chemical shift) (Reproduced by
permission of The Royal Society, London).

NMR nucleus. The CSA interaction increases linearly with the strength of the external
magnetic field B_0 (11).

c) Quadrupolar Interactions

Some of the interested nuclei considered in connection with zeolite science, like ^{27}Al or
^{17}O, possess a quadrupolar moment Q (besides the magnetic moment), that means nuclei
with a spin quantum number I larger or equal to one. The quadrupolar moment is a result of
the non-spherical distribution of nuclear charge (13). Nuclei with a quadrupolar moment
interact with the non-spherical, symmetrical electric field gradient. This **quadrupolar
interaction** is determined by the charge distribution of the surrounding electrons and other
nuclei and can range up to several MHz, completely dominating the entire spectrum. The
quadrupolar powder pattern are mainly affected by the so-called second order quadrupolar
interactions which decrease with increasing magnetic field strength (11).

2.2. Experimental techniques to achieve high-resolution solid-state NMR spectra

NMR spectra of solids with the quality of high resolution can be achieved when the above
mentioned line broadening phenomena are removed or at least considerably reduced. Only in
these cases spectra can be obtained which enable, for example, crystallographically non-
equivalent sites of a zeolite or AlPO$_4$ molecular sieve framework to be resolved as individual
resonance lines. High resolution NMR spectra of solids can be received by use of one or
more experimental techniques described in the following:

a) Dipolar Decoupling (DD)

Heteronuclear dipolar decoupling averages to zero the interactions of the abundant nuclei, like ^{1}H, with the rare spins under observation, as, for example [29]Si. Dipolar decoupling is obtained by irradiating the Larmor frequency of the abundant spin causing the dipolar broadening, while observing the less abundant nucleus to be investigated. During this decoupling, a fast spin flipping of the decoupled nuclei is created, resulting in no influence of the abundant spins on the rare spins.

b) Multiple Pulse Sequences (MPS)

Homonuclear dipolar decoupling can be achieved by using so-called **multiple pulse sequences (MPS)**. These are carefully tailored short and intense pulse cycles averaging the homonuclear dipolar interactions by reorientation of the nuclear spins.

c) Magic Angle Spinning (MAS)

Both, line broadenings caused by dipolar and first order quadrupolar interactions can be removed by **magic angle spinning (MAS)**, discovered by Andrew et al. (16) and Lowe (17). In addition, the anisotropy of the chemical shifts can be reduced to their isotropic values by applying MAS. During magic angle spinning, the sample is rotated quickly about an angle of $\Theta = 54^{\circ}44'$ in relation to the axis of the external magnetic field B_O (see Figure 3). All three phenomena (dipolar and first order quadrupolar interactions, CSA) have a dependence on the second-order Legendre polynomial: $3 \cos^2 \Theta - 1$. That means, if Θ is chosen to be $54^{\circ}44'$ (the so-called magic angle), the expression $3 \cos^2 \Theta - 1$ becomes equal to zero. In consequence, the nuclear spin interactions discussed are reduced to their values in solution: the dipolar and first order quadrupolar interactions are removed, whereas the chemical shift anisotropy is reduced to their isotropic values. In some way, MAS operates as a substitute for the molecular motion in solids. MAS cannot completely average out the second order quadrupolar interaction, but the resonances are narrowed by a factor of 4 (13).

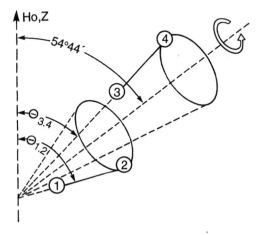

Fig. 3. Effect of magic angle spinning: By rotation about the magic angle the time averaged value of all binding vectors becomes $54^{\circ}44'$ (Reproduced by permission of Bruker GmbH, Karlsruhe).

436

The spinning speed of the rotor during a MAS experiment should be in the range of the line width of the signal recorded under static conditions, otherwise (which means at lower spinning speeds) the main resonance line is accompanied by a series of spinning side bands occuring at integral multiples of the spinning speed. However, even a slow MAS spinning speed provides a considerable narrowing of the lines broadened by chemical shift anisotropy and first order quadrupolar interactions, but still flanked by spinning side bands. Anyway, in order to avoid side bands, also in the case of chemical shift anisotropy, one would recommend a high spinning speed. Broad lines caused by dipolar coupling can only be removed completely if the MAS rotation frequency is higher than the strength of the dipolar decoupling.

Fortunalety, the dipolar and quadrupolar interactions are independing on the magnetic field strength, whereas the chemical shift anisotropy is proportional to the external magnetic field strength B_0. That means, higher spinning speeds must be applied at higher fields. Nowadays, quite high spinning speeds of many kHz can be realized, which are necessary when dealing with complex side band pattern of samples possessing strong dipolar interactions and/or large chemical shift anisotropies.

An extrem example for a nucleous with large chemical shift anisotropy is given in Figure 4. Applying a spin rate which is too low, the resulting spectrum is quite confusing (Figure 4b). Increasing the spinning speed improves the situation (Figure 4c), but does not solve the sideband problem. However, if the spinning speed is the maximum achievable, overlap of sidebands with others lines and assignment may become an important complication. In such cases the isotropic chemical shift may be identified by varying the spinning speed, whereby only the main line will remain invariant (13). Another possibility is the suppression of sidebands by applying **TOSS** pulse sequences (Total Suppression of Sidebands), which can, unfortunately, not be used in a quantitative manner.

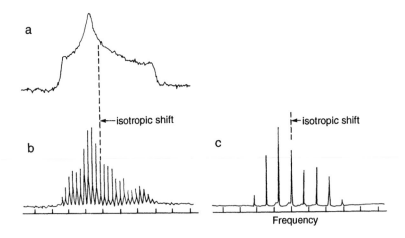

Fig. 4. Effect of spinning rate on the spectrum of a highly anisotropic nucleus: a) static sample, b) slow spinning, c) faster spinning (Reproduced by permission of John Wiley and Sons, New York).

d) Dynamic Angle Spinning (DAS) / Double Orientation Rotation (DOR)

Quadrupolar nuclei interact not only with the magnetic field in which the sample is placed but also with the electric field gradient. The combination of both effects results in an anisotropy that can not longer be removed by the magic-angle spinning alone (18). A detailed analysis of the averaging process of quadrupolar nuclei shows that second-order quadrupolar interactions depend on a fourth-order Legendre polynomial, described by 35 $\cos^4 \Theta$ - 30 $\cos^2 \Theta$ + 3 (18). By introducing only one magic angle, it is never possible to average dipolar interactions, CSA and second-order quadrupolar interactions simultaneously. Pines, Lippmaa and Samoson realized that introduction of two independant angles should average the effects of both tensors (19-21). It should be mentioned that the two magic angles are not unique. There is more than one "magic-angle pair" which can reduce the two terms to zero. In principle, there were two different ways to meet this challenge. One, which is called **dynamic angle spinning (DAS)**, is to rotate the sample sequentially about two different axes, inclined to the magnetic field at angles of 37.38^0 and 79.19^0: the sample is rotated about the first axis, reoriented, rotated about the second axis and so on. The other possibility is **double orientation rotation (DOR)**, in which the sample is spun simultaneously about two axes, the first inclined to the magnetic field at the magic angle and the second at the angle given by zero of the fourth-order Legendre polynomial, that means 30.56^0 (18) (see Figure 5).

To perform DOR in practice is a demanding task from an engineering point of view. The spinning rotors are driven by a flow of gas, which must be delivered to one rotor embedded in another (18). Double-angle rotation will certainly be a major new technique in structure elucidation of zeolites and $AlPO_4$ molecular sieves containing nuclei with quadrupolar moments. Figure 6 shows a ^{27}Al DOR spectrum of hydrated VPI-5 compared to the ^{27}Al MAS NMR spectrum, telling us that the broad single line pattern corresponding to tetrahedreally coordinated Al in the MAS spectrum contains much more structural information after recording the DOR spectrum (22).

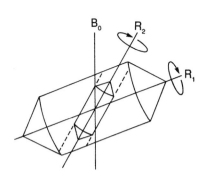

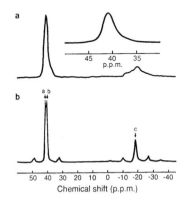

Fig. 5. Represenation of double rotation NMR (Reproduced by permission of Taylor & Francis Ltd.,Basingstoke).

Fig. 6. ^{27}Al NMR spectra of hydrated VPI-5 under conditions of (a) MAS and (b) DOR (Reproduced by permission of Macmillan Magazines Ltd., London).

e) Zero-field NMR Spectroscopy

Pines has developed a method called **zero-field NMR**, in which the sample is placed in a large magnetic field for spin polarization and is then shuttled to a region of low or zero field, where certain field-dependent broadenings disappear (23, 24). Application of zero-field NMR allows the measurement of dipolar couplings, from which distances can be determined. However, zero-field experiments suffer from low sensitivity and can only be applied to materials with reasonably long relaxation times T_1 to allow for the time it makes to shuttle the sample in and out of the field. This field cycling technique measures the evolution of the magnetization at zero-field with the sensitivity of high-field NMR. Zero-field NMR is particularly suited to the investigation of nuclei with quadrupolar moments, since it avoids orientation-dependent broadening.

f) Quadrupole Nutation NMR Spectroscopy

Quadrupole nutation NMR is a technique, introduced by Samoson and Lippmaa (25), where in a two-dimensional way the effect of the quadrupole interaction is separated from other line broadening interactions, simply by allowing the magnetization to evolve during an incremented periode t_1, under the influence of a strong radio frequency field.

As an example, quadrupole nutation NMR of nuclei with half-integer quadrupolar spin in zeolites can distinguish between nuclei of the same chemical element subjected to different quadrupole interactions, the signals of which overlap in conventional spectra. The situation is favourable for half-integer quadrupolar spins since the $m=1/2 <-> m= -1/2$ transition for these nuclei is broadened by the quadrupole interaction only in second-order perturbation theory. The technique can be usefully applied for the determination of the local environment of Al in zeolitic catalysts (26). It allows discrimination between species of similar chemical shift but different quadrupolar coupling constants. The main difficulty in the interpretation is the complex spectrum that results from a nutation experiment since it can consist of many overlapping powder patterns (27).

g) Cross Polarization (CP)

The spin-lattice relaxation times T_1 in solids can be very long and the recording time for a simple NMR spectrum of a solid could be extremely long without any experimental improvement. On the other hand, the transversal relaxation times T_2 are usually short. Since T_1 controls the recycle time between the experiments a method improving the sensitivity (i.e. the signal-to-noise ratio) of the spectra of rare spins (like silicon-29 or carbon-13) has been developed: **Cross polarization (CP)**, which does, however, not influence the resolution of the spectrum.

Cross polarization allows transfer of magnetization (or polarization) from an abundant species (usually ^1H) to a dilute species, which is under observation.

The cross polarization experiment can be described through the following steps:
1. Excitation of the abundant spins by a 90° pulse.
2. Magnetization transfer from the abundant spins to the dilute spins by simultaneously applying "matching" radio frequency fields to both type of spins (spin locking), according to the **Hartmann-Hahn** condition: $\gamma_1 \times B_1 = \gamma_2 \times B_2$ (γ = magnetogyric ratio). During this period both the abundant and dilute spins are in states with equal

fluctuation frequency of the z-magnetization, whereby mutual spin flips of abundant and rare spins are possible. The net magnetization is rapidly transferred from the abundant to the rare spins.

3. Acquisition of the free induction decay of the rare spins by continued irradiation of the ^{1}H field for heteronuclear dipolar decoupling.

4. Repetion of the cycle (see Figure 7).

The benefits are primarily an intensity enhancement of the dilute spin signal by a factor of γ abundant/γ dilute and a reduction of the recycle time between experiments since the rate-determining relaxation time is now that of the abundant species, rather than that of the rare spins. Usually, the relaxation of the abundant spins are much faster than the dilute spin relaxation (13). The cross polarization experiment may thus be repeated with much shorter intervals, leading to a further increase of the signal-to-noise ratio of the rare spin NMR spectrum within a given period of time. The effectiveness of the cross polarization experiment depends on the strength of the dipolar interaction between the abundant and rare spin systems, i.e. on the distance between the actual nuclei (proportional to r^{-3}, where r is the distance between the abundant and the dilute nuclei) (11). The efficiency of magnetization transfer decrease extremely fast as the distance between the abundant and rare spins increase. Bulk Si nuclei in SiO_2, for example, will be relatively distant from any proton species, whereas Si nuclei at the surface will be relatively close to surface OH groups. The application of a cross polarization ^{29}Si MAS NMR pulse sequence will therefore enable us to study Si nuclei at the surface separately from those in the bulk material.

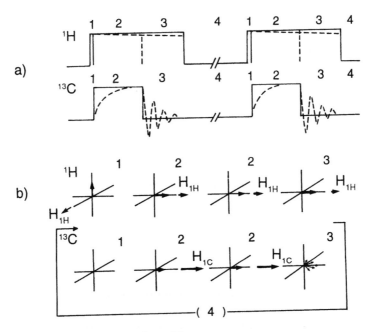

Fig. 7. a) Pulse sequence used for ^{1}H - ^{13}C cross polarization and b) behaviour of the ^{1}H and ^{13}C spin magnetization during the cross polarization sequence (Reproduced by permission of C.F.C. Press, Guelph).

In addition, one may make use of the differences in magnetization transfer rates for the discrimination between chemical species. Rigid or immobilized species experience a more efficient magnetization transfer process than more mobile molecules. In conclusion, the CP technique provides information about close coordinations between protons and dilute nuclei besides motional relationships (11, 13).

Since the cross polarization experiment relies on the availability of dipolar interactions, a problem may be arised when it is used at high fields. Since high MAS spinning speed must be applied, the dipolar interactions may be fully averaged, thereby destroying the CP efficiency (13).

One should emphasis, that under normal conditions, the CP experiment does not provide quantitative results. Finally, the cross polarization sequence does not influence the line width.

2.3. Combination of techniques to obtain high-resolution solid-state NMR spectra

In order to obtain optimum line narrowing and improved sensitivity in a solid-state NMR spectrum of a zeolite or $AlPO_4$ molecular sieve, the experimental techniques discussed in chapter 2.2. may be applied in combination, as, e.g., **CP/MAS, DD/MAS, CP/DOR** or **CRAMPS (Combined rotation and multiple pulse spectroscopy)**. An example of such a combination is shown in Figure 8. However, in the case of many inorganic samples,

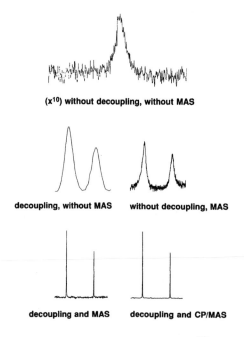

Fig. 8. Effect of different experimental conditions on the ^{13}C spectrum of adamantane (all spectra were measured under identical conditions), (Reproduced by permission of Bruker GmbH, Karlsruhe).

including zeolites and AlPO$_4$ molecular sieves, there are usually no protons (or only few) covalently bonded to the framework. In these cases one cannot apply cross-polarization and dipolar ^{1}H decoupling is not necessary. Therefore, application of MAS with a simple 90° pulse sequence (or DOR/DAS in the case of nuclei with quadrupolar moments) is sufficient to reduce the CSA (or quadrupolar broadenings in the case of spins with I > 1/2). If the CSA is not too large, the experiments can be carried out at higher magnetic fields, since sensitivity and resolution is maximized at high field, whereas the dipolar interactions are not effected and the line broadening due to second order quadrupolar interactions is minimized (11).

The **rotor synchronization** technique provides synchronization of radio frequency pulses to rotor positions for experiments like, for example, CP/MAS, CRAMPS or dynamically adjusted TOSS.

3. RELEVANT NUCLEI FOR HIGH-RESOLUTION SOLID-STATE NMR SPECTROSCOPY

All of the relevant basic nuclei contributing to the framework of zeolites and AlPO$_4$ molecular sieves are detecable to NMR investigations by its naturally isotopes: ^{29}Si, ^{27}Al, ^{31}P and ^{17}O. Both, the ^{31}P (I=1/2) and ^{27}Al isotopes are 100% abundant and spectra are easily detected within reasonable time. However, ^{27}Al has a quadrupole moment which can cause line broadening due to the interaction with the electric field gradient. The ^{17}O isotop has a low natural abundance (0.037%) besides a nuclear quadrupole moment (line broadening!), which in combination makes the registration of ^{17}O NMR spectra without enrichment almost impossible. Interesting investigations of ^{17}O NMR can be done by using enriched zeolite or AlPO$_4$ material.

The ^{29}Si isotope has a natural abundance of 4.7% and no quadrupole moment (I=1/2). The obtained resonance lines for ^{31}P and ^{29}Si are usually narrow, and due to their important role as framework elements (besides ^{27}Al), these nuclei have been widely used in solid-state NMR studies of zeolites and AlPO$_4$ molecular sieves for structural investigations.

Solid-state ^{1}H NMR of protons, OH groups, adsorbed H$_2$O, organic sorbates and probe molecules containing hydrogen in zeolites and AlPO$_4$ molecular sieves has been developed as a capably method for getting information about different kinds of hydrogen in terminal or bridging OH groups, varying environments for hydrogen containing probe molecules and acidity investigations.

Other nuclei which can substitute isomorphously the usual framework elements in zeolites or AlPO$_4$ molecular sieves are observable by solid-state NMR, e.g. ^{11}B, ^{9}Be, ^{73}Ge and ^{69,71}Ga. Charge compensating cations, like ^{7}Li, ^{23}Na, ^{39}K, ^{133}Cs or ^{195}Pt, are suitable for NMR experiments. However, most of those elements possess a quadrupole moment, which limits usually the application.

Furthermore, organic compounds used as templates during hydrothermal synthesis or as sorbates in the zeolite framework can be detected by applying ^{13}C CP/MAS NMR spectroscopy.

Finally, ^{129}Xe is a very suitable and sensitive isotope for probing the pore architecture of zeolites and $AlPO_4$ molecular sieves. The extended Xe electron cloud is easily deformable due to interactions between, e.g. the Xe atoms and the channel wall of a zeolite framework, and deformation results in a large low-field shift of the Xe resonance. In addition, ^{129}Xe NMR can be used to study metal particles in zeolites, while reduction-oxidation reactions can be monitored (13). Table 2 summarizes the NMR properties of a number of nuclei which have been used in NMR investigations of zeolites and $AlPO_4$ molecular sieves (11).

4. TWO-DIMENSIONAL (2D) SOLID-STATE NMR SPECTROSCOPY

High-resolution solid-state NMR spectroscopy has nowadays become a standard technique in many laboratories, whereas two-dimensional solid-state NMR has not yet reached that status. In liquid-state NMR, the use of two-dimensional techniques provides a lot of information on the connectivities between atoms within molecular structures. Although many restrictions apply, 2D NMR techniques allow to get increased information in the solid state as well, like Si-O-Si connectivities in zeolite frameworks (28, 29).

The 2D NMR experiment usually consists of three main periods, which can be described as follows:
1. **Preparation**: The spin system is in a well defined state and spin-lattice relaxation occurs after a 90° pulse (or CP sequence).
2. **Evolution**: The spin system processes depending upon the influence of different nuclear spin interactions. This period is terminated by applying a second radio frequency pulse. **Compared to a one-dimensional (1D) NMR experiment, this period is variable and not fixed** ($t_1 + \Delta t_1$ with n number of experiments).
3. **Detection**: Recording of the free induction decay (FID) followed by processing of the resonance signal.

Figure 9 shows the pulse sequence and the spectrum of a 2D NMR experiment according to the so-called **COSY (Correlation Spectroscopy)** technique (15). The 2D NMR spectrum is processed after successive Fourier transformation (FT) after t_2 and t_1: The receiver signal is a function of both t_1 and t_2, and we obtain two frequency variables, F_1 and F_2 (see Figure 9). The 2D COSY experiment is based on measurements of direct bondings (scalar couplings, through bond couplings) within a molecular structure, which allows to establish bonding connectivities between the lattice elements of a zeolite or $AlPO_4$ molecular sieve framework. The COSY spectrum can be interpretated as follows: After the second 90° pulse we have two different cases, depending on the coupling situation:

1. Non-coupling nuclei (J=0): Only t_1 depending modulation of the magnetization, and we expect only signals on the diagonal, since $F_1 = F_2$.
2. Coupling nuclei (J≠0): Magnetization is now depending also on the scalar interactions, and we expect off-diagonal signals (cross peaks).

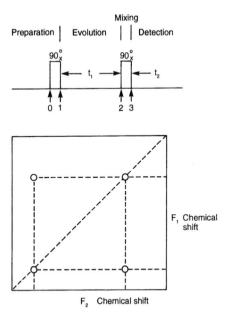

Fig. 9. Pulse sequence and schematic representation of a contour plot of a 2D COSY NMR experiment (Reproduced by permission of C.A. Fyfe, Vancouver).

Fyfe et al. have done pioneering work in connection with the introduction of 2D solid-state NMR spectroscopy with respect to three dimensional connectivities within zeolite lattices (30-38). Application of 2D homonuclear COSY (both frequency axes represent ^{29}Si chemical shifts) enabled Fyfe et al. to yield the correct connectivities for the (known) lattice structure of ZSM-39 (30, 31), see Figure 10. As seen in the Figure, a clear connectivity is established between T_1 and T_2 as well as for T_2 and T_3, but not between T_1 and T_3. This finding is in line with the known structure of ZSM-39 (space group of the high-temperature form is $Fd3$). In this study, the used sample of ZSM-39 was enriched to about 85% in ^{29}Si in order to increase the signal-to-noise ratio and enhance the connectivities that result from the ^{29}Si-O-^{29}Si interactions. Fyfe et al. showed that natural-abundance experiments on these systems are indeed feasible, making the technique quite generally applicable (30-32).

Exactly the same results were obtained with **spin diffusion** experiments performed at ZSM-39 (see Figure 11) (31). Frequencies can be affected by spin diffusion between sites having different NMR parameters, when, for example, magnetization is transported through a solid by means of mutual spin "flip-flops" that can occur even in the absence of atomic or molecular motion (39). By monitoring the correlation among frequencies in the different dimensions of a multidimensional NMR experiment, it is possible to learn about the mechanisms and rates of reorientation and diffusion processes in solids (39).

The three-dimensional connectivities in the zeolites ZSM-12, ZSM-22 (KZ-2), DD3R (deca-dodecasil 3R), ZSM-5 and ZSM-23 have been investigated by Fyfe and coworkers using the 2D COSY and **2D INADEQUATE (Incredible Natural Abundance Double Quantum Transfer Experiment)** sequences (29, 32-37). In the 2D INADEQUATE, only

444

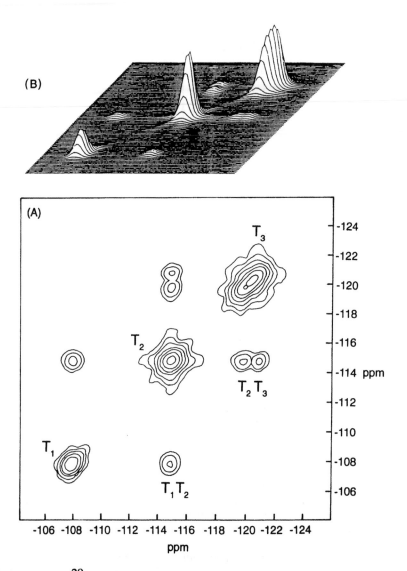

Fig. 10. Results of a ^{29}Si COSY experiment on ZSM-39 carried out at 373 K: A) contour plot, B) stacked plot (Reproduced by permission of the Royal Society of Chemistry, Cambridge). Schematic representation of the zeolite ZSM-39 lattice framework: see Figure 11.

double quantum coherence signals are observed, that means only signals created by coupled ^{29}Si nuclei. The F_2 domain is the normal chemical shift frequency scale and the F_1 domain represents the double-quantum frequencies of the resonances. Connected signals occur equally on either side of the diagonal of the plot at the same frequencies in F_2 as the corresponding resonances from a simple 1D experiment (34).

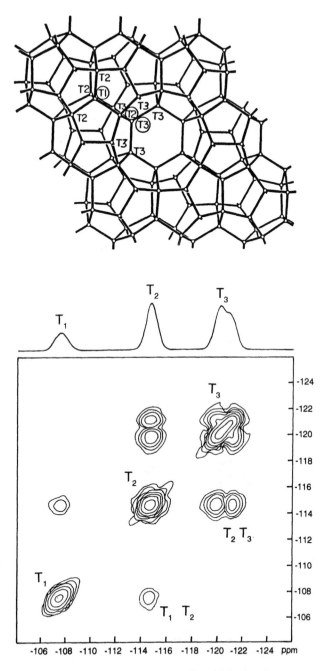

Fig. 11. Schematic representation of the zeolite ZSM-39 lattice framework and the contour plot of a ^{29}Si 2D spin-diffusion experiment on ZSM-39 (Reproduced by permission of the American Chemical Society, Washington).

The COSY experiment is quite easy to carry out, since it does not require detailed knowledge of the coupling constants. However, this experiment was originally designed to indicate homonuclear scalar couplings between abundant nuclei, and its application to the case of dilute nuclei, like ^{29}Si, creates the following difficulty: Most of the ^{29}Si nuclei doesn't show homonuclear scalar coupling, which results in the registration of intense broad signals on the diagonal, hiding cross peaks due to connectivities between spins of small chemical shift differences, like the coupling between T_4 and T_6 in the 2D COSY of ZSM-12 (see Figure 12). Zeolite ZSM-12 is a three-dimensional framework structure in which the asymmetric unit consists of seven crystallographically inequivalent T sites (29, 33).

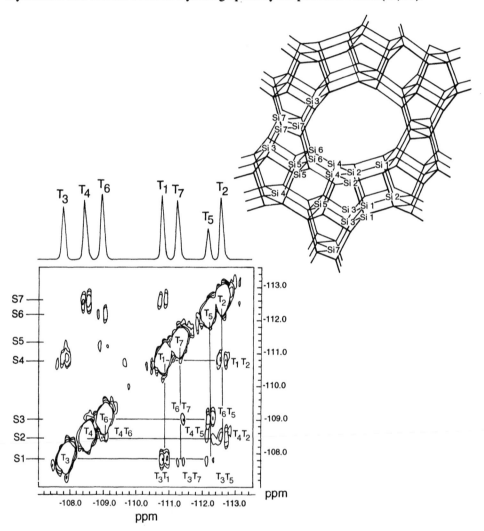

Fig. 12. Schematic representation of the zeolite ZSM-12 lattice framework and the contour plot of a ^{29}Si COSY experiment on ZSM-12 (Reproduced by permission of the American Chemical Society, Washington).

Compared to the 2D COSY experiment, the 2D INADEQUATE experiment has a number of advantages: Those signals due to isolated spins are suppressed (diagonal peaks), which improves the required dynamic range for the connectivities and the detectibility of satellite signals. In addition, a better S/N ratio may be reached and any spinning rates may be used, since spinning side bands will not appear in the spectrum. Therefore, it is much easier to observe connectivities between resonances close in frequency which occur close to the large diagonal signals in the COSY experiment (33). Figure 13 shows the 2D INADEQUATE experiment on ZSM-12, including the connectivities between T_4 and T_6, which is clearly resolved here but was ambiguous in the 2D COSY experiment (see Figure 12) due to the close proximity of the cross-peaks to the diagonal (33).

However, the main drawback of the INADEQUATE experiment is that a reasonable good estimate of the coupling constant must be available, otherwise the experiment will be very inefficient (29). The direct measurement of the coupling constants in solids is normally not possible due to the large linewidths. However, the 2D spectra recorded by Fyfe et al. allow the measurement of the ^{29}Si-O-^{29}Si scalar couplings in zeolite ZSM-12, ranging from 10-15 Hz (33).

In the cases of ZSM-12 and ZSM-22, the results of the natural abundance $^{29}Si/^{29}Si$ COSY and INADEQUATE 2D NMR experiments are in exact agreement with the lattice structures, the INADEQUATE experiment being particularly successful by detecting all of the connectivities (33).

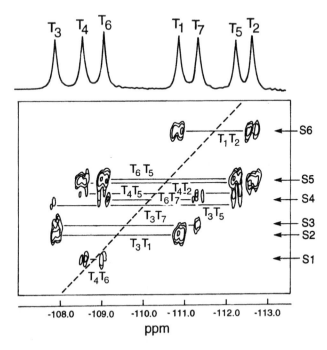

Fig. 13. Contour plot of a ^{29}Si INADEQUATE experiment on ZSM-12 at 300 K (Reproduced by permission of the American Chemical Society, Washington).

The three-dimensional connectivities in the structure of zeolite DD3R have been studied by Fyfe and coworkers (36). 2D COSY experiments on a 85% ^{29}Si enriched sample are in agreement with the expected connectivities. In addition, the results indicate that the structure is of lower symmetry than expected. 2D INADEQUATE experiments confirm these results, including the lower symmetry of the structure, and, in this case, all of the expected connectivities were observed. A high-temperature INADEQUATE experiment shows that the symmetry of the structure increases to that expected from the previous single crystal XRD measurement (36).

One of the most complex zeolite structures known, ZSM-5, has been investigated using 2D solid-state NMR techniques in order to study the three-dimensional bonding connectivities (34). The ZSM-5 structure is difficult to investigate by 2D NMR connectivity experiments since the known structures contain either 24 T sites (monoclinic room temperature form) or 12 T sites (orthorhombic form induced by increasing the temperature or by adding two molecules of p-xylene per unit cell). Therefore, a very large number of ^{29}Si-O-^{29}Si connectivities will occur within the 2D plots, and a substantial number of connectivities between signals of similar resonance frequencies will be obscured in the COSY experiments (34).

The 2D INADEQUATE experiments on ZSM-5 gave substantially better results than 2D COSY experiments on the same highly siliceous samples. In the case of the orthorhombic form (12 T sites), almost all of the expected connectivities were observed, whereas for the monoclinic form (24 T sites), 38 of the total of 48 connectivities were clearly registrated, allowing the assignment of the spectrum (34, 35).

An evaluation of chemical shift-structure correlations from a combination of XRD and 2D ^{29}Si MAS NMR data (INADEQUATE) was done for highly siliceous ZSM-12, -22 and -5, resulting in good correlations with a variety of structure related functions for the best data sets on single crystals of ZSM-5 (38).

Klinowski and coworkers report about a 2D J-scaled ^{29}Si COSY experiment applied to a sample of highly siliceous mordenite, revealing three cross peaks, which were assigned to the four distinct crystallographic sites of mordenite in the following manner: T_1 : T_3 : T_2+T_4 (40). He points out that during the J-scaled COSY pulse sequence an upscaling of the scalar couplings involved in the COSY experiment takes place, leading to enhanced cross peak intensities and consequently improved spectral resolution between adjacent diagonal and cross peaks (40).

So far, the 2D INADEQUATE experiment turns out to be the method of choice with respect to investigations of three-dimensional bonding connectivities within zeolite lattices. This approach is of general interest, since the experiments can be carried out on ^{29}Si natural abundant zeolite samples. However, the studies up to now have been performed on samples of known structure. Fyfe et al. report for the first time the application to a structure, which details are less well established: zeolite ZSM-23 (37). This highly siliceous zeolite is reported with an orthorhombic unit cell (space group *Pmmn*) with 24 T atoms. The proposed structure has a one-dimensional pore system with noncircular channels (41). The NMR results reveal that there are twelve inequivalent T sites with equal occupancies in the unit cell. The 2D INADEQUATE experiment shows a complete ^{29}Si-O-^{29}Si connectivity pattern, but the results are not consistent with the proposed interpretation of the XRD for this zeolite. Even connectivity patterns for different lower-symmetry space groups gave no correspondence. The NMR data must probably be combined with synchrotron X-ray diffraction measurements in order to obtain consistent results (37).

The aluminophosphate VPI-5 has been in the focus for scientists dealing with microporous materials, since this extra-large pore system was discovered a few years ago. The recent NMR results about this aluminophosphate will be discussed in chapter 5.1.2.1., however, interesting solid-state 2D polarization transfer experiments involving quadrupolar nuclei have been reported recently (42-45). As pointed out earlier, the cross-polarization experiments have been used to transfer magnetization from abundant to rare spin systems. Fyfe et al. (42-44) and Veeman and coworkers (45) introduced cross-polarization experiments involving **spin coherence transfer** from quadrupolar spin systems to spin-1/2 nuclei, utilizing the very short T_1 relaxation times of quadrupolar nuclei to detect spectra of spin-1/2 systems within a reasonable time schedule. Both **2D heteronuclear correlation** experiments using cross-polarization (where the ^{27}Al polarization is transferred to the ^{31}P spins during spin-locking and the ^{31}P free induction decay is recorded) and **2D TEDOR (Transferred-echo double resonance)** experiment have been registered for VPI-5 (42-44) (see Figure 14). Both 2D spectra reveal cross peaks between all three ^{31}P resonances and

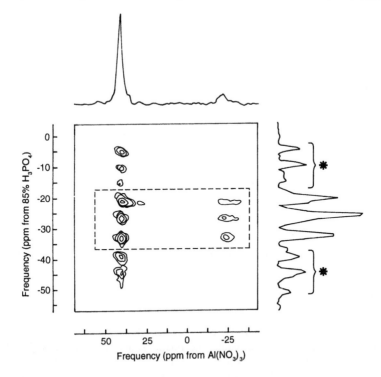

Fig. 14. 2D ^{27}Al to ^{31}P TEDOR experiment, displaying connectivities between the three ^{31}P resonances and the two resolved ^{27}Al resonances (Reproduced by permission of Elsevier Science Publishers B.V., Amsterdam).

both the resonances from the tetrahedrally and octahedrally coordinated ^{27}Al sites are in agreement with the crystal structure of VPI-5. The examples demonstrate the usefulness of TEDOR experiments to observe spin-1/2 nuclei with long T_1 relaxation times which are close to quadrupolar spins, like in zeolites and AlPO$_4$ molecular sieves. These experiments (besides TEDOR another dipolar dephasing double resonance experiment was used: **REDOR (Rotational echo double resonance)**) allow to determine connectivities between quadrupolar and spin-1/2 nuclei in zeolite and AlPO$_4$ molecular sieves (44). REDOR uses rotational echos from MAS spectra and measures directly dipolar couplings between nuclei, thus determining internuclear distances. In rotational resonance, a multiple of the spinning speed is matched to the chemical shift difference between two nuclei to produce information about the dipolar coupling between the two spins. This technique seems to become the solid-state equivalent of the **NOESY (Nuclear Overhauser Enhancement and Exchange Spectroscopy)** experiment, since it provides through space distance information (24).

Veeman et al. (45) applied a REDOR pulse sequence in order to record the 2D heteronuclear ^{27}Al-^{31}P correlation spectrum of VPI-5, confirming the proposed structure of VPI-5 and the results obtained by Fyfe et al. (42-44). In addition, since the intensities of the cross peaks differ, he and his coworkers conclude with an assignment of the three phosphorus sites in hydrated VPI-5 which will be reviewed in chapter 5.1.2.1.

Klinowski et al. showed that ^{31}P NMR spin diffusion in hydrated VPI-5 can be suppressed by fast MAS but not by high-power decoupling (46). In the same paper the authors report on ^{13}C NMR spin diffusion experiments *in situ* on the products of methanol conversion into gasoline over ZSM-5 with respect to spectral assignments. Furthermore, a 2D ^{29}Si J-scaled COSY spectrum of highly siliceous mordenite revealed the connectivities of tetrahedral sites, permitting an unambiguous assignment of all signals (46).

2D nutation NMR spectroscopy has been used to receive detailed information about interactions of quadrupolar nuclei in zeolitic systems. Application of MAS at high magnetic field strength B_0, high power radio frequency irradiation, small spherical sample volumes and careful data processing lead to highly resolved 2D ^{23}Na nutation spectra of sodalite containing different enclathrated sodium salts. Up to six lines have been resolved in the F_2 dimension over a chemical shift range of 20 ppm with minimun shift differences of less than 3 ppm (47).

5. THE STRUCTURE OF ZEOLITES AND ALPO$_4$ MOLECULAR SIEVES

5.1. Framework Characterization
5.1.1. Zeolites
^{29}Si NMR

A number of excellent review papers have been published during the 1980ies, summarizing the structural information and relationships available through the ^{29}Si NMR data on zeolites. At this stage, I would like to mention briefly the most important highlights from this period and draw the readers attention for further details to those review papers (3-5, 7, 8, 10-13).

One of the most important results of Lippmaa, Engelhardt and coworkers (1,2) was to establish the relationship between the ^{29}Si chemical shift sensitivity and the degree of condensation of the silicon-oxygen tetrahedra, i.e. the number and type of tetrahedrally coordinated atoms (T-atoms, with T= Si, Al or other lattice atoms) connected to a given SiO$_4$ unit. The degree of condensation is symbolized by Si (n Al), with n=0, 1, 2, 3 or 4, where n indicates the number of Al atoms sharing oxygens with the SiO$_4$ tetrahedron under consideration. Furthermore, the ^{29}Si chemical shift is influenced by the Si-O-T bond angle and silicon-oxygen bond length, which means, that chemically equivalent but crystallographically inequivalent Si nuclei may have different chemical shifts (3). Chemical shift ranges for the different SiO$_4$ units are given in Figure 15, and an example of the ^{29}Si MAS NMR spectrum of Y zeolite (Si/Al=2.6) is shown in Figure 16. Generally, the ^{29}Si resonance shifts to lower field by ca. 5 ppm per additional aluminium. Therefore, up to five lines per crystallographically inequivalent Si site may be observed in the ^{29}Si NMR spectrum. In conclusion, quite a bit of information about the local environment of the SiO$_4$ tetrahedra forming the zeolite lattice can be obtained from the ^{29}Si chemical shift data (11).

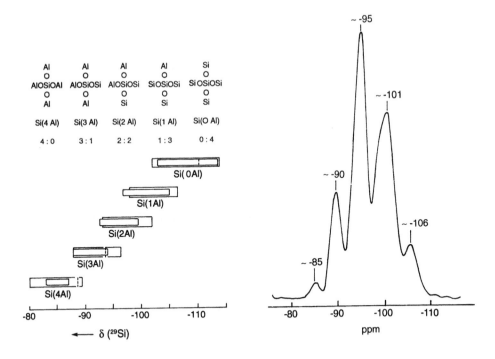

Fig. 15. The five possible local environments of a Si atom in a zeolite framework together with the corresponding chemical shift ranges (Reproduced by permission of C.A. Fyfe, Vancouver).

Fig. 16. ^{29}Si MAS NMR spectrum of Na Y zeolite (Si/Al ratio=2.6).

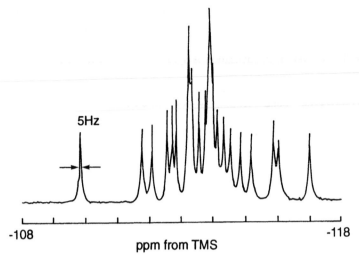

Fig. 17. ^{29}Si MAS NMR spectrum of highly siliceous ZSM-5 (Reproduced by permission of Macmillan Magazines Ltd., London).

One of the most famous examples is the ^{29}Si MAS NMR spectrum of highly siliceous ZSM-5 (Figure 17), which shows at 295 K 20 well resolved lines (linewidths as narrow as 5 Hz), representing the 24 crystallographically distinct Si sites of the monoclinic form of ZSM-5 (48).

Highly resolved ^{29}Si- and ^{13}C MAS NMR spectra of pure silica zeolites of the MFI-type, synthesized according to a new route, reveal new features, which are indicative of the very low defect content and the very high crystallinity of these materials (49). Twenty different lines belonging to the 24 crystallographically nonequivalent Si sites of the monoclinic structure were observed (linewidth of 9 Hz) for the calcined sample (49).

A new CP-technique using fluorine as a magnetization source for silicon was applied to a purely siliceous MFI-type zeolite and octadecasil, synthesized in fluoride medium. ^{19}F-^{29}Si CP/MAS NMR proved to be a useful tool to improve the sensitivity of the Si atoms and also a discriminating technique to specify the location of the fluorine in the zeolitic lattice (50).

Finally, the quantitative ratio of tetrahedral Si and Al in the framework of a zeolite can be directly calculated from the signal intensities (I) according to the following equation:

$$\frac{Si}{Al} = \frac{\sum\limits_{n=0}^{4} I_{Si(n\,Al)}}{\sum\limits_{n=0}^{4} 0.25\, n I_{Si(n\,Al)}} \qquad (1)$$

This method has been used successfully in connection with investigations of dealuminated (ultrastabilized) and realuminated zeolites (see Figure 18). The successful dealumination of zeolite Y by H_4EDTA was demonstrated by Klinowski et al. who confirmed the creation of hydroxyl nests in the zeolite lattice (51). One should emphasize, that **NMR** yields the **framework Si/Al** ratio since only lattice Si and Al are detected, whereas **elemental analysis** provides the **total sample composition**. In addition, equation 1 is independent of the specific structure of the zeolite, but cannot be directly applied to spectra containing overlapping signals from Si (n Al) units of crystallographically non-equivalent Si sites.

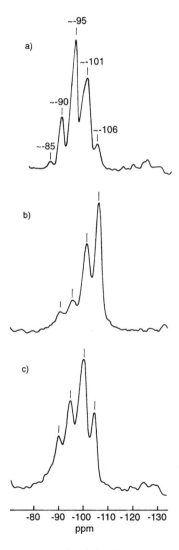

Fig. 18. ^{29}Si MAS NMR spectra of Na Y zeolite: a) parent material, b) dealuminated (ion exchange and steam treatment) and c) realuminated (KOH treatment).

OH groups (or coordinated water molecules) connected to the outer or inner surface of the porous structure of a zeolite or in lattice defects can be detected by the ^{1}H - ^{29}Si cross-polarization technique. Strong intensity enhancements are observed in the CP spectra for the resonances of $(TO)_3SiOH$ or $(TO)_2Si(OH)_2$ but not of $Si(OT)_4$ environments (3).

The differences in the ^{29}Si chemical shifts of Si nuclei located in the same chemical environment are mainly due to changes in the bonding geometry of the corresponding Si (n Al) unit. It has been shown, that the chemical shift of a certain Si (n Al) unit is linearly correlated with the average of the four SiOT bond angles at the central Si atom and also with the average SiO bond length. By means of linear regression analysis quantitative relationships can be established and used to estimate mean SiOT bond angles from ^{29}Si chemical shifts in zeolites, since a chemical shift change of about 0.6 ppm is to be expected for 1^0 change in bond angle, with high-field shifts for larger angles (3).

Usually there are two factors in the solid-state ^{29}Si NMR spectra of zeolites contributing to line broadening: chemical shift anisotropy of the ^{29}Si atoms and heteronuclear dipolar interactions between ^{29}Si and ^{27}Al or other NMR active nuclei (8, 11). The application of the MAS technique is therefore essential, and usually sufficient, for recording highly resolved ^{29}Si NMR spectra of zeolites (11). Removal of Al from the lattice and replacing it with Si reduces the line width to about 1 ppm, indicating that Al is responsible for the line broadening (9).

The high resolution solid-state NMR spectra received for high-silica zeolites indicate that zeolites by no means are rigid structures. Dramatic sorbate-specific spectral changes were observed after absorption of organic compounds (benzene, xylene etc.) in, for example, ZSM-5, and similar effects were observed with temperature variations. These changes reflect alterations in bond angles and lengths in the lattice of zeolites (13).

The ^{29}Si spin-lattice relaxation times (T_1) in zeolites are relatively short. Values in the range of 5-30 seconds have been measured in various zeolites, with only small variations with respect to the number of Al atoms in a Si (n Al) unit or changes in the Si/Al ratio (12). However, T_1 values of crystallographically inequivalent ^{29}Si nuclei often differ significantly within the same sample (12). The influence of oxygen upon the spin-lattice relaxation times is dramatic, whereas water and paramagnetic impurities play a less dominant role in relaxation (10). For example, in dealuminated mordenite T_1 of the three signals is reduced from 236, 210 and 225 seconds under argon to 0.4, 0.26 and 0.27 seconds, respectively, under oxygen, i.e. by three orders of magnitude (52). This (and other reasons) explain why short recycle times can by applied (normally 5 seconds) when recording ^{29}Si MAS NMR spectra of zeolites, still obtaining quantitatively reliable spectra (12). The shortening of the spin-lattice relaxation times is simply due to the accessibility of Si nuclei to oxygen in the porous structure of zeolites and explains why relaxation in compact silicates (like quartz or glasses) is much slower (10).

In the following, we will focus on the recent results from the last few years dealing with the framework characterization of zeolites.

Weller et al. report about new limits of ^{29}Si solid-state NMR chemical shift ranges for the single resonance of sodalites with different compositions and containing various type of metals. The resonance frequency correlates well with the geometry of the $Si(OAl)_4$ tetrahedron and with the lattice parameters, determined by powder neutron diffraction. In particular, they report chemical shift values for these materials for silicon in a $Si(OAl)_4$ environment covering the range from -76 to -93 ppm, compared to former values between -82 and -89 ppm (53). In addition, the authors determined the variation of the ^{29}Si resonance frequency with the Si-O-Al bond angle (Θ) by neutron diffraction, fitting the expression $\delta_{Si} = 1.89 - 0.631\,\Theta$ (53). Thus, the framework geometry in this class of materials can be determined even when the compounds are of poor crystallinity.

Decadodecasil-3R (DD3R), a novel microporous clathrasil, has been studied with respect to a correlation between the ^{29}Si NMR data and the mean Si-Si distances, determined most accurately by XRD. The experimental ^{29}Si MAS NMR spectrum was correlated with a simulated pattern based on XRD data, allowing assignments of the various signal intensities to different Si sites. The presented data consistently contribute to the idea that there indeed exists a general correlation between structural parameters and chemical shifts (54). The most exciting feature of the DD3R structure is the presence of large cavities (19-hedra), which are interconnected by 8-rings of SiO_4 tetrahedra. The DD3R can therefore be considered to be the link between the normally nonporous clathrasils and the open Si rich zeolites (54).

Certain interest has been concentrated upon zeolite beta, since this material can be synthesized with Si/Al ratios higher than 3. This zeolite is of particular interest with respect to cracking and hydrocracking of large hydrocarbon molecules requirering large pore microporous materials. The structure of zeolite beta was reported recently by two independent groups (55-57). This zeolite is a three-dimensional 12-membered ring system, consisting of an intergrowth of one tetragonal (A) and two monoclinic (B and C) closely related crystalline polytype phases (55-57). Pérez-Pariente et al. observed four different signals in the ^{29}Si MAS NMR spectra of beta samples with Si/Al ratios between 7.6 and 110 (58), the lines centered at -104, -111.5, -113 and -115.7 ppm for beta with Si/Al > 100. However, Fyfe et al. showed that the ^{29}Si NMR spectrum of highly siliceous beta consists of a minimum of nine peaks devided in three sets of lines centered at about -111, -113 and -115 ppm, respectively, which supports the existence of nine nonequivalent tetrahedral sites in the unit cell of zeolite beta (59). An attempt to assign the lines recorded by Pérez-Pariente et al. was done: the lines detected above -111 ppm in samples with high Si content could correspond to nonequivalent Si nuclei, with no Al in the second coordination sphere, occupying different structural sites in the zeolite. The presence of the -104 ppm signal in highly siliceous samples was explained by the existence of Si in Si (3 Si, 1 OH) environments (58), whereas Fyfe et al. did not correlate the ^{29}Si NMR data with the proposed structure for zeolite beta (59).

^{29}Si NMR investigations related to the framework of zeolite theta-1 have been carried out by Challoner et al. (60) and Fyfe et al. (59). This unidimensional 10-membered ring zeolite has 24 T-sites per unit cell, with four crystallogaphically distinct positions. These crystallographically nonequivalent sites have been recorded at -110.9 (T_3), -112.8 (T_1), -113.1 (T_2) and 114.3 ppm (T_4), with a relative intensity ratio of 2:1:1:2, respectively (59). This spectrum is in good agreement with structural data of Liebau (61). In addition, Challoner et al. (60) measured the ^{29}Si T_1 relaxation times of the different T sites within the same order of magnitude (ranging from 119 to 138 seconds).

Tuel at al. (62) used variable temperature ^{29}Si MAS NMR spectroscopy to correlate the monoclinic-orthorhombic transition of the titanosilicate TS-1 with the Ti content. TS-1 is a zeolite of the pentasil family where a fraction of the Si ions are replaced by Ti ions, and this material has been found to be a remarkable catalyst for the oxidation of various organic substrates with H_2O_2 (62). The ^{29}Si MAS NMR spectra of TS-1 at temperatures over 275 K exhibit a broad signal at -113 ppm (Si-O-Si connectivities) associated with a more or less pronounced shoulder at -116 ppm (due to distorted Si-O-Si bonds resulting from the presence of Si-O-Ti linkages). Below this temperature, the shoulder disappears and the spectra are similar to those of ZSM-5 recorded at room temperature. The transition temperature for the investigated sample was centered around 275 K (see Figure 19) (62). As shown in the Figure, the transition temperature decreased almost linearly with the Si/Ti ratio. Such a relationship is useful as a tool to investigate the extent of isomorphous substitution.

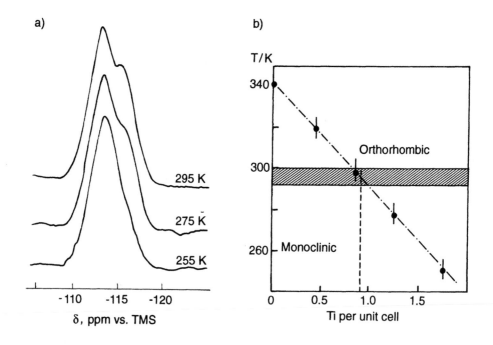

Fig. 19. ^{29}Si MAS NMR spectra of TS-1 recorded at different temperatures (a). Variation of the monoclinic to orthorhombic transition temperature as a function of the Ti content in the zeolite (b) (Reproduced by permission of the Royal Society of Chemistry, Cambridge).

The hydrothermal dealumination of zeolite ZSM-20 (intergrowth of FAU and EMT) was followed by Stöcker et al. using ^{29}Si MAS NMR spectroscopy (63). The starting material with a Si/Al-ratio of 3.9 was steamed at 700°C for 5 days, resulting in a spectrum consisting of a narrow single line at -108 ppm (see Figure 20). The broad signal of low intensity centered around the sharp peak is due to formation of some amorphous material (63). Davis reports ^{29}Si MAS NMR data of another intergrowth of hexagonal and cubic faujasite, called VPI-6, showing only resonances for Si(4Al) at -84.9 and Si(3Al) at -88.5 ppm (64). Weitkamp et al. recorded ^{29}Si MAS NMR spectra of EMT consisting of three intense lines at -93.5 (Si(2Al)), -99.8 (Si(1Al)) and -105.6 ppm (Si(0Al)) along with a small line at -87.5 ppm (Si(3Al)) (65). Only tetrahedrally coordinated Al was observed through the ^{27}Al MAS NMR spectrum (signal at 61.3 ppm).

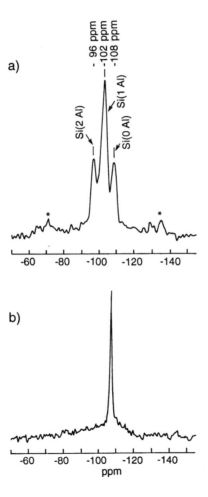

Fig. 20. ^{29}Si MAS NMR spectra of zeolite ZSM-20: a) parent material and b) dealuminated by ion exchange and steam treatment (Reproduced by permission of Munksgaard, Copenhagen (Acta Chemica Scandinavica)).

Zeolites containing three-membered ring systems have been investigated by Annen et al. applying ^{29}Si MAS NMR spectroscopy (66, 67). Lovdarite (beryllosilicate mineral for which a synthetic analogue has been reported) and VPI-7 (novel zincosilicate molecular sieve which is structurally related to lovdarite) exhibit resonances below -80 ppm, which are assigned to the T-sites at the centers of the so-called "spiro-5" units that are shared by two three-membered rings (3 MR). The 3 MR in lovdarite and VPI-7 are in the form of "spiro-5" units, which is composed of five T-atoms that form two fused 3 MR (see Figure 21). This low-field resonance is important, since ^{29}Si NMR chemical shifts are typically in the range of -85 to -110 ppm for zeolites. ZSM-18, which contains isolated 3 MRs does not exhibit ^{29}Si resonances below -86 ppm, which means, that ^{29}Si MAS NMR spectroscopy can be used as a test for the presence of "spiro-5" units in zeolites. The ^{29}Si NMR spectra of VPI-8, VPI-9 and VPI-10 (novel phases with unknown structure) did not show lines below -80 ppm. Therefore, it is unlikely that any of these materials contain "spiro-5" units (66).

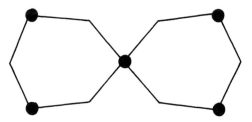

Fig. 21. Schematic illustration of the "spiro-5" building unit (Reproduced by permission of Elsevier Science Publishers B.V., Amsterdam).

Driven equilibrium and spin-echo Fourier Transform spectroscopy (DEFT and **SEFT)** have been used to observe ^{29}Si high-resolution solid-state MAS NMR spectra of the zeolites omega, L, mordenite and ZSM-5. The sequences allow the measurements of T_2 relaxation times, dramatic improvements in signal-to-noise ratios and selective observation of signals with long T_2 relaxation times (68).

The effect of a paramagnetic Cu^{2+} ion exchange at Cu X zeolite on the NMR line broadening was studied by using ^{29}Si MAS NMR (69). The NMR spectrum showed very selective, local line broadening effects for the Si atoms in the SiO_4 framework units.

^{27}Al NMR

The main applications of ^{27}Al NMR spectroscopy in connection with zeolites has been the monitoring of dealumination processes and the detection of extra-framework aluminium, since extra-lattice aluminium is octahedrally coordinated and gives rise to resonance lines at about 0 ppm, whereas framework aluminium is tetrahedrally coordinated and resonates between ca. 50 and 65 ppm.

As mentioned before, ^{27}Al has a nuclear quadrupole moment (I=5/2) which interacts with the electric field gradient caused by the non-spherically symmetric charge distribution around the ^{27}Al nucleus. Both, line broadening/distortion and chemical shift changes may arise from those quadrupolar interactions resulting in more difficult interpretations of the ^{27}Al NMR spectra. On the other hand, the high sensitivity of the ^{27}Al isotope (natural abundance of 100% and generally short relaxation times) allows the determination of very small amounts of aluminium in the samples. Spin-lattice relaxation times (T_1) of ^{27}Al in zeolitic frameworks have been measured by Haase et al. in the range of 0.3 to 70 milliseconds (depending on the temperature) (70). The relaxation is governed by the quadrupolar interaction with the electric field gradients. T_1 dramatically increases upon dehydration of the samples (12).

Usually, the only aluminium transition recorded in zeolite and AlPO$_4$ molecular sieve samples is the central +1/2 <-> -1/2 transition, which is only depending on the second order quadrupolar interaction. This interaction decreases with increasing magnetic field strength B_0 (see Figure 22) and can partly be reduced by MAS. However, complete removal of the second order quadrupolar interaction can be achieved by applying either DOR or DAS (see chapter 2.2.).

Obeying Loewenstein's rule (forbidden Al-O-Al linkages), zeolites and AlPO$_4$ molecular sieves give rise to quite simple ^{27}Al NMR spectra, consisting of signals due to only one type of tetrahedral aluminium environment [Al (OSi)$_4$] besides evtl. octahedrally coordinated aluminium. No relationships between the chemical shifts and the Si, Al ordering or Si/Al ratio have been established, whereas relations between the ^{27}Al chemical shifts and mean Si-O-Al bond angles exist, with shift values carefully corrected for quadrupolar shift contributions (11).

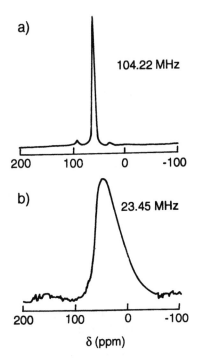

Fig. 22. ^{27}Al MAS NMR spectra of Y zeolite at different magnetic field strenghts:
a) 104.22 MHz and b) 23.45 MHz (Reproduced by permission of C.A. Fyfe, Vancouver).

The quantitative use of ^{27}Al NMR data allows determination of the relative proportions of lattice and extra-lattice Al in zeolite samples, providing that all aluminium is detected in the spectra (no signal intensity loss due to quadrupolar interactions, no "NMR invisible Al").

The distinction between Al species having the same chemical shifts and strongly overlapping lines but different quadrupolar couplings can be made via the two-dimensional quadrupole nutation NMR technique (13). The resolution of ^{27}Al sites with different quadrupolar coupling constants is presented in the F_1 dimension, whereas the regular spectrum is shown in the F_2 dimension (containing combined information about chemical shift and second order quadrupole shifts).

No doubt, ^{27}Al MAS NMR is a valuable tool in probing the coordination, location and quantity of Al in zeolites and AlPO$_4$ molecular sieves, but less useful than ^{29}Si MAS NMR in detailed studies for the lattice structure (3). However, introduction of DAS and DOR will allow much more detailed insight with respect to structural information available by use of ^{27}Al NMR spectroscopy. Since a number of excellent review papers exist, which summarize the structural information obtained by application of ^{27}Al NMR on zeolites, I would like to draw the readers attention for further details again to those reviews (3-5, 7, 8, 10-13) and highlight only results published recently dealing with framework characterization of zeolites by using ^{27}Al NMR spectroscopy.

Klinowski et al. report a considerable advance towards quantitative determination of aluminium in zeolites, and showed that all aluminium can be detected by NMR provided certain experimental conditions are met (10, 71). For the selective excitation of the central transition, maximum line intensity is achieved by very short radiofrequency pulses. In addition, the ^{27}Al NMR spectra were recorded without magic angle spinning in order to avoid spinning sidebands which often coincide with genuine signals (71). The authors also addressed this problem using ^{27}Al quadrupole nutation NMR spectroscopy to monitor the status of Al during, for example, thermal treatment of the sample. Using this technique, a series of free induction decays during the interval t_2 is acquired applying radiofrequency pulses while monotonically increasing the pulse length. The resulting 2D spectra allow ^{27}Al sites with different quadrupolar couplings to be resolved along F_1, which contains only the quadrupolar information, whereas F_2 contains the combined chemical shift and the second order quadrupole shift (10), see Figure 23. In conclusion, quantitative determination of Al in zeolites by NMR requires a careful design of the experimental conditions (10).

The realumination of zeolite Y samples has been followed by ^{27}Al quadrupole nutation NMR spectroscopy by Klinowski et al. (10, 72). The reinsertion of Al was performed by hydrothermal isomorphous substitution using strong basic solutions (KOH). Upon treatment with 0.5 m KOH the non-framework Al signal disappears due to conversion of non-framework Al into lattice aluminium (72).

^{1}H-^{27}Al CP/MAS NMR spectroscopy has been used to monitor selectively the non-framework aluminium species in dealuminated zeolite Y (Si/Al=4 - 14) (73-75). The ^{27}Al MAS NMR spectrum of the parent material (Na Y zeolite) consists of a single line at about 60 ppm (tetrahedrally coordinated lattice Al). The hydrothermally treated samples (Figure 24) reveal two more resonances at about 0 and 30 ppm, associated with non-framework Al. The intensities of those lines, and particularly that at 30 ppm, increases with further degree

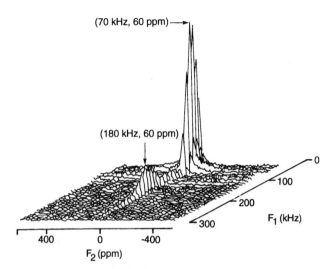

(70 kHz, 60 ppm) ——→

(180 kHz, 60 ppm)

0

100

200

F$_1$ (kHz)

300

400   0   -400

F$_2$ (ppm)

Fig. 23. ^{27}Al quadrupole nutation NMR spectrum of 80% NH$_4$-exchanged Y zeolite (Si/Al=2.5). Numbers in parentheses have units of kHz (for F$_1$) and ppm (for F$_2$), respectively (Reproduced by permission of Elsevier Science Publishers B.V., Amsterdam).

of dealumination, while the intensity of the line at 60 ppm decreases. Furthermore, Figure 24 shows that when the spectra are recorded with CP the intensity of the signals at 0 and 30 ppm increases relative to the signal at 60 ppm. This indicates that the 30 ppm line is a separate ^{27}Al resonance, possibly due to five-coordinated aluminium and not part of a second-order quadrupole lineshape as previously reported (73-75). The possibility that two (or more) ^{27}Al resonances overlap at about 60 ppm was checked by ^{27}Al quadrupole nutation NMR with fast MAS (12 - 13 kHz). The resonance at 56 ppm was identified as four-coordinated non-framework Al, whereas the line at 62 ppm was assigned to framework aluminium. Both resonances at 62 and 56 ppm, which overlap in the ordinary MAS spectrum, were resolved by quadrupole nutation NMR. The assignment of the signals could be done based on quadrupolar coupling constant measurements and second-order quadrupolar shift and lineshape analysis . All in all, the possibility that non-framework species in ultrastable zeolite Y are four-, five- and six-coordinated must be considered (73-75).

The adsorption of acetylacetone on zeolite mazzite lead to a transformation of NMR undetectable aluminium species into visible ones and to an *in situ* transformation of the aluminium coordination inside the zeolite framework (76). The addition of acetylacetone on an acid-leached sample of mazzite produced a decrease in signal intensity of the Al atoms located in the four-member rings of the lattice and an increase of the intensity of the signal due to the octahedrally coordinated atoms (76).

Ray et al. showed that at least four clearly defined types of Al species exist in dealuminated Y zeolite: in addition to framework tetrahedral Al (60 ppm) and non-lattice octahedral Al (0 ppm) the authors demonstrated by both a variation of the magnetic field and application of ^{27}Al DOR NMR that there is a second tetrahedral species (47 ppm) as well as a pentacoordinated Al species at 30 ppm (77).

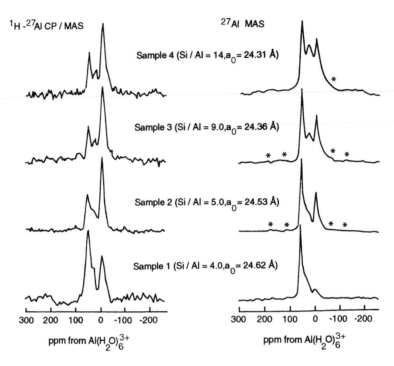

Fig. 24. ^{27}Al MAS and ^{1}H - ^{27}Al CP/MAS NMR spectra of increasing dealuminated zeolite H Y. Asterisks denote spinning side bands (Reproduced by permission of Elsevier Science Publishers B.V., Amsterdam).

The ^{27}Al MAS and DOR NMR spectra of Ca and W containing aluminate sodalites were recorded by Engelhardt and coworkers (78). Dramatic line narrowing was observed in the DOR spectra of the samples. The ^{27}Al DOR NMR spectra of Ca,W aluminate sodalite show seven sharp central lines accompanied by a manifold of spinning sidebands. These lines correspond to the seven crystallographically nonequivalent Al sites of the Ca,W aluminate sodalite framework derived from X-ray analysis. From the difference in the line positions in the 9.4 and 11.7 T spectra the quadrupole coupling constants, quadrupole induced shifts and the isotropic chemical shifts of each Al site have been calculated (78).

A linear relationship has been found to exist between the ^{27}Al and ^{71}Ga NMR chemical shifts of a series of structurally analogous aluminium and gallium compounds (including zeolites):

$$\delta^{71}Ga = 2.83 \ (\delta^{27}Al) - 4.50 \tag{2}$$

This relationship allows the prediction of ^{71}Ga chemical shifts for such gallium compounds from the ^{27}Al values already known for their aluminium analogues, and also a greater understanding of observed ^{71}Ga chemical shifts (79).

^{17}O NMR

Oxygen is part of the zeolite and $AlPO_4$ molecular sieve framework structures, and detailed solid-state ^{17}O NMR studies should be very attractive. However, the ^{17}O isotope has a very low natural abundance (0.037%) and a quadrupole moment which renders the observation of ^{17}O NMR spectra very difficult without expensive ^{17}O enrichment of the samples. Therefore, ^{17}O NMR investigations of zeolites have been carried out exclusively on ^{17}O enriched samples and a summary of former studies has been given by Engelhardt (11). Differences in chemical shifts and quadrupolar coupling constants (QCC) have been found for oxygens bonded to two silicons and those bonded to one silicon and one aluminium: ca 50 / 35 ppm (for the chemical shifts) and ca 5 / 3.2 MHz (for the QCC), respectively (13). ^{17}O NMR studies of a few gallosilicates have identified two chemically distinct oxygen sites, Ga-O-Si and Si-O-Si, in those systems, allowing some interpretation of the bonding nature of these oxygens (80, 81).

^{11}B NMR

Boron and aluminium belong to the same group of the periodic system of elements and, hence, the NMR related properties of the ^{11}B nucleus are similar to those of ^{27}Al. ^{11}B has a nuclear spin quantum number of 3/2, which means a quadrupole moment. The natural abundance of ^{11}B is reasonable high (80.4%). So far, the main applications of solid-state ^{11}B NMR within the field of zeolites have been probing the boron coordination in isomorphously substituted boron containing zeolites like B-ZSM-5 (82, 83), B-silicalite (boralite) and B-ZSM-11 (11, 13, 82). In those systems, boron may occur as tetrahedral BO_4 (framework boron chemical shift range between -1 and -4 ppm), trigonal BO_3 or non-framework BO_4 coordinations (chemical shift range of 1 to 2 ppm). Tetrahedral BO_4 and trigonal BO_3 units have similar isotropic chemical shifts but substantially different quadrupolar coupling constants. However, at high magnetic field and using MAS, tetrahedrally coordinated boron shows usually narrow single resonances, whereas trigonal boron exhibits quadrupolar line pattern due to its high quadrupolar interaction. Anyway, sufficient resolution usually allows the differentiation between those two boron species (11).

Zones and coworkers report ^{11}B MAS NMR studies on boron containing SSZ-24, an all-silica isostructural analog of $AlPO_4$-5, with all boron in tetrahedral lattice positions (84). Different treatments of boron-containing MFI single crystals were followed by ^{11}B MAS NMR spectroscopy checking the tetrahedral coordination of boron atoms (85). BAPO-5, the boron substituted analogue to $AlPO_4$-5 has been studied by Harris et al., showing that boron in the precurser form of the molecular sieve is present in the framework as tetrahedral BO_4 units (86). Qiu and coworkers report on ^{11}B MAS NMR studies on single crystals of BAPO-5, concluding with the observation that boron substitute Al in the framework of the molecular sieve (87).

$^{69/71}$Ga NMR

^{69}Ga and ^{71}Ga NMR spectra of the gallium analogues of zeolites Na-X, Na-Y, Na-sodalite and Na-natrolite have been recorded by Oldfield et al. (88). Both Ga isotopes have a smaller nuclear spin number (I=3/2) than ^{27}Al, which means much broader line widths for Ga relative to Al when the second order quadrupolar interaction is dominant. This indicates that solid-state Ga NMR is expected to be much more difficult than Al NMR, because it requires severe experimental conditions, such as much higher spinning speeds for MAS and/or shorter 90^0 pulses to excite the expected broader spectral widths (88). Of the two Ga isotopes, ^{71}Ga (natural abundance 39.6%) is more attractive because of its smaller quadrupole moment and larger gyromagnetic ratio, which helps to compensate for its lower natural abundance.

The results of Oldfield et al. indicate that Loewenstein's rule holds for gallosilicates as well as aluminosilicates, i.e., no evidence of the existence of either Ga-O-Ga linkages or of a gallium-rich extra-lattice phase were observed in the ^{71}Ga NMR spectra. The measured isotropic chemical shifts are all in the range of 169-184 ppm downfield from an external standard of 1 M Ga(NO$_3$)$_3$. All of the spectra show only single peaks. Since none of the gallium analogue zeolites studied show any resonance near ca 0 ppm (where octahedrally coordinated Ga is expected) the authors suggest that any extra-framework Ga present in their gallosilicates is unlikely to be present as Ga(H$_2$O)$_6^{3+}$ species (88). Furthermore, the results indicate that measurement of the apparent chemical shifts for pairs of isotopes of nonintegral spin quadrupolar nuclei in solids is a useful new approach for determination of both isotropic chemical shifts and nuclear quadrupole couplings constant values (88).

Quantitative determinations of gallium in MFI-type gallo- and galloalumino zeolites have been performed by Gabelica et al. (89, 90) and Diaz et al. (91) applying ^{71}Ga MAS NMR. The amount of tetrahedrally coordinated lattice Ga was directly and quantitatively measured for a wide series of Ga or Ga, Al - MFI zeolites. The substitution of Ga for Al in the framework of ZSM-5 and ZSM-12 was monitored by ^{71}Ga NMR, concluding with the presence of only one type of gallium in these samples (92). Incorporation of gallium into the lattice positions of mordenite was confirmed by ^{71}Ga MAS NMR, showing a single signal at about 150 ppm (93). Novel gallophosphates of the LTA type structure have been synthesized by Kessler and coworkers (94). However, the ^{71}Ga MAS NMR spectra were poorly resolved and the broad signal at about -37 ppm was attributed to Ga atoms in a five- or six-fold coordination (94).

Ga NMR spectra of the GaPO$_4$ based "cloverite" will be reviewed in chapter 5.1.3.

^{51}V NMR

A new vanadium silicalite with the ZSM-48 structure has been synthesized by Tuel at al., and ^{51}V NMR spectroscopy indicated the presence of isolated, well dispersed vanadium atoms (95). Evidence for framework positions of V atoms were suggested on the basis of physico-chemical characterization.

5.1.2. AlPO$_4$ and SAPO Molecular Sieves

Since the first announcement of the synthesis of porous crystalline AlPO$_4$ and SAPO materials (96, 97) a large number of NMR investigations for structural characterization have been performed. AlPO$_4$ and SAPO molecular sieves are interesting molecules for solid-state NMR investigations since they consist of ^{31}P and ^{27}Al nuclei, both of them 100% natural abundant. The ^{31}P MAS NMR spectra generally show symmetrical lines in the shift range of - 14 to - 35 ppm (relative to 85% phosphoric acid = 0 ppm), consistent with tetrahedral PO$_4$ sites in the lattice of these materials (11).

The ^{27}Al NMR spectra are more complex and may exhibit broad and asymmetrical spectral pattern due to severe quadrupolar interactions. The lineshape sometimes show the patterns of unaveraged second-order quadrupolar interaction. The ^{27}Al chemical shifts assigned to tetrahedrally coordinated Al in these materials usually cover the range between 29 to 46 ppm (relative to aqueous aluminium nitrate), whereas lines in the range of - 7 to around - 23 ppm are attributed to octahedrally coordinated aluminium (11). In addition, the ^{27}Al chemical shift and lineshape are strongly dependent on the specific crystalline structure under investigation.

5.1.2.1.   VPI-5, AlPO$_4$-8 and related Aluminophosphates

VPI-5 has received great attention during the last years since it was first reported in 1988 (98, 99). The growing interest in VPI-5 is mainly due to the unusual structure and properties of this material with extra-large pores consisting of 18-membered rings (see Figure 25).

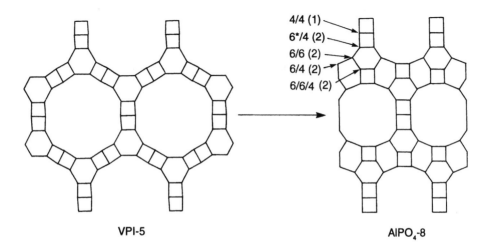

VPI-5   AlPO$_4$-8

Fig. 25. Schematic structures of VPI-5 and AlPO$_4$-8. The different crystallographic sites (relative numbers in parentheses) are indicated for AlPO$_4$-8.

^{31}P NMR

^{31}P NMR spectroscopy is considered as the most sensitive technique performing information on the local structure and structural modifications involving the tetrahedrally coordinated framework elements phosphorus and aluminium in AlPO$_4$ molecular sieves.

The ^{31}P MAS NMR spectrum of VPI-5 is quite unusual and has been (and is still) an item of discussion. The structure of VPI-5 consists of two crystallographically unique phosphorus sites, which are located in the 6-membered rings and in the atomic positions that belong to the two adjacent 4-membered rings and are in the atomic ratio of 2:1, respectively. The ^{31}P MAS NMR spectrum of **dehydrated VPI-5** indeed consists of two lines at about -26/-27 and -31/-32 ppm in an area ratio that closely approximates 2:1 (100-102). However, the ^{31}P MAS NMR spectrum of **hydrated or as-synthesized VPI-5** reveals three lines of equal intensity at about -23, -27 and -33 ppm (101-112). In connection with the assignment of those three lines a number of papers appeared dealing with this topic, and the discussion is still going on, both using ^{31}P NMR spectroscopy and XRD. McCusker et al. investigated the structure of hydrated VPI-5 by synchrotron X-ray, and concluded that two water molecules complete an octahedral coordination sphere around the framework aluminium atom between the fused 4-membered rings (113). Furthermore, all the water molecules are located in the 18- ring channels of VPI-5 forming a triple helix structure.

In the early paper by Davis et al. the authors did not assign the three resonances to phosphorus sites, however they run ^{31}P CP/MAS spectra (Figure 26) in which the line at -23/-24 ppm is enhanced over the two resonances at -26/-27 and -33 ppm, indicating that the water which enhances the ^{31}P NMR resonance is relatively immobile (101). Grobet et al. explained the lines at -23 and -27 ppm by proposing that half of the Al atoms located at the 6-rings in VPI-5 are octahedrally coordinated (two water molecules and four framework oxygens), whereas the line at -33 ppm was assigned to P in the center of the fused 4-memberd rings (103). In fact, a number of authors/groups relate the high-field line at -33 ppm to P in the double 4-membered rings: David et al. (112), J.P. van Braam Houckgeest et al. (107), A. Clearfield et al. (106), J. Klinowski et al. (114, 115) and Veeman et al. (110, 111). The two low-field signals at -23 and -27 ppm were by those authors assigned to the P atoms in the 6-membered rings. However, the number and coordination of the Al atoms connected with the 6-membered ring phosphorous atoms is still a matter of discussion. Sophisticated NMR techniques, like spin diffusion (115) and 2D heteronuclear [27]Al - [31]P correlation spectroscopy (111), were used to solve the assignment question.

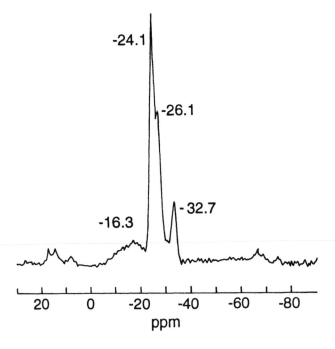

Fig. 26. ^{1}H - ^{31}P CP/MAS NMR spectrum of hydrated VPI-5 (Reproduced by permission of the American Chemical Society, Washington).

Other groups tend to assign the phosphorous in the center of the fused 4-membered rings to the low-field signal at -23 ppm: Maistriau et al. (105) and Akporiaye et al. (109, 116, 117). In tetrahedrally linked frameworks like zeolites and $AlPO_4$ molecular sieves, the chemical shifts of NMR signals usually correlate with the mean T - O - T angles (according to this relation the isotropic chemical shift is expected to decrease with increasing mean T - O - T angle, (118)). However, different conclusions with respect to the validity of this correlation for VPI-5 have been drawn by various authors, and details could be find by consulting the paper of Davis et al. (112).

The crystallization of VPI-5 was monitored by NMR spectroscopy, and a comparison of the spectra of the intermediate products clearly demonstrate the strong influence of the continued hydrothermal heating on the local environment of both P and Al atoms (119).

The stability of VPI-5 and the transformation to **$AlPO_4$-8**, a 14-membered ring $AlPO_4$ molecular sieve (see Figure 25), have been extensively studied by XRD and solid-state NMR spectroscopy. Depending on the quality of the sample and the treatment conditions (evacuation, extremely slow heating rate, careful removal of water etc.), the structrue of VPI-5 can be preserved up to at least 400° C, as followed by ^{31}P MAS NMR (120). Under sealed conditions and sligthly elevated temperatures (70 to 150 ^oC), the hydrated VPI-5 undergoes a reversible dehydration/rehydration effect, which results in a merging of the two low-field signals in the ^{31}P MAS NMR spectrum (109, 114, 121) and splitting again after cooling to room temperature.

Under more drastic conditions (higher temperatures, faster heating rates and unsealed conditions), VPI-5 transforms to $AlPO_4$-8, monitored by recording of a complete different ^{31}P MAS NMR spectrum (see Figure 27, 102). Some authors claim a (partly) reverse process of this transformation under certain conditions (105).

The ^{31}P MAS NMR spectrum of **hydrated $AlPO_4$-8** (see Figure 27) shows three signals at -21, -25 and -30 ppm, with an intensity ration of 1:2:6, which is not incompatible with the five independent crystallographic sites of $AlPO_4$-8 (see Figure 25), if three of the sites are equivalent with respect to the NMR chemical shift (102). The NMR non-equivalence of the 6/4 and 6*/4 sites of hydrated $AlPO_4$-8 has been investigated by Akporiaye et al. (116), which is in contradiction to Maistriau et al. (105).

Parallel to the discussion about the ^{31}P MAS NMR signal assignment of VPI-5/$AlPO_4$-8, closely related $AlPO_4$ materials, designated as H1, H2, H3 and H4, have been investigated with respect to similarities between those compounds and VPI-5 (103, 106, 112, 122). D'Yvoire reported in 1961 the synthesis of those four $AlPO_4$ - H phases (123). **$AlPO_4$ - H1** shows a very similar ^{31}P MAS NMR pattern as hydrated VPI-5, and was considered by Szostak et al. (122) as a template or organic-free VPI-5, whereas Davis et al. (112) and Clearfield et al. (106) don't see evidence to support any claim that VPI-5 and $AlPO_4$-H1 are the same material. **$AlPO_4$ - H2** and VPI-5 share the same building units and have to some extent similar physicochemical properties such as the ^{31}P MAS NMR spectra: three signals at -24, -29 and -33 ppm. However, the relative intensities of these signals are very different compared to VPI-5 (124). The ^{31}P MAS NMR spectrum of **$AlPO_4$ - H3** consists of two

468

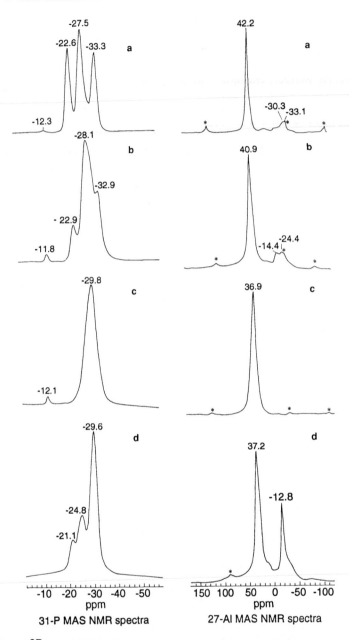

Fig. 27. ^{31}P and ^{27}Al MAS NMR spectra of a) VPI-5 dried at 60 ^{o}C/overnight, b) VPI-5 evacuated at 54 ^{o}C/overnight and calcined at 250 ^{o}C/overnight, c) AlPO$_4$-8 (transformed from VPI-5 by calcination at 400 ^{o}C/overnight) and d) hydrated AlPO$_4$-8. Asterisks denotes spinning side bands (Reproduced by permission of Elsevier Science Publishers B.V., Amsterdam).

signals at -24 and -26 ppm, very similar to the spectrum of **MCM-1** (125), whereas Szostak et al. conclude that both **AlPO$_4$ - H3** and **H4** do not represent microporous phases (122).

^{27}Al NMR

The interpretation of the ^{27}Al NMR spectra of **hydrated VPI-5** and **AlPO$_4$-8** has by no means created the same type of controversial discussion as observed for the assignment of the ^{31}P MAS NMR spectra. This might be due to the fact of the clear resonance assignment of the lines by application of ^{27}Al DOR NMR. As reported by many authors, the ^{27}Al MAS NMR spectrum of hydrated VPI-5 (see Figures 6 and 27) consists of two signals: one sharp line at ca. 40 ppm (tetrahedrally coordinated Al) and a broad line pattern (due to second order quadrupolar interaction) ranging from -15 to -45 ppm (octahedrally coordinated Al), with an intensity ratio of 2:1 (101-106, 108-110, 114, 120, 121, 126). As resolved by ^{27}Al DOR NMR (see Figure 6), the low-field line in the ^{27}Al MAS NMR spectrum at about 40 ppm consists in fact of two signals, representing the 6/4 positions of Al in the structure of hydrated VPI-5, whereas the broad high-field line pattern is assigned to Al in the center of the fused 4-membered rings (22, 127-130).

Especially ^{27}Al DOR NMR has provided new insight into the structural changes of VPI-5 during adsorption of water. In the dehydrated sample of VPI-5, two tetrahedrally coordinated framework aluminium sites were resolved, at 36 and 34 ppm, respectively, and no signal for octahedrally coordinated Al (128, 129). The DOR spectra are reproduced upon successive hydration/dehydration treatment, indicating that the water adsorption process is fully reversible (129). Grobet et al. report on the magnetic field dependence of the isotropic second-order quadrupolar shift for crystallographic site discrimination in VPI-5 by ^{27}Al DOR NMR (127). The ^{27}Al DOR NMR spectrum of hydrated VPI-5 in a field of 4.7 T shows two distinct tetrahedral Al sites due to differences in second-order isotropic quadrupolar shifts, reflecting different crystallographic symmetries in the crystal structure. In a field of 11.7 T the observed shifts of the two tetrahedral Al sites are identical, resulting in a spectrum with only one resonance line (127).

Variable temperature (VT) NMR results on VPI-5 have recently thrown more light on the nature of adsorbed water in hydrated VPI-5. However, the changes in the ^{27}Al MAS NMR VT spectra of VPI-5 are much less dramatic compared to the changes in the ^{31}P MAS NMR VT spectra. Under sealed conditions and careful temperature change (RT to 150 °C and back to RT), the ^{27}Al MAS NMR spectrum of VPI-5 shows a reversible dehydration/hydration effect by decrease/increase of the high-field signal assigned to octahedrally coordinated Al (104, 105, 109, 120). Even up to 400 °C, the VPI-5 structure can be kept stable when the water is carefully removed by evacuation and extremely slow temperature increase (0.1 °C/min): the resulting ^{27}Al MAS NMR shows only tetrahedrally coordinated Al (signal at 35 ppm) (120). However, after calcination overnight at 400 °C with a faster heating rate, the VPI-5 transforms to **AlPO$_4$-8**, revealing a signal at 37 ppm, representing tetrahedrally coordinated Al sites. Upon hydration, an additional broad line for octahedrally coordinated Al sites is observed at -13 ppm (see Figure 27, 102).

Klinowski et al. indicate that hydrated VPI-5 undergoes a high-temperature phase transformation above 80 °C. and a low-temperature structural transformation between -83 and -53 °C., as followed by ^{27}Al quadrupole nutation NMR (114, 126). In parallel to this,

Akporiaye et al. received evidence from the VT experiments on $AlPO_4$-8, which indicate that there are two temperature regimes to be involved: the range below 100 ^oC involves changes in the spectra very similar to VPI-5 related to temperature effects on water in the channels. Above 110 ^oC, differences between the VT behaviour of the two structures were observed and are associated with the complete loss of octahedral signal in the ^{27}Al MAS NMR spectra of $AlPO_4$-8. Considering that the octahedral signal of $AlPO_4$-8 is due to framework Al, it would appear that one third of Al sites formally associated with water in VPI -5 are preserved to some degree in $AlPO_4$-8 after rehydration (116).

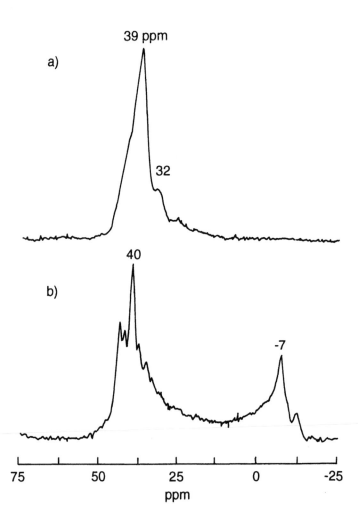

Fig. 28. ^{27}Al DOR NMR spectra at 11.7 T of (a) dehydrated and (b) hydrated $AlPO_4$-8 (Reproduced by permission of Elsevier Science Publishers B.V., Amsterdam).

The ^{27}Al DOR NMR spectra of AlPO$_4$-8 (see Figure 28) are significantly different from those of VPI-5, reflecting major modification of the VPI-5 structure after heating. Moreover, the broad ^{27}Al lines in Figure 28 indicate substantial disorder in the Al environments of the hydrated AlPO$_4$-8 framework. Upon rehydration, sharpened lines appear at 40 and -7 ppm, reflecting a more ordered Al arrangement with both tetrahedrally- and octahedrally-coordinated Al species present. Nevertheless, the broad feature connecting these two lines indicates a continuous distribution of Al environments spanning the range between these two relatively ordered sites (128-130). Rocha et al. studied hydrated AlPO$_4$-8 by fast ^{27}Al MAS NMR at 4.7 and 9.4 T, ^{1}H-^{27}Al CP/MAS NMR and by quadrupole nutation NMR spectroscopy, concluding with the observation of at least four and (possibly) five crystallographically inequivalent sites for aluminium (131).

The ^{27}Al MAS NMR spectrum of **AlPO$_4$ - H1** shows a very similar pattern to the corresponding spectrum of hydrated VPI-5 (122), whereas the ^{27}Al MAS NMR spectrum of **AlPO$_4$ - H2** reveals a resonance at 36 ppm that corresponds to tetrehedral Al and twin lines at -14 and -17 ppm due to octahedral Al species (two water molecules and four framework oxygens) (124). The ^{27}Al CP/MAS NMR spectrum shows that the resonances associated with octahedral Al species are more enhanced than the one for tetrahedral Al. This is expected since water molecules are chemically bonded to the octahedral Al (124). In the case of **AlPO$_4$ - H3**, the following ^{27}Al MAS NMR pattern was found: a sharp signal at about 40 ppm (tetrahedrally coordinated Al) and a group of signals between -10 and -45 ppm, representing octahedrally coordinated Al (103, 122, 125), again similar to MCM-1. From the NMR experiments, Grobet et al. conclude that the framework topology of MCM-1 and AlPO$_4$ - H3 are identical, and that the latter is the Si-free homologue (125). A completely different pattern was observed for **AlPO$_4$ - H4**: From the ^{27}Al MAS NMR, the material appears to contain 4,- 5- and 6-fold coordinated Al, with a substantial disorder of the Al environments (122).

Other nuclei than ^{27}Al and ^{31}P

A few investigations were carried out on Si containing VPI-5 (also refered as MCM-9), and the ^{29}Si MAS or CP/MAS NMR spectra consist of a more or less broad line pattern in the range of -87 to -111 ppm (108, 132, 133), indicating a limit of isomorphous substitution of Si in the framework of VPI-5 (112, 132). The transformation of Si-VPI-5 (MCM-9) to SAPO-8 was monitored by NMR by Derouane et al. (134), following the same transformation pattern as for VPI-5 to AlPO$_4$-8. However, Grobet et al. consider the ^{31}P MAS NMR pattern of MCM-9 as a superposition of the spectra of VPI-5, AlPO$_4$ - H3 and SAPO-11 phases (125).

The dynamics of water molecules in VPI-5 was studied by ^{2}H NMR on D$_2$O adsorbed samples (121, 135, 136). Two distinct types of water molecules were found in VPI-5. The first correspond to water molecules undergoing fast isotropic reorientation within the channels and the second to water molecules bound to framework Al (136, 137).

Finally, ^{1}H NMR of organic template in hydrated VPI-5 were recorded, applying a spin-lock pulse sequence in order to suppress the dominant water signal and to observe (under MAS conditions) very narrow template signals (138).

5.1.2.2.   Other AlPO$_4$ and SAPO Materials

Besides VPI-5/AlPO$_4$-8 and related materials some interest has been concentrated upon other AlPO$_4$- and SAPO-materials, like AlPO$_4$-5, AlPO$_4$-11, AlPO$_4$-14, SAPO-5, SAPO-11, SAPO-17, SAPO-34 and SAPO-37. Therefore, some recent NMR results about these materials will be reviewed here.

The structure of **AlPO$_4$-5** is closely related to that of VPI-5, though the 6-membered rings of the former are separated by single (instead of double) 4-membered rings, resulting in smaller main channel dimensions. The ^{27}Al DOR NMR spectrum of dehydrated AlPO$_4$-5 contains a single line at 36 ppm, a result which corresponds with the ^{27}Al MAS NMR spectrum. The narrow linewidth indicates ordered framework with the lone tetrahedral Al environment expected from the structural configuration. Adsorption of water has a much different effect on Al ordering in AlPO$_4$-5 than in VPI-5. The ^{27}Al DOR NMR spectrum of hydrated AlPO$_4$-5 shows significantly less-ordered Al-environments with broad line features that can be attributed to tetrahedral and octahedral Al species at 39 and -14 ppm, respectively. In addition, a small, broad feature is observed at 7 ppm, a region where lines from penta-coordinated Al species have been observed (128, 129). In conclusion, the highly ordered framework of the dehydrated material is replaced by disordered Al environments following incorporation of water into the pore spaces of the relatively narrow channels of AlPO$_4$-5. The highly ordered VPI-5 framework, which yields narrow NMR lines, can be kept even after adsorption of water, due to the large channels allowing ordering of the adsorbed water molecules.

The dynamics and influence of water molecules in AlPO$_4$-5 have been studied by NMR spectroscopy. Meinhold et al. showed by ^{27}Al CP/MAS NMR that the intensity of the 6-coordinated Al is depending on the water content and is not a measure of the ion-exchange capacity of AlPO$_4$-5 (139, 140). Two distinct types of water molecules were found in AlPO$_4$-5 samples loaded with adsorbed D$_2$O by recording the ^{2}H NMR spectra (135, 136). The average ^{27}Al-^{31}P distances in AlPO$_4$-5 and AlPO$_4$-11/SAPO-11 have been determined to 3.14 Å and 3.13 Å, respectively, by applying the **SEDOR (Spin Echo Double Resonance)** technique (141).

The incorporation of Si into the **AlPO$_4$-11** framework was followed by solid-state NMR, proposing two different mechanisms: first, Si into a hypothetical P position and second, 2 Si for Al + P. Summarizing the experimental results, it can be stated that ^{27}Al MAS NMR investigations did not provide information concerning the Si incorporation (142). Zibrowius et al. studied the incorporation of Si into SAPO-5, SAPO-31 and SAPO-34, and again, neither ^{27}Al nor ^{31}P MAS NMR were suitable to detect the isomorphous substitution of P by Si. However, by means of ^{29}Si MAS NMR, not only the presence of Si, but also the type of incorporation (aluminosilicate (SA) or SAPO domains) could be directly monitored (143). In addition, the amount of Si incorporated on P tetrahedral positions was determined indirectly by applying ^{1}H MAS NMR, measuring the concentration of acidic hydroxyl groups (143).

The structure of AlPO$_4$-11 comprises three crystallographically different tetrahedral sites, and structural changes related to water adsorption have been studied be several authors using NMR. Tapp et al. conclude that adsorbed water results in significant structural changes, which have been shown to be reversible, and no evidence was found for the presence of chemisorbed water (144). Spectral simulations of ^{27}Al MAS and double rotation NMR spectra measured at different magnetic field strengths proved the preferential hydration of one tetrahedral Al site in AlPO$_4$-11, transforming this site reversibly into octahedral Al (Peeters et al., 145, 146). For AlPO$_4$-5, however, the interaction of water with the molecular sieve occurs randomly (147). The interaction of water, methanol, ammonia and acetonitrile with AlPO$_4$-5 and -11 was studied by ^{27}Al MAS and DOR NMR, and both water and ammonia are able to coordinate to part of the lattice aluminium, leading to five- and/or six-coordinated Al (147). The same group distinguished three different crystallogaphic sites (ratio 1:1:1) in the structure of dehydrated AlPO$_4$-5 by ^{31}P MAS and ^{27}Al MAS and DOR NMR spectroscopy (147). Barrie et al. studied hydrated AlPO$_4$-11 using ^{27}Al MAS and DOR NMR at 9.4 and 11.7 Tesla (148). The MAS spectrum at 9.4 T of the hydrated sample (five crystallographically different T-sites) showed three lines in the tetrahedral region with intensity ratios of 21.3 : 60.2 : 18.5, together with an octahedral signal (ca. 20% of the total area) (148). Further ^{27}Al DOR NMR investigations at different field strengths were done by Grobet et al. showing the presence of five tetrahedral Al signals and one octahedral signal all of equal intensity, provided there exists no discrimination among the Al at the different crystallographic sites for coordination with water molecules (149). The importance of ^{27}Al CP/DOR NMR was demonstrated by Wu et al. illustrating the signal enhancement of the octahedral line (Al site with adsorbed water molecules) compared to the signals for the tetrahedral Al species (150).

Vedrine et al. investigated as-synthesized and calcined (550 ^{o}C) AlPO$_4$-11 and SAPO-11 samples by ^{31}P MAS NMR in order to check the effect of hydration on the number of P and Al atoms per asymmetric unit: The as-synthesized and calcined samples revealed three P and three Al atoms per unit, whereas the evacuated and rehydrated samples showed five P and five Al atoms per asymmetric unit (151). In addition, calcination enhances the crystallinity of all samples, resulting in better resolved ^{31}P MAS NMR spectra (151).

The ^{31}P and ^{27}Al MAS NMR spectra of **AlPO$_4$-14A** have been reported by Goepper et al. (152). The two resonances at -19 and -29 ppm (ratio 1 : 2.5) in the ^{31}P-spectrum are indicative for tetrahedral P sites, and the known structure of AlPO$_4$-14A requires four crystallographically unique P positions. Since three of the Al - O - P angle values are close and very different from the fourth one (correlation between chemical shift and the mean T - O - T angle), two NMR signals were expected in the ^{31}P MAS NMR spectrum (152). The ^{27}Al MAS NMR shows in the first approximation two distinct signals around -23 (octahedral Al sites) and in the 20-30 ppm range (tetrahedral Al-sites) (152).

^{27}Al MAS, CP/MAS and DOR NMR spectra of **AlPO$_4$-14** have been recorded by Rocha et al. (130). The AlPO$_4$-14 system is isostructural to GaPO$_4$-14, and the authors detected at least two tetrahedral Al sites (signals at 41 and 31 ppm) besides two other signals at 6 and -7 ppm, indicative for 5- and 6-coordinated Al in hydrated AlPO$_4$ material. The signal assignment for Al was confirmed by CP/MAS experiments, where the two tetrahedral Al sites cross-polarize far less efficiently than the species with additional OH coordination. The ^{27}Al DOR NMR spectrum is much more complicated due to overlap of multiple spinning sidebands, but a signal at 42 ppm is clearly visible. A second signal was observed at 33 ppm. In addition, the authors report the ^{31}P MAS NMR spectrum of AlPO$_4$-14, with three signals in a ratio of 1 : 2 : 1 (130). The results of Rocha et al. are slightly different from those of

Zibrowius et al. (153), and the former authors attribute this difference to overlapping spinning sidebands in their ^{27}Al spectrum (130).

Solid state NMR studies of **AlPO$_4$-18** revealed that this molecular sieve contains three crystallographically independent sites for both P and Al in a 1 : 1 : 1 population ratio (154).

^{27}Al DOR NMR at a magnetic field of 11.7 T distinguishes the extremely distorted 5-coordinated Al sites in the molecular sieve precurser **AlPO$_4$-21**. Upon calcination, AlPO$_4$-21 transforms to **AlPO$_4$-25,** which has two tetrahedral sites with similar isotropic chemical shifts that cannot be resolved in an 11.7 T field. The two tetrahedral environments, however, have different quadrupole coupling constants and are distinguished by DOR at 4.2 T field. The quadrupole coupling constants obtained for these sites indicate that the tetrahedral Al environments are less distorted in the hydrated material (155). According to Akporiaye et al. two badly resolved signals at 41 and 37 ppm could be detected for AlPO$_4$-21 by ^{27}Al MAS NMR (109). However, AlPO$_4$-25 showed only one signal for tetrahedral Al sites at 34 ppm in the ^{27}Al MAS NMR spectrum. The ^{31}P MAS NMR spectrum contains a strong band at -32 ppm and a quite small signal at -20 ppm for AlPO$_4$-25, whereas three P signals were observed for AlPO$_4$-21 at -16, -22 and -26 ppm, respectively (109).

NMR spectra of MeAPO's have been studied by a few authors. The ^{31}P MAS NMR spectrum of **MgAPO-20** consists of two major signals at -21 and -28 ppm which were assigned to P(2Al, 2Mg) and P(3Al, 1Mg) units by Barrie et al. (156). Two additional weak signals at -14 and -35 ppm were attributed to P(1Al, 3Mg) and P(4Al), respectively (156). The strong symmetrical ^{27}Al MAS NMR signal at 34 ppm corresponds to tetrahedral Al(4P) environments (156). ^{31}P spin-lattice relaxation times were followed by Chao et al. on **CoAPO-5** and **CoAPO-34** molecular sieves (157), assuming that the incorporation of paramagnetic cobalt ions in an AlPO$_4$ framework affects the magnetic properties of the P nuclei. Indeed, the ^{31}P relaxation on CoAPO is mainly influenced by framework cobalt ions and not by exchangable non-framework cobalt (157). In addition, the isomorphous substitution of Al by Co in the **CoAPO-5** and **CoAPO-11** molecular sieves lead to partly NMR-invisible phosphorous (158).

A number of papers, including a comprehensive study by Blackwell and Patton (159), appeared in 1988 describing solid-state NMR investigations of **SAPO-5, SAPO-11, SAPO-20, SAPO-34, SAPO-35** and **SAPO-37** (159-162). For further details about those compounds at that time, please consult the mentioned papers. In general, SAPO-n crystals often contain aluminosilicate (SA) domains, where the silicon is concentrated, next to SAPO domains, where all the phosphorus is located. The T-atom compositions of SA and SAPO domains can be determined with ^{29}Si MAS NMR. Brønsted acid sites can be present in the SA parts, in the SAPO parts and at the boundaries between these different domains, and in order to explain the catalytic activity of SAPO-5, SAPO-11 or SAPO-37 materials, the contributions of SA and SAPO crystal domains have been determined using ^{29}Si MAS NMR (163).

The thermal stability and the effect of water coordination of SAPO-34 were investigated by Watanabe et al. (164). ^{27}Al, ^{29}Si- and ^{31}P MAS NMR spectra and XRD as a function of temperature showed the microporous structure of SAPO-34 to be stable up to 1000 ^{o}C. The effect of water coordination on the ^{27}Al- and ^{31}P MAS NMR spectra also showed the adsorptive properties of SAPO-34 to be stable up to 1000 °C. (164). The changes occuring in SAPO-37 upon heating the molecular sieve to 1000 °C. were studied using ^{29}Si-, ^{27}Al- and ^{31}P MAS NMR spectroscopy (165). The removal of the template at 600 °C. leads to a shift of both ^{29}Si and ^{27}Al NMR signals, suggesting an interaction of the template with some SiO$_4$ and AlO$_4$ tetrahedra. A further solid-state transformation was shown to occur between 800 and 900 °C., which involves the formation of siliceous islands, generating silicoaluminate domains. However, the faujasite crystal structure is still maintained (165).

An extensive multinuclear MAS NMR investigation of the erionite-like **AlPO$_4$-17** and **SAPO-17** was performed by Zibrowius et al. (166), demonstrating for the first time a splitting of the ^{29}Si NMR line (resonances at -93 and -97 ppm) caused by crystallographically inequivalent Si sites in a SAPO-type material. Furthermore, AlPO$_4$-17 and SAPO-17 undergo pronounced structural changes during template removal and rehydration. All ^{27}Al NMR lines observed for the different states of AlPO$_4$-17 (as-synthesized (lines at 39, 35 and 16 ppm), calcined and rehydrated) were assigned to the different crystallographic environments present. The well-resolved ^{31}P NMR lines of the calcined (resonances at -29 and -34.5 ppm) and rehydrated AlPO$_4$-17 exhibit the intensity ratio 2 : 1 expected for the ERI structure. Incorporation of Si in the AlPO$_4$-17 framework has only a slight influence on the ^{31}P MAS NMR spectra, whereas the ^{27}Al MAS NMR spectra are changed. For instance, the splitting observed for the tetrahedral line of the as-synthesized AlPO$_4$-17 is not obtained for SAPO-17. All the individual lines seem to be broadened (166).

5.1.3. 20-Membered-Ring GaPO$_4$ and AlPO$_4$ Molecular Sieves

A novel synthetic gallophosphate molecular sieve, called **cloverite** (shape of a four-leafed clover), exhibits very large pore openings (13 Å) and cages (29 Å) (see Figure 29, 167).The

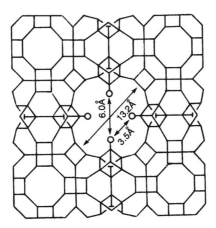

Fig. 29. Schematic structure of the GaPO$_4$ "cloverite". The circles represent terminal hydroxyl groups (Reproduced by permission of Macmillan Magazines Ltd., London).

cubic structure with a three-dimensional channel system consists of five crystallographically distinct phosphorus and gallium sites with the relative multiplicities of 1 : 1 : 2 : 2 : 2. The ^{31}P MAS NMR spectra recorded by several authors show very poor resolution and reveal a spectral pattern as presented in Figure 30. The signals were observed at about -2, -5, -10, -11 and -12/17 ppm and devided by Ozin et al. into three main groups: a (-2 and -5 ppm), b (-10 and -11 ppm) and c (-17 ppm) (168). The two low-field signals were associated with the P sites connected to terminal hydroxyl groups (168, 169). The most intense spectral feature (b) is assigned to the remaining phosphorus T sites, without an exact relation to specific tetrahedral sites (168, 169). Obviously, the exact assignment of the P resonances is extremely complex. A partial assignment of distinct crystallographic P-sites was done by applying spectral editing techniques, like the dipolar-dephasing method, which utilizes differences in dipolar ^{31}P-^{1}H interactions. By use of this technique, five resonance lines could by resolved in the ^{31}P MAS NMR spectrum (169). Under CP/MAS conditions (see Figure 30), only signal (a) is enhanced, confirming its assignment to the hydroxylated P site. Thermally treated cloverite (vacuum, 400 ^{o}C) shows an intense broad line around -17 ppm, a partially resolved, structured weak downfield shoulder extending from around -4 to -12 ppm and a very weak downfield impurity line a 20 ppm (168). The ^{31}P MAS NMR spectra of a dehydrated cloverite recorded by Hanna et al. were interpreted in line with the finding just reported: strong signals at -10 and -11 ppm (P(OGa)$_4$ sites) and a weak signal at -1 ppm (P(OH) sites (170).

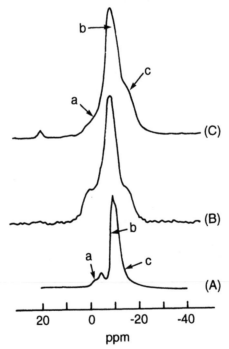

Fig. 30. ^{31}P NMR spectra of cloverite: (A) as-synthesized with MAS, (B) with CP/MAS at 150 ^{o}C and (C) with MAS and proton decoupling at 150 ^{o}C (Reproduced by permission of the American Chemical Society, Washington).

Barr et al. assigned the main signal at -11.2 ppm (with a shoulder at -12.6 ppm) in the ^{31}P MAS NMR spectrum of as-synthesized cloverite to the P(OGa)$_4$ units, whereas the small line at -6.3 ppm corresponds to P-OH phosphorus atoms (171). The observed intensity ratio was close to 7:1, as required by the structural determination. The correctness of their assignment was confirmed by the ^{31}P CP/MAS NMR experiment, which showed that the intensity of the line at -6.3 ppm was significantly increased upon cross-polarization. The P atom in question is therefore near a proton (171).

The ^{1}H MAS NMR spectrum of as-synthesized cloverite displays four signals at 7.6, 6.4, 3.4 and 2.0 ppm superimposed on a very broad background signal. After dehydration at 150 ^{o}C, it is found that a room temperature rehydration process has essentially no effect on the ^{1}H chemical shifts (168). Part of the fluoride from the synthesis was still present in the structure after calcination at 450 ^{o}C, as monitored by ^{19}F MAS NMR (172). However, the flouride was in a slightly different environment.

Since this GaPO$_4$ material has drawn a lot of attention, several authors report on the ^{71}Ga MAS NMR of cloverite (168-170, 173). Merrouche et al. assigned the downfield signal at 103 ppm in the ^{71}Ga MAS NMR spectrum of cloverite to tetrahedral Ga sites from a GaPO$_4$ impurity, whereas the signal at -38 ppm was assigned to the cloverite five coordinated Ga (173). In contradiction to this assignment, Hanna et al. associated the downfield signal at 92 ppm to five coordinated Ga and the second resonance at -24 ppm to six coordinated gallium (170). ^{71}Ga MAS and CP/MAS NMR spectra were run by Ozin et al. (168) and Meyer zu Altenschildesche et al. (169), without any clear conclusion, except indicating that ^{71}Ga DOR and CP/DOR NMR could allow more precise interpretation of the ^{71}Ga NMR measurements.

An extraordinary strong quadrupolar coupling was shown to be the reason for a misleading line shape in the ^{71}Ga MAS NMR spectra of cloverite (174).

^{129}Xe NMR spectra were reported, showing that all sample pretreatments yielded only one resonance in the ^{129}Xe NMR spectra (168). The presence of only one signal might be interpreted to mean that Xe only gains access to one of the channel systems of cloverite, presumably through the 20 T-atom ring window into the 29 Å supercage (168).

^{13}C MAS and CP/MAS NMR spectra were reported on as-synthesized cloverite, mainly with respect to the location, form and distribution of the templating species (quinuclidine) (168, 169). However, no clear conclusion could be drawn so far.

A microporous crystalline AlPO$_4$ molecular sieve possessing a 20-membered ring system and designated as **JDF-20** has been synthesized by Xu et al. (175). The structure consists of elliptically shaped 20-membered ring channels intersected by smaller 10- and 8-membered rings. The ^{27}Al MAS NMR spectrum of JDF-20 consists of one broad resonance at 33 ppm (tetrahedral Al sites), whereas the ^{31}P MAS NMR exhibits two lines at -22 and -29 ppm, indicating that two types of tetrahedral P environments exist in JDF-20 (175). Phase transition of JDF-20 to AlPO$_4$-5 has been reported (175), but so far no NMR data are available about this transition.

5.2. Zeolite and AlPO$_4$ Molecular Sieve Synthesis

Zeolites and AlPO$_4$ molecular sieves are prepared by hydrothermal synthesis using the framework element containing solutions and the appropriate amount of inorganic bases and organic templates, finally adjusted for the correct pH value. Several studies have been carried out in order to understand the mechanism of the hydrothermal synthesis. A number of them were reviewed by Engelhardt (11) and Clague et al. (13) in 1989, and I would recommend the reader to consult those papers with respect to NMR investigations related to zeolite synthesis at that time. ^{29}Si-, ^{27}Al- and ^{31}P-NMR can be applied to both liquid and solid phases formed during the synthesis of zeolites and AlPO$_4$ molecular sieves and is a powerful technique for identifying the structure, distribution and concentration of ions in the starting solution, the gel formation and the transformation of the gel into the crystalline phase.

An excellent review, summarizing the findings concerning the nature and distribution of species present in silicate and aluminosilicate solutions and gels, has been presented by Bell (176). Particular attention was drawn on establishing the effects of pH, cation composition, Si/Al ratio and the solvent composition (176). As an example, Figure 31 illustrates a ^{29}Si NMR spectrum for a sodium silicate solution. The individual resonances comprising this spectrum are grouped into bands each of which is designated as Q^n: Q^0 denotes the monomeric anion SiO$_4$$^{4-}$, Q^1 end-groups of chains, Q^2 middle groups in chains or cycles, Q^3 chain branching sites and Q^4 three-dimensionally cross-linked groups (176). Many of the individual resonances observed in a spectrum like that shown in Figure 31 can be assigned to specific silicate structures (often as single or multiple ring structures, like SBU's) based on spectra of well defined silicate species and detailed 2D-NMR experiments of ^{29}Si enriched silicate solutions (13, 176). Such studies have revealed that the chemical shift of a given Si atom depends on its connectivity, the Si-O bond lengths, the Si-O-Si bond angles, as well as the pH of the medium and the cation type (176). After consulting the work of Bell, there is no doubt that high-resolution ^{29}Si NMR spectroscopy provides considerable insight into the structure and distribution of silicate and aluminosilicate anions present in solutions and gels from which zeolites are formed (176). The narrowness of individual lines and the sensitivity of the chemical shift to details of the local chemical environment makes it possible to identify exact chemical structures. Using ^{29}Si NMR, the concentration of 19 silicate anions have been determined even quantitatively in sodium silicate solutions of a composition suitable for the synthesis of zeolite A and Y (177).

Alkali metal NMR spectroscopy has demonstrated the formation of cation-anion pairs, which is postulated to affect the dynamics of silicate and aluminosilicate formation and the equilibrium distribution of these species (176). ^{27}Al NMR has proven useful in identifying the connectivity of Al to Si. However, due to quadrupolar broadening, this technique cannot be used to define the precise environment of Al atoms (176).

In the following, a few recent papers will be reviewed dealing with the synthesis or preparation conditions of zeolites/AlPO$_4$ molecular sieves.

Sivasanker et al. examined silico-titanium precursors in the synthesis of TS-1 by ^{29}Si NMR spectroscopy and optimized the synthesis procedure resulting in the incorporation of up to eight Ti ions/unit cell corresponding to a Si:Ti ratio of 11 (178). ^{13}C CP/MAS and ^{29}Si MAS NMR have been used to study the crystallization of zeolite ZSM-39 in the

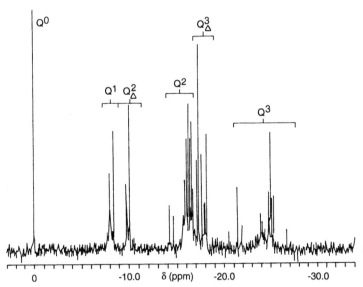

Fig. 31. ^{29}Si NMR spectrum of a 3.0 mol% SiO$_2$ (Reproduced by permission of the American Chemical Society, Washington).

tetramethylenediamine (TMEDA)-SiO$_2$-H$_2$O system (179). With the formation of the precursor from SiO$_2$ species surrounding TMEDA followed by nucleation, the ^{13}C chemical shift of the -C$_2$H$_4$- group moves progressively to higher field (179).

The influence of synthesis conditions and post-synthesis treatments on the nature and quantity of structural defects in highly siliceous MFI-type zeolites have been monitored by Guth et al. using ^{29}Si NMR spectroscopy (180). The formation and annealing of at least two kinds of local defects were detected. High pH values and high concentrations of alkali metal and fluoride ions prevent a complete polycondensation of the silicate species and generate materials that contain significant levels of nonbonding point defects. By contrast, a high crystallization temperature results in the ready hydrolysis of the terminal Si-O$^-$ groups, leading to structures containing fewer defects. A large number of T vacancies (hydroxyl nests) are created in zeolites crystallized at low temperatures in the presence of high Pr$_4$N$^+$ concentrations and in the absence of F$^-$ or Na$^+$ ions. Such conditions favor the formation of double-5-ring silicate anions. The authors propose the hypothesis, that such species may condensate further, leading to frameworks with an important number of empty T- sites. Based on these findings, ideal synthesis conditions can be proposed to produce MFI zeolites which contain a minimum number of defects (180).

Differences in ^{19}F chemical shifts were used to distinguish zeolite samples prepared in OH$^-$ medium from those prepared in F$^-$ medium by applying ^{19}F MAS NMR spectroscopy (181). ^{19}F can be used as a chemical and structural probe in as-synthesized or post-synthesis treated samples (181).

Finally, the possibility of formation and stabilization of tetrahedral Al for the synthesis of AlPO$_4$-5 was investigated in relation to the preparation of the reaction mixture. The Al coordination in the reaction mixtures and products was characterized by means of ^{27}Al MAS NMR. The authors conclude, that the role of the template in the formation of AlPO$_4$-5 is to stabilize the newly created tetrahedral Al and to prevent conversion to octahedral Al by shielding from the attack by water (182).

So far, a lot of understanding about the mechanism of zeolite and AlPO$_4$ molecular sieve synthesis has been possible by using NMR spectroscopy. However, more systematic work still needs to be performed in order to get a deeper insight into the mechanism, since the type of species formed during the synthesis route appears to be highly dependent on the method of preparation (13).

5.3.   Characterization of Cations in Zeolites

The excess negative charge of the lattice, due to the presence of Al or other substituent metal ions, is balanced either by protons or by other exchangeable cations. These ions are located as well defined sites in the zeolite, but may also be mobile (13). A very common balancing cation is the proton, and since this ion is responsible for creation of the Brønsted acid sites, ^{1}H NMR studies will be reviewed in chapter 8, dealing with acidity of zeolites and SAPO's.

Of the large number of cations which balance the charge of zeolite frameworks, only a limited selection is of interest from an NMR point of view. These are in particular ^{23}Na, ^{7}Li and ^{133}Cs, and recent published papers dealing with this subject will be reviewed in the following. However, all three nuclei possess a quadrupolar moment with the associated difficulties in meassurement. Again, double rotation NMR has provided a deeper insight into the understanding of NMR spectra of nuclei with spin numbers larger than 1/2.

A study by Ozin et al. provides a satisfying demonstration of the power of ^{23}Na DOR NMR carried out at 11.7 and 9.6 T magnetic field strengths for elucidating site-specific structure, bonding and dynamical details for Na cations and a range of adsorbed guest molecules, like PMe$_3$, Mo(CO)$_6$ and water, in sodium Y zeolite (183). Quantitative adsorption of those guests affects both the positions and line shape of the ^{23}Na resonances from specific extra-lattice Na$^+$ sites, and the evolution of the ^{23}Na DOR NMR spectra with the progressive introduction of guest molecules allows to probe direct "solvation" effects involving the Na cations in the larger supercages, as well as indirect effects on the Na cations in adjacent smaller sodalite cavities (183). Additional site specific dynamical information was obtained by running ^{23}Na DOR NMR spin-lattice relaxation measurements. The results indicate a lower T$_1$ value for ^{23}Na nuclei at site I, as compared to the site II Na cations, which might be related to the rather small quadrupolar interaction experienced by the Na cations inside the hexagonal prism (183). A similar study dealing with W(CO)$_6$ encapsulated in sodium Y zeolite was reported by the same group, again applying ^{23}Na DOR NMR spectroscopy (184).

Na cations localized at crystallographically distinct sites in dehydrated Y and EMT zeolites were characterized using ^{23}Na DOR, MAS and 2D nutation NMR spectoscopy (185). Two signals for Na cations were identified, a low field (gaussian) line at -12 ppm and a high-field (quadrupole pattern) line, with an isotropic shift of -8 ppm. The former signal was attributed to Na at the S I sites and the latter one to Na at the sites S I' as well as S II in FAU and EMT. The authors emphasize, that rotor-synchronized pulse excitation, variable spinning speeds of the DOR rotor and measurements at at least two different magnetic field strengths are prerequisites for a reliable interpretation of the ^{23}Na DOR NMR spectra (185). The electronic state of Na metal dispersed in the cavities of Na X and Na Y was investigated by ^{23}Na NMR, concluding with the statement that the Na is described as interacting paramagnetic cluster ions rather than as metallic sodium clusters (186).

The ^{23}Na DOR NMR spectra of the low-temperature form of different sodium sodalites reveal the presence of several inequivalent Na sites for this type of zeolites (78). Dramatic line narrowing was observed in the DOR NMR spectra of all samples.

Karge et al. assigned the ^{23}Na MAS NMR resonances of ammonium- and lanthanum-exchanged sodium Y zeolites as follows: The signal at about -9 ppm (referenced to crystalline NaCl) was assigned to Na lattice cations located at S III sites of the large cavities, whereas the line at about -13 ppm is indicative of Na cations in the truncated octahedra (beta cages). A ^{23}Na signal at -5.5 ppm sometimes observed in spectra of La, Na-Y zeolites was tentatively attributed to Na cations located in the hexagonal prism (S I site) (187).

High-resolution solid-state MAS NMR was used to investigate the progressive dehydration of Na X and Na Y zeolites and to study the cation distribution in a series of dehydrated, partially exchanged (Li, Ca) Na X. The ^{23}Na MAS NMR spectrum of totally dehydrated Na X exhibits a line at -40 ppm which differs from that observed in Na Y, demonstrating the influence of the Si/Al ratio. Li^+ has a tendency to substitute the Na^+ in the more constrained sites, leading to narrow ^{23}Na lines, whereas Ca^{2+} seems to force the remaining Na^+ into highly asymmetric environments where it gives broad NMR signals which can only be partially observed. The ^{7}Li MAS NMR signal was quantitatively detected and at least two clearly separated lines at 0.3 and -0.7 ppm are observed in dehydrated samples (188).

^{133}Cs MAS NMR spectroscopy was applied to monitor the local environment of hydrated and dehydrated cesium mordenite by Gerstein et al. (189). The anhydrous sample showed two signals at -191 and -57 ppm (center of signal mass), with relative intensities of roughly 3:1, respectively. The assignment of the resonances to cesium locations was made on the basis of the structural differences of the six-ring coordination site VI from the eight-ring sites II and IV (189). After correction for the second-order quadrupolar shift, the downfield signal was attributed to site VI while sites II and IV gave rise to the high-field shifts. In the fully hydrated sample all three sites possess an identical isotropic shift value of -64 ppm (189). Ahn et al. demonstrated that ^{133}Cs NMR spectra revealed marked changes in the line shape as a function of temperature for a fully hydrated zeolite A sample containing a mixture of Li, Na and Ce cations (190).

The line-shape of solid-state ^{139}La NMR spectra was observed to be quite sensitive to the local symmetry of the La ion sites. The investigation of calcined La-exchanged Y zeolites reveals sharp and broad signals corresponding to La ions located in supercages and in small cages, respectively (191).

5.4. Monitoring the Pore Architecture of Zeolites and AlPO$_4$ Molecular Sieves by ^{129}Xe NMR Spectroscopy

One of the main advantages of application of zeolites and AlPO$_4$ molecular sieves is the shape-selectivity of this type of material, which arises due to differential diffusion of molecules with different sizes and shapes in the zeolite/AlPO$_4$ pores. Therefore, it is very instructive to monitor the pore architecture directly, with a molecule that "observes" the zeolite/AlPO$_4$ structure. ^{129}Xe was shown to be a very suitable and sensitive nucleus for this purpose (13). The extended xenon electron cloud is easyly deformable (e.g. due to collisions), and deformation results in a large low-field shift of the ^{129}Xe-resonance (13). From an NMR point of view, the ^{129}Xe isotope has a spin 1/2, the natural abundance is 26% and its sensitivity of detection relative to proton is 10^{-2}. Hence, assuming fast exchange, the shift of ^{129}Xe absorbed in zeolites/AlPO$_4$ molecular sieves can be regarded as the sum of several additive contributions (13, 192-194). Therefore, the chemical shift δ of xenon adsorbed in a pure zeolite/AlPO$_4$ is:

$$\delta = \delta_0 + \delta_{Xe} + \delta_E + \delta_S \tag{3}$$

where δ_0 is the reference (Xe gas at infinitely low pressure), δ_{Xe} is a contribution from Xe-Xe collisions, δ_E results from electric field gradients (cations) and δ_S is due to collisions between Xe and cage or channel walls of the zeolite.

From equation (3) it is clear that, when extrapolating to vanishing Xe pressure, the xenon chemical shift is determined mainly by the electric field term (which can be neglected for decationized zeolites and those containing only alkali metal cations (194)) and the xenon - zeolite/AlPO$_4$ interaction (13). Indeed, a number of review papers exist, concluding that the chemical shift δ_S of xenon adsorbed on zeolites and extrapolated to zero concentration depends only on the internal void space of the solid (193-196). The smaller the channels or cavities, or the more restricted the diffusion, the greater the δ_S becomes (194). A systematic study of zeolites whose structure is already known demonstrates that the ^{129}Xe chemical shift is linearly related to the pore size of the zeolite (195), and the xenon chemical shift increases with decreasing mean free path for the xenon atoms in the pores (13). A number of different influences, like structure, void space, crystallinity, pore blocking, metal cations (incl. paramagnetic cations), chemisorption of gases etc., have been investigated and reviewed, and I would like to recommend the study of those papers from the late 1980ies, with Ito and Fraissard as the main investigators (193-196). The "^{129}Xe NMR method" has been developed to such a stage where it can be used to study (and predict) the pore architecture of zeolites/AlPO$_4$ molecular sieves with unknown structures. For example, before the structure of zeolite beta was known, Fraissard et al. were able to determine the approximate form of the internal void volume of this zeolite (197).

Anyway, the latest results dealing with the application of ^{129}Xe NMR on zeolites/AlPO$_4$ molecular sieves are reviewed in the following. The sensitivity of ^{129}Xe NMR spectroscopy made it possible for Dybowski et al. to distinguish between two different macroscopic environments for xenon sorbed in silicalite (198). Using this effect, the authors estimated the amount of material with template left in the channels from the ^{129}Xe NMR spectrum. Howe and coworkers investigated the variation of ^{129}Xe NMR chemical shifts with the aluminium content of ZSM-5 and -11 zeolites, and concluded with an increase in chemical shift with increasing number of Al atoms per unit cell (199). This variation is explained by changes in the effective electric field experienced by xenon within the zeolite pores as the Al content and, hence, the proton concentration vary (199). In a way, ^{129}Xe NMR can be used to probe the concentration of Al in the zeolite framework.

The distribution of Na metal particles inside the cavities of Na Y zeolite has been investigated using ^{129}Xe NMR spectroscopy. For a sample prepared by vapor-phase deposition at 247 ^{o}C., the ^{129}Xe NMR spectrum showed three lines which were interpreted in terms of domains of nonuniformly distributed metal particles, oxidized particles and empty cavities. After annealing the sample at 397 ^{o}C., the ^{129}Xe NMR spectrum collapses to one single line, characteristic of a narrow particle size distribution (200). The surface of Pt clusters with average size between 1 and 8 nm supported on Y zeolite was probed by ^{129}Xe NMR as the Pt surface coverage with hydrogen was increased. A distinct change in the structure of the hydrogen overlayer was observed when the chemisorbed hydrogen fills all the next nearest neighbour metal sites and the nearest neighbours start to be occupied (201).

Low-temperature (-129 ^{o}C.) ^{129}Xe NMR studies of steam-dealuminated Y zeolites revealed that the number of micropores with a pore size similar to the supercage could be estimated by associating the ^{129}Xe NMR line-width minimum and the chemical shift at the minimum with a definite physical state of the xenon in the supercage (202). It was shown that ^{129}Xe NMR could be used to determine the number of supercages in the modified H-Y zeolites. Furthermore, Cheung et al. defined the mean free path Δ as a measure of the frequency of collision with the framework and hence measures the free volume available to the xenon atoms (202), according to the dependence of δ_S on the mean free path, Δ, of xenon in a zeolite (203):

$$\delta_S = 243 \, [2.054/(2.054 + \Delta)] \tag{4}$$

In a second paper, Cheung investigated xenon adsorbed in offretite, silicalite, L and A zeolites at -129 ^{o}C (204). Changes of the chemical shift and linewidth of the ^{129}Xe resonance as a function of the xenon loading reflect the size of the pores in zeolites. In zeolites with small pores, the chemical shift increases gradually with loadings, and the resonance is broad, while in zeolites with large pores, the chemical shift undergoes rapid increases in the high loading regime, accompanied by drastic reductions in the line width. The differences in the chemical shift and linewidth may be attributed to a gas-liquid phase transition of the adsorbed xenon in zeolites with large pores (204).

Omega, ferrierite and mordenite structures were studied by Ito et al. applying ^{129}Xe NMR (205). Using this method, the authors recognized a short-range crystallinity in Na-omega, a distribution of Xe atoms in the channels parallel to the c and b axes of Na- and K-ferrierite and the approximate location of cations as well as the intergrowth formation of ferrierite-mordenite structures. They were able to determine the concentration of a mordenite phase included in ferrierite (205).

The ^{129}Xe NMR technique has been successfully used to the study of ion exchange between two zeolites, Rb-Na X and Na Y zeolites (206). The adsorbed water existing in the system greatly favours this exchange, however, solid-state ion exchange under fully dehydrated conditions could also be observed (206). ^{129}Xe NMR has been applied to the study of organic molecules adsorbed in Na-Y zeolite, and Pines at al. found the technique to be sensitive to both the nature and the concentration of the adsorbed species (207). The position of the ^{129}Xe NMR resonance is a function of both the guest species and its concentration, as well as the density of the Xe inside the supercage of the Na-Y zeolite (see Figure 32, 207). Their results showed a difference in the interaction of Xe with aromatic molecules (benzene and trimethylbenzene) as compared to the saturated hydrocarbon chain molecule n-hexane. The authors believe that the aromatic molecules are adsorbed parallel to the zeolite surface while n-hexane molecules may be adsorbed to the surface. In addition, they found evidence for blocking of the supercage windows due to guest molecule size or orientation (207).

Successful monitoring of metals and metal cluster formation in zeolites has been performed by several authors. Radke and coworkers gained new insight into the detailed chemistry of metal-zeolite catalyst preparation by probing the dispersion of Pt on Na Y zeolite by ^{129}Xe NMR (208). The ^{129}Xe chemical shift moves upfield in Pt-Na Y zeolite samples calcined at progressively higher temperatures. The NMR data indicate formation of highly dispersed Pt in Na Y zeolite when the calcination temperature is close to 400 $^{\circ}$C (208). Ryoo et al. confirmed the usefulness of ^{129}Xe NMR for measuring the distribution of Pt species during the preparation of Pt-Na Y catalysts, either by impregnation or cation exchange (209). The ion exchange method resulted in a uniform distribution of Pt species in the zeolite channel, whereas the location of Pt species during the impregnation was found to be affected by the time and drying mode: uniform distribution occurs only after longer thermal treatments in a 100% relative humidity chamber (209).

Parallel findings were observed by van Santen et al. on Ni-Na Y zeolite samples using ^{129}Xe NMR spectroscopy (210). Again, ion exchange resulted in a more homogeneous distribution of the Ni species than impregnation. Ni exchanged Na Y zeolites have also been investigated by Fraissard et al. (211). In spite of the paramagnetism of the Ni^{2+} cations, it was possible to determine quantitatively the electric and magnetic effects for different samples at various levels of exchange and hydration as well as the location of Ni cations in the zeolite (211). The ^{129}Xe NMR resonances of Xe in Cu^{2+} exchanged Na Y zeolites are displaced to higher field compared to Na Y, and this displacement increases with the Cu content (212).

^{129}Xe NMR has been used as a tool for probing zeolite coking by Fraissard et al. (213-215). His group demonstrated that the extra-framework Al atoms of H Y zeolite and H ZSM-5 are involved in the coke formation. ^{129}Xe NMR enabled the authors to detect extra-framework Al remaining in the cages or the channels of both zeolites after dealumination (213, 214). In the case of H Y zeolite, the extra-framework Al atoms are among the first coke formation centres (strong acid sites). As long as coke increases up to 10% the supercages are lined with coke and the remaining internal volume consists of narrow channels. Beyond 10% the coke also effects the external surface of the zeolites, with the formation of coke microcavities between the crystallites (214). Even coking and regeneration of FCC catalysts have been studied by ^{129}Xe NMR, with the possibility to monitor the different stages of regeneration (215). In this way it was shown that relatively mild oxidation caused structure defects leading to a significant loss of microporosity (215).

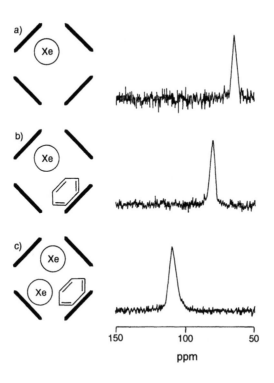

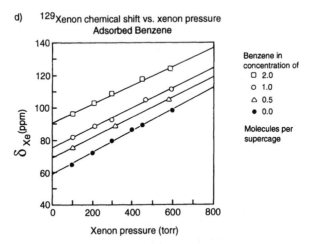

Fig. 32. Schematic illustration of a Na Y supercage with (a) adsorbed xenon (100 Torr), (b) xenon (100 Torr) plus one benzene molecule per cage and (c) xenon (600 Torr) plus one benzene molecule per cage on the left, and the resulting ^{129}Xe NMR spectrum for each on the right. (d) shows the full chemical shift data (in units of ppm versus xenon pressure) for adsorbed benzene in concentrations as indicated (Reproduced by permission of the American Chemical Society, Washington).

The potential of ^{129}Xe NMR to discern domains of different chemical compositions in SAPO molecular sieves was outlined by Davis and coworkers (216, 217). The chemical shift of ^{129}Xe moves downfield as the size of the cavities and channels become smaller for the AlPO$_4$-n and SAPO-n molecular sieves in a way similar to that observed for zeolites (217). In addition, for a particular structure type, the chemical shift moved upfield when going from zeolite to SAPO-n (e.g. Na Y to SAPO-37), and also when going from SAPO-n to AlPO$_4$-n (e.g. SAPO-5 to AlPO$_4$-5). This trend was dramatically illustrated by the data for erionite and AlPO$_4$-17 (217). The authors propose that the deshielding of xenon, when comparing the chemical shift of ^{129}Xe of AlPO$_4$-n molecular sieves to that of zeolites, is due to the greater polarizability of the xenon by the stronger electric fields and field gradients in the zeolites (217). However, Chen at al. claims an exception from this relation for his observation of the chemical shift of Xe adsorbed on SAPO-37 to be identical to that of xenon adsorbed on a Y zeolite (218). Further work is necessary to clarify this point. In addition, the ^{129}Xe chemical shifts were recorded as identical for AlPO$_4$-5, SAPO-5 and MAPO-5 (218).

^{129}Xe NMR of xenon adsorbed in SAPO-37 was recognized as a powerful tool for monitoring the removal of occluded water and organic molecules from the as-synthesized SAPO-37 submitted to different thermal treatments (219). The authors also confirm the sensitivity of moisture to the framework of H-SAPO-37, i.e., its instability and loss of crystallinity (219). Anisotropic chemical shifts were measured for ^{129}Xe sorbed in AlPO$_4$-11 by Ripmeester et al. (220), whereas Maistriau and coworkers investigated a MCM-9 sample containing about 10% SAPO-11 (221). The ^{129}Xe NMR chemical shift of the low-field resonance at 153 ppm has the same pressure dependence as that observed for SAPO-11, and corresponds, therefore, to xenon in the contaminant SAPO-11 phase. The high-field line at 89 ppm evidences a phase possessing much larger pores, and was attributed to xenon in the intracrystalline void volume of MCM-9 (221). Although MCM-9 is the dominant phase, its ^{129}Xe NMR signal has the lowest intensity, which is easily rationalized using the confinement theory that describes the behaviour of molecules in microporous solids (221).

^{129}Xe NMR has been used to characterize the two novel molecular sieves VPI-5 and AlPO$_4$-8 (222). The results are in agreement with the structure of VPI-5, but reveal a second type of pore (about 10.7 Å in diameter), which could correspond to a 16-tetrahedra ring resulting from joining two 6-tetrahedra rings by one 4-tetrahedra ring instead of two 4-tetrahedra rings (222). A difference between the ^{129}Xe NMR and XRD results was found with respect to AlPO$_4$-8, which may arise from the partial collaps of the lattice during the transformation of VPI-5 to AlPO$_4$-8. This collaps may lead to a decrease in the length of the channels and a blockage of some of them, which reduces xenon diffusion and, consequently, the mean free path (222).

The location of residual water and coke deposits in zeolites was the topic of a correlation between the results of ^{129}Xe NMR chemical shifts and ^{1}H NMR pulsed-field-gradient measurements, done by Fraissard and Kärger (223).

An unusual flexibility of the framework of zeolite rho upon xenon loading and temperature change was detected by Smith and coworkers applying ^{129}Xe NMR spectroscopy (224, 225). A cell theory of interacting lattice gas has been introduced by Cheung, interpretating the ^{129}Xe NMR data of xenon adsorbed in Na A zeolite by means of including a repulsive interaction term in case of filling the α-cages with Xe atoms (226).

As mentioned earlier, the influence of monovalent cations in zeolites on the ^{129}Xe chemical shift can be neglected, whereas divalent cations do have an effect on Xe shifts. In fact, the interaction of these cations with xenon is very strong, so that Xe adsorbs preferentially at these cation sites (13).

6. ADSORBED MOLECULES IN ZEOLITES AND AlPO$_4$ MOLECULAR SIEVES

The role of adsorbed molecules in zeolites and AlPO$_4$ molecular sieves can range from the influence of organic templates to frame transitions of zeolite structures due to guest molecules. The structure and position of template species are of great interest for elucidating the process of zeolite/AlPO$_4$ molecular sieve formation, as outlined in the review by Engelhardt (11). The favoured technique in this respect is the ^{13}C CP/MAS NMR spectroscopy, which allows recording of highly resolved NMR spectra of organic templates due to the sensitivity of the isotropic chemical shift to the environment of the carbon atom (11).

^{13}C MAS NMR has been successfully applied for the elucidation of the nature of the interaction of methanol with zeolitic and SAPO-based molecular sieves (12). For example, strongly bound surface CH$_3$-O-Si groups are formed at 250 °C when methanol is adsorbed on the SAPO-5 (12). In addition, the ^{1}H chemical shift of the hydroxyl resonance is very sensitive to the type of zeolite on which methanol is adsorbed (12). Further details and examples are described in the review papers by Engelhardt (11) and Klinowski (12).

The molecular mobility of substrates in zeolites (or catalysts in general) has been an important topic in connection with studies of adsorbed molecules. The most obvious parameters which give insight into molecular mobility are the relaxation times (13), and an excellent overview of the use of relaxation measurements for investigating adsorbed species was given by Pfeifer (227).

Another related parameter is the lineshape of the resonance of a quadrupolar nucleus, which can yield the anisotropy of molecular motion, for example by applying 2D ^{2}H NMR spectroscopy, from which direct information of modes of motion can be derived (13).

A study of sorbate and temperature-induced changes in the lattice structure of ZSM-5 was performed by Fyfe et al. (8) by using highly crystalline and siliceous samples. The change between the monoclinic and orthorhombic forms occurs between 354 and 357 °C and in the case of p-xylene loading at approximately two molecules per 96 T atom unit cell ("low-loaded" form). However, in the case of acetylacetone, the effects are more complex, with two low-temperature monoclinic forms in equilibrium with the high-temperature orthorhombic form (228). Considerable dipolar interaction between the p-xylene protons and the ^{29}Si nuclei in the ZSM-5 lattice were observed in the case of the "high-loaded" form (4-8 molecules of sorbed p-xylene per 96 T unit cell) (229). Conner et al. found that the temperature at which the transition of the monoclinic to the orthorhombic form of ZSM-5 takes place depends on the Si/Al ratio of the zeolite (230). The discrete, reversible transition was followed by ^{29}Si MAS NMR and a similar dependence was found for ZSM-11, too

(230, 231). Phosphorous modification of the ZSM-5 does not change the transition temperature, however, steam treatment of the zeolite does (230).

^{1}H MAS NMR was used to study the chemistry of methanol adsorbed on zeolites H-ZSM-5, Na-ZSM-5 and K-ZSM-5. On H-ZSM-5, methanol forms the $CH_3OH_2^+$ methoxonium ion at low coverages. At higher coverages, large protonated clusters of strongly hydrogen-bonded molecules associated with the Brønsted-acid sites are present. In zeolites Na-ZSM-5 and K-ZSM-5 methanol is coordinatively adsorbed on the cation and forms weakly hydrogen-bonded clusters at higher equilibrium pressures (232). ^{29}Si MAS NMR was used to locate the naphthalene molecules sorbed at room temperature in a saturated silicalite · 4 $C_{10}H_8$ complex, with a placement of the sorbate molecules at the channel intersections (site I) of the MFI structure (233).

^{27}Al and ^{13}C MAS NMR spectroscopy was applied to monitor acetylacetone (AcAc) adsorbed on H-ZSM-5 samples of varying acid-site concentration and extra-lattice Al content. Three types of adsorbed acetylacetone were observed: physisorbed AcAc, AcAc interacting strongly via the carbonyl groups with acid sites within the zeolite structure, and AcAc weakly adsorbed on extra-lattice Al and giving rise to a sharp octahedral Al signal. The authors conclude, that AcAc adsorbs more strongly onto the Brønsted acid sites than onto extra-lattice aluminium. In addition, the use of AcAc as a means of visualizing "NMR invisible" extra-framework Al in H-ZSM-5 should be treated with caution (234).

Topotactic metal organic chemical vapor deposition in zeolite Na Y, using dimethylcadmium and dimethylzinc [$(CH_3)_2M$], was followed by solid-state NMR spectroscopy (235). ^{23}Na MAS NMR revealed the existence of a secondary anchoring interaction between supercage site II Na$^+$ cations and the nucleophilic methylgroups of the $(CH_3)_2M$ guest.

Those examples demonstrate the evidence that the temperature or the adsorption of molecules (whose dimensions approach the pore dimensions) induce changes in the solid structure, i.e. a flexibility of the solid structure (230). Zeolites can, therefore, be regarded as non-rigid solids.

For studying the dynamics of small molecules and the intracrystalline mass transfer in the pores of zeolites and AlPO$_4$ molecular sieves, a varity of approaches can be chosen. However, the groups of Pfeifer and Kärger have done pioneering work with respect to NMR self-diffusion studies in zeolites, especially in connection with the **pulsed field gradient (PFG)** technique (236). Fast translational motions can be studied by applying this method, in which the actual displacement of a molecule is measured (13). Excellent reviews, describing the molecular migration in zeolitic adsorbate-adsorbent systems, were written by Kärger et al. (236-238). The main parameters determining the transport properties of those systems are the coefficients of intracrystalline and long-range self-diffusion and the molecular intercrystalline exchange rates. The authors showed that these quantities may be determined directly by NMR PFG technique in combination with the NMR tracer desorption technique. The varity of conditions to which molecular migration is subjected in the interior of the different types of zeolites/AlPO$_4$ molecular sieves gives rise to characteristic concentration dependences of intracrystalline self-diffusion. The NMR studies yield at least five different patterns of concentration dependence (236). During NMR diffusion studies with labeled molecules, molecular positions are simply recorded by the phase of the Larmor precession of the nuclear spins about the external magnetic field. Since the interaction energy between the nuclear spins and the external magnetic field is much less than the thermal energy of the molecules, this way of labelling does not influence the molecular mobility (236).

A number of papers dealing with this topic have been published since the review paper appeared in 1987, like the experimental evidence for the self-consistency of the NMR diffusion data obtained by using the ^{1}H- and ^{19}F NMR spectroscopy to investigate the self-diffusion of CHF_2Cl in Na X zeolite by varying the intensity of the external magnetic field (239). Conventional ^{1}H NMR signal intensity measurements have been used to monitor macroscopically the kinetics of molecular exchange of deuterium-labeled molecules between the intracrystalline space of zeolite crystallites and the surrounding atmosphere. After reaching equilibrium, a PFG-experiment was performed within the same sample tube. In this way it became possible to compare results for the intracrystalline migration of adsorbed molecules derived from macroscopic (intensity) and microscopic (PFG) measurements on identical samples. For benzene adsorbed on zeolite Na X the intracrystalline diffusivities resulting from these two techniques were found to be in satisfactory agreement. This result proves that molecular exchange between benzene adsorbed in the intracrystalline space of Na X and the surrounding atmosphere is essentially controlled by intracrystalline mass transfer (240).

^{129}Xe NMR has been successfully used to study the self-diffusion of xenon in zeolites (241). In zeolites Na X and ZSM-5, the self-diffusion coefficients were found to decrease with increasing concentration while for zeolite NaCa A they are essentially constant. The highest diffusivities were observed in zeolite Na X. This is in agreement with the fact that due to the internal pore structure the steric restrictions of molecular propagation in zeolite Na X are smaller than those in Na Ca A and ZSM-5 (241). Mass transfer and chemical reaction in zeolite channels in which the individual molecules cannot pass each other (single-file systems) were studied by Monte Carlo simulations, applying a single jump model for the elementary steps of diffusion (242).

The coefficients of intracrystalline self-diffusion of the n-alkanes from propane to n-hexane adsorbed in zeolite ZSM-5 are studied by means of the PFG NMR technique over a temperature range from -20 to 380 °C (243). The diffusivities are found to decrease monotonically with increasing chain lengths. Over the considered temperature range, the diffusivities in ZSM-5 are found to be intermediate between those for Na X and NaCa A zeolite, and the diffusivities of n-alkanes are independent of the Si/Al ratio of the zeolite lattice (243).

Quite a number of PFG NMR measurements have been carried out using ^{1}H NMR spectroscopy with hydrocarbons as probe molecules. Recent progress in the experimental technique of PFG NMR has enabled self-diffusion studies in zeolites using other nuclei than ^{1}H (244). Applying ^{129}Xe and ^{13}C PFG NMR, the temperature dependence of the coefficients of self-diffusion of Xe, CO and CO_2 in zeolites X, NaCa A and ZSM-5 were studied. In all cases the measured diffusivities are found to follow Arrhenius dependence (244). In a second paper dealing with the same zeolites, the authors determined the diffusivities directly from the slope of the echo amplitude vs the field-gradient pulse widths, showing a satisfactory agreement with the corresponding values obtained from an analysis of the NMR tracer desorption curves (245).

The process of adsorption of n-hexane in a bed of zeolite Na X was monitored by NMR imaging in combination with PFG NMR (246).The intracrystalline diffusivities were found to depend exclusively on the given sorbate concentration, independent of the time interval elapsed since the onset of the adsorption process. The authors may conclude that adsorbent accommodation during the process of adsorption is not of significant influence on the molecular mobility (246).

Monitoring the diffusional barriers on the external surface of zeolite crystallites was performed by Kärger et al. applying ^{129}Xe PFG NMR spectroscopy (247). The surface permeability of Na X, NaCa A and ZSM-5 was studied. The passage through the external surface of a Na X crystallite was found to be of minor importance, whereas for NaCa A and ZSM-5 zeolites, this process was found to be significantly retarded. The authors conclude that xenon represents a sensitive tool for probing structural distortion in the surface layer of zeolites (247).

On the basis of the known results one may conclude that this novel PFG NMR technique can be a routine method for characterization of diffusivities in zeolite/AlPO$_4$ molecular sieves.

7.   NMR STUDIES OF CATALYTIC REACTIONS ON ZEOLITES AND SAPO MOLECULAR SIEVES

Solid-state NMR spectroscopy can be used as a powerful tool for getting important information on the structure and properties of intracrystalline guest species. Both ^{1}H and ^{13}C MAS NMR can probe directly the role of the active site in shape-selective catalytic reactions on zeolites and SAPO molecular sieves (12, 248). The catalytic conversion of methanol to hydrocarbons in the gasoline boiling range using, for example, zeolite ZSM-5, SAPO-5 and SAPO-34 has attracted much attention, and review papers covering the work by Klinowski et al. about this topic are available (12, 248). Therefore, I would just like to draw the readers attention to a few examples of their studies and refer to those review papers for further details.

Klinowski and coworkers were able to identify 29 different organic species in the adsorbed phase during the catalytic conversion of methanol to hydrocarbons over ZSM-5 (249). The kind and quantity of the species present inside the particles (olefins, aliphatics and aromatics up to C$_{10}$) were monitored directly by ^{13}C MAS NMR spectroscopy. The authors were able to identify CO as an intermediate in the reaction and could distinguish between mobile and attached species. However, the mechanism of the reactions involved, particularly with respect to the formation of the first carbon-carbon bond, remains still a matter for speculation (249). In addition to chemical shift information, and the monitoring of the number and relative intensity of the various ^{13}C signals, 2 D ^{13}C NMR spectra, like 2 D J-coupled and 2 D spin diffusion experiments, have been used to determine the connectivity of carbons and the number of protons attached to each carbon atom in the various organics as well as the details of ^{13}C-^{1}H couplings (12).

The same authors compare the shape-selective catalytic conversion of methanol to olefins and aliphatics over ZSM-5 with the results obtained for SAPO-34, which has the framework topology of chabazite (12, 250). Figure 33 shows the ^{13}C MAS NMR spectra of SAPO-34 recorded after reaction at different temperatures. After adsorption of methanol and no thermal treatment a single signal at 50 ppm was observed (methanol resonance). After heating to 150 °C, a rather poorly resolved spectrum is obtained consisting of two signals, but only after heating to 300 °C or higher a multitude of narrow resonances were observed in the aliphatic region (-10 to 40 ppm) and no other signals were detected (12, 250). The formation and distribution of the registrated C_1 to C_7 aliphatics are discussed taking into account the framework structure of SAPO-34, concluding with the observation that SAPO-34 converts methanol more selectively to C_3 than to C_2 hydrocarbons (12, 250).

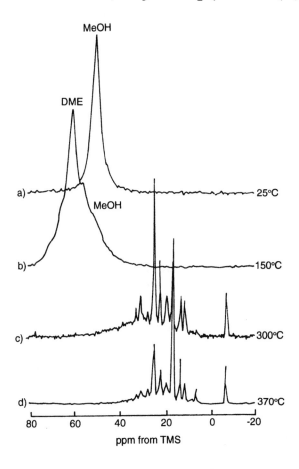

Fig. 33. ^{13}C MAS NMR spectra (with high-power decoupling) of the samples: (a) after adsorption of methanol and no thermal treatment, (b) after heating to 150 °C, (c) after heating to 300 °C for 15 min and (d) after heating to 370 °C for 10 min. Catalyst: SAPO-34 (Reproduced by permission of the American Chemical Society, Washington).

The same authors describe ^{13}C MAS NMR investigations of the nature of the interaction of methanol with ZSM-5 and SAPO-5 by using sealed Pyrex microreactors which fit exactly the MAS rotors and provide an ideal environment for high-temperature catalytic reactions under static reactor conditions (251). Klinowski et al. observed by slow MAS spinning that strongly surface Me-O-Si groups were formed at 250 °C when methanol was adsorbed on SAPO-5, whereas in zeolite ZSM-5, the methanol became strongly bound to the framework at room temperature without formation of methoxy groups (251). Even ^{1}H MAS NMR studies were performed in order to study the chemistry of (deuterated) methanol adsorbed on the zeolites ZSM-5, Y, A, L, silicalite and the SAPO's -5, -11 and -34 (252). Methanol forms the $CH_3OH_2^+$ methoxonium ion and is hydrogen-bonded in the intracrystalline space of the molecular sieves. The authors suggest that clusters of methanol molecules associated with Brønsted acid sites are present at high coverages, whereas at low coverages the methanol molecules are bound at Brønsted acid sites. In Na-ZSM-5, the methanol was coordinately adsorbed on the cation (252). In addition, 2D ^{13}C solid-state NMR spin-diffusion measurements were carried out in connection with the methanol conversion into gasoline over H-ZSM-5, allowing the authors to re-examine former spectral assignments. They received new information on the distribution of C_3-C_7 species adsorbed in the intracrystalline space of ZSM-5 (253).

The reaction mechanism of methanol conversion to hydrocarbons over SAPO-34 was followed by ^{13}C MAS NMR spectroscopy and gas chromatography by Xu et al. (254). The medium acidity of H-SAPO-34 was found to be the determining behaviour of this catalyst towards the formation of light olefins with high selectivities (254).

^{13}C and ^{1}H MAS NMR spectroscopy has been used to monitor the conversion of methanol to light hydrocarbons over intergrowths of the zeolites offretite and erionite with varying contents of the two structures (255). The sample richest in offretite shows the lowest selectivity to light olefins. This was correlated to the ease of formation of long-chain polymeric hydrocarbons within the intracrystalline space of offretite. The presence of those polymers was confirmed by ^{13}C MAS NMR and the authors attribute this observation to the uninterrupted one-dimensional pore system. However, in erionite, where the channel system is constricted at every 15 Å, polymers cannot be formed (255).

^{13}C CP/MAS NMR analysis of the oligomerization of ethene over H-ZSM-5, H-Y and H-mordenite indicated that the degree of branching of the final products increased with increasing pore dimensions of the zeolite (256). The medium-pore zeolite H-ZSM-5 produced more linear products than the large-pore zeolites H-mordenite and H-Y. Ethene sorbed in ZSM-5 was stable below 0 °C, but at higher temperatures it was oligomerized to higher, branched reaction products. The observation of intermediate olefinic oligomers inside the zeolite pores are in line with the traditional model of acid-catalyzed oligomerizations in which the reaction proceeds via carbenium ions. However, in this case the observation was better described in terms of a mechanism involving formation of alkoxide species (256).

2D J-resolved and CP/MAS ^{13}C NMR spectroscopy was applied by Stepanov et al. following the pathway for the transfer of the ^{13}C label from the CH_2 group of isobutyl alcohol into the hydrocarbon skeleton of butene oligomers during the isobutyl alcohol dehydration inside a H-ZSM-5 zeolite (257). First, the label was transferred selectively into the CH_2 group of the isobutyl silyl ether reaction intermediate, and then into the CH and

CH$_3$ groups of the isobutyl fragment of the same intermediate and/or butene oligomers. Finally, it was scrambled over the carbon skeleton of the oligomers. The obtained data suggest that isobutyl carbenium ion was formed as a reaction intermediate or transition state during the transformation of isobutyl silyl ether into butene oligomers (257). Deuterium solid-state NMR was successfully used to monitor the molecular mobility of organic species during the catalytic dehydration of tert. butyl alcohol on H-ZSM-5 zeolite, concluding that the alcohol molecules are predominantly localized at the channel intersections (258). Furthermore, an estimation of the influence of the channel walls on the geometry of adsorbed tert. butyl alcohol was performed.

Investigations of reactions on zeolite catalysts using *in situ* solid-state NMR spectroscopy have been performed by Haw and coworkers applying a special rotor design, abbreviated as **CAVERN (cryogenic adsorption vessel enabling rotor nestling)** (259, 260). *In situ* MAS NMR studies of catalytic reactions face several difficult sample handling requirements, since most catalyst-adsorbate samples are air or moisture sensitive and require preparation on a vacuum line, including loading into the MAS rotor (259). The main advantage of the CAVERN design is that it permits the MAS rotor to be returned to the vacuum line for the adsorption of additional reactants prior to resuming *in situ* NMR studies. The term CAVERN is now used to designate any device for sealing or unsealing a MAS rotor on a vacuum line. Figure 34 shows a drawing of the second generation CAVERN apparatus (259). For details of the CAVERN design please check with the references (259, 260). The use of this design was illustrated by demonstrating the sequential adsorption feature of methanol on H-ZSM-5. Although the first aliquot of methanol exhibited a long induction period in the conversion of methanol to gasoline at 250 °C, none was observed following adsorption of a second aliquot, showing that the products have an influence on the observed kinetics (259).

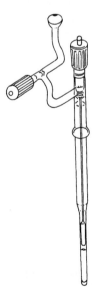

Fig. 34. Diagram of the CAVERN apparatus (Reproduced by permission of Academic Press, Orlando).

A shallow-bed CAVERN design was developed by the same group allowing the catalysts to be activated under shallow-bed conditions at temperatures up to at least 500 °C prior to adsorption, eliminating the need to handle activated catalysts in a glove box in order to transfer it into the MAS rotor (260). The new device was illustrated by ^{133}Cs MAS NMR spectra of samples of zeolite Cs-ZSM-5 loaded with methanol. The investigation clearly showed that a very homogeneous adsorbate loading was achieved with the shallow-bed CAVERN system (260). In a separate paper, *in situ* ^{13}C and ^{77}Se MAS NMR were used to investigate the formation and reactivity of trialkylonium species on zeolite catalysts H-ZSM-5 and H Y (261). The results suggest that onium ions may be useful for measuring the Brønsted acid strength in catalysts.

Conversion of metal halides to ethylene and other hydrocarbons on basic, alkali metal-exchanged zeolites at low temperatures was followed by *in situ* NMR spectroscopy (262). All reactions were performed in a batch mode in sealed MAS rotors while the contents were continuously monitored by *in situ* ^{13}C MAS NMR. Methyl iodide reacted on Cs X below room temperature to form a framework-bound methoxy species in high yield. ^{133}Cs MAS NMR was used to characterize the interactions of methyl iodide and other adsorbates with the cation in zeolite Cs-ZSM-5. Solvation of the alkali metal cation was reflected in large, loading-dependent chemical shifts for ^{133}Cs. Interactions between the cation and adsorbates were also reflected in the ^{13}C chemical shifts of the alkyl halides and ethylene (262).

The change of product selectivity during the synthesis of croton aldehyd from acetaldehyd by coadsorption of water was demonstrated by Haw et al. (263). After water coadsorption, the authors observed a change of catalytic activity (lower acidity) of H-ZSM-5 in connection with this acid catalyzed aldolreaction, resulting in a selective formation of croton aldehyd even at elevated temperature (120 °C). Without coadsorption of water the acetaldehyd reacted to a complex product mixture decomponating at higher temperature (160 °C) (263). The changes were monitored by *in situ* ^{13}C MAS NMR spectroscopy.

8.   CATALYST ACIDITY

The acidic properties of zeolites or SAPO molecular sieves are of considerable importance with respect to catalyzed reactions in heterogeneous catalysis. It is vital to know the concentration, strength and accessibility of the Brønsted and Lewis acid sites and the details of their interaction with adsorbed species (12). For zeolites, for example, ^{29}Si MAS NMR plays a crucial role in the determination of the amount of aluminium which is part of the zeolite lattice as well as the distribution of Al atoms over distinct crystallographic sites (13). Furthermore, ^{27}Al MAS and DOR NMR are additional tools to distinguish between framework and non-framework aluminium. While ^{29}Si and ^{27}Al NMR are suitable for determining the concentration of acid sites, the actual acid strength can be probed either by direct study of the acid protons (Brønsted acid sites) applying ^{1}H NMR or by using probe molecules (13).

Again, a number of review papers exist describing NMR studies on zeolite acidity, and the papers by Pfeifer an coworkers (264) and Klinowski et al. (10, 12 and 248) could be considered as suitable information about this topic of zeolite research.

The Brønsted acidity of zeolites arises from the presence of accessible hydroxyl groups associated with framework aluminium (12). As already mentioned, ^{1}H MAS NMR is an advanced tool for probing the protonic sites in zeolites and SAPO molecular sieves. However, solid-state ^{1}H NMR is an experimentally difficult technique, particularly in cases where a high concentration of protons cause extreme line broadening due to proton-proton dipolar interactions, which must be removed by multiple quantum decoupling (13). In addition, heteronuclear dipolar interactions of protons with aluminium can complicate the NMR spectra, especially taking into account the generally narrow range of proton chemical shifts. Fortunately for dehydrated zeolites/SAPOs, the proton density is often so low that reasonable ^{1}H MAS NMR spectra can be obtained at high magnetic fields and using fast spinning speeds. In this way, four distinct types of protons have been identified and quantified by their chemical shifts:

1. non-acidic, terminal SiOH groups on the surface of zeolite crystallites and crystal defect sites (1.5 - 2 ppm)
2. AlOH groups at non-framework aluminium (2.6 - 3.6 ppm)
3. acidic, bridging hydroxyl groups SiO(H)Al (3.6 - 5.6 ppm)
4. ammonium ions (6.5 - 7.5 ppm)

The most important sites are the bridging SiO(H)Al groups, which represent the catalytically active sites in zeolites. The concentration of the protons in the various sites can be directly determined from the peak intensities. In addition, the deprotonation energy can be measured by chemical shift changes: the chemical shifts increase with increasing Sanderson electronegativity. Moreover, the ^{1}H chemical shift may be correlated with the proton donor ability of the corresponding site and can thus provide information on the acid strength of the protons (3). Measuring the interactions between surface hydroxyl groups and water by ^{1}H NMR can be used to characterize the strength of Brønsted acid sites.

Extensive studies of the acidity of zeolite H Y have been carried out and are summarized in the papers of the following references: 12, 13, 264-266. Recent ^{1}H MAS NMR investigations confirm that extra-framework aluminium present in dealuminated (ultrastable) zeolite Y was reintroduced into the framework by treatment with strongly basic solutions at elevated temperatures (267). The realuminated sample contains twice as many Brønsted acid sites than the ultrastable precursor and (with an accuracy of 20%) the same number of acid sites as the parent as-prepared zeolite. However, not as many hydroxyl groups associated with the framework Al in the product are accessible to pyridine as in the parent sample (267). Harris and coworkers analyzed highly resolved ^{1}H MAS NMR spectra of dried H Y zeolite by Gaussian deconvolution in order to afford estimates of the absolute concentrations of the different types of OH groups present: the non-framework AlOH sites gain intensity at the expense of SiOH and Brønsted sites with time (268).

^{1}H MAS NMR spectroscopy allowed the measurement of the intensities of the signals of two different types of bridging hydroxyl groups in H Y zeolite: bridging hydroxyl groups pointing into the large cavities (at 3.9 ppm) and into the six-membered oxygen rings of the zeolite Y (at 4.9 ppm) (269).

Variable temperature ^{1}H MAS NMR was used to characterize the structure and dynamics of hydrogen bonded adsorption complexes between various adsorbates and the Brønsted acid site in H ZSM-5: the Brønsted proton chemical shift of the active site was found to be extremely sensitive to the amount of type of adsorbate (acetylene, ethylene, CO and benzene) introduced (270).

A study of the geometry and location of bridging OH groups in zeolites and SAPO molecular sieves was performed by Hunger et al. using ^{1}H MAS NMR sideband analysis and ^{29}Si CP/MAS NMR spectroscopy (271). The numerical analysis of the proton sideband pattern in zeolites A, faujasite, erionite, mordenite, pentasil and SAPO's-5, -17, -34 and -37 yields H-Al distances of the hydroxyl protons to the adjacent framework aluminium nucleus which cover a range between 0.234 and 0.252 nm. From the results, a relation between the H-Al distance and the size of the channel rings was derived (271).

Acid sites in zeolites/SAPO molecular sieves can be sensitively probed by the study of adsorbed molecules like pyridine, ammonia (^{15}N NMR) or trialkylphosphines (^{31}P NMR). There is at present only a handful of publications involving ^{15}N NMR of molecules sorbed on zeolites since enriched adsorbents need to be used to improve the poor signal to noise ratio. Valuable data have been obtained for zeolites X, Y and mordenite, which are summarized in the references (12, 13). A new approach was introduced by Rothwell et al. (272), who explored the utilization of the much more sensitive ^{31}P nucleus as a probe for acidity (13). Trimethylphosphine (TMP) was shown to be an effective probe molecule for studying both Brønsted and Lewis acid sites in acidic Y-type zeolites. Interaction of the trimethylphosphine with Brønsted acid sites gives rise to the protonated adduct having ^{31}P MAS NMR chemical shifts in the range -1 to -4 ppm. In a partially oxidized sample the protonated form of TMP oxid exhibits a resonance at 64.6 ppm. TMP also interacts with Lewis acid sites in dealuminated zeolites showing a ^{31}P resonance at -62 ppm. Depending on the treatment conditions additional resonances were observed in the range of -31 to -58 ppm (273, 274).

A new method for measuring the Brønsted acidity of solids (including zeolites) has been proposed by Batamack et al. (275) using wide-line ^{1}H NMR spectroscopy. This technique is based on the interaction of Brønsted acid sites with water molecules when the number of these water molecules is equal to the number of the Brønsted sites. The concentrations and the internal distance parameters of the oxy-protonated species are obtained by simulation of wide-line ^{1}H NMR spectra recorded at 4K in order to rule out motion of the species. Samples whose acid strength is small give only hydrogen-bonded $H_2O...HOS$ groups (S = solid). Formation of H_3O^+ ions occurs when the acid strength is larger (275).

9.   MESOPOROUS MATERIALS

A new family of silicate/aluminosilicate mesoporous molecular sieves designated as M41S and kanemite were introduced a few years ago (276-280), and NMR investigations have been done to characterize those materials.

MCM-41, one member of the M41S family, exhibits a hexagonal arrangement of uniform mesopores whose dimensions may be engineered in the range of about 15 to greater than 100 Å (276). The ^{29}Si MAS NMR spectra of MCM-41 closely resemble those of amorphous silica, that means the spectra can be separated into three very broad signals at -92, -102 and - 111 ppm (276). However, quantitative analysis is very difficult due to the broad and overlapping signals.

The relative number of incompletely condensed (Q^3) and fully condensed (Q^4) silicon atoms in MCM-41 can be determined by ^{29}Si MAS NMR, indicating the degree of hexagonal mesostructures (277).

Japanese researchers characterized kanemite, a highly ordered mesoporous material prepared from layered polysilicates (278-280): the ^{29}Si MAS NMR spectrum showed a sharp signal at -97 ppm due to Q^3 units, indicating a single layered structure. The spectrum of the silicate-organic complex showed two signals due to a Q^3 unit (-99 ppm) and a Q^4 unit (-109 ppm). The appearance of the strong Q^4 unit clearly indicates the formation of a three-dimensional SiO_2 network from the layered kanemite (278, 279).

^1H NMR was used to determine the pore diameter of water-saturated MCM-41 systems by measuring the spin-lattice relaxation rates (281).

10. CONCLUDING REMARKS

It has been shown that solid-state NMR spectroscopy is a powerful tool for characterization of zeolites and $AlPO_4$ molecular sieves. There is no doubt that this field of research has benefited quite a lot from the latest progress in the development of NMR techniques, like higher magnetic field strengths, faster sample spinning, two-dimensional pulse sequences, extended *in situ* capabilities etc. The opportunities to investigate NMR active nuclei with quadrupolar moments by applying, for example DOR, will extent the scope in the still expanding area of zeolites and $AlPO_4$ molecular sieves. In addition, the progress in theoretical understanding and simulation methods indicates a promising outlook for the future.

The author is aware about the fact, that not all contributions published during this period could be taken into account in connection with the preparation of this paper. However, he hopes that this review, which covers the NMR results related to zeolites and $AlPO_4$ molecular sieves during the last four years, can contribute to provide an overview within this part of zeolite and $AlPO_4$ molecular sieve research.

Acknowledgements
The author is indebted to those colleagues and co-workers who contributed to this review. Thanks are due to Tordis Whist for technical assistance in connection with the preparation of the figures, and the relevant publishers are gratefully acknowledged for their permission to reproduce the figures as cited in the text. Financial support by The Research Council of Norway (NFR) and SINTEF is gratefully acknowledged.

REFERENCES

1.   E. Lippmaa, M. Mägi, A. Samoson, G. Engelhardt and A.R. Grimmer, J. Am. Chem. Soc., 102 (1980) 4889.
2.   E. Lippmaa, M. Mägi, A. Samoson, M. Tarmak and G. Engelhardt, J. Am. Chem. Soc.,103 (1981) 4992.
3.   G. Engelhardt, Trends in Anal. Chem., 8 (1989) 343.
4.   G. Engelhardt and D. Michel, High-Resolution Solid-State NMR of Silicates and Zeolites, John Wiley & Sons, Chichester, 1987.
5.   C.A. Fyfe, Solid-State NMR for Chemists, C.F.C. Press, Guelph, 1983.
6.   C.A. Fyfe, L. Bemi, R. Childs, H.C. Clark, D. Curtin, J. Davies, D. Drexler, R.L. Dudley, G.C. Gobbi, J.S. Hartman, P. Hayes, J. Klinowski, R.E. Lenkinski, C.J.L. Lock, I.C. Paul, A. Rudin, W. Tchir, J.M. Thomas and R.E. Wasylishen, Phil. Trans. R. Soc. Lond. A 305 (1982) 591.
7.   C.A Fyfe, J.M. Thomas, J. Klinowski and G.C. Gobbi, Angew. Chem., 95 (1983) 257.
8.   G.T. Kokotailo, C.A. Fyfe, G.C. Gobbi, G.J. Kennedy, C.T. DeSchutter, R.S. Ozubko and W.J. Murphy, Zeolites (Eds.: H. Drzaj, S. Hocevar and S. Pejovnik), Elsevier, Amsterdam, 1985, p. 219.
9.   G.T. Kokotailo, C.A. Fyfe, G.J. Kennedy, G.C. Gobbi, H. Strobl, C.T. Pasztor, G.E. Barlow and S. Bradley, Stud. Surf. Sci. Catal., 28 (1986) 361.
10. J. Klinowski, Colloids and Surfaces, 36 (1989) 133.
11. G. Engelhardt, Stud. Surf. Sci. Catal., 58 (1991) 285.
12. J. Klinowski, Chem. Rev., 91 (1991) 1459.
13. A.D.H. Clague and N.C.M. Alma, Analytical NMR, John Wiley & Sons, Chichester, 1989, p. 115.
14. J.B. Nagy and E.G. Derouane, Perspectives in Molecular Sieve Science, Am. Chem. Soc., 1988, p. 2.
15. C.A. Fyfe, Lecture, 9th International Zeolite Conference - Summer School on Zeolites, Montreal, 1992.
16. E.R. Andrew, A. Bradbury and R.G. Eades, Nature, 182 (1958) 1659 and 183 (1959) 1802.
17. I.J. Lowe, Phys. Rev. Lett., 2 (1959) 285.
18. J. Klinowski, Nature, 346 (1990) 509.
19. A. Samoson, E. Lippmaa and A. Pines, Mol. Phys., 65 (1988) 1013.
20. Y. Wu, B.Q. Sun, A. Pines, A. Samoson and E. Lippmaa, J. Magn. Reson., 89 (1990) 297.
21. A. Samoson and A. Pines, Rev. Sci. Instrum., 60 (1989) 3239.
22. Y. Wu, B.F. Chmelka, A. Pines, M.E. Davis, P.J. Grobet and P.A. Jacobs, Nature, 346 (1990) 550.
23. A. Pines, Proc. 100th School of Physics "Enrico Fermi" (Ed.: B. Maraviglia), North-Holland, Amsterdam, 1988, p. 43.
24. L.W. Jelinski., Anal. Chem., 62 (1990) 212 R.
25. A. Samoson and E. Lippmaa, Chem. Phys. Lett., 100 (1983) 205.
26. H. Hamdan and J. Klinowski, J.C.S. Chem. Commun., (1989) 240.
27. W.S. Veeman, Z. Naturforsch. 47 A (1992) 353.
28. W.S. Veeman, A.P.M. Kentgens and R. Janssen, Fresenius Z. Anal. Chem., 327 (1987) 63.

29. C.A. Fyfe, Y. Feng, G.T. Kokotailo, H. Grondey and H. Gies, Bruker Report 139 (1993) 29.

30. C.A. Fyfe, H. Gies and Y. Feng, J.C.S. Chem. Commun., (1989) 1240.

31. C.A. Fyfe, H. Gies and Y. Feng, J. Am. Chem. Soc., 111 (1989) 7702.

32. C.A. Fyfe, H. Gies, Y. Feng and G.T. Kokotailo, Nature, 341 (1989) 223.

33. C.A. Fyfe, Y. Feng, H. Gies, H. Grondey and G.T. Kokotailo, J. Am. Chem. Soc., 112 (1990) 3264.

34. C.A. Fyfe, H. Grondey, Y. Feng and G.T. Kokotailo, J. Am. Chem. Soc., 112 (1990) 8812.

35. C.A. Fyfe, H. Grondey, Y. Feng and G.T. Kokotailo, Chem. Phys. Lett., 173 (1990) 211.

36. C.A. Fyfe, H. Gies, Y. Feng and H. Grondey, Zeolites, 10 (1990) 278.

37 C.A. Fyfe, H. Grondey, Y. Feng, G.T. Kokotailo, S. Ernst and J. Weitkamp, Zeolites, 12 (1992) 50.

38. C.A. Fyfe, Y. Feng and H. Grondey, Microporous Mater., 1 (1993) 393.

39. B.F. Chmelka and A. Pines, Science, 246 (1989) 71.

40. W. Kolodziejski, P.J. Barrie, H. He and J. Klinowski, J.C.S. Chem. Commun., (1991) 961.

41. A.C. Rohrman, R.B. LaPierre, S.L. Schlenker, J.D. Wood, E.W. Valyocsik, M.K. Rubin, J.B. Higgins and W.J. Rohrbaugh, Zeolites, 5 (1985) 352.

42. K.T. Mueller, C.A. Fyfe, H. Grondey, K.C. Wong-Moon and T. Markus, Bull. Magn. Reson., 14 (1992) 9.

43. C.A. Fyfe, H. Grondey, K.T. Mueller, K.C. Wong-Moon and T. Markus, J. Am. Chem. Soc., 114 (1992) 5876.

44. C.A. Fyfe, K.T. Mueller, H. Grondey and K.C. Wong-Moon, Chem. Phys. Lett., 199 (1992) 198.

45. E.R.H van Eck and W.S. Veeman, J. Am. Chem. Soc., 115 (1993) 1168.

46. W. Kolodziejski and J. Klinowski, Molecular Sieves (Eds.: M.L. Occelli and H.E. Robson), van Nostrand Reinhold, New York, 1992, p. 473.

47. G. Engelhardt, J.- Ch. Buhl, J. Felsche and H. Foerster, Chem. Phys. Lett., 153 (1988) 332.

48. C.A. Fyfe, J.H. O'Brien and H. Strobl, Nature, 326 (1987) 281.

49. J.-M. Chézeau, L. Delmotte, J.-L. Guth and M. Soulard, Zeolites, 9 (1989) 78.

50. F.M. Hoffner, L. Delmotte and H. Kessler, Zeolites, 13 (1993) 60.

51. J. Datka, W. Kolodziejski, J. Klinowski and B. Sulikowski, Cat. Lett., 19 (1993) 159.

52. J. Klinowski, T.A. Carpenter and J.M. Thomas, J.C.S. Chem. Commun., (1986) 956.

53. M.T. Weller and G. Wong, J.C.S. Chem. Commun., (1988) 1103.

54. N.C.M. Alma-Zeestraten, J. Dorrepaal, J. Keijsper and H. Gies, Zeolites, 9 (1989) 81.

55. M.M.J. Treacy and J.M. Newsam, Nature, 332 (1988) 249.

56. J.M. Newsam, M.M.J. Treacy, W.T. Koetsier and C.B. DeGruyter, Proc. R. Soc. Ser. A 420 (1988) 375.

57. J.B. Higgins, R.B. LaPierre, J.L. Schlenker, A.C. Rohrman, J.D. Wood, G.T. Kerr and W.J. Rohrbaugh, Zeolites, 8 (1988) 446.

58. J. Pérez-Pariente, J. Sanz, V. Fornes and A. Corma, J. Catal., 124 (1990) 217.

500

59.   C.A. Fyfe, H. Strobl, G.T. Kokotailo, C.T. Pasztor, G.E. Barlow and S. Bradley, Zeolites, 8 (1988) 132.
60.   R. Challoner, R.K. Harris, K.J. Packer and M. J. Taylor, Zeolites, 10 (1990) 539.
61.   T.V. Whittam, Eur. Pat. Appl. No. 0055046 A1 (1981).
62.   A. Tuel and Y. Ben Taarit, J.C.S. Chem. Commun., (1992) 1578.
63.   M. Stöcker, S. Ernst, H.G. Karge and J. Weitkamp, Acta Chem. Scand., 44 (1990) 519.
64.   M.E. Davis, Molecular Sieves (Eds.: M.L. Occelli and H.E. Robson), van Nostrand Reinhold, New York, 1992, p. 60.
65.   J. Weitkamp and R. Schuhmacher, Proc. 9th IZC Montreal (Eds.: R. von Ballmoos et al.), Butterworth-Heinemann, Stoneham, 1993, p. 353.
66.   M.J. Annen and M.E. Davis, Microporous Mater., 1 (1993) 57.
67.   M.J. Annen and M.E. Davis, Molecular Sieves (Eds.: M.L. Occelli and H.E. Robson), van Nostrand Reinhold, New York, 1992, p. 349.
68.   M.W. Anderson, Magn. Reson. Chem., 30 (1992) 898.
69.   J.H. Kwak and R. Ryoo, J. Phys. Chem., 97 (1993) 11154.
70.   J. Haase, H. Pfeifer, W. Oehme and J. Klinowski, Chem. Phys. Lett., 150 (1988) 189.
71.   P.P. Man and J. Klinowski, J.C.S. Chem. Commun., (1988) 1291.
72.   H. Hamdan and J. Klinowski, J.C.S. Chem. Commun., (1989) 240.
73.   J. Rocha and J. Klinowski, J.C.S. Chem. Commun., (1991) 1121.
74.   J. Rocha, S.W. Carr and J. Klinowski, Chem. Phys. Lett., 187 (1991) 401.
75.   J. Rocha and J. Klinowski, Molecular Sieves (Eds.: M.L. Occelli and H.E. Robson), van Nostrand Reinhold, New York, 1992, p. 70.
76.   W.A. Buckermann, C.B. Huong, F. Fajula and C. Gueguen, Zeolites, 13 (1993) 448.
77.   G.J. Ray and A. Samoson, Zeolites, 13 (1993) 410.
78.   G. Engelhardt, H. Koller, P. Sieger, W. Depmeier and A. Samoson, Solid State Nucl. Magn. Reson., 1 (1992) 127.
79.   S.M. Bradley, R.F. Howe and R.A. Kydd, Magn. Reson. Chem., 31 (1993) 883.
80.   H.K.C. Timken, G.L. Turner, J.P. Gilson, L.B. Welsh and E. Oldfield, J. Am. Chem. Soc., 108 (1986) 7231.
81.   H.K.C. Timken, N. Janes, G.L. Turner, S.L. Lambert, L.B. Welsh and E. Oldfield, J. Am. Chem. Soc., 108 (1986) 7236.
82.   G. Coudurier, A. Auroux, J.C. Vedrine, R.D. Farlee, L. Abrams and R.D. Shannon, J. Catal., 108 (1987) 1.
83.   A. Cichocki, J. Datka, A. Olech, Z. Piwowarska and M. Michalik, J.C.S. Faraday Trans., 86 (1990) 753.
84.   R.A. Van Nordstrand, D.S. Santilli and S.I. Zones, Molecular Sieves (Eds.: M.L. Occelli and H.E. Robson), van Nostrand Reinhold, New York, 1992, p. 373.
85.   R. de Ruiter, A.P.M. Kentgens, J. Grovtendorst, J.C. Jansen and H. van Bekkum, Zeolites, 13 (1993) 128.
86.   I.P. Appleyard, R.K. Harris and F.R. Fitch, Zeolites, 6 (1986) 428.
87.   S. Qiu, W. Tian, W. Pang, T. Sun and D. Jiang, Zeolites, 11 (1991) 371.
88.   H.K.C. Timken and E. Oldfield, J. Am. Chem. Soc., 109 (1987) 7669.
89.   Z. Gabelica, C. Mayenez, R. Monque, R. Galiasso and G. Ciannetto, Molecular Sieves (Eds.: M.L. Occelli and H.E. Robson), van Nostrand Reinhold, New York, 1992, p. 190.

90. Z. Gabelica and M. Wiame, Abstract ZMPC '93, Nagoya, 1993, p. 245.
91. A. Diaz, R. Monque and M. Bussolo, Bull. Magn. Reson., 15 (1993) 112.
92. S.L. Lambert, Proc. 9th IZC Montreal (Eds.: R. von Ballmoos et al.), Butterworth-Heinemann, Stoneham, 1993, p. 223.
93. A.J. Chandwadkar, R.A. Abdulla, S.G. Hegde and J.B. Nagy, Zeolites, 13 (1993) 470.
94. A. Merrouche, J. Patarin, M. Soulard, H. Kessler and D. Anglerot, Molecular Sieves (Eds.: M.L. Occelli and H.E. Robson), van Nostrand Reinhold, New York, 1992, p. 384.
95. A. Tuel and Y. Ben Taarit, Appl. Catal., A 102 (1993) 201.
96. S.T. Wilson, B.M. Lok, C.A. Messina, T.R. Cannan and E.M. Flanigen, J. Am. Chem. Soc., 104 (1982) 1146.
97. B.M. Lok, C.A. Messina, L. Patton, R.T. Gajek, T.R. Cannan and M.E. Flanigen, J. Am. Chem. Soc., 106 (1984) 6092.
98. M.E. Davis, C. Saldarriaga, C. Montes, J. Garces and C. Crowder, Nature, 331 (1988) 698.
99. M.E. Davis, C. Saldarriaga, C. Montes, J. Garces and C. Crowder, Zeolites, 8 (1988) 362.
100. M.E. Davis, C. Montes, P.E. Hathaway and J.M. Garces, Zeolites: Facts, Figures, Future (Eds.: P.A. Jacobs and R.A. van Santen), Elsevier, Amsterdam, 1989, p. 199.
101. M.E. Davis, C. Montes, P.E. Hathaway, J.P. Arhancet, D.L. Hasha ans J.M. Garces, J. Am. Chem. Soc., 111 (1989) 3919.
102. M. Stöcker, D. Akporiaye and K.P. Lillerud, Appl. Catal., 69 (1991) L 7.
103. P.J. Grobet, J.A. Martens, I. Balakrishnan, M. Mertens and P.A. Jacobs, Appl. Catal., 56 (1989) L 21.
104. Z. Gabelica, L. Maistriau and E.G. Derouane, Molecular Sieves (Eds.: M.L. Occelli and H.E. Robson), van Nostrand Reinhold, New York, 1992, p. 289.
105. L. Maistriau, Z. Gabelica, E.G. Derouane, E.T.C. Vogt and J. van Oene, Zeolites, 11 (1991) 583.
106. J.O. Perez, P.J. Chu and A. Clearfield, J. Phys. Chem., 95 (1991) 9994.
107. J.P. van Braam Houckgeest, B. Kraushaar-Czarnetzki, R.J. Dogterom and A. de Groot, J.C.S. Chem. Commun., (1991) 666.
108. Y. Shangqing, Q. Shilem, P. Wenqin and Y. Long, Chem. Res. in Chin. Univ., 7 (1991) 11.
109. D. Akporiaye and M. Stöcker, Zeolites, 12 (1992) 351.
110. G. Engelhardt and W. Veeman, J.C.S. Chem. Commun., (1993), 622.
111. E.R.H. van Eck and W.S. Veeman, J. Am. Chem. Soc., 115 (1993) 1168.
112. H.-X. Li and M.E. Davis, J.C.S. Faraday Trans., 89 (1993) 957.
113. L.B. McCusker, Ch. Baerlocher, E. Jahn and M. Bülow, Zeolites, 11 (1991) 308.
114. J. Rocha, W. Kolodziejski, H. He and J. Klinowski, J. Am. Chem. Soc. 114 (1992) 4884.
115. W. Kolodziejski, H. He and J. Klinowski, Chem. Phys. Lett., 191 (1992) 117.
116. D.E. Akporiaye and M. Stöcker, Proc. 9th IZC Montreal (Eds.: R. von Ballmoos et al.), Butterworth-Heinemann, Stoneham, 1993, p. 563.
117. D.E. Akporiaye and M. Stöcker, Microporous Mater., 1 (1993) 423.
118. D. Müller, E. Jahn, G. Ladwig and U. Haubenreisser, Chem. Phys. Lett., 109 (1984) 332.

502

119. E. Jahn, D. Müller and J. Richter-Mendau, Molecular Sieves (Eds.: M.L. Occelli and H.E. Robson), van Nostrand Reinhold, New York, 1992, p. 248.
120. D.E. Akporiaye, M. Stöcker and K.P. Lillerud, Acta Chem. Scand., 46 (1992) 743.
121. H. He, W. Kolodziejski and J. Klinowski, Chem. Phys. Lett., 200 (1992) 83.
122. B. Duncan, M. Stöcker, D. Gwinup, R. Szostak and K. Vinje, Bull. Soc. Chim. Fr., 129 (1992) 98.
123. F. D'Yvoire, Bull. Soc. Chim. Fr., (1961) 1762.
124. H.X. Li and M.E. Davis, J.C.S. Faraday Trans., 89 (1993) 951.
125. P.J. Grobet, H. Geerts, J.A. Martens and P.A. Jacobs, Stud. Surf. Sci. Catal., 52 (1989) 193.
126. J. Rocha, W. Kolodziejski, I. Gameson and J. Klinowski, Angew. Chem., 104 (1992) 615.
127. P.J. Grobet, A. Samoson, H. Geerts, J.A. Martens and P.A. Jacobs, J. Phys. Chem., 95 (1991) 9620.
128. B.F. Chmelka, Y. Wu, R. Jelinek, M.E. Davis and A. Pines, Stud. Surf. Sci. Catal., 69 (1991) 435.
129. R. Jelinek, B.F. Chmelka, Y. Wu, M.E. Davis, J.G. Ulan, R. Gronsky and A. Pines, Catal. Lett., 15 (1992) 65.
130. J. Rocha, J. Klinowski, P.J. Barrie, R. Jelinek and A. Pines, Solid State Nucl. Magn. Reson., 1 (1992) 217.
131. J. Rocha, X. Liu and J. Klinowski, Chem. Phys. Lett., 182 (1991) 531.
132. B. Kraushaar-Czarnetzki, R.J. Dogterom, W.H.J. Stork, K.A. Emeis and J.P. van Braam-Houckgeest, J. Catal., 141 (1993) 140.
133. E.G. Derouane, L. Maistriau, Z. Gabelica, A. Tuel, J.B. Nagy and R. von Ballmoos, Appl. Catal., 51 (1989) L 13.
134. L. Maistriau, Z. Gabelica and E.G. Derouane, Appl. Catal., 67 (1991) L 11.
135. D. Goldfarb, I. Kustanovich and J.O. Perez, Proc. 9th IZC Montreal (Eds.: R. von Ballmoos et al.), Butterworth-Heinemann, Stoneham, 1993, p. 553.
136. D. Goldfarb, H.X. Li and M.E. Davis, J. Am. Chem. Soc., 114 (1992) 3690.
137. H. He, W. Kolodziejski, J. Rocha and J. Klinowski, Molecular Sieves (Eds.: M.L. Occelli and H.E. Robson), van Nostrand Reinhold, New York, 1992, p. 340.
138. W. Kolodziejski, J. Rocha, H. He and J. Klinowski, Appl. Catal., 77 (1991) L 1.
139. R.H. Meinhold and N.J. Tapp, J.C.S. Chem. Commun., (1990) 219.
140. R.H. Meinhold and N.J. Tapp, Zeolites, 11 (1991) 401.
141. E.R.H. van Eck and W.S. Veeman, Solid State Nucl. Magn. Reson., 1 (1992) 1.
142. E. Jahn, D. Müller and K. Becher, Zeolites, 10 (1990) 151.
143. B. Zibrowius, E. Löffler and M. Hunger, Zeolites, 12 (1992) 167.
144. N.J. Tapp, N.B. Milestone, M.E. Bowden and R.H. Meinhold, Zeolites, 10 (1990) 105.
145. M.P.J. Peeters, J.W. de Haan, L.J.M. van de Ven and J.H.C. van Hooff, J.C.S. Chem. Commun., (1992) 1560.
146. M.P.J. Peeters, J.W. de Haan, L.J.M. van de Ven and J.H.C. van Hooff, J. Phys. Chem., 97 (1993) 5363.
147. M.P.J. Peeters, L.J.M. van de Ven, J.W. de Haan and J.H.C. van Hooff, J. Phys. Chem., 97 (1993) 8254.
148. P.J. Barrie, M.E. Smith and J. Klinowski, Chem. Phys. Lett., 180 (1991) 6.

149. P.J. Grobet, H. Geerts, J.A. Martens and P.A. Jacobs, Proc. 9th IZC Montreal (Eds.: R. von Ballmoos et al.), Butterworth-Heinemann, Stoneham, 1993, p. 545.
150. Y. Wu, D. Lewis, J.S. Frye, A.R. Palmer and R.A. Wind, J. Magn. Reson., 100 (1992) 425.
151. R. Khouzami, G. Coudurier, F. Lefebvre, J.C. Vedrine and B.F. Mentzon, Zeolites, 10 (1990) 183.
152. M. Goepper and J.L. Guth, Zeolites, 11 (1991) 477.
153. B. Zibrowius, U. Lohse and J. Richter-Mendau, J.C.S. Faraday Trans., 87 (1991) 1433.
154. H. He and J. Klinowski, J. Phys. Chem., 97 (1993) 10385.
155. R. Jelinek, B.F. Chmelka, Y. Wu, P.J. Grandinetti, A. Pines, P.J. Barrie and J. Klinowski, J. Am. Chem. Soc., 113 (1991) 4097.
156. P.J. Barrie and J. Klinowski, J. Phys. Chem., 93 (1989) 5972.
157. S.H. Chen, S.P. Sheu and K.J. Chao, J.C.S. Chem. Commun., (1992) 1504.
158. M.P.J. Peeters, L.J.M. van de Ven, J.W. de Haan and J.H.C. van Hooff, Colloids Surf., A 72 (1993) 87.
159. C.S. Blackwell and R.L. Patton, J. Phys. Chem., 92 (1988) 3965.
160. D. Freude, H. Ernst, M. Hunger and H. Pfeifer, Chem. Phys. Lett., 143 (1988) 477.
161. J.A. Martens, M. Mertens, P.J. Grobet and P.A. Jacobs, Stud. Surf. Sci. Catal., 37 (1988) 97.
162. D. Hasha, L.S. de Saldarriaga, C. Saldarriaga, P.E. Hathaway, D.F. Cox and M.E. Davis, J. Am. Chem. Soc., 110 (1988) 2127.
163. J.A. Martens, P.J. Grobet and P.A. Jacobs, J. Catal., 126 (1990) 299.
164. Y. Watanabe, A. Koiwai, H. Takeuchi, S.A. Hyodo and S. Noda, J. Catal., 143 (1993) 430.
165. M. Derewinski, M.J. Peltre, M. Briend, D. Barthomeuf and P.P. Man, J.C.S. Faraday Trans., 89 (1993) 1823.
166. B. Zibrowius and U. Lohse, Solid State Nucl. Magn. Reson., 1 (1992) 137.
167. M. Estermann, L.B. McCusker, C. Baerlocher, A. Merrouche and H. Kessler, Nature, 352 (1991) 320.
168. R.L. Bedard, C.L. Bowes, N. Coombs, A.J. Holmes, T. Jiang, S.J. Kirkby, P.M. Macdonald, A.M. Malek, G.A. Ozin, S. Petrov, N. Plavac, R.A. Ramik, M.R. Steele and D. Young, J. Am. Chem. Soc., 115 (1993) 2300.
169. H. Meyer zu Altenschildesche, H.J. Muhr and R. Nesper, Microporous Mater., 1 (1993) 257.
170. S.M. Bradley, R.F. Howe and J.V. Hanna, Solid State Nucl. Magn. Reson., 2 (1993) 37.
171. T.L. Barr, J. Klinowski, H. He, K. Alberti, G. Müller and J. Lercher, Nature, 365 (1993) 429.
172. J. Patarin, C. Schott, A. Merrouche, H. Kessler, M. Soulard, L. Delmotte, J.L. Guth and J.F. Joly, Proc. 9th IZC Montreal (Eds.: R. von Ballmoos et al.), Butterworth-Heinemann, Stoneham, 1993, p. 263.
173. A. Merrouche, J. Patarin, H. Kessler, M. Soulard, L. Delmotte, J.L. Guth and J.R. Joly, Zeolites, 12 (1992) 226.
174. B. Zibrowius, M.W. Anderson, W. Schmidt, F.F. Schüth, A.E. Aliev and K.D.M. Harris, Zeolites, 13 (1993) 607.

504

175. Q. Huo, R. Xu, S. Li, Y. Xu, Z. Ma, Y.Yue and L. Li, Proc. 9th IZC Montreal (Eds.: R. von Ballmoos et al.), Butterworth-Heinemann, Stoneham, 1993, p. 279.

176. A.T. Bell, Zeolite Synthesis (Eds.: M.L. Occelli and H.E. Robson), Am. Chem. Soc., Symp. Ser. 398 (1989), p. 66.

177. A.V. McCormick, A.T. Bell and C.J. Radke, Zeolites, 7 (1987) 183.

178. A. Thangaraj and S. Sivasanker, J.C.S. Chem. Commun., (1992) 123.

179. X. Tang, Y. Sun, T. Wu, L. Wang, L. Fei and Y. Long, J.C.S. Faraday Trans., 89 (1993) 1839.

180. J.M. Chézeau, L. Delmotte, J.L. Guth and Z. Gabelica, Zeolites, 11 (1991) 598.

181. E. Klock, L. Delmotte, M. Soulard and J.L. Guth, Proc. 9th IZC Montreal (Eds.: R. von Ballmoos et al.), Butterworth-Heinemann, Stoneham, 1993, p. 611.

182. E. Jahn, D. Müller, W. Wieker and J. Richter-Mendau, Zeolites, 9 (1989) 177.

183. R. Jelinek, S. Özkar, H.O. Pastore, A. Malek and G.A. Ozin, J. Am. Chem. Soc., 115 (1993) 563.

184. R. Jelinek, S. Özkar and G.A. Ozin, J. Phys. Chem., 96 (1992) 5949.

185. M. Hunger, G. Engelhardt, H. Koller and J. Weitkamp, Solid State Nucl. Magn. Reson., 2 (1993) 111.

186. G. Schäfer, W.W. Warren Jr., P. Anderson and P.P. Edwards, J. Non. Cryst. Sol., 156-158 (1993) 803.

187. H.K. Beyer, G. Pál-Borbély and H.G. Karge, Microporous Mater., 1 (1993) 67.

188. J.M. Chézeau, A. Saada, L. Delmotte and J.L. Guth, Proc. 9th IZC Montreal (Eds.: R. von Ballmoos et al.), Butterworth-Heinemann, Stoneham, 1993, p. 659.

189. P.J. Chu, B.C. Gerstein, J. Nunan and K. Klier, J. Phys. Chem., 91 (1987) 3588.

190. M.K. Ahn and L.E. Itou, J. Phys. Chem., 95 (1991) 4496.

191. B. Herreros, P.P. Man, J.M. Manoli and J. Fraissard, J.C.S. Chem. Commun., (1992) 464.

192. T. Ito and J. Fraissard, Proc. 5th IZC Napoli, Heyden & Son, London, 1980, p. 510.

193. J. Fraissard and T. Ito, Zeolites, 8 (1988) 350.

194. M. Springuel-Huet, J. Demarquay, T. Ito and J. Fraissard, Stud. Surf. Sci. Catal., 37 (1988) 183.

195. D.W. Johnson and L. Griffiths, Zeolites, 7 (1987) 484.

196. J. Fraissard, Z. Phys. Chem. (Munich) 152 (1987) 159.

197. R. Benslama, J. Fraissard, A. Albizane, F. Fajula and F. Figueras, Zeolites, 8 (1988) 196.

198. C. Tsiao, D.R. Corbin, V. Durante, D. Walker and C. Dybowski, J. Phys. Chem., 94 (1990) 4195.

199. S.M. Alexander, J.M. Coddington and R.F. Howe, Zeolites, 11 (1991) 368.

200. E. Trescos, L.C. de Ménorval and F. Rachdi, J. Phys. Chem., 97 (1993) 6943.

201. M. Boudart, R. Ryoo, G.P. Valença and R. Van Grieken, Catal. Lett., 17 (1993) 273.

202. T.T.P. Cheung and C.M. Fu, J. Phys. Chem., 93 (1989) 3740.

203. J. Demarquay and J. Fraissard, Chem. Phys. Lett., 136 (1987) 314.

204. T.T.P. Cheung, J. Phys. Chem., 94 (1990) 376.

205. T. Ito, M.A. Springuel-Huet and J. Fraissard, Zeolites, 9 (1989) 68.

206. Q.J. Chen and J. Fraissard, Chem. Phys. Lett., 169 (1990) 595.

207. L.C. de Ménorval, D. Raftery, S.B. Liu, K. Takegoshi, R. Ryoo and A. Pines, J. Phys. Chem., 94 (1990) 27.

208. B.F. Chmelka, R. Ryoo, S.B. Liu, L.C. de Ménorval, C.J. Radke, E.E. Petersen and A. Pines, J. Am. Chem. Soc., 110 (1988) 4465.
209. O.B. Yang, S.I. Woo and R. Ryoo, J. Catal., 123 (1990) 375.
210. T.I. Korányi, L.J.M. van de Ven, W.J.J. Welters, J.W. de Haan, V.H.J. de Beer and R.A. van Santen, Catal. Lett., 17 (1993) 105.
211. A. Gedeon, J.L. Bonardet, T. Ito and J. Fraissard, J. Phys. Chem., 93 (1989) 2563.
212. B. Boddenberg and M. Hartmann, Chem. Phys. Lett., 203 (1993) 243.
213. J.L. Bonardet, M.C. Barrage and J. Fraissard, Proc. 9th IZC Montreal (Eds.: R. von Ballmoos et al.), Butterworth-Heinemann, Stoneham, 1993, p. 475.
214. M.C. Barrage, J.L. Bonardet and J. Fraissard, Catal. Lett., 5 (1990) 143.
215. J.L. Bonardet, M.C. Barrage and J. Fraissard, Am. Chem. Soc. - Preprints, 38 (1993) 628.
216. E.G. Derouane and M.E. Davis, J. Mol. Catal., 48 (1988) 37.
217. M.E. Davis, C. Saldarriaga, C. Montes and B.E. Hanson, J. Phys. Chem., 92 (1988) 2557.
218. Q.J. Chen, M.A. Springuel-Huet and J. Fraissard, Chem. Phys. Lett., 159 (1989) 117.
219. N. Dumont, T. Ito and E.G. Derouane, Appl. Catal., 54 (1989) L 1.
220. J.A. Ripmeester and C.I. Ratcliffe, Proc. 9th IZC Montreal (Eds.: R. von Ballmoos et al.), Butterworth-Heinemann, Stoneham, 1993, p. 571.
221. L. Maistriau, E.G. Derouane and T. Ito, Zeolites, 10 (1990) 311.
222. Q.J. Chen, J. Fraissard, H. Cauffriez and J.L. Guth, Zeolites, 11 (1991) 534.
223. J. Fraissard and J. Kärger, Zeolites, 9 (1989) 351.
224. M.L. Smith, D.R. Corbin and C. Dybowski, J. Phys. Chem., 97 (1993) 9045.
225. M.L. Smith, D.R. Corbin, L. Abrams and C. Dybowski, J. Phys. Chem., 97 (1993) 7793.
226. T.T.P. Cheung, J. Phys. Chem., 97 (1993) 8993.
227. H. Pfeifer, NMR: Basic Principles and Progress 7, (Eds.: P. Diehl, E. Fluck and R. Kosfeld), Springer, Berlin (1972), p. 53.
228. C.A. Fyfe, H. Strobl, G.T. Kokotailo, G.J. Kennedy and G.E. Barlow, J. Am. Chem. Soc., 110 (1988) 3373.
229. C.A. Fyfe, Y. Feng, H. Grondey and G.T. Kokotailo, J.C.S. Chem. Commun., (1990) 1224.
230. W.C. Conner, R. Vincent, P. Man and J. Fraissard, Catal. Lett., 4 (1990) 75.
231. C.A. Fyfe, H. Gies, G.T. Kokotailo, C. Pasztor, H. Strobl and D.E. Cox, J. Am. Chem. Soc., 111 (1989) 2470.
232. G. Mirth, J.A. Lercher, M.W. Anderson and J. Klinowski, J.C.S. Faraday Trans., 86 (1990) 3039.
233. B.F. Mentzen, M. Sacerdote-Peronnet, J.F. Bérar and F. Lefebvre, Zeolites, 13 (1993) 485.
234. S.M. Alexander, D.M. Bibby, R.F. Howe and R.H. Meinhold, Zeolites, 13 (1993) 441.
235. M.R. Steele, P.M. Macdonald and G.A. Ozin, J. Am. Chem. Soc., 115 (1993) 7285.
236. J. Kärger and H. Pfeifer, Zeolites, 7 (1987) 90.
237. J. Kärger and D.M. Ruthven, Diffusion in Zeolites and other microporous Solids, John Wiley & Sons, New York, 1992.

506

238. J. Caro, H. Jobic, M. Bülow, J. Kärger and B. Zibrowius, Adv. Catal., 39 (1993) 351.
239. J. Kärger and R.M. Ruthven, Zeolites, 9 (1989) 267.
240. C. Förste, J. Kärger and H. Pfeifer, J. Am. Chem. Soc., 112 (1990) 7.
241. W. Heink, J. Kärger, H. Pfeifer and F. Stallmach, J. Am. Chem. Soc., 112 (1990) 2175.
242. J. Kärger, M. Petzold, H. Pfeifer, S. Ernst and J. Weitkamp, J. Catal., 136 (1992) 283.
243. W. Heink, J. Kärger, H. Pfeifer, K.P. Datema and A.K. Nowak, J.C.S. Faraday Trans., 88 (1992) 3505.
244. J. Kärger, H. Pfeifer, F. Stallmach, N.N. Feoktistova and S.P. Zhdanov, Zeolites, 13 (1993) 50.
245. F. Stallmach, J. Kärger and H. Pfeifer, J. Magn. Reson., A 102 (1993) 270.
246. J. Kärger, G. Seiffert and F. Stallmach, J. Magn. Reson., A 102 (1993) 327.
247. J. Kärger, H. Pfeifer, F. Stallmach and H. Spindler, Zeolites, 10 (1990) 288.
248. J. Klinowski and M.W. Anderson, Magn. Reson. Chem., 28 (1990) S 68.
249. M.W. Anderson and J. Klinowski, Nature, 339 (1989) 200.
250. M.W. Anderson, B. Sulikowski, P.J. Barrie and J. Klinowski, J. Phys. Chem., 94 (1990) 2730.
251. M.W. Anderson and J. Klinowski, J.C.S. Chem. Commun., (1990) 918.
252. M.W. Anderson, P.J. Barrie and J. Klinowski, J. Phys. Chem., 95 (1991) 235.
253. W. Kolodziejski and J. Klinowski, Appl. Catal., A 81 (1992) 133.
254. Y. Xu, C.P. Grey, J.M. Thomas and A.K. Cheetham, Catalytic Science and Technology, Vol.1, Kodansha Ltd., Tokyo, 1991, p. 79.
255. M.W. Anderson, M.L. Occelli and J. Klinowski, J. Phys. Chem., 96 (1992) 388.
256. K.P. Datema, A.K. Nowak, J. van Braam-Houckgeest and A.F.H. Wielers, Catal. Lett., 11 (1991) 267.
257. A.G. Stepanov and K.I. Zamaraev, Catal. Lett., 19 (1993) 153.
258. A.G. Stepanov, A.G. Maryasoo, V.N. Romannikov and K.I. Zamaraev, Proc. 10th International Congress on Catalysis (Eds. L. Guczi al.), Akadémiai Kiadó, Budapest, 1993, p. 621.
259. E.J. Munson, D.B. Ferguson, A.A. Kheir and J.F. Haw, J. Catal., 136 (1992) 504.
260. E.J. Munson, D.K. Murray and J.F. Haw, J. Catal., 141 (1993) 733.
261. E.J. Munson, A.A. Kheir and J.F. Haw, J. Phys. Chem., 97 (1993) 7321.
262. D.K. Murray, J.W. Chang and J.F. Haw, J. Am. Chem. Soc., 115 (1993) 4732.
263. E.J. Munson and J.F. Haw, Angew. Chem., 105 (1993) 643.
264. H. Pfeifer, D. Freude and M. Hunger, Zeolites, 5 (1985) 274.
265. H. Ernst, D. Freude and I. Wolf, Chem. Phys. Lett., 212 (1993) 588.
266. W.P.J.H. Jacobs, J.W. de Haan, L.J.M. van de Ven and R.A. van Santen, J. Phys. Chem., 97 (1993) 10394.
267. J. Klinowski, H. Hamdan, A. Corma, V. Fornés, M. Hunger and D. Freude, Catal. Lett., 3 (1989) 263.
268. P.G. Clarke, K. Gosling, R.K. Harris and E.G. Smith, Zeolites, 13 (1993) 388.
269. E. Brunner, Microporous Mater., 1 (1993) 431.
270. J.L. White, L.W. Beck and J.F. Haw, J. Am. Chem, Soc., 114 (1992) 6182.
271. M. Hunger, M.W. Anderson, A. Ojo and H. Pfeifer, Microporous Mater., 1 (1993) 17.

272. W.P. Rothwell, W.X. Shen and J.H. Lundsford, J. Am. Chem.Soc., 106 (1984) 2452.
273. J.H. Lundsford, P.N. Tutunjian, P.J. Chu, E.B. Yeh and D.J. Zalewski, J. Phys. Chem., 93 (1989) 2590.
274. P.J. Chu, A. de Mallmann and J.H. Lundsford, J. Phys. Chem., 95 (1991) 7362.
275. P. Batamack, C. Doremieux-Morin, R. Vincent and J. Fraissard, J. Phys. Chem., 97 (1993) 9779.
276. J.S. Beck, J.C. Vartuli, W.J. Roth, M.E. Leonowicz, C.T. Kresge, K.D. Schmitt, C.T.W. Chu, D.H. Olson, E.W. Sheppard, S.B. MaCullen, J.B. Higgins and J.L. Schlenker, J. Am. Chem. Soc., 114 (1992) 10834.
277. A. Monnier, F. Schüth, Q. Huo, D. Kumar, D. Margolese, R.S. Maxwell, G.D. Stucky, M. Krishnamurty, P. Petroff, A. Firouzi, M. Janicke and B.F. Chmelka, Science, 261 (1993) 1299.
278. S. Inagaki, Y. Fukushima and K. Kuroda, J.C.S. Chem. Commun., (1993) 680.
279. S. Inagaki, Y. Fukushima, A. Okada, T. Kurauchi, K. Kuroda and C. Kato, Proc. 9th IZC Montreal (Eds.: R. von Ballmoos et al.), Butterworth-Heinemann, Stoneham, 1993, p. 305.
280. T. Yanagisawa, T. Shimizu, K. Kuroda and C. Kato, Bull. Chem. Soc. Japan, 63 (1990) 1535.
281. D.E. Akporiaye, E.W. Hansen, R. Schmidt and M. Stöcker, J. Phys. Chem., 98 (1994) 1926.

J.C. Jansen, M. Stöcker, H.G. Karge and J. Weitkamp (Eds.)
Advanced Zeolite Science and Applications
Studies in Surface Science and Catalysis, Vol. 85
© 1994 Elsevier Science B.V. All rights reserved.

Supported zeolite systems and applications

H. van Bekkum[a], E.R. Geus[a] and H.W. Kouwenhoven[b]

[a] Department of Organic Chemistry and Catalysis, Delft University of Technology, Julianalaan 136, 2628 BL Delft, The Netherlands

[b] Eidgenössische Technische Hochschule, Laboratorium für Technische Chemie, Universitätstraße 6, CH-8092 Zürich, Switzerland.

1. INTRODUCTION

The use of catalysts and adsorbents in the treatment of large gas volumes for removal of minor components and air pollution abatement is rapidly expanding. Generally these applications require a reactor configuration which is tailored for an efficient contact between gas and solid and a low pressure drop over the bed of active material. This configuration is for instance applied in automotive catalytic converters. A thin layer of the active material covers a support, the shape of the latter determines the pressure drop and the efficiency of the active material. In automotive converters the supporting structure consists e.g. of extruded monoliths of a refractory material like cordierite or of corrugated metal sheets. Both types of supporting structures are characterized by a large number of parallel channels having a cross section, with a surface, of about 15 mm^2. The layers of active material are formed by soaking and/or spraying the supporting structure elements with a precursor slurry or solution followed by drying and calcination for removal of solvents and formation of a tightly adhering layer of a high surface area material, which in its turn serves as a support for the catalytically active metals.

A second option for a low pressure drop solid bed reactor is applied in the parallel passage reactor of the Shell process for removal of SO_2 from stackgas. In this reactor the particles of the solid are housed in about 5-10 mm wide metal wire gauze envelopes, which are attached to reactor internals in a configuration forming parallel slid shaped channels, the latter are separated by a layer of reactive solid packed into the envelopes (parallel-passage reactor). Gas components diffuse laterally into the layers of solid and may be removed by adsorption or may be converted by a chemical reaction. The catalyst bed consists of loosely packed extrudates or pellets having a specially designed porous texture, to prevent diffusion problems. Because of the complicated design and the special bed configuration this type of unit is most suitable for large stationary applications.

A third and promising option for low pressure drop reactors consists of the use of zeolites supported on a suitable structured support.

Supported zeolites have been considered in the literature for applications as membranes, adsorbents, catalyst components and in electronic devices. The wide range of well defined pore sizes and the adaptability of the surface chemistry offered by zeolites make them attractive base materials for new developments in these areas.

2. PREPARATION METHODS FOR SUPPORTED ZEOLITES

The two main methods for the preparation of supported zeolites are :
1) Covering an existing support with a zeolite layer by contacting it with a zeolite slurry (dip-coating technique).
2) Growth of an (oriented) layer of zeolite crystals on an existing support by in situ zeolite synthesis.

It will be clear in advance that the envisaged application has a strong bearing on the required crystal orientation of the composite material and on the nature of the support.

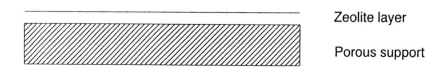
Zeolite layer

Porous support

Figure 1. Continuous zeolite layer on a porous support.

A zeolite based membrane requires a continuous layer of zeolite crystals with properly aligned zeolite channels which should be supported on a porous carrier material., cf. Figure 1. This is best achieved by applying in situ synthesis, method 2). Zeolite slurries as applied in method 1) may also result in zeolitic membranes provided that the slurry contains a dissolved polymeric substance, which forms the continuous phase after removal of the solvent.

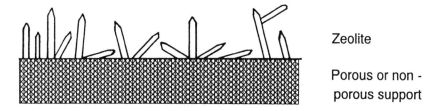

| | Zeolite |
| | Porous or non - porous support |

Figure 2. Axially grown zeolite crystals on a support (schematic).

For the preparation of supported zeolite based catalysts, an axial orientation of densely packed zeolite crystals covering a high fraction of the carrier surface is preferred, cf. Figure 2, and provides the best exposure of the zeolitic surface. Compared with zeolitic membranes this aspect is however much less critical and both method 2) and method 1) may lead to good results. It is obvious that neither zeolitic catalysts nor the analogous zeolitic adsorbents do require a porous support.

On the other hand for sensor applications mutually oriented crystals having a preferred channel orientation must be deposited and fixed on a support having specific properties.

Preparation and applications of zeolites supported on metallic or oxidic carrier materials require a high thermomechanical stability. Since the thermomechanical properties, such as the thermal expansion coefficients, of the components in these composites will usually be different, and moreover frequently temperature-dependent, this aspect needs to be carefully considered. The problem is well known from the enamel- and ceramic industry and an in depth discussion for clay-supported zeolites is presented in the thesis of Geus [1].

2.1. Deposition of zeolites by dip-coating

Dip-coating is an efficient and well proven technique for the deposition of thin layers of an oxidic material on existing supports. The dip-coating technique is very versatile, since it may be applied to most support materials, however complicated their shape. Both the support and the material to be deposited should avail of reactive surface groups. It appears that condensation reactions between SiOH groups on the outer surface of the zeolite crystals and hydroxyl groups on oxidic and metallic supports usually result in the formation of an adhering zeolite layer upon curing of a zeolite coated support. The interaction is not strong however and the thickness of the layer is moreover very limited. Accordingly a precursor of an hydraulic binder is usually added to the zeolite slurry, which upon curing assists in the formation of a strongly bonded zeolite layer on the support. The thickness of this layer depends on the zeolite content of the slurry and the number of times that the dip-coating is repeated. In refs. [2] and [3] the deposition is described of a zeolite layer on a conventional ceramic monolith as applied in automotive muffler catalysts. The manufacturing method is analogous to the conventional dip-coating technique applied for muffler catalysts and the zeolite slurry contains precursors of binders based on either alumina, zirconia, silica, titania or silica-alumina etc. The zeolites may be tailored for a specific application before the coating operation and e.g. Haas et al. [3] describe a coating consisting of a H-mordenite modified by impregnation with aqueous solutions of salts of Pt and Ni.

Supported zeolites prepared by the dip-coating technique are applied in adsorption and in catalysis. Lachman et al. [2] developed a zeolite coated monolith which serves to eliminate or reduce hydrocarbon emissions during start-up of an internal combustion engine. The reduction of the emissions is achieved by the installation of a special muffler system as depicted in Figure 3, which contains downstream from the catalytic converter an additional hydrocarbon adsorber, consisting of the zeolite loaded monolith described above.

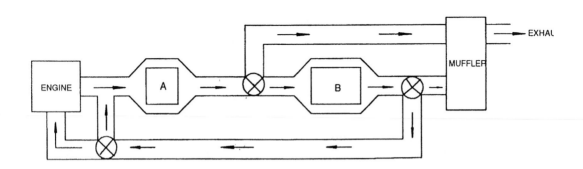

Figure 3. Dual converter engine exhaust system [2]. A: catalytic muffler, B: adsorber.

According to Lachman et al. [2] to prevent preferential water adsorption, the zeolite should be a high silica material, the total hydrocarbon adsorption capacity in the adsorber should be at least 6 g. This capacity is supposedly sufficient to prevent the hydrocarbons from entering the atmosphere during the warm-up period of the catalytic converter to its light-off temperature of 360 C, which is usually reached within about 70 sec after engine start-up. After light-off, the loaded adsorber is regenerated by heating with the hot catalytic converter off gas and the resulting hydrocarbon carrying exhaust gases are either charged to the catalytic converter or to the engine inlet. In this application adsorbents based on activated carbon cannot be used because of their flammability above 150 °C.

Similarly zeolite coated filter elements are described by Haas et al. [3] for removal and catalytic incineration of soot from diesel exhaust gases. A ceramic filter element is dip-coated with a mordenite slurry, the mordenite is in the H-form and is preimpregnated with an aqueous solution of a Pt and a Ni compound. As shown by measurements of the pressure drop over the filter element, the collected diesel soot starts burning at about 420 °C, i.e. well within the normal range of temperatures of diesel exhaust gases and at a temperature which is about 130 °C lower than the initial burning temperature of a soot loaded, non coated filter element. This allows use of a swing reactor system for removal of diesel soot from the exhaust gases, the first filter being loaded with soot, while the other is being regenerated by burning off the soot. As an additional advantage it is claimed [3] that emissions of CO, NOx and hydrocarbons are greatly reduced due to the catalytic activity of the metal loaded zeolite.

Blocki [4] reports the application of hydrophobic zeolitic adsorbents for controlling dilute solvent emissions generated e.g. by paint spraying units. It appears that in this article the term hydrophobic zeolite means silicalite. Presently active carbon is widely applied as sorbent in solvent adsorbers. Compared to active carbons high silica zeolites present some advantages such as

    non flammable material

    thermally stable to about 1000 °C

    hydrophobic surface

    efficient adsorbent at very low contaminant concentrations

    chemically inert surface

The non flammability and thermal stability are specially advantageous in cases where the exhaust gases of high ketone content solvents have to be treated. The effect of the hydrophobic surface on the adsorption capacity for water as a function of relative humidity is shown in Figure 4 for an activated carbon and a hydrophobic zeolite.

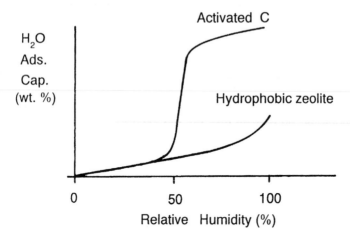

Figure 4. Water adsorption capacity of high silica zeolite and activated carbon. Effect of relative humidity [4].

For removal of entrained solvent vapours the contaminated exhaust gas is for instance contacted with a zeolite coated honeycomb structure in a modular solvent concentrator as depicted in Figure 5. The modular solvent concentrator is a continuous unit, and the adsorber consists of a rotating cylindrical frame containing adsorbent filled elements.

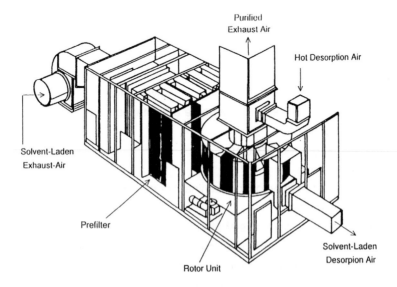

Figure 5. Modular solvent concentrator [4].

The cylindrical unit rotates through three zones: an adsorption zone, where solvents are removed from the contaminated gas; a desorption zone, where the adsorbed solvents are stripped from the laden adsorbent by a hot gas stream and a cooling zone where the adsorbent is cooled before entering the adsorption zone. Some results are presented in Table 1 and show that high silica zeolites may be applied for emission abatement and remove 90-100% of the entrained solvent from the exhaust gas by adsorption at concentrations in the solvent laden gas in the ppm range.

The efficiency for the removal of the individual solvent components depends on the composition of the exhaust gas and on the structure of the high silica zeolite.

Table 1
Adsorption efficiency test

Solvent			Concentration (ppmv)		
Component	vol %	bp (°C)	in	out	efficiency (%)
Butyl cellosolve acetate	5	192	4.3	0	100
Xylenes	80	138	68	3.3	95
Isopropyl alcohol	8	82	7	0.8	88
Ethyl acetate	3	77	3	0	100
Butyl cellosolve	4	171	3.4	0	100

The desorbed organics in the regeneration gas are removed by incineration and thus the total effluent gas vented to the atmosphere contains only minor traces of organic solvents.

2.2. Growth of zeolite layers by in situ synthesis

Active self-supporting units useful in adsorption and catalysis and consisting of layers of randomly oriented zeolite crystals on a large variety of supports may be prepared by the dip-coating technique. Total coverage of an existing support with a closed layer of zeolite crystals will hardly ever be obtained by dip-coating and is not relevant for applications other than in membranes.

Research on the growth of closed layers of zeolite crystals on existing supports by in situ synthesis is presently emphasized because of the promising applications in membrane technology. Porous metallic or oxidic supports consisting of aggregates of powders and/or thin wires, are generally applied and covered by deposition of zeolite crystals from the synthesis mixture. Growth of closed layers of large or small mutually oriented zeolite crystals on supports might allow additional applications in the manufacture of electronic devices. Both in the open literature and in patents many publications concerning the preparation procedure have recently appeared. Main problems are the formation of a dense layer of crystals, their mutual orientation and the prevention of the occurrence of non covered support areas. An early reference to the deposition of zeolites on an existing support is given by Albers and Grant [5] who describe the synthesis of zeolites A, X and Y deposited on alumina, silica, silica-alumina or other oxidic materials, dispersed in the synthesis mixture. The method is straightforward and is characterized by the use of a common seeding mixture for the three zeolites. The compositions of the synthesis mixtures are as usual and the common seeding solution has the composition: $16Na_2O:Al_2O_3:15SiO_2:320H_2O$.

3. STRUCTURED CATALYSTS BY IN SITU GROWTH OF ZEOLITES ONTO SUPPORTS

One of the processes requiring a low pressure drop, dust proof reactor, is the heterogeneously catalyzed selective reduction of nitrogen oxides. Industrial flue gases generally contain considerable amounts of dust and it is often preferable to install the $deNO_x$ unit before the dust removal unit. The requirement of low pressure drop results from the fact that most flue gas streams are at almost ambient pressure, which leaves no room for a large pressure drop across the reactor.

A major technique for removing NO_x from flue gas is Selective Catalytic Reduction (SCR) with ammonia, which is added in a substoichiometric amount before the flue gas is contacted with the $deNO_x$ catalyst. The reaction equation is

$$4 \text{ NO} + 4 \text{ NH}_3 + O_2 \rightarrow 4 \text{ N}_2 + 6 \text{ H}_2O$$

Vanadia, often together with titania, supported onto silica or alumina generally serves as the catalyst in SCR of NO_x, and Shell developed [6] two types of reactor to accommodate this catalyst under low pressure drop conditions.

Zeolites, particularly Cu- (or Co-) exchanged MFI-zeolites, may form an alternative for the vanadia-catalyst. Cu-ZSM-5 showed good NO_x reducing properties in the presence of ammonia [7] as well as with hydrocarbons [8] as reductants. Moreover this catalyst was shown to be able [9] to decompose NO_x to the elements in the absence of added reductants.

Taking the work of Jansen et al. [10] as a starting point two groups at Delft (Van den Bleek et al. and the first author et al.) embarked on growing ZSM-5 zeolites axially onto metal supports and applying these composite systems, after suitable modification, in SCR of NO_x with ammonia [11].

Stainless steel (AISI 316) gauze having 35 μm wire thickness and 44 μm mesh size was used as a support for ZSM-5 crystals. This metal grade is composed of 0.08% C, 1% Si, 2% Mn, 2% Mo, 11.5% Ni, 17% Cr and balance Fe. Onto the outer surface a layer of zeolite crystals can be crystallized which is chemically bonded to the supporting metallic structure. Before crystallization the metal gauze was cut into the desired size and rolled up to fit in the autoclave (for crystallization) and in the reactor (for kinetic experiments). The support was then cleaned in boiling ethyl alcohol for about 1 hour to remove surface contaminants after which it was dried at 120 °C for 2 hours. This thermal treatment causes enhanced chromium diffusion to the freshly cut ends of the gauze and subsequent oxidation promotes healing of the Cr_2O_3 rich surface layer. Before the metal gauze was exposed to the crystallization environment it was accurately weighed. It may be noted that in an industrial application the composite catalyst might be also shaped as stretched, corrugated or folded sheets.

Metal gauze supported ZSM-5 was synthesized by hydrothermal synthesis in dilute aqueous solution containing tetraethyl orthosilicate (TEOS), tetrapropyl-ammonium hydroxide (TPA-OH), sodium aluminate and sodium chloride in a molar ratio: 2.6:1.0:0.06: 1.6 and H_2O 1767.

After preparation of the synthesis mixture, it was homogenized by overnight shaking. Th slurry was subsequently poured in a teflon-lined autoclave containing the vertically positioned metal gauze cylinder. The zeolite was deposited on the metal support during 20 h crystallisation at 170 °C.

Figures 6 and 7 show the starting gauze and the ZSM-5-loaded gauze obtained upon applying the above synthesis formulation, respectively. Average crystal dimensions are 7.3 x 5.1 x 2.2 μm. The crystal loading amounted to 3.5 wt %.

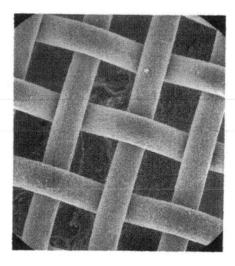

Figure 6. Stainless steel gauze, 35 μm wire thickness.

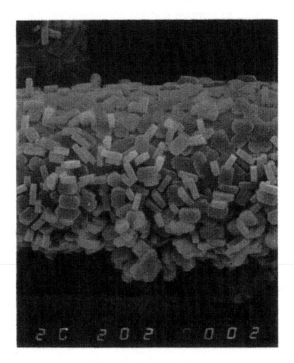

Figure 7. ZSM-5 crystals grown onto metal gauze [11], synthesis formulation see text.

Figure 8. ZSM-5 crystals grown onto metal gauze, synthesis formulation see text.

When a molar ratio TEOS:TPAOH:NaAlO$_2$:NaCl:H$_2$O 2.5:1:1.3:1566 was applied the result was as reflected by Figure 8. In all cases the characteristic morphologies of MFI crystals are observed.

Calcination of the as-synthesized supported ZSM-5 crystals was done at 500 °C for 10 hours in a programmable oven as follows: heating to 80 °C at 1 °C/min; 1 h dwell; heating to 500 °C at 1 °C/min; 10 h dwell; cooling to room temperature at 1 °C/min.

Copper(II) ions were introduced into the ZSM-5 crystals by repeated ion exchange in an 0.017 M copper(II) acetate solution. ICP-AES analysis showed the crystals (of Figure 7) to possess an Si/Al ratio of 44.1. The Al/Cu ratio was found to be 0.88, showing excess ion exchange which is often encountered for ZSM-5 (cf. ref. [12]).

The surface area of the composite material was measured by using a Micromeretics ASAP 2000 M apparatus. As the external surface of the metal wire gauze is small (0.0114 m^2/g) the surface area stemmed essentially entirely from the zeolite crystals. The surface area in micropores was 352 m^2/g zeolite at a micropore volume of 0.17 cm^3/g zeolite. These data are in good agreement with literature data for ZSM-5. The same holds for the sharp pore distribution found, with a median pore diameter of 0.5 nm. The BET surface area was found to be 435 m^2/g, pointing to a rather rough outer surface (83 m^2/g) of the crystals.

The catalyst tested by Calis et al. [11] in the deNO$_x$ reaction had a total weight of 8.4 g and consisted of 96.5 wt % steel support gauze 3.5 wt % ZSM-5 zeolite and 0.1 wt % copper. Experimental deNO$_x$ conditions were 350 °C, 1.1 bar, NO 20-200 ppmv, NH$_3$ 30-1000 ppmv, O$_2$ 0-5 v %, H$_2$O 0-15 v %, N$_2$ carrier gas. A WHSV of 1340 h^{-1} was applied (based on the weight of the Cu-ZSM-5 crystals, thus excluding the weight of the metal gauze support).

The experiments were conducted in a RotoBerty internal recycle reactor which contains a basket in which the catalyst packing is placed. This construction allows any shape or amount of catalyst packing provided it fits in the basket.

The deNO$_x$ reaction turned out to be first order in nitric oxide, zero order in ammonia, positive order in oxygen and negative order in water. The latter compound was found to suppress the ammonia oxidation. The performance of the metal supported Cu on ZSM-5 catalyst was perfectly stable and did not deteriorate as a result of the application of a number of reaction cycles.

The deNO$_x$ activity of the Cu-ZSM-5 catalyst packing was compared with the (extrapolated) activity of a commercial catalyst consisting of 5 wt % vanadia and titania on amorphous silica. Three different methods for comparing the performance are presented in Table 2.

Table 2
The activity of the Cu-ZSM-5 on gauze catalyst packing compared with the extrapolated activity of a vanadia-titania-on-silica catalyst, at 350 °C

Basis of comparison	Act. Cu-ZSM-5/V-Ti-Silica
mol of active metal	30
kg of carrier plus active metal	3.5
kg of catalyst packing	0.13

In conclusion, a catalyst packing consisting of Cu-exchanged ZSM-5 zeolite, grown onto a stainless steel metal wire gauze seems very promising for industrial deNO$_x$-processes.

In connection with the use of metal supported zeolites in catalysis we mention the work of Jianquan et al. Using a special experimental set-up [13] these authors have grown mordenite crystals onto chromium metal. The composite catalyst was active in xylene isomerization.

4. ZEOLITE-BASED MEMBRANES BY IN SITU GROWTH OF ZEOLITES ONTO SUPPORTS

Patent applications [14-17] describe the preparation of membranes by hydrothermal deposition of zeolites on an existing porous support such as a filter having an average pore diameter in the range of about 0.05 to 10 μm. Membranes are prepared by a two or three step synthesis procedure. The latter method starts with a multichannel ceramic filter block, with channels of 4 mm internal diameter and consisting of a composite of alpha-alumina and zirconia, having an average pore size of 50 nm. In a first step the block was three times impregnated with a 0.1 M solution of $Co(NO_3)_2.6H_2O$, followed by drying at 90 °C. It was subsequently calcined at 550 °C for 6 h and as a result the surface of the block was coated with an oxide of cobalt. The thus pretreated block was subsequently covered with a freshly prepared hydrogel of composition:

$$1.83\ Na_2O:Al_2O_3:1.78\ SiO_2:133\ H_2O$$

and heated during 2 h at 90 °C. After cooling the block was removed from the gel and cleaned thoroughly for removal of adhering solids and residual solution, finally the block was dried at 90 °C. The further pretreated block was again covered with a fresh gel of the same composition as above and heated during 24 h at 90 °C. After washing and drying as above examination of the block by SEM showed it to be covered by a continuous surface layer of zeolite A crystals. It appeared that the first short treatment in fresh hydrogel at 90 °C resulted in the formation of silicic acid globules on the surface of the block. The size of these globules was about 300 nm. The zeolite crystals presumably start growing under conditions of high supersaturation in the gel matrix present on the ceramic carrier.

Haag et al. [18] describe the deposition of siliceous MFI on porous and non porous metallic, oxidic or teflon supports aimed at the preparation of membranes. The composition of the synthesis mixture was

$$2.2\ Na_2O:100\ SiO_2:2823\ H_2O:5.22\ TPABr$$

and neither a specific pretreatment of the supports nor a special adaptation of the composition of the synthesis mixture is mentioned. The MFI layer formed on the teflon support was about 250 μm thick, could mechanically be separated off and was strong enough for further experimentation. The layer consisted of an agglomerate formed by loosely packed small, 0.1 μm crystallites at the teflon side and tightly packed intergrown 10-100 μm size crystals at the solution side. This inhomogeneous growth of the zeolite layer is presumably due to the high initial supersaturation in the solution leading to the formation of many crystallisation nuclei. The crystallisation and formation of dense layers of crystals on teflon and other supports is discussed more extensively later in this article. Permeability properties of the membrane grown on teflon were determined in a Wicke-Kallenbach cell in the steady state mode at 569 and 595 °C and 1 bar total pressure on both sides of the membrane. Some permeability coefficients and selectivities calculated for three two component mixtures and measured at 595 °C are presented in Table 3.

Table 3
Permeability coefficients of two component gas mixtures over MFI membrane at 595 °C and 1 bar [18]

Feed composition (mol %)	P1 (cm^2/s)	P2 (cm^2/s)	Selectivity
21% O_2, 79% N_2	1.31×10^{-4}	1.22×10^{-4}	1.05
49.4% H_2, 50.6 CO	2.63×10^{-4}	1.63×10^{-4}	1.60
9.5% nC6, 16.6% 22DMB*	0.33×10^{-4}	0.035×10^{-5}	17.2

* 2,2-Dimethylbutane, balance He.

Haag et al. [18] present other examples of interesting properties of the MFI membrane layers, such as their application as reactor wall which will be discussed later.

Engelen and Van Leeuwen [19] describe the preparation of self-supporting membranes, consisting of aggregates of randomly oriented zeolite A crystals. A membrane is made in several steps by first forming a tube or a plate from a mixture of kaolin and zeolite A. The shaped article, containing additional organic binders to increase its green strength, is calcined at a temperature of about 700 °C. At this temperature the organics are removed, kaolin is converted into the easily reacting metakaolin and the article attains enough strength to be handled. The kaolin used in this process should contain less than 0.3% wt quartz, since at a higher quartz content the surface of the membrane will eventually contain holes, making it less selective. The metakaolin in the shaped articles is converted into zeolite A by hydrothermal treatment with an aqueous NaOH solution. The resulting articles consist mainly of aggregated zeolite A crystals and the porosity of these aggregates is reported to decrease with increasing kaolin content of the zeolite A/kaolin mixture, as is shown in Table 4. The aggregates are further processed in order to seal the non zeolitic channels by an additional hydrothermal deposition of zeolite A.

Table 4

Porosity of zeolite A aggregates

% w kaolin	Average pore size (μm)	Porosity (%)
25	0.37	24
38	0.29	23
50	0.14	14

To this end the aggregates are treated with a solution containing silicate, aluminate and alkali ions while having one of its sides exposed to the synthesis mixture. Preferably the crystallisation is carried out in the presence of an organic amine [19] in order to grow large crystals of zeolite A in a thin toplayer which has preferentially a thickness of 0.01-10 μm. The latter depends on the time that the precursor is subjected to hydrothermal deposition of a zeolite layer and it is claimed that a closed membrane can be made in this way.

Table 5

Effect of closing time on membrane permeability [19]

Membrane	Closing time (h)	H_2 permeability (mol/m^2 s.Pa)
zeolite A aggregate	0	$2.4.10^{-4}$
1	16	$2.9.10^{-5}$
2	48	$8.2.10^{-7}$
3	72	$1.1.10^{-10}$

In Table 5 some data are presented on the hydrogen permeability at 1.5 bar of a thus prepared zeolite A membrane, as a function of the time during which the hydrothermal deposition was continued (closing time). Since the membrane consists of zeolite A, its pore size is dependent on size and number of the exchangeable cations and it is shown in Table 6 that the hydrogen permeability increases dramatically upon Ca exchange.

Table 6
Effect of cation on the permeability of zeolite A based membrane [19]

Cation	Effective pore size (nm)	H_2 permeability.10^{-9} (mol/m^2 s.Pa)
Na^+	0.42	0.11
Ca^{2+}	0.48	290

It is claimed in the patent that membranes consisting of other zeolites may also be prepared according to this general method, provided that in the synthesis steps the composition of the kaolin/zeolite mix is adapted to the composition of the target zeolite, e.g. by addition of silica.

5. SYNTHESIS OF MFI MEMBRANES ON POROUS CERAMIC SUPPORTS

Geus et al. [20] have studied the growth of high silica ZSM-5 onto three macroporous ceramic supports: α-alumina, clay and zirconia. Also two-layer systems were prepared and tested in which e.g. clay or zirconia were covered by a thin mesoporous film of metakaolin.

Zeolite films were grown hydrothermally on disks of the ceramic porous supports with a standard synthesis mixture consisting of silica (Aerosil 200), sodium hydroxide and tetrapropylammonium bromide (TPABr) in water having a molar composition 100 SiO_2:160 Na_2O:150 TPA_2O:16666 H_2O. All experiments were performed in Teflon-lined stainless-steel autoclaves (40 cm^3) at 180 °C for 1-5 days. The substrates were positioned on the bottom of the autoclave, and the synthesis mixture was added. Two-layered supports were always positioned with the mesoporous layer on top.

After hydrothermal treatment, the disks were washed with water and then ethanol, dried and inspected by visible light and scanning electron microscopy. The nature of the films was confirmed by powder X-ray diffraction (XRD). Qualitative chemical analysis of the supports was obtained by EDAX elemental analysis. The chemical composition of the clay material, the separate zeolitic material and the supernatant liquid of various reaction mixtures were determined quantitatively by ICP and AAS analysis.

For some of the supports - α-alumina, metakaolin-on-clay - alkaline leaching took place causing the composition of the synthesis mixture to shift from the range for MFI crystallization. Consequently phases with a higher Al content, like analcime (ANA) were found to crystallize onto these supports. By far the best results were achieved upon growing ZSM-5 on clay and a continuous MFI film was deposited on top of the disk.

The clay support was home-prepared from a standard clay mixture (Royal Delft Ware Factory), consisting of 14.7 wt. % kaolin clays, 17.9 wt. % ball clays, 5.7 wt. % feldspars, 27.1 wt. % quartz, 4.4 wt. % $CaCO_3$ and 30.2 wt. % water. To 100 g of the clay mixture, 15 g of cristobalite, 30 g of $CaCO_3$ and 60 g of water were added. Disk-shaped supports were prepared by pouring the clay suspension into shallow glass rings (diameter 25 mm; height 2-3 mm) on a gypsum surface, followed by drying in air. The supports were calcined at 900 °C in air for at least 1 h, and polished to smooth disks on similar material to avoid contamination. The addition of cristobalite (phase transition at 270 °C) to the original clay mixture resulted in a constant thermal expansion coefficient (TEC) of ca. 10^{-5} C^{-1} (temperature range 20-650 °C).

The MFI-clay composites were carefully calcined at 400 °C (heating rate 1 °C min^{-1}) for at least 20 h. Films were inspected by light and electron microscopy, and the MFI structure was established by XRD. Both an as-synthesized and a calcined film were polished to avoid shadow effects in the XRD experiment.

The top layer of the MFI-clay composite consisted of an array of randomly oriented, intergrown MFI crystals with a layer thickness of ca. 50-80 μm (Figure 9). The film seemed strongly bound to the clay phase (Figure 10). Qualitative elemental analysis by means of EDAX (Figure 10) revealed a film with high-silica content on top of a non-homogeneous support, consisting of mainly silicon, aluminium and calcium oxides. The silicon signal has been depressed by over three orders of magnitude in order to obtain a more detailed image.

The mechanical and thermal stability of the MFI-clay films upon calcination up to 400 °C is excellent. The developing tensile and compressive stresses upon heating/cooling are qualitatively evaluated by Geus [1]. Reference is the synthesis temperature, 180 °C. Important discontinuities are the removal of the template (> 350 °C), which is accompanied by a contraction of the lattice, and the orthorhombic-to-monoclinic phase transition at 77 °C.

10 μm

Figure 9. SEM image of the top layer of the ZSM-5 clay composite membrane (cross section; magnification 1000 x).

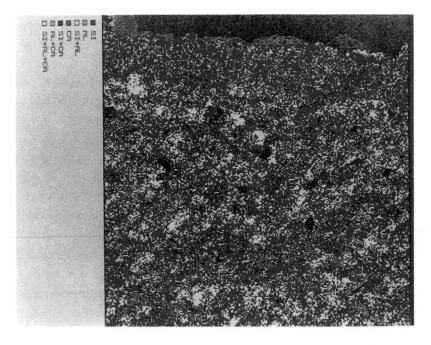

Figure 10. EDAX elemental image of the cross-sectional overview of the composite of Figure 9 [1].

5.1. Permeation measurements

Both as-synthesized and calcined ZSM-5 clay composite films were attached to perforated metal disks using a gas-tight epoxy resin as sealant, leaving a free membrane surface of some 1.5 cm^2. In a membrane cell two Viton O-rings were pressed on both sides of the metal disk. In this way no mechanical stress was directly applied on the membrane.

As-synthesized (template-containing) ZSM-5 clay composite films proved to be gas-tight. The single-component permeation results on a calcined MFI-clay membrane at 21 and 145 °C are summarized in Table 7. Significant and reproducible time lags are found at both temperatures for all gases, but upon variation of the feed pressure, the time lag varies slightly. Note that at 145 °C, all permeabilities decrease, except that for hydrogen.

Table 7

Pure-gas permeation results (transient state) on an MFI-clay membrane

gas	permeability (P) mol m m^{-2} Pa^{-1} s^{-1}	time lag (t$_1$)/s	diffusivity[a] $D(t_1)$/m^2 s^{-1}	D^{ss}/m^2 s^{-1}
temp. = 21 °C; feed pressure = 2 bar; initial permeate pressure = 2 Pa				
H$_2$	8.2×10^{-12}	0.4	4.2×10^{-9}	2.0×10^{-8}
N$_2$	2.7×10^{-12}	11	1.5×10^{-10}	4.0×10^{-10}
CH$_4$	3.8×10^{-12}	18	9.3×10^{-11}	2.2×10^{-10}
CO$_2$	3.2×10^{-12}	60	2.8×10^{-11}	5.0×10^{-11}
CF$_2$Cl$_2$	9.0×10^{-13}	118	1.4×10^{-11}	--
temp. = 145 °C; feed pressure = 2 bar; initial permeate pressure = 15 Pa				
H$_2$	9.4×10^{-12}	0.1	1.7×10^{-8}	3.3×10^{-8}
N$_2$	2.1×10^{-12}	4.7	3.5×10^{-10}	7.4×10^{-10}
O$_2$	2.4×10^{-12}	3.5	4.8×10^{-10}	--
CH$_4$	2.5×10^{-12}	4.5	3.7×10^{-10}	8.3×10^{-10}
CO$_2$	2.7×10^{-12}	42	4.0×10^{-11}	$6.5 \times 1.^{-10}$
CF$_2$Cl$_2$	6.0×10^{-13}	57	2.9×10^{-11}	--

[a] $D(t_1)$ is calculated from the time lag t_1, D^{ss} from the quasi-steady-state permeation rates.

Comparison with pulsed field gradient NMR data on methane diffusion [21] shows that the present values for $D(t_1)$ and D^{ss} are two orders of magnitude lower.

The presence of the relatively low porosity support is supposedly the main cause for the substantially lower diffusivities because: (i) an extra resistance for mass flow is introduced, and (ii) the effective membrane surface area is lower than the total membrane surface area, which is related to the zeolite/support interface porosity. Permeation is presumably obstructed for those parts of the MFI film that are strongly bound to the clay phase.

For a binary mixture of propane-propene at room temperature, no selectivity is obtained (permeability is 3.1×10^{-12} mol m m^{-2} Pa^{-1} s^{-1} for both gases). For n-butane-isobutane, initially a high selectivity for n-butane is observed, which decreases due to an increase in isobutane flow, and concurrently a decrease of n-butane flow. It takes over 1 h to reach steady state.

Recently a membrane layer consisting of a mixture of ZSM-5 and ferrierite was grown by Matsukata et al. [22] onto a porous alumina plate. The membrane layer was ca. 20 μm thick. The permeabilities for N_2 and O_2 were found to be $1.7.10^{-12}$ and $1.2.10^{-12}$ mol m N^{-1} s^{-1}, respectively, these values are comparable with those of Geus et al. [20] given above.

6. POROUS METAL SUPPORTED ZEOLITE MEMBRANES

Porous metal supported continuous films of MFI-type zeolite have been grown by Sano et al. [23] and by Geus et al. [24].

The first mentioned group of investigators used a porous stainless steel disk (5 cm diameter) with an average pore diameter of 2 μm, which was positioned on the bottom of an autoclave containing a silicalite-1 synthesis mixture (TPABr as the template).

After 48 h at 170 °C an aggregate of MFI crystals (20-100 μm) was formed at the surface. The average thickness of the silicalite-1 layer was 460 μm. Upon calcination at 500 °C the layer did not disintegrate. Pervaporation experiments on an ethanol/water mixture (5 v/v % EtOH) showed a high separation factor: α EtOH/H_2O = 60. Fluxes amounted to 0.2-0.8 kg m^{-2} h^{-1} over the temperature range 30-60 °C.

Within a close collaboration between the groups of Moulijn and one of the present authors, Geus and Bakker [24] prepared continuous layers of MFI on two-layer porous, sintered stainless (AISI 316) steel support disks (diameter 25 mm; thickness 3 mm) provided with a thin (50-150 μm) top layer of metal wool (R/F 1) which were obtained from Krebsöge (Radevormwald, Germany). For the construction of high-temperature membrane modules, similar disks (diameter 20 mm) were enclosed in non-porous stainless steel (AISI 316) rings (cf. Figure 11). Tight contact between both parts was achieved by thermal assemblage (porous support at liquid nitrogen temperature; non-porous disk at 400 °C).

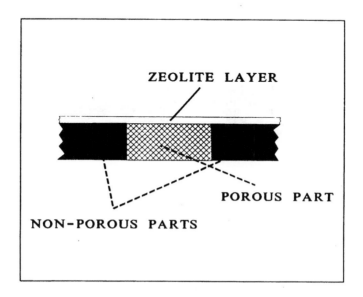

Figure 11. Schematic view of a zeolite layer on top of a porous/non-porous substrate.

Two stainless steel cylinders, provided with commercial flange connections (Leybold-Heraeus), were subsequently connected to the non-porous part of the disks with a gold alloy. In Figure 12 a photograph of the complete high-temperature module is shown. During hydrothermal treatment, the top section of the module (with the smooth top layer) is filled with the synthesis mixture, and the lower section is effectively closed by a Teflon cylinder (also shown in Figure 12). Both sides of the module are closed with flange connections using Teflon sealing rings.

Prior to the formation of MFI layers within the above high-temperature modules, several hydrothermal syntheses were performed on separate porous supports within Teflon lined (30 ml) autoclaves. In these experiments the chemical composition of the synthesis mixture was optimized for the formation of a continuous MFI layer, fully covering the porous metal substrate.

Figure 12. High temperature permeation cell as developed by Geus and Bakker [24].

Especially rather diluted synthesis mixtures at pH 13.5-14.0 were applied. Aerosil 200 (Degussa) was used as the silicon source and the presence of alkali metal ions was kept as low as possible in order to ensure the maximal incorporation of TPA in the framework. TPA to SiO_2 ratio was relatively high.

After ageing the synthesis mixtures while stirring for 5 h at ambient conditions, the hydrothermal syntheses were performed at 180 °C.

Two formulations leading to a continuous polycrystalline MFI film were (molar ratio):

$$100 \; SiO_2:100 \; TPA:50 \; OH^-:11000 \; H_2O \; (\text{method A})$$

and

$$100 \; SiO_2:230 \; TPA:75 \; OH^-:14000 \; H_2O \; (\text{method B})$$

Synthesis times were 48 and 45 h, respectively.

SEM pictures of the layer obtained when using the first mentioned gel composition are shown in Figure 13a (top view) and Figure 13b (cross section).

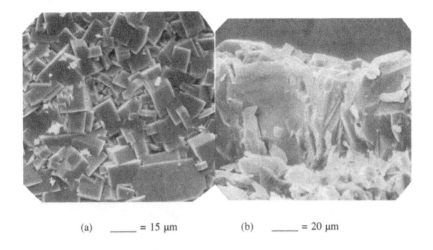

(a) _____ = 15 μm   (b) _____ = 20 μm

Figure 13a. SEM picture of a continuous MFI layer (top view), prepared using method A.

Figure 13b. SEM picture of a cross section of the same two-layer stainless steel supported MFI layer (magnification 600 x).

Figures 14a and b show the cross section of the above three-layer membrane system as observed and analyzed by SEM and EDAX, respectively.

A pure silica MFI layer (red) is supported by the (blue) stainless steel substrate, consisting mainly of iron, chromium, and nickel. Some small MFI crystals can be observed on the stainless steel within the wide pore section. The thin metal top layer with substantially smaller pores has retained the greater part of its porosity. This is rationalized by the low applied silica concentration in connection with the limited hold-up of the synthesis mixture within the macropores of the support. During the crystallization process, the pores within the metal support become isolated from the bulk solution by the developing MFI layer. On the outer surface, on the other hand, the crystal growth can continue from nutrients in the bulk solution.

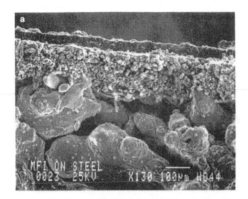

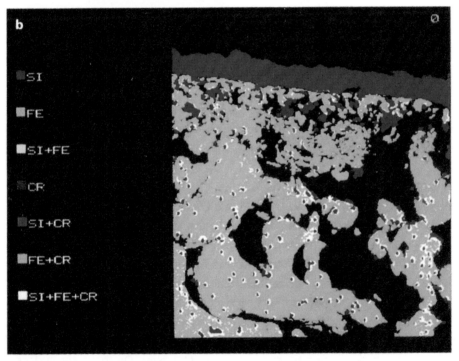

Figure 14a. Cross-sectional overview (SEM) of the MFI composite membrane, prepared according to formulation A (magnification 130 x).

Figure 14b. EDAX elemental image of the cross-sectional overview of the same two-layer stainless steel supported MFI membrane [1].

Unless the crystallization is favoured in a specific direction as on the extremely smooth and non-porous Si-wafers [10], total coverage of the porous support requires a relatively thick layer. Inevitably, the growth on a macroporous support will lead to a randomly grown crystalline layer, because the crystal growth proceeds from nuclei with a large variation in orientation. Hence, the minimal layer thickness is expected to be correlated to the maximal pore size of the porous support. For this reason two-layer stainless steel supports have been used, thus combining a high porosity support and a smooth top layer with a smaller pore size (ca. 10 μm).

6.1. Permeation experiments

Permeation experiments were performed according to the Wicke-Kallenbach concept (concentration gradient as the driving force), using helium as a purge gas. The permeate side was continuously flushed with helium, and analyzed by mass spectrometry. In all experiments the zeolite layer was exposed to the feed side, because the inner surface of this compartment was fully covered with MFI material. High purity gases were used: helium (99.996%), methane (99.5%), neon, n-butane, and isobutane (99.95%). The mass spectrometer was calibrated by self-made mixtures in helium within the same range as the permeate concentrations.

The as-synthesized metal supported MFI layers proved to be gas-tight for a pure neon feed. In Figure 15 the pure gas (30% in helium) permeation behaviour of all gases through the calcined (400 °C, 1 °C/min, dwell 16 h) MFI layer (prepared according to method B) is shown. The time to detect methane (34 s) and neon (38 s) is somewhat higher than the delay time of the experimental set-up (26 s). The detection of the butane fluxes depends strongly on the isomer: n-butane after 42 s, and isobutane after ca. 2 min. For methane, neon, and n-butane the steady state fluxes are reached within 1-2 min. In the case of isobutane it takes over 10 min to reach steady state completely, and the permeation rate is 1-2 orders of magnitude lower. Under the applied conditions, the steady state permeabilities (in $mol.m.m^{-2}.s^{-1}.Pa^{-1}$) amount to $1.2*10^{-11}$ (methane), $1.5*10^{-12}$ (neon), $4.5*10^{-12}$ (n-butane), and $7.0*10^{-14}$ (isobutane).

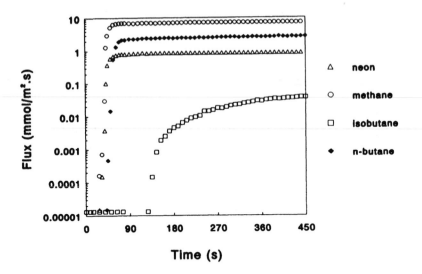

Figure 15. Pure gas transient permeation behaviour (30% in helium; 1 bar total pressure) for methane, neon, n-butane, and isobutane at 25 °C.

Figure 16 shows the permeation behaviour of a (50/50) methane/n-butane mixture. The methane flux starts out similar to the pure gas measurement in Figure 15, but levels off after a few seconds to reach a temporary maximum. The permeation behaviour of n-butane is hardly affected by the presence of methane instead of helium on the feed side. Steady state is reached within 2 minutes, similar to the pure gas permeation experiment for n-butane. The selectivity (α), however, amounts to approximately 50 in favour of n-butane, whereas the ideal separation factor, based on pure gas permeabilities, is in favour of methane. In this case, separation is apparently governed by a difference in adsorption strength, and by the reduced mobility (vide infra) of methane within the zeolite micropores in the presence of strongly adsorbing molecules. Owing to the strong adsorption of n-butane, hardly any methane may enter the zeolite micropores, and the driving force for methane permeation is substantially reduced. The methane permeation is equally reduced in the presence of isobutane. On other supported MFI layers a similar behaviour of binary mixtures of weakly and strongly adsorbing molecules has been observed [24].

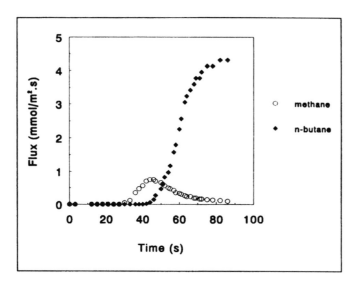

Figure 16. Transient permeation behaviour of a binary methane/n-butane mixture at 25 °C [24].

The permeation behaviour (steady state) of a (50/50) methane/n-butane mixture (pure) as a function of temperature is shown in Figure 17. A steady increase of the n-butane flux is observed (up to approximately 160 °C). The methane permeation rate remains low up to approximately 140 °C, and the selectivity is in favour of n-butane. At higher temperatures, the methane flow becomes substantial and even exceeds the decreasing n-butane flow at approximately 230 °C. At still higher temperature (up to 350 °C) the rise in methane permeation levels off. The same permeation behaviour for both gases is observed upon cooling, and has been found to be reproducible for several subsequent cycles. Thus, the metal supported MFI membrane remains thermomechanically stable up to at least 350 °C.

All observed features in Figure 17 may be related to the fact that both diffusion and adsorption are temperature dependent. For the n-butane permeation rate, the initial rise with temperature may arise from the higher intrinsic diffusivity, as diffusion is an activated process. Concurrently, however, the n-butane adsorption shifts out of the saturation area, first on the permeate side, so the driving force for n-butane permeation is also increased. For still higher temperatures, the driving force for n-butane decreases, as the adsorbate concentration on the feed side is no longer near saturation. Still higher n-butane permeation rates may be reached by applying higher n-butane (partial) feed pressures.

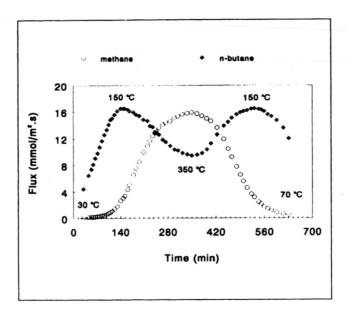

Figure 17. Steady state permeation rates of a 50/50 methane/n-butane mixture as function of temperature (heating rate 1 °C/min; cooling rate 1.5 °C/min) [24].

In conclusion the permeation selectivity is mainly governed by adsorption, provided both molecules have access to the zeolite pores. Strongly adsorbing molecules are favoured over weakly adsorbing species.

Additional examples of single and multi-component transport through metal-supported MFI membranes can be found in another paper of Bakker and Geus et al. [25].

Regarding the permeation of binary organic mixtures through MFI membranes we also mention results of Caro et al. [32] using ZSM-5 crystals embedded in a metal foil. Upon permeation of a 1:1 n-heptane/toluene mixture in the beginning almost pure heptane passed the membrane. In the steady state, at 95 °C, however, the permeate is substantially enriched in toluene (separation factor 10). This may be understood in terms of preferential toluene adsorption. At room temperature the enrichment was, however, much smaller.

Table 8 summarizes the results obtained on permeation of binary mixtures over MFI-based membranes.

Table 8

Permeation of binary mixtures over MFI-membranes

Membrane configuration	Binary mixture	Temp. °C	Selectivity	Ref.
continuous intergrown crystal layer	2,2-DMB n-C_6	23-49	17 (23 °C)	18
crystals embedded in metal foil	$PhCH_3$ n-C_7	23-95	10 (95 °C)	32
continuous intergrown layer on porous support	n-C_4/CH_4 i-C_4/CH_4	23-400	> 100 (23 °C)	24

The results of Haag et al. are attributed to shape-selective discrimination by the zeolite, whereas the results of the two other groups are ascribed to differences in adsorption strength.

A further refinement has been achieved by Jansen et al. [10] by growing MFI layers onto silicon wafers (non-porous) in an oriented way. The straight channels are (within 4°) oriented perpendicular to the support surface. When attained in a membrane configuration this would, of course, be of great advantage.

7. CATALYTIC ZEOLITE MEMBRANES

In many processes water or other small molecules are generated during reaction, which should be removed because the reaction is prevented from going to completion by e.g. thermodynamic limitations. In such systems reactors might be applied having a membrane consisting of a narrow pore zeolite. In case water has to be removed membranes based on one of the forms of zeolite A seem to be adequate. The advantage of a membrane over in situ drying (batch drying) or continuous drying (column drying) is that a continuous water removal takes place without saturating the adsorbent. On the other hand organics may be removed form aqueous solutions by applying a high silica membrane of a suitable pore size. An application might be the continuous removal of ethanol from carbohydrate fermentation liquids. The micro-organisms used are deactivated at an alcohol content > 12%. By continuous removal of the alcohol produced its concentration may be kept at a level below 12% enabling a higher final conversion of the carbohydrate feed.

Zeolite membranes may play either a passive or an active role in catalytic (organic) conversion reactions and the potential applications of zeolite membrane reactors are quite promising. Both liquid phase and gas phase reactions may advantageously be carried out in a membrane reactor, and transport from the reaction zone is promoted by continuous removal of the permeating molecules. Selective removal of product molecules is beneficial in equilibrium limited conversion reactions, since the conversion per pass in enhanced and the downstream product purification is simplified.

The membrane plays a passive role in case its main function is to:
- selectively remove product molecules,
- supply an active component to the reaction zone.

The membrane plays an active role in case its function is to
- catalyse the reaction
- function as a carrier for the active catalyst component.

Few examples of conversion reactions in a zeolite membrane reactor have been reported in the literature. Haag et al. [18] described a reactor consisting of an inner porous alumina tube, carrying on its interior wall a 12 μm ZSM-5 membrane. The alumina tube was mounted in an outer steel tube and the inter-tube annulus had a width of about 0.5 cm, both tubes were separately fitted with feed and product lines. This type of reactor was applied in some reactions and the synthesis conditions of the ZSM-5 membrane were adapted to obtain the appropriate Si/Al ratio for the envisaged application. Some of the applications described [18] are:

a) The membrane of the reactor consisted of ZSM-5 having a Si/Al ratio > 20.000, which is essentially inactive in catalysis. The inner tube of the reactor was loaded with H-ZSM-5 extrudates having a Si/Al ratio of 70. A feed stream of vapourized cumene was passed over the catalyst at 350 °C, one bar and a WHSV of 10 h^{-1}. Propene and benzene were withdrawn from the annular space which was swept with N_2. Under comparable conditions in a conventinal reactor the reaction product consisted of disproportionation products of cumene and only traces of propene were formed.

b) The same membrane reactor as in example a) above was charged with a 0.6% Pt on alumina catalyst mixed with alumina. A stream of 40 g/h isoprene containing about 2% of the undesirable 1,3-pentadiene was charged in the outer annular space reactor at 100 °C. Hydrogen in a molar ratio of 0.1 H_2/hydrocarbon was charged to the inner and the outer reactor. The effluent consisted of isoprene contaminated with less than 1% 1,3-pentadiene and some n-pentane.

c) A membrane reactor having the same configuration as described before, was equipped with a ZSM-5 membrane having a Si/Al ratio of 700. The inner reactor was loaded with a mixture of 10 g molybdenum oxide and 20 g quartz particles. At 200 °C, 25 bar and a WHSV of 2 h^{-1}, cyclohexane was passed through the inner reactor while 40 ml of oxygen was passed through the outer reactor. The effluent of the inner reactor contained cyclohanol and cyclohexanone and only little CO_2. Compared to a conventional reactor system, where cyclohexane and oxygen are cofed to the reactor, the selectivity to the desired product of the membrane reactor system is much higher.

d) A membrane reactor having the same configuration as described before, was equipped with a K-exchanged ZSM-5 membrane having a Si/Al ratio of 220. The membrane was impregnated with chloroplatinic acid to give 0.001 wt % Pt based on total quantity of zeolite, and reduced at 500 °C in hydrogen to form platinum metal supported on the zeolite covering the inner surface of the membrane tube. The catalyst was applied in the dehydrogenation of isobutane at 560 °C and atmospheric pressure. The product from the inner reactor contained isobutene and the product of the annular reactor was hydrogen.

The first three examples mentioned above illustrate the effect of a membrane playing a passive role and the last is an example of a membrane actively participating in the reaction.

Another use of zeolite membranes is in the controlled release of a reactant towards the reaction phase. In the absence of zeolite consecutive reactions are inevitable here. This principle can enhance selectivity as has been shown in the (batch) use of bromine-loaded CaA for para-selective bromination of aniline and the toluidines [26]. The addition can be controlled by varying the type of zeolite, the thickness of the membrane layer, the pressure difference etc.

For reviews on catalytically active (ceramic) membranes the reader is referred to Zaspalis [27] and Armor [28].

8. MISCELLANEOUS

In view of the accelerating effects of microwave radiation on zeolite crystallization, as found by Arafat and Jansen et al. [29], some zeolite synthesis experiments were carried out in the presence of metal and ceramic supports. Thus an activated Cu metal support platelet was immersed in a NaA synthesis mixture. After fast heating to 120 °C followed by 10 min at 95 °C in the microwave oven, zeolite NaA crystals, firmly attached to the metal oxide surface but not fully grown, with smoothed edges and corners were observed. When using a ceramic support and under the above described conditions, fully grown cube-type crystals were found on a module of cordierite. Apparently microwaves stimulated the interaction between the growing zeolite crystal and the copper surface more than with the cordierite.

Various potential applications of supported zeolites can be envisaged in the field of sensors. Zeolites can be designed and tuned so as to become selective towards a given molecule or a group of molecules [30].

Finally we mention the potential use of metal-supported zeolites in heat pumps [31]. Here, heats of adsorption and desorption (e.g. water in zeolite A) are applied, and improvement of the thermal conductivity of the adsorption beds is an essential factor.

9. CONCLUSIONS

Supported zeolites are a new type of industrial inorganic composites having promising properties for application in catalysis, sorption, separation and probably as electronic components.

The adaptability of both the chemical composition and the porous texture of the zeolite structure allows the development of materials which are tailored for specific applications.

Specially for problems related to the environmental compatibility of processes supported zeolites offer interesting options for unconventional problem solving.

REFERENCES

1. E.R. Geus, "Preparation and Characterization of Composite Inorganic Zeolite Membranes with Molecular Sieve Properties", Thesis Delft University of Technology, 1993.
2. I.W. Lachman, M.D. Patil and L.S. Socha, EPA 0 460 542, 1991, to Corning Inc.
3. J. Haas, C. Plogand and J. Steinwandel, DEOS 3 716 446, 1988, to Dornier System GmbH.
4. S.W. Blocki, Env. Progress, 12, 1993, 226.
5. E.W. Albers and C.E. Grant, US Pat. 3.730.910, 1973, to W.R. Grace & Co.
6. A.F. Woldhuis, Procestechnologie 9 (1992) 16.
7. E.g. W. Held, A. König and L. Puppe, DEP 4.003.515, 1991.
8. E.g. J.L. d'Itri and W.M.H. Sachtler, Catal. Lett. 15 (1992) 289.
9. E.g. W.K. Hall and J. Valyon, Catal. Lett. 15 (1992) 311.
10. J.C. Jansen, W. Nugroho and H. van Bekkum, in Proc. 9th Int. Zeolite Conf., Montreal, 1992, eds. R. von Ballmoos et al., Butterworth-Heineman, U.S.A., 1993, 247.
11. H.P. Calis, A.W. Gerritsen, C.M. van den Bleek, C.H. Legein, J.C. Jansen and H. van Bekkum, Can. J. Chem. Eng., in the press.
12. J. Sarkany, J.L. d'Itri and W.H.M. Sachtler, Catal. Lett. 16 (1992) 241.
13. X. Wenyang, J. Dong, L. Jinping, L. Jianquan and W. Feng, J. Chem. Soc., Chem. Commun. 1990, 755.
14. G.J. Bratton and T.Naylor, EPA 0 481 658, 1992, to Britt. Petr. Int. Ltd.

15. S.A.I. Barri, G.J. Bratton, T. Naylor and J.D. Tomkinson, EPA 0 481 659, 1992, to Brit. Petr. Int. Ltd.
16. S.A.I. Barri, G.J. Bratton and T. Naylor, EPA 0 481 659, 1992, to Brit. Petr. Int. Ltd.
17. G.J. Bratton and T. Naylor, PCT Int. Appl. WO 93/19840, 1993, to Brit. Petr. Int. Ltd.
18. W.O. Haag, E.W. Valyocsik and J.G. Tsikoyiannis, EPA 0460512, 1991, to Mobil Oil Co. J.G. Tsikoyiannis and W.O. Haag, Zeolites 12 (1992) 126.
19. C.W.R. Engelen and W.F. van Leeuwen, PCT Int. Appl. WO 93/19841, 1993, to ECN.
20. E.R. Geus, M.J. den Exter and H. van Bekkum, J. Chem. Soc., Faraday Trans. 88 (1992) 3101.

21. J. Caro, M. Buelow, W. Schirmer, J. Kärger, W. Heink, H. Pfeifer and S.P. Zdanov, J. Chem. Soc., Faraday Trans. 81 (1985) 2541.

22. M. Matsukata, N. Nishiyama and K. Ueyama, J. Chem. Soc., Chem. Commun. 1994, 755.

23. T. Sano, H. Yanagishita, Y. Kiyozumi, D. Kitamoto and F. Mizukami, Chem. Lett. 1992, 2413.

24. E.R. Geus, H. van Bekkum, W.J.W. Bakker and J.A. Moulijn, Microporous Mat. 1 (1993) 131.

25. W.J.W. Bakker, G. Zheng, F. Kapteijn, M. Makkee, J.A. Moulijn, E.R. Geus and H. van Bekkum, in Precision Process Technology, eds. M.P.C. Weijnen and A.A.H. Drinkenburg, Kluwer Academic Publishers, 1993, p. 425.

26. M. Onaka and Y. Izumi, Chem. Lett. 1984, 2007.

27. V.T. Zaspalis, Thesis University Twente, 1990.

28. J.N. Armor, Chemtech 1992, 557; Appl. Catal. 49 (1989) 1.

29. J.C. Jansen, A. Arafat, A.K. Barakat and H. van Bekkum, in Molecular Sieves, Synthesis of Microporous Materials, eds. M.L. Occelli and H.E. Robson, Vol. 1 (1992) p. 507. Cf. also A. Arafat, J.C. Jansen, A.R. Ebaid and H. van Bekkum, Zeolites 13 (1993) 162.

30. J.H. Koegler, H.W. Zandbergen, J.L.N. Harteveld, M.S. Nieuwenhuizen, H. van Bekkum and J.C. Jansen, contribution 10th IZC, Garmisch-Partenkirchen, 1994.

31. G. Cacciola and G. Restuccia, Heat recovery systems & CHP, in the press.

32. J. Caro, P. Kölsch, E. Lieske, M. Noack and D. Venzke, Congress Abstracts, New Directions in Separation Technology, Noordwijkerhout, 1993, p. 6. The authors thank Dr. J.C. Jansen for drawing their attention to this work.

J.C. Jansen, M. Stöcker, H.G. Karge and J. Weitkamp (Eds.)
Advanced Zeolite Science and Applications
Studies in Surface Science and Catalysis, Vol. 85
Elsevier Science B.V.

The intersection of electrochemistry with zeolite science

Debra R. Rolison

*Naval Research Laboratory, Surface Chemistry Branch; Code 6170
Washington, DC 20375-5342 USA*

CONTENTS

1. INTRODUCTION

Why have zeolites been of interest in electrochemical science, especially over the past decade? That such an interest in the intersection of electrochemistry and zeolite science does exist can be discerned by the, perhaps surprising, number of reviews on this topic in the past five years [1-5]. Figure 1 demonstrates that the intersection of electrochemistry and zeolite science is not the empty set, but yields many shared attributes--not the least being the heterogeneous nature of reactions at the surface of either an electrode or a zeolite. But even greater interest arises from the perception that examining the commonalities between electrochemistry and zeolite science and then exploring beyond them will lead us to new science and the discernment of new reactive schemes. Some aspects of the new possibilities will be discussed in this chapter.

From the electrochemical perspective, zeolites offer a microstructured domain where the physical structure and the chemical nature of the zeolite may (but not necessarily will) affect electron-transfer reactions and may influence known chemical steps that couple with electron transfers at the electrode-solution interface. In keeping with the above perception of new possibilities when zeolite crystals are added to the electrode-solution interphase, new chemical steps may arise from the interaction of the electron-transfer reactants or products with zeolite that are not usually seen at the electrode-solution interface in the absence of zeolite. In the shorthand used in electrochemistry to discuss sequential elementary reactions, one would then seek to study electron-transfer reactions (E) to electroactive molecules or ions which are in turn coupled to chemical reactions (C) that may precede (CE) or follow single (EC) or multiple (ECE, $ECEC$...) electron-transfer step(s) [6].

The following well-known zeolitic characteristics offer the physical and chemical means to influence electrochemical reactions:

(i) **molecular sieving**—which affects sorption into the zeolite crystal based on the size and shape of the sorbate and arises from the nanometer-scale dimensions of the pore openings and channels in the zeolite crystalline lattice;

(ii) **cation-exchange capacity**—which can affect the distribution of cations in the solution bathing the zeolite through cation-exchange reactions and arises from the extraframework, charge-compensating mobile cations present in the aluminosilicate lattice to counter the equivalence of negatively charged, tetrahedrally bonded aluminum centers; and

(iii) **catalytic effects**—which alter the rate of chemical processes and can arise from either strong-acid sites in the zeolite or extrinsic centers synthesized in or supported on the lattice.

ZEOLITE SCIENCE   ELECTROCHEMISTRY

A   X,Y

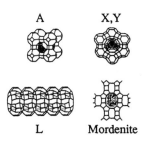

L   Mordenite

- *dc* electronic insulators
- molecular sieving based
  on size, shape, charge
- catalyst and catalyst support

- e⁻ source or sink
- mV control of the free energy
  of reaction $\Delta G^o = -nFE^o$
- material-dependent e'catalysis
  of non-reversible reactions

- diffusion of reactants and products
- local electrostatic fields of V/Å
- charge neutrality maintained by mobile ions
- heterogeneous reactions

Figure 1. The intersection of electrochemistry with zeolite science.

The wish to modify the physical and chemical environment surrounding the electron-transfer zone at the surface of an electrode, so as to control and alter electrode processes, has been the guiding rationale underlying the twenty years of research in chemically modified electrodes [7-9]. The use of zeolites to modify electrodes has been an active subset of this field [1-5].

From the perspective of zeolite science, electrochemistry offers experimental and conceptual means to explore issues arising should externally controlled voltage or current be imposed during chemical reactions normally inherent to or affected by zeolites. Obvious reactions that would be candidates for electrochemical influences are any zeolite chemistry known to be ionic, oxidative, or reductive in nature. Electrochemistry can also provide a measurement of the rate of fundamental processes inherent to zeolites, such as the kinetics of the ion-exchange process in zeolites and the intracrystalline diffusion of electroactive reactants. The liquid electrolytic conditions common to electrochemistry can afford new opportunities for catalytic and electrocatalytic chemistry with zeolites, especially as the use of zeolites in liquid-phase (non-electrolyte) synthesis and catalysis has become of increased interest of late [10-13]. Coupling zeolites to electrochemistry also provides openings in advanced applications of zeolites in electrochemical devices, including power sources and sensors.

My goal for this chapter will be to offer a view (rather than yet another review) with the addition of a perspective on the intersection of these two sciences. This chapter will also emphasize, using results from the literature and my group, a consideration of the questions that arise when electrochemical experiments are performed in the presence of zeolites and a discussion of the possible pitfalls awaiting the unwary (and the wary!) who interpret electrochemical results once the electrolytic environment intersects the microstructured domain of zeolites.

2. ELECTROCHEMICAL CONSIDERATIONS *(for the zeolite scientist)*

2.1. Electrodics *vs.* Ionics

Electrodics is a term encompassing the heterogeneous nature of most (but not all) electrochemical phenomena, and as seen in Figure 2, it involves the study of electrochemical processes in which transfer of an electron occurs across a two-dimensional boundary (the interface) between two phases—that of the electrode and that of an ionically conducting solution or solid (*i.e.*, the electrolyte) which contains the mobile ions [14]. Zeolite-modified electrodes are inherently electrodic and are doubly heterogeneous, once due to electron transfer events at an electron-conducting material and then again as reactants and products interact with the intracrystalline or extracrystalline surface of the zeolite or as zeolite-supported reactants interact with the electrode.

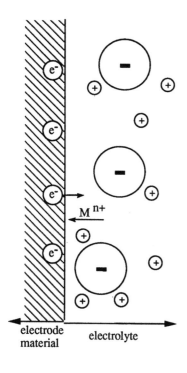

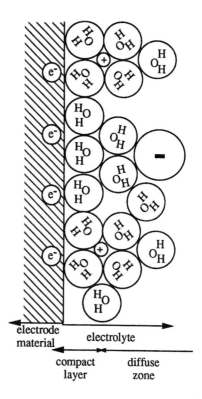

Figure 2. Electrodics: Charge transfer across a boundary between an electronic phase and an ionic phase.

Figure 3. Classical model of the double layer of charge at an electrified surface negative of the potential of zero charge.

Ionics concerns the study of ionically conducting phases (liquid or solid) with no concern for electrodic events. Zeolites are *dc* electronic insulators and, consequently, are not candidates to be direct electrode materials, *i.e.*, a material capable of acting as an electron bank to contribute electrons to or garner electrons from a reactant. Aluminosilicate zeolites are ionic conductors, however, and are capable of either solution-like ionic conduction, when hydrated, or solid-state ionic conduction when dry and taken to temperatures above 200°C [15]. Because of the link in common to the field of ionics, electrochemical reactions and the chemical and catalytic reactions of the aluminosilicate zeolites irrevocably occur in a charge-balanced yet ionic environment.

2.2. Interface *vs.* Interphase

In photoemission, when an electron crosses from an electron-conducting phase to vacuum, the boundary is sharp, as sharp as is implied in Figure 2. In a liquid electrolyte, the zone at the surface of an electrode consists of a highly structured region with dipole-oriented solvent and a double layer of charge-segregated ions which resembles a capacitive dielectric. Moving out from the electrode surface into solution there lies a less highly structured ionic zone (the diffuse layer) which finally transitions to the randomness of the bulk electrolyte solution. This classic model of the electrode-liquid electrolyte junction is summarized in Figure 3 for an electrode poised negative of its potential of zero charge. Because the electrode-liquid electrolyte intersection cannot rigorously be called one-dimensional, some electrochemists refer to this boundary region as an interphase rather than an interface—but both terms can be found in electrochemical discussions.

When an electrode is charged (or electrified) by poising the material at a potential positive or negative of its potential of zero charge (a quantity dependent on the type of electrode material, crystal-plane orientation, and the identity of the electrolyte in which it is studied), the charge on the metal is mirrored in solution by movement and restructuring of the ions in the double layer and diffuse layer. The movement of charge (q) in time yields current (I) through Eqn. (1):

$$q = \int I \cdot dt \tag{1}$$

The net motion of ions in solution needed to counterbalance the charge assumed by the electrode (q_m) creates a capacitive (or charging) current such that the charge in solution at the electrode interface (q_s) equals $-q_m$. When electron transfer to a chemical species occurs (generating a faradaic charge, and, thus, a faradaic current), the capacitive current also flowing (as the ionic charge re-structures at the electrode surface with changes in applied potential) becomes the noise to the signal of the faradaic current. These two types of current need to be separated to derive information about the electrochemical transformation. Faraday's Law relates the faradaic charge to the number of electrons exchanged (n) and the amount of material involved in the reaction (m in moles) through the Faraday constant ($F = 96,495$ coulombs per equivalent of electronic charge):

$$q = nFm \tag{2}$$

2.3. Charge Balance

The transfer of an electron is inherently a process requiring electrostatic compensation. If an electron is donated or accepted by the electrode material, the electron gas of the metal compensates by pushing or pulling an electron from the electrical circuit and mirroring the process at the second electrode in the circuit. A second electrode is always present because electrochemistry is inherently a coupled process: if something is oxidized, something else in the system must be reduced.

When a chemical species accepts an electron from a cathode (by definition, the electrode at which reductions occur), its additional negative charge can be electrically balanced by the mobile ions in the ionically conductive phase either by a cation moving toward or a mobile anion moving away from the newly charged species. This ionic dance is mirrored at the anode for the chemical species which gives up an electron and becomes more positively charged. The movement of charged species is intrinsic and necessary for electrochemistry: without ionic mobility to balance the charge conferred by the electron-transfer reaction, the electron transfer event can be completely shut down.

Similarly, the fixed anionic site associated with the aluminum of an aluminosilicate zeolite lattice, as schematically depicted in Figure 4, must be balanced by an equivalence of cationic charge: as stated previously, these cations confer a cation-exchange capacity to the aluminosilicate zeolites. The mobility of these compensating cations within the zeolite is influenced by their physical location in the zeolite, their size and charge, and the content of water in the zeolite [15-17].

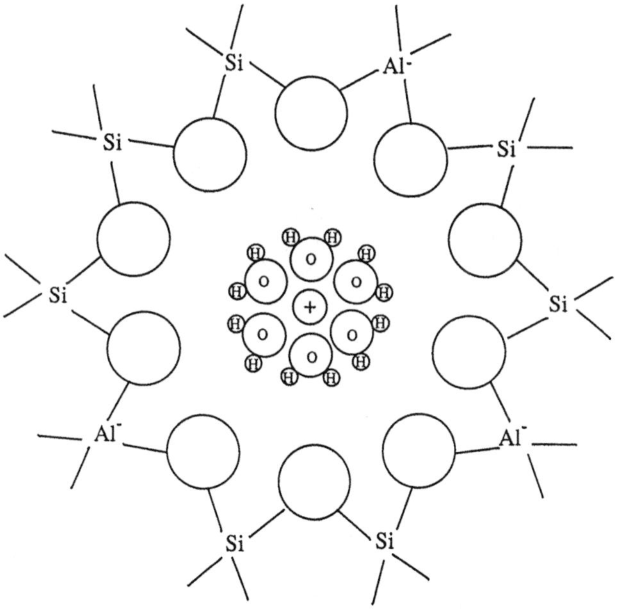

Figure 4. The tetrahedral network of Si, Al, and O atoms comprising the framework structure of aluminosilicate zeolites with a representation of the innate electrostatics and ion chemistry imposed by the presence of four-coordinate Al.

When aluminosilicate zeolites, with their ionic character, are placed at the electrode-liquid electrolyte interface, the complexity of the ionics of the system has been increased, as illustrated in Figure 5. The zeolite has charge-compensating cations present in the small cages or channels of the lattice structure, which may or may not be size excluded from ionic communication with the mobile charge-compensating cations present in the large supercages or channels of the zeolite lattice. These latter cations are available to exchange with cations present in the liquid electrolyte *if* the solvated electrolyte cations are not size excluded from the zeolite. Also present are extracrystalline cations compensating for fixed anionic lattice sites at the extracrystalline surface of the zeolite. These extracrystalline cations are readily available for ion exchange with electrolyte cations as the issues of size exclusion, ionic mobilities, and solvation energies are less critical factors for this process. Baker and co-workers have used Ag-exchanged zeolite Y (Ag^{I}Y) as a means to explore the exchange of Ag$^+$ between the small and large cages of zeolite Y and with electrolyte cation [18-20]; this groundbreaking work investigates the ionic bookkeeping in the zeolite-electrochemistry intersection and will be described more fully later.

Liquid electrolytes may be aqueous or non-aqueous, so that considerations of solvation may vary greatly when zeolite-modified electrodes are studied in non-aqueous media. Although the understanding of ion-exchange equilibria in zeolites

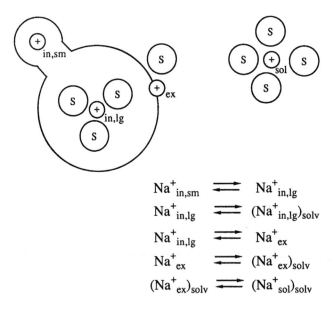

$$Na^+_{in,sm} \rightleftharpoons Na^+_{in,lg}$$

$$Na^+_{in,lg} \rightleftharpoons (Na^+_{in,lg})_{solv}$$

$$Na^+_{in,lg} \rightleftharpoons Na^+_{ex}$$

$$Na^+_{ex} \rightleftharpoons (Na^+_{ex})_{solv}$$

$$(Na^+_{ex})_{solv} \rightleftharpoons (Na^+_{sol})_{solv}$$

Figure 5. Zeolite ionics meet liquid electrolyte ionics—Legend: in: intracrystalline; ex: extracrystalline; sol: solution; sm: small channel; lg: large channel; S: solvent.

in non-water solvents is still a much understudied area [21], it is known that selectivities and exchange orders determined in water for a range of zeolites are not identical for the same zeolites and exchanging species in alcohols [22-24].

The understanding of the ion-exchange kinetics and equilibria characteristic of zeolite cation-exchangers in non-alcohol organic solvents suitable for electrochemistry (typically those with a reasonable dielectric constant [25]) may not yet be established, but it is a necessary component of the studies of zeolite-modified electrodes (ZME) in non-aqueous electrolytes. A data base is being incidentally established as the electrochemical characterizations of ZMEs in non-aqueous electrolytes are undertaken, yet this is one area where electrochemistry focused on the issues of ion exchange will be critical for application of zeolites in electrochemical devices requiring non-aqueous electrolytes. More work on this matter is crucial.

Furthermore, the effect of trace water in the non-aqueous solvent or of water present in incompletely dehydrated zeolites may markedly affect the ionic interactions between extraframework cations in the zeolite and cations in a liquid, non-aqueous electrolyte. Senaratne and Baker have used $Ag^{I}A$-modified electrodes in aqueous and dimethylformamide (DMF) electrolytes to show that in the size-excluded solvent (DMF), the electrochemical response due to Ag^{+} reduction and oxidation of the electroformed Ag^{0} is minimal relative to the response in water [26]. Trace additions of water (at ppm levels) to the DMF electrolyte are sufficient to increase the electrochemical signal. The greater the water content in the DMF, the higher are the currents measured for the stripping of the electroformed Ag^{0} to Ag^{+}—this sensitivity to trace water permits the $Ag^{I}A$-modified electrode to be used as an analytical sensor for water in non-aqueous solvents at levels below that typical of Karl Fischer determinations. With inert handling of the $Ag^{I}A$-modified electrode (i.e., dry-box conditions), 100 ppb detection of water in DMF was achieved [26]. The handling necessary to know and control the water content of a zeolite when one desires to study a modified zeolite or a zeolite-modified electrode in non-aqueous electrolyte should not be overlooked and assumed to be an unimportant variable.

2.4. Mass Transport

Another commonality to zeolites and electrochemistry lies in the importance of diffusion as the means to transport reactant to (and product from) the heterogeneous reacting surface. Because the flow of current arising from electron-transfer reactions to solutional reactants (i.e., faradaic current) is a direct expression of the rate of the electron-transfer reaction at the electrode, the rate of mass transport of reactant to the electrode surface also controls the magnitude of the faradaic current (and therefore the reaction rate). As the reactant becomes consumed at the electrode-electrolyte interface, concentration gradients arise relative to the bulk concentration of the reactant in the electrolyte. Fick's First Law (for a one-dimensional (linear) approach to an electrode of surface area, A) expresses the reaction rate (faradaic current, I) as a function of the flux of the reactant concentration (C) and its diffusion coefficient (D):

$$I = -nFAD(\partial C/\partial x) \tag{3}$$

Other means of mass transport are possible in electrochemistry; these are electromigration (which is relevant only for charged species and describes movement induced by an electric field) and convection. To minimize electromigration effects on the reactant/product couple to be studied, it is customary to use supporting electrolyte concentrations of an as-inert-as-possible dissociating salt that are 100-times greater than the reactant concentration [27]. These electrolyte ions will now carry >99% of the ionic current through the bulk electrolyte, so the faradaic current measured arises almost exclusively from diffusive transport of the reactant to the surface of the electrode.

Convection is achieved by stirring the solution (or by unwanted room vibrations) or more controllably by flowing the electrolyte past the electrode or by rotating the electrode (as in hydrodynamic voltammetry [28]) and is an effective way to increase contact of a solution-phase reactant with the electrode surface. When preparative-scale changes in an electroactive solute are desired, as in bulk electrolysis, the electrolyte is stirred. When diffusion is desired as the dominant mode of mass transport, care needs to be taken to keep the electrochemical solution (and system) quiescent. The worst case, from an analytic mathematical viewpoint, is when the mass transport is not controlled, but has contributions to the mass flux (and, hence, to the current) from two or all three forms of mass transport—diffusion, convection, and electromigration.

The electrochemical characterization of chemically modified electrodes has engendered an appreciation of the varied diffusive regimes that can exist when the electroactive reactant no longer approaches the surface of the electrode from the bulk electrolyte via semi-infinite (planar) diffusion. Modification of electrode surfaces with covalently bonded or chemisorbed monomolecular layers of electroactive reactant creates a physical situation in which the electroactive moieties can be exhaustively electrolyzed (i.e., completely reacted) in the course of the electrochemical experiment; in this respect, the electrochemical results resemble those obtained with electrochemical cells containing only thin-layers (on the order of micrometers) of electrolyte [29]. In these circumstances the amount of material reacted per unit electrode area is on the order of 10^{-10} mol/cm^2— which is why all the molecules can be completely electrolyzed in seconds or minutes rather than hours.

With the electroactive reagent sited at the electrode surface, semi-infinite diffusion from the bulk solution is not an option. As the modifying layers become polymeric [30] or multi-integrated [31], the mass transport is described by finite diffusion [32], also analogous to conditions operative in thin-layer electrochemistry, but at short times [33]. When the electroactive centers are immobilized at an electrode surface by enmeshing or binding them to a polymer, the ability of electrons to transfer through the layer is described by a mixed-valent, self-exchange-based electron-hopping mechanism [34]. Electron hopping in these layers is not a result of tunneling from a fixed electron donor and fixed electron acceptor site. The elasticity, plasticity, and segmental motion characteristic of polymers works to bring redox centers sufficiently close for electron exchange to occur. This physical contribution of the polymer to the resulting faradaic process —and the need to have mobile, counterbalancing, solvated ions in the polymer— can markedly slow the apparent "diffusion" of the redox species, so that the current eventually detected at the electrode surface yields an apparent diffusion coefficient with values that can approach those of solid-state diffusion (10^{-8}-10^{-11} cm^2 s^{-1}).

A zeolite is not an organic polymer capable of segmental motion, so that electron transfer between redox centers sorbed in zeolite cannot be physically equated to the polymeric motion-assisted processes as described for polymer-modified electrodes [34]—however, the use of the equations derived for the finite-diffusion conditions operative in polymeric cases may be valid as a global measure of the charge-transport limited currents measured for redox-modified zeolites.

Another aspect of diffusion needs to be considered that pertains for reactions occurring at the surface of micrometer-sized objects. Much of the recent interest in ultramicroelectrodes, or electrodes in which one dimension of the electrode material is sized on the scale of micrometers or less [35-39], is due to the rapidity at which semi-infinite, planar diffusion of a species to an electrode of this dimension converts to non-planar diffusion once the electron-transfer reaction is initiated at the electrode surface—this yields an increase in the amount of reactant transported to (and of product transported from) the microscale surface, so that when faradaic current is scaled to electrode area, much higher current densities for the same reaction can be achieved at an ultramicroelectrode than at a larger electrode.

A layer (or layers) of zeolites on the surface of a normally sized planar electrode creates micrometer-scale pathways past the micrometer-sized zeolite crystals to the underlying electrode surface. These pathways more likely represent a stagnant, unstirred pool of electrolyte, much as can be observed for arrays of micrometer-sized sunken hole electrodes created in an insulating surface [40]. An assumption should not be made that semi-infinite, planar diffusion controls the mass transport of electroactive reactant past the zeolite to the surface of the electrode just because that condition prevails at the unmodified electrode.

2.5. Thermodynamics *vs*. Kinetics

Electrochemistry offers the experimental means to characterize the state of a reversible system at equilibrium, *i.e.*, thermodynamic information can be readily obtained. Electrodics also offers the experimental means to study electrode kinetics[1]—the reaction kinetics of electron-transfer processes. When thermo-dynamic information is desired, the rate of the forward reaction must be equal to the rate of the reverse reaction—in an electrochemical system (either an electrodic or ionic electrochemical system) this means that no **net** current is flowing at the measuring (or working or indicator) electrode. Under these circumstances, and when the species taking part in the electrochemical reactions are at unit activity, the measured electrode potential (or to use the historical term: electromotive force) of the system is the standard value, E°, for the specific electrochemical reaction. The electrochemical cell, consists of two half-cell reactions—one reduction and one oxidation—where the standard potential for the half-cell reaction is a measure of the oxidizing or reducing power of the electroactive species. The standard potential has great significance as it can be related to the standard free energy of the reaction (ΔG°) and the equilibrium constant (K) of the reaction through Eqn. (4).

$$-\Delta G^\circ = nFE^\circ = RT \ln K \tag{4}$$

These relationships, and the relative ease by which the E° of an electrochemical cell could be determined, was exploited in the late 19th and early 20th century to establish the quantitative basis of thermodynamics and the free energy of reactions [41].

When the components of the electrochemical system are not at unit activities, for example:

$$a\ Ox_1 + b\ Red_2 \longrightarrow c\ Red_1 + d\ Ox_2 \tag{5}$$

the measured equilibrium potential, E, relates to the free energy change by $-\Delta G = nFE$, and since the equilibrium constant for the overall reaction in an electrochemical cell can be described by the same mass-action accounting used for chemical equilibria, the expression, still describing equilibrium (zero net current) conditions, becomes the well-known Nernst equation, Eqn. (8):

$$-\Delta G = RT \ln K - RT \ln \left[\frac{(Red_1)^c (Ox_2)^d}{(Ox_1)^a (Red_2)^b} \right] \tag{6}$$

$$nFE = nFE^\circ - RT \ln \left[\frac{(Red_1)^c (Ox_2)^d}{(Ox_1)^a (Red_2)^b} \right] \tag{7}$$

---

[1] The kinetics of the faradaic reaction at the electrode is not described as electrokinetics, but as electrode kinetics —electrokinetics describes a class of electric-field-driven phenomena, such as electrophoresis, electroosmosis, and streaming potentials.

$$E = E° - \frac{RT}{nF} \ln \left[\frac{(Red_1)^c (Ox_2)^d}{(Ox_1)^a (Red_2)^b} \right] \tag{8}$$

The Nernst equation is strictly written for activities of reactants and products, rather than concentrations, but under the high ionic-strength conditions typical of most electrochemical reactions (due to the presence of excess supporting electrolyte), the activity coefficients, γ_i (defined for $a_i = \gamma_i c_i$), remain constant despite the charge transfer occurring throughout the redox reactions [42]. This condition allows the standard potential to be modified to the formal potential, $E°'$, where:

$$E°' = E° - \frac{RT}{nF} \ln \left[\frac{(\lambda_{Red_1})^c (\lambda_{Ox_2})^d}{(\lambda_{Ox_1})^a (\lambda_{Red_2})^b} \right] \tag{9}$$

and the Nernst equation can then be written for species' concentrations when using the $E°'$ obtained for specific experimental conditions (including the nature and concentration of supporting electrolyte).

One of the earliest electrochemical uses of zeolites was to measure equilibrium potentials of a natural zeolite membrane undergoing cation exchange [43]. This type of potentiometric determination [44] provides information on the activities of and selectivities for the species exchanging between the membrane and the ionic solution. Because zeolites are cation exchangers, the reaction of interest is that for the cations, but the overall electrical neutrality of the system maintains because the equilibrium potential is measured against a reference electrode, and at that interface a balancing flow of anions can occur.[2] Traditional potentiometry has recently been used in the guise of a sodium-ion-selective electrode to characterize in situ the kinetics of ion exchange of NaY slurries with aqueous solutions of Mg^{2+}, Ca^{2+}, Sr^{2+}, and Ba^{2+} (as well as the reverse displacement of the alkaline-earth cations by Na^+) [45]. Half-times of 10-35 s for forward exchange and 5-65 min for the reverse exchange were obtained. The response time of the sodium-ion-selective electrode to 90% of the full value was found to be 5 s and was deemed sufficient for the kinetics studies.

The greater interest in electrochemical characterization of zeolite-modified electrodes has been in the use of time-dependent measurements where the current flowing at the working electrode is monitored as a function of an applied potential. In electrodics, the shape of the current-voltage curve or the time-dependent current or coulombic charge measured for an applied voltage waveform are diagnostic of the electron-transfer reactions that give rise to the signal.

The fundamental equation relating current (and, thus, electron-transfer kinetics) and potential (and, thus, driving force) is the Butler-Volmer equation, Eqn.

---

[2] Thou shalt not forget electroneutrality in electrochemical situations is the minor theme of this chapter, as should be apparent by now.

(11), which defines the heterogeneous reaction kinetics for the net electron-transfer reaction as based on the rates (*i.e.*, the faradaic currents) of the forward and back reactions (reduction and oxidation, respectively) of the general electroreaction:

$$Ox + n\ e^- \Leftrightarrow Red \tag{10}$$

$$I = nFAk°\{c_{red}[e^{(1-\alpha)nF(E-E°')/RT}] - c_{ox}[e^{-\alpha nF(E-E°')/RT}]\} \tag{11}$$

where c_{ox} and c_{red} are the respective **surface** concentrations of the oxidized and reduced forms of the electroactive species; $k°$ is the standard rate constant for the heterogeneous electron-transfer reaction (at the standard potential) in SI units of m s^{-1} (although this quantity is often reported in cm s^{-1}); the quantity $(E-E°')$—also known as the overvoltage, η—is a measure of the extra energy imparted to the electrode beyond the equilibrium potential for the reaction; and α describes the symmetry factor for the forward reaction (and $(1-\alpha)$ that for the back reaction)—the symmetry factor can be considered comparable to the order of the redox reaction with respect to electrons and has been further defined in Marcus theory [46]. Inspection of Eqn. (11) shows that when the net current is zero (*i.e.*, at equilibrium), the Butler-Volmer description collapses to the Nernst equation (Eqn. (8)). If the electrode kinetics for the reaction are very fast (*i.e.*, $k°'$ approaches infinity), the Butler-Volmer equation can also be shown to reduce to the Nernst equation—even when current flows: this is the condition of electrochemical reversibility [46].

2.6. Electroanalytical Techniques of Interest

As expressed in the Butler-Volmer equation (Eqn. (11)), current has an exponential dependency on potential. Electroanalytical methods that poise the potential of the system more than a few tens of mV away from the equilibrium potential—where the current is linearly dependent on the potential, as can be seen by expanding the exponential terms of Eqn. (11)—are called large-amplitude techniques [47] and include the potential-sweep method of choice (to-date) when studying zeolite-modified electrodes: cyclic voltammetry. In cyclic voltammetry, the current at the working electrode is measured as the potential (versus a reference electrode) is varied linearly in time to a chosen potential (E_{sw}, the switching potential) and then reversed with the entire waveform repeated as desired—this cycling of the potential in time permits both the forward and reverse electron-transfer steps to be observed. The speed at which the potential is swept (v, or scan rate) can be varied to study the kinetics of the electron-transfer reaction and to gauge the electrochemical and chemical stability (or reactivity) of the reactant and the electrogenerated product. The cyclic voltammogram for an electrochemically reversible reaction at a planar electrode under linear (semi-infinite) diffusion conditions (schematically depicted in Figure 6A) yields the characteristic shape seen in Figure 6B, where the current at the top of the peak (I_p) is:

$$I_p = 0.4463nFA(\frac{RT}{nF})^{1/2} D^{1/2}v^{1/2}C^*$$ (12)

where C^* represents the bulk concentration of the electroactive species (for current in amperes, A is in cm^2, D is in cm^2 s^{-1}, v is in V s^{-1}, and C^* is mol cm^{-3}). The peak maximum and tailing arise from the interplay of the exponential dependence of current on potential and the depletion of the electroreacting species at the surface of the electrode. Also shown in Figure 6 are voltammograms characteristic of ideal, electrochemically reversible electron transfer for a surface-confined species (where I_p is proportional to v rather than $v^{1/2}$), Figure 6C, and when non-linear diffusion predominates—as at an ultramicroelectrode—where the maximum current is a plateau rather than a peak and is essentially invariant with scan rate, Figure 6D.

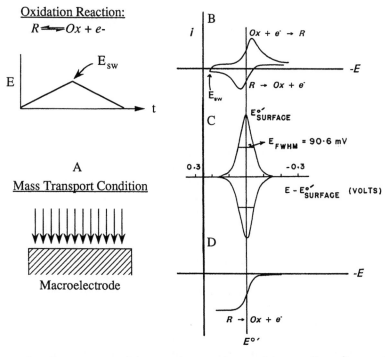

Figure 6. Current-potential waveshapes observed by cyclic voltammetry for a simple electron-transfer reaction: A, B: mass-transport condition of semi-infinite diffusion to the electrode surface (A) and the ideal voltammogram for this condition (B); C: ideal voltammogram for a surface-confined reactant; and D: ideal voltammogram at an ultramicroelectrode.

Cyclic voltammetry offers a powerful survey of the electrochemical response of a system and provides qualitative insight into the electrochemical and chemical reversibility of an electron-transfer species and has proven to be a guide to environmental effects on the voltammetric response of electroactive solutes present in chemically or physically modified or multi-integrated domains [9]. For quantitative insight into the physical parameters underlying the electrochemical response (*e.g.*, the diffusion coefficients, the number of electrons being transferred) or even the analytical concentration of the reacting substance, other electroanalytical methods which permit a readier separation of the charging (capacitive) response from the faradaic information are preferred.

Transient potential-step techniques such as chronocoulometry ($q(t)$) or chronoamperometry ($I(t)$) - in which the potential is stepped from a value where no faradaic reaction occurs to one beyond the standard potential - allows the rapidly decaying capacitive charge or current to be discriminated against the slower temporal evolution of faradaic charge or current; when the potential step is reversed (as in double-step methods) information about both the forward and reverse electron-transfer steps is obtained [47,48]. Baker and co-workers have recently used chronoamperometric characterization of Ag^IA-modified electrodes as a function of electrolyte temperature to determine the activation energy for intracrystalline Ag^+ ion-exchange in the A lattice as well as the diffusion coefficient (and activation energy) for exchange of zeolitic Ag^+ and electrolytic Na^+ [20].

Another manner in which capacitive information can be dissociated from the faradaic is to use potential-time waveforms in which the current is sampled after the capacitive current has been established (and decayed to near-zero). The simplest forms of this approach are embodied in the pulse polarographic methods (such as differential pulse and normal pulse polarography). The greater accuracy of the faradaic information obtained by the pulse methods make them preferred to cyclic voltammetry when determining concentrations of electroactive species—and because of the inherently better signal-to-noise, lower concentrations of electroactive solute can be measured [49]. Even greater sophistication in the choice of pulsed waveform leads to the method referred to as square-wave voltammetry [49,50]. Derouane and co-workers have taken advantage of square-wave voltammetry to analyze for the amount of methyl viologen (MV^{2+}) displaced from zeolite Y during the cyclic voltammetric characterization of their (viologen-modified zeolite)-modified electrodes. They were able to detect MV^{2+} in aqueous NaCl electrolyte by this method at concentrations as low as 50 nM [51]. Although unnecessary for the purpose of the cited study, there is no reason why the advantages of square-wave voltammetry cannot be applied to the characterization of the zeolite-modified electrode itself—especially under conditions of dilute concentrations of electroactive species in the zeolite bulk.

3. INTERSECTING ELECTROCHEMISTRY AND ZEOLITES

3.1. Confined Zeolites at the Electrode Surface: Summary of Preparatory Methods for Zeolite-Modified Electrodes

Because zeolites are *dc* electronic insulators, the study of zeolites in electrochemical interphases requires close physical contact of the zeolite crystal to an electronic conductor. Thus, the first step in creating and studying the intersection of zeolite science and electrochemistry lies in deciding how to create an interphase containing an electrode and a zeolite. This construction is often achieved by fixing the zeolite in a layer at a solid electrode surface or by forming an electrode particle-zeolite particle composite. In my earlier review [2], I listed the procedures developed to fabricate zeolite-modified electrodes as a means to provide a quick guide to the specific literature. For this chapter, I have brought the table, identically organized, but updated as a means to provide a capsule summary of the pre-1989 and post-1989 research in the area of zeolite-modified electrodes; see Table 1. As this chapter is primarily a discussion of zeolite + electrochemical issues rather than a discussion of every new (and old) piece of work, this table will serve to present research that will not necessarily be elaborated further in the text.

The considerable expansion of research since 1989 in the combinatory area of zeolites and electrochemistry can be gauged by the doubling of entries for zeolite-modified electrodes in Table 1, but this increase (including the work of researchers completely new to the area since 1989) is primarily due to increasing the variety of electroactive species studied with zeolite-modified electrodes rather than new ways to solve the means to intersect a zeolite with an electrode. Carbon-zeolite composites (whether pressed dry or held together with organic binders) and zeolite-polymer coatings (created by evaporation of an aliquot of a zeolite-polystyrene suspension on the electrode surface[3]) remain the principle methods used to fabricate zeolite-modified electrodes. As the generic how-to aspects of the various procedures were previously described [2], it is now time to look at the limiting aspects of these methods of preparation on the durability of the zeolite-modified electrode (ZME) in use and the quality of the electrochemical - mostly voltammetric - signals obtained with it.

In aqueous electrolytes, either the zeolite-modified carbon pastes or the zeolite/polystyrene coatings (Z/PS) can be adequately durable as the water does not immediately undermine the coherence provided by the organic binder or polymer. Long-term durability must not be assumed, however. When Derouane *et*

---

[3] I would like to point out that the first reference to coating an electrode surface with a zeolite-polystyrene overlayer was by deVismes, *et al.* [64] and not, as cited by so many papers on zeolite-modifed electrodes, the work of Pereira-Ramos, Messina, and Perichon [62], who instead pressed a powder composite of Ag^{I}A zeolite with graphite to form their modified electrode.

Table 1 Strategies for the Preparation of Zeolite-Modified Electrodes [a]

Method	*Zeolite Type*	*Electron-transfer Co-factor*	*Ref.*
Z/Conductive Composite			
A. Z/Carbon Paste			
1. 1 Z:10 (1 g of graphite/1 mL of vaseline)	natural Z from volcanic rock (Canary Islands)	Hg^{2+}	52
2. 40-50% Z:C paste [Type S, Bioanalytical Systems]	NaA, NaY	*(i)* $Ru(NH_3)_6^{3+}$; MV^{2+}; PV^{2+}; HV^{2+}; Cu^{2+}	53
	NaA, NaY	*(ii)* MV^{2+}	54
3. 1.1 g of hydrated Z + 2.1 g of graphite [Acheson 38] + 1.9 g of mineral oil	NaA	Ag^+	55
4. 9 mg of $Cu^{II}Z$ + 67 mg of C powder + 24 mg of paraffin oil	Na mordenite	Cu^{2+}	56
5. 700 mg of (MVZ + C powder) + 300 mg of mineral oil	NaY (LZ-Y-54)	MV^{2+}	51,57
6. 80:20 (Z:C powder) + unknown amount of unspecified oil	TS-1	framework Ti(III)	58
7. 40:50 (Z:graphite) + 10 wt% unspecified wax, melted	NaA; NaY (LZ-Y-52)	HQ; 1,3-DHB; MeOH, EtOH; N_2H_2	59
B. Z-Carbon-Polymer Composite			
1. 4.4 wt% {7.2 wt%} Z/4.8 wt% {7.6 wt%} Ketjenblack C [Akzo-Chemie]/ 1 wt% {2.2 wt%} AIBN/89.7 wt% {83.1 wt%} 60:40 styrene:DVB	NaA {NaY}	{$Ru(NH_3)_6^{3+}$; MV^{2+}}	60
C. Z-Conductive Powder Mixture			
1. 0.05% Z/100 g of Pb powder (30 μm)/10 mL of H_2O/10 mL of H_2SO_4	general	$Pb/PbSO_4$	61
2. 15 mg Ag^IZ/50 wt% graphite	mordenite-type	Ag^+/Ag^0	62
3. [(100 g of $Na_2SiO_3\cdot9H_2O$/100 mL of triethanolamine/700 mL of H_2O)/graphite [Aquadag] (or MoS_2 or TiS_2)] + 40 g of $NaAlO_2\cdot3H_2O$/ 1.25 g of Z/(100 mL of triethanolamine/700 mL of H_2O)]	13X	general	63
4. 30 mg of (ML)Z/30 mg of graphite, pressed pellet	NaY	*(i)* $TMPyP^{4+}$; $M(TMPyP^{4+})$ (M = Mn(III); Co(III); or Fe(III))	64
	NaX-type	*(ii)* $Co^{II}Pc$	65
	NaY (LZ-Y-52)	*(iii)* Co^{II}(salen)	66
	NaY	*(iv)* M(salen) (M = Mn(III) or Fe(III))	67
	NaY	*(v)* $MPcF_{16}$ (M = Co(II); Cu(II))	68
5. 50 mg of (ML)Z/50 mg of graphite pressed pellet	NaY (LZ-Y-52)	$M(bpy)_3^{2+}$; $M(phen)_3^{2+}$ (M = Co(II); Ni(II); or Fe(II))	69

a. Abbreviations: MV^{2+}: methylviologen; PV^{2+}: pentylviologen; HV^{2+}: heptylviologen; HQ: hydroquinone; DHB: dihydroxybenzene; AIBN: 2,2'-azobis(2-methylpropinonitrile); DVB: divinylbenzene; $TMPyP^{4+}$: tetra(*N*-methyl-4-pyridyl)porphyrin; Pc: phthalocyanine; salen: *N,N'*-bis(salicylaldehyde)ethylenediimine; bpy: 2,2'-bipyridine; phen: *o*-phenanthroline.

Table 1. Strategies for the Preparation of Zeolite-Modified Electrodes[a]

Method	*Zeolite Type*	*Electron-transfer Co-factor*	*Ref.*
Co-Electrodeposition			
A. *Z/organic salt (ratio dependent on organic)*	NaA, KA	1,4-DNB; TCNQ; 1,3-DNB; $Ru(bpy)_3^{2+}$	70,71
B. *Z/Conducting polymers*	NaA	pyrrole	3
Evaporation of Z/Polymer Suspension			
A. *Z/Polystyrene (PS)*			
1. 100 mg of Z/10 mg of PS/1 mL of THF	NaY	$TMPyP^{4+}$; $M(TMPyP^{4+})$ (M = Mn(III); Co(III); and Fe(III))	64
2. 80 mg of Z/5 mg of PS/10 mL of CH_2Cl_2	NaA; NaY	MV^{2+}	53,72
3. 100 mg of (ML)Z/10 mg of PS/1 mL of THF	NaY	$M(bpy)_3^{2+}$ $(M'(CpR)_2$ or $M'(Cp)CpR)$ (M = Ru,Os; M' = Co,Fe)	73,74
4. 100 mg of (V/P)Z/10 mg of PS/1 mL of THF	NaY	$M(TMPyP^{4+})$/Viologen; M = Zn(II),Co(II)	74,75
5. 12.7 mg of Fe^{II}Z/3 mg of PS/0.2 mL of THF	NaX; NaCaA; NaKA; Na mordenite	Fe^{2+}/Fe^{3+}	76
6. 100 mg of Z/50 mg of PS/1 mL of $CHCl_3$	NaA	Cd^{2+}; Al^{3+} [b]	3
7. 100 mg of Ag^IZ/10 mg of PS/2 mL of THF	NaA / NaA; NaY; CsY	(i) Ag^+ / (i) Ag^+	26 / 77
8. 80 mg of Hg^{II}Z/7 mg of PS/10 mL of 1,2-dichloroethane	NaA	Hg^{2+}	78
9. 80 mg of Hg^{II}Z/7 mg of PS/20 mL of CH_2Cl_2	NaA	Hg^{2+}	79
10. 100 mg of Fe^{III}Z/10 mg of PS/? mL of THF	NaY	Fe^{3+}/Fe^{2+}; Prussian Blue	80
11. 100 mg of modified Z/10 mg of PS/1 mL of THF	NaY	(i) $(Alk)M^{II}[Fe^{3+/2+}(CN)_6]$ (Alk = Na^+ or K^+), (M = Ni(II),Co(II), Fe(II))	81,82
		(ii) $C_6H_5NH_3^+$	83
B. *Z/C/Polystyrene (PS)*			
1. (100 mg of (M)Z + 100 mg of graphite)/10 mg of PS/2 mL of THF	NaY (LZ-Y-52) / NaA; NaY (LZ-Y-72) / NaY	(i) Ag^+/Ag^0 / (ii) Cu^{2+}; Ag^+ / (ii) Co^{2+}	18,20 / 19 / 77
C. *Z Layer/Polystyrene (PS) Overcoat*			
1. Evaporate 5 μL of 4 mg of Z/mL H_2O on electrode surface; evaporate 5 μL of 3 mg of PS/25 mL of THF over Z layer	NaA	Ag^+/Ag^0	84
D. *Z/Polyethylene Oxide (PEO)*			
1. 40 mg of Z/30 mL of PEO gel	Na mordenite	framework Al(III) $[H_2O$ oxid$^n/O_2$ red^n (?)]	85
2. x mg of Z/x mL (PEO + $LiBF_4$)	unknown Z	none (ionic electrolyte)	86

a. Abbreviations: DNB: dinitrobenzene; TCNQ: tetracyanoquinodimethane; bpy: 2,2'-bipyridine; THF: tetrahydrofuran; $TMPyP^{4+}$: tetra(*N*-methyl-4-pyridyl)porphyrin; MV^{2+}: methylviologen; Cp: cyclopentadienyl. **b.** Z layer used as an ion-exchange membrane for the potentiometric determination of the metal cations listed.

Table 1 Strategies for the Preparation of Zeolite-Modified Electrodes [a]

Method	*Zeolite Type*	*Electron-transfer Co-factor*	*Ref.*
Silane-linked Z Overlayer			
1. $SnO_2(O)_2$-Si⌢N⌣⌢N⌣Si-$(O)_2$-Z	NaY (LZ-Y-72)	$Fe(CN)_6^{4-}/Os(bpy)_3^{2+}/Fc^+$	74,87
Photopolymerized Z Layer			
1. 5 μL of 15 mg of Z (silanized with $(EtO)_3SiCH=CH_2$)/2 mL of CH_3CN evaporated on Pt; illuminated with 254-nm light	NaA	Ag^+	88
Z/Epoxy Composite			
1. Z pellet infused with low-viscosity epoxy resin (10 g of vinylchlorohexane dioxide + 6 g of propylene glycol diglycidyl ether + 26 g of nonenylsuccinic anhydride + 0.4 g of (diemethylamino)ethanol)	Cs mordenite	Cs^+; Ag^+; K^+; Na^+; Li^+; Ba^{2+}; Ca^{2+}; Cu^{2+} [b]	89
2. Z pellet infused with low-viscosity resin (unspecified resin)	NaA; NaCaA; NaY	Cu^{2+}; Fe^{3+} [b]	90
Z Only			
A. Pressed Pellets—Mechanical Contact to Electrodes			
1. Two electrodes: Current collector and Zn anode	NaX	$(Cu^{2+}$; Ag^+; $Hg^{2+})/Zn$	91
2. Two or three Pt electrodes (T=210-430 °C)	NaA; NaA·NaNO₃ [c]	Na^+ $(O_2?)$	92,93
3. WE: Ag; CE: PtSn; RE: Pt (T=355 °C)	NaA	Ag^+/Ag^0, Na^+ $(O_2?)$	92
4. Two or three Pt electrodes (T=250-425 °C)	$[Cd(NO_3)_2]_nNa_{12-n}A$ ·$xNaNO_3$ [c]	Cd^{2+}/Cd^0, Na^+ $(O_2?)$	94,95
5. WE: Pt or Ag; CE: Pt or Ag; RE: Pt (T=23 °C; 255-425 °C)	NaA; $[Pb(NO_3)_2]_n$ $Na_{12-n}A·xNaNO_3$ [c]	Pb^{2+}/Pb^0; Na^+ $(O_2?)$; Ag^+/Ag^0	93
6. WE: Pt; CE: Ag; RE: Ag (T=315-400 °C)	Ag^IA [d]	Ag^+/Ag^0 $(O_2?)$	96
7. WE/RE: Pt/Ag or Ag/Pt; CE: Ag (T=RT; 325 °C)	LiA; NaA	$H^+/OH^-/H_2O/H_2/O_2$, Li(hydr)	97
8. Two Cu electrodes (T=273; 298; 300K)	Na faujasite	$H^+/OH^-/H_2O$; $Cu^0/Cu^+/Cu^{2+}$	98
B. Pressed Pellet—Electrodeposition of Electrode Contact			
1. porous Au coating electrodeposited on 1 face of Z pellet	NaY	$AuCl_4^-$; $H^+/OH^-/H_2O$; NH_3; [e] EtOH [e]	99
C. Floated or Compacted Film—Mechanical Contact to Electrodes			
1. Cu, stainless steel, or graphite electrodes	NaX	H_2O, EtOH, acetone	100
2. Two Pt electrodes (T=RT; 350-500 °C)	NaA; NaY	O_2; UO_2^{2+}; Co^{2+}; Ni^{2+}; Cu^{2+}	101

a. Abbreviations: bpy: 2,2'-bipyridine; Fc^+: ferricenium; WE: working electrode; CE: counter electrode; RE: reference electrode. **b.** Z layer used as an ion-exchange membrane for the potentiometric determination of the metal cations listed. **c.** Z modified by the inclusion complex listed via exposure to the appropriate molten nitrate salt(s). **d.** Z heated to 1273K before electrochemistry—a temperature high enough to cause structural damage to the crystalline lattice of the zeolite. **e.** Porous Au-coated Z pellet used as a gas-phase/liquid electrolyte separator to study the electro-oxidation of the listed gases at the Au face.

al. studied a MV^{2+}-exchanged NaY-carbon paste composite, they observed currents for the reduction of MV^{2+} that were dependent on the number of previous cycles of the modified electrode in the electrolyte or for how long the paste had been exposed to aqueous media; these observations indicate electrolyte penetration into μm-thick regions of the paste. Stable measurements could be made, but only after 50 sweeps or more through the first reduction wave for MV^{2+} at 0.2 V s^{-1} [57].

Some groups rely on only the first or the first few voltammetric sweeps before the carbon or polymer composite becomes compromised by the electrolyte —even in water. When pressed powder composites are used as the form of the zeolite-modified electrode, one must accept that significant, and possibly complete, perfusion of the pressed electrode by the electrolyte has occurred and that the electroreaction zone is no longer the face of the pressed electrode, but the body of the electrode. This latter problem, at least in aqueous electrolytes, can be minimized by using the hard carbon-polymer-zeolite composites devised by Shaw and co-workers [60], but at the expense of a more cumbersome electrode preparation.

Pressed carbon-zeolite electrodes and the Z/PS-coated electrodes have been used in non-aqueous solvents such as DMF [19,26,65], dimethylsulfoxide (DMSO) [19,65-68], acetonitrile [19,66,67,69,72,77], and methanol [19], but the durability of such composites in these solvents can be lessened under such stresses as stirring of the electrolyte (or rotation of the electrode) or during large current excursions typical of solvent/electrolyte electroreactions, which may be gas evolving. The molecular weight (and MW distribution) of the polystyrene (or any other polymer) used to cohere a zeolite layer at the electrode surface should be routinely listed in the experimental details (and unfortunately has not been in the past)—especially when Z/PS systems are studied in non-aqueous solvents which will swell the polymer even if dissolution of the polymer binder does not immediately occur. Similarly for the zeolite-carbon composites, the specific carbon or graphite powder used to prepare the composite should be detailed as the specific surface area of conductive carbon can vary significantly and this will affect the electrochemical characteristics of the ZME.

Many of the fabricated forms of a zeolite-modified electrode yield generally unaesthetic-*looking* voltammetry for very practical reasons. The Z/PS composite combines two resistive elements in the coating and this not unexpectedly can lead to a sloping voltammogram—as is characteristic of a resistive element in the electrical circuit with its ohmic (linear) dependence of current on voltage; this is shown in Figure 7A for a stylized example of a resistive component distorting the shape of a cyclic voltammogram. Some authors then seek to improve the conductivity of the ZME by grinding the zeolite with carbon powder to increase the number of zeolite-conductor junctions, *i.e.*, the use of carbon powder increases the effective surface area of the electrode—but as capacitance (and,

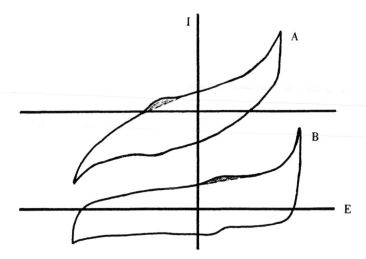

Figure 7. Cyclic voltammetry to avoid—distortion of the I-E wave due to resistive and/or capacitive effects: A: a sample voltammogram showing the effect of a high resistivity component at the electrode surface; B: a sample voltammogram showing the effect of high capacitance (noise) dwarfing the faradaic signal (shaded gray).

thus, charging current) is directly proportional to electrode surface area, the voltammetry of the (Z + C)-modified electrode will now have an increased background—as is shown in Figure 7B for a stylized example of a capacitive component (noise) dwarfing the faradaic information (signal) from an electroactive species.

In aqueous electrolytes and when no size limitations exist on the ingress of solvent or cations into the zeolite bulk, the large capacity of electroactive species associated with the zeolite bulk can provide ample signal (faradaic current) to minimize the bothersome nature of any resistive or capacitive components on the I-V wave shape, but when the desired faradaic response is minimal (due to low loadings or limitations in solvent/cation access to the zeolite bulk), these features can dominate the I-V response. In these cases, characterization of the ZME via classical cyclic voltammetry should be augmented by electroanalytical techniques which discriminate against the charging current, as discussed in Section 2.6.

The ideal method and means to fabricate a zeolite-modified electrode that is sufficiently conductive, amply durable, and reasonably easy-to-prepare does not yet exist. Approaches seeking to modify planar electrode surfaces with a layer of zeolite monograins continues and may offer more rigor in defining the reactive portion of a zeolite-modified electrode over that (not) achieved with the composite methods. Mallouk *et al.* started this approach by devising a multi-step synthetic procedure to silanize zeolite Y to a SnO_2 surface—much of the complexity of their synthetic procedure arose because of their desire to build-in anion-exchange sites in the molecule covalently linking the SnO_2 to NaY by silane linkages [87].

Calzaferri and co-workers have also devised two approaches to monograin modification of a planar electrode surface [84,88]. The simpler version relies on floating single-to-several layers of zeolite on the electrode surface from an aqueous suspension of the zeolite; after drying, the zeolites are coated with an aliquot of polystyrene dissolved in tetrahydrofuran which dries to a thin polymer film enfolding the zeolite particles to the electrode. By placing the zeolites first at the electrode surface, rather than evaporating an aliquot of a zeolite + polystyrene suspension, greater contact of the relevant material (the zeolite) with the electrode results [84]. In the second procedure, which avoids the use of polystyrene with its attendant difficulties as discussed above, Calzaferri and co-workers first silanized NaA by stirring the dehydrated powder at room temperature for four days in an acetonitrile solution of triethoxyvinylsilane. The silanized zeolite was floated onto a platinum surface (from dry acetonitrile), the solvent evaporated, and then exposed to 254-nm irradiation to polymerize the vinyl groups; this mechanically durable coating could then be further chemically modified to introduce an electroactive guest [88].

A recent promising approach involves the synthesis of thin, continuous (and, ultimately, adherent), films of pure zeolite on a metallic support from the starting precursor gel—*i.e.*, the film is achieved without embedding zeolite particles in a polymeric or ceramic matrix. Initial attempts by Suib and co-workers [102] have produced 1-μm to 1-mm thick films of zeolite Y crystallites on copper and other metal substrates (including platinum). One relevant and important observation for these films is that they were quite conductive as evidenced by minimal charging during the SEM analysis—this is a critical attribute if the electrochemical possibilities of such films are to be pursued. More recently, Jansen *et al.*, have grown oriented (mainly axially oriented) MFI zeolite crystals at a thickness of 1 μm on a range of metallic supports, including copper, nickel, titanium, and aluminum [103]. The application of these pure zeolite films to electrochemical objectives has yet to be reported.

3.2. Unconfined Zeolites in the Electrolyte: Dispersions and Slurries

One quick way to sidestep the intricacies involved in capturing the zeolite on or in the electron conductor is not to even try but instead to suspend the zeolite particles in the electrolyte. The size of commercially available zeolite powder has a particle size on the order of 1 μm, which is too large to sustain the particles as a long-term colloidal suspension. Mechanical means or gas dispersion can be used to maintain suspension of the zeolite particles in the liquid. In this manner the zeolite particles are free to contact the electrode physically, but ephemerally. This approach is analogous to the use of fluidized or dispersed electrodes in practical electroreactors, only the zeolite is not a millimeter-scale or micrometer-scale metallic or graphitic particulate electrode as would be used in fluidized [104-106] or dispersed electroreactors [107-109], respectively.

The various uses of monopolar or bipolar dispersions of modified zeolites to affect electrocatalytic reactions are listed in Table 2; also included in this table are some of the literature examples of photoelectrochemical reactivity at dispersed modified zeolites as driven by electron donor-acceptor chemistry. Most of the electrocatalytic uses have involved supporting non-metallic modifiers on the zeolite that, in turn, can catalyze selective organic oxidations.

Table 2. Electrochemistry and Photoelectrochemistry of Zeolite Dispersions [a]

Zeolite—Electron-transfer Modifier	*Dispersion Density g of Z/L*	*Substrate*	*Ref.*
NaX; NaA—Ru(bpy)$_3^{2+}$ [b]	unspecified (unstirred)	TMPD; 10-PP	110
NaY; Na mordenite—Ag$^+$ [c]	7.1	Ag$^+\rightarrow$Ag^0; synthesis of Cu^{II}Z, Zn^{II}Z, and Mn^{II}Z	111
NaY—Ru(bpy)$_3^{2+}$ [b]	unspecified (unstirred)	O$_2\rightarrow{}^1$O$_2$: MCH; (Me)$_2$C=C(Me)$_2$	112
faujasite—Mn^{III}(TMPyP^{4+}) [c]	53.3	2,6-di-*tert*butylphenol	113
Pt-KL—EDTA^{2-}/Zn^{II}(TMPyP^{4+})/MV^{2+} [b]	20.0	H$_2$O$\rightarrow$H$_2$	114
KL; NaY—Ru(bpy)$_3^{2+}\sim$2DQ^{2+}/BV^{2+} [b]	50.0	none	115
Pt-NaY [c]	1.25-4.0	H$_2$O/H$_2$/O$_2$; Fc; Fe(cp$\sim$R)$_2$	116, 117
NaA—Ag$^+$ [b]	4.0	Cl$^-\rightarrow$Cl$_2$	118
KL; NaY, Na mordenite— Ru(bpy)$_3^{2+}$/MV^{2+} [b]	6.3	none [d]	119
NaY—Pd^{II}Cu^{II} [c]	2.6	propene	120-3
NaY; LiY; KY; RbY; CsY, CaY [c,e]	2.6	H$_2$O$\rightarrow$H$^+$	124-5
NaY [c]	2.6	Aroclor®6050; Cl$_2$FC-CF$_2$Cl chlorobenzene	126-8
NaY—Ru(bpy)$_3^{2+}$/MV^{2+}; NaY—Ru(bpy)$_3^{2+}$/Os(bpy)$_3^{2+}$ [b]	10.0	none [d]	129, 130

a. Abbreviations: TMPD: *N,N,N′N′*-tetramethyl-*p*-phenylenediamine; 10-PP: 10-phenylphenothiazine; MCH: 1-methyl-1-cyclohexene; bpy: 2,2′-bipyridine; TMPyP^{4+}: tetra(*N*-methyl-4-pyridyl)porphyrin; EDTA^{2-}: ethylenediaminetetraacetic acid; MV^{2+}: methyl viologen; 2DQ^{2+}: diquat (*N,N′*-ethenyl-2,2′bipyridinium); BV^{2+}: benzyl viologen; Fc: ferrocene; cp: cyclopentadienyl; Aroclor®6050: mixture of polychlorinated terphenyls. b. Photoelectrochemical study. c. Electrochemical study. d. Charge and energy transfer were studied between the listed electron donors and acceptors. e. Ion exchange between the zeolitic mobile cation and electrogenerated proton in the electrified dispersion was studied.

4. ELECTROCHEMISTRY OF ZEOLITES MODIFIED WITH ELECTROACTIVE GUESTS

4.1. Charge Balance and Electrochemical Reactions in Zeolitic Interphases

Two mechanistic models, as originally proposed by Shaw et al. [53], have been postulated to describe the electron-transfer event for a zeolite-associated electroactive guest—and inextricably linked (yet again) is the issue of electroneutrality.

$$E_{(Z)}^{m+} + n\ e^- + n\ C_{(s)}^+ \Leftrightarrow E_{(Z)}^{m-n} + n\ C_{(Z)}^+ \tag{13}$$

$$E_{(Z)}^{m+} + m\ C_{(s)}^+ \Leftrightarrow E_{(s)}^{m+} + m\ C_{(Z)}^+ \tag{14a}$$

$$E_{(s)}^{m+} + n\ e^- \Leftrightarrow E_{(s)}^{m-n} \tag{14b}$$

where E^{m+} is the electroactive probe, C^+ is the electrolyte cation (univalent in this formulation), and (z) and (s) refer to zeolite and solution phase, respectively. The charge-transport mechanism described by Eqn. (13) implies that electron transfer to a zeolite-bound (or entrained) electroactive species can occur at/on/in the zeolite environs as long as a cation from the electrolyte solution can enter the zeolitic system to balance the extra negative charge brought in by the electron. This scheme has been designated as that most characteristic of intracrystalline redox. The two steps of Eqn. (14) describe a charge-transport mechanism which first postulates cationic displacement of the electron-transfer reactant from the zeolite by electrolyte cations, followed by traditional electron transfer at the electrode-solution interface to the no-longer zeolite-associated redox species. In this scheme the electron transfer occurs externally to the bulk of the zeolite and is referred to as an extracrystalline redox process.

4.2. Extracrystalline Redox Processes

Much of the published voltammetric characterization of metal-ion modified zeolites can be deciphered in a manner consistent with Eqn. (14) in which electrolyte cation displaces the zeolite guest cation, so that the zeolite-free guest undergoes electroreaction at the electrode surface. The strongest evidence for the extracrystalline redox mechanism has been and continues to be the manner in which the electroactivity of the zeolite-modified electrode can be shut down when the electrolyte consists of size-excluded cations and how electroactivity can be restored by additions of non-size-excluded cations. Creating electrolyte conditions to shut down the voltammetry by using size-excluded cations (or more precisely, size-excluded solvated cations) should be part of every voltammetric characterization of a zeolite-modified electrode.

In acetonitrile electrolyte and using a Co^{II}Y-modified electrode (a coating formed from Z+C dispersed in a polystyrene solution), Baker and co-workers showed that the voltammetry for reduction of Co^{II} could be essentially completely blocked when the electrolyte cation was the nominally size-excluded tetrabutylammonium cation (TBA$^+$) but freely seen when Li$^+$ was used as the cation [77]. The concept of *complete* suppression of the voltammetric response needs to be further qualified since:

(i) an ion-exchanged electroactive guest present on the external (extracrystalline) surface of the zeolite is free to exchange with any electrolyte cation and provide a (low) level of reactant in the electrolyte for direct electron-transfer reaction with the electrode surface—please refer to Figure 5 and consider the various locations of cation as depicted therein to now be the environments available to the electroactive guest;

(ii) ion-exchanged electroactive guest present on the extracrystalline surface of the zeolite can, if in contact with the electrode surface (be it planar or particulate), also directly react with the electrode without an exchange step—but, based on the relative surface areas involved, this can not be considered to be a high frequency occurrence;

(iii) even highly pure organic solvents can have enough adventitious impurities present to provide a roughly μM level of ion carriers (this has been observed repeatedly in the studies with ultramicroelectrodes using "salt-free electrolyte" conditions [36,38]) and these adventitious ions may not be size excluded, even though nominally size-excluded electrolyte cations are deliberately added—and, of course, even the purest water is 10^{-7} M in protons; and

(iv) given sufficient driving force (such as high equilibrium constants for complex or precipitate formation of the guest with an anion or ligand present in the electrolyte) and sufficient time, even nominally size-excluded electrolyte cations may distort sufficiently to displace a cationic electroactive guest from the zeolite bulk.

The first three possibilities above will define low levels of faradaic current due to either the small concentration of reactants involved (either from exchange of the extracrystalline electroactive guest or that exchanged by adventitious non-size-excluded cations) or the low number of appropriate junctions of zeolite-associated, but extracrystalline, guest with the electrode surface. The last possibility, item *(iv)*, which may lead to substantial levels of current, has been demonstrated by Baker *et al.*—again for the Co^{II}Y-modified electrode in acetonitrile —when TBA$^+$ salts were added to the TBA$^+$-blocking electrolyte, but with I$^-$ or SCN$^-$ as the anion, rather than the non-complexing BF$_4^-$ anion, thereby shifting the equilibria to the formation of the metal complex in solution [77]. These authors have pointed out, and I wish to re-emphasize here, that although Barrer found TBA$^+$ to be blocking for faujasite [131], he prescribes a kinetic diameter of ≥ 10 Å as necessary to achieve total size-exclusion [132]—and TBA$^+$ is smaller than that.

Tetrahexylammonium salts have been used as non-aqueous electrolyte salts, but THA$^+$ also has surfactant/micellar character and may not provide the ideal replacement for TBA$^+$ in the electrochemical characterization of zeolite-modified electrodes. The caution must be not to rely unequivocally on TBA$^+$ as a blocking cation for zeolites X and Y in all circumstances.

As detailed in previous reviews on zeolite-modified electrodes [2,3,5], such systems as Ag^{I}A [55,62,26,77,84], Hg^{II}A, [78,79] and Cu^{II}Y [53] have shown voltammetry in aqueous electrolytes that is fully consistent with extracrystalline redox, *i.e.*, the displaced (zeolite-free) metal ion is reduced to metal and then reoxidized back to the metal ion during the subsequent anodic stripping sweep. In their original paper postulating the two mechanisms for electron transfer in zeolite-modified electrodes [53], Shaw and co-workers could not fully ascribe the extracrystalline redox mechanism to the voltammetry they observed with (MV^{2+})-exchanged NaY. However in recent work by Derouane and co-workers with MV^{2+}-exchanged NaY (prepared as a ZME in a carbon-paste composite) [51,57], they deduced that if intracrystalline redox occurred, the voltammetric current for reduction of MV^{2+} (to MV$^{+\cdot}$) in the zeolite bulk should vary linearly with the exchange level (loading) of MV^{2+} in NaY—rather than show a dependence on the concentration of free MV^{2+} in the electrolyte solution, as extracrystalline redox would demand. Their results showed that the solution concentration of free MV^{2+} (in initially MV^{2+}-free electrolyte) was the important observable—a result again consistent with electrochemistry arising from an extracrystalline redox process [51].

4.3. Intracrystalline Redox Processes

At the time of my first review on zeolites and electrochemistry [2], there were neither interpretations of the voltammetric data from zeolite-modified electrodes which supported the possibility of non-mediated intracrystalline redox process nor were there even any claims of **non-mediated** intracrystalline redox. As of this review, there are now a number of reports of intracrystalline redox, including non-mediated intracrystalline redox. There should first be a recognition of the difference between direct and mediated electron transfer. It would seem best to use the convention customary in electrodics, where **direct** electron transfer is an event between the redox species and an electrode surface (*i.e.*, an electronic conductor), while **mediated** electron transfer requires a second electroactive species which has facile, direct electron transfer with the electrode surface and then acts to shuttle electrons to the redox species of interest (which often has slow electron-transfer kinetics with the electrode surface at its standard (thermodynamic) potential).

A claim of **direct** electron transfer to an intracrystalline species is important as the desire for a supracommunal response from intracrystalline redox species would fuel many of the advanced applications anticipated of zeolite-modified electrodes and electrode-modified zeolites. However, as the claim of direct intracrystalline redox is so important, it is incumbent on any claimants to demonstrate unequivocally that direct intracrystalline redox has occurred.

As formulated by Eqn. (13), intracrystalline redox implies that electrons from a distant electrode (where distant is defined on the scale of electron transfer where the rate of tunneling events falls to negligible when the centers are separated by more than 15 Å) can transition (somehow) to a redox species sited in the zeolite interior. Implicit in the intracrystalline redox mechanism of Eqn. (13) is the physical siting of the redox species as an encapsulated or "ship-in-the-bottle" guest in the zeolitic host. The location is implicit because if the intracrystalline redox species can migrate within the interior of the zeolite, little reason remains not to postulate migration to the exterior of the zeolite where the extracrystalline electron-transfer mechanism of Eqn. (14) holds sway.

Does intracrystalline redox occur? Can intracrystalline redox occur? The critical issues are those of charge transport: electronic communication and ionic balance. The electrochemical communication of an encapsulated redox species with the outside world—such as the electrode, which by garnering or providing electrons provides a global antenna to monitor and transmit the electrochemical conversation—can be readily imagined to require one or some combination of the following scenarios:

(1) a direct electronic path through the electron-insulating zeolite matrix to electrically connect an extracrystalline conductive electrode and the intracrystalline electron donor or acceptor—the ever-desired "molecular wire"; or

(2) a suitably small, suitably charged, suitably oxidative (or reductive), **and** suitably mobile electron-transfer mediator to enter the microporous zeolite, visit the encaged redox centers, discuss their relative electronic states, and then return to the external world (*i.e.*, the electrode) to report; or

(3) an intracrystalline redox site population of sufficient density such that electron self-exchange can occur between encaged species as electrons tunnel/hop/transfer from one encaged species to its nearest neighbor and so on until the electron passes across (or is carried by) an extracrystalline-sited redox species to the electrode.

In the absence of one of the above three conditions, assigning the electrochemical response to direct, intracrystalline electrochemical communication requires strong evidence and proof that the most likely short circuit - an adventitiously or deliberately added electron-transfer mediator, as described by Scenario *(2)* above - is not disguising the results. As diffusion in the zeolite can be liquid-like, the timescale of voltammetry dovetails all-too-neatly with possible diffusive events in the zeolite interior. The question now becomes: what is diffusing?—an electron-transfer mediator and the solvated ions necessary to compensate for shifting charges as the encapsulated redox species undergo electron transfer?—or just the solvated ions necessary to compensate for shifting charges as the encapsulated redox species undergo *direct* electron transfer?

Great effort by Bedioui and Balkus and colleagues has been spent creating encapsulated organometallic complexes in zeolites X and Y [65-69]. In addition to spectroscopic characterization of these modified zeolites showing formation of the expected complex(es), the electrochemistry of these materials as studied with pressed powder composites of graphite + modified zeolite has been pursued to address the possibility of intracrystalline electron transfer. Their most recent work has shown minimal electroactivity for ZMEs prepared from faujasite-encapsulated M^{II}hexadecafluorophthalocyanine (M = Cu or Co) after air-handling precautions were taken during preparation of the material and after multiple Soxhlet extractions were made to remove extracrystalline material [68]. Some electroactivity persists, but at a level estimated to be 0.5% of the total complexes encapsulated—the authors propose that these reactive centers are those near the external surface allowing sufficient intimacy for electron transfer to an occluded graphite particle to occur [68]. This observation does not bode well for the important goal of supracommunal electronic communication throughout the zeolite lattice and to the external world. The "ship-in-the-bottle" approach does seem to have benefits for site isolation of the encapsulated complexes during electroreaction, *e.g.*, dimerization of the metallophthalocyanines, as is common in homogeneous solution, was minimized for the zeolite-encapsulated form as gauged by the voltammetry for the zeolite-modified electrode [68].

Although Bein and his co-workers continue to study intracrystalline polymerization of materials that could act as conducting paths (*i.e.*, as molecular wires) [133-136], the conductivity of the zeolite-encapsulated polymer composite remains disappointing. When and if direct intracrystalline redox can be demonstrated at practical levels, the next issue becomes one of leaving sufficient room in the zeolite (stuffed as it is with encapsulated catalyst, unreacted ligands or monomer, and/or molecular wires) so as to be able to do electrocatalytic chemistry with a desired substrate!

4.4. Ionic Bookkeeping

In the continuing study by the Baker group of the electrochemical behavior of Ag^I-modified zeolites, the most recent report [20] deals with a thorough dissection of the silver ion environments in Ag^I-Y-modified electrodes and the coupled ionic equilibria operative when this modified electrode is used in aqueous electrolyte. Their ability to discriminate against ion exchange between the small and large cages in zeolite Y was aided by the extensive knowledge of the siting of Ag^+ in zeolite Y as a function of the number of silver ions per unit cell—so that for less than 4 silver ions per unit cell ($Ag_4Na_{52}Y$), the silver ions will reside in the small channel system (the sodalite cage/hexagonal prism, *s*) and only occupy the large channel (supercage, *l*) sites above that value; also studied was the fully exchanged form ($Ag_{56}Y$). A further distinction was made between the bound or electrostatically sited ions (*b*) and the unbound or mobile (*m*) ions. The only species capable of approaching the electrode-solution interface (*esi*) will be the mobile ions and their mobility will be diffusive (thus, a physical situation exists that can suitably be studied by electrochemical techniques). Baker *et al.*, define the following equilibria using the above siting notation, (*M* is the electrolyte cation):

$$Ag^+_{b,l} \xleftrightarrow{k_1} Ag^+_{m,l} \xleftrightarrow{D_l} Ag^+_{esi}$$
$$M^+_{m,l} \xleftrightarrow{} M^+_{b,l} \qquad M^+_{esi} \xleftrightarrow{} M^+_{m,l} \tag{15}$$

where D_l is the diffusion coefficient for the mobile silver ions in the large channel.

These coupled equilibria (as denoted by the connecting $\cap$) imply that if the intracrystalline exchange between bound and mobile ions is rate limiting (*i.e.*, k_1 is small), then diffusion control as measured by the electrochemistry will not be observed. If this step is fast, as would be likely, then the concentration of the cations in the large channel between bound and mobile forms will be in equilibrium. These authors describe the small-large channel exchange, thusly:

$$Ag^+_{b,s} \xleftrightarrow{k_2} Ag^+_{b,l} \xleftrightarrow{k_3} Ag^+_{m,l} \xleftrightarrow{D_l} Ag^+_{esi}$$
$$M^+_{b,l} \xleftrightarrow{} M^+_{b,s}$$
$$M^+_{m,l} \xleftrightarrow{} M^+_{b,l}$$
$$M^+_{esi} \xleftrightarrow{} M^+_{m,l} \tag{16}$$

In the coupled equilibria represented by Eqn. (16) any communication of silver ions in the small channel system of zeolite Y beyond the zeolite bulk can only occur through the intercession of the silver ions in the large channel system. Although these authors also wrote the equations for the equilibria relevant for direct communication of mobile silver ions in the small channel with silver ion at the electrode-solution interface, this pathway was ignored based on the kinetics model suggested by Brown *et al.* [137].

The chronoamperometric response operative for an extracrystalline electroactive solute, such as Ag^+_{esi}, should follow the equations derived for (semi-infinite) diffusion control. This is known as Cottrell behavior and faradaic current will vary linearly with $t^{-1/2}$. Baker and his co-workers showed that for silver ions associated only with the small channel system (using $Ag_2Na_{54}Y$-modified electrodes), both cyclic voltammetry and chronoamperometry described kinetics that were not under diffusion control, confirming that the slow step for the silver ions in the small channel system is to exchange into the large channel system (*i.e.*,

k_2—see Eqn. (16)—is «k_3». The faradaic current (and, therefore concentration) of Ag_{esi}^+ as a function of temperature for $Ag_2Na_{54}Y$-modified electrodes yielded Arrhenius plots with an activation energy for small-large channel exchange of 35 ± 1 kJ mol^{-1}. Once the Ag loading reached a level where ions populated the large channel system, as for $Ag_xNa_{56-x}Y$ ($x \geq 5.9$)- and $Ag_{56}Y$-modified electrodes, the kinetics, as measured by either cyclic voltammetry or chronoamperometry, were diffusion-controlled [20]. The electrochemistry of this system must have a negligible contribution from the intracrystalline redox mechanism. In hindsight, it seems odd to postulate intracrystalline redox (in the absence of molecular wires) if any intra-extra mobility can exist for the electroactive guest in the zeolite bulk.

5.   ADVANCED APPLICATIONS OF ZEOLITES IN ELECTROCHEMISTRY

As fascinating as are the electrochemical studies of zeolite-modified electrodes, the ultimate benefit of such studies to electrochemistry will be when practical devices or processes are achieved that fully (or at least partially) exploit the zeolite characteristics described in the introduction. Some of the initial efforts to move beyond the promise implicit in the intersection of electrochemistry and zeolite science are described below.

5.1.   Electroanalysis and Sensors

The efforts to use zeolite-modified electrodes for electroanalysis are among the most abundant examples of applications residing in the literature on ZMEs. In electroanalysis the measurement of charge, current, or equilibrium potential is related to the concentration of the unquantified analyte. I have previously reviewed the use of zeolite-modified electrodes for electroanalysis [3], and will not repeat that discussion here. A common demonstration of electroanalytical suitability of a zeolite-modified electrode is to:

(1)   use the zeolite present in the electrode composite or coating to sequester the analyte (often via cation exchange with mobile extraframework cations);

(2)   follow the pre-concentration step by sweeping the potential of the electrode to react the harvested analyte while measuring the amount of faradaic current that flows; and

(3)   use calibration curves, devised using known concentrations of the analyte (preferably in the same solution environment (matrix) as the unknown, to determine the concentration of the analyte.

The potential of the electroreaction is diagnostic of the identity of analyte and the faradaic current (or faradaic charge) permits quantitative analysis.

Traditional electroanalysis does not always require direct electroreaction of the analyte [138] and many indirect analytical methods exist in the literature. As previously discussed, Senaratne and Baker have used the nearly null response of

Ag^{I}A-modified electrodes in DMF electrolyte as a sensor for trace water—the water preferentially solvates the electrolyte cations which can then exchange for zeolitic Ag$^+$. The reduction of this species to silver metal and subsequent re-oxidation (anodic stripping) to Ag$^+$ provides the detectable electrochemical signal. Analogously to normal protocol for anodic stripping voltammetry, the signal that is used for quantitation is that due to the re-oxidation to metal ion rather than the current due to the reduction of the metal ion to metal.

Indirect approaches to the electroanalysis of non-electroactive cations (Li$^+$, Na$^+$, K$^+$, and Cs$^+$) at 10 ppm levels have also been demonstrated by Baker and Senaratne [19] with Ag^{I}Y-modified electrodes in methanolic TBA$^+$ electrolyte. The low levels of non-size-excluded alkali cations displaced zeolitic Ag$^+$ into the electrolyte which again provided the quantifying signal. This scheme echoes that achieved with the hexacyanometallate films related to Prussian Blue (M^{II}[Fe^{III}(CN)$_6$]$^-$) which can be electrodeposited on electrode surfaces and which exhibit zeolite-like exchange of cations based on size. Indirect electroanalyses for the alkali cations have also been achieved with the cyanometallate films [139, 140].

The concept of sensor invokes the absence of the human analyzer and the presence of a small, often thin device that can monitor the analyte in real-world, *in-situ* conditions. These requirements could be met if some of the direct and indirect electroanalyses already demonstrated can be made with the thin zeolite films [102,103] discussed in Section 3.1. Although not devised as electrochemical sensors, Bein and co-workers have grown thin zeolite composite films on piezoelectric devices (such as surface acoustic wave devices or quartz crystal microbalances where the mass due to sorbed gases is monitored as a change in the frequency of the piezoelectric device) and they have achieved molecular sieving of gas-phase organic analytes [141,142]. Gas-phase analyses will be less convoluted than those in electrolyte due to the absence of the ionic and solvation issues accompanying electrochemistry—as depicted schematically in Figure 5 and illustrated by example in the above text. This is not to diminish the contribution that zeolite-modified electrode surfaces can (and have) brought to electroanalysis—but because of the multiple processes, equilibria, and kinetics involved in the intersection of zeolite science and electrochemistry (perhaps collision rather than intersection would be the better description), the methods developed for analysis require a thorough understanding of what truly gives rise to the electrochemical signal.

5.2. Electrocatalysis

Devynck and co-workers have used zeolite Y to support Mn^{III}porphyrin on the extracrystalline surface via ion exchange. A slurry of the Mn^{III}porphyrin/Y in non-aqueous electrolyte (with 0.1 *M* tetraethylammonium perchlorate) could then catalyze the oxidation of an organic substrate (2,6-di-*tert*butylphenol) when the catalytic (Mn(II)) form of the complex was electrogenerated by reduction of the

Mn(III) centers through collisional contact of the modified zeolite particle with the cathode of the cell. The platinum cathode in this experiment functions as a feeder (or collector) electrode would in a practical electroreactor. The reducing potential applied at this electrode also regenerates the catalytic Mn(II) centers after substrate oxidation to 2,6-di-*tert*butyl-p-benzoquinone and diphenoquinone occurs [113]. No reaction was observed until the potential was applied to the platinum feeder cathode. Collisional contact of dispersed particles to the electrode surface obviously will provide fewer intersection opportunities of electrochemistry with zeolite—these authors ran their dispersed electroreactions for 70 hours in order to exhaustively react the substrate.

Sarradin *et al.* have also used monopolar dispersions of Ag^l-modified zeolites (mordenite and NaY) in basic electrolyte and at a Pt collector electrode to synthesize other transition metal-ion-exchanged forms of the zeolite—but to do so in one hour, rather than by the slower process of equilibrium ion exchange—by driving the silver ions from the zeolite via reduction at the cathode of the electroreactor [111]. The determination of the amount of silver reduced permitted a quantitative measurement of the amount of transition metal ion (*e.g.*, Cu^{2+}, Mn^{2+}, and Zn^{2+}) that exchanged into the zeolite from the electrolyte. At temperatures of 60°C, nearly 100% of the silver ion originally present in the modified zeolite could be removed to the cathode as an electrodeposited layer.

In work in my group [116,117,120-128], we have turned to the study of dispersed zeolites in electroreactors that contain pure liquids and to which no electrolyte salts are added. The advantages of removing the electrolyte salt consist of removing a high concentration of electroreactant—and potential chemical reactant—as electrolyte salts are rarely as ideally inert as desired. The electroreactivity of the ions added to provide the high ionic strength typical of electrochemistry limits the magnitude of potential that can be applied to an electrochemical cell—and so limits the amount of thermodynamic (and kinetic) driving force that can be applied to do chemistry.

A serious disadvantage of removing the electrolyte salt is the loss of information and control cherished in electrochemistry: the potential of the anode and cathode in the system and the ability to control that potential *vs.* a thermodynamic reference electrode. A further complication arises because while the amount of voltage applied to the system may be known, the voltage gradients in the cell and throughout the low ionic strength medium will be variable in ways not necessarily foreseen. Because of the high ionic strengths typical of most electrochemical reactions, the potential dropped across the electrolyte is essentially zero, but with the infinitely dilute/low ionic strength conditions operative in our electroreactors, voltage gradients can exist in the bulk of the medium rather than just at the electrode-solution interface. We feel this loss of potential control is worth forgoing in order to study electroreactions under atypical conditions and environments—particularly in the presence of zeolites, so as to explore beyond the commonalities so apparent in the intersection of electrochemistry and zeolite science.

Our first work [116,117] borrowed from Gallezot's procedures to prepare Pt nanocrystallites in zeolite Y [143]. These nanocrystallites were then shown to behave as supported ultramicroelectrodes for the electrolysis of water in the absence of deliberately added electrolyte salt [116,117] and for the oxidation of ferrocenyl derivatives in pure acetonitrile [117]. Under the low electric field conditions of these experiments (<300 V cm⁻¹), the intracrystalline-sited Pt nanocrystallites were shown to be electrochemically silent by the use of a size-excluded ferrocenyl derivative as the electroactive probe [117]. The size of the electrodic extracrystalline-sited Pt structures were then varied as a function of the weight loading of Pt and allowed us to demonstrate an electrode-size dependence for the effectiveness of water oxidation and reduction, where the smallest particles (≤2.5 nm) were found to be the most effective [117].

A key recognition of these ultramicroelectrode studies at Pt^0-NaY for water electrolysis was the ionic role of the zeolite, so that sub-mM levels of Na^+ would be observed in the 18 MΩ-cm water after several polarization curves (i.e., apply V, measure I) [117]. As protons are generated at the feeder anode during water electrolysis (albeit balanced by OH^- production at the feeder cathode), the zeolite particles can undergo proton-Na^+ exchange upon contacting the anode via Eqn. (17):

$$Na^+_{(z)}Z + H^+_{(sol)} \rightarrow H^+_{(z)}Z + Na^+_{(sol)} \qquad (17)$$

The concentration of Na^+ in the dispersion stabilizes in a matter of minutes (as seen by attaining steady-state current at a constant applied voltage and by ex-situ analysis of Na^+ in the liquid phase of the dispersion) and does not further deplete Na^+ from the zeolite bulk over the course of an hour of applied voltage to the aqueous dispersion of NaY.

We have continued to study this time-dependent conductivity phenomenon of the electrified zeolite dispersions [124,125] and have recently reported that the evolution in time of the current (including a maximum which relaxes to the steady-state current, as seen in Figure 8) is mirrored by the bulk conductivity of the dispersion as a function of time [125]. This latter result was obtained by ac impedance at frequencies where the contribution to the equivalent circuit of the electrochemical cell by the interfacial (i.e., electrode-dispersion interface) impedances was minimal [125]. The importance of this observation lies in ruling out any faradaic contribution at the electrode surfaces to the current maximum observed before the system achieves steady-state. The maximum reflects a higher, but short-lived, conductance in the bulk dispersion—as we remain unable to tie this maximum to a maximum in the concentration of plausible solutional ions

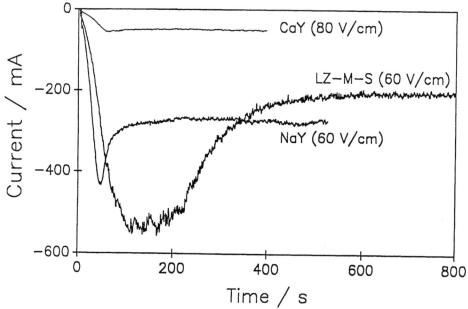

Figure 8: Current-time profiles for electrifying the listed aqueous zeolite dispersions (2.5 g/L suspension density) at the listed applied voltage gradients.

(*e.g.*, Na^+, aluminates, or silicates), we are left with the surface conductance of the dispersed zeolite as the most likely cause of this phenomenon.

The time to reach a current/conductivity maximum (or to reach current/conductivity steady state) at a given applied voltage gradient (on the order of tens of V cm^{-1}) is also a function of the mobile, charge-compensating cation in the zeolite, as well as the structure of the zeolite [124]. Based on studying the alkali series of zeolite Y: in general, for a *common zeolite composition and structure*, the more mobile the hydrated cation (in infinitely dilute homogeneous solution), the faster will steady-state conditions be reached—for example, CsY dispersions achieved steady state (and a peak maximum) far faster than did LiY dispersions. For CaY, roughly twice the voltage gradient had to be applied to reach a steady-state current relative to dispersions of NaY (and hydrated Ca^{2+} is more mobile than is Na^+ in infinitely dilute homogeneous solution), see Figure 8. With the same extraframework cation, but different zeolite structures, common current-time profiles are not obtained—synthetic sodium mordenite dispersions were roughly twice as slow to achieve steady-state as NaY dispersions as shown in Figure 8 [124]. All of these observables have led us to conclude that in addition to the importance of the ionic mobility of the exchangeable cations, the local electrostatic field at the bound zeolite site (which will be a function of the valency and size of the bound cation as well as the zeolite composition (Al/Si ratio) and structure) controls the ionicity of zeolite dispersions when electrified in pure water [124].

When aqueous zeolite dispersions are electrified in the presence of organic substrates (a process we term: electrified microheterogeneous catalysis, EMC), we have observed two general classes of reactivity:

(1)   if a molecular catalyst is supported on the zeolite, selective synthesis occurs; and

(2)   if the native zeolite is used, oxidation occurs, often resulting in molecular degradation to water-soluble fragments.

The selective synthesis we chose to study was the Wacker oxidation of propene in pure water with dispersed Pd(II)- and Cu(II)-modified NaY zeolite (Pd^{II}Cu^{II}NaY). The traditional selective partial oxidation product for this bifunctional catalyst is acetone—whether reacted in homogeneous solution or when supported on NaY for gas-phase reaction at elevated temperatures. Based on the preparation of the catalyst (primarily calcining the Pd(II)-exchanged NaY before exchanging in Cu^{2+}), we were able to prepare acetone as the selective partial oxidation product (non-calcined catalyst) or propylene oxide (calcined catalyst) [120-122]. Of great note: propylene oxide is not an oxidation product normally observed with this molecular catalyst.

The electrifying force is critical to achieve reaction as in the absence of an applied *dc* voltage or current, no oxidation product forms. We run the reaction at 0°C to avoid any thermal driving force for the reaction and the need for an applied electrifying force shows that the electrifying force drives the reaction rather than temperature. Selectivity for propylene oxide over other C_3 oxygenates is > 80%. Any applied voltage >10 V is sufficient to drive the reaction (higher voltages neither improve the reaction yield nor lower the selectivity).

If the reaction is run in dilute NaOH (1 mM), to act as a control for the Na$^+$ ions which move into the aqueous phase upon electrification of the zeolite dispersion, all oxidation selectivity is lost [120-122]. Attempts to achieve propylene oxide production by additions of a chemical oxidant (H_2O_2) to the dispersed catalyst, in the absence of an applied voltage or current, produced no oxidation product at all. The key features to create propylene oxide via EMC are the application of voltage (or current) *and* the ability to run the reaction in essentially salt-free conditions. The prime drawback with our current reactor design is the negligible concentration of propene dissolved in water at 0°C—as we use a one-pass reactor, most of the propene gas passes through unreacted [120-122].

Our control experiments for the partial selective oxidation of propene via EMC of dispersed Pd^{II}Cu^{II}NaY implied that the use of unmodified NaY led to oxygenation and degradation of the organic substrate. We are now turning our attention to the degradative capabilities of EMC as the means to devise an ambient temperature, liquid-phase process to destroy or detoxify polychlorinated organic molecules. Using chlorobenzene and chloronaphthalene as model compounds, as well as a commercial sample of polychlorinated terphenyls and CFC-113, we have observed degradation of these molecules to water-soluble fragments, including chloride (and, if applicable, fluoride) ion. The best success has been achieved with the alkali-ion compensated zeolite Y dispersions. Other non-faujasite zeolites and other oxides (such as alumina, silica, and kaolinite) lead to oxygenation of the aromatic ring rather than degradation of the aromatic ring [123,126-128]. Our preliminary data implicate a reactive oxygen-based species that is created during the concomitant electrolysis of water at the platinum feeder electrodes during the EMC process [128]—but since such a species would be generated with the non-faujasite, non-zeolitic oxides, additional chemistry must be occurring unique to the alkali-ion-compensated Y zeolites.

5.3. Power Sources

The global desire for clean sources of energy has revitalized the research into electrochemical-based power sources: batteries and fuel cells. Issues of ion transport and the morphology of the active electrode materials are fundamental and potentially limiting in batteries. Issues of ion transport and water management (including transport of water to prevent water starvation or flooding of the electrodes) are key problems in fuel cells. The ionics and size/shape selectivity of zeolites are natural attributes that should confer design advantages in batteries, while the ionics, size/shape selectivity, and profound hydrophilicity of the aluminosilicate zeolites should offer design advantages in fuel cells. The ability of zeolites to sequester electroreactants while providing a high capacity matrix are also of interest in fuel cells [144-146], as has been previously discussed [2], however new efforts with zeolites in fuel cells do not seem to have appeared.

An early and prominent example of zeolites as applied in electrochemical systems—and one that seriously predates the modern (mid-1980's onward) use of zeolites in electrochemistry—was the patent by Freeman/Union Carbide for a solid-state battery, which, in the customary shorthand for an electrochemical cell, consisted of:

$$Zn \mid \mid Zn^{2+} \mid NaX \mid M^{II}X \mid \mid Au; \quad M = Cu^{2+}; \; Ag^{+}; \; or \; Hg^{2+}$$

so that zeolite X acted as both a solid-state electrolyte (and physical separator of the cathode (Au) from the anode (Zn) and as the host for the catholyte (the transition metal cations).

A more recent use of zeolites in solid-state batteries was by Thackeray and Coetzer who also used zeolite (NaA) as a catholyte host, but where the catholyte was the neutral species iodine and the solid-state electrolyte was a silver-ion conductor:

580

$$Ag \,|\, Ag + Ag_{44}I_{53}(C_{11}H_{30}N_3)_3 \,|\, Ag_{44}I_{53}(C_{11}H_{30}N_3)_3 \,|\, (I_2)NaA + C + Ag_{44}I_{53}(C_{11}H_{30}N_3)_3 \,|\, C$$

The ability of NaA to sorb ~46 wt% of iodine was an additional attraction to its use in the solid-state cell. One drawback to using a neutral catholyte was found in the temperature dependence of the discharge characteristics of the cell: as the temperature of the battery was elevated, iodine desorbed. The authors also specifically recommended the desirability of using a thin film of zeolite to minimize internal resistance.

Also of interest in solid-state batteries are the ion-conducting polymers. Recent work has described the fabrication of a composite containing polyethylene oxide (PEO, a Li^+ conductor), a lithium salt, and zeolites [85,86]. The zeolite-containing films have higher conductivity than do the Li^+-doped PEO films and this improvement was attributed to the presence of the zeolite interrupting crystalline domains of the PEO; these domains are known to limit the ionic conductivity of such systems.

While zeolites have been used in both the practical electrochemical devices of batteries and fuel cells, significant work can still be envisioned. The possibilities afforded by the fabrication of thin zeolite films on electrode materials of interest in batteries or fuel cells provide a promising place to start.

6. CONCLUSION

The interest of researchers in combining the domains of zeolite science and electrochemistry has further blossomed in the past five years as the range of electroactive guests supported in and on zeolites has increased. The frustration ensuing from the intersection has been that the complexity of the issues of ionics, solvation, and electron transfer that underpin any electrochemical process are multiplied when the process occurs in and around the microstructured environment provided by zeolites. Such complexity can be refracted to become opportunity once the interactions and interplay between the zeolitic and electrochemical domains is more fully understood. A prime example of the information that need to be gathered is the work by the Baker group which has addressed the issues of extracrystalline *vs.* intracrystalline electron transfer—and the coupled ionics— when the electroactive guest has not been created to be an encapsulated guest. Their understanding of the Ag^I-modified zeolite system in electrochemical circumstances has led to the use of Ag^I-zeolite-modified electrodes as effective sensors for trace levels of water in organic solvents and for non-electroactive cations.

Unfortunately, as discussed in Section 4.3, the implications for *direct* intracrystalline electron transfer remain grim. Electrocatalysis with near-extracrystalline encapsulated species may still offer advantages to minimize catalyst deactivation and to retain the advantage of ease of separation of catalyst from product that is inherent to heterogeneous catalysts. The use of dispersed zeolites to affect liquid-phase catalysis as driven by electrochemical, rather than thermal, driving forces also promises to contribute to the contemporary interest in the zeolite community to use zeolites for the synthesis of fine chemicals in the liquid phase. As our group has shown—by electrifying liquid-phase zeolite dispersions in salt-free environments, unanticipated reaction chemistry can be pursued.

ACKNOWLEDGMENTS

Our work with zeolites in the electrochemical milieu has been funded by the Office of Naval Research. I would like to thank my new (post-1989) research associates in the zeolite-electrochemistry intersection for their efforts and intellectual stimulation: Joe Stemple, Dave Blauch, Dave Curran, Liz Hayes, and Carol Bessel. Significant assistance from Liz Hayes on the figures in this chapter is gratefully acknowledged.

REFERENCES

1. G.A. Ozin, A. Kuperman, and A. Stein, *Angew. Chem. Int. Ed. Engl.*, 28 (1989) 359-376.
2. D.R. Rolison, *Chem Rev.*, 90 (1990) 867-878.
3. D.R. Rolison, R.J. Nowak, T.A. Welsh, and C.G. Murray, *Talanta* 38 (1991) 27-35.
4. A.J. Bard and T.E. Mallouk, in *Molecular Design of Electrode Surfaces*; R.W. Murray (ed.), Wiley, New York, 1992, pp. 271-312.
5. M.D. Baker and C. Senaratne, in *Electrochemistry of Novel Materials*, J. Lipkowski and P.N. Ross (eds.), VCH, Amsterdam, 1994, in the press.
6. G. Cauquis in *Organic Electrochemistry: An Introduction and a Guide*, M.M. Baizer and H. Lund (eds.), Marcel Dekker, New York, 1983, pp. 38-48.
7. R.W. Murray, in *Electroanalytical Chemistry*; Vol. 13, A.J. Bard (ed.), Marcel Dekker, New York, 1984, pp. 191-368.
8. J.S. Miller (ed.), *Chemically Modified Surfaces in Catalysis and Electrocatalysis*, ACS Symposium Series No. 192, American Chemical Society, Washington, DC, 1988.
9. R.W. Murray, in *Molecular Design of Electrode Surfaces*; R.W. Murray (ed.), Wiley, New York, 1992, pp. 1-48.
10. W. Hölderich, M. Hesse, and F. Näumann, *Angew. Chem. Int. Ed. Engl.*, 27 (1988) 226-246.
11. R.A. Sheldon, in *Heterogeneous Catalysis and Fine Chemicals II*, M. Guisnet et al. (eds.), Elsevier, Amsterdam, 1991, pp. 33-54.
12. W.F. Hölderich, *Catal. Sci. Tech.*, 1 (1991) 31-45.
13. W.F. Hölderich and H. van Bekkum, in *Introduction to Zeolite Science and Practice*, H. van Bekkum, E.M. Flanigen, and J.C. Jansen (eds.), Elsevier, Amsterdam, 1991, pp.631-726.

14. J.O'M. Bockris and A.K.N. Reddy, *Modern Electrochemistry*, Vol. 1, Plenum Press, New York, 1976, pp. 2-12.

15. D.W. Breck, *Zeolite Molecular Sieves*, Wiley-Interscience, New York, 1974, pp. 397-410.

16. D.C. Freeman, Jr., and D.N. Stamires, *J. Chem. Phys.*, 35 (1961) 799-806.

17. D.N. Stamires, *J. Chem. Phys.*, 36 (1962) 3174-3181.

18. M.D. Baker and J. Zhang, *J. Phys. Chem.*, 94 (1990) 8703-8708.

19. C. Senaratne and M.D. Baker, *Anal. Chem.*, 64 (1992) 697-700.

20. M.D. Baker, C. Senaratne, and J. Zhang, *J. Phys. Chem.*, 98 (1994), 1668-1673.

21. R.P. Townsend, *Pure & Appl. Chem.*, 58 (1986) 1359-1366.

22. D.W. Breck, *Zeolite Molecular Sieves*, Wiley-Interscience, New York, 1974, pp. 578-585.

23. P.C. Huang, A. Mizany, and J.L. Pauley, *J. Phys. Chem.*, 68 (1964) 2575-2578.

24. A. Dyer and R.B. Gettins, *J. Inorg. Nucl. Chem.*, 32 (1970) 2401-2410.

25. D.T. Sawyer and J.L. Roberts, Jr., *Experimental Electrochemistry for Chemists*, J. Wiley, New York, 1974, pp. 203-210.

26. C. Senaratne and M.D. Baker, *J. Electroanal. Chem.*, 332 (1992) 357-364.

27. A.J. Bard and L.R. Faulkner, *Electrochemical Methods: Fundamentals and Applications*, Wiley, New York, 1980, pp. 125-127.

28. Yu.V. Pleskov and V.Yu. Filinovskii, *The Rotating Disc Electrode*, Consultants Bureau (Plenum), New York, 1976.

29. A.T. Hubbard and F.C. Anson, in *Electroanalytical Chemistry*, Vol. 4, A.J. Bard, (ed.), Marcel Dekker, New York, 1970, pp. 121-214.

30. R.W. Murray, *Ann. Rev. Mater. Sci.*, 14 (1984) 145-169.

31. R.W. Murray, A.G. Ewing, and R.A. Durst, *Anal. Chem.*, 59 (1987) 379A-390A.

32. P. Daum, J.R. Lenhard, D. Rolison, and R.W. Murray, *J. Am. Chem. Soc.*, 102 (1980) 4649-4653.

33. D.M. Oglesby, S.H. Omang, and C.N. Reilley, *Anal. Chem.*, 37 (1965) 1312-1316.

34. N.A. Surridge, J.C. Jernigan, E.F. Dalton, R.P. Buck, M. Watanabe, H. Zhang, M. Pinkerton, T.T. Wooster, M.L. Longmire, J.S. Facci, and R.W. Murray, *Faraday Discuss. Chem. Soc.*, 88 (1989) 1-17.

35. R.M. Wightman, *Anal. Chem.*, 53 (1981) 1125A-1134A.

36. *Ultramicroelectrodes*, M. Fleischmann, S. Pons, D.R. Rolison, and P.P. Schmidt, (eds.), Datatech Systems, Morganton, NC, 1987.

37. R.M. Wightman and D.O. Wipf, *Electroanalytical Chemistry*, Vol. 15, A.J. Bard (ed.), Marcel Dekker, New York, 1989, pp. 267-353.

38. *Microelectrodes: Theory and Applications*, M.I. Montenegro, M.A. Queirós, J.L. Daschbach (eds.), Kluwer Academic, Amsterdam, 1991.

39. J. Heinze, *Angew. Chem. Int. Ed. Engl.*, 30 (1993) 1268-1288.

40. Y. Shimizu and K.-I. Morita, *Anal. Chem.* 62 (1990) 1498-1501.

41. G.N. Lewis, M. Randall, K.S. Pitzer, and L. Brewer, *Thermodynamics*, 2nd Ed., McGraw-Hill, New York, 1961, pp. 3-5.

42. K.B. Oldham and J.C. Myland, *Fundamentals of Electrochemical Science*, Academic Press, San Diego, 1994, pp. 120-126.

43. C.E. Marshall, *J. Phys. Chem.*, 43 (1939) 1155-1164.

44. R.P. Buck, "Potentiometry: pH Measurements and Ion-Selective Electrodes" in *Techniques in Chemistry*, Vol. 1, Pt. 2A, A. Weissberger and B.W. Rossiter (eds.), Wiley Interscience, New York, 1971, pp. 61-162.

45. F. Shunong, F. Yilu, and L. Peiyan, *Talanta*, 41 (1994) 155-157.

46. K.B. Oldham and J.C. Myland, *Fundamentals of Electrochemical Science*, Academic Press, San Diego, 1994, pp. 167-175.

47. W.R. Heineman and P.T. Kissinger in *Laboratory Techniques in Electroanalytical Chemistry*, P.T. Kissinger and W.R. Heineman (eds.), Marcel Dekker, New York, 1984, pp. 51-161.

48. A.J. Bard and L.R. Faulkner, *Electrochemical Methods: Fundamentals and Applications*, Wiley, New York, 1980, Ch. 5, pp. 136-212.

49. J. Osteryoung, *Acc. Chem. Res.*, 26 (1993) 77-83.

50. J.G. Osteryoung and R.A. Osteryoung, *Anal. Chem.*, 57 (1985) 101A-110A.

51. A. Walcarius, L. Lamberts, and E.G. Derouane, *Electrochim. Acta*, 38 (1993) 2267-2276.

52. P. Hernández, E. Alda, and L. Hernández, *Fresenius Z. Anal. Chem.*, 327 (1987) 676-678.

53. B.R. Shaw, K.E. Creasy, C.J. Lanczycki, J.A. Sargeant, and M. Tirhado, *J. Electrochem. Soc.*, 135 (1988) 869-876.

54. K.E. Creasy and B.R. Shaw, *Electrochim. Acta*, 33 (1988) 551-556.

55. J. Wang and T. Martinez, *Anal. Chim. Acta*, 207 (1988) 95-102.

56. N. El Murr, M. Kerkeni, A. Sellami, and Y. Ben Taârit, *J. Electroanal. Chem.* 246 (1988) 461-465.

57. A. Walcarius, L. Lamberts, and E.G. Derouane, *Electrochim. Acta*, 38 (1993) 2257-2266.

58. S. de Castro-Martins, S. Khouzami, A. Tuel, Y. Ben Taârit, N. El Murr, and A. Sellami, *J. Electroanal. Chem.*, 350 (1993) 15-28.

59. M. Xu, W. Horsthemke, and M. Schell, *Electrochim. Acta*, 38 (1993) 919-925.

60. B.R. Shaw and K.E. Creasy, *J. Electroanal. Chem.*, 243 (1988) 209-217.

61. Furukawa Battery Co., Ltd., Jpn-Patent 83 12,263 (1983) [CA(99:25460g)].

62. J.-P. Pereira-Ramos, R. Messina, and J. Périchon, *J. Electroanal. Chem.*, 146 (1983) 157-169.

63. R.C. Galloway, Ger. Offen. DE 3,311,461 (1983).

64. B. deVismes, F. Bedioui, J. Devynck, and C. Bied-Charreton, *J. Electroanal. Chem.*, 187 (1985) 197-202.

65. F. Bedioui, E. De Boysson, J. Devynck, and K.J. Balkus, Jr., *J. Electroanal. Chem.*, 315 (1991) 313-318.

66. F. Bedioui, E. De Boysson, J. Devynck, and K.J. Balkus, Jr., *J. Chem Soc., Faraday Trans.*, 87 (1991) 3831-3834.

67. L. Gaillon, N. Sajot, F. Bedioui, J. Devynck, and K.J. Balkus, Jr., *J. Electroanal. Chem.*, 345 (1993) 157-167.

68. K.J. Balkus, Jr., A.G. Gabrielov, S.L. Bell, F. Bedioui, L. Roué, and J. Devynck, *Inorg. Chem.*, 33 (1994) 67-72.

69. K. Mesfar, B. Carre, F. Bedioui, and J. Devynck, *J. Mater. Chem.*, 3 (1993) 873-876.

70. C.G. Murray, R.J. Nowak, and D.R. Rolison, *J. Electroanal. Chem.*, 164 (1984) 205-210.

584

71.   D.R. Rolison, manuscript in preparation.

72.   H.A. Gemborys and B.R. Shaw, *J. Electroanal. Chem.*, 208 (1986) 95-107.

73.   Z. Li and T.E. Mallouk, *J. Phys. Chem.*, 91 (1987) 643-648.

74.   J.S. Krueger, C. Lai, Z. Li, J.E. Mayer, and T.E. Mallouk, in *Inclusion Phenomena and Molecular Recognition*, J.L. Atwood (ed.), Plenum Press, New York, 1990, pp. 365-378.

75.   Z. Li, C.M. Wang, L. Persaud, and T.E. Mallouk, *J. Phys. Chem.*, 92 (1988) 2592-2597.

76.   C. Iwakura, S. Miyazaki, and H. Yoneyama, *J. Electroanal. Chem.*, 246 (1988) 63.

77.   M.D. Baker, C. Senaratne, and J. Zhang, *J. Chem. Soc., Faraday Trans.*, 88 (1992) 3187-3192.

78.   J. Cassidy, E. O'Donoghue, and W. Breen, *Analyst*, 1145 (1989) 1509-1510.

79.   J. Cassidy, W. Breen, E. O'Donoghue, and M.E.G. Lyons, *Electrochim. Acta*, 36 (1991) 383-384.

80.   K.L.N. Phani and S. Pitchumani, *Electrochim. Acta*, 37 (1992) 2411-2414.

81.   S. Bharathi, K.L.N. Phani, J. Joseph, S. Pitchumani, D. Jeyakumar, G. Prabhakara Rao, and S.K. Rangarajan, *J. Electroanal. Chem.*, 334 (1992) 145-132.

82.   S. Bharathi, K.L.N. Phani, J. Joseph, S. Pitchumani, D. Jeyakumar, and G. Prabhakara Rao, *J. Electroanal. Chem.*, 360 (1993) 347-349.

83.   K.L.N. Phani, S. Pitchumani, and S. Ravichandran, *Langmuir*, 9 (1993) 2455-2459.

84.   J. Li and G. Calzaferri, *J. Chem. Soc., Chem. Commun.*, (1993) 1430-1432.

85.   G. Shi, G. Xue, W. Hou, J. Dong, and G. Wang, *J. Electroanal. Chem.*, 34 (1993) 363-366.

86.   N. Munichandraiah, L.G. Scanlon, R.A. Marsh, B. Kumar, and A.K. Sircar, *Batteries and Fuel Cells for Stationary and Electric Vehicle Applications*, 93-8, A. Langrebe and Z. Takehara (eds.), The Electrochemical Society, Pennington, NJ, 1993, pp. 97-111.

87.   Z. Li, C. Lai, and T.E. Mallouk, *Inorg. Chem.*, 28 (1989) 178-182.

88.   G. Calzaferri, K. Hädener, and J. Li, *J. Chem. Soc., Chem. Commun.*, (1991) 653-654.

89.   *(a)* G. Johansson, L. Risinger, and L. Faelth, L. *Hung. Sci. Inst.*, 49 (1980) 47-51; *(b) Anal. Chim. Acta*, 119 (1977) 25-32.

90.   M. Demertzis and N. P. Evmiridis, N.P. *J. Chem. Soc., Faraday Trans. I*, 82 (1986) 3647-3655; N. P. Evmiridis, M.A. Demertzis and A.G. Vlessidis, *Fresenius J. Anal. Chem.*, 340 (1991) 145-152.

91.   D.C. Freeman, Jr., *(a)* US-Patent 3,186,875 (1965); *(b)* Brit-Patent 999,948 (1965).

92.   M. Šušic and N. Petranovic, *Electrochim. Acta*, 23 (1978) 1271-1274.

93.   M.V. Šušic and N. Petranovic, *Bull. - Acad. Serbe Sci. Arts, Cl. Sci. Nat. Math., Sci. Nat.*, 68 (1979) 21-30.

94.   M.V. Šušic, *Electrochim. Acta*, 24 (1979) 535-540.

95.   M.V. Šušic, *Sci. Sintering.*, 11 (1979) 141-153.

96.   N. Petranovic and M. Šušic, *Zeolites*, 3 (1983) 271-273.

97.   M. Šušic, *Bull. - Acad. Serbe Sci. Arts, Cl. Sci. Nat. Math., Sci. Nat.*, 90 (1985) 69-78.

98.   O. Vigil, J. Fundora, H. Villavicencio, M. Hernandez-Velez, and R. Roque-Malherbe, *J. Mater. Sci. Lett.*, 11 (1992) 1725-1727.

99.   O. Enea, *Electrochim. Acta*, 34 (1989) 1647-1651.

100. R. Stockmeyer, *Spez. Ber. Kernforschunganlage Juelich*, 165 (1982) 23 pp.

101. K.E. Creasy and B.R. Shaw, *J. Electrochem. Soc.*, 137 (1990) 2353-2354.

102. S.P. Davis, E.V.R. Borgstedt, and S.L. Suib, *Chem. Mater.*, 2 (1990) 712-719.

103. J.C. Jansen, W. Nugroho, and H. van Bekkum, in *Proceedings from the Ninth International Zeolite Conference, Montreal, 1992*, Vol. I, R. von Ballmoos, J.B. Higgins, M.M.J. Treacy (eds.), Butterworth-Heinemann, Stoneham, MA, 1993, pp. 247-254.

104. M. Fleischmann, J.W. Oldfield, and D.F. Porter, *J. Electroanal. Chem.*, 29 (1971) 241-253.

105. F. Goodridge, C.J.H. King, and A.B. Wright, *Electrochim. Acta*, 22 (1977) 1087-1091.

106. B.J. Sabacky and J.W. Evans, *J. Electrochem. Soc.*, 126 (1979) 1180-1187.

107. M. Fleischmann, J. Ghoroghchian and S. Pons, *J. Phys. Chem.*, 89 (1985) 5530-5536.

108. M. Fleischmann, J. Ghoroghchian, D. Rolison and S. Pons, *J. Phys. Chem.*, 90 (1986) 6392-6400.

109. J. Ghoroghchian, S. Pons, and M. Fleischmann, *J. Electroanal. Chem.*, 317 (1991) 101-108.

110. L.R. Faulkner, S.L. Suib, C.L. Renschler, J.M. Green, and P.R. Bross, in *Chemistry in Energy Production*, R.G. Wymer and O.L. Keller (eds.), Wiley, New York, 1982, pp. 99-113.

111. J. Sarradin, J.-M. Louvet, R. Messina, and J. Périchon, *Zeolites*, 4 (1984) 157-162.

112. T.L. Pettit and M.A. Fox, *J. Phys. Chem.*, 90 (1986) 1353-1354.

113. B. deVismes, F. Bedioui, J. Devynck, C. Bied-Charreton, and M. Perrée-Fauvet, *Nouv. J. Chim.*, 10 (1986) 81-82.

114. L. Persaud, A.J. Bard, A. Campion, M.A. Fox, T.E. Mallouk, S.E. Webber, and J.M. White, *J. Am. Chem. Soc.*, 109 (1987) 7309-7314.

115. J.S. Krueger, J.E. Mayer, and T.E. Mallouk, *J. Am. Chem. Soc.*, 110 (1988) 8232-8234.

116. D.R. Rolison, R.J. Nowak, S. Pons, J. Ghoroghchian, M. Fleischmann, in *Molecular Electronic Devices III*; F.L. Carter, R.E. Siatkowski, H. Wohltjen, (eds), Elsevier, Amsterdam, 1988, 401-410.

117. D.R. Rolison, E.A. Hayes, W.E. Rudzinski, *J. Phys. Chem.* 93 (1989) 5524-5531.

118. R. Beer, G. Calzaferri, and W. Spahni, *Chimia*, 42(1988) 134-137.

119. Y.I. Kim and T.E. Mallouk, *J. Phys. Chem.*, 96 (1992) 2879-2885.

120. D.R. Rolison and J.Z. Stemple, *J. Chem. Soc., Chem. Commun.*, (1993) 25-27.

121. J.Z. Stemple and D.R. Rolison, manuscript in preparation.

122. D.R. Rolison and J.Z. Stemple, US-Patent 5,296,106 (1994).

123. D.R. Rolison and J.Z. Stemple, US-Patent 5,288,371 (1994).

124. D.R. Rolison, J.Z. Stemple, and D.J. Curran, in *Proceedings from the Ninth International Zeolite Conference, Montreal, 1992*, Vol. II, R. von Ballmoos, J.B. Higgins, M.M.J. Treacy (eds.), Butterworth-Heinemann, Stoneham, MA, 1993, pp. 699-705.

125. D.N. Blauch and D.R. Rolison, *J. Electroanal. Chem.*, in the press.

126. D.R. Rolison and J.Z. Stemple, US-Patent 5,282,936 (1994).

127. J.Z. Stemple and D.R. Rolison, *Decompostition of Halogenated Organic Materials by Electrified Microheterogeneous Catalysis*, NRL Memorandum Report, NRL/MR/6170-92-7165, 1992.

128. E.A. Hayes, J.Z. Stemple, and D.R. Rolison in *Water Purification by Photocatalytic, Photoelectrochemical, and Electrochemical Processes*, T.L. Rose, The Electrochemical Society, Pennington, N.J., in the press.

129. E.S. Brigham, P.T. Snowden, Y.I. Kim, and T.E. Mallouk, *J. Phys. Chem.*, 97 (1993) 8650-8655.

130. J.S. Krueger and T.E. Mallouk, in *Kinetics and Catalysis in Microheterogeneous Systems*, M. Grätzel and K. Kalyanasundaram (eds.), *Surfactant Sci. Ser.*, 38, Marcel Dekker, New York, 1991, pp. 461-490.

131. R.M. Barrer, *Proc. Chem. Soc.*, (1958) 99-112.

132. R.M. Barrer, *J. Incl. Phenom.*, 1 (1983) 105-123.

133. P. Enzel and T. Bein, *J. Chem. Soc., Chem. Commun.*, (1989) 1326-1327.

134. T. Bein and P. Enzel, *(a) Angew. Chem., Int. Ed. Engl.*, 28 (1989) 1692-1694; *(b) Synth. Met.als*, 29 (1989) E163-E168; *(c) J. Phys. Chem.*, 93 (1989)6270-6272; *(d) Chem. Mater.*, 4 (1992) 819-824.

135. P. Enzel, J.J. Zoller, and T. Bein, *J. Chem. Soc., Chem. Commun.*, (1992) 633-635; *(e) Synth. Metals*, 55 (1993) 1238-1245.

136. L. Zuppiroli, F. Beuneu, J. Mory, P. Enzel, and T. Bein, *Synth. Metals*, 57 (1993) 5081-5087.

137. L.M. Brown, H.S. Sherry, and F.J. Krambeck, *J. Phys. Chem.*, 75 (1971) 3846-3855.

138. J.J. Lingane, *Electroanalytical Chemistry*, 2nd Ed., Interscience, New York, 1958.

139. L.J. Amos, A. Duggal, E.J. Mirsky, P. Ragones, A.B. Bocarsly, and A.P. Fitzgerald-Bocarsly, *Anal. Chem.*, 60 (1988) 245-249.

140. K.N. Thompson and R.P. Baldwin, *Anal. Chem.*, 61 (1989) 2594-2598.

141. T. Bein, K. Brown, G.C. Frye, and C.J. Brinker, *(a) J. Am. Chem. Soc.*, 111 (1989) 7640-7641; *(b)* U.S.-Patent, 5,151,110.

142. Y. Yan and T. Bein, *(a) J. Phys. Chem.*, 96 (1992) 9387-9393; *(b) Chem. Mater.*, 4 (1992) 975-977; *(c)* Y. Yan and T. Bein, *Microporous Mater.*, 1(1993) 401-411.

143. P. Gallezot, A. Alarcon-Diaz, J.-A. Dalmon, A.J. Renouprez, and B. Imelik, *J. Catal.*, 39 (1975) 334.

144. A.M. Moos, Ger-Patent 1,302,003, (1969) [CA(73:126473g)].

145. Toray Industries, Inc., Jpn-Patent 87 241,265 (1987) [CA(108:24622d)].

146. M. Cruceanu, E. Popovici, and A. Vasile, Rom-Patent 78,511 (1982).

J.C. Jansen, M. Stöcker, H.G. Karge and J. Weitkamp (Eds.)
Advanced Zeolite Science and Applications
Studies in Surface Science and Catalysis, Vol. 85
© 1994 Elsevier Science B.V. All rights reserved.

Past, Present and Future Role of Microporous Catalysts in the Petroleum Industry

S.T. Sie

Delft Technical University, Faculty of Chemical Technology and Materials Science, P.O. Box 5045, 2600 GA Delft, The Netherlands

ABSTRACT

Zeolitic catalysts are currently widely used in a variety of processes of the petroleum industry and represent a factor of very great economic importance. The present paper reviews the current trends in petroleum refining and examines the role of zeolitic catalysts against this background.

The historical development and future perspectives of three major processes of the oil industry that apply zeolitic catalysts, viz., catalytic cracking, hydrocracking and paraffin isomerization are discussed in more detail within this context.

Microporous materials offer many possibilities as catalysts to meet future demands in the catalytic process technology in petroleum refining and synthetic fuels production of the future.

1. INTRODUCTION

Microporous materials, notably zeolites, are presently widely applied as adsorbents and catalysts in the hydrocarbon processing (oil refining) industry. Because of their high and tailorable acidity and their specific pore structures with channels and cavities of molecular dimensions which allow shape-selective conversions of organic molecules, zeolites occupy a prominent place as catalysts in the hydrocarbon processing industry. In more traditional oil refining, zeolite catalysis is involved in the processing of almost every fraction of the crude oil barrel, as is shown in Fig. 1. In the emerging industry of synfuels manufacture, zeolite catalysts have already found a place and will become more important as this industry assumes a greater role in the future. Some major commercial processes in the oil refining and synfuels industries are listed in Table 1.

With respect to the future of zeolites and other microporous materials in the above industries, the question is not whether their important role will be sustained in the future since this can without doubt be answered in an affirmative sense, but how the trends in oil refining and synfuels manufacture are likely to affect the choices and the developments of zeolite-based catalytic processes. The present paper attempts to examine the main trends in hydrocarbon processing and will discuss the possibilities, limitations and desirable developments in zeolite catalysis in this light. Rather than reviewing the multi-

Table 1
Zeolite-based catalytic processes commercially applied in the hydrocarbon processing
industries (oil refining, synfuels)

Catalytic cracking
Hydrocracking
Isomerization of light paraffins
Reformate upgrading
Distillate dewaxing
Luboil dewaxing
Gasoline from methanol
Light olefins from methanol
Gasoline and middle distillates from light olefins
Deep hydrogenation of diesel fuel
Isobutene production from normal butenes

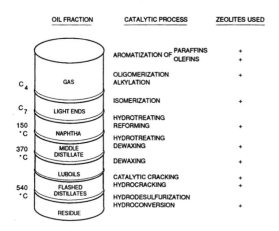

Figure 1. Catalytic processes for different parts of the crude oil barrel [1].

tude of zeolite-based catalytic processes already available and under development at
present, only a limited number of the major processes which are affected by refining
trends will be discussed. An extensive review of zeolite-catalysed processes in the
hydrocarbon processing industry has been given at the previous Summer School on
Zeolites [1].

2. TRENDS IN OIL REFINING

Present and foreseeable main pressures on the hydrocarbon processing industry are listed in Table 2. Economic pressures mainly caused by overcapacity in traditional regions and growth of capacity in developing regions causing strong intra-industry competition call for improved processes with lower investment and operating costs per unit output of product. The change in demand barrel towards a higher output of transportation fuels coupled with a growing supply of heavier feeds including tar sands oils and heavy oils which are more contaminated (sulfur, nitrogen, asphtalthenes, metals) calls for improved conversion capabilities. While meeting the increased demand for the lighter transportation fuels, product quality in a technical sense (octane quality of gasoline, cetane quality for automotive gasoil, smoke point for aviation jet fuel) has to be maintained or improved. Last but not least, environmental factors give rise to additional quality specifications: absence of lead in most grades of gasoline, presence of oxygenated components in "green" or "reformulated" gasoline, lower vapour pressure, lower sulfur content, less benzene and lower total aromatics content. For automotive gasoil (diesel), there is a trend to reduce sulfur content and to lower aromatics concentrations so as to reduce diesel emissions from urban driving (city diesel").

Table 2
Challenges for the hydrocarbon processing industry

Change in demand barrel:	More light and middle distillates as transportation fuels
Change in supply barrel:	Heavier crudes and unconventional heavy oils
	More contaminants: S, N, C, V, Ni
Product quality:	Technical: Octane, Cetane, etc.
	Environmental: Lead, RVP, Benzene, Aromatics, Oxygenates, S
Economic:	Lower investment and operating costs
Environmental:	Lower refinery emissions, less energy usage

The historical trends and future projections for the size and cut of the demand barrel are shown in Fig. 2. The historical and projected breakdown of the demand for transportation fuels is also shown in this figure. It will be evident that the increased demand for light and middle distillate transportation fuels, as contrasted with the reduced demand for residual products imply an increased mobilization of cracking processes which have to convert heavier feedstocks, in addition to increased application of processes that can build up the desired molecules from smaller fragments produced as byproducts from the cracking processes (alkylation, oligomerization), see Fig. 3. Of these processes, the cracking processes that employ zeolitic catalysts (catalytic cracking, hydrocracking) will be further discussed in more detail in this paper.

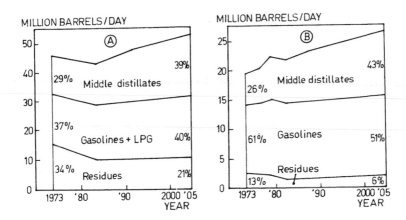

Figure 2. Historical and projected trends in size and cut of the demand barrel (A) and volume and distribution of transport fuels (B). After [2].

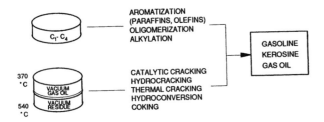

Figure 3. Feedstocks and processes leading to distillate transportation fuels.

In addition to the quantitative aspects of oil products demand, the qualitative aspects also have a great bearing on the choice and desired performance of the conversion processes. Apart from quality aspects which relate to the technical application of the product, product specifications are increasingly influenced by environmental considerations. The increased environmental concern in recent years is illustrated by Fig. 4, which shows an almost explosive growth in environmental legislature in the United States.

Of the environmentally driven changes in product specifications the phasing out of lead and the restriction in aromatics content of gasoline have important implications for the process routes for gasoline manufacture. Maintaining the octane quality of gasoline whilst complying with these specifications has favoured the application of light paraffin isomerization as a means to enhance octane quality of light naphtha or tops, a process that may be carried out with a zeolitic catalyst. The call for oxygenates in gasoline have

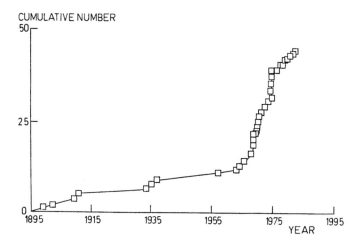

CUMULATIVE NUMBER

Figure 4. Growth in the number of U.S. environmental laws [3].

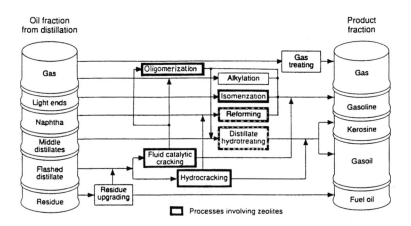

Figure 5. Simplified scheme of a refinery showing the place of some zeolite-based catalytic processes. Adapted from ref. 4.

spurred production of methyl tertiary butyl ether (MTBE) to the extent that availability of feedstock isobutene has become limiting. In this context, a recently developed zeolite-based catalytic route for isomerizing normal butenes to isobutene becomes of interest. Recent focus on clean burning diesel fuel has stimulated the development of catalytic routes for lowering the aromatics content in diesel fuel featuring the use of a zeolite-based catalyst. Of these processes, zeolite-catalysed isomerization of light paraffins will be a subject of further discussion in this paper.

Figure 5 shows a simplified diagram in which the place of zeolite-based processes in a typical refinery scheme becomes apparent.

3.   CATALYTIC CRACKING

3.1.   Historical trends in catalytic cracking

Catalytic cracking is at present a major process for the production of gasoline and probably the most widespread conversion process in the world, being applied in some 300 commercial plants worldwide. Figure 6 shows the growth in fluid catalytic cracking capacity (FCC) in the United States.

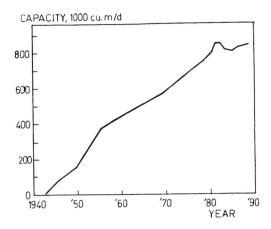

Figure 6. Growth of FCC capacity in the United States [5].

Since the start-up of the first commercial Houdry fixed-bed plant in 1936, the technology of catalytic cracking has seen many innovations: from moving-bed cracking, fluid-bed catalytic cracking (dense bed) till the present riser and residue FCC processes. Extensive reviews of these developments have been published by Avidan et al. [5] and by Biswas and Maxwell [20].

Closely linked with the development in reactor technology are the developments in the cracking catalyst: amorphous silica-alumina catalysts with relatively low alumina content have been superseded by catalysts with high alumina content, followed by the introduction of zeolites as partial replacement for amorphous silica-alumina in the mid sixties.

The success of zeolites in FCC is mainly based on their higher intrinsic activity, their better selectivity towards gasoline as compared with amorphous catalysts, and their good activity retention. The higher activity has allowed the introduction of riser cracking, in which contact times are of the order of seconds rather than minutes, as in dense bed FCC. Due to reduced backmixing and thermal cracking contributions in riser reactors, further selectivity improvements have been achieved. These improvements are illustrated in Figure 7.

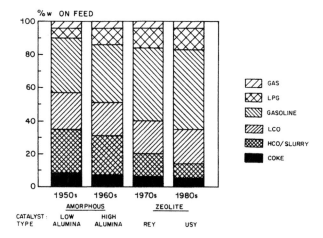

Figure 7. Improvements of FCC selectivity over the years [20].

3.2. Impact of more recent refining trends on FCC technology

3.2.1. Product quality

The zeolitic component in all of the above-mentioned catalysts is of the faujasite type (X, and now practically only Y). Developments in Y-zeolite in the past have been very much focussed on improving the hydrothermal stability by exchanging with rare earth ions and on improving gasoline yield. Due to the higher hydrogen transfer activity of these rare earth exchanged zeolites (REY), high octane olefinic components are transformed into high octane aromatics and low octane paraffins, the latter causing a drop in gasoline octane number. Figure 8 shows that prior to the eighties the increased yield of gasoline has been accompanied by a drop in gasoline octane number.

The drop in gasoline octane quality was not much of a problem as long as lead could be added to boost the octane number. However, with the ban on lead in more recent years, there has been a drive to regain FCC gasoline quality. This has led to the reversal of the declining trend of research octane quality, as shown in Figure 8.

The improvement of gasoline octane quality in the past decade is mainly due to the increased usage of highly dealuminated Y zeolites. The lower alumina content of Y (ultrastable Y or USY) leads to a better maintenance of crystallinity under hydrothermal conditions, but also to a more olefinic gasoline of higher octane number. This is because the acid sites in USY become more isolated so that the rate of bimolecular hydrogen

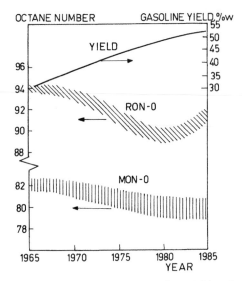

Figure 8. Trends in FCC gasoline yield and octane quality [6].

transfer reactions becomes lower. The effect of dealumination on the unleaded research octane number of the gasoline product is shown in Figure 9. The reduced hydrogen transfer activity of USY also leads to a decreased formation of polyaromatics and coke, as can be seen from Figure 10 . Because of the latter beneficial effect, gasoline yield is not much affected although the cracking activity has been reduced.

The improvement of the octane quality/ gasoline selectivity in successive generations of FCC catalysts is shown in Figure 11.

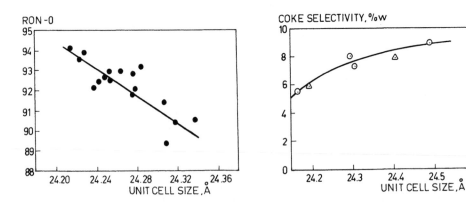

Figure 9 (left). Effect of dealumination of Y zeolite on the unleaded research octane number of the FCC gasoline. The unit cell size is a measure of Al in the zeolite framework; a lower cell size means less Al [7].

Figure 10 (right). Effect of dealumination of Y zeolite on coke selectivity [8].

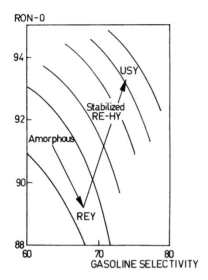

Figure 11. Research octane number versus gasoline selectivity from commercial FCCU data, showing the trend in catalyst type. Adapted from ref. 9.

The trend to replace REY by USY-type FCC catalysts is clearly shown by the increased market share of the latter type of catalysts in North American refineries, see Figure 12.

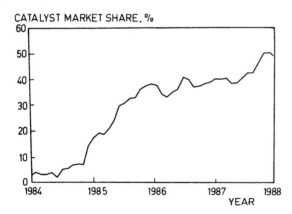

Figure 12. Growth in market share of ultrastable Y-type FCC catalysts in North America [1].

Another way of recovering lost octane quality in the lead-free gasoline era is the application of other zeolites, notably ZSM-5, as octane boosting additive in FCC. The narrower pore ZSM-5 zeolite allows the preferential cracking of less branched, low octane components such as n-paraffins from the FCC gasoline, thereby enhancing its octane number. This shape-selective cracking leads to an increased production of light hydrocarbons, which need not be very unfavourable, however, since light olefins are also valuable components. They can add to distillate production via alkylation and oligomerization processes or they may find an outlet as base chemicals in the petrochemical industry. Because the pore texture of ZSM-5 cannot accommodate polyaromatics, efficient aromatization of acyclic, low branched hydrocarbons to monoaromatics in the gasoline range is possible without extensive coproduction of cycle oils and coke. Figures 13 and 14 show the effect of ZSM-5 addition on the yields of gasoline and lower olefins, illustrating the removal of paraffins from gasoline with formation of light olefins, and the increase in gasoline aromaticity. Figure 15 shows the consequences of these compositional changes on octane quality of the FCC gasoline.

ZSM-5 can be applied as an additive, i.e., separate catalyst particles that contain a high concentration of this zeolite. However, it can also be integrated in a composite catalyst, i.e., the particles of FCC catalyst contain both Y and ZSM-5 zeolites.

Although so far ZSM-5 addition has been practiced mainly for the purpose of octane boosting whilst accepting some increased production of gaseous hydrocarbons, interest may develop in further enhancing the production of light olefins by increasing the contribution of ZSM-5 in catalytic cracking at higher temperatures, shorter contact

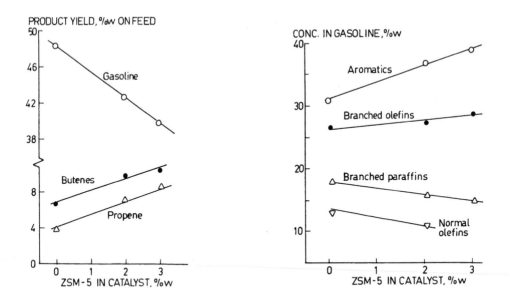

Figure 13 (left). Effect of ZSM-5 addition on yields of gasoline and light olefins in FCC [10].

Figure 14 (right). Effect of ZSM-5 addition on the composition of FCC gasoline [10].

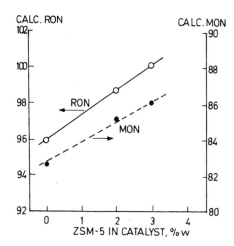

Figure 15. Effect of ZSM-5 addition on the octane quality of FCC gasoline [10].

times and higher catalyst/feed ratios. Such "deep cracking" may be of interest in the future for opening up routes to middle distillates of high quality by oligomerization, and to provide an additional source of light olefins (particularly of propene) for the petrochemical industry. Thus the current primary role of FCC as a gasoline producer may well shift towards that of a light olefins generator in the future.

3.2.2. Processing of heavier feeds

The desire to process heavier feeds so as to comply with the refinery trend for deeper conversion of the oil barrel has led to several developments in both FCC hardware and catalyst to cope with some of the difficulties of processing such feeds.

Accessibility of cracking sites
One problem to be addressed is that of catalyst accessibility to molecules of increasing molecular weight and size. As shown in Figure 16, the proportion of multiring components (polynaphthenes, polyaromatics) tends to increase as the oil fraction becomes heavier.

The limited accessibility of zeolites to bulkier molecules, which is an aspect which is exploited in shape-selective conversions, becomes a hindrance for the cracking of multiring compounds present in heavier oils. Figure 17 shows the strong effect of molecular size on the mobility of molecules in zeolite Y. Whereas larger molecules should theoretically be more easily cracked than similar molecules of lower molecular weight, the reverse is true for the cracking of polynaphthenes over Y zeolite. Figure 18 shows that 3 and 4 ring naphthenes are less rapidly cracked than a mononaphthene, whereas the reverse is true for an amorphous silica-alumina with much larger pores.

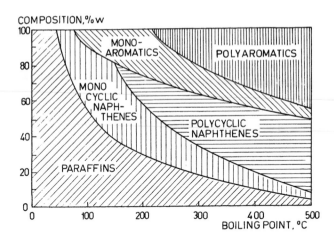

Figure 16. Distribution of hydrocarbon types in a typical crude oil.

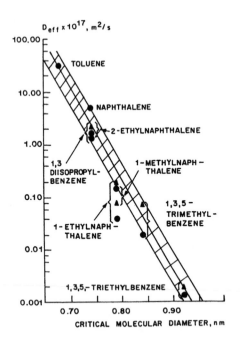

Figure 17. Effect of molecular size on diffusivity of hydrocarbons in zeolite Y [11,12].

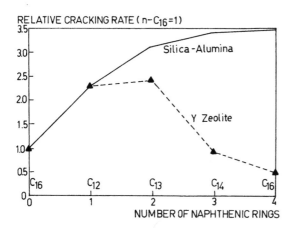

RELATIVE CRACKING RATE (n-C_{16}=1)

Silica-Alumina

Y Zeolite

C_{16} C_{12} C_{13} C_{14} C_{16}

NUMBER OF NAPHTHENIC RINGS

Figure 18. Cracking rates of hydrocarbons over zeolite Y and amorphous silica-alumina [12].

Practical FCC catalysts therefore are composite catalysts, combining an amorphous silica-alumina matrix with relatively large pores (pore diameters of the order of 100 A) with zeolites having pores around 7 A diameter. The former component takes care for the precracking of the larger molecules, whereas the latter serves to further crack the smaller fragments formed. An optimized FCC catalyst should therefore have a pore size distribution which is balanced for a given feed at specific process conditions.

The present trend of processing residual feedstocks in addition to the conventional vacuum gasoil feeds imply that very bulky, carbon-rich molecular complexes called asphalthenes are present in the feed. These complexes cannot easily enter the pores of 50 to 200 A diameter of a usual silica-alumina matrix. For residue FCC catalysts the addition of much larger macropores has therefore been advocated. This addition is claimed to have a beneficial effect on gasoline yield, while coke formation is reduced, as shown in Figures 19 and 20.

Carbon

The increase in cyclic components, in particular polyaromatics, in heavier feedstocks, imply a higher carbon and lower hydrogen content. Therefore, more coke is produced in the FCC process. This results in a need for a higher regeneration capacity and the installation of catalyst coolers to dispose of excess heat in residue FCC. Minimizing coke make is important, and has been realized by improvements in hardware (improved feed inlet devices, devices for rapid and efficient gas/solid separation and improved stripping) but also by improvements in catalyst design, such as the introduction of ultrastable Y zeolite and macroporosity mentioned earlier.

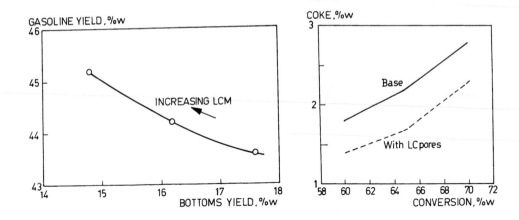

Figure 19 (left). Effect of "liquid catching" macroporosity (LCM) on gasoline and bottoms yield in residue FCC [13].

Figure 20 (right). Effect of "liquid catching" macropores on coke make in residue FCC [14].

The progress in the capabilities to convert feeds of increasing heaviness and higher carbon content during the last decade is shown in Figure 21.

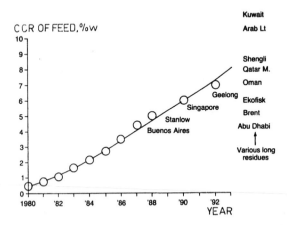

Figure 21. Increase in the Conradson Carbon Content of the feeds of FCC units of Shell with time [15].

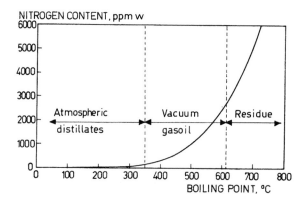

Figure 22. Nitrogen content of fractions of Arabian light crude oil.

Contaminants

As feeds become heavier, the concentration of contaminants increases. Apart from a higher sulfur content, heavier fractions of a crude oil generally also contain larger amounts of organic nitrogen compounds, see Figure 22. Due to their basicity, nitrogen compounds tend to suppress the acidic cracking reaction, as is illustrated in Figure 23.

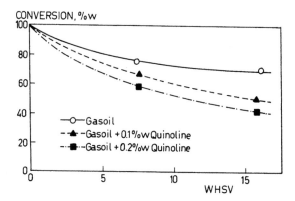

Figure 23. Effect of Quinoline addition on catalytic cracking of gasoil. T = 538 C. [16].

As follows from Figure 23, the activity suppression by nitrogen compounds is not very dramatic, presumably because the coverage of active sites is only moderate at the relatively high temperature. The activity reduction can easily be compensated by an increase in reaction temperature. Alternatively, hydrotreating of catalytic cracker feed which mainly serves to reduce the aromatics, sulfur and metals content, may also alleviate the problem of activity suppression by nitrogen.

Metals (V, Ni) present in residual feedstocks pose a greater problem in residue FCC. Nickel deposited on the catalysts acts as a dehydrogenation catalyst under the prevailing conditions, enhancing the formation of coke and light gases at the expense of gasoline yield. Vanadium also deposits on the catalyst and tends to migrate through the catalyst, forming vanadates and low melting eutectics in the presence of sodium, thus destroying the zeolite.

Figure 24 shows a typical build-up of metals on catalyst during processing of a residual feed in a FCC unit. The deterioration of catalyst performance as a result of this metals build up is shown in Figure 25.

The detrimental effect of nickel can be suppressed by adding antimony or bismuth compounds to the feed. These metals tend to combine with nickel, thus poisoning the dehydrogenation activity. The solution to the problem of vanadium poisoning is generally sought in the incorporation of basic oxides such as MgO or SrO which trap vanadium oxide formed under regeneration conditions.

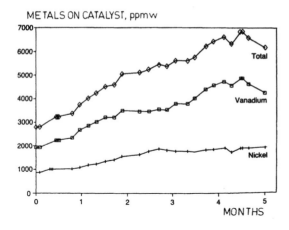

Figure 24. Build up of metals during processing of a residual feed in a residue FCC unit [15].

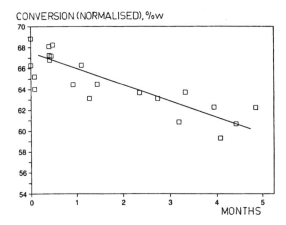

Figure 25. Decline of catalyst performance as a result of metals accumulation on the catalyst [15].

3.3.   Future challenges for FCC

From the developments discussed above it will be evident that catalytic cracking is far from a mature technology, notwithstanding the fact that it was conceived more than half a century ago. Considering the refinery trends discussed before, there can be little doubt that FCC will remain a key process for gasoline manufacture, and that there will be an increased demand for processing heavier, in particular residual, feeds in the future. Technological progress will shift the limits of processable feeds, as shown in Figure 26.

The trend towards higher temperature, shorter residence time cracking may well continue and lead to deeper cracking with enhanced yields of light olefins. Such a development calls for innovations in both hardware and catalysts, and in the latter field zeolites will continue to play a key role.

As contact times become shorter, the cracking reactions on the catalyst will proceed under transient, rather than steady-state conditions and penetration of reactants into the particles will become less and less complete, as is shown in Figure 27. The combination of zeolitic micropores with meso- and macropores of the amorphous matrix as already applied today may in optimal catalysts have to be realized in a more sophisti-cated architecture of the catalyst particle: macroporosity with progressively smaller pores toward the interior, as depicted in Figure 28.

Although significant gains have been achieved in gasoline yield which represent very large amounts of money, the ultimate yield is still far off. In the current, once-through mode of cracking, products are subjected to secondary cracking with consequent losses. Although the transition from the older dense bed cracking (representing more or less a backmixed system) to riser cracking (representing a better approach to a plugflow reactor) has been a great improvement in this respect, further yield gains are theoretically

604

possible. As can be seen from Figure 29, running at low conversions per pass with intermediate removal of gasoline product results in much higher gasoline selectivity than

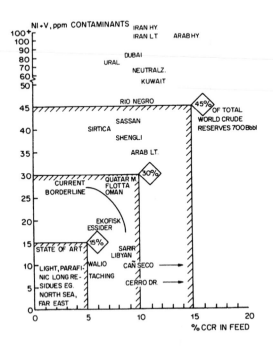

Figure 26. Limits in processable feeds for FCC [17].

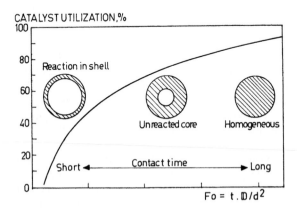

Figure 27. Catalyst utilization as a function of the Fourier number under non-steady state conditions [18].

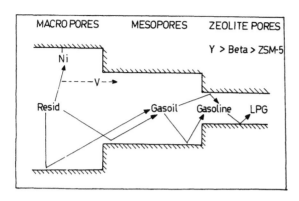

Figure 28. Conceptual pore architecture of a FCC catalyst [19].

single-pass conversion. The latter mode of operation calls for a cross-current, rather than a stirred (dense fluidized bed) or cocurrent (riser) type of reactor, see Figure 30.

The different reactor and other process choices will have their impact on the design of optimal catalysts. Even today, hardly any of the approximately 300 FCC units is identical to another in its design and operating conditions. In principle, this calls for customized catalysts for each unit. The trend towards customizing is reflected in the growth of the number of commercial catalysts offered in the market to a present day value which is as great as the number of FCC units, as shown in Figure 31. With respect to tailorability, zeolites with their great variability may have much to offer for the future.

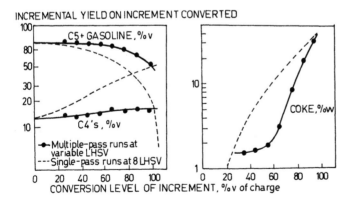

Figure 29. Comparison of gasoline, gas and coke yields in once-through conversion and yields in incremental conversions with intermediate product removal [5].

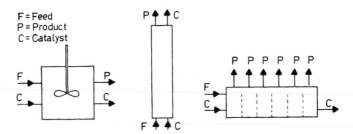

Figure 30. Catalytic cracking in different types of reactor.
From left to right: stirred tank reactor symbolizing a dense fluid bed, cocurrent plug flow reactor symbolizing a riser reactor and a hypothetical cross-flow reactor.

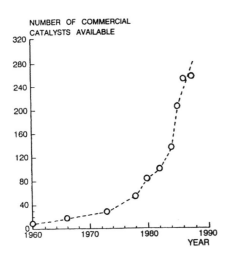

Figure 31. Growth in the number of FCC catalysts in the market [20].

4.   HYDROCRACKING

Hydrocracking is another major catalytic process besides catalytic cracking for converting heavy oil fractions into lighter distillates. In contrast to catalytic cracking, catalyst deactivation is not coped with by very frequent regeneration, but by operating under sufficiently high hydrogen pressure in the presence of a hydrogen activating function on the catalyst so as to prevent formation of polyaromatics and coke. Thus, in principle catalyst activity is stable; i.e. under practical conditions catalyst life can be a year or longer.

4.1. Past developments

The hydrocracking process dates back from the period before world war II, when the Badische Anilin und Soda Fabrik as part of the IG Farben trust developed the so-called Vapour Phase Process for the conversion of oil fractions boiling up to 325°C into gasoline by hydrogen treatment under severe conditions [21]. The feedstock, which may be derived from crude oil or from coal, was processed at temperatures of about 450°C and at pressures of about 300 bar in two steps: a primary prehydrogenation step over a catalyst consisting of unsupported Ni/W sulfide to remove nitrogen, followed by a cracking step over a catalyst consisting of WS_2 supported on acidic, fluorided clay.

The abundance of cheap crude oil from the Middle East and the good market outlet for residual fuel oil made this process uneconomic in the period directly following the second world war. However, the growth in gasoline consumption in the US and the concomitant increase in FCC capacity caused a surplus in FCC cycle oils which provided an incentive for hydrocracking to convert this oil into naphtha. The advent of improved, more active catalysts allowed more economic hydrocracking at lower temperatures and pressures (350-420°C, 100-150 bar) whilst still complying with the requirement to stay out of the aromatics forming region, see Figure 32. The growth in hydrocracking capacity since that time is depicted in Figure 33.

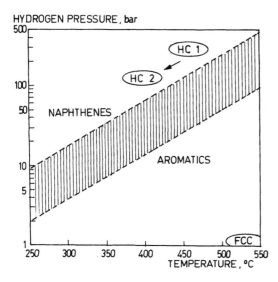

Figure 32. Thermodynamic equilibrium for aromatics hydrogenation. HC 1 = classical hydrocracking before WW II, HC 2 = modern hydrocracking.

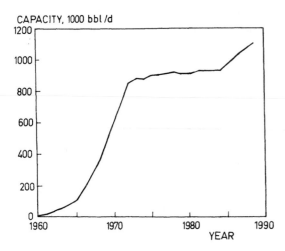

CAPACITY, 1000 bbl /d

Figure 33. Growth in hydrocracking capacity in the US. Data till 1980 from ref. 22.

Whereas in the US the main interest in hydrocracking has traditionally been in the area of gasoline production, hydrocracking in other parts of the world (Europe, Asia) with relatively lower gasoline demand has been more directed towards production of middle distillates (kerosine and gas oil for jet and diesel fuel) from heavier feedstocks. This is because the hydrogen rich middle distillate fractions from hydrocracking have favourable properties as transport fuels, in contrast with the corresponding fraction from a catalytic cracker, see Table 3.

Table 3.
Comparison of the properties of middle distillates from catalytic cracking and hydrocracking [23].

Fraction	Property	FCC	HC
Kerosine (150-200°C)	H, %w	10.4	14.0
	S, ppmw	1660	20
	Smoke point, mm	8	23
Gasoil (250-370°C)	H, %w	9.4	13.6
	S, ppmw	10000	40
	Cetane number	27	55

A review of hydrocracking processes and catalysts has been presented recently [24].

4.2. Zeolites in hydrocracking

Shortly after their succesful introduction in catalytic cracking in the mid sixties, zeolites also entered the hydrocracking scene to replace the amorphous silica-alumina used as acidic cracking component in the second stage catalyst. The advantages of zeolites over amorphous catalysts are listed in Table 4.

Table 4
Relative performance of zeolitic over amorphous catalysts.

Performance characteristic	Zeolite over amorphous
Activity	+
Stability	+/-
Distillate yield	-
Distillate yield decline	+/-
Distillate quality	+
Hydrogen consumption	+
Processibility of heavy feeds	-/+

Some of these performance characteristics will be discussed in more detail below.

Activity
The higher activity and better activity maintenance of zeolitic catalysts compared to amorphous ones is illustrated by the comparative experiments shown in Figure 34. The higher activity of zeolitic catalysts is not only due to their higher intrinsic acidity, but also to their higher tolerance for basic compounds, such as ammonia formed by hydrodenitrogenation of organic compounds in the first stage. This can be seen from the data listed in Table 5.

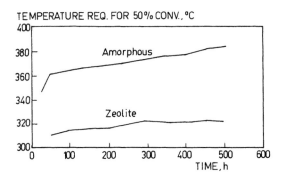

Figure 34. Comparison of typical zeolitic and amorphous silica-alumina hydrocracking catalysts [23].

Table 5
Effect of ammonia on hydrocracking activity of an amorphous and zeolitic catalyst. Activities expressed as first-order rate constant for cracking to C_6-. Ammonia generated by addition of t-butyl amine to the feed.

Feed	ppm NH_3 in H_2	Ni/W/Silica-Alumina	Ni/W/Y
n-C_7	0	7.0	13.6
n-C_7	270	0.07	0.9
Naphtha	0	5.0	11.2
Naphtha	300	0.07	1.0

The higher residual activity of zeolitic catalysts allows to dispense with the usual removal of ammonia from the first-stage effluent in two-stage hydrocracking. Thus, the first-stage effluent can be passed directly over the second-stage catalyst which obviates the need for intermediate cooling down, scrubbing, and reheating. This so-called series-flow operation made possible by zeolitic catalysts is at present still less widely applied than the older single-stage and two-stage hydrocracking modes, but the higher conversions possible compared to single-stage hydrocracking and the cost savings compared with two-stage hydrocracking may well increase its popularity in the future. Figure 35 compares the basic flow schemes of the different hydrocracking process types, while Figure 36 presents a breakdown of current units according to these process types.

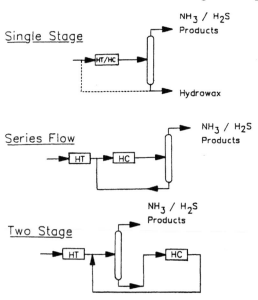

Figure 35. Basic schemes of a single stage hydrocracker with partial conversion and two-stage and series flow hydrocrackers with bottoms recycle to extinction.

611

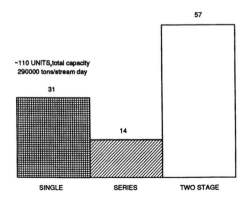

Figure 36. Distribution of existing hydrocrackers according to process type [24].

4.3. Selectivity to middle distillates

The trend towards increasing production of transport fuels, in particular middle distillates, as discussed before imply an increasing need to exploit the capability of hydrocracking to produce these fuels of good quality. In this respect zeolites are at a disadvantage compared with amorphous silica-aluminas, as their higher acidity and narrower pores give rise to lower selectivities to liquid products. This is because the primary fragment molecules of intermediate size have greater difficulty in escaping the pore structure and will therefore be more subject to further cracking. A comparison of the activity and selectivity of typical representatives of these catalyst classes is shown in Figure 37.

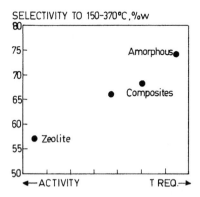

Figure 37. Comparison of typical zeolitic and amorphous hydrocracking catalysts in terms of activity and selectivity for middle distillate production [25].

The higher activity of zeolitic catalysts allows some trade-off of activity to increase middle distillate selectivity. Activity can be moderated with ammonia in the hydrogen gas, which indeed results in a better selectivity to middle distillates. Another way is to reduce the alumina content in Y zeolite by ultrastabilization. With the lower acid site density, a better balance is obtained between the rate of primary cracking and rate of escape of cracked fragments. Catalysts based on very ultrastable Y zeolite (VUSY) with a low alumina content in the framework as evidenced by a low unit cell constant can indeed achieve a much higher selectivity to middle distillates, as shown in Figure 38.

A third possibility is to combine zeolites and amorphous silica-alumina's in composites to obtain catalysts that are reasonably active with acceptable middle distillate selectivity, as shown in Figure 37.

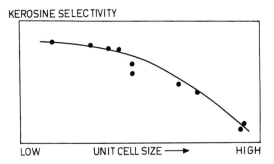

Figure 38. Relation between unit cell size as a measure for framework aluminium and middle distillate selectivity of hydrocracking catalysts based on Y zeolite [23].

4.4. Processing of heavy feeds

The refinery trend toward deeper conversion of the barrel calls for hydrocracking of feeds with higher end point. The main problems associated with processing heavier feeds are the increasing concentration of nitrogen (see Figure 22) and difficult access of bulkier molecules to the acid sites.

Nitrogen in feed

Like in catalytic cracking, organic nitrogen compounds reduce the cracking activity, as can be seen from the data in Table 6. However, the inhibition effect is stronger due to the lower temperatures. Moreover, in contrast to catalytic cracking, the reduction in intrinsic activity can in most cases not be simply compensated by an increase in temperature (by, e.g., 70 °C, see Table 6) since this may require an uneconomic increase in operating pressure (see Figure 32). Thus, coping with the higher nitrogen content of heavier feeds generally calls for more effective nitrogen removal in the first stage, i.e., more severe conditions and/or more active first-stage catalysts.

Table 6.
Effect of nitrogen in feed and temperature in hydrocracking.

N addition to feed	Relative conversion	
	T = 300°C	T = 370°C
None *)	1	--
1000 ppmw N as Quinoline	0.05	1.10
1000 ppmw as Pyrrole	0.07	1.15

*) Feed contained 2 ppmw N

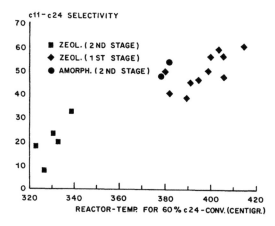

Figure 39. Comparison of activity and middle distillate selectivity of zeolitic and amorphous catalysts under second and first-stage hydrocracking conditions [26].

The high intrinsic activity of zeolites, however, allows to somewhat relax the denitrogenation requirement in the first stage as compared with amorphous catalysts. Although a large proportion of the acid sites in zeolites will be covered by nitrogen species under operating conditions, there is still a reasonable number of active sites available for cracking. This can be seen from Figure 39, which shows that the performance of a zeolitic catalyst under first-stage conditions with moderate concentrations of nitrogen compounds can be rather similar to that of amorphous catalysts under second-stage, essentially nitrogen-free conditions. This allows the use of zeolitic catalysts in so-called stacked beds in the first stage, i.e., the filling of the first-stage reactor consists of a combination of a hydrotreating catalyst upstream and a zeolitic hydrocracking catalyst downstream. Thus, in addition to removing nitrogen from the feed, the first stage now also gives a significant contribution to cracking. This contribution can be maximized by judicious choice of the relative volumes of the two catalysts, as can be seen from Figure 40.

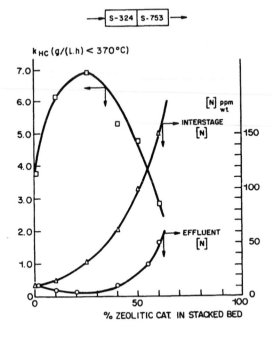

Figure 40. Overall cracking activity and interstage organic nitrogen levels of stacked beds with different ratios of a Ni/Mo/Alumina catalyst (S-324) and a USY zeolite catalyst (S-753) [27].

Accessibility of cracking sites

As discussed under catalytic cracking, accessibility of cracking sites for the bulkier molecules present in the tail of a heavy feed is a problem with zeolitic catalysts. The increased diffusional control of hydrocracking of naphthenes with more rings can be seen from the reduced conversion rates and lowered activation energies as given in Table 7.

Table 7.

Effect of the number of rings on the reactivity for hydrocracking over a Ni/W/USY catalyst. T=371°C [28].

Feed	Rate constant, g/(g.h)	Activation energy, kcal/mol
Tetralin	14.5	44.0
Perhydrophenanthrene	6.2	35.1
Perhydropyrene	2.8	35.1

Figure 41 shows a comparison of hydrocracking rates as a function of feed boiling point for a zeolitic and an amorphous silica-alumina catalyst. Whereas the amorphous catalyst shows the expected increase of rate with molecular size, the Y zeolite shows a decreasing rate. The size exclusion effect in the latter case is clearly illustrated by field ionization mass spectroscopic analysis of the unconverted fractions of the products obtained. As figure 42A shows, paraffins and monoring naphthenes are less easily cracked than multiring naphthenes over an amorphous catalyst, whereas the reverse is true with a hydrocracking catalyst based on Y zeolite, see Figure 42B.

The consequence of insufficient cracking of heavy molecules present in the tail of high end-point feeds is that in operation which aims at recycle to extinction these molecules build up in the recycle stream. To maintain per-pass conversion at the desired level, reactor temperatures have to be increased more rapidly than usual to compensate for the increased refractiveness of the combined feed with time, as can be seen in Figure 43. The build up of heavy molecules in recycle operation is depicted in Figure 44. Figure 45 shows the decrease in selectivity caused by the increase in operating temperature required to maintain conversion rates.

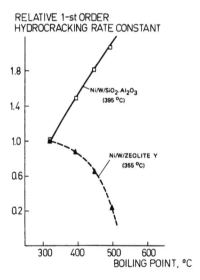

Figure 41. First-order rate constant of amorphous silica-alumina and Y zeolite catalysts as a function of the boiling point of fractions from a Middle East vacuum gasoil [29].

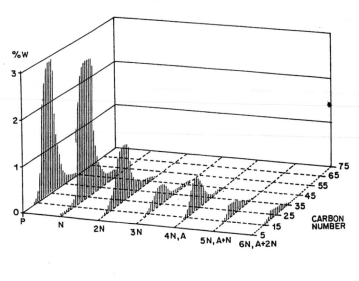

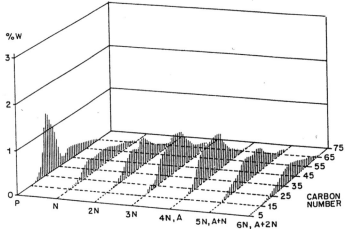

Figure 42. Field ionization mass spectra of the unconverted fractions from two-stage hydrocracking of a vacuum gasoil [30].
A(bove): catalyst based on amorphous silica-alumina.
B(elow): catalyst based on Y zeolite.

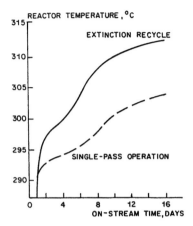

Figure 43. Temperature required for 60% per-pass conversion as a function of on stream time in hydrocracking a gasoil blend over a Ni/W/RE-X catalyst [31].

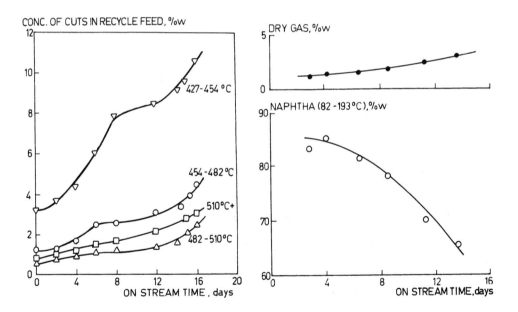

Figure 44 (left). Increase of the concentration of higher boiling components in the recycle stream during hydrocracking over a Ni/W/RE-X catalyst [31].

Figure 45 (right). Deterioration of selectivity during recycle-to-extinction hydrocracking over a Ni/W/RE-X catalyst [31].

618

To improve the capability of hydrocracking to process feedstocks with high boiling end points, the accessibility of cracking sites for bulky molecules can be improved in similar ways as discussed for catalytic cracking: dealumination of Y zeolite improves the balance between diffusion and cracking rates, while a combination of amorphous silica-alumina and a zeolite in composite catalysts can be applied so as to profit from the high cracking rates of multiring naphthenes over the amorphous component (see Figures 18, 41 and 42A) while utilizing the high activity of the zeolite component for smaller size molecules with relatively low intrinsic reactivity: paraffins, smaller naphthenes. In principle, reducing the size of zeolite crystals should also increase accessibility, but without sacrificing activity.

Improved designs of hydrocracking catalysts in the last decade have resulted in improved capabilities to process heavy feeds, as is illustrated by Figure 46.

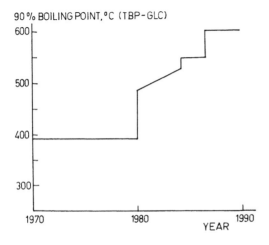

Figure 46. Increase of heaviness of feedstocks processed in hydrocrackers of Shell [29,32].

4.5. Future challenges for hydrocracking

From the viewpoint of the trend of maximizing the production of transport fuels, the main challenge for hydrocracking will be its further extension to heavier residual feedstocks. Such feedstocks which contain asphaltenes and metals, are now converted by thermal processes or by catalytic hydroconversion to products which are still relatively heavy and which need further upgrading and conversion in separate processes including hydrocracking.

In the hydroprocessing of residual feedstocks, hydrodemetallization and hydroconversion take place over specially tailored catalysts based on amorphous carriers. The separate trends in residue hydroconversion towards higher conversions and in hydrocracking of non-residual feedstocks in the direction of heavier feedstocks would seem to logically converge towards an integrated process in which zeolite based hydrocracking catalysts are used in a final stage in the hydroconversion of residual feedstocks, rather than in a separate hydrocracking process.

As discussed before under catalytic cracking, more optimal catalysts in hydrocracking may perhaps be obtained by a sophisticated design of catalyst particles. The pore architecture may be designed so as to provide an optimal match with the different sizes of the molecules to be cracked. In addition, catalyst particles may be designed so as to take advantage of diffusion limitations of cracking and denitrogenation reactions. Thus, denitrogenation and cracking duties are not only varied in the direction of the process stream by applying different catalysts in separate reactors and catalyst beds, but also inside the catalyst particles themselves, see Figure 47.

For the above and other innovations in hydrocracking in the future, the zeolite catalysts have to fulfill specific demands, but this challenge may well be met by taking advantage of the great variability and other features of microporous materials.

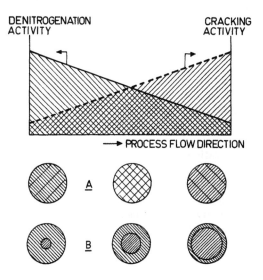

Figure 47. Conceptual distributions of denitrogenation and cracking activities in hydrocracking.
A. With homogeneous catalyst particles.
B. With tailored inhomogeneity in catalyst particles.

5. PARAFFIN ISOMERIZATION

5.1. Historical development

Skeletal isomerization of light paraffins for octane enhancement of light naphtha or "tops" dates back to the period before world war II, when processes were applied that used monofunctional acid catalysts of the Friedel Crafts type. Aluminium choride was the preferred acid, either applied onto a solid support by sublimation (vapour phase isomerization process) or dissolved in a molten eutectic mixture with antimony chloride (liquid phase process). Rapid catalyst deactivation requiring continuous replenishment of fresh acid and the need to carry out the isomerization in the presence of significant amounts of HCl (to obtain Brønsted acidity) were major complications in these type of processes.

The use of bifunctional catalysts which combined the acid function with a noble-metal (Pt) hydrogenation function was a major breakthrough since these catalysts can be operated stably under moderate hydrogen pressures. In the first examples of these bifunctional catalysts, developed in the fifties, the acidic component consisted of chlorided alumina and this type of catalyst is still in use today in the Penex process of UOP. In the late sixties, bifunctional isomerization catalysts with zeolites as the acidic component were developed. The Hysomer process of Shell, which uses H-Mordenite as the acid component, saw its first commercial application in 1970 [33].

Isomerization of light naphtha has only recently enjoyed much interest. In the era of leaded gasoline without restrictions on aromatics in gasoline, the desired octane quality of gasolines could be economically achieved by addition of lead and more severe reforming. However, the environmentally driven changes in gasoline specifications has considerably increased the importance of paraffin isomerization.

A review of paraffin isomerization innovations has been published recently [34].

5.2. Features of zeolites in paraffin isomerization

The preferred zeolite at present is the hydrogen form of mordenite. Partial dealumination by acid leaching to a silica/alumina ratio of about 20 has been found to give maximum activity, see Figure 48. The activity allows isomerization of pentanes and hexanes at attractive rates at temperatures around 250°C. While this is significantly below the temperature required with amorphous silica-alumina based catalysts (about 400°C), it is higher than the temperature required with catalysts based on chlorided alumina, which can operate between 120 and 170°C.

Since the isomerization of paraffins is an equilibrium-limited reaction with less branching according as temperatures are higher, zeolitic catalysts are at a disadvantage compared with catalysts based on chlorided alumina. At the higher operating temperatures required for zeolitic catalysts the octane ceilings are lower, see Figure 49.

Since the thermodynamic equilibrium sets a maximum to the conversion of n-paraffins achievable, removing unconverted normal paraffins from the isomerized product by selective adsorption becomes of interest if higher octane values are aimed for. Selective removal of normal paraffins is possible using 5A type molecular sieves by a gas phase adsorption/desorption process (Isosiv process of Union Carbide) or a liquid phase adsorption/desorption process (Molex process of UOP).

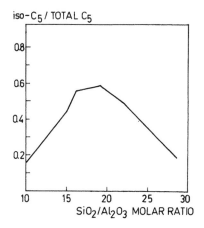

iso-C$_5$ / TOTAL C$_5$

SiO$_2$/Al$_2$O$_3$ MOLAR RATIO

Figure 48. Effect of the silica/alumina ratio of mordenite on pentane isomerization activity [35].

In such a recycle mode of operation the octane quality disadvantage of zeolite-based isomerization processes largely disappears, as can be seen in Figure 50.

In the case of isomerization over a bifunctional, zeolite-based catalyst as in the Hysomer process the operating temperature, pressure and hydrogen flow rate can be chosen so as to fit the requirements of a gas phase selective adsorption process such as the Isosiv process with desorption by purging with hydrogen. Thus, rather than applying two separate processes, the isomerization and selective adsorption processes can be closely integrated into a single process with significant savings in equipment and costs. This is an advantage of a zeolite-based isomerization process, since the operating conditions with a catalyst based on chlorided alumina do not permit such a close integration. The integrated Hysomer/Isosiv combination is known as the Total Isomerization Package (TIP) process. Depending upon the composition of the feed, it is advantageous to install the isomerization section either upstream of the adsorption section (for feeds with high n-paraffins content), or downstream (for feeds with high isoparaffins and cyclic contents). These alternative configurations of the TIP process are shown in Figure 51.

The most important advantage of zeolitic catalysts, however, is their robustness. As can be seen in Figures 52 and 53, a Pt/Mordenite catalyst can be operated in the presence of moderate amounts of sulfur and quite appreciable amounts of water. The tolerance for moisture obviates the need for feed driers, which are generally required in the case of catalysts based on chlorided alumina. Even with the latter precaution to ensure dry operation, stripping of HCl from the chlorided alumina catalyst cannot be entirely avoided, necessitating continuous replenishment of chloride during operation and coping with some HCl slip in the product. The zeolite-based Hysomer catalyst has demonstrated excellent stability in commercial practice, with catalyst lifes of up to 7 years. A catalyst batch deactivated by operational mishaps can usually be regenerated easily.

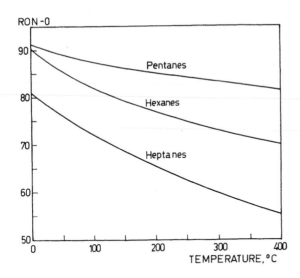

Figure 49. Octane values of isomeric pentanes, hexanes and heptanes at equilibrium in the gas phase (calculated from API project 44 data).

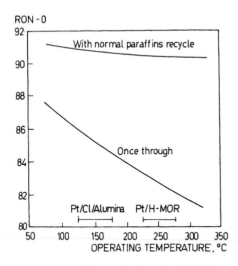

Figure 50. Effect of normal paraffins recycle and temperature on the octane numbers achievable in isomerization. Feed: 60% pentanes, 30% hexanes, 10% cyclic C6 [36].

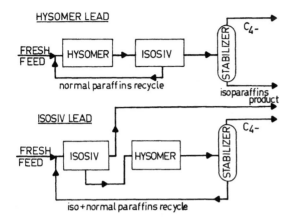

Figure 51. Alternative configurations of the TIP process.

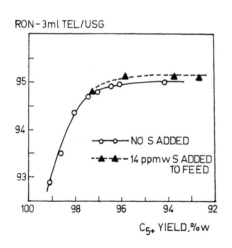

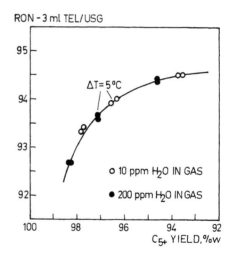

Figure 52 (left). Effect of sulfur on the yield/octane performance of a Pt/Mordenite catalyst in isomerization. Feed: hydrotreated light naphtha A, dry operation.

Figure 53 (right). Effect of water on the yield/octane performance of a Pt/Mordenite catalyst in isomerization. Feed: hydrotreated light naphtha B.

5.3. Future of paraffin isomerisation

Against the background of the lead ban from gasoline and the pressure on aromatics in gasoline it can be expected that paraffin isomerization will remain important in the future, and this also pertains to the use of zeolites for this purpose. Zeolite-catalysed paraffin isomerization capacity has enjoyed a rapid growth particularly in the past decade, as can be seen in Figure 54. At present the share of zeolitic catalysts in the total isomerization capacity of light naphtha is about as large as that of the older amorphous catalysts, as can be seen in Figure 55.

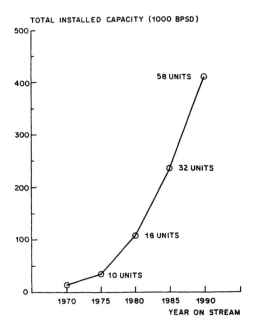

Figure 54. Growth of total installed Hysomer capacity [24].

A challenge for the future is the isomerization of lighter and heavier paraffins than pentane and hexane with zeolite catalysts. Mechanistic considerations indicate, however, that it will difficult to isomerize n-butane by the normal mechanism involving a protonated cyclopropane intermediate, while isomerization of n-heptane and higher paraffins will be accompanied by significant cracking [38,39]. Formation of isobutane from n-butane may occur by a dimerization-isomerization- cracking path and for such a route zeolites with a specially favourable pore texture, such as ferrierite, may offer advantages, as has been recently demonstrated in the skeletal isomerization of normal butenes [40].

Although isomerization of n-heptane results in a considerable octane gain, the octane number of the isomerate with largely monomethyl hexane isomers is still low (see Figure 49). A molecular sieve adsorption process that discriminates between dibranched alkanes and less branched paraffins could boost the octane number of the isomerate in a similar way as in the TIP process.

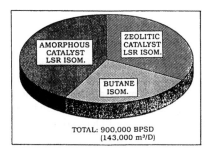

Figure 55. Shares of isomerization of light straight-run naphtha (LSR) with zeolitic and amorphous catalysts and isomerization of butane [34,37].

The equilibrium-limited nature of paraffin isomerization makes the concept of further integration of reaction and separation than in the TIP process of interest, i.e. application of reactive distillation or membrane catalysis. Conceptual schemes for membrane catalysis in paraffin isomerization are shown in Figure 56 for the production of

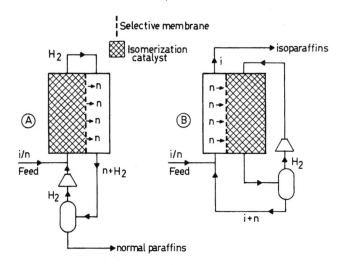

Figure 56. Conceptual schemes for membrane catalysis.
A: For production of normal paraffins.
B: For production of isoparaffins.

either normal paraffins or isoparaffins. In these schemes zeolites are involved in isomerization as well as separation. Realization of this concept requires specially adapted zeolites for both duties.

6. OTHER ZEOLITE-CATALYSED PROCESSES IN PETROLEUM REFINING

The processes discussed above represent the most widespread applications of zeolites in oil refining. It will be clear from the foregoing discussions that zeolites are not unique in these applications since catalysts based on amorphous materials are available as alternatives. Notwithstanding this and the fact that zeolites even have some drawbacks, e.g., with respect to access of molecules of large size, they offer sufficient advantages to secure a firm position in these applications, which undoubtedly will further increase in strength.

There are other, smaller scale applications of zeolites in oil refining, however, where zeolites are more unique. These include processes in which shape selectivity is involved, such as in the upgrading of reformate where residual low-octane paraffins are selectively removed by hydrocracking (Selectoforming) or by a combination of cracking and alkylation (M-forming). Shape selectivity of zeolitic catalysts also features in catalytic dewaxing processes for distillates (MDDW) and luboils (MLDW) and in oligomerization of light olefins to middle distillates of good quality (MOGD, SPGK). Zeolites can also have specific advantages as carrier for metals, e.g., noble metals. Cases in point are the high dehydrocyclization activity of Pt for light paraffins when supported on Ba/K/L-zeolite in the absence of sulfur (Aromax), and the high tolerance of noble metals for nitrogen and sulfur in the feed when they are supported on a zeolite. The latter finding forms the basis for a new process (SMDH) for deep hydrogenation of aromatics in diesel oil, which was succesfully commercialized recently by Shell [41].

These and other processes using microporous catalysts yet to be developed will enlarge the capabilities of the petroleum industry in the future to meet increasingly strict specifications on product quality.

7. MANUFACTURE OF SYNTHETIC FUELS

The present easy availability and low cost of crude oil have been deterrents to the large scale production of synthetic fuels or fuel components. Exceptions are the production of synthetic fuels from coal by Sasol in South Africa, gasoline from natural gas derived methanol by the MTG proces in New Zealand, and middle distillates from natural gas by the SMDS process in Malaysia, where special economic circumstances (cheap raw materials with limited market outlets, desire for self-suffiency in transport fuels) prevail. Another exception is the production of MTBE from isobutene and methanol, which has grown rapidly in the past 5 years because of lead elimination from gasoline and the required presence of oxygenates in "reformulated" gasoline. The growth in MTBE production is shown in Figure 57.

However, there can be little doubt that in the more distant future synthetic fuels manufacture will become important. Proven reserves of natural gas are increasing rapidly relative to oil and current estimates predict a gas reserve equal to or exceeding that of oil. Many gas finds at remote locations without nearby markets will provide an incentive for synfuels production from gas. Ultimately, coal will remain as a very large source of energy which may also be utilized as feedstock for synfuels production on a large scale. Table 8 lists some estimates of the ultimate economically recoverable reserves of different sources of fossil energy. These data may be compared with a current energy demand of about 0.1 billion barrels/day oil equivalent (worldwide, but excluding centrally planned economies).

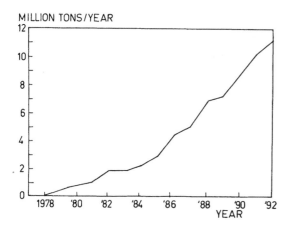

Figure 57. Growth in world production of MTBE [42].

Table 8.
Estimated ultimate economic recoverable sources of energy [43].

Source	Reserve, billion* bbl. oil equivalent
Oil	2000
Tar Sands	1500
Shale oil	2500
Gas	3000
Coal	53000

*) 1 billion = 1000 million

628

Many of the available process routes for the production of synfuels involve the use of zeolitic catalysts, as can be seen in Figure 58. A discussion of these processes shown in this figure is beyond the scope of the present paper. A survey of the application of zeolites as catalysts in synfuels production has been given in another publication [24].

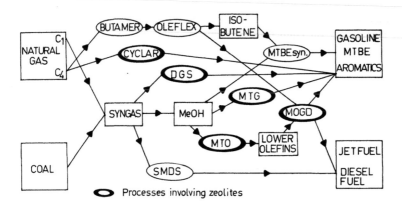

Figure 58. Catalytic process routes for the production of synthetic fuels from natural gas and coal.

8. CONCLUDING REMARKS

From the foregoing discussion it will be clear that the application of zeolites is now well entrenched in oil refining. The economic benefit derived from the use of zeolites in this area, especially in catalytic cracking and hydrocracking, is very large and may well reach dollar figures of ten to eleven digits per annum.

The drive towards more transport fuels from the oil barrel and the tighter specifications of oil products because of growing environmental concerns give rise to a need for more extensive and more precise chemical transformation of the molecules in oil fractions and in this respect zeolites have an increasingly important role to play as catalysts for these transformations.

Environmental considerations with respect to the manufacturing processes themselves also tend to favour clean and selective catalytic processes over others, e.g., thermal processes such as coking that are less selective and less environmentally friendly. The generally lower energy consumption of catalytic processes, as exemplified by a comparison of catalytic dewaxing with a non-catalytic alternative such as solvent dewaxing also tends to favour catalytic routes. Finally, catalytic routes for cleaning waste streams such as flue gases increase in importance and in this area too zeolites have a role to play.

The manufacture of synthetic fuels from natural gas and coal will ultimately become important in the future and this will further strengthen the role of microporous catalysts.

Up to now, only a limited number of zeolites, notably Y, mordenite and ZSM-5 have found large scale commercial use as catalysts. Even within this limited choice of basic structures, it has been found possible to adapt the properties of these zeolites by various ways of modification to suit particular needs. The very large and still increasing range of available microporous materials includes not only silica-alumina zeolites but also the large and growing family of aluminiumphosphates and other mixed oxides. New materials are being discovered and further developed, such as large-pore crystalline materials. The large and growing reservoir of available new materials implies that there is still quite some untapped potential in microporous catalysts. One may therefore be confident that catalysis with microporous catalysts is well placed to face the challenges that may arise in the hydrocarbon industry in the future.

LIST OF ABBREVIATIONS

CCR   Conradson Carbon Residue
DGS   Direct Gasoline Synthesis
FCC   Fluid Catalytic Cracking
LCM   Liquid Catching Macropores
MDDW   Mobil Distillate Dewaxing
MLDW   Mobil Luboil Dewaxing
MOGD   Mobil Olefin to Gasoline and Distillates
MTBE   Methyl Tertiary Butyl Ether
MTG   Methanol To Gasoline
MTO   Methanol To Olefins
RVP   Reid Vapour Pressure
SMDH   Shell Middle Distillate Hydrogenation
SMDS   Shell Middle Distillate Synthesis
SPGK   Shell Poly Gasoline and Kero
TIP   Total Isomerization Package

REFERENCES

1. I.E. Maxwell and W.H.J. Stork, Hydrocarbon Processing with Zeolites, in H. van Bekkum, E.M. Flanigan and J.C. Jansen (eds.) Introduction to Zeolite Science and Practice, Elsevier, Amsterdam, 1991, p.571.
2. H. May, H.J. Tausk, K. Oblӓnder and A.A. Reglitzky, Automotive Diesel Fuel - Future Supply and Demand, in Proc. 12th World Petroleum Congr., J. Wiley, Chichester, 1987, Vol 4, p. 187.
3. J.A. Kusumano, Chem. Tech., Aug. 1992, 482.

4. I.E. Maxwell, C. Williams, F. Muller and B. Krutzen, Zeolite Catalysis - for the Fuels of Today and Tomorrow, presentation for the Royal Society Science into Industry Exhibition and Soiree, London, May 1992.

5. A.A. Avidan, M. Edwards and H. Owen, Oil & Gas J., Jan. 8 (1990) 33.

6. J. Rabo, in F.A. Ribeiro, A.E. Rodrigues, L.D. Rollman and C. Naccache (eds.), Zeolite Science and Technology, Martinus Nijhoff, The Hague, 1984, p. 291.

7. L.A. Pine, P.J. Maher and W.A. Wachter, Ketjen Catalyst Symposium, Amsterdam, 1984. J. Catal., 85 (1984) 466.

8. J. Magnussen and R. Pudas, Activity and Product Distribution Characteristics of the Currently used FCC Catalyst Systems, Katalistics 6th Annual Fluid Catalytic Cracking Symposium, Munich, 1985.

9. B. de Kroes, C.J. Groenenboom and P. O'Connor, New Zeolites in FCC, Ketjen Catalysts Symposium 1986, Scheveningen, The Netherland, 1986.

10. J. Biswas and I.E. Maxwell, Appl. Catal. 58 (1990) 1.

11. R.M. Moore and J.R. Katzer, A.I.Ch.E. J., 18 (1972) 816.

12. D.M. Nace, Ind. Eng. Chem. Product Res. Dev. 9 (1970) 203.

13. P. O'Connor, L.A. Gerritsen, J.R. Pearce, P.H. Desai, S. Yanik and A. Humphries, Hydrocarbon Processing, Nov. 1991, p.76.

14. P. O'Connor, A.W. Gevers, A. Humphries, L.A. Gerritsen and P.H. Desai, Concepts for Future Residuum Catalyst Development, in M.L. Ocelli (ed.), Fluid Catalytic Cracking II. Concepts in Catalyst Design, ACS Symposium Series 452, Am. Chem. Soc., Washington D.C., 1991, p. 318.

15. M.J.P.C. Nieskens, M.J.H. Borley, K.-H. Roebschlaeger and F.H.H. Khouw, The Shell Residue Fluid Catalytic Cracking Process, paper presented at the National Petroleum Refiners Association Meeting, San Antonio, Texas, March 1990.

16. S.E. Voltz, D.M. Nace, S.M. Jacob and V.W. Weekman Jr., Ind. Eng. Chem. Process Des. Dev. 11 (1972) 261.

17. J.E. Naber and M. Akbar, Hydrocarbon Technology International 1989/1990, p. 37.

18. P. O'Connor, The Role of Diffusion in Bottoms Conversion, paper presented at the Ketjen Catalysts Symposium, Amsterdam, 1986.

19. P. O'Connor and A.P. Humphries, Accessibility of Functional Sites in FCC, paper presented before the Div. of Petroleum Chem., 206th National Meeting, Am. Chem. Soc., Chicago, Ill., Aug. 22-27, 1993.

20. J. Biswas and I.E. Maxwell, Appl. Catal. 63 (1990) 259.

21. E.E. Donath, Fuel Proc. Technology, 1 (1977) 3.

22. R.F. Sullivan and J.W. Scott, The Development of Hydrocracking, in B.A. Davis and W.P. Hettinger (eds.), Heterogeneous Catalysis, Selected American Histories. ACS Symposium Series 222, Am. Chem. Soc., Washington D.C., 1983, p. 293.

23. A. Hoek, T. Huizinga, A.A. Esener, I.E. Maxwell, W.H.J. Stork, F.J. van de Meerakker and O. Sy, Oil & Gas J., April 22, 1991, p. 77.

24. P.M.M. Blauwhoff, E.P. Kieffer, S.T. Sie and W.H.J. Stork, Zeolites as Catalysts in Industrial Processes, in L. Puppe and J. Weitkamp (eds), Catalysis and Zeolites - Fundamentals and Applications, Springer, Berlin, in the press.

25. J.E. Naber, W.H.J. Stork, P.M.M. Blauwhoff and K.J.W. Groeneveld, Technological Response to Environmental Concerns, Product Quality Requirements and Changes in Demand Pattern, Lecture at DGMK Meeting, Sept. 1990. Oil-Gas European Magazine 1 (1991) 27.

26. P.J. Nat, Erdoel und Kohle, Erdgas, Petrochemie, 42 (1989) 447.

27. A.A. Esener and I.E. Maxwell, Improved Hydrocracking Performance by Combining Conventional Hydrocracking and Zeolites in Stacked Bed Reactors, in M.L. Ocelli and R.G. Anthony (eds.), Advances in Hydrotreating Catalysts, Annual A.I.Ch.E. Meeting, Nov. 27-Dec. 2, 1988, Washington D.C., 1988, Elsevier, Amsterdam, p. 263.

28. H.W. Haynes, J.F. Parcher and N.E. Helmer, Ind. Eng. Chem. Process Des. Dev., 22 (1983) 401.

29. I.E. Maxwell, Catal. Today, 1 (1987) 385.

30. A. van Dijk, A.F. de Vries, J.A.R. van Veen, W.H.J. Stork and P.M.M. Blauwhoff, Catal. Today, 11 (1991) 129.

31. T.-Y. Yan, Ind. Eng. Chem. Process Des. Dev., 22 (1983) 154.

32. J.W. Gosselink, W.H.J. Stork, A.F. de Vries and C.H. Smit, in B. Delmon and G.F. Froment (eds.), Catalyst Deactivation 87, Elsevier, Amsterdam, 1987, p. 279.

33. H.W. Kouwenhoven and W.C. van Zijll-Langhout, Chem. Eng. Progress, 67 (1971) 65.

34. P.J. Kuchar, J.C. Bricker, M.E. Reno and R.S. Haizmann, Fuel Proc. Technology, 35 (1993) 183.

35. P.B. Koradia, J.R. Kiovski and M.Y. Asim, J. Catal., 66 (1980) 290.

36. A. Hennico and J.-P. Cariou, Hydrocarbon International 1990/1991, p. 68.

37. M.E. Reno, R.S. Haizmann, B.H. Johnson, P.P. Piotrowski and A.S. Zarchy, Hydrocarbon International 1990/1991, p. 73.

38. S.T. Sie, Ind. Eng. Chem. Research, 31 (1992) 1881.

39. S.T. Sie, Ind. Eng. Chem. Research, 32 (1993) 403.

40. H.H. Mooiweer, K.P. de Jong, B. Kraushaar-Czarnetski, W.H.J. Stork and P. Grandvallet, Oil & Gas J., submitted for publication.

41. J.P. van den Berg, J.P. Lucien, G. Germaine and G.L.B. Thielemans, Fuel Proc. Technology, 35 (1993) 119.

42. Shell Briefing Service, Issues in Refining, Shell Int. Petroleum Co. Ltd., Group Public Affairs, London 1989.

43. K. Davies, The Oil/Chemical Interface, Shell Int. Petroleum Co. Ltd., Group Public Affairs, London, December 1988.

J.C. Jansen, M. Stöcker, H.G. Karge and J. Weitkamp (Eds.)
Advanced Zeolite Science and Applications
Studies in Surface Science and Catalysis, Vol. 85
© 1994 Elsevier Science B.V. All rights reserved.

Application of molecular sieves in view of cleaner technology. Gas and liquid phase separations

F. Fajula[a] and D. Plee[b]

[a]Laboratoire de Chimie Organique Physique et Cinétique Chimique Appliquées, URA 418 CNRS. School of Chemistry, 8 Rue de l'Ecole Normale, 34053 Montpellier Cedex, France

[b]CECA Adsorption, Groupement de Recherches de Lacq, B.P. 34, 64170 Lacq, France

1. INTRODUCTION

Because of their unique selectivities displayed in the fields of catalysis, adsorption and ion-exchange, zeolite molecular sieves have provided, since the very beginning of their commercialization, effective solutions for controlling pollution of our environment. Zeolite-based technologies can indeed address the environmental issue following the two general strategies i) by minimizing the production of pollutants and ii) by a secondary treatment of the effluents. The introduction of the EniChem hydroxylation process on Ti-MFI and the use of zeolites as sequestring agents in detergents to substitute phosphates are examples, among many others, of the first approach. Examples of the second approach are found in the nuclear industry, in radioactive waste storage and cleanup, or in the removal of SO_2 and NOx from industrial flue gases.

Law regulations restricting pollutant emissions exist indeed for long. In the past, however, the economic feasability was the main consideration in determining the best available control technology. Today, the legislation strengthenspollution control while the obvious solutions have been tried. More elaborate or creative technologies will be needed therefore , in which zeolites will, undoubtedly, play a role.

In this brief overview we present some of the challenges where the use of zeolites as selective adsorbents may offer one effective solution.

Adsorption/separation processes have led to the first commercial application of zeolites A, X and Y (structural types LTA and FAU) some fourty years ago. These zeolites have gained now a major technological importance throughout the petroleum and chemical industries and their dominant role will probably not be challenged in a near future. Actually, the environmentally forced new developments involving zeolite molecular sieves that emerge today concern essentially the use of hydrophobic/organophilic high-silica zeolites (hydrophobic molecular sieves or HMS) for the adsorption of volatile organics.

Despite a large number of scientific papers and patents - mostly based on laboratory work - on the subject, only few examples can be found in the public and technical literature demonstrating the feasability, or drawbacks, of such a technology. In this chapter, the potential and limitations of these materials are discussed and some examples that the authors have considered as representative of the current trends, are presented.

2. ADSORPTION OF VOCs ON HYDROPHOBIC ZEOLITES

2.1 What are VOCs ?

Since VOCs (volatile organic compounds) are of high concern in most developped countries, a great deal or research is devoted to find out solutions to reduce or suppress their emission. This can be achieved either by designing new cleaner processes, where emissions are drastically reduced, by reusing the emitted compounds, solvent recovery being a good example, or by destroying them.

According to the United States Environmental Protection Agency (US EPA), VOCs correspond to stable products exhibiting a vapor pressure above 0.1 mm Hg under normal temperature and pressure conditions. Molecules of the VOC's family are found therefore in almost every branch of the chemical industry and belong to the different groups of organic chemistry (alcohols, ketones, aromatics, chlorinated hydrocarbons,..)

Several environmental problems are associated with the emission of VOCs in the atmosphere.

One of the most widely discussed issue is the greenhouse effect which is believed to raise the average earth temperature. VOCs are charged with the tropospheric ozone production and, more generally speaking, with the photochemical reactions involved in smog formation in urban areas and in the forest decay.

VOCs are sometimes toxic, carcinogenic, irritating and/or flammable compounds (1,2).

The Centre Interprofessionnel Technique d'Etude de la Pollution Atmospherique (CITEPA) has estimated (3) the total EEC VOC's emission in 1985 to 20 million tons per year. The main sources are :

Transportation	31 %
Solvent uses	19 %
Industrial processes	3 %
Refineries	1 %
Combustion	4 %
Miscellaneous[*]	42 %

[*]Gas distribution, coal mining, domestic wastes.

The commitment of the French governement is to reduce the VOC's emissions by 30 % at the turn of the century and stringent regulations are also implemented in other countries.

VOC's control is indispensable for either economical or environmental reasons. It is for example a rather common operation today to recover solvents instead of incinerating them. Recovery for reuse is not always considered as the more economical solution but it is currently considered as a more and more acceptable option by the public opinion (4, 5).

The most commonly recovered solvents are:
- toluene
- heptane
- hexane
- carbon tetrachloride
- acetone
- ethyl acetate

- methyl ethyl ketone (MEK)
- naphthalene
- methylene chloride

but many other organics are also suitable for recovery by adsorption.

2.2 Processes and technology

Adsorption/separation processes are based on a difference in either the adsorption equilibrium or the intracrystalline diffusitivities. In the first case the selectivity is determined by the thermodynamics of the adsorption whereas in the second the kinetics dominate. A very large body of scientific literature, books and reviews analyzes the influences of the composition and the pore architecture on both the energetics and kinetics of adsorption and diffusion in zeolites (see for example 6-9). The way in which the two factors superimpose during the process of adsorption is still matter of intensive research.

With respect to the applications currently in operation at an industrial scale, the basic principle of an adsorption/separation process is that the selectively adsorbed species is adsorbed and desorbed in sequence. Technological differences are found essentially in the type of contact between the mobile and sorbent phases (gas/solid or liquid/solid, co or counter-current, fixed or moving bed) and in the desorption method (PSA, TSA, VSA or use of a desorbent) (10-12).

Adsorption of volatile organics is most generally performed using fixed beds of adsorbents or rotor concentrators, using steam or hot air for the desorption step (13).

- Fixed bed technology

This technology is based on a two adsorbers scheme (Figure 1) for processing gas or wastewater streams. One bed is operating on the adsorption mode and the flux is shifted to the second one once the breakthrough occurs. The first bed is regenerated by steam, the consumption of which depends on the volume and nature of the chemicals to be desorbed.

It may be considered as a rule of thumb that more steam or higher temperature will be necessary the higher the boiling point of the adsorbed molecules. For activated carbon adsorbents, typical steam consumptions are in the range of 0.25 to 0.5 kg of steam per kg of adsorbent.

The stripped gas is then condensed and the VOCs recovered by gravity. Water soluble solvents require distillation instead of decantation.

This process description holds for a typical adsorption/desorption cycle where the clean effluent coming off the adsorbent bed is vented (open loop design). Another process (close loop design), employed for wastewater treatment, uses the clean gas in a counter-current stripper which separates VOCs from influent wastewater and sends the vapor stream to the adsorbent beds.

- Rotor concentrator

This technology is used in cases where short contact times must be achieved, for small VOCs concentrations and large volumes of gas streams. The gas to be processed flows through a corrugated wheel whose honeycomb structure has been coated with powdered adsorbent . The continuously rotating wheel is divided into three zones, one for adsorption, one for desorption and the third for cooling the regenerated zone in order to maintain the operating temperature (Figure 2). The VOCs stream may be flown axially or radially throuh

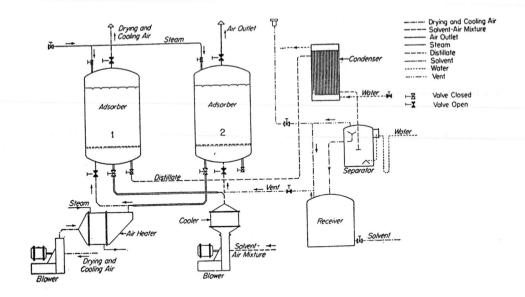

Figure 1. Flow scheme of a two adsorbers separation unit.

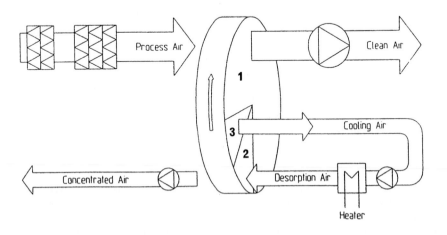

Figure 2. Operation principle of a rotor concentrator. (1 : Adsorption zone ; 2 : Regeneration zone ; 3 : Cooling zone).

the wheel. The desorption effluent contains stripped VOCs at 5 to 15 times the concentration of the influent gas.

Such a design allows for very short contact times and the limiting parameter becomes the gas diffusitivity in the pores of the adsorbent.

An alternative option exists which comprises a rotating wheel loaded with granular adsorbent. The process conditions lead to longer contact times since, in this case, mass transport is also limited by inter-particle diffusion. Adsorption capacities are usually much higher than in the honeycomb design since the adsorbent weight is several orders of magnitude greater.

The operating and installation costs of this technology are relatively low. Rotor concentrators are widely used today to control solvent vapors in painting cabins or plastics industries.

2.3 Hydrophobic molecular sieves adsorbents

The emergence of highly siliceous zeolites (14, 15) opened the way to new adsorbents for organics in water and moist streams. This is readily explainable since neutral adsorbents are not expected to adsorb molecules with a low electron density, such as water, to a great extent.

Chen first reported this phenomenon and proposed the use of HMS to remove organics from water (16, 17). This result stimulated numerous researches aiming to applications in the field of depollution. More generally, it must be pointed out that separations can be achieved on highly siliceous zeolites according to the molecule polarity, even in the absence of water. For example, aluminium containing faujasites adsorb preferentially aromatics from aromatics/paraffins mixtures (18). The opposite behaviour is observed on solids with Si/Al ratios greater than 30 (9). These examples are worth mentioning but it seems clear that the main applications of HMS we may foresee will be in the depollution of air and water streams.

At present, two main zeolitic types, available at a commercial scale, are receiving attention. The first one belongs to the MFI type (silicalite, silica-rich ZSM-5) and can be obtained directly by synthesis (14, 15). The second is prepared by dealumination of FAU type zeolite (Y) in order to increase the silicon to aluminium ratio above *ca.* 50. Two main dealumination procedures - by combined hydrothermal/acid treatments and by reaction with silicon tetrachloride - allow to reach this goal and are well established now (19-21). The properties of the resulting adsorbents depend strongly on the preparation method (*vide infra*).

Uses of powdered hydrophobic molecular sieves do exist but, generally, the products have to be formed as beads, extrudates, full or hollow cylinders. Binders, when used, can significantly affect the hydrophobic properties of the adsorbent. Alumina containing binders should be avoided because of their proper water affinity and of the migration of aluminium atoms into the zeolite framework upon high temperature treatment (22, 23). Cellulose acetate, aluminium-poor sepiolite, colloïdal silica and tetraethoxysilane constitute possible binders providing good mechanical strength without loss of hydrophobic properties (24).

2.4 Compared properties of activated carbons and hydrophobic molecular sieves

Both activated carbons and HMS present advantages and drawbacks. The former are rather cheap materials, readily available from many companies. Moreover, they have high initial adsorption capacities. HMS show, generally, limited initial adsorption capacities, they are thermally and chemically very stable products and do not lead to side reactions. In most cases they are more easily regenerated than active carbons and exhibit better retention of performances. Their relative high cost prevents their extensive use for the moment. HMS are particularly useful for treating VOCs in small concentrations in streams with high relative humidity.

- Hydrophobicity

The definition of the hydrophobic character of HMS in terms of a separation factor or hydrophobicity index is not easy. The hydrophobic/organophilic nature of an adsorbent can be qualitatively revealed by single component isotherms, as shown in Table 1, for example, where the adsorption capacities for water and trichloroethylene (TCE) on activated carbon (Pinewood type) and dealuminated faujasite are reported for different sorbate concentrations in air (25).

Table 1
Adsorption capacities at 25 °C of activated carbon and dealuminated Y zeolite for two concentrations of water and TCE in air

	Water concentration (%, w/w)		TCE concentration (%, w/w)	
	0.2	1	0.2	1
Activated carbon	1.5	40	41	60
Dealuminated Y	0.5	4	23	26

It appears clearly from these data that the HMS has a smaller affinity for water than activated carbon. However, dealuminated Y is expected to perform better than carbon only for a high water concentration.

A better picture of the hydrophobic properties can be gained from the breakthrough curves determined in competitive adsorption experiments. Only this type of measurement allows the true adsorption capacities to be determined. It has to be understood that the terms hydrophobic or organophilic do not mean that water is excluded from the adsorbent. These terms only mean that at comparable pressures (or concentrations) of water and organic, the latter is preferentially adsorbed. However, at a very low coverage, if the water pressure is much higher than the organic pressure, a significant fraction of the total adsorption capacity will be taken up by water. This is actually the situation encountered in the adsorption of trace amounts of organics present in waste streams.

The competitive gas-phase adsorption measurements of water/ethanol (26,27), water/hydrocarbons (28,29) and water/chlorinated hydrocarbons (30,22) mixtures reported recently demonstrate that the shift from hydrophilic to hydrophobic of the surface of zeolites as a function of the Si/Al ratio is a well established phenomenon but the crossover point is strongly dependent on the nature of the molecule adsorbed, the structure type of the zeolite, the adsorption temperature and the final calcination temperature of the adsorbent. This latter effect is attributed to the presence of silanol groups, that constitute hydrophilic sites and require high temperature treatments to be annealed. Moreover, extrapolation of gas-phase results to the separation of organics from liquid water is not straightforward. A recent report by Radeke et al. (31) compared the phenol adsorption isotherms obtained on two carbons (Fi 400, 0.4 cm^3/g, 880 m^2/g and AG 3, 0.36 cm^3/g, 660 m^2/g), two polymeric resins (Wolfatites Y 77, 1 cm^3/g, 1250 m^2/g and EP 61, 1 cm^3/g, 350 m^2/g,) and two dealuminated faujasites with Si/Al ratios of 95 obtained, one by steam plus acid dealumination (DY 1, 0.22 cm3/g, 690 m²/g) and the other by SiCl$_4$ treatment (DY 2, 0.19 cm^3/g, 500 m²/g). The results, shown in Figure 3, reveal a very poor hydrophobicity of the two zeolites, in spite of their high Si/Al. These results contest some of the conclusions drawn from the gas-phase experiments and point out the importance of the mode of preparation of the zeolite adsorbents and the limits of model studies.

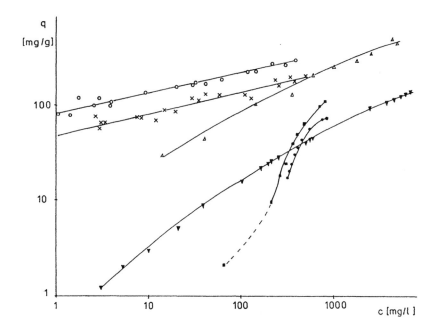

Figure 3. Phenol adsorption isotherms on (o) Fi 400, (x) AG 3, (▼) Y 77, (▲) EP 61, (·) zeolite DY 1 and (■) zeolite DY 2.

Another factor that can also affect the efficiency of HMS is the presence of competing species such as salts, high boiling organics or other impurities. The latter can induce chromatographic effects on the adsorbent which lead to premature breakthrough.

- Inertness and regenerability
    HMS constitute an interesting alternative to activated carbons when the VOC stream presents a fire risk with the latter. This can particularly occur when solvents such as oxygenates have to be removed. In addition, impurities as well as ashes present on the carbon surface can act as catalysts that promote either polymerization (styrene) or oxidation (MEK, cyclohexanone). Both risks are avoided on siliceous adsorbents.
    Another advantage of HMS lies in their easier regenerability. Steam regeneration of activated carbon leaves a solid heel of adsorbate very difficult to remove without causing damage to the adsorbent. Furthermore, steam hydrolysis may occur, giving rise to hydrochloric acid and corrosion. In most applications, the use of steam is not necessary

with HMS as complete desorption can be achieved with air at moderate temperatures. However, as discussed under section 3.2 below, the easier regeneration of HMS may constitute a severe drawback in some specific applications.

To date, no general guide can be proposed to choose either hydrophobic zeolites or activated carbons. The real systems to be processed are much more complex than the ones generally handled in laboratory experiments. Several parameters interfere, the importance of which being not necessarily related directly to the technical problem to be solved (location of the pollutant source, permanent or temporary pollution, constancy of the emission composition). The advantages of using HMS adsorbents may come from energy savings and better recovery rates, but must be ascertained by a case by case study. This is illustrated by the three examples presented in the next section.

3. EXAMPLES OF APPLICATIONS

3.1 Solvent recovery

Early solvent recovery plants have been installed essentially for economic reasons, the value of the reusable solvent justifying the investment in very short term. Today, as environmental regulations tighten and treatment of more and more dilute air streams is required, the economic decision becomes one of recovery versus destruction methods, such as incineration (4). In any case, solvent separation and concentration from the process air constitutes an intermediate stage. This is usually achieved using activated carbon adsorbents.

Though widely applied, the latter present several problems (flammability, strong retention of high-boiling solvents, surface reactivity, humidity control) that could be readily solved by using hydrophobic zeolite adsorbents. As the cost of hydrophobic zeolite is still very high, its use will be economically limited to applications for which activated carbon is not well-suited. The two examples (32) developed below present case studies showing that activated carbon and hydrophobic zeolites (MFI type) are complementary technologies rather than competing solutions.

The figures given in Table 2 illustrate the influence of the composition of the process air on the performance of zeolite adsorbent. The adsorption device was a rotor concentrator designed to handle air flows up to 79 000 Nm^3/h. The process air was flown radially into the center of the cylinder on which the honeycomb elements coated with zeolite had been attached.

As seen in the table, the zeolite removal efficiency is largely independent of solvent concentration within a large range, suggesting the possibility of handling important concentration fluctuations. In addition, high efficiencies are obtained for solvents with differing polarities and boiling points. However, unlike activated carbon, the hydrophobic zeolite shows little efficiency for solvents like solvesso 100 (a C_9 aromatic cut) and xylene, when treating solvent composition one. This has probably to do with the size of the molecules whose diameters are of the same order, or even greater, than the zeolite pore size. With solvent composition two, the removal efficiency of xylene is, by contrast, much higher and an overall abatement over 95 % is obtained.

This first example confirms the performaces of zeolite adsorbents for treating emission streams. However, such a technology could be successfully applied only in the case of solvent composition two.

Table 2
Efficiency of hydrophobic zeolite as a function of effluent composition.

Technology : Rotor concentrator

Adsorbent : Silicalite

Test conditions :

Inlet gas temperature(°C)	32
Relative humidity (%)	60
Face velocity (m/sec)	1.8
Reactivation temperature (°C)	160
Concentration ratio	10

Solvent composition one

Solvent	Vol.(%)	BP(°C)	Concentration (ppm) In	Out	Efficiency
Solvesso 100	20	160	14	1.72	87.7
Ethanol	21.9	78	15.3	3.37	78
MAK	16.1	150	11.3	N.D.	100
MEK	8.3	80	5.9	0.05	99.1
MIBK	7.6	118	5.3	N.D.	100
But. Acet.	9.9	126	6.9	N.D.	100
Xylene	9.7	138	6.7	1.15	82.8
Oct. Acet.	6.5	200	4.6	N.D.	100
	Average	127	70	6.29	91

Solvent composition two

Solvent	Vol.(%)	BP(°C)	Concentration (ppm) In	Out	Efficiency
But. Cell. Acet.	5	192	4.3	0	100
Xylene	80	138	68	3.3	95.5
IPA	8	66	6.8	0.8	88
Et. Acet.	3	77	2.6	0	100
But. Cell.	4	168	3.4	0	100
	Average	134	85	4.1	95.2

The second case study concerns the elimination of solvents from an exhaust stream of 560 000 Nm^3/h containing about 90 ppm of polluants and requiring > 95 % abatement. Because of the fire hazard presented by the stream containing 30 - 40 % ketones, adsorption on activated carbon was immediately eliminated as an option.

Two possible systems were then evaluated; direct regenerative thermal oxidation and hydrophobic zeolite concentration/thermal oxidation. Each system was capable of meeting the performance criteria. Table 3 summarizes the costs of each option. The installation cost for the zeolite-based technology was slightly lower but, above all, the utility costs made the choice obvious.

Table 3

Comparative costs for the direct regenerative thermal oxidation (option A) and hydrophobic zeolite concentration/thermal oxidation (option B) processes (10^3$)

	Option A	Option B
Installation cost	8 200	7 800
Fuel	574.6	79.2
Electricity	496.1	134.9
Dry Filters	-	60.2
Annual cost	1 070.7	274.3

3.2 Cold-start hydrocarbon collection form automobile exhausts

Catalytic converters designed to remove air polutants from automobile exhausts are fully operational only at sufficiently high temperature (catalyst light-off temperature). Emissions that exit the engine while the catalyst is still cold pass therefore the exhaust system intact and represent an appreciable fraction of the air pollutants released into the atmosphere. Cold-start hydrocarbons (HCs), in particular, can account for more than 60% of the integrated HC's emission. Typical HC's concentrations stand in the range 1000 - 1500 ppm. Water, carbon monoxide, carbon dioxide and NOx constitute the major exhaust components.

One strategy to control hydrocarbon cold-start emissions is to collect and store them in an adsorbent until they can be released to the activated catalyst (33,34). The use of a high silica zeolite trap as adsorbent presents several advantages due to its hydrophobicity and high thermal stability. Thus the zeolite trap could be positioned near the engine and be self-generating in such a configuration; hydrocarbons adsorbed during the cold start would desorb and react as the catalyst warms up and becomes operational. Activated carbons, in contrast, should be positioned far behind the exhaust catalyst to prevent combustion of the carbon.

The concept of the zeolite adsorption trap has been tested by Heimrich et al.(35) on a laboratory prototype adapted to the exhaust system of a commercial gasoline-fueled vehicle (1986 Honda Accord LXi). The adsorbent was a ZSM-5 zeolite with a Si/Al ratio of 100-150 which was coated on three cordierite honeycomb substrates. Each substrate measured 8.3 cm in diameter and 12.5 cm in length and had a square cell geometry with 62 cells per square centimer. The design of the adsorbent device was based on the expected zeolite adsorption capacity and the quantity of cold-start hydrocarbons emitted from the vehicle.

Such an arrangement led to a total volume of two liters which corresponded to 1.4 times the volume of the original catalytic converter.

Due to the low desorption temperature of the zeolite (around 80°C) as compared to the working temperature of the catalyst (typically automotive catalysts light-off around 400°C and routinely operate in the range 400 to 600°C), the adsorbent cannot be installed directly upstream of the catalytic converter. This limitation implies a rather complicated valving and piping arrangement to isolate the trap from the exhaust stream during regeneration and to purge the adsorbed HCs into either the intake air or the exhaust stream in front of the catalyst after warm-up. Two of the possible configurations are represented in Figure 4 and the corresponding management of the main valves is given in Table 4. Alternate configurations consider the possibility of using hot exhaust gas instead of air for hydrocarbon desorption. The adsorbent is installed downstream of the catalyst, in a cooler location in order to improve its sorption capacity. The catalyst will therefore receive directly the engine exhaust heat to bring it to light-off temperature as soon as possible, thereby lowering total emissions.

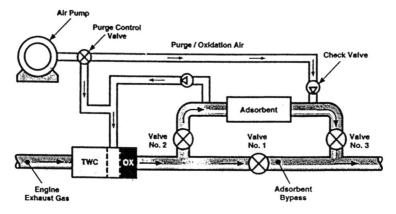

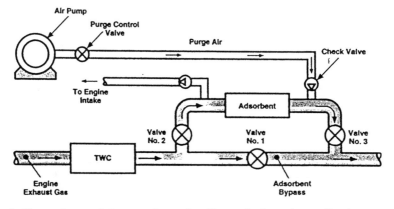

Figure 4. Flow scheme of the experimental cold-start hydrocarbon collection

Table 4.
Exhaust management valve positions

Valve number	Hydrocarbon Collection	Hydrocarbon Storage & Purge
1	closed	open
2	open	closed
3	open	closed

Results of one of the tests conducted over a period of time of 600 seconds are presented in Figure 5. In this experiment the vehicle was started, allowed to idle for 20 seconds and then accelerated to a constant speed of 48.3 km/h (30 mi/h). Net hydrocarbon collection was observed until the adsorption bed temperature reached 83°C. Desorption occurred then, as illustrated by a higher downstream hydrocarbon concentration. This desorption continued beyond the point where catalyst light-off occurred (when its temperature reached 390°C) due to the high temperature of the adsorbent bed.

Other experiments conducted using different standardized driving schedules are described in the work of Heimrich et al.(35). They demonstrate that a 35 % reduction in cold-start hydrocarbon emissions can be achieved using zeolite adsorbents. Such an improvement is however small compared to that of electrically-preheated catalysts. The latter, which currently represents the leading technology, have shown reductions in total hydrocarbon cold-start emissions of 80%. Several shortcomings of the zeolite-based adsorption trap technology have been identified. Firstly, the low sorption capacity of the zeolite imposes too large adsorbent volumes to be practical. The primary reason for insufficient HCs storage is the competitive adsorption of the high concentrations of water vapor and other components present in automotive exhaust.

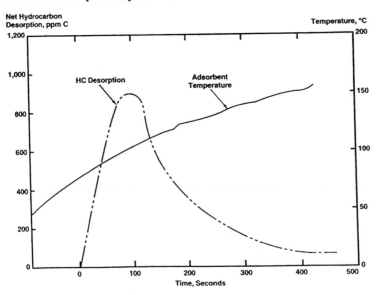

Figure 6. Hydrocarbon desorption with hot exhaust gas.

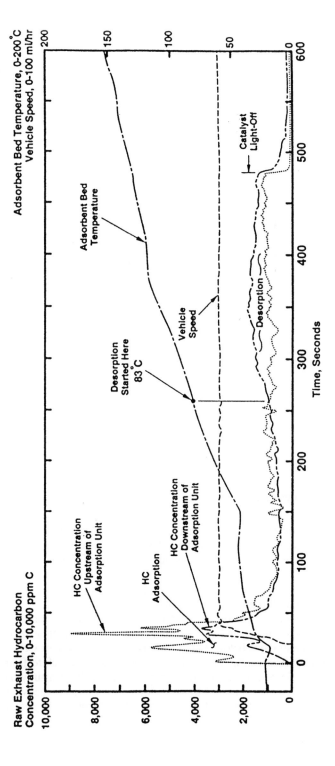

Figure 5. Hydrocarbon adsorption and desorption for a cold-start experiment.

Secondly, the concept of the purging of the collected hydrocarbons must be reconsidered. Figures 6 and 7 show desorption profiles using hot exhaust gas and secondary air (from an air pump), respectively. Using hot exhaust gas (Fig. 6), desorption appeared to be nearly complete after 400 seconds but the adsorbent bed temperature was 155°C. This is significantly too high for starting a new adsorption cycle. An air purge (140 l/min) allowed to heat slowly the adsorbent, retarding desorption, but after 600 seconds the element had not completely desorbed (Fig. 7). The above two points suggest that potential adsorbents need to be selected on the basis of their desorption characteristics as well as their ability to adsorb components from the exhaust. Finally one main problem to be solved is the management of a (cost effective) complex system of valves, sensors and heat exchangers in order to control adsorption and desorption temperatures, desorption completeness, space velocities.

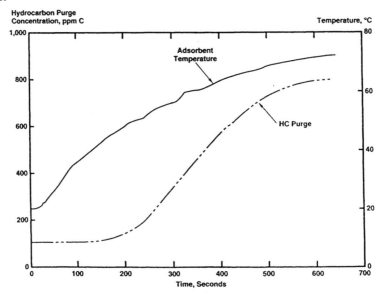

Figure 7. Hydrocarbon desorption with air.

A significant improvement could be gained by designing an adsorbent able to retain selectively the most harmful constituents emitted. The future use of zeolites in exhaust post-treatment systems will then rely on the efforts and creativity of the automotive engineers and zeolite scientists.

3.3 Removal of chlorinated hydrocarbons from water

Reduction of chlorinated hydrocarbons (CHCs) discharges into air and water is a major challenge for the chemical and dry cleaning industries (36). New regulations will only allow CHC's concentrations below 3 ppm, depending on the type of CHC, in aqueous effluent streams. The CHC's concentrations have to be reduced therefore from saturation levels down to sub ppm levels. Water purification treatments will be mainly based on two processes: air or steam stripping followed by adsorption. Adsorption on activated carbon or polystyrene resins has been widely used in the past (37). Most of the required effluent treatment plants will have to deal with flow rates in the range 50 to 200 m^3/h. An in-situ regeneration is therefore needed to cut the cost of the replaced adsorbent. The limited options and difficulties of regenerating activated carbon and resin systems on a large scale pose major problems to their use for this type of separation. In that respect, HMS, which present a good compromise between sorption capacity, specificity, regenerability and mass transfer, appear then as potentially interesting adsorbents (22,30).

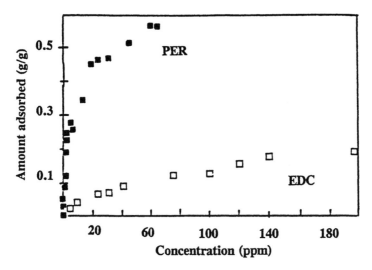

Figure 8. Adsorption isotherms of DCE and PER on activated carbon.

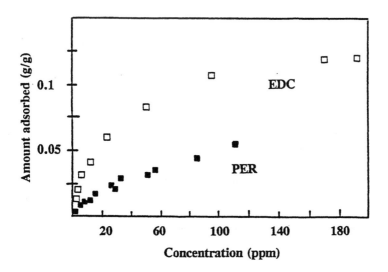

Figure 9. Adsorption isotherms of DCE and PER on silicalite.

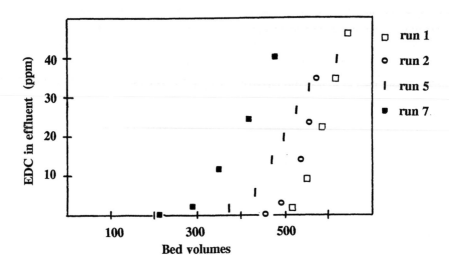

Figure 10. DCE breakthrough curves on carbon in real effluent treatment.

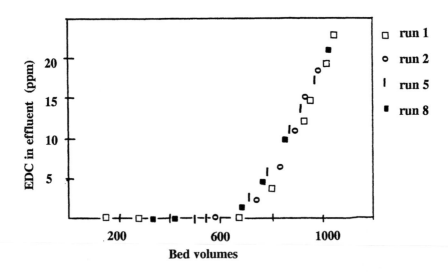

Figure 11. DCE breakthrough curves on silicalite in real effluent treatment.

Figures 8 and 9 show single component isotherms for 1,2-dichoroethane (DCE) and tetrachloroethylene (PER) on activated carbon and silicalite, respectively (38). PER is more strongly adsorbed than DCE on activated carbon with nearly twice the capacity. The reverse is true on silicalite. Both adsorbents show a relatively high loading at very low concentrations, which is a prerequisite for the removal of trace impurities, but the sorption capacity of silicalite is definitely much smaller than that of activated carbon.

The regenerability of the two adsorbents has been investigated with model solutions (500 to 1500 ppm of EDC and 80 to 90 ppm of PER) using a temperature swing process with preheated (150°C) nitrogen as purge gas. EDC was readily desorbed from both adsorbents, at a temperature around 100°C. For PER, the desorption was complete at the final regeneration temperature only in the case of silicalite. After 10 cycles, the solid heel for activated carbon was 0.05 g/g of adsorbent and the adsorption capacity decrease by 15 % from its original value. The solid heel on silicalite was 20 times less and the cyclic sorption capacity was reduced by less than 2 %.

Breakthrough curves obtained in studies on real plant effluents containing 40 ppm of DCE as main impurity and trace amounts of high boiling CHCs and hydrocarbons are shown in Figures 10 and 11. With activated carbon no cyclic capacity could be observed due to the loss of adsorption sites through adsorption of high boiling impurities. Regeneration temperatures above 270°C would be needed to regenerate completely the bed. After 10 runs the capacity dropped by 56 %.

Silicalite, by contrast, reached a stable cyclic capacity after 4 runs. Its capacity was decreased by only 16%. The exclusion of high boiling molecules such as hexachlorobenzene from the micropores by a molecular sieving effect and the easier regeneration explain such a behaviour.

4. CONCLUSIONS

Among the new zeolite-based adsorption technologies with the aim of a better control of pollutant emissions that emerge today, the removal of volatile organic compounds on hydrophobic molecular sieves is the most documented issue, in both the scientific and technical literature. This new technology appears as complementary to the one based on activated carbons, in niche applications.

In most of the applications developed to date, HMS are incorporated into carbon-based existing processes, without major modifications. In those cases they show high efficiency for treating small pollutant concentrations present in large streams.

HMS offer net advantages because of their chemical inertness and high thermal stability. They are particularly useful in solvent recovery operations where a fire risk exists when using activated carbon.

Besides their price, the main disadvantage of HMS is their relatively small initial adsorption capacity. A better exploitation of their unique shape selective properties could overcome this limitation.

Because of the complexity of the effluents to be processed, only case studies can decide on the introduction of such a technology in actual installations.

650

REFERENCES

1. M. Maës, Revue "Eau, Industrie, Nuisances", 154 (1992) 33.
2. G. Martin, Biofutur, p. 22 Sept. 1993.
3. Special Report on VOCs, Energie Plus, p. 19, July 1993.
4. M.J. Ruhl, Chem. Eng. Progress, 123 (1993) 37.
5. M.H. Stenzel, Chem. Eng. Progress, 89 (1993) 36.
6. R.M. Barrer, "Zeolites and Clay Minerals as Sorbents and Molecular Sieves", Academic Press, 1978.
7. J. Kärger and D.M. Ruthven, "Diffusion in Zeolites", Wiley, 1992.
8. H. Stach, U. Lohse, H. Thamm and W. Schrimer, Zeolites, 6 (1986) 74.
9. R.M. Dessau, "Adsorption and Ion Exchange with Synthetic Zeolites", W.H. Frank, Ed., ACS Symp. Ser. 135 (1980) 123.
10. "Fundamentals of Adsorption", A.L. Myers and G. Belfort, Eds., AIChE, 1983.
11. "Adsorption, Science and Technology", A.E. Rodrigues, M.D.LeVan and D. Tondeur, Eds., NATO ASI Series, Ser. E Vol. 158, 1988.
12. "Zeolite Technology and Applications, Recent Advances", J. Scott, Ed., Noyse Data Corp. USA, p. 260, 1980.
13. J.R. Graham and M. Ramaratnam, Chem. Eng., p. 6, Feb. 1993.
14. R.J. Argauer and G.R. Landolt, US Pat. 3 702 886 (1972).
15. E.M. Flanigen, J.M. Bennett, R.W. Grose, J.P. Cohen, R.L. Patton, R.M. Kirchner and J.V. Smith, Nature, 271 (1978) 612.
16. N.Y. Chen, US Pat. 3 732 326 (1973).
17. N.Y. Chen, J. Phys. Chem.,80 (1976) 60.
18. C.N. Satterfield, C.S. Chen and J.K. Smeets, Aiche, 20 (1974) 612.
19. J. Scherzer, J. Catal., 54 (1978) 285.
20. U. Lohse, E. Alsdorf and H. Stach, Z. Anorg. Allg. Chem., 447 (1978) 64.
21. H.K. Beyer and I. Belenykaja, Stud. Surf. Sci. Catal., 5 (1980) 203.
22. B. Lledos, Thesis, University of Montpellier, 1993.
23. D. Anglerot, Personnal communication.
24. E. Sextl, E. Roland, P. Kleinschmit and A. Kiss, Eur. Pat. 516 949 A1 (1992).
25. D. Plee, Proc. IInd Intern. Symp. Characterization and Control of Odours and VOC in the Process Industry, Louvain la Neuve, Belgium, Nov. 1993, in press.
26. B. Günzel, J. Weitkamp, S. Ernst, M. Neuber and W.D. Deckwer, Chem. Ing. Tech., 61 (1989) 66.
27. J. Weitkamp, S. Ernst, B. Günzel and W.D. Deckwer, Zeolites, 11 (1991) 314.
28. C.H. Berke, A. Kiss, P. Kleinschmit and J. Weitkamp, Chem. Eng. Tech.,63 (1991) 623.
29 J. Weitkamp, P. Kleinschmit, A. Kiss and C.H. Berke, Proc. 9th Intern. Zeolite Conf., Montreal 1992, R. von Ballmoos, J.B. Higgins, M.M.J. Treacy, Eds., Butterworth-Heinemann, Vol. II, 79 (1992).
30. R. Schumacher, S. Ernst and J. Weitkamp, Proc. 9th Intern. Zeolite Conf., Montreal 1992, R. von Ballmoos, J.B. Higgins, M.M.J. Treacy, Eds., Butterworth-Heinemann, Vol. II, 89 (1992).
31. K.H.Radeke,U. Lohse, K. Struve, E. Weiss and H. Schröder, Zeolites, 13 (1993) 69.
32. S.W. Blocki, Environ. Progress, 12 (1993) 226.

33. K. Otto, C.N. Montreuil, O. Todor, R.W. McCabe and H.S. Gandhi, Ind. Eng. Chem. Res.,30 (1991) 2333.
34. H. Von Bluecher and E. De Ruiter, Ger. Offen. 4 039 951; 4 039 952 (1992).
35. M.J. Heimrich, L.R. Smith and J. Kitowski, Soc. Auto. Eng., 920847 (1992).
36. M. Helou, Actualité Chimique, 69, Jul. - Sept. 1993.
37. M. Schäfer, H.J. Schröter and G. Peschel, Chem. Eng. Technol., 14 (1991) 59.
38. L. Utiger, A.F. Gordon, D.L. Cresswell and L.S. Kershenbaum, NSF/CNRS Workshop "Adsorption Processes for Gas Separations" Gif sur Yvette, 249, 1991.

J.C. Jansen, M. Stöcker, H.G. Karge and J. Weitkamp (Eds.)
Advanced Zeolite Science and Applications
Studies in Surface Science and Catalysis, Vol. 85
© 1994 Elsevier Science B.V. All rights reserved.

Crystalline Microporous Phosphates: a Family of Versatile Catalysts and Adsorbents

J.A. Martens and P.A. Jacobs

Centrum voor Oppervlaktechemie en Katalyse
KU Leuven
Kard. Mercierlaan 92, B-3001 Heverlee, Belgium

1. INTRODUCTION

The development of crystalline microporous phosphates was initiated by Wilson et al. in 1982, with the synthesis of a series of materials with aluminophosphate composition [1]. In the relatively short period of time since that elapsed invention, the diversity of structure types and compositions of phosphate based crystalline microporous oxides has become comparable to that of the silicate based zeolites and molecular sieves. The incorporation of silicon next to phosphorus and aluminium in crystalline microporous oxide frameworks resulted in the synthesis of silicoaluminophosphates in 1984 [2]. For various structure types, the relative amounts of Si, Al and P in the framework can be varied such that gradual transitions from phosphate based to silicate based frameworks are possible. Actually, 19 different transition metals and main group elements with valencies ranging from I to V, viz. (Li(I), Co(II), Fe(II), Mg(II), Mn(II), Zn(II), Be(II), Ni(II), Sn(II), B(III), Al(III), Ga(III), Fe(III), Cr(III), Si (IV), Ge(IV), Ti(IV), As(V) and V(V) have been successfully combined with the P(V) framework element in at least one framework type [3]. A newcomer in the field may get the impression that any combination of elements is possible, which is certainly not the case.

Rather than enumerating catalytic conversion reactions and sorptive processes for which crystalline microporous phosphate based materials have shown to possess interesting properties, in this chapter some concepts are developed allowing to rationalize the physico-chemical properties of this family of materials. Special attention has been payed to concepts, based on which the feasibility and the degree of incorporation of individual elements in the different structure types can be predicted.

2. AlPO$_4$'s AND GaPO$_4$'s

2.1. The 'AlPO$_4$' and 'GaPO$_4$' topological concept

In all microporous crystalline phosphate structures that have been solved, phosphate has a valency of five and a coordination number of four. A one to one combination of P(V) and Al(III) or Ga(III) in a tridimensional network of corner sharing oxygen tetrahedra centered by these elements results in an electroneutral framework. In the regular microporous crystalline aluminophosphate and gallophosphate structures, no oxo bridges between two Al

654

(Ga) atoms, or between two P atoms are present. The strict alternation of Al (Ga) and P in the tetrahedral nodes of the framework precludes the occurrence of odd membered rings of corner sharing tetrahedra. This framework concept is reflected in the code name for crystalline microporous aluminophosphates and gallophosphates, which is based on the idealized chemical composition formula of the frameworks, viz. 'AlPO$_4$' and 'GaPO$_4$', respectively [4,5]. The majority of microporous crystalline phosphate-based structures fit with the AlPO$_4$ (GaPO$_4$) concept (Table 1). Materials containing other elements can be thought of as being derived from an imaginary, isostructural AlPO$_4$ (or GaPO$_4$), in which phosphorus and/or the trivalent element have undergone partial or even complete substitution. Such approach is useful to rationalize the isomorphic substitution behaviour of the individual elements as well as the physico-chemical properties of the material, as will be discussed later in this chapter.

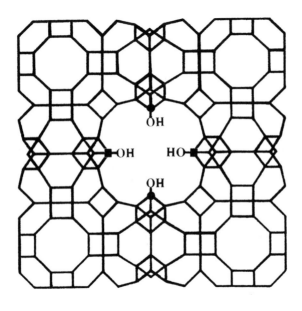

● P

■ Ga

Figure 1 A. Examples of interrupted framework: the aluminophosphate JDF-20 (after ref.7)

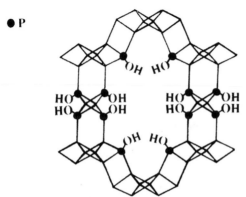

● P

Figure 1 B. Examples of interrupted framework: the gallophosphate with -CLO topology (after ref.6)

Table 1. Phosphate-based crystalline microporous framework types and their matching with the 'AlPO$_4$' or 'GaPO$_4$' concept

Framework structure type (IUPAC CODE)[a]	Type species	AlPO$_4$ concept	window size[c]
ABW	ZnPO-ABW	obeyed	8
AEI	AlPO$_4$-18	obeyed	8
AEL	AlPO$_4$-11	obeyed	10
AET	AlPO$_4$-8, MCM-37	obeyed	14
AFI	AlPO$_4$-5	obeyed	12
AFO	AlPO$_4$-41	obeyed	10
AFR	SAPO-40	obeyed	12
AFS	MAPSO-46	obeyed	12
AFT	AlPO$_4$-52	obeyed	8
AFY	CoAPO-50	obeyed	12
ANA	AlPO$_4$-24	obeyed	8
APC	AlPO$_4$-C, AlPO$_4$-H3, MCM-1	obeyed	8
APD	AlPO$_4$-D	obeyed	8
AST	AlPO$_4$-16	obeyed	6
ATN	MAPO-39	obeyed	8
ATO	AlPO$_4$-31	obeyed	12
ATS	MAPO-36	obeyed	12
ATT	AlPO$_4$-33, AlPO$_4$-12-TAMU	obeyed	8
ATV	AlPO$_4$-25	obeyed	8
AWW	AlPO$_4$-22	obeyed	8
BPH	Beryllophosphate-H	obeyed	12
CHA	SAP0-34, CoAPO-44, CoAPO-47, ZYT-6	obeyed	8
-CLO	cloverite	interrupted framework	20
ERI	AlPO$_4$-17	obeyed	8
FAU	SAPO-37	obeyed	12
GIS	MAPSO-43	obeyed	8
LEV	SAPO-35	obeyed	8
LOS	BePO-LOS	obeyed	6
LTA	SAPO-42	obeyed	8
SOD	AlPO$_4$-20	obeyed	6
RHO	BeAsP0-RHO	obeyed	8
VFI	VPI-5, MCM-9, AlPO$_4$-54	obeyed	18
_[b]	AlPO$_4$-12	interrupted framework	8
-	AlPO$_4$-14	obeyed	8
-	AlPO$_4$-14A	contains Al-O-Al bonds	8
-	AlPO$_4$-15	obeyed	8
-	AlPO$_4$-21	obeyed	8

a, approved by the IZA Structure Commission by the end of 1991 [17];
b, no structure code assigned; c, number of T-atoms in smallest ring circumscribing channels

Table 1. Continued.

Framework structure type (IUPAC CODE)[a]	Type species	AlPO$_4$ concept	window size[c]
_[b]	AlPO$_4$-H1	obeyed	18
-	AlPO$_4$-EN3	obeyed	8
-	GaPO$_4$-C$_7$	obeyed	8
-	JDF-20	interrupted framework	20
-	AlPO$_4$-JDF	obeyed	8
-	Ga$_9$P$_9$O$_{36}$OH.HNEt$_3$	obeyed	10
-	AlPO$_4$-CJ2	obeyed	8
-	ZnPO-dab	interrupted framework	8

a, approved by the IZA Structure Commission by the end of 1991 [17];
b, no structure code assigned; c, number of T-atoms in smallest ring circumscribing channels.

Exceptions to the AlPO$_4$ (GaPO$_4$) concept are the interrupted frameworks found in cloverite (-CLO) [6], JDF-20 [7], AlPO$_4$-12 [8] and ZnPO-Dab [9] and the peculiar AlPO$_4$-14A structure containing Al-O-Al linkages [10]. In the -CLO framework, interruptions occur at an equal number of Ga and P atoms, whereas in JDF-20, only P atoms are missing a fourth linkage (Fig.1). At the framework interruptions, the coordination of P, Al and Ga atoms is saturated with an hydroxo ligand (Fig.1).

2.2. Permanent dipoles

Because each oxygen of the framework is connected to two T-atoms, a tridimensional (3-D) network of corner-sharing tetrahedra can be represented by a line drawing, in which the line segments represent the oxygen atoms and the T atoms ly at the nodes. The framework structure of several AlPO$_4$'s and GaPO$_4$'s can be constructed starting from a parallel stack of simple two-dimensional (2-D) nets, in which each node is three-connected [4,11,12]. A 2-D net is denoted with the size of the three circuits that meet at the different types of nodes. For instance, in the (4.4.18)(4.6.18)$_2$ net on which the VFI framework is based (Fig.2), two 4-rings and an 18-ring meet at one type of node, a 4-ring, a 6-ring and an 18-ring at the second one. The abundance of the (4.6.18) type of nodes is twice that of the (4.4.18) nodes. The complete description of a 3-D net requires the specification of the fourth linkage of the nodes in the 2-D nets. In a stack of 2-D nets, for each node there are two possibilities for the formation of the fourth linkage, viz. binding with a node in the previous net or in the subsequent net. This variability explains the diversity of structures that can be constructed starting from a stacking of one type of 2-D nets.

Several AlPO$_4$ structures with tubular pores can be rationalized by the 2-D net approach (Table 2). In the ATV, AEL, AFO, AFI, AET and VFI topologies, interlinking of the 2-D nets occurs through a same pattern: in each 2-D net, the nodes around the framework circuits are alternately bound with the previous and subsequent 2-D net. This strict alternation of the linking has two major effects. The first is that the structures possess a monodimensional tubular micropore system, the pores running in the direction perpendicular to the 2-D nets. The second effect is that in these structures, each 2-D net is linked with the previous 2-D net through Al atoms, and with the subsequent 2-D net through P atoms, as shown for the VFI topology in Fig.2. Such linkage should give rise to particular dipolar properties.

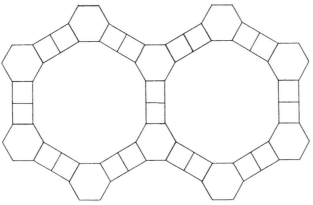

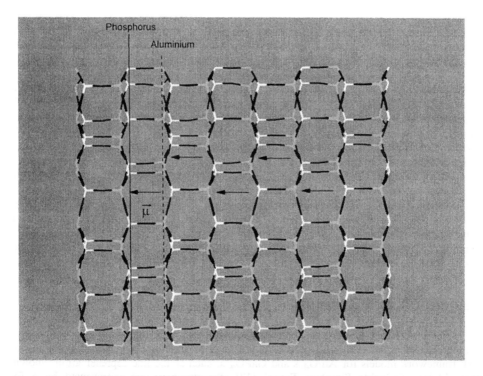

Figure 2. Upper part: **(4.4.18)(4.6.18)₂** 2-D net of the VFI framework. Lower part: VFI framework viewed perpendicular to the c axis. Layers of Al atoms (grey) and P atoms (white) alternate in the structure giving rise to permanent dipoles, oriented according to the channel direction.

Table 2. AlPO$_4$ molecular sieve structures based on the alternate linkage mode of 2-D nets

2-D net[a]	framework topology	window size
(4.6.8)$_2$(6.8.8)	ATV	8-ring
(4.6.10)$_4$(6.6.10)	AEL	10-ring
(4.6.6)(4.6.10)(4.10.10)	AFO	10-ring
(4.6.12)	AFI	12-ring
(4.6.6)(4.4.14)(4.6.14)$_2$(6.6.14)	AET	14-ring
(4.4.18)(4.6.18)$_2$	VFI	18-ring

a, a 2-D net is denoted with the numbers of nodes present in the three framework circuits that meet at the different types of nodes.

The positive electronic charge on the Al atoms of the framework is higher than on the P atoms, owing to the lower electronegativity [13]. Each pair of adjacent Al,P atoms represents a permanent dipole. In the projection of the VFI framework shown in Fig.2, the 2-D nets are perpendicular to the plane of the drawing. The P and Al atoms of the individual 2-D nets are positioned in discrete parallel planes. Al --> P dipoles are systematically present and oriented according to the channel direction. In the other crystallographic directions, the permanent dipoles have a variety of orientations. This polar property is unique for the group of AlPO$_4$ frameworks listed in Table 2. In other AlPO$_4$ topologies, which are not based on 2-D nets and the alternate linkage mode, the polar properties are less pronounced.

The polar nature of the micropore wall was observed first in AlPO$_4$-5 [14]. The polar property can be exploited to orient the crystals in a strong electric field [15]. Since the framework dipoles have the same orientation as the micropores, the parallel orientation of crystals with monodimensional channel systems can be advantageously used, e.g. for the preparation of molecular sieve membranes [15]. The Al --> P dipoles may play an important role in many processes, including adsorption and catalytic conversion reactions with polar molecules.

2.3. The Al and Ga coordination concept

Microporous crystalline phosphates are crystallized from aqueous or alcoholic media under hydrothermal conditions, in presence of organic structurizing agents called templates. In framework models for AlPO$_4$'s and GaPO$_4$'s, such as the line representations shown in the 'Atlas of Zeolite Structure Types' [17], the structures are represented by a four-connected network of corner sharing tetrahedra. In the majority of as-synthesized materials, specific framework Al and Ga atoms interact with extra-framework species such as water molecules, hydroxide, fluoride or phosphate anions giving rise to coordination numbers for these elements that are higher than four. The negative electronic charges of the framework-bound anions are balanced by the positive charges of organic and/or inorganic cations that are encapsulated in the micropores. In the early literature on AlPO4's [1,4,19], it was thought that aluminium with a coordination number of four was essential in order to obtain molecular sieving properties. Meanwhile most of the structures have been solved and the aluminium and gallium coordinations determined based on diffraction data, ^{27}Al and [71]Ga NMR. From the detailed structure determinations, it appears that there is no such correlation of Al or Ga coordination with pore size. Four-, five- and six-coordinated Al and Ga atoms occur in materials with any pore size and in materials with regular and interrupted frameworks.

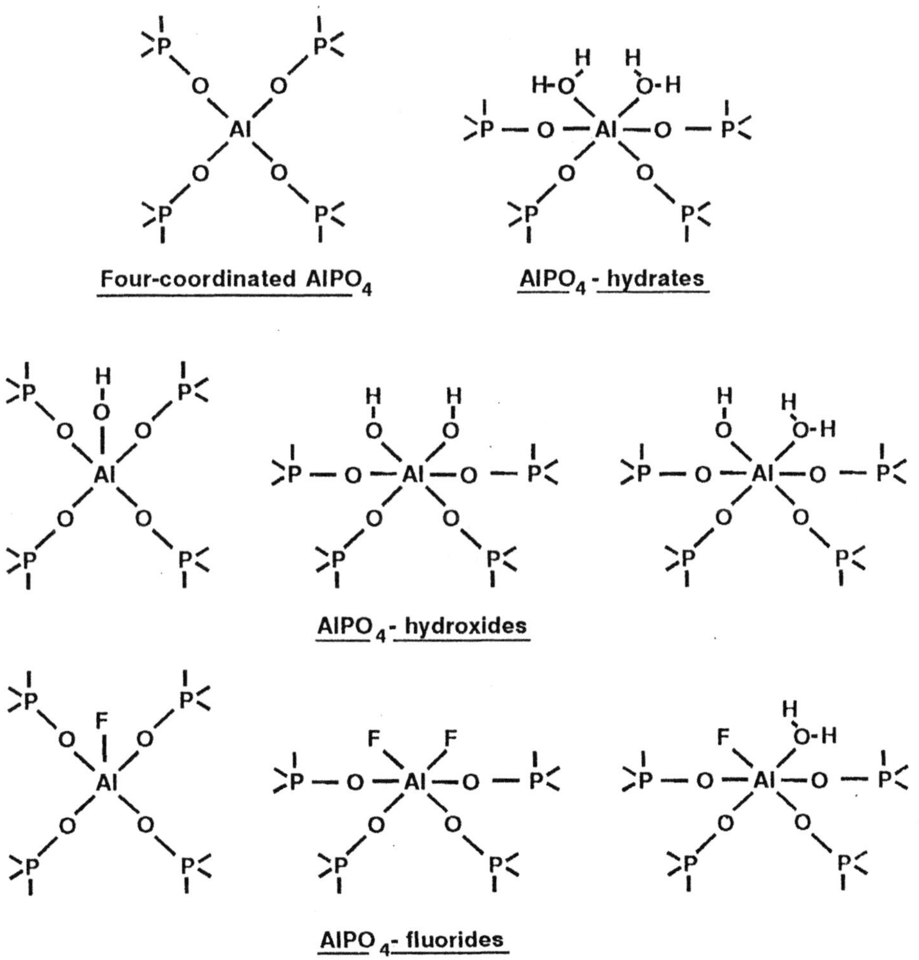

Figure 3. Al coordinations occurring in as-synthesized four-coordinated AlPO4's, AlPO4 hydrates, AlPO4 hydroxides and AlPO4 fluorides.

Table 3. Classification of as-synthesized $AlPO_4$ materials according to aluminium coordinations

	Al coordinations	bridges between Al atoms[a]
Four coordinated $AlPO_4$'s		
$AlPO_4$-5	AlO_4	-
$AlPO_4$-11	AlO_4	-
$AlPO_4$-12 TAMU	AlO_4	-
JDF-20	AlO_4	-
$AlPO_4$ hydrates		
variscite	$AlO_4(H_2O)_2$	-
metavariscite	$AlO_4(H_2O)_2$	-
$AlPO_4$-8	AlO_4 ; $AlO_4(H_2O)2$	-
$AlPO_4$-H3	AlO_4 ; $AlO_4(H_2O)2$	-
VPI-5	AlO_4 ; $AlO_4(H_2O)2$	-
AlPO4-H1	AlO_4 ; $AlO_4(H_2O)2$	-
$AlPO_4$ hydroxides		
$AlPO_4$-15	$AlO_4(OH)_2$; $AlO_4(OH)(H_2O)$	$3(OH)_1$; $2(OH)_2$
$AlPO_4$-14	AlO_4 ; $AlO_4(OH)$; $AlO_4(OH)_2$	$3(OH)_1$; $2(OH)_2$
$AlPO_4$-14A	AlO_4 ; $AlO_2(OH)_2$; $AlO_5(OH)$	$2(OH)_1$
$AlPO_4$-12	AlO_4 ; $AlO_4(OH)$	$2(OH)_1$
$AlPO_4$-21	AlO_4 ; $AlO_4(OH)$	$2(OH)_1$
$AlPO_4$-EN3	AlO_4 ; $AlO_4(OH)$	$2(OH)_1$
$AlPO_4$-17	AlO_4 ; $AlO_4(OH)$	$2(OH)_1$
$AlPO_4$-18	AlO_4 ; $AlO_4(OH)$	$2(OH)_1$
$AlPO_4$-31	AlO_4 ; $AlO_4(OH)$	
$AlPO_4$-20	AlO_4 ; $AlO_4(OH)$	
$AlPO_4$ fluorides		
$AlPO_4$-CJ2	$AlO_4(OH)$; AlO_4F ; $AlO_4(OH)_2$; $AlO_4F(OH)$; AlO_4F_2	
$AlPO_4$-CHA	AlO_4 ; AlO_4F_2	$2(F)_2$
$AlPO_4$ phosphates		
$AlPO_4$-22	AlO_4 ; $AlO_4(PO_3(OH))$	$4(PO_3(OH))_1$

a, the terminology $_x(A)_y$ defines the number, x, of Al atoms bridged by a single A species and the number, y, of bridges between the x Al centers.

Table 4. Classification of as-synthesized materials according to Ga coordinations

	Ga coordinations	bridges between Ga atoms[a]
GaPO$_4$ hydroxides		
GaPO$_4$.2H$_2$0	GaO$_4$(OH)$_2$; GaO$_4$(OH)(H$_2$0)	3(OH)$_1$; 2(OH)$_2$
GaPO$_4$-C$_7$	GaO$_4$(OH)$_2$; GaO$_4$(OH)(H$_2$0)	3(OH)$_1$
GaPO$_4$-CJ2	GaO$_4$(OH) ; GaO$_4$F ;	
	GaO$_4$(OH)$_2$; GaO$_4$F(OH) ; GaO$_4$F$_2$	
GaPO$_4$-12	GaO$_4$; GaO$_4$(OH)	2(OH)$_1$
GaPO$_4$-21	GaO$_4$; GaO$_4$(OH)	2(OH)$_1$
GaPO$_4$-14	GaO$_4$; GaO$_4$(OH) ; GaO$_4$(OH)$_2$	3(OH)$_1$; 2(OH)$_2$
Ga$_9$P$_9$O$_{36}$OH.HNEt$_3$	GaO$_4$; GaO$_4$(OH)	3(OH)$_1$
GaPO$_4$ fluorides		
cloverite	GaO$_4$F	4(F)$_1$
GaPO$_4$-LTA	GaO$_4$F	4(F)$_1$

a, $_x$(A)$_y$ representation explained in Table 3.

The family of *as-synthesized* AlPO$_4$ structures can be subdivided according to the aluminium coordination as follows (Table 3, Fig.3). In *four coordinated AlPO$_4$'s*, all framework Al atoms are four-coordinated (Al^{IV}). In *AlPO$_4$ hydrates*, Al atoms at specific crystallographic framework positions are six-coordinated (Al^{VI}) and have two aquo ligands in addition to the four oxo bridges to framework P atoms. This type of aluminium is denoted as 'aluminium dihydrate'. In *AlPO$_4$ hydroxides*, a number of specific Al atoms are five- or six-coordinated depending on whether these atoms are linked either to one hydroxo ligand (Al^V), one hydroxo ligand and one aquo ligand or to two hydroxo ligands (Al^{VI}). Fluor exhibits a similar chemistry in the coordination sphere of Al atoms of *AlPO$_4$ fluorides*. In *AlPO$_4$ phosphates* a phosphate anion not belonging to the framework is trapped in a systematic way in specific cages. The oxygens of these phosphate anions occupy a fifth coordination site of framework Al^V atoms. Examples of the different types of AlPO$_4$'s are given in Table 3. Whereas aquo ligands bind to one Al atom, hydroxo, fluoro and phosphate ligands are always bound to two Al atoms at least (Table 3). It has to be emphasized that the Al coordination in the as-synthesized material may be different from the one after calcination and readsorption of water or other polar molecules, as will be discussed in section 2. The classification of as-synthesized materials according to Al coordination is essential for understanding isomorphic substitution, as will be explained in section 3.

The as-synthesized GaPO$_4$'s are either *GaPO$_4$ hydroxides* or *GaPO$_4$ fluorides* (Table 4). Examples of GaPO$_4$ hydrates and four coordinated GaPO$_4$'s have not been reported.

2.4. High temperature stability of AlPO$_4$'s

Activation of the phosphate based molecular sieves involves the evacuation of the pore filling species including the organic templates and the framework ligands. This is currently done by calcination. The high temperature stability of AlPO$_4$'s (and GaPO$_4$'s) is dependent on the Al (Ga) coordination. When the interactions of Al (Ga) atoms with their fifth and sixth ligand are removed, displacive transformations, involving changes of relative atom positions, topotactic and disruptive transformations, involving bond breaking and relinking, or even structural collapse are observed.

2.4.1. Four-coordinated AlPO$_4$'s

The four coordinated AlPO4's are thermally and hydrothermally very stable structures. These structures withstand heating at temperatures exceeding 1300 K. Only minor structural changes are observed upon calcination and rehydration. For instance, as-synthesized and calcined, dehydrated AlPO$_4$-11 has a body-centered orthorhombic unit cell (parameters a = 1.35 nm; b = 1.85 nm; c = 0.83 nm) [19]. Upon hydration of calcined AlPO$_4$-11, compression of the framework takes place along the b and c axes, while expansion occurs along the a axis to yield a primitive orthorhombic cell (parameters a = 1.38 nm; b = 1.80 nm; c = 0.81 nm) [20]. As-synthesized and calcined, dehydrated AlPO$_4$-11 samples exhibit ^{27}Al MAS NMR resonance lines in the chemical shift region typical of tetrahedral Al coordination [21,22]. The calcined, rehydrated form shows an additional ^{27}Al MAS NMR resonance line in the region of octahedrally coordinated Al, assigned to AlO$_4$(H$_2$O)$_2$ coordinations [21]. The presence of water in the coordination sphere of Al is confirmed by a strong enhancement of the Al^{VI} line when using (^{1}H-^{27}Al) CP MAS NMR [21].

The formation of AlO$_4$(H$_2$O)$_2$ environments has also been observed upon rehydration of calcined AlPO$_4$-5 [23]. From the literature [20-23] on rehydration of calcined AlPO$_4$-5 and AlPO$_4$-11, it appears that the formation of aluminium hydrate sites is (i) reversible, (ii) occurring to an extent that is structure dependent and (iii) altering the topological space group.

2.4.2. AlPO$_4$ hydrates

Water molecules play an important structural role in AlPO$_4$ hydrates. In AlPO$_4$-H3 [24] and VPI-5 [25,26], water is the only micropore filling molecule. Upon dehydration, the coordination number of Al is gradually decreased from six to four by removal of the aquo ligands, resulting in dramatic changes of the bond angles in the framework. AlPO$_4$-H$_3$ exhibits complex topotactic and displacive phase transitions upon dehydration-rehydration treatments, related to the removal and reinsertion of the two aquo ligands of aluminium [27].

The phase transitions of VPI-5 upon dehydration and rehydration have been the subject of several investigations [28-33].

A VPI-5 structure with aluminium monohydrate sites can be obtained by careful dehydration [28]. The phase transitions of VPI-5 dihydrate, VPI-5 monohydrate, and anhydrous VPI-5 upon dehydration and rehydration are of the displacive type and are reversible [28]. Depending on temperature and atmosphere, a disruptive transition can occurs, yielding the AET topology [29-33] (Fig.4). The VFI into AET transition reduces the pore openings from 18- to 14-rings and is sometimes reversible [29]. In VPI-5 (VFI topology) and AlPO$_4$-8 (AET topology), one third of the framework aluminium is aluminium dihydrate [33,34]. In VPI-6, aluminium dihydrate is located on the (4.4.18) nodes of the 2-D nets [26]. In $AlPO_4$-8, the (4.4.14) and one of the two types of (4.6.14) nodes are the locus of aluminium dihydrate [29,33] (Fig.4). During the VFI --> AET phase transformation, aluminium dihydrate is preserved (Fig.3).

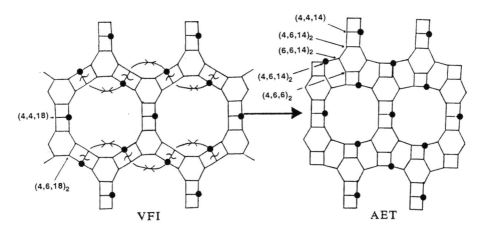

Figure 4. VFI --> AET phase transition with preservation of aluminium dihydrate sites.

2.4.3. $AlPO_4$ hydroxides

The thermal and hydrothermal stability of $AlPO_4$ hydroxides and $GaPO_4$ hydroxides depends on the structure type. Hydroxo ligands bridge between two or three Al or Ga atoms and distort the framework. This distortions vanishes upon removal of the hydroxo ligands. For example, in $AlPO_4$-21, one third of the framework Al is four coordinated; two thirds are five coordinated due to the presence of an hydroxo ligand [35,36]. Calcination of $AlPO_4$-21 leads to a topotactic transformation and yields the $AlPO_4$-25 material, which is a four-coordinated $AlPO_4$ [37]. Displacive transformations are observed in $AlPO_4$'s with ERI topology [38].

2.5. Extra-large-pores: $AlPO_4$'s and $GAPO_4$'s and the 12-ring barrier

Zeolites and molecular sieves are often classified according to pore size. The pore size is conveniently measured as the number of T atoms in the smallest window a guest molecule has to cross in order to reach the largest void volumes. In the common classification, large-pore materials have windows composed of 12-membered rings of T atoms. In extra-large-pore materials the windows are composed of higher membered rings.

Several extra-large-pore phosphate-based molecular sieves have been synthesized. One class of phosphates with extra-large-pores are the materials with VFI [39] and $AlPO_4$-H1 [40] framework types, possessing 18-membered ring pore openings, and the molecular sieve $AlPO_4$-8, having a pore system with 14-membered ring windows [41,42]. These materials are $AlPO_4$ hydrates. 20-Membered rings are present in the interrupted frameworks of JDF-20 [7], a four-coordinated $AlPO_4$ (Fig.1) and in cloverite (-CLO topology) [6], a $GaPO_4$ with five-coordinated Ga. The strategies for the synthesis of extra-large-pore materials that have been applied successfully are summarized in Fig.5. For four-coordinated frameworks, the upper limit of the number of T-atoms in micropore windows is 12. Increasing the window size necessitates an unfavourable enlargement of the O-T-O angles in the ring far beyond the ideal tetrahedral angle of 109°. One way to facilitate the formation of O-T-O angles approaching 180° is by introducing framework elements with a coordination number of VI, such as dihydrate aluminium. The two water molecules occupy cis positions in the coordination sphere of aluminium and protrude into the channel. The alternative approach is to interrupt the framework at positions in the wall of the extra-large-pore. This situation is encountered in the -CLO and JDF-20 structures.

2.6. Adsorption properties

The surface selectivity of $AlPO_4$-based molecular sieves is reported to be weakly to moderately hydrophilic [5]. The introduction of framework charges in substituted analogues generates strong adsorption centra for polar molecules [38]. The substituted analogues encompass a range of moderately to highly hydrophilic surface properties, comparable to aluminosilicate zeolites [2].
Adsorption of permanent gases and hydrocarbons on $AlPO_4$'s follows a micropore volume-filling mechanism resulting in type I isotherms, which is the common isotherm shape for microporous sorbents [42-44]. Unusual sorption properties regarding polar molecules have been found with many $AlPO_4$ materials.
The adsorption of water in $AlPO_4$'s gives rise to a type V isotherm shape, as observed with AFI, AEL, AEI, ERI, SOD and VFI framework types [44-48] (Fig.6). Adsorption of small amounts of water at low pressure is followed by a number of isobar rises and steps, depending on the structure type. All $AlPO_4$'s exhibit hysteresis extending to very low pressures. This water sorption behaviour is explained by chemisorption on specific framework aluminium sites, that are transformed into aluminium dihydrate sites. The remark has to be made here that aluminium dihydrate formation upon water adsorption occurs on all $AlPO_4$'s, irrespective of the Al coordination in the as-synthesized form. The low water uptake at low pressures suggests that water enters the coordination shell of Al only after a certain amount is adsorbed. The hysteresis is not due to capillary condensation, but to the specific water structure at the aluminium dihydrate sites, which persists up to very low pressures during desorption (Fig.6).
The behaviour of calcined $AlPO_4$'s in water adsorption seems to be dependent on the aluminium coordination in the as-synthesized form. In $AlPO_4$'s with AFI and AEL framework type, which are four-coordinated $AlPO_4$'s in the as-synthesized form, and in $AlPO_4$'s of the AEI and ERI type, which are $AlPO_4$ hydroxides in the as-synthesized form, there is only one water uptake step, corresponding to the direct transition from four-coordinated aluminium to aluminium dihydrate. In VPI-5, which is an $AlPO_4$ hydrate in its as-synthesized form, the formation of aluminium monohydrate and dihydrate upon readsorption of water on evacuated samples occurs in consecutive steps (Fig.6). The Al atoms involved in the coordinative adsorption processes and in aluminium dihydrate formation during the crystallization are the same [48]. The heat of water sorption on $AlPO_4$'s is intermediate between the value found in NaX zeolite and silica molecular sieves [48].

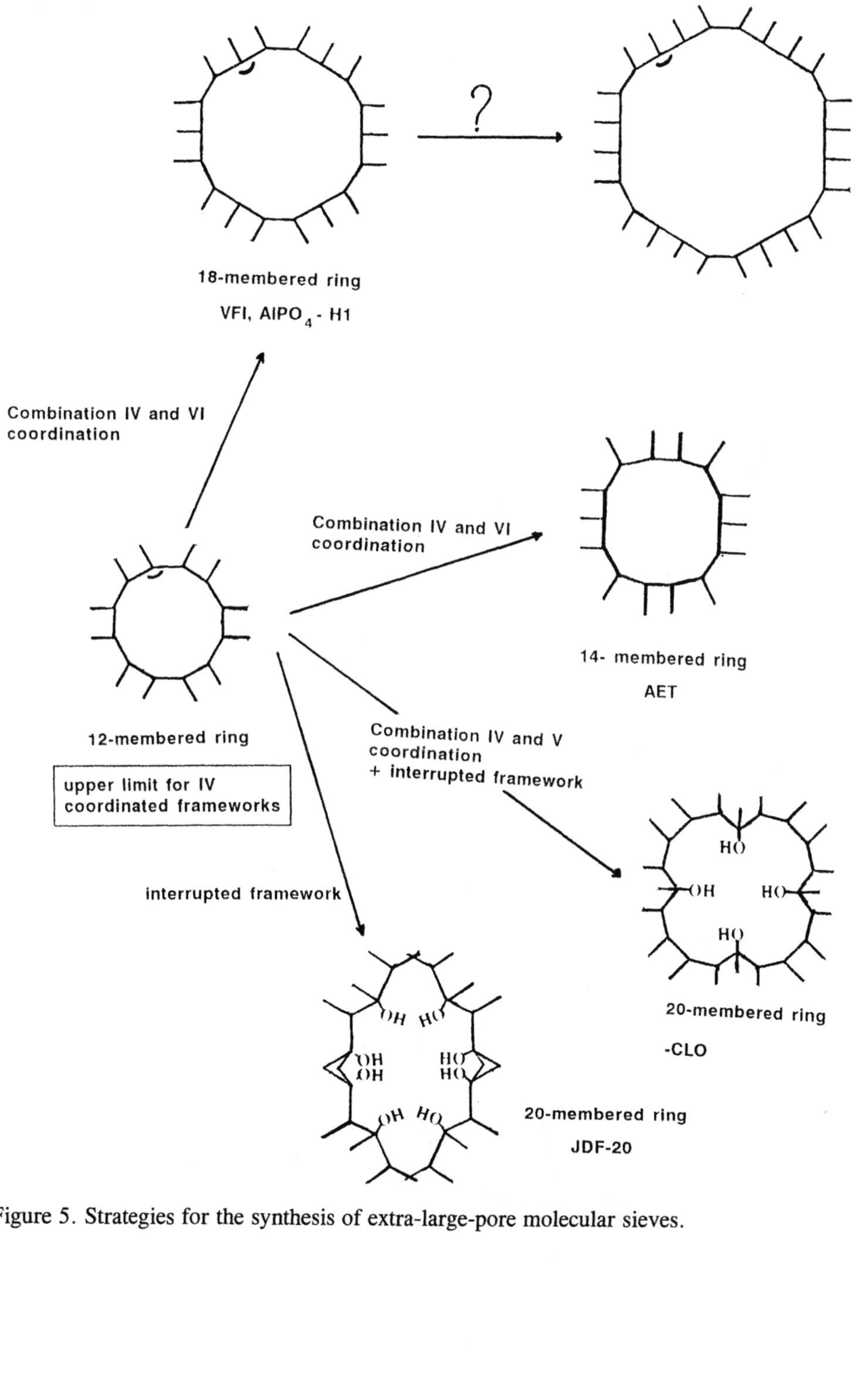

?
18-membered ring
VFI, AlPO4 - H1
Combination IV and VI coordination
Combination IV and VI coordination
14- membered ring
AET
12-membered ring
upper limit for IV coordinated frameworks
Combination IV and V coordination + interrupted framework
interrupted framework
HO
OH
HO
HO
20-membered ring
-CLO
OH HO
OH HO
OH HO
OH HO
20-membered ring
JDF-20

Figure 5. Strategies for the synthesis of extra-large-pore molecular sieves.

The sorption behaviour of other polar molecules such as ammonia, methanol and ethanol in $AlPO_4$'s is similar to that of water [47,48]. These molecules are coordinatively adsorbed, with one or two molecules per Al site, depending on their nature and the $AlPO_4$ framework type. In $AlPO_4$-18 (AEI topology), methanol creates Al^V environments, whereas water and ammonia create Al^{VI} coordinations. Methanol starts generating Al^V coordinations only after ca. one molecule of methanol is sorbed per cavity. Jänchen et al. suggested that the sorbate structure effectively interacting with the framework aluminiums is a methanol dimer [47]. Based on ^{27}Al DOR NMR, it could be shown that the Al atoms that adopt the Al^V coordination in the as-synthesized material and after methanol adsorption are the same atoms [47]. Ammonia is coordinated to the same Al atoms as water [47]. In $AlPO_4$-5, the adsorption of ammonia creates five and six coordinated Al, while adsorbed methanol does not interact with the framework aluminiums [49].

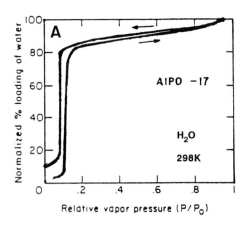

Figure 6. Water adsorption isotherms on $AlPO_4$ molecular sieves. A, Example of hysteresis on $AlPO_4$-17 (adapted from ref.45); B, Adsorption isotherms on VFI-5 and $AlPO_4$-5 compared to NaX (adapted from ref.48).

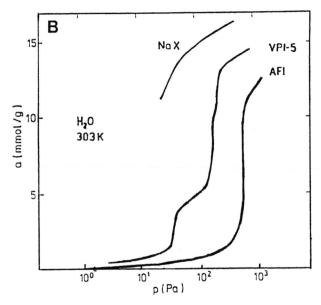

3. Isomorphic substitution

Per definition, isomorphic substitution corresponds to a replacement of an element in a crystalline lattice by another element with similar cation radius and coordination requirements. Isomorphic substitution is currently achieved during crystallization, by adding the element to be incorporated to the synthesis mixture. Many types of isomorphic substitutions are possible in microporous crystalline aluminophosphates and gallophosphates. The complexity arises from the presence of two types of atoms, Al (or Ga) and P, which are both susceptible to substitution, and from the diversity of coordinations of the framework Al (or Ga) atoms. To elements substituting Al (or Ga), microporous crystalline phosphates offer the possibility of adopting coordination numbers IV, V or VI. For phosphates obeying the $AlPO_4$ ($GaPO_4$) concept or having interrupted frameworks with strict Al,P or Ga,P alternation, the isomorphic substitution mechanisms can be classified into substitutions of Al (Ga) atoms (SM I), substitutions of P atoms (SM II), and substitutions of pairs of adjacent Al (Ga) and P atoms (SM III) [3,5] (Fig.7). SM III has been observed only with silicon. In phosphates, the valency rather than the cationic radius dertermines whether an element replaces Al (Ga) or P. The SM I and SM II mechanisms can be subdivided according to the valency of the element introduced: monovalent elements (SM Ia), bivalent elements (SM Ib), trivalent elements (SM Ic), tetravalent elements (SM IIa), and pentavalent elements (SM IIb). The configurations of the elements in frameworks that are subjected to the different types of substitutions are shown in Fig.7. The same scheme applies to $GaPO_4$'s and $AlPO_4$'s. Other types of substitutions not shown in Fig.7 are unlikely [5]. They would lead either to positively charged frameworks, or to a too high negative charge density.

Elements occurring in the synthesis medium under two oxidation states (e.g. Fe(II) and Fe(III) may be incorporated according to two substitution modes simultaneously (SM Ib and SM Ic in this instance). In materials in which more than one element is incorporated, each element exhibits its specific substitution behaviour, depending on its valency and coordination requirement.

For materials obeying the $AlPO_4$ ($GaPO_4$) concept, the degree and type of substitution can be derived, in principle, from the normalized oxide composition, expressed as $M_xAl_yP_zO_2$, in which the sum of the x + y + z fractions equals 1 [5]. The degree of substitution corresponds to the value of 0.5 - y for Al; 0.5 - z for P. In practice, this approach often does not turn out to be very useful. The deduction of SM of a material based on the experimentally determined oxide composition can be misleading, e.g. when trace amounts of impurities with a deviating chemical composition are present.

Isomorphic substitutions of the types SM Ia, SM Ib and SM IIa lead to negatively charged frameworks. After removal of the organic template from the micropores, these negatively charged frameworks have the potential of Brönsted acidity and cation exchange, provided the incorporated elements are stable in the framework.

With respect to isomorphic substitution, $GaPO_4$'s have received much less attention in literature compared to $AlPO_4$'s. Since the principles of substitution are essentially the same, in the following sections, the discussion on isomorphic substitution will be limited to $AlPO_4$'s.

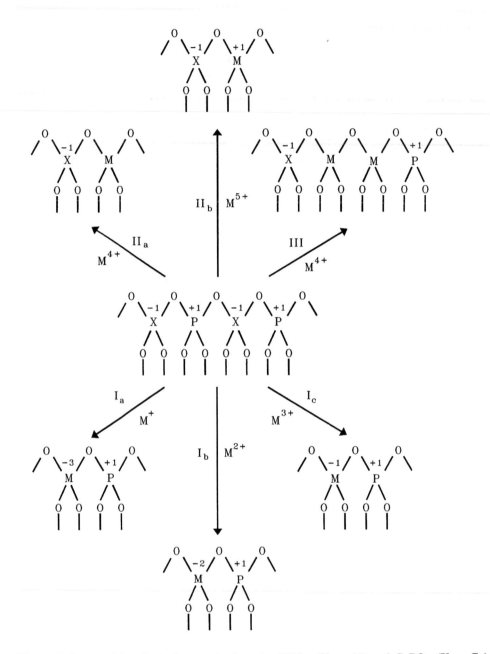

Figure 7. Isomorphic substitution mechanisms in $AlPO_4$ (X = Al) and $GaPO_4$ (X = Ga) based frameworks.

Table 5. Isomorphic substitutions in AlPO$_4$ based materials generating framework charges

Substitution Mechanism:	SM Ia	SM Ib	SM IIa
Elements:	Li(I)	Co(II), Fe(II) Mg(II), Mn(II) Zn(II)	Si(IV)

Four coordinated AlPO$_4$'s

Framework topologies:	probably all reported for AFI and JDF-20	all not reported for ATT	all not reported for ATT
Sustitution degree[a]:	unknown	0...S...ES	0...S

AlPO$_4$ hydrates

Framework topologies:	none	specific ? reported for VFI	specific ? reported for VFI
Substitution degree:	-	traces	traces

AlPO$_4$ hydroxides, fluorides and phosphates

Framework topologies:	specific: SOD CHA AlPO$_4$-14	specific: SOD CHA AlPO$_4$-14 ERI ATO	specific: SOD CHA ERI ATO
Substitution degree:	unknown	0...S	0... < S or 0...S

No AlPO$_4$ analogue

Framework topolgies:	none	specific: AFS AFY ATN ATS FAU GIS LEV	specific: AFS AFR FAU GIS LEV
Substitution degree:	-	S	S

Remarks:	incorporation of extra-framework Li	sometimes incorporation of extra-framework Me(II)	no incorporation of extra-framework Si

a, S: stoichiometric substitution level; ES: substitution level exceeding stoichiometric substitution.

3.1. Isomorphic substitutions generating framework charges

For obvious reasons, isomorphic substitutions generating framework charges have received most attention in literature. The behaviour of the different classes of $AlPO_4$'s with respect to isomorphic substitutions generating framework charges (SM Ia, Ib and IIa) is summarized in Table 5.

3.1.1. Four-coordinated $AlPO_4$'s

The four-coordinated $AlPO_4$'s are susceptible to the three types of substitutions introducing framework charges, SM Ia, Ib and IIa [3,5]. In these $AlPO_4$'s, organic templates act as pore fillers, with only weak van der Waals interactions with the framework atoms [50]. Four-coordinated $AlPO_4$ materials such as $AlPO_4$-5 and $AlPO_4$-11 can be synthesized using a diversity of templates. For instance, the prerequisite of an organic molecule of being capable of templating $AlPO_4$-5 crystallization is that the diameter of the molecule does not exceed the pore diameter of ca. 0.8 nm [18]. The templates are either unprotonated amines or protonated amines or tetraalkylammonium cations, encapsulated as ion pairs, e.g. as organic ammonium hydroxide and tetraalkylammonium hydroxide molecules. For each template, the template:framework stoichiometry in the as-synthesized material is fixed [5]. These relationships between the encapsulated template and the framework explain why elements generating negative framework charges can be incorporated into the frameworks of four coordinated $AlPO_4$'s. The negative framework charge can effectively be balanced by the organic template, which becomes positively charged by protonation (in the case of amines), or omission of the anion (in the case of ion pairs). Specimens in which the framework charge generated by isomorphic substitution corresponds to the maximum charge that can be balanced with the organic template cations are conveniently called the 'stoichiometrically substituted' compositions.

The template:framework stoichiometry dictates the maximum degree of substitution in the instance of Si incorporation according to SM IIa [5]. The key parameter for achieving a high degree of substitution is to select a small template molecule, exhibiting a high template:framework stoichiometry.

The template:framework stoichiometry is less stringent with respect to the incorporation of bivalent elements according to SM Ib [3,5]. For several structure types and bivalent elements, SM Ib type substitutions covering the range from the pure $AlPO_4$ to beyond the stoichiometric substitution level have been reported [51]. When the stoichiometric degree of substitution is exceeded, extra-framework M^{2+} cations function as additional charge compensating species [51]. With Co(II), Fe(II) and Mn(II), part of the metal occurs frequently in extra-framework positions already from low levels of substitution on [52-58].

The incorporation of Li(I) according to the SM Ia mechanism creates two formal negative framework charges at the Li sites. The framework charge is balanced partly by the monovalent organic cations, partly by extra-framework Li^+ cations [3,59].

3.1.2. $AlPO_4$ hydroxides, fluorides and phosphates

In their as-synthesized form, this type of $AlPO_4$'s has negatively charged frameworks due to the binding of anionic ligands with framework Al atoms. The negative framework charges originating from the hydroxo and fluoro ligands are removed upon calcination. They should not be confused with the framework charges introduced by isomorphic substitution. In uncalcined $AlPO_4$ hydroxides, fluorides and phosphates, the negative framework charges of the anionic framework ligands are balanced with encapsulated organic cations. Isomorphic substitutions of the type SM Ia, Ib and IIa disturb this electrostatic charge coupling. With increasing degree of incorporation of the foreign element, the anionic framework ligands are progressively removed such that the framework-template charge balance remains satisfied. It has to be stressed that the omission of the anionic framework ligand must always take place in order to maintain the charge balance, even when a framework atom that is not coordinated to an anionic ligand (Al^{IV} or P^{IV} atom) is replaced by an element of a lower valency. The success of the substitution depends on the stability of the framework in absence of the interactions with the hydroxo, fluoro and phosphate species.

For the structure types listed in Table 5, the anionic ligand can at least partly be removed upon isomorphic substitution [5]. For other framework types such as AEI, AFT, AWW, and those of the $AlPO_4$-15, $AlPO_4$-21 and $AlPO_4$-EN3 type, the binding of the anionic ligands seems to be essential, since substituted specimens have not been synthesized. Isomorphic substitution may nevertheless become possible by altering the nature of the anionic framework ligand. For the hydroxides this could eventually be achieved by crystallization in fluoride medium.

3.1.3. $AlPO_4$ hydrates

$AlPO_4$ hydrates are rather unreactive towards isomorphic substitutions generating framework charges (Table 5). Several explanations can be advanced to explain this behaviour. (i) For some materials like VPI-5 and $AlPO_4$-H3, there is almost no incorporation of organic molecules in the micropores during crystallization. Isomorphic substitutions generating framework charges are unlikely since there is no species in the micropores capable of balancing eventual framework charges. (ii) The introduction of charged species in the micropores is expected to disturb the precious water structures, such as the triple helix structure of water molecules present in the micropores of VPI-5. (iii) The significant phase transformations of $AlPO_4$ hydrates upon dehydration suggest that the dihydrate sites are essential for the structural integrity. A replacement of framework aluminium atoms coordinated with two aquo ligands with elements not capable of adopting such coordination is unlikely. (iv) With respect to Si incorporation according to SM IIa, the generation of Si(4Al) environments in which some of the Al atoms are hexa-coordinated is probably unfavoured given the analogy with aluminosilicate zeolites, where Al coordinated to more than four oxygen atoms does not occur.

3.1.4. Compounds without $AlPO_4$ analogue

Several compounds do not have an $AlPO_4$ analogue [5,60] (Table 5). Most of the time these compounds have stoichiometric compositions (maximum degree of substitution tolerated by framework:template stoichiometry) and the framework Al atoms are four coordinated. Lowering of the degree of substitution is difficult to achieve. The decrease of the framework charge at constant template:framework stoichiometry would have to be compensated by one of the following mechanisms: (i) introduction of anions forming ion pairs with the cationic template molecules, (ii) binding of anions to framework Al atoms or (ii) deprotonation of amines. For the compounds without $AlPO_4$ analogue, it seems that none of these changes can be realized, and that is not possible to decrease the substitution level significantly.

3.1.5. Compounds without Al or Ga

Microporous crystalline beryllophosphates (BePO's) and zincophosphates (ZnPO's) are one to one combinations of Be(II) or Zn(II) with P(V) in tetrahedral oxide frameworks. Harvey and Meier reported BePO's with RHO, BPH, GIS, EDI and ANA topologies [61]. Gier and Stucky synthesized BePO's and ZnPO's with FAU, LOS, ABW and SOD framework types [9,62-64]. The large negative framework charge is balanced with Li^+, Na^+ and tetramethylammonium cations. Stability problems arise when these compounds are heated at moderate temperatures, limiting potential applications to mild temperature conditions.

3.2. Isomorphic substitutions not generating framework charges

3.2.1. Substitutions of type Ic

Literature on SM Ic type isomorphic substitution according to which Al is replaced by other trivalent elements is rather scarce [5]. However, this type of chemistry may lead to interesting catalytic properties. Cr(III) has been incorporated into the framework of $AlPO_4$-14, an $AlPO_4$ hydroxide [65]. The siting of Cr in the framework could be determined. Cr adopts a sixfold coordination and replaces selectively $AlO_4(OH)_2$ sites in the framework.

The occurrence of $GaPO_4$'s that are isostructural with $AlPO_4$ hydroxides containing Al in four-, five- and sixfold coordination (Table 3 and 4) shows that Ga(III) is able to substitute Al(III) in its three types of coordination. Due to its small cation radius, B(III) can not mimic Al^V and Al^{VI} coordinations and substitutes Al^{IV} selectively. Boron incorporation is typically found in four coordinated $AlPO_4$'s [5].

3.2.2. Substitutions of type IIb

Partial replacement of P with As(V) has been reported to be possible for $AlPO_4$'s with AEL and AEI topologies, and for BePO's and ZnPO's. The incorporation of traces of V(V) in $AlPO_4$-5 seems possible. Its siting in the framework is complex and is probably not in agreement with the SM IIb concept [66,67].

3.2.3. Substitutions of the type SM III

Silicon is the only element exhibiting SM III. The various possibilities to generate silicoaluminophosphate (SAPO) frameworks by substitution of P and Al atoms with Si are shown in Fig.8. The substitution is conveniently explained using two-dimensional grid representations of T-atom configurations (Fig.8). Such two-dimensional representations are meaningful only for the determination of the first neighbours of the individual T atoms. For the determination of second neighbours, the exact framework circuits (instead of the 4-rings of the two-dimensional grids) have to be considered.

In SAPO's, the formation of Si-O-P linkages is unlikely [5]. The substitution of an isolated pair of adjacent Al and P atoms in an $AlPO_4$ framework with two Si atoms inevitably generates Si-O-P linkages (SM III *ho*, see Fig.8) and does not occur. One way to avoid this linkage is by applying the SM III substitution in an heterogeneous instead of homogeneous way (SM III *he*, Fig.8). By replacing systematically all Si and P atoms in a certain section of the crystal starting from the external surface, no P-O-Si linkages nor framework charges are formed (Fig.8 d). An electroneutral framework is generated, comprising $AlPO_4$ layer(s) and topotactic SiO_2 overlayer(s). At the bounday of the two crystal domains, Si(3Si,1Al) and Si(1Si,3Al) environments are present. It has to be stressed that in this instance, no framework charges are associated with these Si environments.

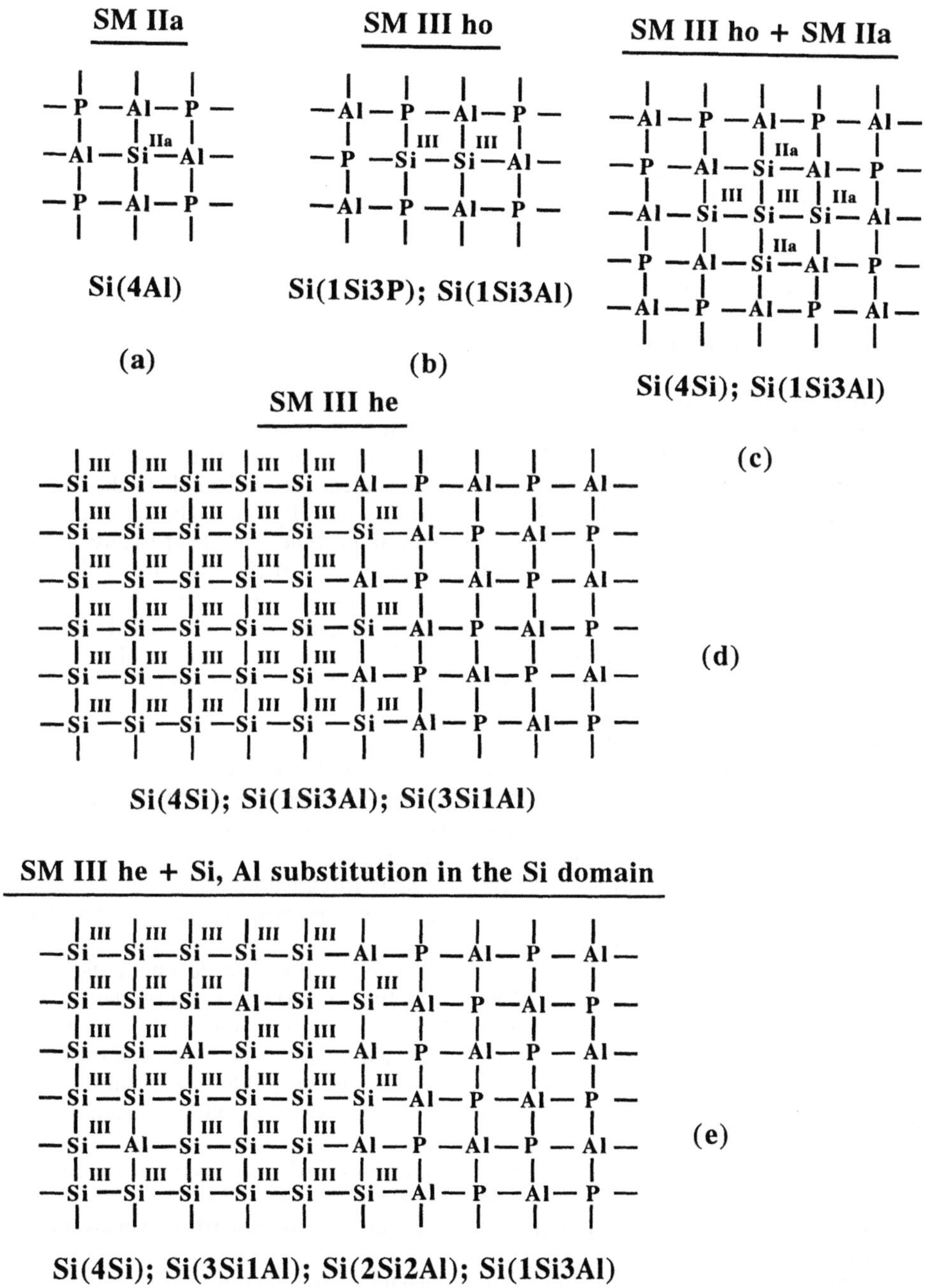

Figure.8. Two-dimensional representation of different possible T atom configurations arising after Si substitution in an AlPO$_4$ framework. The exponents indicates the substitution mechanism of the individual Si atoms (adapted from ref.68).

AEL is the only framework topology for which extensive and virtually pure SM III *he* has been realized using dipropylamine as organic template [68-70]. The highest reported level of Si incorporation amounts to 46% of the T atoms [68]. The maximum intensity of the ^{29}Si MAS NMR signal of these SAPO-11 samples is at -112 ppm (relative to TMS), which is the chemical shift for the Si(4Si) environment in the siliceous layers [68]. The virtual absence of SM IIa is apparent from the very weak ^{29}Si NMR signal intensity in the chemical shift region from -92 to -95 ppm.

A limited amount of silicon can be substituted according to SM III *he* in the AlPO$_4$ hydrates with APC (MCM-1, ref.71) and VFI [72-75] topology. The extent of substitution does not exceed a few percentages of the T-atoms and the siliceous crystal domain is located near the crystal surface [75].

3.3. Combinations of substitution mechanisms

3.3.1. SM IIa and SM III: generation of silicon patches

In the majority of reported SAPO materials, Si is incorporated according to a combination of SM IIa and SM III mechanisms. The contribution of SM IIa and SM III is sensitive to many synthesis parameters, e.g. to the Si content [68,70,76-84], to the nature of the organic template [79,85,86], the amine/Al$_2$O$_3$ and P$_2$O$_5$/Al$_2$O$_3$ ratio in the synthesis mixture [69], pH [68] and the crystallization time and temperature [84]. Generally, the first silicon atoms are incorporated according to SM IIa. Beyond a critical Si concentration, which may be far below the stoichiometric SM IIa substitution level, SM IIa and SM III start to occur simultaneously, and extensive regions in the individual crystals become siliceous. Situations may be encountered where some Si atoms of the silicon patches are replaced with Al atoms, thus generating negative framework charges (Fig.8e).

3.3.2. SM III ho versus SM III he

The SM III substitution pattern can be homogeneous (SM III ho) or heterogeneous (SM III he). As explained in section 3.2.3, SM III he does not introduce lattice charges. The introduction of a pair of Si atoms according to SM III ho results in the formation of three Si-O-P linkages (Fig.8b) and is unlikely [5]. In order to avoid Si-O-P bonds, a combination of SM III ho and SM IIa is necessary. The smallest SiO$_2$ patch with only Si-O-Al bonds at its boundaries contains 5 Si atoms, three of which are incorporated according to SM IIa, two according to SM III ho (Fig.8c).

For some of the SAPO materials, it has been possible to discriminate among the rival models (SM III he + SM IIa versus SM III ho + SM IIa). In SAPO-5 samples, the SM III substitutions are of the heterogeneous type [68,87]. SAPO-5 crystals containing large layers of SiO$_2$ composition, representing e.g. up to 25% of the oxide lattice have been obtained [70]. The silicon is concentrated at the surface of the crystals [80,84]. The substitution behaviour of Si in SAPO-31 appears to be similar to that of SAPO-5, but the amount of Si incorporated is lower [88].

Whereas the use of dipropylamine in the synthesis of SAPO-11 gives rise to almost pure SM III he (see section 3.2.3), samples with SM IIa and SM III combinations crystallize in presence of diisopropylamine [89]. The nature of the SM III (he versus ho) is not known.

A SAPO-34 sample synthesized in a fluoride medium containing triethylamine yielded crystals in which Si was substituted by a combination of SM IIa and SM III mechanisms [90]. The nature of the SM III (he versus ho) is not known.

The framework of some SAPO-42 and SAPO-20 samples corresponds to the situation of Fig.8 e [76,91]. The individual crystals contain a small amount of silicoaluminophosphate intergrown with aluminosilicate. The atomic arrangements of Si, Al and P can be rationalized by a combination of SM IIa and SM III mechanisms [76,91]. Sodalite cages in SAPO-20 and SAPO-42 contain occluded phosphate anions [76,91].

3.3.3. The SAPO-37 debate: rival models for the same material or different materials ?

The stoichiometric SAPO-37 species corresponding to Si incorporation according to SM IIa has a framework T-atom composition of 12.5% Si, 50% Al and 37.5% P [92]. Each type of framework atom has only one type of environment. Each Si tetrahedron is linked to 4 Al tetrahedra (Si(4Al) environment), each P tetrahedron to 4 Al tetrahedra (P(4Al) environment) and each Al tetrahedron to 3 P and 1 Si tetrahedron (Al(3P,1Si) environment). The negative framework charge is balanced with 2 tetrapropylammonium cations in the FAU cages, and 1 tetramethylammonium cation in each TOC cage [92].

Stoichiometric SAPO-37 samples have been prepared by several research teams [80,82,92,93]. In specimens with higher levels of Si incorporation, the framework charge is not increased with respect to the stoichiometric species, as determined e.g. using ^{29}Si NMR [83] or from the TO_2 composition formula [81]. Some silicon-rich SAPO-37 samples have even a lower framework charge [81]. The SM III ho + SM IIa and the SM III he + SM IIa model have been proposed in literature to account for Si substitution in SAPO-37. A two-dimensional grid representation of the T-atom configurations proposed in the two models is shown in Fig.9. In Model 1 [81-83,93], Si patches are created by a combination of SM IIa and SM III ho mechanisms. The smallest SiO_2 patch avoiding Si-O-P linkages contains 5 Si atoms, three of which are incorporated according to SM IIa, two according to SM III ho (Fig.8c, Fig.9). This ensemble of Si atoms generates a formal framework charge of -3. Consequently, in the neighbourhood of the silicon patch, the number of isolated Si atoms (incorporated according to SM IIa) has to be reduced, in order to respect the template:framework charge balance (Fig.9). In Model 1, Al(4P) environments are present (Fig.9). In Model 2 [70,94], the stoichiometric SAPO-37 framework is proposed to be modified by Si incorporation according to the SM II *he* mechanism, e.g. by the generation of siliceous domains near the surface of the crystals (Fig.9).

Fundamentally different Si,Al,P configurations arise from the two rival Models presented in literature to account for Si incorporation in SAPO-37 (Fig.9). The question remains whether actual SAPO-37 samples all adopt the same Si,Al,P configuration, or whether depending on the synthesis conditions, different materials are obtained. In this respect, a comparison of literature data on Si,Al,P compositions of the synthesis mixture and the SAPO-37 crystals obtained could be elucidating (Fig.10). In the crystallizations of SAPO-37 by Franco et al. [95], the crystallization product tends to be enriched with P. These materials are interpreted as combinations of SM III ho + SM IIa. In the crystallizations by Martens et al. [94], the products are enriched with Si and Al. The latter materials are proposed to be generated based on SM III he + SM IIa.

Derewinski et al. studied the behaviour of a stoichiometric SAPO-37 sample at temperatures between 1073 and 1173 K with ^{29}Si MAS NMR [101] and concluded that at these temperatures, the atoms start to diffuse in the lattice resulting in the formation of silicon patches and $AlPO_4$ domains. The resulting framework is proposed to be in agreement with Model 1.

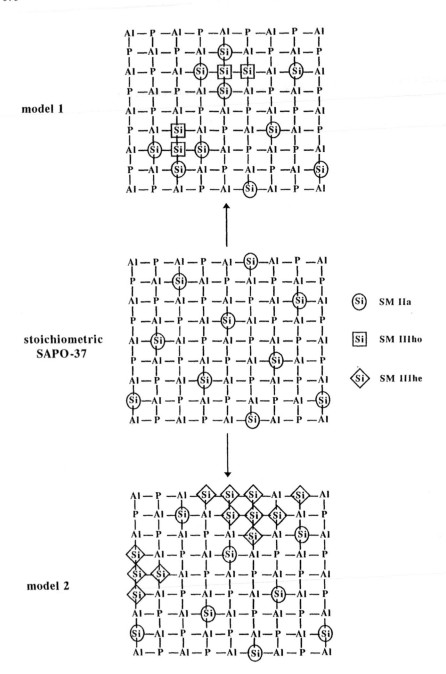

Figure 9. Models of Si,Al,P configurations in SAPO-37 proposed in literature.

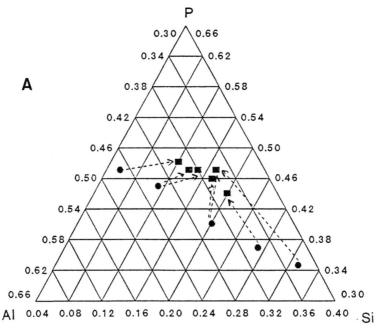

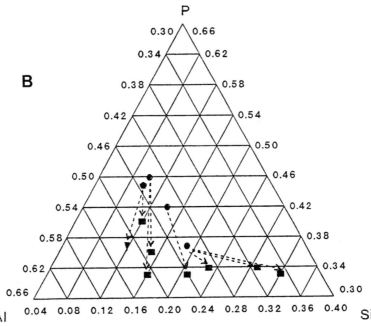

Figure 10. Literature data on Si,Al,P compositions of synthesis mixtures (squares) and SAPO-37 products (circels). A, data from ref.95; B, data from ref.94 and unpublished results.

3.3.4. Interpretation of the ^{29}Si MAS NMR spectra of SAPO-37 according to Model 2

^{29}Si MAS NMR spectra of SAPO-37 samples are composed of six resonance lines, that are assigned as follows: -86 ppm (Si(4Al)^{SA}, -90 ppm (Si(4Al)^{SAPO}, -94 ppm Si(3Al)^{SA}, -98 ppm Si(2Al)^{SA}, -102 ppm Si(1Al)^{SA}, -106 ppm Si(OAl)^{SA} (Fig.11). The Si(4Al)^{SA} signal is not always present [81-83,93]. According to Model 2, the framework can be subdivided into silicoaluminophosphate (SAPO) regions with a fixed Si,Al,P composition, corresponding to that of the stoichiometric SAPO-37, and aluminosilicate (SA) regions. The composition of the SA fraction of the sample can be derived from the Si(nAl) distribution obtained from the ^{29}Si MAS NMR spectrum. Since the framework composition of the SAPO part is fixed (12.5%Si, 37.5%P, 50% Al), the intensity of the ^{29}Si MAS NMR signal of the Si(4Al)^{SAPO} line allows to calculate the fraction of the framework occupied by the SAPO domains [70,94] (Table 7).

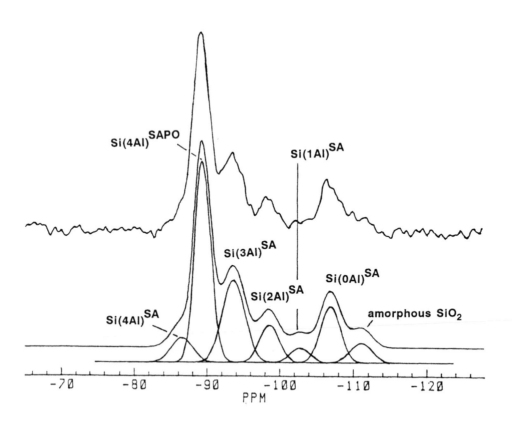

Figure 11. Decomposition of a ^{29}Si MAS NMR spectrum of a typical SAPO-37 sample and interpretation of the resonance lines according to Model 2 (adapted from ref.94).

3.3.5. Brönsted acidity of SAPO's

From the discussion on the modes of Si incorporation in SAPO's, it follows that attempts to rationalize the catalytic activity of SAPO's by concepts such as overall Si content or lattice charge are meaningless. In SAPO crystals containing SA and SAPO domains, Brönsted acid sites are located in the SAPO domains, in the SA domains and at the domain interphases (Fig.12).

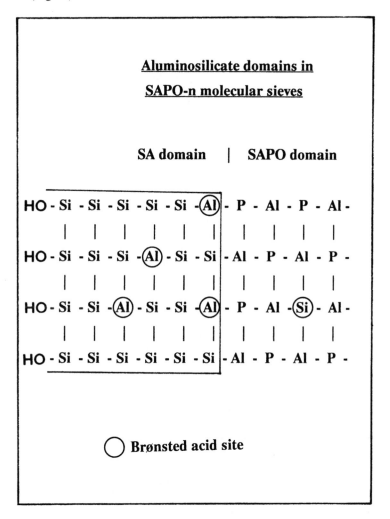

Figure 12. Location of negatively charged framework sites giving rise to Brönsted acidity in SAPO's generated by SM III he + SM IIa (adapted from ref.70).

From a study on the catalytic conversion of decane on SAPO-5, SAPO-11 and SAPO-37 samples [70], it was concluded that the contribution of the different types of acid sites to the catalytic activity is strongly dependent on the structure type.

In SAPO-5, the SA domains have essentially an SiO_2 composition and do not contain catalytically active Brönsted acid sites. The catalytic activity of SAPO-5 is situated in the SAPO domains, where Si atoms give rise to the Brönsted acidity. During the crystallization, the negative lattice charge that is created by the incorporation of Si atoms is balanced with organic cations. Therefore, the Brönsted acid sites in the SAPO domains are isolated from each other and homogeneously distributed. In the cracking of butane over a series of SAPO-5 samples with different Si contents, it was found that the catalytic activity per strong acid site (quantified by ammonia desorption above 580 K) was constant [96]. Similar acid strenghts were found for samples with pure SM IIa and samples with SM III and SM IIa combinations [97,98].

In SAPO-11 synthesized with dipropylamine, the SA domains do not contain Al and the catalytic activity is due to acid sites located at SAPO-SA interphases [70]. In SAPO-37, Brönsted acid sites are present in the SA and SAPO domains as well as on the phase boundaries. The strongest acid sites exhibiting the highest catalytic turnover numbers are located at the SA - SAPO interphases. This conclusion is supported by the results of molecular orbital calculations (CNDO/2) performed by Ojo et al. [99]. Charges on the hydrogen atoms of bridging Si-OH-Al groups in double 4-ring units were calculated for several Si,Al,P configurations, some of which are shown in Fig.13. The most acidic proton is found in configurations generated by a combination of SM III and SM IIa, at the borderline of the two types of domains. Although the double 4-ring unit does not occur in the FAU framework of SAPO-37, the result highlights the importance of framework heterogeneity. In decane hydrocracking, silicon-rich SAPO-37 samples proposed to be generated based on SM III he + SM IIa exhibit a catalytic activity that is comparable to H-Y zeolites [70]. The catalytic activity of stoichiometric SAPO-37 is much lower, suggesting that the few protons at SA-SAPO interphase exhibit unusually high turnover frequencies [70]. Based on TPD of ammonia, the acid strength in SAPO-37 proposed to be generated based on SM III ho + SM IIa is independent of Si content [102].

The thermal stability of the hydroxyls of SAPO-37 materials is comparable to that of ultrastable Y zeolites [100].

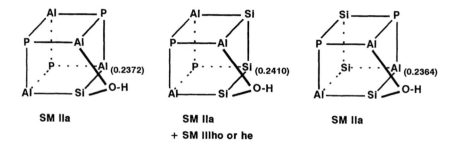

Figure 13. Charges on hydrogen atoms of Si-OH-Al bridges in double 4-ring units (adapted from ref.99).

4. Acknowledgements

JAM acknowledges the Flemisch National Fund for Scientific Research for a Research position as Senior Research Associate.

5. References

1. S.T. Wilson, B.M. Lok, C.A. Messina, T.R. Cannan and E.M. Flanigen, J. Am. Chem. Soc., 104 (1982) 1146.
2. B.M. Lok, C.A. Messina, R.L. Patton, R.T. Gajek, T.R. Cannan and E.M. Flanigen, J. Am. Chem. Soc., 106 (1984) 6092.
3. E.M. Flanigen, B.M. Lok, R.L. Patton and S.T. Wilson, in 'New Developments in Zeolite Science and Technology', Proceed. 7th Int. Zeolite Conf., Eds. Y. Murakami, A. Lijima and J.W. Ward, Kodansha, Elsevier, Amsterdam, Oxford, New York, Tokyo, 1986, p.103.
4. J.M. Bennett, W.J. Dytrych, J.J. Pluth, J.W. Richardson, Jr., and J.V. Smith, Zeolites, 6 (1986) 349.
5. E.M. Flanigen, R.L. Patton and S.T. Wilson, in 'Innovation in Zeolite Materials Science', Eds. P.J. Grobet, W.J. Mortier, E.F. Vansant and G. Schulz-Ekloff, Stud. Surf. Sci. Catal. No.37, Elsevier, Amsterdam, 1988, p.13.
6. M. Estermann, L.B. McCusker, C. Baerlocher, A. Merrouche and H. Kessler, Nature, 352 (1991) 320.
7. Q. Huo, R. Xu, S. Li, Y. Xu, Z. Ma, J.M. Thomas and A.M. Chippendale, J. Chem. Soc., Commun. (1992) 875.
8. J.B. Parise, J. Chem. Soc., Chem. Commun. (1984) 1449.
9. W.T.A. Harrison, T.E. Gier, T.M. Nenoff and G.D. Stucky, in Proceed. 9th International Zeolite Conference Vol.I, Eds. R. von Ballmoos, J.B. Higgins, M.M.J. Treacy, Butterworth-Heinemann, Stoneham, 1993, p.399.
10. J.J. Pluth and J.V. Smith, Acta Crystallogr., C43 (1987) 866.
11. J.V. Smith, Chem. Rev., 88 (1988) 149.
12. J.V. Smith, in 'Zeolites: Facts, Figures, Future', Eds. P.A. Jacobs and R.A. van Santen, Studies in Surface Science and Catalysis Vol.49A, Elsevier, Amsterdam, Oxford, New York, Tokyo, 1989, p.29.
13. L. Uytterhoeven, W.J. Mortier and P. Geerlings, J. Phys. Chem., 16 (1989) 50.
14. J.M. Bennett, J.P. Cohen, E.M. Flanigen, J.J. Pluth and J.V. Smith, in 'Intrazeolite Chemistry', Eds. G.D. Stucky and F.G. Dwyer, Am. Chem. Soc. Symp. Ser. Vol.218, 1983, p.109.
15. J. Caro, B. Zibrowius, G. Finger, M. Bülow, J. Cornatowski and W. Hübner, German Patent DE 41 09 038 (1992).
16. J. Caro, G. Finger, E. Jahn, J. Kornatowski, F. Marlow, N. Noack, L. Werner and B. Zibrowius, in Proceed. 9th International Zeolite Conference Vol. II, Eds. R. von Ballmoos, J.B. Higgins, M.M.J. Treacy, Butterworth-Heinemann, Stoneham, 1993, p.683.
17. W.M. Meier and D.H. Olson, Zeolites 12 (1992) 449.
18. S.T. Wilson, B.M. Lok, C.A. Messina and E.M. Flanigen, in Proceed. Sixth International Zeolite Conf., Eds. D. Olson and A. Bisio, Butterworths, Guildford, 1984, p.97.

19. J.M. Bennett, J.W. Richardson, Jr., J.J. Pluth and J.V. Smith, Zeolites, 7 (1987) 160.

20. R. Khouzami, G. Coudurier, F. Lefebvre, J.C. Vedrine and B.F. Mentzen, Zeolites, 10 (1990) 183.

21. M. Goepper, F. Guth, L. Delmotte, J.L. Guth and H. Kessler, in 'Zeolites: Facts, Figures, Future', Eds. P.A. Jacobs and R.A. van Santen, Studies in Surface Science and Catalysis Vol. 49B, Elsevier, Amsterdam, 1989, p.857.

22. C.S. Blackwell and L. Patton, J. Phys. Chem., 88 (1984) 6135.

23. R.H. Meinhold and N.J. Tapp, J. Chem. Soc. Chem. Commun. (1990) 219.

24. J.J. Pluth and J.V. Smith, Nature, 318 (1985) 165.

25. P.R. Rudolf and C.E. Crowder, Zeolites, 10 (1990) 163.

26. L.B. McCusker, Ch. Baerlocher, E. Jahn and M. Bülow, Zeolites, 11 (1991) 308.

27. E.B. Keller, W.M. Meier and R.M. Kirchner, Solid State Ionics, 43 (1990) 93.

28. J.A. Martens, E. Feijen, J.L. Lievens, P.J. Grobet and P.A. Jacobs, J. Phys. Chem., 95 (1991) 10025.

29. E.T.C. Vogt and J.W. Richardson, Jr., J. Solid State Chem., 87 (1990) 469.

30. M.J. Annen, D. Young, M.E. Davis, O.B. Cavin and C.R. Hubbard, J. Phys. Chem., 95 (1991) 1380.

31. K. Vinje, J. Ulan, R. Szostak and R. Gronsky, Appl. Catal., 72 (1991) 361.

32. W. Schmidt, F. Schüth, H. Reichert, K. Unger and B. Zibrowius, Zeolites, 12 (1992) 2.

33. D.E. Akporiaye and M. Stöker, in Proceed. 9th International Zeolite Conference Vol. I, Eds. R. von Ballmoos, J.B. Higgins, M.M.J. Treacy, Butterworth-Heinemann, Stoneham, 1993, p.563.

34. J.A. Martens, H. Geerts, P.J. Grobet and P.A. Jacobs, in 'Zeolite Microporous Solids: Synthesis, Structure and Reactivity', Eds. E.G. Derouane, F. Lemos, C. Naccache and F.R. Ribeiro, NATO ASI Ser.C, Vol.352, Kluwer Academic Publishers, Dordrecht, Boston, London, 1992, p. 477.

35. J.M. Bennett, J.P. Cohen, G. Artioli, J.J. Pluth and J.V. Smith, Inorg. Chem., 24 (1985) 188.

36. J.B. Parise and C.S. Day, Acta Crystallogr., C41 (1985) 515.

37. J.W. Richardson, Jr., J.V. Smith and J.J. Pluth, J. Phys. Chem., 94 (1990) 3365.

38. U. Lohse, E. Löffler, K. Kosche, J. Jänchen and B. Parlitz, Zeolites, 13 (1993) 549.

39. M.E. Davis, C. Saldarriaga, C. Montes, J. Garces and C. Crowder, Nature, 331 (1988) 698.

40. D.M. Poojary and A. Clearfield, Zeolites, 13 (1993) 542.

41. R.M. Dessau, J.L. Schlenker and J.B. Higgins, Zeolites, 10 (1990) 522.

42. S.T. Wilson, B.M. Lok, C.A. Messina, T.R. Cannan and E.M. Flanigen, in 'Intrazeolite Chemistry', Eds. G.D. Stucky and F.G. Dwyer, Am. Chem. Soc. Symp. Ser. No. 218, 1983, p.79.

43. J. Jänchen, H. Stach, P.J. Grobet, J.A. Martens and P.A. Jacobs, Zeolites, 12 (1992) 9.

44. H. Stach, H. Thamm, K. Fiedler, B. Grauert, W. Wieker, E. Jahn and G. Öhlmann, in 'New Developments in Zeolite Science and Technology', Proceed. 7th Int. Zeolite Conf., Eds. Y. Murakami, A. Lijima and J.W. Ward, Kodansha, Elsevier, Amsterdam, Oxford, New York, Tokyo, 1986, p.539.

45. P.B. Malla and S. Komarneni, Mat. Res. Soc. Symp. Proceed. Vol.233, Material Research Society, 1991, p.237.

46. M.B Kenny, K.S.W. Sing and C.R. Theocharis, J. Chem. Soc. Chem. Commun. (1991) 974.

47. J. Jänchen, M.P.J. Peeters, J.W. de Haan, L.J.M. van de Ven, J.H.C. van Hooff, I. Girnus and U. Lohse, J. Phys. Chem., 97(46) (1993) 12042.

48. J. Jänchen, H. Stach, P.J. Grobet, J.A. Martens and P.A. Jacobs, in Proceed. 9th International Zeolite Conference Vol. II, Eds. R. von Ballmoos, J.B. Higgins, M.M.J. Treacy, Butterworth-Heinemann, Stoneham, 1993, p.22.

49. I. Kustanovich and D. Goldfarb, J. Phys. Chem., 95 (1991) 8818.

50. J.J Pluth and J.V. Smith, in 'Zeolites: Facts, Figures, Future', Eds. P.A. Jacobs and R. van Santen, Studies in Surface Science and Catalysis, Vol. 49B, Elsevier, Amsterdam, 1989, p.835.

51. S.T. Wilson and E.M. Flanigen, in 'Zeolite Synthesis', Eds. M.L. Occelli and H.E. Robson, Am. Chem. Soc. Symp. Ser. No. 398, 1989, p.329.

52. B. Kraushaar-Czarnetzki, W.G.M. Hoogervorst, R.R. Andréa, C.A. Emeis, and W.H.J. Stork, in 'Zeolite Chemistry and Catalysis", P.A. Jacobs, N.I. Jaeger, L. Kubelkova and B. Wichterlova, Studies in Surface Science and Catalysis No.69, Elsevier, Amsterdam, 1991, p.231.

53. S. Ernst, L. Puppe and J. Weitkamp, in 'Zeolites: Facts, Figures, Future', Eds. P.A. Jacobs and R.A. van Santen, Studies in Surface Science and Catalysis, Vol.49A, Elsevier, Amsterdam, 1988, p.447.

54. R.A. Schoonheydt, R. De Vos, J. Pelgrims and H. Leeman, in 'Zeolites: Facts, Figures, Future', Eds. P.A. Jacobs and R.A. van Santen, Studies in Surface Science and Catalysis, Vol.49A, Elsevier, Amsterdam, 1988, p.559.

55. S. Schubert, H.M. Ziethen, A.X. Trautwein, F. Schmidt, H. Li, J.A. Martens, and P.A. Jacobs, in 'Zeolites as Catalysts, Sorbents and Detergents Builders', Eds. H.G. Karge and J. Weitkamp, Studies in Surface Science and Catalysis Vol.46, Elsevier, Amsterdam, 1989, p.735.

56. A.F. Ojo, J. Dwyer and R.V. Parish, in 'Zeolites: Facts, Figures, Future', Eds. P.A. Jacobs and R.A. van Santen, Studies in Surface Science and Catalysis, Vol.49A, Elsevier, Amsterdam, 1988, p.227.

57. P. Wenqin, Q. Shilun, K. Qiubin, W. Zhiyun and P. Shaoyi, in 'Zeolites: Facts, Figures, Future', Eds. P.A. Jacobs and R.A. van Santen, Studies in Surface Science and Catalysis, Vol.49A, Elsevier, Amsterdam, 1988, p.281.

58. G. Brouet, X. Chen and L. Kevan, in Proceed. 9th International Zeolite Conference Vol. I, Eds. R. von Ballmoos, J.B. Higgins, M.M.J. Treacy, Butterworth-Heinemann, Stoneham, 1993, p.489.

59. S. Qiu, W. Tian, W. Pang, T. Sun and D. Jiang, Zeolites, 11 (1991) 371.

60. J.M. Bennett and B.K. Marcus, in 'Innovations in Zeolite Materials Science', Eds. P.J. Grobet W.J. Mortier, E.F. Vansant and G. Schulz-Ekloff, Studies in Surface Science and Catalysis Vol.37, Elsevier, Amsterdam, 1988, p.269.

61. G. Harvey and W.M. Meier, in 'Zeolites: Figures, Facts, Future', Eds. P.A. Jacobs and R.A. van Santen, Studies in Surface Science and Catalysis, Vol. 49A, Elsevier, Amsterdam, 1989, p.179.

62. T.E. Gier and G.D. Stucky, Nature, 349 (1991) 508.

63. W.T.A. Harrison, T.E. Gier and G.D. Stucky, Zeolites 13 (1993) 242.

64. T.M. Nenoff, W.T.A . Harrison, T.E. Gier, J.M. Nicol and G.D. Stucky, Zeolites, 12 (1992) 770.

65. M. Helliwell, V. Kaucic, G.M.T. Cheetham, M.M. Harding, B.M. Kariuki and P.J. Rizkallah, 9th; International Zeolite Conference, Montreal 1992, Recent Research Report No.203.

66. C. Montes, M.E. Davis, B. Murray and M. Narayana, J. Phys. Chem., 94 (1990) 6431.

67. M.S. Rigutto and H. van Bekkum, J. Mol. Catal., 81 (1993) 77.

68. M. Mertens, J.A. Martens, P.J. Grobet and P.A. Jacobs, in 'Guidelines for Mastering the Properties of Molecular Sieves - Relationship between the Physicochemical Properties of Zeolitic Systems and Their Low Dimensionality', Eds. D. Barthomeuf, E.G. Derouane and W. Höldrich, NATO ASI, Ser. B, Vol.221. Plenum Press, New York, London, 1990, p.1.

69. E. Jahn, D. Müller and K. Becker, Zeolites, 10 (1990) 151.

70. J.A. Martens, P.J. Grobet and P.A. Jacobs, J. Catalysis, 126 (1990) 299.

71. J.A. Martens, B. Verlinden, M. Mertens, P.J. Grobet and P.A. Jacobs, in 'Zeolite Synthesis', Eds. M.L. Occelli and H.E. Robson, Am. Chem. Soc. Symp. Ser. No. 398, 1989, p.305.

72. H. Cauffriez, L. Delmotte and J.L. Guth, Zeolites, 12 (1992) 121.

73. J.A. Martens, H. Geerts, L. Leplat, G. Vanbutsele, P.J. Grobet and P.A. Jacobs, Catalysis Letters, 12 (1992) 367.

74. J.A. Martens, I. Balakrishnan, P.J. Grobet and P.A. Jacobs, in 'Zeolite Chemistry and Catalysis', Eds. P.A. Jacobs, N.I. Jaeger, L. Kubelkova and B. Wichterlova, Studies in Surface Science and Catalysis, Vol.69, Elsevier, Amsterdam, 1991, p.135.

75. M.E. Davis, C. Montes, P.E. Hathaway and J.M. Garces, in 'Zeolites: Facts, Figures, Future', Eds. P.A. Jacobs and R.A. van Santen, Studies in Surface Science and Catalysis, Vol.49A, Elsevier, Amsterdam, 1989, p.199.

76. D. Hasha, L.S. de Saldarriaga, C. Saldarriaga, P.E. Hathaway, D.F. Cox and M.E. Davis, J. Am. Chem. Soc., 110 (1988) 2127.

77. C.S. Blackwell and R.L. Patton, J. Phys. Chem., 92 (1988) 3965.

78. N.J. Tapp, N.B. Milestone and D.M. Bibby, in 'Innovation in Zeolite Materials Science', Eds. P.J. Grobet, W.J. Mortier, E.F. Vansant and G. Schulz-Ekloff, Studies in Surface Science and Catalysis, No.37, Elsevier, Amsterdam, 1988, p.393.

79. J. Jänchen, V. Penchev, E. Löffler, B. Parlitz and H. Stach, Collect. Czech. Chem. Commun., 57 (1992) 826.

80. N. Rajic, D. Stojakovic, S. Hocevar, V. Kaucic, Zeolites 13, (1992) 384.

81. P.P. Man, M. Briend, M.J. Peltre, A. Lamy, P. Beaunier and D. Barthomeuf, Zeolites, 11 (1991) 563.

82. A. Corma, V. Fornes, M.J. Franco, F. Melo, J. Perez-Pariente and E. Sastre, Proceed. 9th International Zeolite Conference Vol.II, Eds. R. von Ballmoos, J.B. Higgins and M.M.J. Treacy, Butterworth-Heinemann, Stoneham, 1993, p.343.

83. M.A. Makarova, A.F. Ojo, K.M. Al-Ghefaili and J. Dwyer, Proceed. 9th International Zeolite Conference Vol.II, Eds. R. von Ballmoos, J.B. Higgins and M.M.J. Treacy, Butterworth-Heinemann, Stoneham, 1993, p.259.

84. D. Young and M.E. Davis, Zeolites, 11 (1991) 277.

85. R. Wang, C.F. Lin, Y.S. Ho, L.J. Leu and K.J. Chao, Appl. Catal., 72 (1991) 39.

86. A.I. Biaglow, A.T. Adamo, G.T. Kokotailo and R.J. Gorte, J. Catalysis, 131 (1991) 252.

87. J.A. Martens, M. Mertens, P.J. Grobet and P.A. Jacobs, in 'Innovation in Zeolite Materials Science', Eds. P.J. Grobet, W.J. Mortier, E.F. Vansant and G.Schulz-Ekloff, Studies in Surface Science and Catalysis, Vol.37, Elsevier, Amsterdam, Oxford, New York, Tokyo, 1988, p.97.
88. H-L. Zubowa, E. Alsdorf, R. Fricke, F. Neissendorfer, J. Richter-Mendau, E. Schreier, D. Zeigan and B. Zibrowius, J. Chem. Soc. Faraday Trans., 86(12) (1990) 2307.
89. L. Yang, Y. Aizhen and X. Qinhua, Appl. Catal. 67 (1991) 169.
90. Y. Xu, P. Maddox and J.W. Couves, J. Chem. Soc. Faraday Trans., 86(2) (1990) 425.
91. G.H. Kühl and K.D. Schmitt, Zeolites, 10 (1990) 2.
92. L. Sierra de Saldarriaga, C. Saldarriaga and M. Davis, J. Am. Chem. Soc., 109 (1987) 2686.
93. L. Maistriau, N. Dumont, J.B. Nagy, Z. Gabelica and E.G. Derouane, Zeolites, 10 (1990) 243.
94. J.A. Martens, C. Janssens, P.J. Grobet, H.K. Beyer and P.A. Jacobs, in 'Zeolites, Facts, Figures and Future', Eds. P.A. Jacobs and R.A. van Santen, Studies in Surface Science and Catalysis, Vol.46, Elsevier, Amsterdam, Oxford, New York, Tokyo, 1989, p. 215.
95. M.J. Franco, J. Perez-Pariente, A. Mifsud, T. Blasco and J. Sanz, Zeolites 12 (1992) 386.
96. C. Halik, S.N. Chaudhuri and J.A. Lercher, J. Chem. Soc., Faraday Trans. I, 85(11) (1989) 3879.
97. C. Halik and J.A. Lercher, J. Chem. Soc., Faraday Trans. I, 84 (1988) 4457.
98. J. Meusinger, H. Vinek, G. Dworeckow, M. Goepper and J.A. Lercher, in 'Zeolite Chemistry and Catalysis, Eds. P.A. Jacobs, N.I. Jaeger, L. Kubelkova and B. Wichterlova, Studies in Surface Science and Catalysis, No.69, Elsevier, Amsterdam, 1991, p.373.
99. A.F. Ojo, J. Dwyer, J. Dewing, P.J. O'Malley and A. Nabhan, J. Chem. Soc. Faraday Trans., 88(1) (1992) 105.
100. S. Dzwigaj, M. Briend, A. Shikholeslami, M.J. Peltre and D. Barthomeuf, Zeolites 10 (1990) 157.
101. J.M. Peltre, M. Briend, A. Lamy and D. Barthomeuf, J. Chem. Soc. Faraday Trans. 86(22) (1990) 3823.
102. M. Briend, M.J. Peltre, A. Lamy, P.P. Man and D. Barthomeuf, J. Catal. 138 (1992) 90.

Keyword Index

STUDIES IN SURFACE SCIENCE AND CATALYSIS

Advisory Editors:
B. Delmon, Université Catholique de Louvain, Louvain-la-Neuve, Belgium
J.T. Yates, University of Pittsburgh, Pittsburgh, PA, U.S.A.

Printed and bound by CPI Group (UK) Ltd, Croydon, CR0 4YY

03/10/2024

01040328-0012